Mathematical Foundations of Quantum Information and Computation and Its Applications to Nano- and Bio-systems

Theoretical and Mathematical Physics

The series founded in 1975 and formerly (until 2005) entitled *Texts and Monographs in Physics* (TMP) publishes high-level monographs in theoretical and mathematical physics. The change of title to *Theoretical and Mathematical Physics* (TMP) signals that the series is a suitable publication platform for both the mathematical and the theoretical physicist. The wider scope of the series is reflected by the composition of the editorial board, comprising both physicists and mathematicians.

The books, written in a didactic style and containing a certain amount of elementary background material, bridge the gap between advanced textbooks and research monographs. They can thus serve as basis for advanced studies, not only for lectures and seminars at graduate level, but also for scientists entering a field of research.

For other titles published in this series, go to
http://www.springer.com/series/720

Masanori Ohya · Igor Volovich

Mathematical Foundations of Quantum Information and Computation and Its Applications to Nano- and Bio-systems

Springer

Masanori Ohya
Information Sciences
Tokyo University of Science
Yamazaki 2641
278-8510 Noda
Japan
ohya@rs.noda.tus.ac.jp

Igor Volovich
Mathematical Physics
Steklov Mathematical Institute
Gubkin St 8
119991 Moscow
Russia
volovich@mi.ras.ru

ISSN 1864-5879
e-ISSN 1864-5887
ISBN 978-94-007-0170-0
e-ISBN 978-94-007-0171-7
DOI 10.1007/978-94-007-0171-7
Springer Dordrecht Heidelberg London New York

Cover design: eStudio Calamar S.L.

Printed on acid-free paper

Springer is part of Springer Science+Business Media (www.springer.com)

Preface

Matter consists from atoms and elementary particles. It is desirable to get information about atoms and use it in various fields of technology. It was discovered in the twentieth century that elementary particles, atoms and molecules are governed by quantum laws which are different from the classical Newton mechanics. The information treated at the level of atom is called quantum information which is based on the description by quantum states of various quantum systems.

Achievements of the fundamental research are going to be useful in the technological industrial applications. Nanotechnology or quantum technology is the ability to engineer at atomic or molecular levels, it is the creation of materials, devices and systems through the control of matter on the nanometer (one-billionth (10^{-9}) meter) scale. Nanotechnology requires quantum mechanics and quantum information theory to reach its goals.

Quantum information theory is a rather old subject. Quantum mechanics was created by Heisenberg, Schrödinger, Dirac and others in the 1920s. Quantum information theory is based on quantum mechanics, and it has its origins in the 1930s when von Neumann introduced quantum entropy and Einstein, Podolsky and Rosen considered a sort of information gained as a result of measurement on the entangled particle.

In 1950s, in particular after Shannon, the sciences named "information" and "communication" were extensively developed. In the same vein, the processing ability of computers has been increasing each year, and now information technology (IT) expands to almost all our daily life. However, one looks for more accurate, speedier computers and safer communication, and the present stage of computers and IT is not in fullness. Further development will be possible in some conceptually different category, whose candidate is due to quantum mechanics. Mathematical foundations of quantum information were considered by Umegaki, Stratonovich, Ingarden and others in the early 1960s and developed by many other researchers. Important contributions to the modern development of quantum computing and nanoscience was made by Feynman. Information theory and technology based on quantum mechanics are called quantum information theory and nanotechnology, respectively.

In this book, we will discuss some mathematical features of quantum information and quantum computer (mainly computation) as mentioned above. Moreover, we will discuss the following topics: (1) basics of classical and quantum probability, functional analysis, stochastic analysis; (2) mathematical foundations of classical and quantum communication theory; (3) quantum entropy, relative entropy, mutual entropy, information and adaptive dynamics and their use in information communication and description of chaos; (4) Bell's type inequalities, quantum entanglement and their dependence on spatial variables; (5) various quantum algorithms and a new algorithm for quantum computations which goes beyond the quantum Turing machine paradigm, with an application to NP complete problems; (6) classical and quantum cryptography; (7) quantum teleportation; (8) some topics in quantum measurements, quantum electrodynamics and quantum fields; (9) applications to life science.

Further, we will consider applications of these mathematical methods to quantum dots and some topics in nanoscience which have the potential to be useful in future technology. We will also discuss some bio-systems in terms of mathematics treated in this book.

These areas are being quickly developed, and the book does not pretend to give a complete description of these topics. We try to describe the mathematical foundation of these fields.

Some of these topics have been already discussed in several textbooks; see the bibliography. This book contributes to the consideration of these topics by treating them from the mathematical points of view. For instance, (1) our description is basically done in an infinite dimensional Hilbert space which is essential for studying quantum systems; (2) the dynamics of systems and its observation are considered from their modes of existence; (3) we investigate some basic notions of quantum information from the point of view of the quantum field theory, e.g., we investigate the spatial dependence of the entangled states which leads to a new perspective to the so-called quantum non-locality and Bell's theorem.

Various applications of mathematics developed in quantum and quantum-like methods to bio-systems will be considered, including the recent experimental discovery of quantum effects in photosynthesis.

The immensely long DNA (sequence of the four bases in the genome) contains information on life, and decoding or changing this sequence is involved in the expression and control of life. In quantum information, meanwhile, we produce various "information" by sequences of two quantum states, and think of ways of processing, communicating and controlling them. It is thought that the problems we can process in time T using a conventional computer can be processed in time nearly $\log T$ using a quantum computer. However, the transmission and processing of information in the living body might be much faster than of quantum information. Seen from this very basic viewpoint, developing the mathematical principles that have been found in quantum information should be useful in constructing mathematical principles for life sciences which have not been established yet. The mechanism of processing information in life is also expected to be useful for the further growth of quantum information.

So it will be very natural to try to find a new bridge between quantum information and life science [41–43]. We believe that the mathematics or mathematical notions discussed in this book are or will become important for bioscience and further developments of quantum information and nanotechnology.

Interrelation between mathematics, information and life science as it is presented in this book is shown in the following figure.

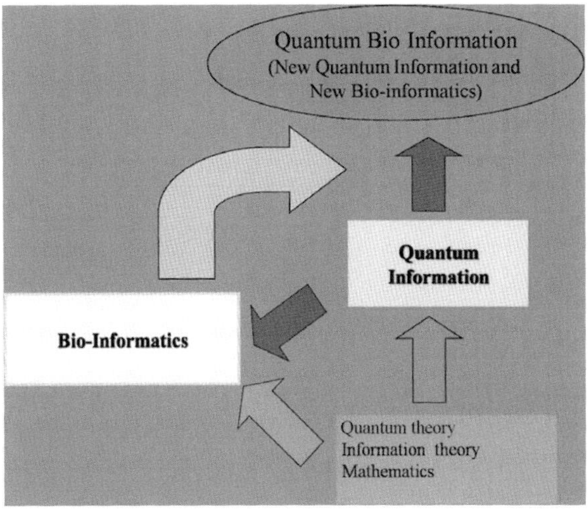

Acknowledgements

We are very grateful to Professors L. Accardi, I.Ya. Aref'eva, V. Belavkin, K.-H. Fichtner, W. Freudenberg, T. Hida, A. Khrennikov, A. Kossakowski, A. Jamiołkowski, W.A. Majewski, I. Ojima, D. Petz, I. Yamato for fruitful discussions, and we also thank N. Watanabe, S. Akashi, S. Kozyrev, T. Matsuoka, K. Sato, K. Inoue, S. Iriyama, A. Pechen, A. Truschechkin, M. Asano, T. Miyadera, T. Hara, Y. Hirota, Y. Tanaka for reading the draft of this book, giving comments and making the bibliography.

Noda, Japan — Masanori Ohya

Moscow, Russia — Igor Volovich

Contents

Chapter 1
Introduction

"My greatest concern was what to call it. I thought of calling it 'information', but the word was overly used, so I decided to call it 'uncertainty'. When I discussed it with John von Neumann, he had a better idea. Von Neumann told me, 'You should call it entropy, for two reasons. In the first place, your uncertainty function has been used in statistical mechanics under that name, so it already has a name. In the second place, and more important, nobody knows what entropy really is, so in a debate you will always have the advantage."—From a conversation between Claude Shannon and John von Neumann regarding what name to give to the "measure of uncertainty" or attenuation in phone-line signals (1949).

Quantum theory, i.e., quantum mechanics and quantum field theory, is a modern fundamental physical theory. Information theory and technology based on quantum mechanics is called quantum information theory or technology, and its first step was made by von Neumann introducing quantum entropy of a density operator 20 years ahead of Shannon. Quantum information theory constitutes a part of quantum theory which deals with the notions of quantum state, channel, measurement, and entropy.

In this book, we present the mathematical foundations of quantum information and quantum computation in various aspects such as quantum entropy, algorithm, entanglement, communication, cryptography, teleportation, and nanosystems. We describe in some details not only the material which has already become standard in the field, but also the results related with the authors' interests such as quantum mutual entropy, the space–time dependence of quantum entangled states, a new paradigm for quantum computations which goes beyond the quantum Turing machine, mathematical aspects of theory of nano- and bio-systems, relativistic quantum information, adaptive dynamics.

It should be pointed out here that many researchers recently discussed topics of quantum information and quantum computation in finite dimensional Hilbert spaces. However, the essential feature of quantum mechanics is based on the infinite dimensionality of a Hilbert space (for example, the spatial dependence of the system), so that it is desirable for the theory of quantum information and computation to be described in infinite dimensional Hilbert spaces. Thus the present book treats things in an infinite dimensional Hilbert space as much as possible by anticipating future development.

M. Ohya, I. Volovich, *Mathematical Foundations of Quantum Information and Computation and Its Applications to Nano- and Bio-systems*,
Theoretical and Mathematical Physics,
DOI 10.1007/978-94-007-0171-7_1, © Springer Science+Business Media B.V. 2011

This introduction is devoted to explain some fundamentals. Furthermore, we will describe some aspects of quantum information and quantum computer, historically, conceptually and intuitively in this introduction.

1.1 Entropy and Information Communication

Physics is considered as the "theory of matter", or equivalently, the "theory of existence in itself". *Information theory* (entropy theory) is considered as the "theory of events" as the probability theory, so that it will be considered as the "theory of changes". *Quantum information* can be regarded as a synthesis of these two theories. The key concept of quantum information bridging between matter and event, so between two modes of being, is "entropy", which was introduced by Shannon in classical systems and by von Neumann in quantum systems.

According to Shannon, information is related to uncertainty in a way that dissolution of uncertainty can be regarded as acquisition of information, so that information is described by entropy:

$$\textit{Information} = \textit{Uncertainty} = \textit{Entropy}.$$

Historically, the concept of entropy was introduced to describe the flow of heat, then it was recognized that the entropy describes chaos or uncertainty of a system. A system is described by a state. A state is described by a probability distribution or a measure in classical systems (Chap. 3) and a density operator in quantum systems (Chap. 5), which is a rather abstract concept not belonging to an object (observable) to be measured, but a means to get measured values. The entropy defined through a state of a system implies that the entropy is not an object considered in the usual "objective" physics, and it is an existence coming along the action of "observation".

The channel is a key concept for the description of any state change both in classical and quantum systems, which will be extensively discussed and used in several chapters of this book.

Communication of information and computer processing are expressed in the following figure.

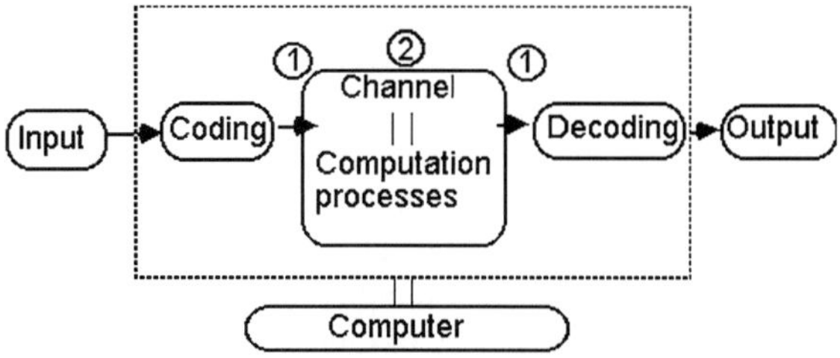

Information is described by a state with a suitable coding, and it is sent through a channel, whose outcome is decoded to receive original information.

Now let us mention briefly the mathematical expression of a quantum state and entropies. Quantum-mechanical description starts from assigning a Hilbert space $\mathcal{H}$ to the physical system. A *state* of the system is described by a density operator, i.e., a positive operator ρ of trace one, $\operatorname{tr}\rho = 1$. Pure states are represented by unit vectors in $\mathcal{H}$. An observable of the system is represented by a self-adjoint operator A in $\mathcal{H}$. Its expectation value is given by $\operatorname{tr}(A\rho)$. We consider here the operator A with the eigenvalues λ_n and the resolution of unity $E_n, n = 1, 2, \ldots$; we consider here the discrete spectrum. In the more general case, one speaks about a positive operator-valued measure (POVM) as an observable. The time evolution of states is described by the Schrödinger equation and the corresponding unitary operators.

The results of the measurements of the system are given by the probability distribution $\operatorname{Prob}(\lambda_n) = \operatorname{tr}(E_n\rho)$. According to the von Neumann–Luders *projection postulate*, the state ρ transforms after the measurement to the state $E_n\rho E_n / \operatorname{tr}(E_n\rho)$. This state transformation and also a unitary time evolution are examples of *channels*. A channel Λ^* is a linear mapping on the algebra of observables; its dual mapping Λ which sends states to states, is also called a channel. Any channel can be given in the Kraus–Sudarshan form $\Lambda^*\rho = \sum_k A_k\rho A_k^*$, $\sum_k A_k^* A_k \le 1$, where A_k are some operators.

An important characterization of information about a state is the von Neumann entropy $S(\rho) = -\operatorname{tr}\rho\log\rho$. It might decrease under the channel transform transformation. The uncertainty of a state ρ with respect to another state σ is represented by the *relative entropy* $S(\rho, \sigma) = \operatorname{tr}\rho(\log\rho - \log\sigma)$ which was introduced by Umegaki in 1962 [763]. A further development was performed by several researchers, and it will be discussed in Chap. 7. The theory of quantum information is now very much based on quantum probability which has been started by von Neumann and then by Umegaki in the early 1960s and developed by many other researchers. The relative entropy is a kind of distance between two states. The von Neumann entropy may be expressed in terms of the relative entropy:

$$S(\rho) = \sup\left\{\sum_i \mu_i S(\rho_i, \rho);\ \rho = \sum_i \mu_i\rho_i\right\}$$

where the least upper bound is taken over all finite convex orthogonal decompositions of the state ρ. Based on this formula and on the ideas of Kolmogorov, Gelfand and Yaglom in the classical information theory, Ohya in 1983 introduced the *quantum mutual entropy* with a channel Λ^* sending a quantum state to another quantum state

$$I(\rho; \Lambda^*) = \sup\left\{\sum_i \mu_i S(\Lambda^*\rho_i, \Lambda^*\rho);\ \rho = \sum_i \mu_i\rho_i\right\}.$$

The mutual entropy represents the amount of information correctly transmitted through the channel.

It seems that the notion of information in the classical information theory is clear enough intuitively, i.e., it has two aspects: one of which is the quantitative aspect, the amount of information described by the entropy, and the other is a message represented as a sequence of symbols (letters). The notion of quantum information is more subtle because in quantum theory the role of the observer is important. In a broad sense, *quantum information is the theory of entropy and dynamics for quantum states*. We will carefully discuss such various aspects of quantum information in this book. The details of quantum entropy and its use are discussed in Chaps. 7 and 9.

1.2 Classical Computer Versus Quantum Computer

Let us compare a classical computer with a quantum computer.

Classical computer (CC)

- Input is expressed by a sequence of 0s and 1s, and each digit is respectively translated into "OFF" and "ON" of electrical signals.
- Turing machine (we call it the classical Turing machine (CTM) here), equivalently circuits: It reads electrical signals 0, 1 and changes the signals according to processing commands, and it saves the results in the memory and makes outputs. Here the processing commands are running by the fundamental gates such as `NOT`, `AND`, `OR`, and `XOR`.

Quantum computer (QC)

- Input is expressed by a sequence of 0s and 1s, which are translated into two different quantum states.
- Quantum Turing machine (QTM), equivalently quantum circuits: It reads quantum states and changes the states according to quantum dynamics.

The basic difference between a classical computer and a quantum computer is that there exists a coherence among states in the course of processing, whose precise meaning will be discussed in Chaps. 2 and 11.

In a classical computer there exist inevitable demerits such as

(1) Strong effect from thermal noise due to electric current.
(2) Information loss by an irreversible gate that is actually used in a classical computer.

1.2.1 Logical Irreversibility and Energy Loss

Let us discuss information loss by irreversible gates. Although it is possible to make reversible gates even for a classical computer, such a reversible gate needs more resources for computation, so that such gates are not used in a usual classical computer. On the other hand, quantum computation is, in principle, done by unitary transformations, so that it is reversible.

Now, we show how much information will be lost in classical gates.

Let us take an `AND/OR` gate Fig. 1.1 to discuss this problem.

In a `NOT` gate, the information is conserved, but in an `AND/OR` gate information is lost.

I_1	I_2	O_{AND}	O_{OR}
0	0	0	0
0	1	0	1
1	0	0	1
1	1	1	1

Fig. 1.1 AND/OR gate and truth table

1.2.2 Thermodynamic Interpretation of Irreversible Computation

Landauer and von Neumann considered that this information loss is related to heat generation. They explained it as follows: The information of input and that of output are respectively given by

$$S_{\text{in}} = k \log 4, \qquad S_{\text{out}} = k \log 2$$

where k is the Boltzmann constant. Hence the loss of information turns out to be

$$\Delta S = -k \log 2 = \frac{\Delta Q}{T},$$

and Landauer and von Neumann considered that it causes generation of heat ΔQ given by

$$\Delta Q = -kT \log 2 \quad \text{(energy loss)}$$

where T is the temperature of the system.

1.2.3 Information Theoretic Interpretation

After Shannon, the information-theoretic explanation of the above information loss should be:

$$S_{\text{in}} - I \text{ (mutual entropy)} = \frac{3}{4} \log 3$$
$$\Rightarrow \text{information loss.}$$

The classical entropy S_{in} and the mutual entropy I above will be discussed in Chap. 6. Reversible classical computation was discussed by Bennett, and he showed that the computational complexity is much more than irreversible one.

1.2.4 Resolution of Demerits (1), (2) by Quantum Computer

We can resolve the demerits of a classical computer by a quantum computer: Demerit (1) can be avoided because quantum noise is very small. Demerit (2) is re-

Fig. 1.2 NOT gate

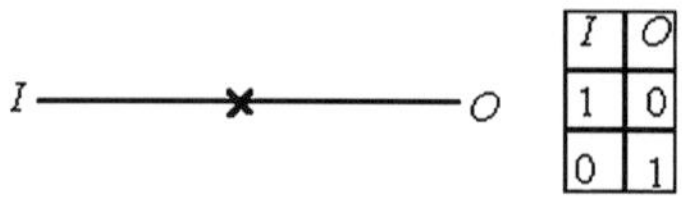

solved because state is, in principle, changed by unitary dynamics in a quantum computer, so that information (entropy) is preserved in the process of computation as

$$I(\rho; U^* \cdot U) = S(\rho),$$

where $S(\rho)$ is the von Neumann entropy of an input state ρ, and I is the quantum mutual entropy, both discussed in Chap. 7.

1.3 What Are Quantum Gates and Computation?

1.3.1 Gates

Quantum computation is, in principle, done by reversible quantum gates (Fredkin–Toffoli gate, Feynman gate, and other gates), that is, unitary transformations.

We here explain the fundamental three gates, originally given by Feynman. A more general discussion on gates is given in Chaps. 2 and 10. Let 0 be described by a state vector $|0\rangle = \binom{1}{0} \in \mathbb{C}^2$ and 1 by a vector $|1\rangle = \binom{0}{1} \in \mathbb{C}^2$ by Dirac's bra–ket notation

(1) NOT gate is the gate to compute the function

$$f_{\mathrm{NOT}}(x) = 1 - x \ (\mathrm{mod}\ 2), \quad x = 0, 1$$

so that the corresponding truth table is as in Fig. 1.2. The unitary operator describing this gate is written by

$$U_{\mathrm{NOT}} = |0\rangle\langle 1| + |1\rangle\langle 0|$$

where, for instance, $|0\rangle\langle 1| = \binom{1}{0}(0\ 1) = \begin{pmatrix} 0 & 1 \\ 0 & 0 \end{pmatrix}$ and so

$$U_{\mathrm{NOT}} = \begin{pmatrix} 0 & 1 \\ 1 & 0 \end{pmatrix}.$$

(2) Controlled NOT gate (Exclusive OR (XOR)) is the gate to compute the function

$$f_{\mathrm{XOR}}(x, y) = (x, x \oplus y), \quad x \oplus y \equiv x + y \ (\mathrm{mod}\ 2).$$

Its truth table and the unitary operator are given as

$$U_{\mathrm{CNOT}} = |0\rangle\langle 0| \otimes I + |1\rangle\langle 1| \otimes \big(|0\rangle\langle 1| + |1\rangle\langle 0|\big).$$

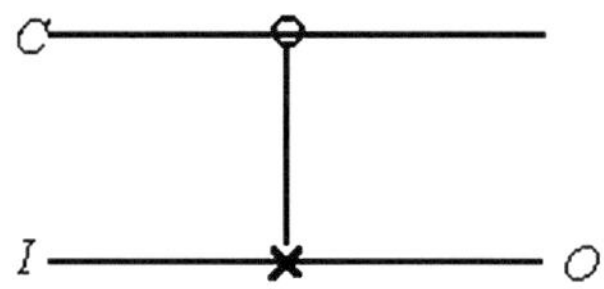

C	I	O
1	1	0
1	0	1
0	1	1
0	0	0

CNOT gate

(3) Controlled–Controlled NOT gate is the gate to compute the AND function

$$f_{\text{AND}}(x, y, 0) = (x, y, x \wedge y)$$

and the OR function

$$f_{\text{OR}}(x, y, 0) = (\bar{x}, \bar{y}, x \vee y),$$

where $\bar{x} = 1 - x$, $x \wedge y = \min\{x, y\}$, and $x \vee y = \max\{x, y\}$. Its truth table and the unitary operator are given below:

$$U_{\text{CCN}} = |0\rangle\langle 0| \otimes I \otimes I + |1\rangle\langle 1| \otimes |0\rangle\langle 0| \otimes I$$
$$+ |1\rangle\langle 1| \otimes |1\rangle\langle 1| \otimes \big(|0\rangle\langle 1| + |1\rangle\langle 0|\big).$$

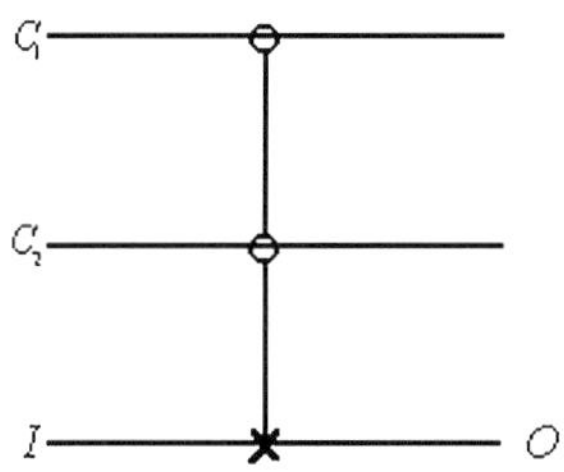

C_1	C_2	I	O
1	1	1	0
1	1	0	1
1	0	1	1
1	0	0	0
0	1	1	1
0	1	0	0
0	0	1	1
0	0	0	0

C-C-N gate

AND gate can be realized by exchanging the role of the gates as follows:

	C_1	C_2	I	0
AND	Input 1	Input 2	0	Output

(4) Other gates to be used for computation can be constructed by the above three gates.

The details on gates and algorithms are discussed in several successive chapters (e.g., Chaps. 2, 11–15).

1.3.2 Quantum Algorithm and Computation

Let us explain the basic idea of quantum computation by comparing it with classical computation. In a classical computer, a state consists of the signals 1 or 0, i.e., a condenser is physically charged or not. In a quantum computer, a state symbolically is 1 or 0, or a superposition of these two; physically, such a state (vector) is expressed by

$$|\xi\rangle = \alpha|0\rangle + \beta|1\rangle; \qquad |\alpha|^2 + |\beta|^2 = 1,$$

which is called a qubit.

When the quantum vectors $|1\rangle$ and $|0\rangle$ are represented by spin-up and spin-down, the superposition of these two vectors is written as

$$\cos\theta|0\rangle + \sin\theta|1\rangle.$$

An input (signal) in a classical computer is described by a sequence $(a_1, \ldots, a_N)$ with each $a_k = 1$ or 0, so that we have 2^N such vectors altogether. On the other hand, an input of a quantum computer can be generally written as $|a_1\rangle \otimes \cdots \otimes |a_N\rangle = \bigotimes_{k=1}^{N}(\alpha_k|0\rangle + \beta_k|1\rangle)$ with $a_k = 1$ or 0 and $|\alpha_k|^2 + |\beta_k|^2 = 1$ for every k.

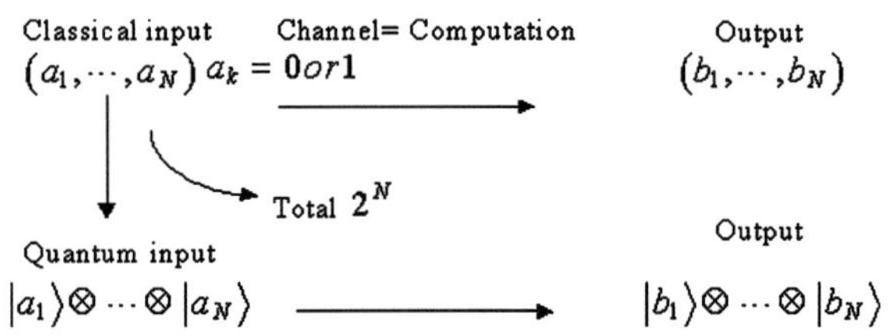

After passing through a channel (computation), the output will be $(b_1, \ldots, b_N)$ and $|b_1\rangle \otimes \cdots \otimes |b_N\rangle$ in classical and quantum settings, respectively.

Here we take the input as

$$|a_1\rangle \otimes \cdots \otimes |a_N\rangle = \bigotimes^{N}\left(\frac{|0\rangle + |1\rangle}{\sqrt{2}}\right) = \frac{1}{\sqrt{2^N}} \bigotimes_{k=0}^{2^N-1} |\xi_k^{(N)}\rangle,$$

where

$$|\xi_k^{(N)}\rangle = |a_1^{(k)}\rangle \otimes \cdots \otimes |a_N^{(k)}\rangle.$$

Let us now consider changing the signs of all a_j^k, that is, the transformation $a_j^k \to -a_j^k$ for all j and k. We have the following complexity.

(1) In the classical case, we need to determine the sign of $(a_1^{(k)}, \ldots, a_N^{(k)})$ for each k. Therefore, the complexity will be 2^N in CC.
(2) In the quantum case, we have only to operate the unitary (N-tuple tensor product of it)

$$\begin{pmatrix} 1 & 0 \\ 0 & -1 \end{pmatrix} \implies |0\rangle \to |0\rangle, \qquad |1\rangle \to -|1\rangle.$$

Thus the complexity is N (the times operating the unitary) in QC.

A mathematical model of a quantum computer as the quantum Turing machine or a quantum circuit was proposed by Deutsch [197]. A quantum Turing machine is given by a quadruplet $\{Q, \Sigma, \mathcal{H}, U\}$. Here Q is a set of states, Σ is an alphabet space, $\mathcal{H}$ is a Hilbert space of complex functions on the configuration space of the classical Turing machine. The unitary operator U is constructed from elementary unitary operators, the so-called quantum gates. The form of this construction defines a quantum program.

Although the ability of a computer has highly progressed, there are several problems which may not be solved effectively, that is, in polynomial time. Among such problems, the NP problem and the NP-complete problem are fundamental. It is known that all NP-complete problems are equivalent, and an essential question is "whether there exists an algorithm to solve an NP-complete problem in polynomial time". Quantum computations were considered by several pioneers like Feynman, Manin, Benioff, Deutsch in 1980. After their pioneering works, several important works have been done on quantum algorithms by Shor, Grover, Bennett, and many others. It was shown by Shor [710] that by means of a quantum computation one can solve the problem of factoring of integers in the polynomial time. Ohya, Masuda and Volovich showed an algorithm solving an NP-complete problem in polynomial time, which goes beyond the usual quantum Turing machine, and it uses, instead of the unitary transformation, more general channels with a chaotic amplifier. This new computational model is called a generalized quantum Turing machine [373], which contains both classical and quantum Turing machine. The computation in a quantum computer is performed on a tensor product Hilbert space, and its fundamental point is to use quantum coherence (entanglement) of states. Therefore, the realization of a quantum computer is strongly based on maintenance of this coherence whose technology will be very hard to realize. Various quantum algorithms such as Shor, Grover, Ohya–Masuda–Volovich are discussed in Chaps. 11–14.

1.4 Locality and Entanglement

The entanglement or the entangled state is an old and important concept of quantum mechanics coming from interference of two different states.

Let φ, φ_j $(j = 1, 2)$ be the state vectors of the coupled system with the assigned Hilbert space $\mathcal{H}_A \otimes \mathcal{H}_B$ given as $\varphi \equiv \alpha\varphi_1 + \beta\varphi_2 \equiv \alpha\psi_A \otimes \psi_B + \beta\omega_A \otimes \omega_B$ with $|\alpha|^2 + |\beta|^2 = 1$. This state φ is entangled if the state of the system A being ψ_A (resp., ω_A) implies that the state of the system B is ψ_B (resp., ω_B) with probability 1. Otherwise, the state φ is said not to be entangled. For instance, the state $\varphi = \alpha|0\rangle \otimes |1\rangle + \beta|1\rangle \otimes |0\rangle$ is an entangled state, but the state $\varphi = \alpha|0\rangle \otimes |1\rangle + \beta|1\rangle \otimes |1\rangle$ is not. If Alice and Bob can share such an entangled state, then they can use it as a key because one of them can know the other's hand when he (or she) reads his (or her) hand. The entangled states and the reduction postulate are essential for discussing quantum cryptography and quantum teleportation.

One of the sources of quantum information is the article of Einstein, Podolsky, and Rosen [218] "Can quantum-mechanical description of physical reality be considered complete?". In this article, paradoxical properties of entangled states of two particles were analyzed. The two particle Hilbert space $\mathcal{H}$ is the tensor product of the one-particle Hilbert spaces $\mathcal{H}_1$ and $\mathcal{H}_2, \mathcal{H} = \mathcal{H}_1 \otimes \mathcal{H}_2$. An *entangled* pure state ψ in $\mathcal{H}$ is such a unit vector which is not the tensor product of vectors from $\mathcal{H}_1$ and $\mathcal{H}_2$, that is, a sum of at least two terms, $\psi = \varphi_1 \otimes \chi_1 + \varphi_2 \otimes \chi_2$.

The correlation function of spins of two particles in an entangled spin state ψ_{spin} can be computed as

$$\langle \psi_{\text{spin}} | s(\alpha) \otimes s(\beta) | \psi_{\text{spin}} \rangle = -\cos(\alpha - \beta).$$

Here $s(\alpha)$ and $s(\beta)$ are the operators of spins of the first and the second particles depending on the angles α and β, respectively. It was shown by Bell (1965) that $\cos(\alpha - \beta)$ cannot be represented in the form $\int \xi(\alpha, \lambda)\eta(\beta, \lambda)\, dP(\lambda)$ where $dP(\lambda)$ is the probability measure on the space of the "hidden variables" λ, and $\xi(\alpha, \lambda)$ and $\eta(\beta, \lambda)$ are appropriately bounded functions. It follows from the so-called Bell's inequalities for the classical correlation functions. This result and appropriate experiments were interpreted as non-compatibility of quantum mechanics with local realism and as manifestation of quantum non-locality.

However, in the previous formula for the quantum mechanical correlation function of spins of the particles, the spatial dependence of the wave function, which is relevant for the consideration of locality, was neglected. A modification of the quantum mechanical correlation function which takes into account the space–time dependence was obtained by Volovich [790]. If one makes quantum mechanical computation with the wave function $\psi = \psi(x, y)$ depending on the spatial variables x and y then, in the simplest case, one obtains the following relation

$$\langle \psi | s(\alpha)E(x) \otimes s(\beta)E(y) | \psi \rangle = g(x, y) \cos(\alpha - \beta),$$

where $E(x)$ and $E(y)$ are the projection operators on the locations of the detectors for the first and the second particle, and the factor $g(x, y)$ describes the location of the detectors of the particles at the points x and y in a three-dimensional space. The function $g(x, y)$ vanishes with a large separation between the detectors. One can prove that if the distance between the detectors is large enough then there exists a representation of this quantum-mechanical correlation function of the form

$$\langle \psi | s(\alpha)E(x) \otimes s(\beta)E(y) | \psi \rangle = \int \xi(\alpha, x, \lambda)\eta(\beta, y, \lambda)\, dP(\lambda).$$

This result shows that quantum mechanics is compatible with an appropriate understanding of local realism at large distances. It is important that in this representation the classical functions (*random fields*) $\xi(\alpha, x, \lambda)$ and $\eta(\beta, y, \lambda)$ depend on the spatial variables x and y, and hence they are local functions. These random fields are classical analogues of quantum fields.

Let us emphasize that the central notion in modern fundamental physics is the notion of a *quantum field* (and its extension to string theory). Therefore, quantum information theory should be based on quantum field theory, first of all on quantum electrodynamics. One of the aims of this book is an attempt to develop quantum information theory starting from quantum field theory. A quantum field is an operator valued function on the Minkowski space–time. In the quantum field theory, there is the fundamental *property of locality* (local commutativity) which for the scalar field $\Phi(x)$ reads

$$\left[\Phi(x), \Phi(y)\right] = 0,$$

if the points x and y are space-like separated. The above mentioned representation which includes classical random fields $\xi(\alpha, x, \lambda)$ and $\eta(\beta, y, \lambda)$ is consistent with the property of locality in the quantum field theory. We will discuss implications of these results to the Bell type experiments and quantum cryptography in Chaps. 8 and 17.

Various studies of separable and entangled states have been recently done by many researchers, and are discussed in Chap. 8.

1.5 What Are Quantum Cryptography and Teleportation?

Let us discuss the basic idea of quantum cryptography and quantum teleportation as a conceptual introduction to Chaps. 17 and 18.

1.5.1 Cryptography

A well known example of a cryptosystem is the Caesar cipher. Julius Caesar allegedly used a simple letter substitution method. Each letter of Caesar's message was replaced by the letter that followed it alphabetically by 3 places. This method is called the Caesar cipher. The size of the shift (3 in this example) should be kept secret. It is called the *key* of the cryptosystem. It is an example of the traditional cryptosystem. It is also called the *private key cryptography*. Anyone who knows the enciphering key can decipher the message. A mathematical theory of classical cryptography has been developed by Shannon.

There is a problem in the private key cryptography which is called the *problem of key distribution*. To establish the key, two users must use a very secure channel. In the *classical world*, an eavesdropper, in principle, can monitor the channel without the legitimate users being aware that eavesdropping has taken place.

In 1976, Diffie and Hellman discovered a new type of a cryptosystem and invented public key cryptography, by which the problem of key distribution was solved. The well-known public key cryptosystem is called RSA, and it is widely used on the Internet and other businesses.

Wiesner and Bennett–Brassard have proposed the idea of quantum cryptography. They used the sending of single quantum particles, eigenstates of noncommutative observables, and their quantum cryptography can solve the key distribution problem. Moreover, it can detect the presence of an eavesdropper, so that we have a secure protocol for information communication.

Non-classical properties of entangled states and the projection postulate are used in quantum cryptography and quantum teleportation first considered by Bennett et al. in the 1990s and implemented in recent experiments. In 1991, using entanglement between Alice (system A) and Bob (system B) such that one can make cryptography read a signal between them, Ekert proposed using such a phenomena of entanglement and Bell's inequalities.

1.5.2 Teleportation

Quantum teleportation is sending a quantum state itself containing all information of a certain system from one place to another. If such an information transmission is realized, then this transmission can be one of ultimate methods because quantum state will not be stolen due to its fragility under observation; that is, quantum state is easily destroyed by observation, so one can easily notice that a state sent from a sender was stolen or not. The problem of quantum teleportation is whether there exist a physical means and a key (or a set of keys) by which a quantum state attached to a sender (Alice) is completely transmitted through an entangled device between Alice and Bob, and a receiver (Bob) can reconstruct the state sent. Bennett and his coworkers showed such a teleportation is possible through a device (channel) made from proper (EPR) entangled states. The basic idea behind their discussion is to divide the information encoded in the state into two parts, classical and quantum, and send them through different channels, a classical channel and the EPR channel. The classical channel is nothing but a simple correspondence between a sender and a receiver, and the EPR channel is constructed by Bell's entangled state. Accardi and Ohya generalized their scheme and discussed the necessity of maximal entangled states and the existence of unique key. A crucial role of the spatial dependence in quantum teleportation schemes was pointed out by Volovich. It is known that the EPR channel is not stable due to decoherence of the entanglement. Fichtner and Ohya studied the quantum teleportation by means of a general beam splitting with coherent states such that it contains the EPR channel as a special case, and we proposed a more stable teleportation model in a mathematically rigorous framework. In this model, the so-called Hida's derivative plays a fundamental role, so that the quantum teleportation is linked to white noise analysis. Furusawa and Kimble discussed the case of continuous variables. Kossakowski and Ohya introduced a new treatment of quantum teleportation, in which any teleportation process becomes linear with respect to an input state and can be perfect even for non-maximal entangled state between Alice and Bob. The quantum teleportation is discussed in Chap. 18.

1.6 Nanosystems and Biosystems

Nanotechnology is the control of matter on an atomic and molecular scale. It deals with structures of the size 100 nanometers or smaller, (1 nm $= 10^{-9}$ m). Nanotechnology became popular after the works of Feynman, Taniguchi and Drexler.

Nanotechnology and nanoscience got started in the 1980s with the birth of cluster science, the invention of the scanning tunneling microscope and the discovery of fullerenes and carbon nanotubes. One should mention the semiconductor nanocrystals, quantum dots and giant magnetoresistance.

Important examples of nanosystems are *quantum dots* which are quasi-zero-dimensional systems of dimensions of the order of 10–1000 nm that contain a small and controllable number of electrons. One has the possibility of controlling the properties of quantum dots, for example, their shape, their dimensions, and the number of confined electrons. Quantum dots are similar to atoms and often referred to as the artificial atoms. However, quantum dots do not have nuclei, and the lateral potential of a quantum dot differs from the Coulomb potential binding electrons in an atom. They have the shape of flat discs. In a good approximation, an electron in a quantum dot can be described by the Hamiltonian

$$H = \frac{1}{2m}\left(\mathbf{p} - \frac{e}{c}\mathbf{A}\right)^2 + \frac{k}{2}\mathbf{r}^2.$$

Here m is the effective mass, $\mathbf{r} = (x, y)$ is the position of the electron on the plane, $\mathbf{p} = (p_x, p_y)$ is the momentum, $\mathbf{A}$ is the vector potential of the magnetic field B, $\mathbf{A} = \frac{1}{2}B(y, -x, 0)$, and e, c, and k are the electric charge, the velocity of light, and a coupling constant. The eigenstates and the energy levels of the Hamiltonian were determined by Fock and Darwin in 1927 by using a transformation to a pair of harmonic oscillators. Quantum dots and some other nanosystems were recently proposed for possible implementations of quantum computers. Some nanosystems are discussed in Chap. 19.

The study of biosystems such as protein, DNA, other biomolecules and brain as such is very important for the genetics and life science. One looks for basic mathematics to describe "life". However, such mathematics is not yet found, so that it is important for us to try using various mathematics and their modifications to understand the mystery of "life". We show some attempts of building mathematical models of a brain and cognitive phenomena in Chap. 21 within the framework stated in the preface. In this chapter, we also discuss recent discoveries in quantum photosynthesis.

Chapter 2
Algorithms and Computation

In this chapter, we first discuss the principles of algorithm and computation in general framework, common both in classical and quantum computers, then we go to the fundamental topics of a Turing machine, algorithm, computation, circuits, and NP-complete problems.

2.1 General Algorithms

An algorithm is a precise formulation of doing something. Algorithms play an important role in mathematics and in computers, and they are employed to accomplish specific tasks using data and instructions. The notion of an algorithm is old; there is, for example, the well known Euclid's algorithm for finding the greatest common divisor of two numbers. Let us exhibit Euclid's algorithm here.

Euclid's algorithm. Given two positive integers m and n, the task is to find their greatest common divisor, i.e., the largest positive integer which divides both m and n. Here m and n are interpreted as variables which can take specific values. Suppose that m is greater than n. The algorithm consists of three steps.

Step 1. Divide m by n and let r be the remainder.
Step 2. If $r = 0$, the algorithm halts; n is the answer.
Step 3. Replace the value of m by the current value of n, also replace the value of n by the current value of r and go back to Step 1.

An algorithm has *input*, i.e., some quantity which is given to it initially before the algorithm begins. In Euclid's algorithm, the input is a pair of two positive integers m and n. An algorithm also has *output*, i.e., some quantity which has a specified relation to the input. In Euclid's algorithm, the output is n in Step 2, which is the greatest common divisor of two given integers.

Exercise 2.1 Prove that the output of Euclid's algorithm is indeed the greatest common divisor.

M. Ohya, I. Volovich, *Mathematical Foundations of Quantum Information and Computation and Its Applications to Nano- and Bio-systems*,
Theoretical and Mathematical Physics,
DOI 10.1007/978-94-007-0171-7_2,

There are various approaches to precise formulation of the concept of an algorithm. There exist classical and quantum algorithms. One of modern precise formulations of the notion of a classical algorithm can be given by using Turing machines. Another approach to algorithms is based on the notion of circuits. Classical circuits and classical Turing machines are used as mathematical models of a classical computer. Quantum circuits and quantum Turing machines are mathematical models of a quantum computer.

The concept of the Turing machine was introduced by A.M. Turing in 1936 [757] for the study of limits of human ability to solve mathematical problems in a formal way. *Any classical algorithm can be implemented on a Turing machine* (this is the so-called A. Church's thesis).

A Turing machine has two main parts: a *tape* and a central unit with a *head* ∇ (see figure below).

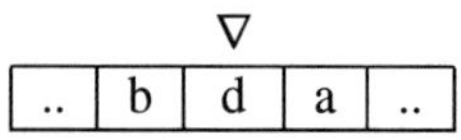

The tape is infinite in both directions and is divided into squares. Each square of the tape holds exactly one of the symbols from a finite set of symbols (this finite set of symbols is called an alphabet). The central unit with the head is in one of the states from a finite set of states (the precise meaning of a state will be fixed later). The head sees at any moment of time one square of the tape and is able to read the content of the square as well as to write on the square. The input is written as a string (sequence) of symbols on the tape. The head starts in a prescribed state. In a single move, the Turing machine can read the symbol on the one square seen by its head, and based on that symbol and its current state, replace the symbol by a difference one, change its state, and move the head one square to the left, or one square to the right, or stay on the same square as before.

A sequence of moves is called a computation. For some pairs of states and symbols on the tape, the machine halts. In this case, symbols remaining on the tape form the output corresponding to the original input. A Turing machine accepts some input strings if it halts on it. The set of all accepted strings is called a language accepted by the Turing machine. Such languages are called recursively enumerable sets.

The Turing machine is a suitable model for the computational power of a classical computer. Its usefulness follows from the Church's thesis which may be reformulated as follows: *The computational power of the Turing machine represents a limit for any realizable classical computer.*

Let us indicate now one method which is general enough to include classical as well as quantum algorithms. Let us take two sets I and O. The set I will represent the input, and the set O represents the output of our computation. Suppose the sets I and O are parts of a larger set $\mathcal{X}$ which will represent configurations of computation. Let $G = \{g_1, \dots, g_r\}$ be a finite set of functions g_i from $\mathcal{X}$ to $\mathcal{X}$. Such functions are called *gates* in computing and G is called the basis of gates. They form the primitive elements from which we will design an algorithm. For example, the gates can represent the basic logical operations such as AND, OR, and NOT as discussed in Chap. 1. Now let us be given a function f which maps the input set I to the output set O. Our problem is to find a sequence of gates $A_G = \{g_{i_1}, g_{i_2}, \dots, g_{i_k}\}$ which computes the function f in the sense that the function can be represented as a composition of gates, i.e., for any input $x \in I$ one has $f(x) = g_{i_1} \circ g_{i_2} \circ \cdots \circ g_{i_k}(x)$. The sequence A_G is called the algorithm or the program of computation using G.

Each input x in the set I defines a computational sequence, $x_0, x_1, \dots,$ as follows: $x_0 = x$, $x_1 = g_{i_1}(x_0), \dots, x_m = g_{i_m}(x_{m-1}), \dots$. One says that the computational sequence terminates in n steps if n is the smallest integer for which x_n is in O, and in this case it produces the output $y = x_n$ from x. One says that the algorithm computes the function $y = f(x)$.

A more general approach would be if one admits that the functions g_i and the function f are not defined everywhere (such functions are called partial functions) and that not every computational sequence terminates. Moreover, one can assume that the transition $x_m = g_{i_m}(x_{m-1})$ takes place with a certain probability (random walk) and that the output space O is a metric space with a metric τ. Then one says that the algorithm makes an approximate computation, in the order of ε, of a function $f(x)$ with a certain probability if one gets a bound $\tau(f(x), x_k) < \varepsilon$ for some $x_k \in O$.

To summarize, *the algorithm for the computation of the function f by using the prescribed set of gates is given by the following data $\{\mathcal{X}, I, O, G, A, f\}$ described above.*

In the considerations on this book, the set $\mathcal{X}$ for the classical Turing machine will be the set of all configurations of the Turing machine, and the gates g_i will form the transition function. For a classical circuit, the gates might, for example, be basic logical operations AND, OR and NOT. For a quantum circuit and for a quantum Turing machine, the set $\mathcal{X}$ might be the Hilbert space of quantum states and the gates g_i could be some unitary matrices and projection operators.

An important issue in computing is the computational complexity. One would like to minimize the amount of time and memory needed to produce the output from a given input. For an input x, let $t(x)$ be the number of steps until the computational sequence terminates. The computational time T of the algorithm is defined by

$$T(n) = \max_x \{t(x) : |x| = n\}$$

where $|x|$ is the length of the description of x. The actual length of the description depends on the model of computation. We are interested, of course, in minimizing the computational time $T(n)$.

2.2 Turing Machine

Now let us give a precise definition of the (classical) Turing machine. An *alphabet* A is a finite set of symbols (letters). For example, $A = \{0, 1\}$ or $A = \{\#, a_1, \dots, a_m\}$. Let A, Q and Γ be three different alphabets. The alphabet $A = \{\#, a_1, \dots, a_m\}$ is often called the tape alphabet. It should include the symbol # which is called the blank symbol.

Definition 2.2 A Turing machine M is a triple $M = \{Q, A, \delta\}$. Here $Q = \{q_0, q_1, \dots, q_r, q_F\}$ is called the set of states. It should include the initial state q_0 and the final state q_F. δ is the transition function defined by $\delta : Q \times A \to Q \times A \times \Gamma$, where the set Γ consists of three symbols $\Gamma = \{L, S, R\}$.

Remark 2.3 In Γ, the letters L, S and R mean that the tape head goes to the left, stays, and moves right, respectively. It is useful to use the notation $\{-1, 0, +1\}$ instead of $\{L, S, R\}$, so that we may use both in the sequel when the later is considered useful.

The above function δ is not necessarily defined on all elements of Q. Such not everywhere defined functions are called partial functions. Normally, δ is defined on all elements of A and on all elements of Q except q_F.

The tape of the Turing machine is a sequence of squares, i.e., the one-dimensional lattice which can be enumerated by integers. Each square of the tape contains a symbol of A. The central unit of the machine is at each moment of time in one state $q \in Q$.

At the beginning, the input $x = (x_1, \dots, x_n)$, $x_i \in A$, is contained in the squares labeled $0, 1, \dots, n-1$, all other squares contain the blank symbol #. The head starts in the state q_0. If the head is in the state q and reads a on the tape, and if $\delta(q, a) = (q', a', \gamma)$ then the machine replaces the contents of the considered square by a', the new state of the head is q', and the head moves one step in direction γ. Here if $\gamma = R$ (or $+1$) then the head moves to the right, if $\gamma = L$ (or -1) then the head moves to the left, and if $\gamma = S$ (or 0) then the head stays on the same square. The computation stops if the head is in the final state q_F. The result of computation $y = (y_1, \dots, y_s)$, where $y_i \in A$, can be read consecutively from the square starting at 0 until the first square containing the blank symbol #.

The Turing machine M transforms the input x into the output y. One can write $y = M(x)$. The computation does not necessarily halt on a given input x. Therefore, the function $M(x) = y$ is not necessarily defined on all inputs x. The following terminology is used. A *word* is a finitestring of symbols of the alphabet A, i.e., an ordered sequence $x_1 x_2 \cdots x_n$ where $x_i \in A$. The *length* $|w|$of the word $w = x_1 x_2 \cdots x_n$ is $|w| = n$. One can write $M(w) = v$ if the word w is an input, and the word v is the output for the Turing machine M. The function $v = M(w)$ is a partial function $M : A^* \to A^*$ from the set A^* of all words over the alphabet A to A^*.

Example 2.4 Let us design an adder, i.e., a Turing machine which makes addition of two natural numbers.

One takes $M = \{Q, A, \delta\}$ where the head alphabet $Q = \{q_0, q_1, q_F\}$ has three states, and the tape alphabet $A = \{\#, 1, +\}$ also has three symbols. If we want to compute the sum, say, of 2 and 3, then we will write the input word as $11 + 111$. The output should be 11111, i.e., 5. Our machine in this case acts as $M(11+111) = 11111$, and similarly for addition of arbitrary natural numbers n and m. The program is given by the following transition function δ:

$$\begin{aligned} \delta(q_0, \#) &= (q_1, \#, L), \qquad & \delta(q_0, 1) &= (q_0, 1, R), \\ \delta(q_0, +) &= (q_0, 1, R), \qquad & \delta(q_1, 1) &= (q_F, \#, S). \end{aligned} \tag{2.1}$$

The program can be given also by the table below.

	#	1	+
q_0	$q_1, \#, L$	$q_0, 1, R$	$q_0, 1, R$
q_1		$q_F, \#, S$	

(2.2)

The machine starts in the state q_0 and looks at the left symbol 1. Then it moves to the right, finds the symbol + and prints instead of it the symbol 1. Then it goes up to the end, i.e., until reaches #, and finally moves one step to the left, prints # instead of 1 and halts.

Let us stress that the Turing machine is finite, it has only 9 symbols and the program is also finite, but the machine can make addition of two arbitrary natural numbers n and m.

Exercise 2.5 Design a Turing machine for the subtraction of two integers.

At every moment of time, the state of the Turing machine can be described as a *configuration* $C = (q, i, w)$ where q is the state of the head, the integer i is the number of the square on the tape to which the head looks, and w is the word on the tape formed by non-blank symbols. The program δ transforms one configuration into another. Computation on the Turing machine is a sequence of configurations.

The set of all words over A is denoted by A^*. A *language* L is a subset of A^*. A Turing machine *decides a language* L if it computes 1 for every input $A \in L$ and 0 if $A \notin L$. This means that the Turing machine computes the characteristic function of the language L. Here we assume that 0 and 1 belong to the alphabet A. A language is called *decidable* if there exists a Turing machine which computes its characteristic function. If a mathematical problem can be formulated as a decidable language then it is called a *solved problem in the sense of Turing*. Otherwise, the problem is called *unsolvable in the sense of Turing*. After the next section, we shall discuss an example of an unsolvable problem (the halting problem).

2.2.1 Gödel Encoding

Let us study the computational power of Turing machines. To this end, it is useful to enumerate all Turing machines in a special way. We have described the program of the Turing machine as a table or a matrix $(\delta(q_i, a_j))$ for all $q_i \in Q$ and $a_j \in A$. Now we will represent the program as a string. If $\delta(q, a) = (q', a', \gamma)$ then we will write the quintuplet (q, a, q', a', γ) where the first two symbols denote the arguments of the function δ and the last three symbols denote its value. The program δ can be written as a string P of quintuplets.

For example, the program of the Turing machine from Example 2.4 can be written as

$$P = (q_0, \#, q_1, \#, L), (q_0, 1, q_0, 1, R), (q_0, +, q_0, 1, R), (q_1, 1, q_F, \#, S).$$

One can remove the brackets. Then we get

$$P = q_0, \#, q_1, \#, L, q_0, 1, q_0, 1, R, q_0, +, q_0, 1, R, q_1, 1, q_F, \#, S.$$

Now if one encodes the symbols of our Turing machine by natural numbers, for example, as

$$\begin{array}{lllll} q_0 \to 1, & q_1 \to 2, & q_F \to 3, & \# \to 4, & \\ 1 \to 5, & + \to 6, & R \to 7, & L \to 8, & S \to 9, \end{array}$$

then one can represent the program as a sequence of natural numbers:

$$P = 1, 4, 2, 4, 8, 1, \ldots, 9. \tag{2.3}$$

We would like to be able to represent or encode individual Turing machines as natural numbers. This would make it possible for one Turing machine to take another Turing machine (in encoded form) as its input. Of course, we assume a certain fixed encoding for the alphabets Q and A. Otherwise, the set of Turing machines will be uncountable because there is an uncountable number of encoded alphabets.

We will need some facts about prime numbers. *Prime numbers* are such natural numbers which are not divisible to other numbers except 1 and themselves.

Exercise 2.6 Prove that there are infinitely many prime numbers.

Every natural number $n \geq 2$ has a unique prime decomposition, $n = p_1^{i_1} p_2^{i_2} \cdots$, where $p_1 = 2, p_2 = 3, \ldots$. This decomposition leads to a one-to-one correspondence between finite sequences of natural numbers $(i_1, i_2, \ldots)$ and natural numbers n. In particular, one can encode the program as the following number:

$$2^1 3^4 5^2 7^4 11^8 \cdots = n.$$

Such an encoding is called the *Gödel encoding*, and n is called the *Gödel number* of the Turing machine M. It is evident that in this way we can enumerate all Turing machines $\{M_n\}$. The set of all Gödel numbers $\{n\}$ is a subset of natural numbers. Note that from this remark it follows that the set of all Turing machines is a countable set. Given a natural number, there is an algorithm for determining whether it is the Gödel number of a Turing machine.

Exercise 2.7 Prove that the number 49 is not a Gödel number.

2.2.2 The Halting Problem

Let us prove that there are uncomputable functions. Let $\{M_n\}$ be the set of all Turing machines where n is a Gödel number. Let us define the following function h: $h(n)$ is defined only for such Gödel numbers n for which M_n does not halt on input n, and for such n on has

$$h(n) = 0.$$

Theorem 2.8 *The function h is not Turing computable.*

Proof Let us assume that the function h is computable. Then there exists a Turing machine M_k with a Gödel number k that computes h. Let us consider what happens when M_k is presented with input k. The machine M_k either halts on the input k or it does not halt. Suppose first that M_k halts on input k, i.e., the machine computes the value of the function $h(k)$. But according to the definition of the function h, the value $h(k)$ is defined only if M_k does not halt on input k. We get the contradiction. Suppose now that M_k does not halt on input k. Then from the definition of h, one gets that $h(k)$ is defined, and moreover $h(k) = 0$. But since M_k does not halt on input k, this means that M_k does not compute the value $h(k)$. However, according to our assumption the function h has to be computed by M_k. Again we get the contradiction. Therefore, our assumption that the function h is computable cannot be true. □

2.2.3 Universal Turing Machine

A universal Turing machine is a machine which can simulate the behavior of any Turing machine whose programs are provided as input data to it. A universal Turing machine is comparable to a general purpose computer which can compile and execute any given input program. The universal Turing machine U will receive as input the description of a particular machine M in addition to a copy of an input x and will simulate the behavior of M when M starts on x. In other words, U receives (δ_M, x) as input, where δ_M is the string of quintuples of M and x is an arbitrary input for M. The universal Turing machine U then simulates the behavior of M when M starts on x. One can prove that there exists a universal Turing machine [197].

2.2.4 Partially Recursive Functions

Which functions are computable on Turing machines? This is the class of partially recursive functions. They are generated from basic functions by using generating rules. The basic functions are defined on the natural numbers. There are three basic functions:

(1) The successor function

$$s(x) = x + 1.$$

(2) The zero function

$$\mathbf{z}(x) = 0.$$

(3) The identity function

$$i(x) = x.$$

Similarly, the basic functions of several variables are defined by $n(x_1, \ldots, x_m) = 0$ and $i_j^m(x_1, \ldots, x_m) = x_j$.

Generating rules include composition, primitive recursion and minimization. For example, the function $f(x) = 1$ can be constructed from the basic functions by using the composition as follows: $f(x) = s(\mathbf{z}(x))$.

Let $g(x, y)$ be a given function of two variables and k be a natural number. We define a new function $r(x)$ by using g and the primitive recursion as:

$$r(0) = k,$$
$$r(x+1) = g\big(x, r(x)\big), \quad x \geq 0.$$

Similarly for functions of several variables, let us take two functions $g(x_1, \ldots, x_{n+1})$ and $h(x_1, \ldots, x_{n-1})$. We define a new function $f(x_1, \ldots, x_n)$ by using g, h, and the primitive recursion as:

$$f(x_1, \ldots, x_{n-1}, 0) = h(x_1, \ldots, x_{n-1}),$$
$$f(x_1, \ldots, x_{n-1}, y+1) = g\big(x_1, \ldots, x_{n-1}, y, f(x_1, \ldots, x_{n-1}, y)\big), \quad y \geq 0.$$

Example 2.9 The addition function $f(x, y) = x + y$ can be defined through the use of primitive recursion and composition as

$$f(0, y) = i(y) = y,$$
$$f(x+1, y) = s\big(f(x, y)\big).$$

Exercise 2.10 Define the multiplication function $f(x, y) = xy$ from the basic functions by using composition and the primitive recursion.

The minimization produces a new function $f(x_1, \ldots, x_n)$ from a function $g(x_1, \ldots, x_{n+1})$ by the relation

$$f(x_1, \ldots, x_n) = \min\big\{y \mid g(x_1, \ldots, x_n, y) = 0\big\}.$$

Partially recursive functions are those functions which are generated from the basic functions by using a finite number of generating rules.

The following theorem describes the class of functions which are computable on Turing machines (for a proof, see, e.g., [757]).

Theorem 2.11 *The class functions computable on Turing machines is equivalent to the class of partially recursive functions.*

2.3 Boolean Functions, Gates and Circuits

2.3.1 Boolean Functions and Gates

In computations, it is often convenient to use the binary system and to reduce a problem to the study of Boolean functions. A Boolean function $f(x_1, \ldots, x_n)$ is a function of n variables where each variable takes values $x_i = 0$ or 1 and the function also takes values 0 or 1. If we denote $B = \{0, 1\}$ then the function f is a map $f : B^n (= \underbrace{B \times \cdots \times B}_{n}) \to B$.

There are three important Boolean functions which represent logical operations AND, OR and NOT. They are also called *gates* as before. The functions are:

(1) The conjunction

$$g_{\mathrm{AND}}(x_1, x_2) = x_1 x_2 = \min(x_1, x_2).$$

(2) The disjunction

$$g_{\mathrm{OR}}(x_1, x_2) = \max(x_1, x_2) = \begin{cases} 1, & \text{if } x_1 = 1 \text{ or } x_2 = 1, \\ 0, & \text{if } x_1 = 0 = x_2. \end{cases}$$

(3) The negation

$$g_{\mathrm{NOT}}(x) = \begin{cases} 1, & \text{if } x = 0, \\ 0, & \text{if } x = 1. \end{cases}$$

One also uses the notations $g_{\mathrm{AND}}(x_1, x_2) = x_1 \wedge x_2$, $g_{\mathrm{OR}}(x_1, x_2) = x_1 \vee x_2$ and $g_{\mathrm{NOT}}(x) = \overline{x}$. $\wedge$ and $\vee$ are often called *meet* and *join*, respectively. One has also the disjunction of several variables which is the composition of the binary disjunctions:

$$g_{\mathrm{OR}}(x_1, x_2, \ldots, x_n) = g_{\mathrm{OR}}\big(x_1, g_{\mathrm{OR}}\big(x_2, \ldots, g_{\mathrm{OR}}(x_{n-1}, x_n)\big)\big).$$

One has

$$g_{\mathrm{OR}}(x_1, x_2, \ldots, x_n) = \max(x_1, x_2, \ldots, x_n) = x_1 \vee x_2 \vee \cdots \vee x_n.$$

Similarly, for the conjunction of several variables one has

$$g_{\mathrm{AND}}(x_1, x_2, \ldots, x_n) = g_{\mathrm{AND}}\big(x_1, g_{\mathrm{AND}}\big(x_2, \ldots, g_{\mathrm{AND}}(x_{n-1}, x_n)\big)\big)$$

and

$$g_{\mathrm{AND}}(x_1, x_2, \ldots, x_n) = \min(x_1, x_2, \ldots, x_n) = x_1 x_2 \cdots x_n = x_1 \wedge x_2 \wedge \cdots \wedge x_n.$$

All logical circuits may be described in terms of the three fundamental elements.

If one denotes

$$x^\sigma = \begin{cases} \overline{x}, & \text{if } \sigma = 0, \\ x, & \text{if } \sigma = 1 \end{cases}$$

then it follows

$$x^\sigma = 1 \iff x = \sigma. \tag{2.4}$$

Exercise 2.12 Prove the above relation (2.4).

Now let us prove that every Boolean function can be represented as a superposition of the fundamental elements.

Theorem 2.13 *Every Boolean function $f(x_1, \dots, x_n)$ can be represented in the following form*

$$f(x_1, \dots, x_n) = \bigvee_{\sigma_1, \dots, \sigma_n} x_1^{\sigma_1} x_2^{\sigma_2} \cdots x_n^{\sigma_n} f(\sigma_1, \dots, \sigma_n). \tag{2.5}$$

Proof For a given $(x_1, \dots, x_n)$ one has either $f(x_1, \dots, x_n) = 1$ or 0. First, let us assume $f(x_1, \dots, x_n) = 1$. Then (2.4) leads to

$$x_1^{x_1} \cdots x_n^{x_n} f(x_1, \dots, x_n) = 1,$$

and one gets (2.5).

Next let us assume $f(x_1, \dots, x_n) = 0$. Note that $x_1^{\sigma_1} x_2^{\sigma_2} \cdots x_n^{\sigma_n} = 1$ only if $x_1 = \sigma_1, \dots, x_n = \sigma_n$, which means that the right-hand side $x_1^{x_1} x_2^{x_2} \cdots x_n^{x_n} f(x_1, \dots, x_n) = 0$ form the assumption $f(x_1, \dots, x_n) = 0$. □

The basic logical operations AND, OR and NOT are not independent. For example, the conjunction function g_{AND} can be represented as the superposition of the functions g_{OR} and g_{NOT} because one has

$$g_{\text{AND}}(x, y) = g_{\text{NOT}} \circ g_{\text{OR}}\big(g_{\text{NOT}}(x), g_{\text{NOT}}(y)\big)$$

or

$$xy (\equiv x \wedge y) = \overline{\overline{x} \vee \overline{y}}. \tag{2.6}$$

Exercise 2.14 Prove the above relation (2.6).

Analogously the disjunction is represented by conjunction and negation:

$$x \vee y = \overline{\overline{x} \wedge \overline{y}}.$$

It follows that not only the set

$$\mathcal{A}_1 = \{g_{\text{AND}}, g_{\text{NOT}}, g_{\text{OR}}\}$$

is a complete universal system (i.e., every Boolean function can be represented by this set $\mathcal{A}_2$) but also the set

$$\mathcal{A}_2 = \{g_{\mathrm{NOT}}, g_{\mathrm{AND}}\}$$

is a complete universal system for the Boolean system.

There are other complete universal systems. The following gates are important. They are the CNOT gate (*controlled* NOT *gate*) $g_{\mathrm{CNOT}} : B^2(= B \times B) \to B^2$,

$$g_{\mathrm{CNOT}}(x, y) = (x, x \oplus y)$$

and the *Toffoli gate* $g_{\mathtt{Toff}} : B^3(= B \times B \times B) \to B^3$:

$$g_{\mathtt{Toff}}(x, y, z) = (x, y, z \oplus xy).$$

Here $x \oplus y$ means addition in modulo 2:

$$0 \oplus 0 = 0, \qquad 0 \oplus 1 = 1 \oplus 0 = 1, \qquad 1 \oplus 1 = 0.$$

One can prove that the Toffoli gate is a universal gate for reversible computations.

2.3.2 Circuits

A representation is called a disjunctive normal form. It gives an example of a circuit. Every Boolean function can be written in such a form. However, this representation is not necessary the simplest representation of a function in terms of the fundamental ones. Some functions admit simpler representations.

Sometimes instead of the fundamental logical operations it is convenient to fix another set of functions as basic functions and represent any function in terms of these basic functions. A representation of a function using a given set of functions is called a *circuit*. We introduce some more notations before we give a formal definition of the circuit. For example, we shall write the function

$$f(x_1, x_2, x_3) = f_{\mathrm{AND}}\big(f_{\mathrm{OR}}(x_1, x_3), x_2\big)$$

as

$$f(x_1, x_2, x_3) = f_{\mathrm{AND}}^{(12)} \circ f_{\mathrm{OR}}^{(13)}(x_1, x_2, x_3)$$

where the superscript (13) in $f_{\mathrm{OR}}^{(13)}$ indicates that this function acts on the first and third arguments in (x_1, x_2, x_3) and the superscript (12) in $f_{\mathrm{AND}}^{(12)}$ indicates that the function acts on the first and second arguments.

In the more general case, if g is a function of k variables, $g : B^k \to B^m$ and $n \geq k$, then we define a function $g^{(i_1,\ldots,i_k)} : B^n \to B^{m+n-k}$ as

$$g^{(i_1,\ldots,i_k)}(x) = \big(g^{(i_1,\ldots,i_k)}(x_{i_1}, \ldots, x_{i_k}), x \setminus \{x_{i_1}, \ldots, x_{i_k}\}\big).$$

Here $x = (x_1, \ldots, x_n)$.

More precisely, if we introduce auxiliary variables $t_1 = x_1$ and $t_2 = f_{\mathrm{OR}}(x_2, x_3)$ then we can write $f(x_1, x_2, x_3) = f_{\mathrm{AND}}^{(12)}(t_1, t_2)$.

The functions $\{g_{\mathrm{AND}}, g_{\mathrm{OR}}, g_{\mathrm{NOT}}\}$ form a *universal basis* among Boolean functions in the sense that every Boolean function can be represented as a superposition of these functions. We reformulate Theorem 2.13 as

Theorem 2.15 *Every Boolean function $f(x_1, \ldots, x_n)$ can be written as a superposition of the three fundamental elements as*

$$f(x_1, \ldots, x_n) = g_{i_1}^{(\alpha_1)} \circ g_{i_2}^{(\alpha_2)} \circ \cdots \circ g_{i_k}^{(\alpha_k)}(x_1, \ldots, x_n).$$

Here g_i is one of the three fundamental elements, and the superscript (α) of $g_i^{(\alpha)}$ indicates on which arguments the element acts.

A circuit is a sequence of gates of the form where the functions g_i are some fixed Boolean functions.

Definition 2.16 A circuit C is the following set of data

$$C = \left\{G, f; f = g_{i_1}^{(\alpha_1)} \circ g_{i_2}^{(\alpha_2)} \circ \cdots \circ g_{i_k}^{(\alpha_k)}\right\},$$

where $G = \{g_1, \ldots, g_r\}$ is a finite set of Boolean functions (gates), f is a Boolean function of n variables, and one has the representation

$$f(x_1, \ldots, x_n) = g_{i_1}^{(\alpha_1)} \circ g_{i_2}^{(\alpha_2)} \circ \cdots \circ g_{i_k}^{(\alpha_k)}(x_1, \ldots, x_n).$$

2.4 Computational Complexity, P-problem and NP-problem

There are several problems which may not be solved effectively, namely, in polynomial time. The most fundamental problems are NP-problems and NP-complete problems. It is known that all NP-complete problems are equivalent, and an essential question is *whether there exists an algorithm to solve an NP-complete problem in polynomial time*. Such problems have been studied for decades now, and for each such problem all known algorithms have an exponential running time in the length of the input so far. A P-problem and an NP-problem are defined as follows:

Let n be the size of input.

Definition 2.17

(1) A P-problem is a problem with a time needed for solving it at worst a polynomial in n. Equivalently, it is a problem which can be recognized in a time polynomial in n by a deterministic Turing machine.

(2) An NP-problem is a problem that can be solved in polynomial time by a nondeterministic Turing machine (there must be a verifier that runs in polynomial time). This can be reexpressed as follows: Let us consider a problem of finding a solution of $f(x) = 0$. We can check in time polynomial in n whether x_0 is a solution of $f(x) = 0$, but we do not know whether we can find the solution of $f(x) = 0$ in time polynomial in n.
(3) An NP-complete problem is such an NP-problem to which any other NP-problem can be reduced in polynomial time.

Examples of a P-problem and an NP-complete problem

(1) *Euler closed path: P-problem.* Let $G = (V, E)$ be a graph, V be a set of vertices of the graph, and E be a set of edges of the graph. The problem of whether there exists a closed path from an arbitrary point v of V to v itself passing through all edges of E is called the *Euler closed path problem.* Let n be the number of all edges. Then it is known that the problem can be solved in polynomial time of order $O(n^3)$.
(2) *Hamilton closed path: NPC-problem.* In the above mentioned graph $G = (V, E)$, the problem of whether there exists a closed path from an arbitrary point v of V to v itself passing through all vertices of V is called the *Hamilton closed path problem.* It is not known whether an algorithm to solve this problem in polynomial time exists or not.
(3) *Traveling Salesman problem: NPC-problem.* Let $C = \{c_1, \ldots, c_n\}$ be the set of n cities, and let M be a given natural number. The distance $d(c_i, c_k)$ between two cities c_i, c_k is given, and the order of visiting all cities in $C = \{c_1, \ldots, c_n\}$ is denoted by π, namely, $c_{\pi(1)} \to \cdots \to c_{\pi(n)}$. Then the problem of whether there exists a π satisfying the following inequality

$$\sum_{i=1}^{n-1} d(c_{\pi(i)}, c_{\pi(i+1)}) + d(c_{\pi(n)}, c_{\pi(1)}) \leq M$$

is called the *traveling salesman problem.*
(4) *SAT problem: NPC-problem.* Let $X \equiv \{x_1, \ldots, x_n\}$ be a set. Then x_k and its negation $\bar{x}_k$ $(k = 1, 2, \ldots, n)$ are called *literals*, and the set of all such literals is denoted by $X' \equiv \{x_1, \bar{x}_1, \ldots, x_n, \bar{x}_n\}$. The set of all subsets of X' is denoted by $\mathcal{F}(X')$, and an element $C \in \mathcal{F}(X')$ is called a *clause*. We take a truth assignment to all Boolean variables x_k. If we can assign the truth value to at least one element of C, then C is called *satisfiable*. When C is satisfiable, the truth value $t(C)$ of C is regarded as true, otherwise as false. Take the truth values as "true $\leftrightarrow$ 1", "false $\leftrightarrow$ 0". Then C is satisfiable iff $t(C) = 1$.

Let $B = \{0, 1\}$ be a Boolean lattice with the usual join $\vee$ and meet $\wedge$, and let $t(x)$ be the truth value of a literal x in X. Then the truth value of a clause C is written as $t(C) \equiv \bigvee_{x \in C} t(x)$.

Moreover, the set $\mathcal{C}$ of all clauses C_j $(j = 1, 2, \ldots, m)$ is called *satisfiable* iff the meet of all truth values of C_j is 1; $t(\mathcal{C}) \equiv \bigwedge_{j=1}^{m} t(C_j) = 1$. Thus the SAT problem is written as follows:

Definition 2.18 (SAT problem) Given a Boolean set $X \equiv \{x_1, \ldots, x_n\}$ and a set $\mathcal{C} = \{C_1, \ldots, C_m\}$ of clauses, determine whether $\mathcal{C}$ is satisfiable or not.

That is, this problem is whether there exists a truth assignment to make $\mathcal{C}$ satisfiable. It is known in usual algorithms that it takes polynomial time to check the satisfiability only when a specific truth assignment is given, but we cannot determine the satisfiability in polynomial time when an assignment is not specified.

"The SAT problem is an NP-complete" (Cook's theorem).

It is not known whether every problem in NP can be solved quickly—this is called the P = NP problem (a Millennium problem).

As special representations of the algorithms discussed in this chapter, we will discuss a quantum algorithm, generalized quantum algorithm and a chaotic quantum algorithm in Chaps. 11 and 14.

2.5 Notes

Conventional discussions on a Turing machine and an algorithm can be read in the book [188], in which there are several examples illustrating the moves of Turing machines in computation processes. For a more recent exposition, see [723]. The original paper discussing Gödel encoding, from the point of view of undecidability, is [293]. Fredkin–Toffoli gates used for the construction of computational circuits were introduced in [268]. A popular textbook introducing computational complexity and NP-problems is [278].

Chapter 3
Basics of Classical Probability

In this chapter, we discuss the basics in the classical probability theory. Probability theory is based on measure theory, so that we start by reviewing measure theory.

3.1 Classical Probability

Here we present Kolmogorov's [438] axiomatization of the classical probability theory which is generally accepted.

Definition 3.1 A triple $(\Omega, \mathcal{F}, \mu)$, where Ω is a set (of points ω), $\mathcal{F}$ is a σ-field of subsets of Ω, and μ is a probability measure (or probability) is called a probabilistic model or a probability space.

Here Ω is the sample space or the space of elementary events (outcomes), the set A in $\mathcal{F}$ is called an *event* and $\mu(A)$ is the probability of the event A.

A system $\mathcal{F}$ of subsets of Ω is a σ*-field* if the following conditions are satisfied:

1. $\Omega \in \mathcal{F}$
2. $A \in \mathcal{F} \Rightarrow \bar{A} = \Omega \setminus A \in \mathcal{F}$
3. $A_n \in \mathcal{F}, n = 1, 2, \ldots \Rightarrow \bigcup_n A_n \in \mathcal{F}$.

Exercise 3.2 Prove that the system of all subsets of a finite set is a σ-field.

Finally, a function $\mu : \mathcal{F} \to [0, \infty)$ is a *probability measure* if it has the following properties

1. $\mu(\varnothing) = 0$
2. $\mu(\Omega) = 1$
3. $A_n \in \mathcal{F}, n = 1, 2, \ldots$, and $A_n \cap A_m = \varnothing, m \neq n \Rightarrow \mu(\bigcup_n A_n) = \sum_n \mu(A_n)$.

For example, the ordinary (Lebesgue) measure dx on the unite segment $[0, 1]$ on the real line is a probability measure. Let us consider an example of a probability space.

M. Ohya, I. Volovich, *Mathematical Foundations of Quantum Information and Computation and Its Applications to Nano- and Bio-systems*,
Theoretical and Mathematical Physics,
DOI 10.1007/978-94-007-0171-7_3,

Example 3.3 (Binomial distribution. Coin tossing) Let us discuss a model for the experiment in which a coin is tossed N times independently with probability p (where $0 \le p \le 1$) of heads (for a symmetrical coin $p = 1/2$). Record the results as an ordered set $(a_1, \dots, a_N)$ where $a_i = 1$ for heads ("success") and $a_i = 0$ for tails ("failure"). The sample space of all outcomes is

$$\Omega = \{\omega; \omega = (a_1, \dots, a_N), a_i = 0, 1\}.$$

To each sample point $\omega = (a_1, \dots, a_N)$, we assign the probability

$$\mu(\omega) = p^{\sum a_i} q^{N - \sum a_i},$$

where $p + q = 1$, $p, q \ge 0$. We define the σ-field $\mathcal{F}$ as the set of all subsets of Ω and the measure of an event $A = \{\omega\}$ from $\mathcal{F}$ as the sum of the probabilities of the sample events

$$\mu(A) = \sum_{\omega \in A} \mu(\omega).$$

The events

$$A_m = \{\omega; \omega = (a_1, \dots, a_N), a_1 + \dots + a_N = m\}, \quad m = 0, 1, \dots, N$$

consist of exactly m successes. The set of probabilities $(\mu(A_0), \dots, \mu(A_N))$ is called the *binomial distribution.*

Exercise 3.4 Prove that $\mu(A_m) = {}_N C_m p^m q^{N-m}$, where the binomial coefficient is

$${}_N C_m = \frac{N!}{(N-m)!m!}.$$

We have obtained the triple $(\Omega, \mathcal{F}, \mu)$ which is a probability space. In particular, for $N = 1$ one has $\Omega = \{\omega; \omega = a, a = 0, 1\}$, $\mu(1) = p$ and $\mu(0) = q$.

Exercise 3.5 Prove that the constructed triple $(\Omega, \mathcal{F}, \mu)$ is a probability space.

If $(\Omega, \mathcal{F}, \mu)$ is a probability space then the integral with respect to the measure μ is defined. In particular, an important example of the probability space is given by the real line $\mathbb{R}$. The Lebesgue measure ν on $\mathbb{R}$ is defined on intervals by $\nu((a, b]) \equiv b - a$.

Example 3.6 Let us consider a triple $(\Omega, \mathcal{F}, \mu)$ where Ω is the real line, $\Omega = \mathbb{R}$, the σ-field $\mathcal{F}$ is generated by intervals on the line (such a σ-field is called the Borel σ-field) and $f(\omega)$ is a positive real valued function on the real line such that $\int f(\omega)\, d\omega = 1$. Here $d\omega$ is the Lebesgue measure. Then $d\mu(\omega) = f(\omega)\, d\omega$ is a probability measure and the triple $(\Omega, \mathcal{F}, \mu)$ is a probability space.

Let $(\Omega, \mathcal{F}, \mu)$ be a probability space. A *random variable* ξ is a real valued measurable function $\xi : \Omega \to \mathbb{R}$ (i.e., $\xi^{-1}(\Delta) \in \mathcal{F}$ for any set $\Delta \in \mathcal{B}(\mathbb{R})$, the Borel σ-field on $\mathbb{R}$) on the sample space. The expectation of ξ is the integral

$$E_\mu \xi = \int_\Omega \xi(\omega)\, d\mu(\omega).$$

The set of all measurable functions is denoted by $M(\mathcal{F})$. For any $\xi \in M(\mathcal{F})$,

$$F_\xi(t) \equiv \mu\big(\xi^{-1}(-\infty, t]\big) = \mu\big(\{\omega \in \Omega; -\infty < \xi(\omega) \le t\}\big)$$

is called the distribution function with respect to ξ and $(\Omega, \mathcal{F}, \mu)$, and

$$\mu_\xi(\Delta) \equiv \mu\big(\xi^{-1}(\Delta)\big), \quad \Delta \in \mathcal{B}(\mathbb{R})$$

is called a probability distribution, so that

$$\mu_\xi\big((-\infty, t]\big) = F_\xi(t).$$

One has

$$E_\mu \xi = \int_{\mathbb{R}} t\, dF_\xi(t).$$

Example 3.7 The random variable $\xi = 1_A(\omega)$ is called the indicator (characteristic function) of the event $A \subset \Omega$. Here

$$1_A(\omega) \equiv \begin{cases} 1, & \text{if } \omega \in A, \\ 0, & \text{if } \omega \notin A. \end{cases}$$

It has the properties of the projection:

$$1_A 1_A = 1_A, \qquad 1_A 1_B = 0 \quad \text{if } A \cap B = \varnothing. \tag{3.1}$$

The probability of the event A can be written as the expectation of its indicator

$$\mu(A) = E_\mu 1_A. \tag{3.2}$$

Exercise 3.8 Check the relations (3.1) and (3.2).

The *conditional probability* $\mu(B \mid A)$ of the event B given the event A is defined as

$$\mu(B \mid A) = \frac{\mu(A \cap B)}{\mu(A)}.$$

Here $\mu(A \cap B)$ is the probability of the intersection of the events A and B, and we assume $\mu(A) \neq 0$. It follows the Bayes's formula

$$\mu(A \mid B) = \mu(B \mid A) \frac{\mu(A)}{\mu(B)}.$$

Generally, we have the following theorem.

Theorem 3.9 (Bayes's Theorem) *For any exclusive n events $A_1, \dots, A_n$ with $\bigcup_{k=1}^n A_k = \Omega$ and any event B, we have*

$$\mu(A_k \mid B) = \frac{\mu(B \mid A_k)\mu(A_k)}{\sum_{j=1}^n \mu(B \mid A_j)\mu(A_j)}.$$

Proof It is enough to prove the equality: $\mu(A_k \mid B)\sum_{k=1}^n \mu(B \mid A_j)\mu(A_j) = \mu(B \mid A_k)\mu(A_k)$.

LHS $= \mu(A_k \mid B)\sum_{j=1}^n \mu(B \cap A_j) = \mu(A_k \mid B)\mu(\bigcup_{j=1}^n (B \cap A_j)) = \mu(A_k \mid B)\mu(B \cap \Omega) = \mu(A_k \mid B)\mu(B) = \mu(A_k \cap B) = \mu(A_k)\mu(B \mid A_k) =$ RHS. □

Events A and B are called *independent* if

$$\mu(A \cap B) = \mu(A)\mu(B).$$

We have discussed real-valued random variables. A complex-valued variable is a pair of real-valued random variables.

Finally, it is noted that if a random variable depends on a parameter t, then it is called a *random process*, that is, a random process is a measurable map $\xi : \mathbb{R} \times \Omega \to \mathbb{R}$. We write $\xi(t, \omega) = \xi_t(\omega)$.

3.1.1 Radon–Nikodym Theorem

Let us use the notation $\mu, \nu, p, \dots$ for probability measures on $(\Omega, \mathcal{F})$ in the sequel. An observable on the probability space $(\Omega, \mathcal{F}, \mu)$ is a measurable function f from Ω to $\mathbb{R}$. For the above measure μ and a finite or σ-finite measure ν (i.e., $\nu(\Omega) < +\infty$, or $\nu(\Omega) = +\infty$ and there exists $\{A_n\} \subset \mathcal{F}$ with $\Omega = \bigcup_n A_n$, $\nu(A_n) < +\infty$), μ is said to be absolutely continuous with respect to ν (denoted by $\mu \ll \nu$) if $\nu(A) = 0$, $A \in \mathcal{F}$, implies $\mu(A) = 0$.

Theorem 3.10 (Radon–Nikodym Theorem) *For the above measures μ and ν, if $\mu \ll \nu$, then there exists $f \in L^1(\Omega, \mathcal{F}, \nu)$ in almost everywhere sense such that*

$$\mu(A) = \int_A f(\omega)\nu(d\omega) \left(= \int_A f\, d\nu\right), \quad A \in \mathcal{F}.$$

Remark 3.11

1. Two random variables f and g being equal in almost everywhere sense w.r.t. a measure μ ($f = g\,\mu$-a.e.) means that

$$\mu(\{\omega \in \Omega; f(\omega) \neq g(\omega)\}) = 0.$$

2. $L^1(\Omega, \mathcal{F}, \nu)$ is the equivalence class of all absolutely integrable functions (i.e., $\|f\|_1 \equiv \int_\Omega |f|\,d\nu < +\infty$) on $(\Omega, \mathcal{F}, \nu)$.

Let t be the Lebesgue measure on $\mathcal{B}(\mathbb{R})$ (i.e., $t((a,b]) \equiv b-a$). If $\mu_f \ll t$, then there exists a density distribution $p_f(t)$ such that

$$\mu_f(\Delta) = \int_\Delta p_f(t)\,dt,$$

so that

$$F_f(x) = \int_{-\infty}^x p_f(s)\,ds.$$

We call the triple $(\mathbb{R}, \mathcal{B}(\mathbb{R}), \mu_f)$ the probability space induced from f.

For any $f, g \in M(\mathcal{F})$

$$\mu_{f,g}(\Delta_1, \Delta_2) \equiv \mu\big(f^{-1}(\Delta_1) \cap g^{-1}(\Delta_2)\big), \quad \Delta_1, \Delta_2 \in \mathcal{B}(\mathbb{R})$$

is called the joint probability distribution with respect to f and g. An important property of the joint probability distribution is the following marginal conditions

$$\mu_{f,g}(\Delta_1, \mathbb{R}) = \mu_f(\Delta_1),$$
$$\mu_{f,g}(\mathbb{R}, \Delta_2) = \mu_g(\Delta_2).$$

In the classical probability theory, the expectation, the variance and the covariance of random variables f and g with respect to a measure μ are summarized as follows:

1. (Expectation) $E_\mu(f) = \int_\Omega f\,d\mu = \int_\mathbb{R} t\,dF_f(t)$
2. (Variance) $V_\mu(f) = E_\mu(\{f - E_\mu(f)\}^2)$
3. (Covariance) $C_\mu(f,g) = E_\mu(\{f - E_\mu(f)\}\{g - E_\mu(g)\})$.

3.2 Conditional Expectation and Martingale

For a probability space $(\Omega, \mathcal{F}, \mu)$, let $\mathcal{G}$ be a subfield of $\mathcal{F}$ and $f \in L^1(\Omega, \mathcal{F}, \mu)$. Put

$$\nu(A) = \int_A f\,d\mu, \quad A \in \mathcal{G},$$

then $\nu \ll \mu$ holds. Thus the Radon–Nikodym theorem tells us that there exists a $\mathcal{G}$-measurable function g such that

$$\nu(A) = \int_A g\,d\mu, \quad A \in \mathcal{G}.$$

This g is called the *conditional expectation* of f with respect to $\mathcal{G}$, and it is denoted by $\mathcal{E}(f;\mathcal{G})$. The conditional expectation has the following properties:

1. $f \geq 0 \Rightarrow \mathcal{E}(f; \mathcal{G}) \geq 0$.
2. $\mathcal{E}(f; \mathcal{G}) : L^1(\Omega, \mathcal{F}, \mu) \to L^1(\Omega, \mathcal{G}, \mu)$ is linear.
3. $\mathcal{E}(\mathcal{E}(f; \mathcal{G}); \mathcal{G}) = \mathcal{E}(f; \mathcal{G})$: projection.
4. $\mathcal{G}_1 \subset \mathcal{G}_2 \Rightarrow \mathcal{E}(\mathcal{E}(f; \mathcal{G}_2); \mathcal{G}_1) = \mathcal{E}(f; \mathcal{G}_1)$.
5. $\mathcal{E}(f; \{\varnothing, \Omega\}) = \int_\Omega f \, d\mu$.
6. $f_n \uparrow f \Rightarrow \mathcal{E}(f_n; \mathcal{G}) \uparrow \mathcal{E}(f; \mathcal{G})$.
7. For a finite measure ν, $L^\infty(\Omega, \mathcal{F}, \nu)$ is the equivalence class of all essentially bounded functions (i.e., $\|f\|_\infty \equiv \inf\{\alpha \in \bar{R};\ |f(\omega)| < \alpha, \nu\text{-a.e.}\} < +\infty$). For any $g \in L^\infty(\Omega, \mathcal{G}, \mu)$ and $f \in L^1(\Omega, \mathcal{F}, \mu)$, one has $\mathcal{E}(gf; \mathcal{G}) = g\mathcal{E}(f; \mathcal{G})$.
8. $\|\mathcal{E}(f; \mathcal{G})\|_1 \leq \|f\|_1$.
9. Let ζ be a map from $L^1(\Omega, \mathcal{F}, \mu)$ to $L^1(\Omega, \mathcal{G}, \mu)$ such that $\zeta(\zeta(f)) = \zeta(f)$ and $\int_A \zeta(f) \, d\mu = \int_A f \, d\mu$, $A \in \mathcal{G}$, then $\zeta(f) = \mathcal{E}(f; \mathcal{G})\mu$-a.e.
10. (Jensen's inequality) Let φ be a lower-bounded convex function from $\mathbb{R}$ to $\mathbb{R}$ and $\varphi(f) \in L^1(\mathbb{R}, \mathcal{F}, \mu)$ for a random variable f. Then

$$\varphi\big(\mathcal{E}(f; \mathcal{G})\big) \leq \mathcal{E}\big(\varphi(f); \mathcal{G}\big).$$

Now put $p(A; \mathcal{G}) \equiv \mathcal{E}(1_A; \mathcal{G})$. This $p(A; \mathcal{G})$ is the probability (conditional probability) of the event $A \in \mathcal{F}$ for a given $\mathcal{G}$, which has the following properties:

1. $0 \leq p(A; \mathcal{G}) \leq 1$
2. $A \in \mathcal{G} \Rightarrow p(A; \mathcal{G}) = p(A)$
3. $A \subset B \Rightarrow p(A; \mathcal{G}) \leq p(B; \mathcal{G})$
4. $p(\bar{A}; \mathcal{G}) = 1 - p(A; \mathcal{G})$
5. $p(A; \{\varnothing, \Omega\}) = p(A)$
6. $A \cap B = \varnothing \Rightarrow p(A \cup B; \mathcal{G}) = p(A; \mathcal{G}) + p(B; \mathcal{G})$.

When a sequence $\{\mathcal{G}_n\}$ is increasing ($\mathcal{G}_n \uparrow$), a sequence $\{f_n\}$ of $\mathcal{G}$-measurable functions is called a *martingale* if $f_n \in L^1(\Omega, \mathcal{F}, \mu)$ and $\mathcal{E}(f_{n+1}; \mathcal{G}_n) = f_n$. For a martingale $\{f_n\}$ with respect to increasing σ-fields $\{\mathcal{G}_n\}$, we have:

Theorem 3.12

(i) $\sup\|f_n\|_1 < +\infty \Rightarrow$ *there exists a function* $f \in L^1(\Omega, \mathcal{G}, \mu)$ *such that* $f_n \to f$ *a.e.. Moreover, if* $\{f_n\}$ *is uniformly integrable (i.e.,* $\int_{\{|f_n| \geq a\}} |f_n| \, d\mu \to 0$ *as* $a \to +\infty$*), then there exists a function* $f \in L^1(\Omega, \mathcal{G}, \mu)$ *such that* $f_n \to f$*, a.e. and in* L^1 *sense, that is,* $\mu\{\omega \in \Omega; \lim_{n\to\infty} f_n(\omega) \neq f(\omega) = 0\}$ *and* $\lim_{n\to\infty} \|f_n - f\|_1 = 0$.

(ii) *(Increasing martingale) For* $\mathcal{G}_n \uparrow \mathcal{G}$ *and any* $f \in L^1(\Omega, \mathcal{F}, \mu), \mathcal{E}(f; \mathcal{G}_n) \to \mathcal{E}(f; \mathcal{G})$ *a.e. and in* L^1.

(iii) *(Decreasing martingale) For* $\mathcal{G}_n \downarrow \mathcal{G}$ *and any* $f \in L^1(\Omega, \mathcal{F}, \mu), \mathcal{E}(f; \mathcal{G}_n) \to \mathcal{E}(f; \mathcal{G})$ *a.e. and in* L^1.

3.3 Algebraic Formulation of Classical Probability

Let $(\Omega, \mathcal{F}, \mu)$ be a probability space. The set of all bounded (complex) random variables forms an algebra $\mathcal{A}$. Here algebra means that it is closed under addition and multiplication properly defined. The expectation with respect to a probability measure μ is a linear functional on the algebra $\mathcal{A}$. We denote the functional $\phi(f)$ as

$$\phi(f) = E_\mu(f), \quad f \in \mathcal{A}.$$

It is a linear functional with the following properties:

$$\phi(f^* f) \geq 0 \quad \text{(positivity)},$$
$$\phi(1) = 1 \quad \text{(normalization)}.$$

A linear positive normalized functional on an algebra is called a *state*. Here "$*$" is the complex conjugation. If $\mathcal{A}$ is an arbitrary algebra, then the conjugate linear mapping $* : \mathcal{A} \to \mathcal{A}$ such that $(AB)^* = B^* A^*$ for any elements A, B of the algebra $\mathcal{A}$ is called an *involution*.

We have described how to a probability space one can associate a commutative algebra with involution and a state.

Conversely, to any commutative algebra $\mathcal{A}$ with involution and a state ϕ one can associate a probability space $(\Omega, \mathcal{F}, \mu)$. Therefore, there is a one-to-one correspondence between probability spaces and commutative algebras with involution and state:

$$(\Omega, \mathcal{F}, \mu) \quad \Longleftrightarrow \quad (\mathcal{A}, \phi).$$

Many notions of the classical probability theory are generalized to the non-commutative case when one takes a non-commutative algebra A and a state ϕ, which will be discussed in Chaps. 5 and 7.

3.4 Notes

The basic probability theory was formulated by several mathematicians such as J. Bernoulli, Moivre, Bayes, Legendre, Lagrange, Laplace around the nineteenth century. The probability theory by means of a measure-theoretic formulation was done by Kolmogorov [438]. Before his work, the law of large numbers was studied by Borel and Cantelli, and the Brownian motion was discussed by Wiener following Einstein and Smoluchowski. The classical probability theory is discussed in [234, 355, 378, 438, 457, 479, 546, 631, 658, 766, 768]. One of the important concepts in probability theory is that of a stochastic process which was studied by Doob [207], Bartlett, Ito, Gelfand [280], Kolmogorov, Yaglom, and many others. The white noise analysis was discussed by Volterra, Hadamard, Levy [478, 479], Hida [335]; we will discuss this analysis in Chap. 19. Various interpretations of a probability are considered by Khrennikov [405]. An algebraic formulation of the classical probability theory is discussed in non-commutative (quantum) probability, see [9, 34].

Chapter 4
Basics of Infinite-Dimensional Analysis

We will quickly review the basic facts of infinite-dimensional (functional) analysis needed to describe quantum mechanics and quantum information theory.

4.1 Hilbert Space

Let $\mathcal{H}$ be a vector space over a scalar field $\mathbb{K}$ (real number field $\mathbb{R}$ or complex number field $\mathbb{C}$). For the elements of $\mathcal{H}$ we will use notations such as $x, y, \ldots$ or $\psi, \varphi, \ldots$. We call $\mathcal{H}$ an inner product space if it has an inner product $\langle \cdot, \cdot \rangle$ which is defined as a map from $\mathcal{H} \times \mathcal{H}$ to $\mathbb{K}$ satisfying the following conditions for any $x, y, z \in \mathcal{H}$ and any $\lambda \in \mathbb{K}$:

(i) $\langle x, x\rangle \geq 0$, $\langle x, x\rangle = 0 \Leftrightarrow x = 0$
(ii) $\langle x, y\rangle = \overline{\langle y, x\rangle}$
(iii) $\langle x, \lambda y + z\rangle = \lambda\langle x, y\rangle + \langle x, z\rangle$.

Now put $\|x\| = \sqrt{\langle x, x\rangle}$. One has

Theorem 4.1 (Schwarz inequality) *We have*

$$\left|\langle x, y\rangle\right| \leq \|x\|\|y\|, \quad \forall x, y \in \mathcal{H},$$

where the equality holds iff x and y are linearly dependent.

Proof From (ii) and (iii), one has

$$\langle x + \alpha y, x + \alpha y\rangle = \langle x, x\rangle + \overline{\alpha}\langle y, x\rangle + \alpha\langle x, y\rangle + |\alpha|^2\langle y, y\rangle \geq 0,$$

for any complex number α. Now if $y = 0$ then the inequality is proved. If $y \neq 0$ then let us set $\alpha = -\langle y, x\rangle / \langle y, y\rangle$ in the above equality. We get

$$\langle x, x\rangle - \frac{|\langle y, x\rangle|^2}{\langle y, y\rangle} - \frac{|\langle y, x\rangle|^2}{\langle y, y\rangle} + \frac{|\langle y, x\rangle|^2}{\langle y, y\rangle} \geq 0,$$

which is equivalent to the stated inequality. □

M. Ohya, I. Volovich, *Mathematical Foundations of Quantum Information and Computation and Its Applications to Nano- and Bio-systems*,
Theoretical and Mathematical Physics,
DOI 10.1007/978-94-007-0171-7_4,

From this inequality, one easily see that $\|\cdot\|$ defines a norm on $\mathcal{H}$. That is, for arbitrary $x, y, z \in \mathcal{H}, \lambda \in \mathbb{K}$, $\|\cdot\|$ satisfies:

(i) $\|x\| \geq 0$, $\|x\| = 0 \Leftrightarrow x = 0$
(ii) $\|\lambda x\| = |\lambda|\|x\|$
(iii) $\|x + y\| \leq \|x\| + \|y\|$.

In general, a vector space with norm $\|\cdot\|$ is called a *normed space*. In a normed space, a sequence $\{x_n\}$ is called a *Cauchy sequence* if it satisfies $\|x_n - x_m\| \to 0$ as $m, n \to \infty$. A normed space $\mathcal{H}$ is called complete if all Cauchy sequences converge in $\mathcal{H}$, namely, there exists $x \in \mathcal{H}$ such that $\|x_n - x\| \to 0$. Moreover, we call a complete normed space a *Banach space*. An inner product space is called a (real or complex according to $\mathbb{K} = \mathbb{R}$ or $\mathbb{C}$) *Hilbert space* if it is complete with respect to the norm $\|x\| = \langle x, x\rangle^{1/2}$.

The simplest example of a real Hilbert space is $\mathbb{R}^n$ and a complex one is $\mathbb{C}^n$: For two vectors $z = (z_1, \ldots, z_n)$ and $w = (w_1, \ldots, w_n)$ in $\mathbb{C}^n$, its scalar product is defined as

$$\langle z, w\rangle = \sum \bar{z}_i w_i.$$

A sequence $\{x_n\}$ in $\mathcal{H}$ might converge to $x \in \mathcal{H}$ in one of the following two ways: The convergence is

(i) in the weak sense (denoted by $x_n \to x(w)$) if $\langle x_n, y\rangle \to \langle x, y\rangle, \forall y \in \mathcal{H}$, and
(ii) in the strong sense (denoted by $x_n \to x(s)$) if $\|x_n - x\| \to 0$.

A subset $\mathcal{K}$ of $\mathcal{H}$ is said to be a *subspace* if it is a linear subspace and a *closed subspace* if it is also closed with respect to the norm $\|\cdot\|$ of $\mathcal{H}$. The set $\mathcal{K}^\perp$ defined by $\{x \in \mathcal{H}; \langle x, y\rangle = 0, \forall y \in \mathcal{K}\}$ is a closed subspace and is called the *orthogonal complement* of $\mathcal{K}$. For any closed subspace $\mathcal{K}$ of $\mathcal{H}$, every element $x \in \mathcal{H}$ can be uniquely decomposed as $x = y + z$ $(y \in \mathcal{K}, z \in \mathcal{K}^\perp)$ (*projection theorem*), so that $\mathcal{H}$ can be expressed as the direct sum of $\mathcal{K}$ and $\mathcal{K}^\perp$, that is, $\mathcal{H} = \mathcal{K} \oplus \mathcal{K}^\perp \equiv \{y + z;\ y \in \mathcal{K}, z \in \mathcal{K}^\perp\}$.

Example 4.2 The set of the equivalence classes of square integrable functions;

$$L^2(\mathbb{R}^3) \equiv \left\{\psi; \int_{\mathbb{R}^3} |\psi(\mathbf{x})|^2 \, d\mathbf{x} < +\infty\right\}, \qquad \psi \sim \phi \overset{\text{def}}{\Longleftrightarrow} \int_{\mathbb{R}^3} |\psi - \phi|^2 \, dx = 0$$

is a Hilbert space with the inner product given by

$$\langle \varphi, \psi\rangle = \int_{\mathbb{R}^3} \overline{\varphi(\mathbf{x})}\psi(\mathbf{x}) \, d\mathbf{x}.$$

Example 4.3 Let $\mathcal{S} \equiv \{x \equiv (x_1, x_2, \ldots, x_n, \ldots); x_n \in \mathbb{C}, n \in \mathbb{N}\}$. The set $l^2 \equiv \{x \in \mathcal{S}; \langle x, x\rangle = \sum_n |x_n|^2 < +\infty\}$ is a Hilbert space with the inner product $\langle x, y\rangle \equiv \sum_n \overline{x}_n y_n$.

4.2 Linear Operators in Hilbert Space

Let $\mathcal{H}$ be a Hilbert space and $\mathcal{D}$ be a subspace of $\mathcal{H}$. An $\mathcal{H}$-valued map A defined on $\mathcal{D}$ is called a *linear operator*, or simply an *operator*, if it satisfies

$$A(\lambda x + \mu y) = \lambda Ax + \mu Ay \quad (\forall x, y \in \mathcal{D}, \forall \lambda, \mu \in \mathbb{C}).$$

Example 4.4 Consider the Hilbert space $\mathcal{H} = \mathbb{C}^n$. It has a basis

$$e_1 = \begin{pmatrix} 1 \\ 0 \\ \vdots \\ 0 \end{pmatrix}, \dots, e_n = \begin{pmatrix} 0 \\ 0 \\ \vdots \\ 1 \end{pmatrix}.$$

If A is an operator in $\mathcal{H}$ then one can assign to it a matrix (a_{ij}) given by

$$Ae_i = \sum_j a_{ij} e_j.$$

$\mathcal{D}$ is said to be the *domain* of A and denoted as $\mathcal{D}(A)$, or $\operatorname{dom} A$. A set $\{Ax; x \in \mathcal{D}\}$ is said to be the *range* of A and denoted $\mathcal{R}(A)$ or $\operatorname{ran} A$. An operator A is called a *bounded operator* if its domain is $\mathcal{H}$ itself ($\operatorname{dom} A = \mathcal{H}$) and there exists a constant $M > 0$ satisfying

$$\|Ax\| \le M \|x\| \quad (\forall x \in \mathcal{H}).$$

We denote the set of all bounded operators by $\mathbf{B}(\mathcal{H})$. Note that the norm of the operator A is given by

$$\|A\| = \sup\bigl\{\|Ax\|; \|x\| = 1, x \in \mathcal{H}\bigr\}.$$

An operator which is not an element of $\mathbf{B}(\mathcal{H})$ is called an *unbounded operator*.

For $A \in \mathbf{B}(\mathcal{H})$, there exists a unique operator $A^* \in \mathbf{B}(\mathcal{H})$ such that $\langle x, Ay \rangle = \langle A^*x, y \rangle$ $(\forall x, y \in \mathcal{H})$. A^* is called the *conjugate operator* of A. In particular, A is called a *self-adjoint operator* if $A = A^*$ holds. A is called a *positive operator* (written as $A \ge 0$) if $\langle x, Ax \rangle \ge 0$ holds for any $x \in \mathcal{H}$. For a positive operator $A \ge 0$, there exists a unique positive operator such that $A = B^2$. Thus defined B is written as $\sqrt{A}$. All positive operators are self-adjoint. (Note that this fact does not hold in the case of a real Hilbert space.)

Exercise 4.5 Prove that if an operator A in $\mathbb{C}^n$ has the matrix representation (a_{ij}) with respect to the basis (e_i) then its conjugate operator A^* has the matrix representation (a^*_{ji}) where $a^*_{ji} = \overline{a}_{ji}$.

The following operators A are important:

1. A is a *normal operator* $\Leftrightarrow A^*A = AA^*$

2. A is a *projection operator* $\Leftrightarrow A = A^* = A^2$
3. A is an *isometry operator* $\Leftrightarrow \|Ax\| = \|x\|$ $(\forall x \in \mathcal{H})$
4. A is a *unitary operator* $\Leftrightarrow A^*A = AA^* = I$ (I is an identity operator on $\mathcal{H}$)
5. A is a *partial-isometry operator* $\Leftrightarrow$ A is an isometry operator on $(\mathrm{Ker}\, A)^{\perp}$, where $\mathrm{Ker}\, A \equiv \{x \in \mathcal{H};\ Ax = 0\}$.

Here we note that (i) if A is a *Hermitian operator* then $U = e^{iA}$ is a unitary operator, and (ii) a projection operator P is a self-adjoint operator, $P = P^*$, which satisfies the relation $P^2 = P$. As an example if x is a vector in a Hilbert space $\mathcal{H}$ of unit length, $\|x\| = 1$, then the operator P_x which acts as $P_x y = \langle x, y\rangle x$ is a projection operator, where $\forall y \in \mathcal{H}$. In Dirac's notations (see Chap. 5 for the definition), P_x is written as $|x\rangle\langle x|$.

Let A be an unbounded operator on $\mathcal{H}$ such that $\overline{\mathcal{D}(A)}$ (the closure of $\mathcal{D}(A)$ w.r.t. the norm, i.e., $\mathcal{D}(A) \cup \{$all limit points of Cauchy sequences in $\mathcal{D}(A)\}$) coincides with $\mathcal{H}$ itself. Such an operator A is said to be *densely defined* on $\mathcal{H}$. If there exists $y' \in \mathcal{H}$ for $y \in \mathcal{G} \subset \mathcal{H}$ such that $\langle y, Ax\rangle = \langle y', x\rangle$ for any $x \in \mathcal{D}(A)$ then putting $y' = A^*y$ one obtains an operator A^* on $\mathcal{H}$ with domain $\mathcal{D}(A^*) = \mathcal{G}$. This operator A^* is called the *adjoint operator* of A. The operator A is called a *symmetric operator* (or *Hermitian operator*) if $A \subset A^* (\Leftrightarrow \mathcal{D}(A) \subset \mathcal{D}(A^*))$ holds. In particular, when $\mathcal{D}(A) = \mathcal{D}(A^*)$ holds, we write as $A = A^*$ and call A self-adjoint.

Now we summarize certain properties of spectra. Recall that an operator A on $\mathcal{H}$: $\mathrm{dom}\, A \to \mathrm{ran}\, A$ has its inverse A^{-1}: $\mathrm{ran}\, A \to \mathrm{dom}\, A$ whenever A is one-to-one. We define the *resolvent set* of a densely defined closed operator A as $\mathrm{Re}(A) = \{\lambda \in \mathbb{C};\ (A - \lambda I)^{-1} \in \mathbf{B}(\mathcal{H})\}$, that is, the set of all complex numbers $\lambda \in \mathbb{C}$ such that $A - \lambda I$ has the bounded inverse. The set $\mathrm{Sp}(A) = \mathbb{C} \setminus \mathrm{Re}(A)$ (the complement of $\mathrm{Re}(A)$) is called the *spectrum of* A. In addition, in $\mathrm{Sp}(A)$, the set $\mathrm{Sp}^{(p)}(A) = \{\lambda \in \mathbb{C};\ (A - \lambda I)^{-1}$ does not exist$\}$ is called the set of *point spectra*, and $\mathrm{Sp}^{(c)}(A) = \{\lambda \in \mathbb{C};\ (A - \lambda I)^{-1}$ exists but is unbounded with $\overline{\mathrm{ran}(A - \lambda I)} = \mathcal{H}\}$ is called the set of *continuous spectra*.

An element λ of $\mathrm{Sp}^{(p)}(A)$ is called an *eigenvalue* of A and $x \in \mathcal{H}$ is called an *eigenvector* corresponding to λ if $Ax = \lambda x$. The set $\mathcal{H}_\lambda$ of all eigenvectors corresponding to λ, which can be a subspace, is called the *eigenspace* corresponding to λ. Its dimension, $\dim \mathcal{H}_\lambda$, is called the *multiplicity* of λ. An eigenvalue with multiplicity 1 is called *simple or nondegenerate spectrum.*

Every finite Borel measure $d\mu$ of the real line has the unique decomposition with respect to the Lebesgue measure

$$d\mu = d\mu_{ac} + d\mu_s,$$

where $d\mu_{ac}$ is absolutely continuous with respect to the Lebesgue measure and $d\mu_s$ is singular with respect to the Lebesgue measure (i.e., $d\mu_s$ is supported on a set of Lebesgue measure zero). The singular part $d\mu_s$ can be further decomposed into a (singularly) continuous and a pure-point part,

$$d\mu_s = d\mu_{sc} + d\mu_{pp},$$

where μ_{sc} is continuous on the real line and μ_{pp} is a step function. Accordingly, one defines an absolutely continuous, singularly continuous and pure-point spectrum of a self-adjoint operator A. Note that, in general, the pure point spectrum is not equal to the set of the eigenvalues. The discrete spectrum of the operator A is the set of all eigenvalues which are isolated points of the spectrum and whose corresponding eigenspaces are finite-dimensional. The complement of the discrete spectrum is called the *essential spectrum.*

When A has only discrete spectrum, $\mathrm{Sp}(A) = \mathrm{Sp}^{(p)}(A) = \{\lambda_n\}$. Let P_n be the projection operator onto the eigensubspace $\mathcal{H}_{\lambda_n}$ corresponding to λ_n, then $P_n P_m = \delta_{nm} P_n$ holds, and A can be expressed as $A = \sum_n \lambda_n P_n$. Here δ is the Kronecker's delta which is defined by $\delta_{nm} = 1$ $(n = m)$, $\delta_{nm} = 0$ $(n \neq m)$. More generally, for any self-adjoint operator A, *the spectral theorem* states that there is a family $\{E(\lambda)\}$ of projection operators depending on a real parameter λ which forms the spectral measure such that in particular:

(i) If $\lambda < \mu$, then $E(\lambda)E(\mu) = E(\lambda)$
(ii) $A = \int \lambda \, dE(\lambda)$ $(= \int \lambda E(d\lambda))$.

This $\{E(\lambda)\}$ is often called a *projection valued measure.*

A family of vectors $\{x_n\}$ in $\mathcal{H}$ is called an *orthonormal system* (*ONS*) if $\langle x_k, x_j \rangle = \delta_{jk}$ holds. Moreover, a family of vectors $\{x_n\}$ is called a *completely orthonormal system* (*CONS*)*, or orthnormal basis* (*ONB*), if it is an ONS and $\langle x, x_n \rangle = 0$ for all n implies $x = 0$.

By means of Dirac's notation, various operators are expressed as follows:

1. A is a projection operator $\Leftrightarrow A = \sum_j |x_j\rangle\langle x_j|$ with ONS $\{x_j\}$ of ran A.
2. A is a unitary operator $\Leftrightarrow$ There exist two CONS's $\{x_i\}, \{y_j\}$ such that $A = \sum_j |x_j\rangle\langle y_j|$.
3. A is an isometry operator $\Leftrightarrow$ There exist an ONS $\{x_j\}$ and a CONS $\{y_i\}$ such that $A = \sum_j |x_j\rangle\langle y_j|$.
4. A is a partial-isometry operator $\Leftrightarrow$ There exist two ONSs $\{x_j\}$ and $\{y_j\}$ such that $A = \sum_j |x_j\rangle\langle y_j|$.

4.2.1 Spaces $\mathbf{F}(\mathcal{H}), \mathbf{C}(\mathcal{H}), \mathbf{S}(\mathcal{H})$*, and* $\mathbf{T}(\mathcal{H})$

Let us discuss important subspaces of $\mathbf{B}(\mathcal{H})$.

Definition 4.6

1. An operator $A \in \mathbf{B}(\mathcal{H})$ is said to be a *finite rank* operator if $\dim(\mathrm{ran}\, A) < +\infty$; the set of all finite rank operators is denoted by $\mathbf{F}(\mathcal{H})$.
2. An operator $A \in \mathbf{B}(\mathcal{H})$ is said to be a *compact* operator if $\|Ax_n - Ax\| \to 0$ for any $\{x_n\}$ such that $x_n \to x(w)$; the set of all compact operators is denoted by $\mathbf{C}(\mathcal{H})$.

3. An operator $A \in \mathbf{B}(\mathcal{H})$ is said to be a *Hilbert–Schmidt* operator if for any CONS $\{x_n\}$ the sum $\sum_n \|Ax_n\|$ converges; the set of all Hilbert–Schmidt operators is denoted by $\mathbf{S}(\mathcal{H})$.
4. An operator $A \in \mathbf{B}(\mathcal{H})$ is said to be a *trace class* operator if for any CONS $\{x_n\}$ the sum $\sum_n \langle x_n, Ax_n\rangle$ absolutely converges; the set of all trace class operators is denoted by $\mathbf{T}(\mathcal{H})$.

Remark 4.7 In part 4 above, the *trace* of A is given by $\operatorname{tr} A = \sum_n \langle x_n, Ax_n\rangle$ for any CONS $\{x_n\}$ in the Hilbert space $\mathcal{H}$. The trace does not depend on the choice of CONS.

These subspaces have the following useful properties:

1. $A \in \mathbf{F}(\mathcal{H})$ iff there exist $n \in \mathbb{N}$, $\{\lambda_k; k = 1, \dots, n\} \subset \mathbb{R}$ and ONS $\{x_k\}, \{y_k\}$ such that $A = \sum_{k=1}^{n} \lambda_k |x_k\rangle\langle y_k|$.
2. $A \in \mathbf{C}(\mathcal{H})$ iff there exist a unique $\{\lambda_n\} \subset \mathbb{R}$ with $\lambda_n \downarrow 0$ and ONS $\{x_n\}, \{y_n\}$ such that $A = \sum_n \lambda_n |x_n\rangle\langle y_n|$, where λ_n are the eigenvalues of $|A|$.
3. $A \in \mathbf{S}(\mathcal{H})$ iff $A \in \mathbf{C}(\mathcal{H})$ and $\sum_n |\lambda_n|^2 < +\infty$, where λ_n are the eigenvalues of A.
4. $A \in \mathbf{T}(\mathcal{H})$ iff $A \in \mathbf{S}(\mathcal{H})$ and $\sum_n |\lambda_n| < +\infty$, where λ_n are the eigenvalues of A.
5. $\mathbf{F}(\mathcal{H}) \subset \mathbf{T}(\mathcal{H}) \subset \mathbf{S}(\mathcal{H}) \subset \mathbf{C}(\mathcal{H}) \subset \mathbf{B}(\mathcal{H})$.
6. $\mathbf{S}(\mathcal{H})$ becomes a Hilbert space if given the inner product defined by $\langle\langle A, B\rangle\rangle \equiv \sum_n \langle Ax_n, Bx_n\rangle$ for any CONS $\{x_n\}$. The norm $\|A\|_2 \equiv \sqrt{\langle\langle A, A\rangle\rangle} = \sum_n |\lambda_n|^2$ is larger than the usual norm $\|A\|$, with which $\mathbf{S}(\mathcal{H})$ is a Banach space.
7. $\mathbf{T}(\mathcal{H})$ becomes a Banach space if endowed with the *trace norm* defined by $\|A\|_1 \equiv \sum_n |\lambda_n|$ which is larger than the *Hilbert–Schmidt norm* $\|A\|_2$.

4.3 Direct Sum and Tensor Product of Hilbert Spaces

Let us consider two quantum systems described by $\mathcal{H}$ and $\mathcal{K}$. We, hereafter, need the composite system of these two systems, which are defined on the *direct sum Hilbert space* $\mathcal{H} \oplus \mathcal{K}$ or the *tensor product Hilbert space* $\mathcal{H} \otimes \mathcal{K}$.

The set of pairs of elements of two Hilbert spaces $\mathcal{H}$ and $\mathcal{K}$,

$$\mathcal{H} \oplus \mathcal{K} \equiv \{x \oplus y; x \in \mathcal{H}, y \in \mathcal{K}\},$$

becomes a vector space, where the operation (sum) $\oplus$ is defined by

$$x_1 \oplus y_1 + x_2 \oplus y_2 \equiv (x_1 + x_2) \oplus (y_1 + y_2), \qquad \lambda(x_1 \oplus y_1) \equiv \lambda x_1 \oplus \lambda y_1, \quad \lambda \in \mathbb{K}.$$

In addition,

$$\langle x_1 \oplus y_1, x_2 \oplus y_2\rangle \equiv \langle x_1, x_2\rangle + \langle y_1, y_2\rangle$$

satisfies the conditions of the inner product and makes $\mathcal{H} \oplus \mathcal{K}$ a Hilbert space, which is called the direct sum Hilbert space of $\mathcal{H}$ and $\mathcal{K}$. Identify $x \in \mathcal{H}$ with $x \oplus \mathbf{0}$ and $y \in \mathcal{K}$ with $\mathbf{0} \oplus y$, then $\mathcal{H}$ and $\mathcal{K}$ are closed subspaces of $\mathcal{H} \oplus \mathcal{K}$, and the relations

$\mathcal{H}^\perp = \mathcal{K}$ and $\mathcal{K}^\perp = \mathcal{H}$ hold. A direct sum of an infinite number of Hilbert spaces is defined in the same manner.

Next we define the tensor product Hilbert space of $\mathcal{H}$ and $\mathcal{K}$. On the direct product of Hilbert spaces $\mathcal{H}$ and $\mathcal{K}$

$$\mathcal{H} \times \mathcal{K} \equiv \{(u, v); u \in \mathcal{H}, v \in \mathcal{K}\},$$

define a bilinear functional $x \otimes y$ ($x \in \mathcal{H}$, $y \in \mathcal{K}$) by

$$x \otimes y(u, v) \equiv \langle x, u\rangle\langle y, v\rangle.$$

Denote the set of all linear combinations of such bilinear functionals by $\mathcal{H} \circ \mathcal{K}$. On this set, we define

$$\langle f, g\rangle \equiv \sum_{i=1}^{n}\sum_{k=1}^{m} \lambda_i^* \mu_k \langle x_i, u_k\rangle\langle y_i, v_k\rangle,$$

for any $f = \sum_{i=1}^{n} \lambda_i x_i \otimes y_i$, $g = \sum_{k=1}^{m} \mu_k u_k \otimes v_k \in \mathcal{H} \circ \mathcal{K}$, which gives an inner product on $\mathcal{H} \circ \mathcal{K}$. A Hilbert space obtained as the completion of $\mathcal{H} \circ \mathcal{K}$ w.r.t. this inner product is called the tensor product Hilbert space of $\mathcal{H}$ and $\mathcal{K}$, and it is denoted as $\mathcal{H} \otimes \mathcal{K}$. One can easily see that for any CONS $\{x_i\}$ of $\mathcal{H}$, and any CONS $\{y_k\}$ of $\mathcal{K}$, $\{x_i \otimes y_k\}$ becomes a CONS of $\mathcal{H} \otimes \mathcal{K}$.

Example 4.8 Take $\mathcal{H} = \mathbb{C}^3$ and $\mathcal{K} = \mathbb{C}^2$. Then for $x = (a_1, a_2, a_3)^t \in \mathcal{H}$, $y = (b_1, b_2)^t \in \mathcal{K}$,

$$x \oplus y = (a_1, a_2, a_3, b_1, b_2)^t, \qquad x \otimes y = (a_1b_1, a_2b_1, a_3b_1, a_1b_2, a_2b_2, a_3b_2)^t.$$

Exercise 4.9 Prove that

$$\mathbb{C}^m \otimes \mathbb{C}^n = \mathbb{C}^{mn}.$$

Notice the difference between the tensor product and the direct sum. For example, we have $\mathbb{C}^m \otimes \mathbb{C}^n = \mathbb{C}^{mn}$ but for the direct sum one has $\mathbb{C}^m \oplus \mathbb{C}^n = \mathbb{C}^{m+n}$. In particular, one can prove the relation

$$L^2(\mathbb{R}^m) \otimes L^2(\mathbb{R}^n) = L^2(\mathbb{R}^{m+n}).$$

Finally, we discuss operators on $\mathcal{H} \oplus \mathcal{K}$ and $\mathcal{H} \otimes \mathcal{K}$, and the partial trace on $\mathcal{H} \otimes \mathcal{K}$. For A and B, operators on $\mathcal{H}$ and $\mathcal{K}$, respectively, the operator $A \oplus B$ on $\mathcal{H} \oplus \mathcal{K}$ and the operator $A \otimes B$ on $\mathcal{H} \otimes \mathcal{K}$ are defined for any $x \in \mathcal{H}$ and $y \in \mathcal{K}$ by

$$A \oplus B(x \oplus y) \equiv Ax \oplus By, \qquad A \otimes B(x \otimes y) \equiv Ax \otimes By.$$

For an arbitrary $Q \in \mathbf{B}(\mathcal{H} \otimes \mathcal{K})$, one can define an operator Q' in $\mathcal{K}$ by

$$\langle x, Q'y\rangle \equiv \sum_k \langle x \otimes z_k, Qy \otimes z_k\rangle$$

for any $x, y \in \mathcal{H}$ and a CONS $\{z_n\} \subset \mathcal{K}$. This Q' is a bounded operator on $\mathcal{H}$ and is called the partial trace of Q on $\mathcal{K}$, denoted by $Q' \equiv \operatorname{tr}_{\mathcal{K}} Q$.

4.4 Fock Space

4.4.1 Bosonic Fock Space

Fock space is a tensor algebra over a Hilbert space. Let $\mathcal{H}$ be a Hilbert space (in this context it will be called the one-particle Hilbert space). We denote by $\mathcal{H}^{(n)}$ the n-fold tensor product $\mathcal{H}^{(n)} = \mathcal{H} \otimes \cdots \otimes \mathcal{H}$, set $\mathcal{H}^{(0)} = \mathbb{C}$ and define the Hilbert space

$$F(\mathcal{H}) = \mathbb{C} \oplus \mathcal{H}^{(1)} \oplus \mathcal{H}^{(2)} \oplus \cdots = \bigoplus_{n=0}^{\infty} \mathcal{H}^{(n)}.$$

$F(\mathcal{H})$ is called the (*total*) *Fock space* over $\mathcal{H}$. A vector $\Psi \in F(\mathcal{H})$ is written as $\Psi = (\psi^{(0)}, \psi^{(1)}, \ldots)$ with $\psi^{(n)} \in \mathcal{H}^{(n)}$.

For example, if $\mathcal{H} = L^2(\mathbb{R})$, then $\mathcal{H}^{(n)} = L^2(\mathbb{R}^n)$ and an element $\Psi \in F(\mathcal{H})$ is a sequence of functions $\Psi = (\psi^{(0)}, \psi^{(1)}(x_1), \psi^{(2)}(x_1, x_2), \ldots)$ so that

$$\|\Psi\|^2 = \left|\psi^{(0)}\right|^2 + \sum_{n=1}^{\infty} \left|\psi^{(n)}(x_1, \ldots, x_n)\right|^2 dx_1 \cdots dx_n < \infty.$$

There are two important subspaces of the total Fock space $F(\mathcal{H})$ which are called the *Boson* (or *symmetric*) Fock space and the *Fermion* (or *antisymmetric*) Fock space. Let $\mathcal{P}_n$ be the permutation group on n elements, and let $\{f_k\}$ be a basis for $\mathcal{H}$. For each $\sigma \in \mathcal{P}_n$ we define an operator, which we also denote by σ, on the basis elements of $\mathcal{H}^{(n)}$ by

$$\sigma(f_{k_1} \otimes \cdots \otimes f_{k_n}) = f_{k_{\sigma(1)}} \otimes \cdots \otimes f_{k_{\sigma(n)}}.$$

σ extends by linearity to a bounded operator on $\mathcal{H}^{(n)}$, so we can define

$$S_n = \frac{1}{n!} \sum_{\sigma \in \mathcal{P}_n} \sigma.$$

The operator S_n is an orthogonal projection:

$$S_n^2 = S_n, \qquad S_n^* = S_n.$$

The range of S_n is called the *n-fold symmetric tensor product of* $\mathcal{H}$. We denote $\mathcal{H}_s^{(n)} = S_n \mathcal{H}^{(n)}$. In the case where $\mathcal{H} = L^2(\mathbb{R})$ and $\mathcal{H}^{(n)} = L^2(\mathbb{R}^n)$, $\mathcal{H}_s^{(n)}$ is the subspace of $L^2(\mathbb{R}^n)$ of all functions left invariant under any permutation of the variables. We define the Hilbert space

$$F_s(\mathcal{H}) = \bigoplus_{n=0}^{\infty} \mathcal{H}_s^{(n)}.$$

$F_s(\mathcal{H})$ is called the *bosonic Fock space* over $\mathcal{H}$. We call $\mathcal{H}_s^{(n)}$ the n-particle subspace of $F_s(\mathcal{H})$. There is an operator S on $F(\mathcal{H})$ such that $S \restriction \mathcal{H}^{(n)} = S_n \mathcal{H}^{(n)}$. The vector $\Psi_0 \in F_s(\mathcal{H})$,

$$\Psi_0 = (1, 0, 0, \dots),$$

plays a special role; it is called the *vacuum vector*.

4.4.2 Annihilation and Creation Operators

Let us fix $f \in \mathcal{H}$. For vectors in $\mathcal{H}^{(n)}$ of the form $f_1 \otimes \cdots \otimes f_n$ we define a map $b(f) : \mathcal{H}^{(n)} \to \mathcal{H}^{(n-1)}$ by

$$b(f)(f_1 \otimes \cdots \otimes f_n) = \langle f, f_1 \rangle (f_2 \otimes \cdots \otimes f_n).$$

For $n = 0$ we define

$$b(f)\Psi_0 = 0.$$

$b(f)$ extends by linearity to a bounded operator from $F(\mathcal{H})$ to $F(\mathcal{H})$. The adjoint operator $b(f)^*$ takes each $\mathcal{H}^{(n)}$ to $\mathcal{H}^{(n+1)}$ with the action

$$b(f)^*(f_1 \otimes \cdots \otimes f_n) = f \otimes f_1 \otimes \cdots \otimes f_n.$$

The map $f \to b(f)^*$ is linear, but $f \to b(f)$ is antilinear. We have the following free commutation relation on $\mathcal{H}^{(n)}$:

$$b(f)b(g)^* = \langle f, g \rangle, \quad f, g \in \mathcal{H}.$$

A vector $\Psi = \{\psi^{(n)}\}_{n=0}^{\infty}$ for which $\psi^{(n)} = 0$ for all but finitely many n is called a *finite particle vector*. The set of finite particle vectors will be denoted F_0.

Let A be a self-adjoint operator on $\mathcal{H}$ with domain D_A. Let $D = \{\Psi \in F_0 \mid \psi^{(n)} \in \bigotimes_{i=1}^n D_A$ for each $n\}$ and define $\Gamma(A)$ on $D \cap \mathcal{H}_s^{(n)}$ as

$$\Gamma(A) = A \otimes I \otimes \cdots \otimes I + I \otimes A \otimes \cdots \otimes I + \cdots + I \otimes \cdots \otimes I \otimes A.$$

The operator $\Gamma(A)$ is called the *second quantization* of the operator A; it is essentially self-adjoint on D. For example, for $A = I$ the operator $N = \Gamma(I)$ is called the *number operator*; for $\psi^{(n)} \in \mathcal{H}_s^{(n)}$ one has $N\psi^{(n)} = n\psi^{(n)}$.

For the second quantization of the unitary operator $e^{\mathrm{i}tA}$, we have the formula

$$\Gamma\big(e^{\mathrm{i}tA}\big) = e^{\mathrm{i}t\Gamma(A)}, \quad t \in \mathbb{R}.$$

We define the annihilation operator $a(f)$ on $F_s(\mathcal{H})$ with domain F_0 by

$$a(f) = \sqrt{N+1}\, b(f).$$

It takes each n-particle subspace into $(n-1)$-particle subspace. We have

$$a(f)\Psi_0 = 0.$$

The adjoint operator $a(f)^*$ is called a *creation operator*,

$$a(f)^* \restriction F_0 = S b(f)^* \sqrt{N+1}.$$

One has the following commutation relations on F_0:

$$\big[a(f), a(g)^*\big] = (f, g), \quad f, g \in \mathcal{H},$$
$$\big[a(f), a(g)\big] = 0, \qquad \big[a(f)^*, a(g)^*\big] = 0.$$

Example 4.10 If $(X, d\mu)$ is a measure space and $\mathcal{H} = L^2(X, d\mu)$, then on the Fock space $F_s(\mathcal{H})$ the annihilation and creation operators $a(f)$ and $a(f)^*$ are given by

$$\big(a(f)\psi\big)^{(n)}(p_1, \ldots, p_n) = \sqrt{n+1} \int_X \bar{f}(p)\psi^{(n+1)}(p, p_1, \ldots, p_n)\, d\mu(p),$$

$$\big(a(f)^*\psi\big)^{(n)}(p_1, \ldots, p_n) = \frac{1}{\sqrt{n}} \sum_{i=1}^{n} f(p_i)\psi^{(n-1)}\big(p_1, \ldots, \hat{p}_i, \ldots, p_n\big)$$

where $\hat{p}_i$ means that the p_i is omitted.

Let H be a self-adjoint operator on $F_s(\mathcal{H})$ such that the unitary operator $e^{\mathrm{i}tH}$ maps F_0 to itself for any $t \in \mathbb{R}$. For real-valued functions $f \in L^2(X, d\mu)$, we define on F_0 the *quantum field* operator $A(f, t)$ by

$$A(f, t) = e^{\mathrm{i}tH}\big(a(f) + a(f)^*\big)e^{-\mathrm{i}tH}.$$

4.4.3 Fermionic Fock Space

Now we define the *Fermionic* (*antisymmetric*) Fock space. Let $\varepsilon : \mathcal{P}_n \to \{-1, 1\}$ be the function which is 1 on even permutations and -1 on odd permutations. Define

$$A_n = \frac{1}{n!} \sum_{\sigma \in \mathcal{P}_n} \varepsilon(\sigma)\sigma.$$

A_n is an orthogonal projection on $\mathcal{H}^{(n)}$. The space $\mathcal{H}_a^{(n)} = A_n\mathcal{H}^{(n)}$ is called the n-fold antisymmetric tensor product of $\mathcal{H}$. In the case where $\mathcal{H} = L^2(\mathbb{R})$, $\mathcal{H}_a^{(n)}$ is the subspace of $L^2(\mathbb{R}^n)$ consisting of the odd functions under the interchange of two coordinates. The Hilbert space

$$F_a(\mathcal{H}) = \bigoplus_{n=0}^{\infty} \mathcal{H}_a^{(n)}$$

is called the Fermionic Fock space over $\mathcal{H}$.

4.4.4 Weyl Representation

Let s be a symplectic form on a linear space $\mathcal{H}$, that is, the mapping $s : \mathcal{H} \times \mathcal{H} \to \mathbb{C}$ which is not degenerated and satisfies $s(f, g) = -s(g, f)$. Typically, $\mathcal{H}$ is a complex Hilbert space and $s(f, g) = -\operatorname{Im}\langle f, g\rangle$. The *Weyl algebra*, denoted by CCR($\mathcal{H}$) (CCR; canonical commutation relation), is generated by unitaries $\{W(f); f \in \mathcal{H}\}$ satisfying the *Weyl form* of the canonical commutation relation:

$$W(f)W(g) = e^{\mathrm{i}s(f,g)}W(f+g) \quad (f, g \in \mathcal{H}).$$

Since the linear hull of the unitaries $W(f)$ is dense in CCR($\mathcal{H}$), any state is determined uniquely by its values taken on the Weyl unitaries. The most important state of the Weyl algebra is the Fock state which is given by

$$\varphi\big(W(f)\big) = e^{-\|f\|^2/2} \quad (f \in \mathcal{H}).$$

The GNS Hilbert space (see Sect. 4.6) corresponding to the Fock state is called the *(bosonic) Fock space* $\Gamma(\mathcal{H})$ and the cyclic vector Φ is named *vacuum*. The states

$$\varphi_f(\cdot) = \varphi\big(W(f)^* \cdot W(f)\big)$$

are called *coherent states*, and they are induced by the coherent vectors

$$\pi_F\big(W(f)\big)\Phi = \Phi_f$$

in the *Fock representation* π_F:

$$\varphi\big(W(f)\big) = \big\langle\Phi, \pi_F\big(W(f)\big)\Phi\big\rangle.$$

One has

$$\begin{aligned}\langle\Phi_f, \Phi_g\rangle &= \varphi\big(W(f)^*W(g)\big)\\ &= e^{-\frac{1}{2}\|g-f\|^2}e^{-\mathrm{i}s(f,g)}\\ &= e^{-\frac{1}{2}(\|f\|^2+\|g\|^2)+\langle f,g\rangle}\end{aligned}$$

and

$$\begin{aligned}\varphi_f\big(W(g)\big) &= e^{-\frac{1}{2}\|g\|^2-2\mathrm{i}s(f,g)}\\ &= e^{-\frac{1}{2}\|g\|^2+2\mathrm{i}\mathrm{Im}\langle f,g\rangle} \quad (f,g\in\mathcal{H}).\end{aligned}$$

The field operators are obtained as the generators of the unitary groups $t \mapsto \pi_F(W(tf))$ in the Fock representation. In other words, $B(f)$ is an unbounded self-adjoint operator on $\Gamma(\mathcal{H})$ such that

$$B(f) = \frac{d}{dt}\pi_F\big(W(tf)\big)\big|_{t=0}$$

with an appropriate domain. The creation and annihilation operators are defined as

$$\begin{aligned}a(f)^* &= \frac{1}{2}\big(B(\mathrm{i}f) - \mathrm{i}B(f)\big),\\ a(f) &= \frac{1}{2}\big(B(\mathrm{i}f) + \mathrm{i}B(f)\big).\end{aligned}$$

The positive self-adjoint operator

$$n(f) = a(f)^*a(f)$$

has spectrum Z^+, and it is called the *particle number operator* (for the "f-mode").

4.5 Hida White Noise Analysis

4.5.1 White Noise

White noise W_t in communication theory is defined as a stationary stochastic process whose power density spectrum is constant at all frequencies. Another definition of white noise is that it is the time derivative of the Brownian motion process B_t,

$$W_t = \frac{d}{dt}B_t = \dot{B}_t.$$

Therefore, $\{W_t\}$ should satisfy the following conditions:

$$\begin{aligned}E(W_t) &= 0,\\ E(W_tW_s) &= \delta(t-s).\end{aligned}$$

It means that the white noise should be treated not as usual random variable but a generalized function.

Such a measurable stochastic process does not exists, but there is a generalized stochastic process with these properties. Let $S(\mathbb{R})$ be the Schwartz space of rapidly decreasing functions on $\mathbb{R}$ with the usual topology, and let $S^*(\mathbb{R})$ denote its dual space of tempered distributions. We denote by $\langle \omega, f \rangle$ the action of $\omega \in S^*(\mathbb{R})$ on $f \in S(\mathbb{R})$. Also we denote $\Omega = S^*(\mathbb{R})$. Since $S^*(\mathbb{R})$ is a nuclear, countably Hilbert space, there exists, by the Bochner–Minlos theorem, a probability measure μ on the measurable space $(\Omega, \mathcal{B})$ where $\mathcal{B}$ is the σ-algebra of the Borel subsets of Ω such that its characteristic functional is

$$E\left(e^{\mathrm{i}\langle \cdot, f \rangle}\right) = \int_\Omega e^{\mathrm{i}\langle \omega, f \rangle} \, d\mu(\omega) = e^{-\frac{1}{2}\|f\|^2},$$

where $\|f\|^2 = \int |f(t)|^2 \, dt$. The triple $(\Omega, \mathcal{B}, \mu)$ is called the *white noise probability space* and the (generalized) *white noise process* is the map

$$W : S(\mathbb{R}) \times \Omega \to \mathbb{R}$$

given by

$$W(f, \omega) = \langle \omega, f \rangle.$$

By using the relation

$$E\left(\langle \cdot, f \rangle^2\right) = \|f\|^2,$$

one can extend the definition $W(f, \omega)$ from $f \in S(\mathbb{R})$ to all $f \in L^2(\mathbb{R})$. If we take $f(s) = 1_{[0,t]}(s)$ for all $t > 0$, we can define

$$X_t(\omega) = W(1_{[0,t]}, \omega).$$

One can check that X_t is a Gaussian stochastic process with mean 0 and covariance

$$E(X_t X_\tau) = \min(t, \tau).$$

A continuous version of X_t is a standard Brownian motion B_t. The relation $\dot{B}_t = W_t$ is valid in the sense of generalized functions.

4.5.2 Hida Distributions

According to Hida, the right starting point for the theory of stochastic processes is not the Brownian motion B_t, but the white noise $W_t = \dot{B}_t$. In particular, one studies functionals of the white noise

$$F\left(\dot{B}_t(\omega), t \in T\right),$$

where T is an interval and $\dot{B}_t$ is interpreted as a family of independent identically distributed random variables, a kind of an uncountable system of coordinates

$x_t = \dot{B}_t$. An appropriate infinite-dimensional calculus (white noise calculus) is constructed on a suitable space $(S)^*$ of generalized white noise functionals, *Hida distributions*. It is constructed as a generalization of the *Gel'fand triple*

$$S \subset L^2 \subset S^*.$$

One defines $S = S(\mathbb{R})$ as a countable intersection of Hilbert spaces of operator domains ($p = 1, 2, \ldots$)

$$S = \bigcap_p S_p, \quad S_p = D(A^p),$$

where A is a self-adjoint operator in $L^2 = L^2(\mathbb{R}, dt)$,

$$A = -\frac{d^2}{dt^2} + t^2 + 1.$$

Hida's Gel'fand triple is defined as:

$$(S) \subset (L^2) \subset (S)^*.$$

Here (L^2) is the Fock space $(L^2) = L^2(S^*(\mathbb{R}), d\mu)$, and the space (S) again is obtained as an intersection of operator domains, but now one uses a second quantization $\Gamma(A)$ of the operator A in the Fock space (L^2):

$$(S) = \lim_p (S)_p, \quad (S)_p = D(\Gamma(A^p)).$$

The operator $\Gamma(A)$ acts on the *normal* (or *Wick*) *ordered* monomials $:W(f_1) \cdots W(f_n):$ by

$$\Gamma(A)(:W(f_1) \cdots W(f_n):) = :W(Af_1) \cdots W(Af_n):.$$

Here $W(f) = W(f, \omega)$ is the white noise and the Wick ordered monomials are defined by the recurrent procedure:

$$:W(f_1) \cdots W(f_n): = (1 - P_{n-1}) W(f_1) \cdots W(f_n),$$

where P_n projects onto the subspace of polynomials of order at most n.

Elements form the dual space $(S)^*$ are called Hida distributions. This construction permits defining important functionals such as local Wick powers $:\omega^n(t):$, Donsker's δ-function $\delta(B_t - \alpha)$, and others. There are many applications of the white noise analysis in quantum probability and infinite-dimensional analysis [34, 335–337].

In particular infinite-dimensional differential operators ∂_t and ∂_t^* are introduced, which play the role of annihilation and creation operators and satisfy the commutation relations

$$[\partial_t, \partial_s^*] = \delta(t - s).$$

An important tool in the white noise analysis is the infinite-dimensional rotation group $O(E)$ (we take $E = S(\mathbb{R})$) which consists of those orthogonal linear transformations of the real Hilbert space $L^2(\mathbb{R})$ which are continuous transformations of E. The white noise measure μ is invariant under the natural action of the group $O(E)$.

4.5.3 Lie Algebra of the Hida Rotation Group

Let us discuss the Lie algebra for the Hida rotation group.

Let us consider the Gel'fand triple of real spaces

$$E \subset L^2(\mathbb{R}) \subset E^*.$$

Let us take $E = S(\mathbb{R})$ as the space of test functions (C^∞ and fast decreasing). The *Hida rotation group* $O(E)$ consists of orthogonal transformations of $L^2(\mathbb{R})$ which are homeomorphisms of E. Denote $G = L^2(\mathbb{R})$. The group $O(E)$ is a subgroup of $O(G)$.

We want to describe the Lie algebra of $O(E)$. Since $O(E)$ is a subgroup of $O(G)$ the Lie algebra of $O(E)$ should be a subalgebra of the Lie algebra of $O(G)$. Therefore, we first describe the Lie algebra of $O(G)$.

The Lie algebra of $O(G)$ consists of all anti-self-adjoint operators on the real Hilbert space $G = L^2(\mathbb{R})$. Indeed, if L is an element of the Lie algebra of $O(G)$ then the element of the group $O(G)$ can be written formally as $e^{\varepsilon L} = 1 + \varepsilon L + \cdots$, and from the infinitesimal relation

$$\big\langle (1+\varepsilon L) f, (1+\varepsilon L) g\big\rangle = \langle f, g\rangle$$

we get $L^* = -L$. Examples of such operators are integral operators of the form

$$Lf(t) = \int K(t,s) f(s)\, ds, \quad K(t,s) = -K(s,t)$$

and differential operators

$$L = \frac{d}{dt}, \frac{d^3}{dt^3}, \ldots;$$

$$L_{nm} = t^n \frac{d^m}{dt^m} - (-1)^m \frac{d^m}{dt^m} t^n, \quad m, n = 0, 1, 2, \ldots.$$

Now it would be interesting to investigate which anti-self-adjoint operators belong to the Lie algebra of the Hida rotation group $O(E)$, that is, for which anti-self-adjoint operators L the following condition holds: If $f \in S$ then $e^{\varepsilon L} f \in S$.

Two Elemental Stochastic Processes

There are two basic *elemental* stochastic processes: the *Wiener process* or *white noise* and the *Poisson process* $P(t)$ or the *Poisson noise* $\dot{P}(t)$. The characteristic functional of the Poisson noise has the form

$$C_P(\xi) = \exp\left[\lambda \int \left(e^{i\xi(t)} - 1\right) dt\right]$$

where $\xi \in E$ and $\lambda > 0$ is the intensity. It defines a probability measure μ_P on E.

4.6 Elements of Operator Algebras

Any self-adjoint operator in $\mathbf{B}(\mathcal{H})$ is usually considered as a physical observable, but this is not a trivial fact: When $\dim \mathcal{H} = \infty$, $\mathbf{B}(\mathcal{H})$ contains an infinite number of self-adjoint operators and not all of them may be physically realizable. Therefore, it is natural to think of a smaller subset $\mathcal{N}$ of $\mathbf{B}(\mathcal{H})$ which is physically more realistic. This subset $\mathcal{N}$ should satisfy the following conditions

(i) $\mathcal{N}$ is a *-algebra (i.e., for any $A, B \in \mathcal{N}$ and $\lambda \in \mathbb{C}$, $A + \lambda B \in \mathcal{N}, AB \in \mathcal{N}, A^* \in \mathcal{N}$ (A^* is the adjoint of A)).
(ii) If $\{A_\alpha\} \subset \mathcal{N}$ converges to A in the weak sense (i.e., $\langle x, A_\alpha y\rangle \to \langle x, Ay\rangle$ for any $x, y \in \mathcal{H}$), then $A \in \mathcal{N}$.

Condition (i) is imposed for mathematical convenience in operations of addition and multiplication; and condition (ii) is imposed for physical convenience because the expectation value is computed by taking an inner product and the subset should be closed under this operation.

When $\mathcal{N}$ contains a unit operator I, the above conditions are identical to

$$\mathcal{N}'' = \mathcal{N},$$

where $\mathcal{N}'' = (\mathcal{N}')'$ and

$$\mathcal{N}' \equiv \left\{A \in \mathbf{B}(\mathcal{H}); AB = BA, B \in \mathcal{N}\right\}.$$

The algebra $\mathcal{N}$ satisfying the above condition is called a *von Neumann algebra*. A quantum system is now considered in terms of a von Neumann algebra $\mathcal{N}$. A physical observable is a self-adjoint element of $\mathcal{N}$ and a *state* is a positive linear normalized continuous functional on $\mathcal{N}$; that is, for any $A, B \in \mathbf{B}(\mathcal{H})$ and $\lambda \in \mathbb{C}$,

(i) $\varphi(A^*A) \geq 0$
(ii) $\varphi(A + \lambda B) = \varphi(A) + \lambda\varphi(B)$
(iii) $\varphi(I) = 1$
(iv) if $\|A_n - A\| \to 0$, then $|\varphi(A_n) - \varphi(A)| \to 0$.

Next we discuss some topologies on $\mathbf{B}(\mathcal{H})$. Let $\{A_\lambda; \lambda \in J\}$ be a *net* (that is, J is an ordered set, λ converges to some element in J with this order; refer to [534]) in $\mathbf{B}(\mathcal{H})$, and $A_\lambda \xrightarrow{\tau} A$ means that A_λ converges to A as $\lambda \to \infty$ with respect to the (operator) topology τ:

1. Uniform (operator) topology $\tau^u \overset{\text{def}}{\Longleftrightarrow} \|A_\lambda - A\| \to 0$
2. Strong (operator) topology $\tau^s \overset{\text{def}}{\Longleftrightarrow} \|(A_\lambda - A)x\| \to 0, \forall x \in \mathcal{H}$
3. Weak (operator) topology $\tau^w \overset{\text{def}}{\Longleftrightarrow} \langle x, (A_\lambda - A)y\rangle \to 0, \forall x, y \in \mathcal{H}$
4. Ultrastrong (operator) topology $\tau^{us} \overset{\text{def}}{\Longleftrightarrow}$ For any $\{x_n\} \subset \mathcal{H}_F \equiv \{\{x_n\} \subset \mathcal{H}$: $\sum_{n=1}^{\infty} \|x_n\|^2 < +\infty, \sum_{n=1}^{\infty} \|(A_\lambda - A)x_n\|^2 \to 0\}$
5. Ultraweak (operator) topology $\tau^{uw} \overset{\text{def}}{\Longleftrightarrow} \sum_{n=1}^{\infty} |\langle x_n, (A_\lambda - A)y_n\rangle| \to 0, \forall \{x_n\}, \{y_n\} \in \mathcal{H}_F$
6. Strong $*$ (operator) topology $\tau^{s*} \overset{\text{def}}{\Longleftrightarrow} \|(A_\lambda - A)x\|^2 + \|(A_\lambda^* - A^*)x\|^2 \to 0, \forall x \in \mathcal{H}$
7. Ultrastrong $*$ (operator) topology $\tau^{us*} \overset{\text{def}}{\Longleftrightarrow} \sum_{n=1}^{\infty} \{\|(A_\lambda - A)x_n\|^2 + \|(A_\lambda^* - A)x_n\|^2\} \to 0, \{x_n\} \in \mathcal{H}$.

These topologies are related as follows:

$$\begin{array}{ccccccc} \tau^u & > & \tau^{us*} & > & \tau^{us} & > & \tau^{uw} \\ & & \vee & & \vee & & \vee \\ & & \tau^{s*} & > & \tau^{s} & > & \tau^{w} \end{array}$$

(strong topology > weak topology). When $\dim \mathcal{H} < +\infty$, all topologies coincide.

Denote by $\mathcal{B}$ a uniformly bounded subset of $B(\mathcal{H})$, for example, the unit ball $\mathcal{B}_1 \equiv \{A \in B(\mathcal{H}) : \|A\| \le 1\}$. On this ball, $\tau^{uw} = \tau^w, \tau^{us} = \tau^s, \tau^{us*} = \tau^{s*}$.

When $A_\lambda \to A, B_\lambda \to B$ in a certain topology τ,we shall show whether the following proposition holds or not: (a) $A_\lambda^* \to A^*$; (b) $A_\lambda Q \to AQ$ and $QA_\lambda \to QA, \forall Q \in B(\mathcal{H})$; (c) $A_\lambda B_\lambda \to AB$; (d) $A_\lambda B_\lambda \to AB$ for $\{A_\lambda\}, \{B_\lambda\} \subset \mathcal{B}$.

We write ◯ if the proposition holds, and write × if it does not.

τ	(a)	(b)	(c)	(d)
τ^u	◯	◯	◯	◯
τ^s	×	◯	×	◯
τ^w	◯	◯	×	×
τ^{us}	×	◯	×	◯
τ^{uw}	◯	◯	×	×
τ^{s*}	◯	◯	×	◯
τ^{us*}	◯	◯	×	◯

Note that $\tau_1 < \tau_2$ implies $\mathfrak{M}^{\tau_2}$ (the closure of τ_2-topology) $\subset \mathfrak{M}^{\tau_1}$ for an algebra $\mathfrak{M} \subset \mathcal{B}(\mathcal{H})$.

Let $\mathfrak{M}$ be a $*$-algebra such that $\mathfrak{M}^{us} = \mathfrak{M}$, so that $\mathfrak{M}$ is von Neumann algebra. It is known that if $\mathfrak{M}$ contains unity I, then $\mathfrak{M}^{us} = \mathfrak{M}''$. A linear functional φ

is τ-continuous on $\mathfrak{M}$ if $A_\lambda \to A(\tau)$ implies $\varphi(A_\lambda - A) \to 0$. When $\tau_1 < \tau_2$, τ_1-continuity of φ implies τ_2-continuity of φ.

Denote by $\mathfrak{M}^*$ the set of all τ^u-continuous linear functionals. Put $\mathfrak{M}^*_+ \equiv \{\varphi \in \mathfrak{M}^*; \varphi(A^*A) \geq 0, \forall A \in \mathfrak{M}\}$, $\overline{\mathfrak{M}_\sim} \equiv \{\varphi \in \mathfrak{M}^*; \tau^w\text{-continuous}\}$, $\mathfrak{M}_* \equiv \{\varphi \in \mathfrak{M}^*; \tau^{uw}\text{-continuous}\}$.

We have the following properties:

1. $\overline{\mathfrak{M}_\sim} = \mathfrak{M}^*$, where the closure "$-$" is taken w.r.t. the norm $\|\cdot\|$ on $\mathfrak{M}$.
2. $\varphi \in \mathfrak{M}_\sim \Leftrightarrow \exists\{x_n\}_{n=1}, \{y_n\}_{n=1} \subset \mathcal{H}$ and $N \in \mathbb{N}$ such that $\varphi(A) = \sum_{n=1}^N \langle x_n, Ay_n\rangle$.
3. $\varphi \in \mathfrak{M}^* \Leftrightarrow \exists\{x_n\}_{n=1}, \{y_n\}_{n=1} \subset \mathcal{H}$ such that $\sum_{n=1}^\infty \|x_n\|^2 < +\infty$, $\sum_{n=1}^\infty \|y_n\|^2 < +\infty$, $\varphi = \sum_n \langle x_n, Ay_n\rangle$.

When $\mathfrak{M}$ is identical with $\mathbf{B}(\mathcal{H})$, we have $\mathfrak{M}_* = T(\mathcal{H})$ (trace class). This identity is particularly important.

Theorem 4.11 *$\mathfrak{M}_*$ is a Banach space with respect to the norm of $\mathfrak{M}^*$ and* $(\mathfrak{M}_*)^* = \mathfrak{M}$.

Definition 4.12 $\varphi \in \mathfrak{M}_*$ is normal if $0 \leq A_\lambda \uparrow A$ implies $\varphi(A_\lambda) \uparrow \varphi(A)$.

Let $\mathcal{N}$ be a von Neumann algebra and $\mathfrak{S}$ be the set of all states on $\mathcal{N}$; let $\mathfrak{S} \equiv \mathcal{N}^*_+$ and $\mathfrak{S}_*$ be the set of all normal states on $\mathcal{N}$.

Let us list some properties of a state φ and the space $\mathfrak{S}$.

1. The norm of a state φ is given by $\|\varphi\| = \sup\{|\varphi(A)|; A \in \mathcal{N}, \|A\| \leq 1\} = \varphi(I)$ (when $I \in \mathcal{N}$).
2. A state φ is said to be *faithful* if $\varphi(A^*A) = 0$ implies $A = 0$.
3. A state φ is said to be *pure* if φ cannot be decomposed into a convex combination of other states. (Here we assume that $\mathcal{N}$ contains I.)
4. A state φ is said to be a *mixture* if φ is not pure.
5. Two different topologies are often used in $\mathfrak{S}$:
   (i) Uniform topology—ε-neighborhood $N_\varepsilon(\varphi)$ of φ is
   $$N_\varepsilon(\varphi) = \left\{\psi \in \mathfrak{S}_*; \|\varphi - \psi\| < \varepsilon\right\}.$$
   (ii) Weak $*$ topology—ε-neighborhood $N_\varepsilon(\varphi)$ of φ is
   $$N_\varepsilon(\varphi) = \left\{\psi \in \mathfrak{S}_*; \left|\varphi(A_k) - \psi(A_k)\right| < \varepsilon, A_1, \ldots, A_n \in \mathcal{A}\right\}.$$
6. $\mathfrak{S}$ is the weak $*$ closed convex hull of $\operatorname{ex}\mathfrak{S}$, the set of all extreme points (pure states) of $\mathfrak{S}$; $\mathfrak{S} = {}^{w^*}\overline{\operatorname{co}}\operatorname{ex}\mathfrak{S}$.
7. The expectation value of a self-adjoint element A of $\mathcal{N}$ in a state φ is given by $\varphi(A)$.
8. GNS (Gelfand–Naimark–Segal construction theorem) For any state φ, there exist
   (i) a Hilbert space $\mathcal{H}_\varphi$

(ii) a representation π_φ (i.e., $\pi_\varphi : \mathcal{N} \to B(\mathcal{H}_\varphi)$ such that $\pi_\varphi(AB) = \pi_\varphi(A)\pi_\varphi(B), \pi_\varphi(A^*) = \pi_\varphi(A)^*$)

(iii) and a cyclic vector x_φ (i.e., $\{\pi_\varphi(A)x_\varphi; A \in \mathcal{N}\}^- = \mathcal{H}_\varphi$) such that

$$\varphi(A) = \langle x_\varphi, \pi_\varphi(A)x_\varphi \rangle.$$

The above triple $(\mathcal{H}_\varphi, \pi_\varphi, x_\varphi)$ is uniquely determined up to unitary equivalence. Note that the Hilbert space $\mathcal{H}_\varphi$ is different from the original Hilbert space $\mathcal{H}$.

9. For a state φ, the following are equivalent:
   (i) φ is normal
   (ii) There exists a density operator ρ on $\mathcal{H}$ such that $\varphi(\cdot) = \operatorname{tr} \rho \cdot$.
10. Let $\mathcal{N}$ be a commutative von Neumann algebra on a separable Hilbert space $\mathcal{H}$. Then there exists a Hausdorff space Ω and a probability measure μ such that

$$\mathcal{N} = L^\infty(\Omega, \mu).$$

This means that a von Neumann algebraic description is an extension of the classical probability theory.

In some physical models with infinite degrees of freedom, it is not convenient to fix a Hilbert space from the beginning. The C^*-algebraic expression is useful in such cases. A *C^*-algebra* $\mathcal{A}$ is a Banach $*$-algebra (i.e., a norm $\|\cdot\|$ is defined in $\mathcal{A}$ and it is a complete $*$-algebra w.r.t. $\|\cdot\|$) with $\|A^*A\| = \|A\|^2$, $\|A^*\| = \|A\|$. A state φ on $\mathcal{A}$ is a positive linear normalized continuous functional on $\mathcal{A}$. The properties (1)–(8) of a von Neumann algebra are satisfied in $\mathcal{A}$. In particular, the usual Hilbert space formalism is recovered through the GNS construction theorem above. That is, for any state $\varphi \in \mathfrak{S}$, the state space of a C^*-algebra $\mathcal{A}$, there exists a triple $(\mathcal{H}_\varphi, \pi_\varphi, x_\varphi)$ satisfying the above property 8. Remark that any von Neumann algebra is a C^*-algebra, and for a C^*-algebra $\mathcal{A}$, $\{\pi_\varphi(A); A \in \mathcal{A}\}'' =: \pi_\varphi(\mathcal{A})''$ is a von Neumann algebra. Moreover,

Theorem 4.13 *For general C^*-algebra $\mathcal{A}$, there exists a Hilbert space $\mathcal{H}$ such that a C^*-algebra $\mathcal{B}$ ($\subset \mathbf{B}(\mathcal{H})$) on $\mathcal{H}$ is isometrically isomorphic to $\mathcal{A}$.*

In the sequel, we assume that every von Neumann algebra or C^*-algebra contains a unit operator I (i.e., $IA = AI$ for any element of algebra). Let $P(\mathcal{N})$ be the set of all projections in $\mathcal{N}$. For any $E, F \in P(\mathcal{N})$, E dominates F (denoted by $F \le E$) if $\operatorname{ran} F \le \operatorname{ran} E$, and E is equivalent to F (denoted by $E \cong F$) if there exists a partial isometry W such as $E = W^*W$, $F = WW^*$.

Definition 4.14

(i) E is *finite* if any $F \le E \cong F$ implies $E = F$.
(ii) E is *semifinite* if any $F \le E$ has no finite subprojection.
(iii) E is *infinite* if E is not finite.

(iv) E is *purely infinite* if there does not exist nonzero finite projection F such that $F \leq E$.

There exist some important types of von Neumann algebra:

Definition 4.15

(i) A von Neumann algebra is finite if I is finite.
(ii) A von Neumann algebra is semifinite if I is semifinite.
(iii) A von Neumann algebra is infinite if I is infinite.
(iv) A von Neumann algebra is purely infinite if I is purely infinite.

Moreover,

Definition 4.16 A von Neumann algebra is said to be *σ-finite* if any family of mutually orthogonal projections is countable, equivalently, there exists a faithful normal state on this algebra.

4.7 KMS States and Tomita–Takesaki Theory

Let α be a mapping from a topological group G with an operation $\bullet$ to the set of all automorphisms of the C^*-algebra $\mathcal{A}$ (denoted by $\mathrm{Aut}(\mathcal{A})$) satisfying the following conditions:

1. $\alpha_{t\bullet s} = \alpha_t \alpha_s$ for any $t, s \in G$.
2. $\lim_{t\to 0} \|\alpha_t(A) - A\| = 0$ (strong continuity) for any $A \in \mathcal{A}$, where "0" is the unit in G for the operation $\bullet$.
3. $\alpha_t(A^*) = \alpha_t(A)^*$ for any $t \in G$ and $A \in \mathcal{A}$.

The triple $(\mathcal{A}, \alpha, G)$, or $(\mathcal{A}, \mathfrak{G}, \alpha(G))$, is sometimes called a C^*-dynamical system. For an *α-invariant* state φ (i.e., $\varphi(\alpha_t(A)) = \varphi(A)$ for any $A \in \mathcal{A}$), there exists an important theorem.

Theorem 4.17 *For a C^*-dynamical system $(\mathcal{A}, \alpha, \mathbb{R}; \bullet = +)$ and an α-invariant state φ, there exists a representation π_φ in a Hilbert space H_φ and a strongly continuous one-parameter unitary group $\{u_t^\varphi; t \in \mathbb{R}\}$ such that* (i) $u_t^\varphi x_\varphi = x_\varphi$, *and* (ii) $\pi_\varphi(\alpha_t(A)) = u_t^\varphi \pi_\varphi(A) u_t^{\varphi *}$.

Let us discuss fundamentals of KMS (Kubo–Martin–Schwinger) state and Tomita–Takesaki theory [741].

The KMS condition gives a formulation of the Gibbs state under some temperature which has a meaning in the infinite volume. The equilibrium state of a quantum system with Hamiltonian H in a finite Hilbert space $\mathcal{H}$ with temperature T is described by the Gibbs state density operator ρ

$$\rho = e^{-\beta H}/\mathrm{tr}\, e^{-\beta H}.$$

Here $\beta = 1/kT > 0$ is the inverse temperature, and k is the Boltzmann constant. Let $\mathcal{A}$ be an algebra of operators in $\mathcal{H}$ (for example, the algebra of all linear operators on a finite dimensional Hilbert space). We define the Gibbs state φ on $\mathcal{A}$ at inverse temperature β by

$$\varphi(A) = \operatorname{tr}(\rho A), \quad A \in \mathcal{A}.$$

Let

$$\alpha_t(A) = A(t) = e^{\mathrm{i}tH} A e^{-\mathrm{i}tH}, \quad A \in \mathcal{A}, t \in \mathbb{R}$$

be the Heisenberg dynamics associated with the Hamiltonian H. One can prove that for any pair of operators $A, B \in \mathcal{A}$ the map $t \to \varphi(A(t)B)$ can be analytically continued, and the state φ satisfies the following condition, called the *KMS condition at inverse temperature* $\beta > 0$:

$$\varphi\bigl(A(t - \mathrm{i}\beta)B\bigr) = \varphi\bigl(BA(t)\bigr), \quad t \in \mathbb{R}.$$

Indeed, denoting $Z = \operatorname{tr} e^{-\beta H}$, one has

$$\begin{aligned}\varphi\bigl(A(t - \mathrm{i}\beta)B\bigr) &= Z^{-1}\operatorname{tr}\bigl(e^{-\beta H} e^{\mathrm{i}tH + \beta H} A e^{-\mathrm{i}tH - \beta H} B\bigr) \\ &= Z^{-1}\operatorname{tr}\bigl(e^{\mathrm{i}tH} A e^{-\mathrm{i}tH} e^{-\beta H} B\bigr) = \varphi\bigl(BA(t)\bigr).\end{aligned}$$

Moreover, one can prove that the KMS property of the state is equivalent to the Gibbs formula.

The state φ is α_t-invariant, since

$$\varphi\bigl(A(t)\bigr) = Z^{-1}\operatorname{tr}\bigl(e^{-\beta H} e^{\mathrm{i}tH} A e^{-\mathrm{i}tH}\bigr) = \varphi(A), \quad A \in \mathcal{A}, t \in \mathbb{R}.$$

Let $(\mathcal{A}, \mathfrak{S}, \alpha)$ be a C^*-dynamical system. A state $\varphi \in \mathfrak{S}$ is a *KMS state* with respect to a real constant β and the automorphism α_t if for any $A, B \in \mathcal{A}$ there exists a complex function $F_{A,B}(z)$ such that

1. $F_{A,B}(z)$ is analytic for any $z \in D_\beta \equiv \{z \in \mathbb{C}; -\beta < \operatorname{Im} z < 0\}$ if β is taken as a positive constant. If $\beta < 0$, $D_\beta \equiv \{z \in \mathbb{C}; 0 < \operatorname{Im} z < -\beta\}$.
2. $F_{A,B}(z)$ is bounded and continuous for any $z \in \bar{D}_\beta \equiv \{z \in \mathbb{C}; -\beta \le \operatorname{Im} z \le 0\}$.
3. $F_{A,B}(z)$ satisfies the following boundary conditions: (i) $F_{A,B}(t) = \varphi(\alpha_t(A)B)$ and (ii) $F_{A,B}(t - \mathrm{i}\beta) = \varphi(B\alpha_t(A))$.

The KMS state with respect to the constant β and α_t is called (β, α_t)-KMS state. We denote the set of all (β, α_t)-KMS states by $K_\beta(\alpha)$.

One can easily prove the following two facts:

1. φ is a $(0, \alpha_t)$-KMS state $\Leftrightarrow$ φ is a tracial state.
2. φ is a (β, α_t)-KMS state $\Leftrightarrow$ φ is $(-1, \alpha_{\beta t})$-KMS state.

Theorem 4.18 *(β, α_t)-KMS state φ is α-invariant.*

Theorem 4.19 *The following statements are equivalent:*

(i) $\varphi \in K_\beta(\alpha)$.

(ii) *$\varphi(AB) = \varphi(B\alpha_{i\beta}(A)), \forall A \in \mathcal{A}_\alpha, \forall B \in \mathcal{C}$, where $\mathcal{A}_\alpha$ is the set of all α-analytic elements in $\mathcal{A}$ (i.e., $A \in \mathcal{A}$ is α-analytic if there exists an $\mathcal{A}$-valued analytic function $A(z)$ such that $A(z) = \alpha_z(A)$).*
(iii) *For any f whose Fourier transform is a rapidly decreasing function with a compact support in $C^\infty(\mathbb{R})$ and for any $A, B \in \mathcal{A}$ the following relation is satisfied: $\int f(t)\varphi(\alpha_t(A)B)\,dt = \int f(t - \mathrm{i}\beta)\varphi(B\alpha_t(A))\,dt$, where $f(t - \mathrm{i}\beta) = \int \hat{f}(w)e^{\mathrm{i}w(t-\mathrm{i}\beta)}\,dw$.*

Theorem 4.20 *If φ is faithful, then for a certain β, there exists a unique one-parameter automorphism group α_t such that φ is a (β, α_t)-KMS state.*

Let $\{\mathcal{H}_\varphi, \pi_\varphi, x_\varphi, u_t^\varphi\}$ be the GNS representation of $\varphi \in \mathfrak{S}$. The state $\tilde{\varphi}$ is defined by $\tilde{\varphi}(Q) = \langle x_\varphi, Qx_\varphi\rangle$ for any $Q \in \pi_\varphi(\mathcal{A})''$, which is called the natural extension of φ to $\pi_\varphi(\mathcal{A})''$. The natural extension $\tilde{\alpha}_t$ of α_t is defined by $\tilde{\alpha}_t(Q) = u_t^\varphi Q u_{-t}^\varphi$ for any $Q \in \pi_\varphi(\mathcal{A})$. Then

1. For any $\varphi \in K_\beta(\alpha)$, $\tilde{\alpha}_t(Q) = Q, \forall Q \in Z_\varphi \equiv \pi_\varphi(\mathcal{A})' \cap \pi_\varphi(\mathcal{A})''$.
2. $K_\beta(\alpha) = \{\varphi\} \Rightarrow Z_\varphi = CI$.
3. $\varphi \in \mathrm{ex}\, I(\alpha) = \{\text{extreme points in } I(\alpha)\} \Rightarrow \pi_\varphi(\mathcal{A}) \cap u_t^\varphi(R)' = CI$.
4. $\varphi \in K_\beta(\alpha) \cap \mathrm{ex}\, I(\alpha) \Rightarrow \varphi \in \mathrm{ex}\, K_\beta(\alpha) \Leftrightarrow Z_\varphi = CI$.
5. Since $K_\beta(\alpha)$ is compact in weak $*$-topology and convex, $K_\beta(\alpha)$ is equal to the closure of the set of all convex combinations of extreme points in $K_\beta(\alpha)$ in the weak $*$-topology.

Let us consider a von Neumann algebra $\mathcal{N}$ having a cyclic and separating vector x in a Hilbert space $\mathcal{H}$. Here separating means that $Ax = 0$, $A \in \mathcal{N}$ iff $A = 0$. Define conjugate linear operators S_o and F_o by

$$S_o Ax = A^*x \quad (\forall A \in \mathcal{N}),$$
$$F_o A'x = A'^*x \quad (\forall A \in \mathcal{N}).$$

Their domains contain $\mathcal{N}x$ and $\mathcal{N}'x$, respectively, and they are closable operators: $\bar{S}_o = S$, $\bar{F}_o = F$. We have

1. $S_o^* = F, F_o^* = S$. Let $S = J\Delta^{1/2}$ be the polar decomposition of S. J is called the *modular conjugate operator*. It is conjugate unitary (i.e., $\langle Jx, Jy\rangle = \langle y, x\rangle$, $J^2 = I$ for any $x, y \in \mathcal{H}$). Δ is an unbounded positive self-adjoint operator called the *modular operator*. All these operators depend on $\{\mathcal{N}, x\}$, so that they should be denoted by S_x, F_x, etc., but we omitted x here. The following important equalities hold:
2. $\Delta = FS$ and $\Delta^{-1} = SF$.
3. $F = JSJ = \Delta^{1/2}J = J\Delta^{-1/2}$.

Theorem 4.21 (Tomita theorem)

(i) $J\mathcal{N}J = \mathcal{N}'$, $J\mathcal{N}'J = \mathcal{N}$.
(ii) $JAJ = A^*$ $(\forall A \in \mathcal{N} \cap \mathcal{N}')$.

(iii) $\Delta^{\mathrm{i}t}\mathcal{N}\Delta^{-\mathrm{i}t} = \mathcal{N}, \forall t \in \mathbb{R}$.

Take $\phi(A) = \langle x, Ax\rangle$, $A \in \mathfrak{N}$ and define $\sigma_t^{\phi}(A) = \Delta^{\mathrm{i}t} A \Delta^{-\mathrm{i}t}$, $A \in \mathfrak{N}$.

Theorem 4.22 (Takesaki theorem) *The state ϕ satisfies the KMS-condition w.r.t. σ_t^{ϕ} at $\beta = -1$.*

Remark 4.23 If we define $\sigma_t^{\phi}(A) = \Delta^{-\mathrm{i}t} A \Delta^{\mathrm{i}t}$ then in the above theorem we get $\beta = +1$.

4.8 Notes

For the basic theorems concerning Hilbert spaces and the operators on Hilbert spaces, we refer to [314, 826]. Especially, the mathematical structure of tensor product Hilbert spaces is discussed in [684]. The Hida white noise calculus is discussed in [335–337]. For Theorem 4.11 and Theorem 4.13 concerning the theory of operator algebras, we refer to [330, 740]. In particular, we refer to [131] for Theorem 4.17, Theorem 4.19 and Theorem 4.20 concerning C^*-dynamical systems and the KMS-condition. The proofs of Theorem 4.21 and Theorem 4.22 in Tomita–Takesaki theory are given in [330, 741].

Chapter 5
Basics of Quantum Mechanics

Quantum mechanics was started after works by Heisenberg [325], Schrödinger [692] and Dirac around 1925. Together with relativity theory it is the most fundamental theory in physics today. There are two important points in quantum mechanics:

(1) Quantum mechanics is a statistical theory.
(2) Every quantum system assigns a Hilbert space.

In this chapter, we quickly review the mathematical structure of basic principles quantum mechanics and discuss the fundamentals of quantum probability. In particular, we discuss the notions of a quantum state and quantum operation, conditional probability and joint probability in quantum probability, theory of measurements, positive operator-valued measures and instruments. Finally, we summarize the discussion by presenting the seven principles of quantum mechanics where the fundamental role of the physical space $\mathbb{R}^3$ is emphasized.

5.1 Quantum Dynamics

Let H be a self-adjoint operator in $\mathcal{H}$ called the Hamiltonian of a certain quantum system. In quantum mechanics, in the simplest cases, the Hamiltonian has the same form as in classical mechanics, but the canonical variables become operators.

The *Heisenberg equations* in quantum mechanics have the same form as in classical mechanics. The Hamiltonian for a nonrelativistic particle with mass $m > 0$ on the real line in an external potential $V(x)$ has the same form as in classical mechanics

$$H = \frac{p^2}{2m} + V(q),$$

where p is the momentum and q is the coordinate of the particle, but in quantum mechanics p and q are operators in $L^2(\mathbb{R})$ which satisfy the *commutation relations*

$$[p, q] = -\mathrm{i}\hbar.$$

M. Ohya, I. Volovich, *Mathematical Foundations of Quantum Information and Computation and Its Applications to Nano- and Bio-systems*,
Theoretical and Mathematical Physics,
DOI 10.1007/978-94-007-0171-7_5, © Springer Science+Business Media B.V. 2011

The operators can be defined as

$$q\psi(x) = x\psi(x), \qquad p\psi(x) = -\mathrm{i}\hbar\frac{d}{dx}\psi(x)$$

on some vectors ψ in the Hilbert space $L^2(\mathbb{R})$.

The Heisenberg equations of motion are the same as in classical mechanics:

$$\frac{dq(t)}{dt} = \frac{p(t)}{m},$$
$$\frac{dp(t)}{dt} = -\frac{\partial}{\partial q}V\big(q(t)\big),$$

but here $q(t)$ and $p(t)$ are operator-valued functions:

$$q(t) = U(t)^* qU(t), \qquad p(t) = U(t)^* pU(t)$$

where the unitary operator

$$U(t) = \exp(-\mathrm{i}tH/\hbar)$$

is called the *evolution operator*.

Let $\psi(t) = U(t)\psi(0)$ be a vector in $\mathcal{H}$ describing the state of the system at time t. Then one has the Schrödinger equation on an appropriate domain

$$\mathrm{i}\hbar\frac{\partial\psi(t)}{\partial t} = H\psi(t).$$

In the sequel, we take Planck's constant $\hbar = 1$ for simplicity. For a stationary state, when $\psi(t) = e^{-\mathrm{i}Et}\varphi$, E being a constant, the Schrödinger equation can be written as

$$H\varphi = E\varphi$$

with time-independent vector φ, which is an eigenequation.

5.1.1 Schrödinger Picture

The time evolution of a Hamiltonian system is given by a unitary operator $U(t) = \exp(-\mathrm{i}tH)$; that is, suppose that the system is in a state φ at $t = 0$, then at time t the system is in a state $\psi(t) = U(t)\varphi$. This $\psi(t)$ satisfies the Schrödinger equation (on an appropriate domain)

$$\mathrm{i}\frac{\partial\psi(t)}{\partial t} = H\psi(t).$$

The state explicitly depends on time but the operator does not, which is called the *Schrödinger picture* (or *representation*).

The Schrödinger equation for the nonrelativistic particle with mass $m > 0$ on the real line in an external potential $V(x)$ reads

$$\mathrm{i}\frac{\partial\psi(x,t)}{\partial t} = -\frac{1}{2m}\frac{\partial^2\psi(x,t)}{\partial x^2} + V(x)\psi(x,t).$$

Here we set the Planck constant $\hbar = 1$. If we insert it back then the Schrödinger equation reads

$$\mathrm{i}\hbar\frac{\partial\psi(x,t)}{\partial t} = -\frac{\hbar^2}{2m}\frac{\partial^2\psi(x,t)}{\partial x^2} + V(x)\psi(x,t).$$

5.1.2 Heisenberg Picture

The time evolution of an observable A is determined by

$$A(t) = U(t)^* A U(t),$$

which is a solution of the Heisenberg equation

$$\frac{dA(t)}{dt} = \mathrm{i}\big[H, A(t)\big] \equiv \mathrm{i}\big(HA(t) - A(t)H\big).$$

The operators explicitly depend on time but the states do not, which is called the *Heisenberg picture* (or *representation*).

The Heisenberg picture and the Schrödinger picture are in dual relation to each other in the following sense

$$\big\langle\psi(t), A\psi(t)\big\rangle = \big\langle\psi, A(t)\psi\big\rangle.$$

Let us discuss the description of a free particle as an example, and compare the description of the free particle in quantum mechanics with classical mechanics.

5.1.3 Free Particle

Classical Mechanics

The free particle on the real line in classical mechanics moves with a constant velocity. Its trajectory is

$$x(t) = x_0 + vt. \tag{5.1}$$

Here $x(t)$ is the position of the particle at time t, x_0 is the initial position, and v is the velocity. A state of the particle is described by two real numbers x_0 and v. The momentum p of the particle is equal to the product of the mass of the particle, m, and the velocity, v, i.e., $p = mv$. One describes a state of the particle in the phase space by two canonical variables: the position x and the momentum p.

Quantum Mechanics

In quantum mechanics, the canonical variables become operators. There are no trajectories in the classical sense. A state of the particle on the line in quantum mechanics is described by a complex-valued function $\psi(x,t)$ where x is the coordinate on the line and t is the time variable. The basic postulate of quantum mechanics is that there exists a function $\psi(x,t)$ such that the probability to observe the particle in an interval $\mathbb{I} \subset \mathbb{R}$ at time t is given by

$$\int_{\mathbb{I}} |\psi(x,t)|^2 \, dx.$$

Note that this postulate does not determine the position of the particle, it only determines the *probability* that the particle is located in the interval $\mathbb{I}$. Since it is certain that the particle must be somewhere along the line, we should have

$$\int_{\mathbb{R}} |\psi(x,t)|^2 \, dx = 1 \tag{5.2}$$

at each time.

A *free* particle on the line is described by the function of the form

$$\psi(x,t) = \int \exp\left\{-\mathrm{i}\frac{p^2}{2m}t + \mathrm{i}px\right\} \varphi(p) \, dp, \tag{5.3}$$

where x is the coordinate on the line, and p represents the momentum. Here $\varphi(p)$ is an arbitrary function which satisfies the condition

$$\int |\varphi(p)|^2 \, dp = 1. \tag{5.4}$$

The function (5.3) is a solution of the Schrödinger equation for a free particle

$$\mathrm{i}\frac{\partial \psi(x,t)}{\partial t} = -\frac{1}{2m}\frac{\partial^2 \psi(x,t)}{\partial x^2}. \tag{5.5}$$

Exercise 5.1 Prove that the function (5.3) is a solution of the Schrödinger equation (5.5).

We see that the descriptions of the free particle in classical and in quantum mechanics are quite different. The function $\psi(x,t)$ is called the wave function. Its physical interpretation is that $|\psi(x,t)|^2$ is the density of the probability to observe the particle at time t and the point x.

Note that even in the case when at time $t = 0$ the particle is localized, i.e., $|\psi(x,0)|^2$ does not vanish only on a finite interval, at time $t > 0$ there is a nonvanishing probability to observe the particle near an arbitrary point on the line.

5.1.4 Scattering

Most of the information about interaction between particles is derived from scattering experiments. Suppose that a flux of N particles per unit area per unit time is incident on the target. The number of particles per unit time scattered into a narrow cone of solid angle $d\tau = \sin\theta\, d\theta\, d\phi$, centered about the direction whose polar angles with respect to the incident flux are θ and ϕ, will be proportional to the incident flux N and to the angular opening $d\tau$ of the cone, hence it may be written as $\sigma(\theta,\phi)N\,d\tau$. The proportionality factor $\sigma(\theta,\phi)$ is called the *differential cross-section*. One can compute the differential cross-section from the solution of the stationary Schrödinger equation as follows. We seek a solution of the Schrödinger equation in $\mathbb{R}^3$ (Δ is the Laplace operator)

$$-\frac{\hbar^2}{2m}\Delta\psi(\mathbf{x}) + V(\mathbf{x})\psi(\mathbf{x}) = E\psi(\mathbf{x})$$

that satisfies the asymptotic boundary condition

$$\psi(\mathbf{x}) \sim A\left(e^{i\mathbf{kx}} + \frac{f(\theta,\phi)}{r}e^{ikr}\right)$$

in the limit of large r. Here $f(\theta,\phi)$ is some function, A is a constant, $r = |\mathbf{x}|$, $k = |\mathbf{k}|$. Then the differential cross-section is given by

$$\sigma(\theta,\phi) = \left|f(\theta,\phi)\right|^2.$$

The function $f(\theta,\phi)$ is called the *scattering amplitude*.

The time dependent scattering theory deals with the scattering operator, the S-matrix, which is given by the limit of the evolution operator

$$S = \lim_{\substack{t\to\infty \\ \tau\to-\infty}} e^{i(t-\tau)H_0}e^{-i(t-\tau)(H_0+V)}.$$

Here H_0 is a free Hamiltonian and $H_0 + V$ is the total Hamiltonian including the term V which describes interaction. For the potential scattering, the S-matrix leads to the same scattering amplitude $f(\theta,\phi)$.

5.1.5 Description of General State: Density Operator

Let $\{\varphi_n\}$ be the normalized eigenvectors of an observable A. Using φ_n with probability p_n and observing another observable Q, we compute the expectation as

$$E(Q) = \sum_n p_n\langle\varphi_n, Q\varphi_n\rangle.$$

Define the operator $\Xi_{\varphi,\psi}$ for any $\varphi, \psi \in \mathcal{H}$ as

$$\Xi_{\varphi,\psi} z \equiv \langle \psi, z \rangle \varphi, \quad \forall z \in \mathcal{H}.$$

Following Dirac, this operator $\Xi_{\varphi,\psi}$ is written as

$$\Xi_{\varphi,\psi} \equiv |\varphi\rangle\langle\psi|.$$

Put

$$\rho = \sum_n p_n |\varphi_n\rangle\langle\varphi_n|.$$

Then the above expectation value of Q in the state ρ is given by

$$E_\rho(Q) = \operatorname{tr} \rho Q,$$

where "tr" is the trace (i.e., $\operatorname{tr} A = \sum_n \langle \varphi_n, A\varphi_n \rangle$ for any CONS $\{\varphi_n\}$ in the Hilbert space as was explained in Chap. 4).

This ρ is called a *density operator* or a *mixed state*. Mathematically, a density operator is a certain element of $B(\mathcal{H})$, the set of all bounded linear operators on $\mathcal{H}$, that is, the set of all density operators is

$$\mathfrak{S}(\mathcal{H}) \equiv \left\{\rho \in B(\mathcal{H});\, \rho^* = \rho (\text{self-adjoint}), \rho \geq 0, \operatorname{tr} \rho = 1\right\}.$$

In the terminology of Chap. 4, the density operator is a normalized self-adjoint positive trace-class operator. An element ρ of $\mathcal{S}(\mathcal{H})$ may be called a (general) state and the expectation value of A in a state ρ is given by $E_\rho(Q) = \operatorname{tr} \rho Q$. This ρ is a generalization of a probability distribution for a classical system and a vector state φ.

Now we explain more about Dirac's Ket $|\cdot\rangle$ and Bra $\langle\cdot|$ and discuss their uses. An operator $|\varphi\rangle\langle\psi|$ ($\forall\varphi, \psi \in \mathcal{H}$) is in $B(\mathcal{H})$. The norm of the operator is $\||\varphi\rangle\langle\psi|\| = \|\varphi\|\|\psi\|$, so that $\operatorname{ran}(|\varphi\rangle\langle\psi|) = \{\lambda\varphi;\, \lambda \in \mathbb{C}\}$ is a one-dimensional closed subspace if $\varphi \neq 0$. The inner product of $\varphi, \psi \in \mathcal{H}$ is often written in Dirac's notation as $\langle \varphi, \psi \rangle = \langle \varphi | \psi \rangle$.

Example 5.2 $\mathcal{H} = \mathbb{C}^n$

$$|\varphi\rangle = \begin{pmatrix} t_1 \\ \vdots \\ t_n \end{pmatrix}, \qquad \langle\psi| = (s_1, \ldots, s_n) \quad \Longrightarrow \quad |\varphi\rangle\langle\psi| = \begin{pmatrix} t_1 s_1 & \cdots & t_1 s_n \\ \vdots & \ddots & \vdots \\ t_n s_1 & \cdots & t_n s_n \end{pmatrix}.$$

Exercise 5.3 Let $a = (a_1, a_2, a_3)$ be a vector in the 3-dimensional real space $\mathbb{R}^3$. Prove that the matrix

$$\rho = \frac{1}{2} \begin{pmatrix} 1 + a_1 & a_2 - \mathrm{i}a_3 \\ a_2 + \mathrm{i}a_3 & 1 - a_1 \end{pmatrix}$$

represents a density operator in the two dimensional Hilbert space $\mathbb{C}^2$ if and only if the vector a belongs to the unit ball,

$$|a| = \left(a_1^2 + a_2^2 + a_3^2\right)^{1/2} \leq 1.$$

5.2 States and Observables

5.2.1 *Quantum Parallelism (Duality)*

To see the reason why a description of the quantum mechanics is so different from that of the classical mechanics, let us remind the phenomena of quantum parallelism. One of the most striking properties of quantum particles is their interference properties shown in the double slit experiment (Fig. 5.1).

In the double slit experiment which is referred to also as Young's experiment, a coherent light source illuminates a thin plate with two parallel slits cut in it, and the light passing through the slits strikes a screen behind them. The light waves passing through both slits interfere, creating an interference pattern of bright and dark bands on the screen. There is the following problem here. At the screen, the light is always found to be absorbed as though it were made of discrete particles, called photons. If the light travels from the source to the screen as particles, then according to the classical reasoning the number of particles (or probability) that strike any particular point on the screen is expected to be equal to the sum of those particles (or probabilities) that go through the first slit and those that go through the second slit.

So, according to the classical particle physics, the brightness at any point should be the sum of the brightness when the first slit is blocked and the brightness when the second slit is blocked. In other words, the probability $P(x)$ of finding a particle at the point x on the screen should be the sum of probability $P_1(x)$ when the second slit is blocked and the probability $P_2(x)$ when the first slit is blocked, $P(x) = P_1(x) + P_2(x)$. However, it is found in experiments that unblocking both slits makes some points on the screen brighter, and other points darker. The experiment tells us that $P(x) \neq P_1(x) + P_2(x)$. This can only be explained by the alternately additive and subtractive interference of waves, not the exclusively additive nature of particles, so we obtain that light must have some particle–wave duality. There is the similar property also for electrons and other quantum particles.

It is important to note that the concept of probability is not altered in quantum mechanics. When one says the probability of a certain outcome of an experiment under certain conditions is p, this means that if the experiment is repeated many times under the same conditions, one expects that the fraction of those which give the outcome in question is roughly p. What is changed in quantum mechanics is the method of calculating probabilities. One can say that instead of *exclusive* alternatives, which one uses in classical world, quantum mechanics uses *interfering* alternatives. When we speak about quantum probability, this means, in fact, only special rules for computation of ordinary ("classical") probability.

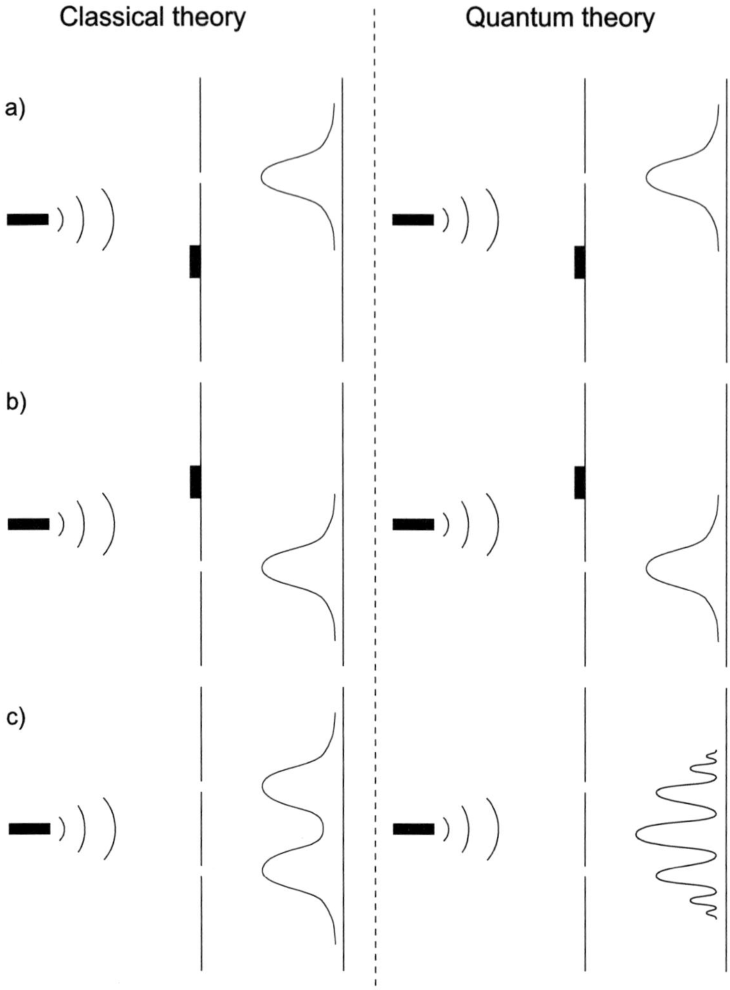

Fig. 5.1 Quantum interference

To state the correct law for the probability $P(x)$, quantum mechanics postulates that $P(x)$ is the absolute square of a certain complex-valued function $\psi(x)$ which is called the *probability amplitude* of arrival at x (it is also called the *wave function*). More generally, to every (pure) state of any quantum system one assigns a unit vector from a Hilbert space. Furthermore, $\psi(x)$ is the sum of two contributions describing the alternatives:

$$\psi(x) = c\big[\psi_1(x) + \psi_2(x)\big]$$

where c is the normalization constant (the total probability must be equal 1). Here $\psi_1(x)$ is the amplitude of arrival through hole 1, $P_1(x) = |\psi_1(x)|^2$, and $\psi_2(x)$ the amplitude of arrival through hole 2, $P_2(x) = |\psi_2(x)|^2$. Then we should have

$$\begin{aligned} P(x) = \left|\psi(x)\right|^2 &= |c|^2\big[\left|\psi_1(x)\right|^2 + \left|\psi_2(x)\right|^2 \\ &\quad + \psi_1(x)^*\psi_2(x) + \psi_1(x)\psi_2(x)^*\big] \\ &= |c|^2\big[P_1(x) + P_2(x) + I(x)\big] \end{aligned}$$

where $I(x) = \psi_1(x)^*\psi_2(x) + \psi_1(x)\psi_2(x)^*$ is the *interference term.*

An explicit form of the wave functions is derived in quantum mechanics by solving the Schrödinger equation (see below). In our case, the wave functions $\psi_1(x)$ and $\psi_2(x)$ can be given as follows. Let x be the distance from the center of the screen to the point on the screen we consider. Furthermore, let r_1 be the distance between the first slit and the point x on the screen, and r_2 the distance between the second slit and the point x. Then the wave functions $\psi_1(x)$ and $\psi_2(x)$ can be approximately written as spherical waves

$$\psi_1(x) = \frac{e^{ikr_1}}{r_1}, \qquad \psi_2(x) = \frac{e^{ikr_2}}{r_2},$$

where we neglect the normalization constants. The same expressions we get in classical optics by solving the Maxwell equations. Here in classical optics k is the wave vector, $k = 2\pi/\lambda$, where λ is the wavelength of light. In quantum mechanics, if we consider electrons or other particles, then $k = p/\hbar$, where p is the momentum of an electron and $\hbar$ is the Planck constant. Now the probability $P(x)$ is proportional to

$$\left|\psi_1(x) + \psi_2(x)\right|^2 = \frac{2}{r_1 r_2}\big[1 + \cos k(r_2 - r_1)\big].$$

A bright fringe on the screen is located at the places x where

$$k(r_2 - r_1) = 2\pi n, \quad n = 0, \pm 1, \ldots.$$

In classical optics, the formula says that a bright fringe on the screen is located at the places x where the difference of distances between the place and the slits equals an integer multiple of the wavelength: $r_2 - r_1 = \lambda n$.

We can simplify this condition as follows. Let a be the slit separation and L distance from slits to screen. Then for large L and small a and x one has

$$r_{1,2} = \sqrt{L^2 + \left(x \mp \frac{a}{2}\right)^2} \simeq L + \frac{1}{2L}\left(x^2 + \frac{a^2}{4}\right) \mp \frac{xa}{2L}$$

and $r_2 - r_1 \simeq xa/L$. Therefore, we obtain the famous Young formula

$$\frac{a}{L} = \frac{\lambda}{x}n, \quad n = 0, \pm 1, \ldots.$$

The interference of particles is essential for quantum physics, which is not in classical physics. This interference (a character of a "wave") is a fundamental property of matter, and it is used in several aspects of quantum information as will be explained in the subsequent chapters. On the other hand, according to several other fundamental experiments such as the photo-effect and Compton-effect, every matter has the property of a "particle". Thus every matter has a duality, namely, the property of a wave and that of a particle, so that quantum mechanics is based on the principle that every existence has the duality.

5.2.2 States

A (pure) state in a quantum system is described by a normalized vector ψ in a certain Hilbert space $\mathcal{H}$. This vector ψ is called the *state vector* or just *state*. Let H be a self-adjoint operator in $\mathcal{H}$, called the Hamiltonian of a certain quantum system. Then the Schrödinger equation is

$$\mathrm{i}\hbar\frac{\partial\psi(t)}{\partial t} = H\psi(t),$$

where $\psi(t)$ is a vector in $\mathcal{H}$ describing the state of the system at time t. In the sequel, we take Planck's constant $\hbar = 1$ for simplicity. For a stationary state, when $\psi(t) = e^{-\mathrm{i}Et}\psi$, the Schrödinger equation can be written as

$$H\psi = E\psi$$

with time independent vector ψ, which is an eigenequation.

For a particle in $\mathbb{R}^3$, the above Hamiltonian depends on $x \in \mathbb{R}^3$, so that the vector also depends on x, and the Hilbert space is given by

$$L^2(\mathbb{R}^3) \equiv \left\{\psi;\ \int_{\mathbb{R}^3} |\psi(\mathbf{x})|^2\, d\mathbf{x} < +\infty\right\}$$

with the inner product given by

$$\langle\varphi, \psi\rangle = \int_{\mathbb{R}^3} \overline{\varphi(\mathbf{x})}\psi(\mathbf{x})\, d\mathbf{x}.$$

A state of the system is a normalized vector ψ of $L^2(\mathbb{R}^3)$, namely,

$$\|\psi\| = \sqrt{\langle\psi,\psi\rangle} = 1.$$

5.2.3 Observables

An observable of a quantum system is described by an Hermite (symmetric) operator A on $\mathcal{H}$:

$$\langle\varphi, A\psi\rangle = \langle A^*\varphi, \psi\rangle = \langle A\varphi, \psi\rangle \quad \text{for any } \varphi, \psi \in \mathcal{D}(A).$$

Remark that most of physically interesting Hermite operators can be extended to self-adjoint operators, and a bounded Hermite operator is self-adjoint so that one often says that an observable is described by a self-adjoint operator.

Note that a general notion of observables is associated with positive operator-valued measures which is discussed below.

5.2.4 Expectation Values

An expectation value of an observable A in a (pure) state φ is given by

$$\langle \varphi, A\varphi \rangle$$

which is a real number because A is Hermitian:

$$\langle \varphi, A\varphi \rangle = \overline{\langle A\varphi, \varphi \rangle} = \overline{\langle \varphi, A^*\varphi \rangle} = \overline{\langle \varphi, A\varphi \rangle}.$$

Only the eigenvalue of A is actually measured in an eigenstate (eigenvector) of A, that is,

$$A\varphi_k = a_k\varphi_k \quad \Longrightarrow \quad \langle \varphi_k, A\varphi_k \rangle = a_k \langle \varphi_k, \varphi_k \rangle = a_k.$$

Moreover, when a state vector φ measuring A is not an eigenstate of A and A is observed by the state φ several times, the eigenvalue a_k is obtained with the probability $p_k = |c_k|^2$. Then the state vector φ is expressed as

$$\varphi = \sum_k c_k \varphi_k$$

using the normalized eigenvectors φ_k $(k = 1, 2, \dots)$ of A. The expectation value of A in this φ becomes

$$\begin{aligned}\langle \varphi, A\varphi \rangle &= \sum_{k,j} \langle c_k\varphi_k, Ac_j\varphi_j \rangle = \sum_{k,j} \overline{c_k} c_j \langle \varphi_k, A\varphi_j \rangle \\ &= \sum_{k,j} \overline{c_k} c_j \langle \varphi_k, a_j\varphi_j \rangle = \sum_{k,j} a_j \overline{c_k} c_j \langle \varphi_k, \varphi_j \rangle \\ &= \sum_k a_k |c_k|^2 = \sum_k a_k p_k.\end{aligned}$$

Example 5.4 When $\mathcal{H} = \mathbb{C}^n$ and an observable A is given by a diagonal matrix, that is,

$$A = \begin{pmatrix} a_1 & & 0 \\ & \ddots & \\ 0 & & a_n \end{pmatrix} : \mathbb{C}^n \to \mathbb{C}^n,$$

and when a state φ is given by

$$\varphi = \left(\sqrt{p_1}, \ldots, \sqrt{p_n}\right)^t \in \mathbb{C}^n, \quad p_k \geq 0 \text{ with } \sum_{k=1}^{n} p_k = 1,$$

the expected value of A shows

$$\langle \varphi, A\varphi \rangle = \sum_{k=1}^{n} a_k p_k \equiv \langle A \rangle_\varphi,$$

which coincides with the expectation value of a classical system.

5.3 Quantum Oscillator, Creation and Annihilation Operators

The Hamiltonian of a *quantum oscillator* has the form

$$H = \frac{1}{2m} p^2 + \frac{\omega^2}{2} q^2.$$

Here ω is a real parameter, $\omega > 0$, and p and q are formal operators which are defined as

$$q\psi(x) = x\psi(x), \qquad p\psi(x) = -\mathrm{i}\hbar \frac{d}{dx}\psi(x)$$

on some vectors ψ in the Hilbert space $L^2(\mathbb{R})$. They satisfy the Heisenberg canonical commutation relations (CCR), by taking $m = \hbar = 1$ in the sequel for simplicity,

$$[q, p] = \mathrm{i}.$$

These relations in terms of operators

$$a = \frac{1}{\sqrt{2\omega}}(\omega q + \mathrm{i}p), \qquad a^* = \frac{1}{\sqrt{2\omega}}(\omega q - \mathrm{i}p)$$

take the form

$$[a, a^*] = I. \tag{5.6}$$

The operators a, a^* are called the annihilation and creation operators, respectively. The Hamiltonian now will take the form

$$H = \omega\left(a^* a + \frac{1}{2}\right). \tag{5.7}$$

One can prove that the complete orthonormal system of eigenvectors of the operator $N = a^* a$ has the form

$$\psi_n(x) = \frac{1}{\sqrt{n!}} a^{*n} \psi_0(x), \quad n = 0, 1, 2, \ldots,$$

where $\psi_0(x) = (\omega/\pi)^{1/4}\exp(-\omega x^2/2)$. One can prove that

$$\psi_n(x) = \sqrt{\frac{\omega}{\pi 2^n n!}} H_n\big(\sqrt{\omega}x\big)\exp\left(-\frac{\omega x^2}{2}\right),$$

where H_n are the Hermite polynomials such that

$$H_n(z) = (-1)^n \exp\big(z^2\big)\left(\frac{d^n}{dz^n}\exp\big(-z^2\big)\right).$$

One has

$$N\psi_n(x) = n\psi_n(x), \quad n = 0, 1, 2, \ldots$$

and

$$H\psi_n(x) = \omega\left(n + \frac{1}{2}\right)\psi_n(x), \quad n = 0, 1, 2, \ldots.$$

This vector $\psi_n(x)$ is equivalently expressed by the vector $|n\rangle$ in Dirac notations and satisfies

$$a|n\rangle = \sqrt{n}|n-1\rangle, \qquad a^*|n\rangle = \sqrt{n+1}|n+1\rangle$$

from algebraic computation by means of (5.6) and (5.7).

Let us define the family of unitary operators $W_{\xi,v}$ in $L^2(\mathbb{R})$ by the relation

$$W_{\xi,v}\varphi(x) = \exp\left[\mathrm{i}v\left(x - \frac{\xi}{2}\right)\right]\varphi(x-\xi)$$

for any $\varphi \in L^2(\mathbb{R})$. Here ξ, v are real parameters. Then one can check the relation

$$aW_{\xi,v}\psi_0(x) = zW_{\xi,v}\psi_0(x),$$

where $z = (\omega\xi + \mathrm{i}v)/\sqrt{2\omega}$. The state denoted by $|z\rangle \equiv W_{\xi,v}|0\rangle$ is called the *coherent* vector. There is the completeness relation for the coherent vectors

$$\int_{\mathbb{C}} |z\rangle\langle z| \frac{d^2z}{2\pi} = I,$$

where $d^2z = d\xi\, dv/2$.

5.4 Symmetries

It is the Galilean relativity principle that the statistics of any measurement is the same in any inertial frame of reference. Galilean group is an important example of a symmetry group. In quantum theory, a symmetry group is described by the unitary or projective representations of the group in a Hilbert space of quantum states of the system.

If G is a group then a family of unitary operators $g \to U_g$, $g \in G$ in a Hilbert space $\mathcal{H}$ is called a *projective unitary representation* of G in $\mathcal{H}$ if

$$U_{g_1} U_{g_2} = \omega(g_1, g_2) U_{g_1 g_2}, \quad \forall g_1, g_2 \in G.$$

Here $\omega(g_1, g_2)$ is a complex-valued function with $|\omega(g_1, g_2)| = 1$. If $\omega \equiv 1$, then the representation is called *unitary*. We shall consider only topological groups G and representations which are weakly continuous.

5.4.1 Representations of the Rotation Group: Spin

Consider rotations in the 3-dimensional Euclidean space: $x \to Rx$. The rotation matrices R form the rotation group G. One can prove that in a Hilbert space of any finite dimension $d = 2, 3, \ldots$ there exists exactly one irreducible projective unitary representation of the rotation group. The number $s = (d-1)/2 = 1/2, 1, 3/2, \ldots$ is called the *spin* of the representation. This spin s is obtained from the eigenvalues of the operator J given below: $J^2 (\equiv \sum_{k=1}^3 J_k^2) \psi = s(s+1)\psi$. In the case of the trivial representation ($d = 1$), one says that spin $s = 0$. We describe here representations of various spins.

Any rotation is a rotation around the unit vector n through an angle θ. There are three infinitesimal generators of rotations around the three coordinate axes. The rotation matrix can be written in the form $R = \exp[-\sum_{k=1}^3 \theta_k L_k]$, where L_k are infinitesimal generators (3-dimensional matrices) satisfying the relations

$$[L_1, L_2] = -L_3, \qquad [L_2, L_3] = -L_1, \qquad [L_3, L_1] = -L_2,$$

and θ_k are real parameters. It is said that the matrices L_k generate the Lie algebra of the rotation group.

Any unitary representation of the rotation group in a d-dimensional Hilbert space can be written in the form

$$U_R = \exp\left[-\mathrm{i} \sum_{k=1}^{3} \theta_k J_k \right],$$

where J_k are Hermitian matrices satisfying

$$[J_1, J_2] = \mathrm{i} J_3, \qquad [J_2, J_3] = \mathrm{i} J_1, \qquad [J_3, J_1] = \mathrm{i} J_2.$$

For the spin $s = 1/2$ ($d = 2$), the solution of the last commutation relations $J_k = \sigma_k / 2$ is given by the Pauli matrices

$$\sigma_1 = \begin{pmatrix} 0 & 1 \\ 1 & 0 \end{pmatrix}, \qquad \sigma_2 = \begin{pmatrix} 0 & -\mathrm{i} \\ \mathrm{i} & 0 \end{pmatrix}, \qquad \sigma_3 = \begin{pmatrix} 1 & 0 \\ 0 & -1 \end{pmatrix}.$$

In this case, one can compute the exponents as

$$U_R = \exp\left[-\frac{i}{2}\sum_{k=1}^{3}\theta_k\sigma_k\right] = I\cos\frac{\theta}{2} - i\left(\sum_{k=1}^{3}\theta_k\sigma_k\right)\frac{1}{\theta}\sin\frac{\theta}{2},$$

where $\theta = \sqrt{\theta_1^2+\theta_2^2+\theta_3^2}$. In particular, the rotation by 2π for $\theta_1 = 2\pi, \theta_2 = \theta_3 = 0$ gives $U_R = -I$.

The Galilean group is given by the following Galilei transformation. The coordinates (x,t) and (x',t') of two inertial frames of reference are related by the *Galilei transformation*

$$x' = Rx + \xi + vt, \qquad t' = t + \tau.$$

Here R is a rotation, v is the velocity, and τ is the time shift. The time shift describes dynamics of the system, and it is represented by a Hamiltonian. Any projective unitary representation of the group of kinematical transformations $x' = Rx + \xi + vt$, $t' = t$, is characterized by two parameters d and μ where $d = 1,2,3,\dots$ (instead of d, the spin $s = (d-1)/2$ is often used), and μ is a real number. This representation is given in the Hilbert space $\mathcal{H} = \mathbb{C}^d \otimes L^2(\mathbb{R}^3)$ by

$$W(\xi, v, R)\psi(x) = \exp\bigl[i\mu v\cdot(x-\xi/2)\bigr]U_R\psi\bigl(R^{-1}(x-\xi)\bigr).$$

Here U_R acts in the space $\mathbb{C}^d$, i.e., on components of the vector-function $\psi(x)$.

Remark 5.5 Spin properties of an electron are described by the relativistic covariant Dirac equation, see Chap. 16.

5.5 Quantum Probability

In this section, we discuss some fundamental facts of quantum probability. Quantum probability in a simple setting is defined by a Hilbert space $\mathcal{H}$ with the set of all projections $P_{\mathcal{H}}$ which represent events, the set of self-adjoint (or Hermitian) operators representing observables, and the set of density operators representing states.

Let φ_k $(k = 1,2,\dots)$ be the eigenvectors of an observable A with the eigenvalues a_k (i.e., $A\varphi_k = a_k\varphi_k$). Prepare a state $\varphi = \sum_k c_k\varphi_k$ with $\sum_k |c_k|^2 = 1$ and observe A by this φ. The probability of obtaining a_k is $|c_k|^2$, and it is denoted by

$$P(A = a_k) = |c_k|^2.$$

The physical interpretation for the change $\varphi \to \varphi_k$ is called *a reduction of the wave packet* (see the subsequent section) through the observation of A in a state φ. Let $\mathcal{H}$ be the Hilbert space spanned by $\{\varphi_k\}$, and let its subspace $\mathcal{H}_\Delta$ be given by

$$\mathcal{H}_\Delta = \text{closure of the linear span of } \{\varphi_k;\, a_k \in \Delta\},$$

where Δ is a Borel set of $\mathbb{R}$ (i.e., $\Delta \in \mathcal{B}(\mathbb{R})$). Let P_Δ be the projection from $\mathcal{H}$ to $\mathcal{H}_\Delta$. Then the probability of obtaining the eigenvalue contained in Δ by observing A in a state φ is

$$P(A \in \Delta) = \|P_\Delta \varphi\|^2.$$

As is explained in the previous section, the expectation value of A in φ is

$$E_\varphi(A) = \langle \varphi, A\varphi \rangle = \sum_k P(A = a_k) a_k = \sum_k |c_k|^2 a_k,$$

and the variance of A in φ is

$$V_\varphi(A) = E_\varphi\left(A^2\right) - E_\varphi(A)^2.$$

When an observable A has a continuous spectrum and its spectral decomposition is

$$A = \int_R a E_A(da),$$

the expectation value is denoted as

$$E_\varphi(A) = \int_R a \big\langle \varphi, E_A(da)\varphi \big\rangle.$$

Here $E_A(\cdot)$ is the spectral measure of A and φ is a vector in $\operatorname{dom} A$ (domain of A).

Again as in the previous section, let us consider a state ρ given by

$$\rho = \sum p_n E_n, \quad E_n = |\varphi_n\rangle\langle\varphi_n|.$$

Then the expectation value of A in a state ρ is

$$E_\rho(A) = \operatorname{tr} \rho A,$$

which becomes

$$\sum_k a_k p_k$$

only when $A\varphi_k = a_k \varphi_k$.

So far, the state φ or ρ is prepared through an observable A, and we have considered the observation of A. Generally, we observe some other physical observable B in this prepared state φ or ρ. Then the expectation value is expressed as

$$E_\varphi(B) = \langle \varphi, B\varphi \rangle,$$
$$E_\rho(B) = \operatorname{tr} \rho B.$$

A comparison of the classical probability theory (CPT) with the quantum probability theory (QPT) is shown in the following table:

	CPT	QPT
Fundamental space	Ω	$\mathcal{H}$
Event	σ-field $\mathcal{F}$	set of all projections $\mathcal{P}_{\mathcal{H}}$
Observable	real random var. f	self-adjoint op. A
State	prob. measure μ	density operator ρ

Note that a probability distribution is a special case of μ and a vector state φ is that of ρ:

$$\langle \varphi, B\varphi \rangle = \operatorname{tr} B|\varphi\rangle\langle\varphi|.$$

There exists the following correspondence between an event of CPT and that of QPT:

CPT	QPT
For any $A, B \in \mathcal{F}$	For any $E, F \in \mathcal{P}_{\mathcal{H}}$
$A \subset B$	$E \leq F$ (i.e., $\operatorname{ran} E \subset \operatorname{ran} F$)
$A \cap B$	$E \wedge F$ (projection to $\operatorname{ran} E \cap \operatorname{ran} F$)
$A \cup B$	$E \vee F$ (projection to $\overline{\operatorname{ran} E \cup \operatorname{ran} F}$)
$A^c \equiv \Omega \setminus A$	$E^{\perp} \equiv I - E$

A map $\varphi : \mathcal{P}_{\mathcal{H}} \to [0, 1]$ such that $\varphi(I) = 1$, $\varphi(\bigcup_j E_j) = \sum_j \varphi(E_j)$ for any $\{E_j\} \subset \mathcal{P}_{\mathcal{H}}$ with $E_i \perp E_j$ $(i \neq j)$ is called a *probability measure on* $\mathcal{P}_{\mathcal{H}}$.

The following theorem may be a witness of the above correspondence.

Theorem 5.6 (Gleason's theorem) *If* $\dim \mathcal{H} \geq 3$ *and* φ *is a probability measure on* $\mathcal{P}_{\mathcal{H}}$, *then there exists a unique density operator* ρ *such that*

$$\varphi(E) = \operatorname{tr} \rho E$$

for any $E \in \mathcal{P}_{\mathcal{H}}$.

5.5.1 Conditional Probability and Joint Probability in QP

The conditional probability and the joint probability do not generally exist in a quantum system, which is an essential difference from a classical system. In the classical probability, the joint probability for two events A and B is

$$\mu(A \cap B),$$

and the conditional probability is defined by

$$\frac{\mu(A \cap B)}{\mu(B)}.$$

In the quantum probability, after a measurement of $F \in \mathcal{P}_{\mathcal{H}}$, a state ρ is changed to

$$\rho_F = \frac{F\rho F}{\operatorname{tr} \rho F}.$$

When we observe an event $E \in \mathcal{P}_{\mathcal{H}}$, the expectation value becomes

$$\operatorname{tr} \rho_F E = \frac{\operatorname{tr} F\rho F E}{\operatorname{tr} \rho F} = \frac{\operatorname{tr} \rho F E F}{\operatorname{tr} \rho F}. \tag{5.8}$$

This expectation value may be interpreted as the *conditional probability in QPT.*

There is also another candidate for the conditional probability in QPT. Is it possible to match well the conditional probability in quantum probability with that in classical probability?

By a direct application of the chart between CPT and QPT, the joint probability and the conditional probability of QPT are expected to be expressed as

$$\varphi(E \wedge F)$$

and

$$\frac{\varphi(E \wedge F)}{\varphi(F)}, \tag{5.9}$$

where we used the notation

$$\varphi(\cdot) = \operatorname{tr} \rho(\cdot).$$

We ask when the above two expressions (5.8) and (5.9) in QPT are equivalent. We will see that actually $\varphi(\cdot \wedge F)/\varphi(F)$ is not a probability measure on $\mathcal{P}_{\mathcal{H}}$, but the expression (5.8) can be a candidate for a conditional probability in quantum probability.

Proposition 5.7

(i) *When E commutes with F, the above two expressions are equivalent, namely,*

$$\frac{\varphi(FEF)}{\varphi(F)} = \frac{\varphi(E \wedge F)}{\varphi(F)}.$$

(ii) *When $EF \neq FE$, $\frac{\varphi(\cdot \wedge F)}{\varphi(F)}$ is not a probability on $\mathcal{P}_{\mathcal{H}}$, so that the above two expressions are not equivalent.*

Proof (i) $EF = FE$ implies $E \wedge F = EF$ and $FEF = EFF = EF^2 = EF$, so that

$$\frac{\varphi(E \wedge F)}{\varphi(F)} = \frac{\varphi(FEF)}{\varphi(F)} = \frac{\varphi(EF)}{\varphi(F)}.$$

(ii) Put $K_\varphi(E \mid F) \equiv \frac{\varphi(E \wedge F)}{\varphi(F)}$ and put $z \in \operatorname{linsp}\{x, y\}$, $z \neq x, y$ for any $x, y \in \mathcal{H}$. Take the projections $P_x = |x\rangle\langle x|$, $P_y = |y\rangle\langle y|$, $P_z = |z\rangle\langle z|$ such that $(P_x \vee P_y) \wedge P_z = P_z$ and $P_x \wedge P_z = 0 = P_y \wedge P_z$.

Then

$$K_\varphi\big((P_x \vee P_y) \wedge P_z \mid F\big) = K_\varphi(P_z \mid F) \neq 0,$$
$$K_\varphi(P_x \wedge P_z \mid F) + K_\varphi(P_y \wedge P_z \mid F) = 0.$$

Therefore,

$$K_\varphi\big((P_x \vee P_y) \wedge P_z \mid F\big) \neq K_\varphi(P_x \wedge P_z \mid F) + K_\varphi(P_y \wedge P_z \mid F),$$

so that $K_\varphi(\cdot \mid F)$ is not a probability measure on $\mathcal{P}_\mathcal{H}$. □

In CPT, the joint distribution for two random variables f and g is expressed as

$$\mu_{f,g}(\Delta_1, \Delta_2) = \mu\big(f^{-1}(\Delta_1) \cap g^{-1}(\Delta_2)\big)$$

for any Borel sets $\Delta_1, \Delta_2 \in B(\mathbb{R})$. The corresponding quantum expression is either

$$\varphi_{A,B}(\Delta_1, \Delta_2) = \varphi\big(E_A(\Delta_1) \wedge E_B(\Delta_2)\big)$$

or

$$\varphi\big(E_A(\Delta_1) \cdot E_B(\Delta_2)\big)$$

for two observables A, B and their spectral measures $E_A(\cdot)$, $E_B(\cdot)$ such that

$$A = \int aE_A\,(da), \qquad B = \int bE_B\,(da).$$

It is easily checked that neither one of the above expressions satisfies either a condition of a probability measure or the marginal condition unless $AB = BA$, so that they cannot be the joint quantum probabilities in the classical sense.

Therefore, we conclude that there do not exist both conditional and joint probabilities in a quantum system in the sense of a classical system.

Let us explain the above situation, as an example, in a physical measurement process. When an observable A has a discrete decomposition like

$$A = \sum_k a_k F_k, \qquad F_i \perp F_j \quad (i \neq j),$$

the probability of obtaining a_k by measurement in a state ρ is

$$p_k = \operatorname{tr} \rho F_k$$

and the state ρ is changed to a state ρ_k such that

$$\rho_k = \frac{F_k \rho F_k}{\operatorname{tr} \rho F_k} \equiv P_\rho(\cdot | F_k).$$

After the measurement of A, we will measure a similar type observable B (i.e., $B=\sum_j b_j E_j$ $(E_i \perp E_j$ $(i \neq j))$ and the probability of obtaining b_j after we have obtained the above a_k for the measurement of A is given by

$$\begin{aligned} p_{jk} &= (\operatorname{tr} \rho F_k)(\operatorname{tr} \rho_k E_j) \\ &= \operatorname{tr} \rho F_k E_j F_k \\ &= P_\rho(E_j | F_k) \operatorname{tr} \rho F_k. \end{aligned} \tag{5.10}$$

This p_{jk} satisfies

$$\begin{aligned} &\sum_{j,k} p_{jk} = 1, \\ &\sum_{j} p_{jk} = \operatorname{tr} \rho F_k = p_k, \end{aligned} \tag{5.11}$$

but not

$$\sum_{k} p_{jk} = \operatorname{tr} \rho E_j$$

unless $E_j F_k = F_k E_j$ $(\forall j, k)$, so that p_{jk} is not considered as a joint quantum probability distribution.

The above discussion shows that the order of the measurement of two observables A and B is essential, and it gives us a different expectation value, hence the state change.

5.5.2 Probability Positive Operator-Valued Measure (POVM)

We consider an example of a positive operator-valued measure on the lattice $\mathbb{Z}^2$. A family of bounded operators $\{M(\Delta),\ \Delta \in \sigma\text{-algebra on } \mathbb{Z}^2\}$ in a Hilbert space is called a probability *positive operator-valued measure (POVM)* if the following conditions are satisfied:

(i) $M(\Delta) \geq 0, M(\varnothing) = 0$
(ii) $M(\bigcup_k \Delta_k) = \sum_k M(\Delta_k)$ for any $\{\Delta_k\}$ with $\Delta_i \cap \Delta_j = \varnothing$ $(i \neq j)$
(iii) $M(\mathbb{Z}^2) = I$.

Let us consider an operator

$$M(\Delta) \equiv \sum_{i,j \in \Delta} F_i E_j F_i$$

for any spectral resolutions $\{F_i\}, \{E_j\} \subset \mathcal{P}_\mathcal{H}$ and any subset $\Delta \subset \mathbb{Z} \times \mathbb{Z} \equiv \mathbb{Z}^2$.

Theorem 5.8

1. *$M(\Delta)$ is a positive operator-valued measure.*
2. *$M(\cdot)$ is a* projection-valued measure *(PVM) (i.e., $M(\cdot)$ is POVM and $M(\cdot)^2 = M(\cdot)$) iff $F_i E_j = E_j F_i$ for any i, j.*

Proof (Part 1) Properties (i) and (iii) of POVM are obvious. Let us prove property (ii). The relation $\Delta_k \cap \Delta_l = \varnothing$ $(k \neq l)$ implies

$$M\left(\bigcup_k \Delta_k\right) = \sum_{i,j \in \bigcup_k \Delta_k} F_i E_j F_i$$
$$= \sum_k \sum_{i,j \in \Delta_k} F_i E_j F_i = \sum_k M(\Delta_k).$$

(Part 2) $[F_i, F_j] = 0$ $(\forall i, j)$ implies

$$F_i E_j F_i = E_j F_i F_i = E_j F_i.$$

Therefore,

$$M(\Delta) = \sum_{i,j \in \Delta} F_i E_j F_i = \sum_{i,j \in \Delta} E_j F_i,$$
$$M(\Delta)^* = \sum_{i,j \in \Delta} F_i^* E_j^* = \sum_{i,j \in \Delta} F_i E_j$$
$$= \sum_{i,j \in \Delta} E_j F_i = M(\Delta).$$

Similarly, we have $M(\Delta)^2 = M(\Delta)$. So $M(\Delta)$ is a projection. Conversely, when $M(\Delta)^2 = M(\Delta)^* = M(\Delta)$ is satisfied, $F_i E_j F_i = E_j F_i E_j$ holds for any $(i, j) \in \Delta \subset Z^2$. Hence $E_j = \sum_i E_j F_i E_j = \sum_i F_i E_j F_i$, which implies

$$F_k E_j = \sum_i F_k F_i E_j F_i = F_k F_j F_k = \sum_i F_i E_j F_i F_k = E_j F_k.$$

□

When we measure $A = \sum_i a_i F_i$ and $B = \sum_j b_j E_j$ in this order, and obtain the value in $\Delta \subset \mathbb{Z} \times \mathbb{Z}$, the state ρ is changed to

$$\rho_\Delta = \frac{\sum_{i,j \in \Delta} E_j F_i \rho F_i E_j}{\operatorname{tr} M(\Delta) \rho}.$$

The above correspondence

$$\rho \to \rho_\Delta$$

is not linear in general.

In order to partially solve the difficulty of the nonexistence of a joint quantum distribution, the notion of a compound state satisfying two conditions (5.10) and (5.11) was introduced by Ohya, and it is discussed in Chap. 7.

5.6 General Frameworks of Quantum Probability

Here we point out briefly that a von Neumann algebra and a C^*-algebra explained in Chap. 4 contain both classical and quantum systems and play essential role in studying some physical models with infinite degrees of freedom.

A quantum system is generally described by the above von Neumann algebraic framework or the C^*-algebraic framework by the following correspondence.

	CPT	v.N. or C^*-QPT
State	μ	$\varphi \in \mathfrak{S}$
Observable	real r.v. $f \in L^\infty(\Omega, \mu)$	s.a. $A \in \mathcal{N}$
Expectation value	$\langle f \rangle = \int_\Omega f \, d\mu$	$\varphi(A)$

5.7 Uncertainty Relation

As was discussed, the expected value of an observable A with respect to a state vector $\psi \in \mathcal{H}$ was

$$E_\psi(A) = \langle \psi, A\psi \rangle,$$

and its dispersion (variance) was given by

$$V_\psi(A) = \langle A^2 \rangle_\psi - \langle A \rangle_\psi^2.$$

Hence one can see that if a vector ψ is an eigenvector of some observable, its dispersion in the state ψ vanishes.

Heisenberg considered that any observation must have an effect on a system if the system is microscopic, and he derived an important inequality for dispersions of two observables. For arbitrary two operators A and B, the inequality

$$V_\psi(A) V_\psi(B) \geq \frac{1}{4} \left| \langle [A, B] \rangle_\psi \right|^2$$

holds if A, B and the commutator are defined on the vector state ψ. If the operator A is the position operator and B is the momentum operator with the commutation relations, i.e.,

$$[q_k, p_j] = \left[q_k, -i\hbar \frac{\partial}{\partial q_j} \right] = i\hbar \delta_{kj} I,$$

then one gets the *Heisenberg's uncertainty principle*.

Note that for a quantum system in a bounded region Heisenberg's uncertainty principle is generally not valid. It is clear from the following remark. One could have a discrete spectrum for the momentum operator; therefore, there are states in which the dispersion of the momentum is zero while the uncertainty of the position cannot be larger then the volume of the region, see [804].

Instead of taking the vector state ψ, we can use a general quantum state ρ. In such a case, the expectation and the variance are given respectively by

$$E_\rho(A) = \operatorname{tr} \rho A \quad \text{and} \quad V_\rho(A) = E_\rho\left(A^2\right) - E_\rho(A)^2.$$

Then the uncertainty relation is written as

$$V_\rho(A) V_\rho(B) \geq \frac{1}{4} \left| E_\rho\left([A, B]\right)\right|^2.$$

These uncertainty relations are easily proved by applying Schwarz inequality.

5.8 Principle of Quantum Measurements

Quantum mechanics has resolved fundamental difficulties of classical physics in describing atoms and elementary particles, and it also tremendously contributed to the whole science in the twentieth century. In spite of this success, quantum mechanics has not yet succeeded in clarifying all fundamental problems. In this section, we address one such problem, namely the *measurement problem*. The problem of measurement comes from the following belief. "Let us consider the process of measurement in a quantum system to obtain the measured results (we call such a process the measurement process). If quantum theory is the ultimate theory of nature, the measurement process itself should be described within the quantum theory."

There exists as of yet no satisfactory answer to this question although there have been several efforts by many physicists. Some physicists even claim that the complete resolution of the problem cannot be expected within quantum theory but needs some new theory beyond quantum mechanics.

In quantum information and quantum computation, one has to treat transmitted information or computation result, so that one should consider the process of quantum measurement. In this section, we discuss what a quantum measurement is from a practical point of view.

As discussed in the previous section, a state of a quantum system is represented by a vector φ in a Hilbert space $\mathcal{H}$ or a density matrix ρ in $\mathcal{H}$. For φ or ρ, the expectation value of an observable A in $\mathcal{H}$ is given by

$$\langle \varphi, A\varphi \rangle \quad \text{or} \quad \operatorname{tr} \rho A, \quad \text{respectively.}$$

Suppose there are two observables A and B to be measured. We must first determine in what state we make their measurements. Namely, it is needed to prepare a state φ or ρ. We assume that this state preparation is done by the observable A. Let

A have discrete eigenvalues (e.g., A is a compact operator), and let a_k and φ_k be its eigenvalues and eigenvectors, respectively (i.e., $A\varphi_k = a_k\varphi_k$). Then from the fundamental principle of quantum mechanics, the so-prepared state has the following form:

$$\varphi = \sum c_k\varphi_k \quad \text{or} \quad \rho = |\varphi\rangle\langle\varphi|,$$

where $\{c_k\}$ is a family of constants satisfying $\sum |c_k|^2 = 1$ and $\{\varphi_k\}$ is an orthonormal set of the eigenvectors.

According to von Neumann, there are two kinds of measurement. In one we measure the same observable immediately after we have measured an observable, and we always obtain the same value for these two successive measurements. Such a measurement process is called *the first kind of measurement*. Otherwise, a process is called *the second kind of measurement*. Since the former process is thought to be physical, we only consider this type of measurement hereafter.

Having prepared a state, we measure two observables A and B. When we measure A in the state ρ and obtain its measured result a_k with probability $|c_k|^2$, the state changes as

$$\rho \to \rho_k = E_k \equiv |\varphi_k\rangle\langle\varphi_k|.$$

5.8.1 Measurement Procedure

We consider a situation where we measure the observable B after the state ρ evolved for a time interval t by some unitary transformation U_t. Namely, the state at t is of the form $\rho(t) = U_t\rho U_t^*$. Assuming B is a compact operator whose spectral decomposition is

$$B = \sum_i b_i F_i \quad \text{with } F_i \equiv |\psi_i\rangle\langle\psi_i|,$$

the probability of getting a measured result b_i in the state $\rho(t)$ is given by

$$P(B = b_i; t) = \operatorname{tr} \rho(t) F_i.$$

Before the measurement of the observable B, the state has two different forms:

1. If we measured the observable A at $t = 0$ and its measured value a_k was read (*selected measurement*), then the state ρ changed to the state $\rho_k = E_k$ at $t = 0$, hence at t we have

$$P(B = b_i; t \mid A = a_k; t = 0) = \operatorname{tr} U_t\rho_k U_t^* F_i.$$

2. If we measured the observable A at $t = 0$ but the measured result was not read, then the state ρ had changed at $t = 0$ into a state $\rho' = \sum E_k\rho E_k = \sum |c_k|^2 E_k$, so that at t

$$P(B = b_i; t) = \operatorname{tr} U_t\rho' U_t^* F_i.$$

That is, depending on the way of measurement of A, the state ρ suffers from one of the following changes:

$$\rho \to \rho_k = E_k$$

or

$$\rho \to \rho'.$$

The state ρ changes to a pure state according to form 1, or it changes to a mixed state according to form 2. In any case, if quantum mechanics can describe all natural phenomena, the above changes (forms 1 and 2) should be described within quantum dynamics. That is, there should exist a Hamiltonian $H^{(i)}$ such that the unitary evolution $U_t^{(i)} = \exp(-\mathrm{i}tH^{(i)})$ provides such a dynamical change of state, where $i = 1$ and 2 correspond to the state changes $\rho \to \rho'$ and $\rho \to \rho_k$, respectively. More precisely, there exists a certain time T (including the case $T = \infty$) such that

$$\rho' = U_T^{(1)} \rho U_T^{(1)*} \quad \text{or} \quad \rho_k = U_T^{(2)} \rho U_T^{(2)*}.$$

However, it is known that such a unitary operator does not exist in general (Fell–Wigner's theorem).

Since reading a measured result should be done after a preparation for the reading, it is natural to consider that the change of form 1 occurs after the change of form 2. Consequently, we may think that the state change takes place as

$$\rho \to \rho' \to \rho_k,$$

where the change $\rho' \to \rho_k$ is made by means of a third unitary evolution $U_t^{(3)} = \exp(-\mathrm{i}tH^{(3)})$.

5.8.2 Reduction of State

The change of a state $\rho' \to \rho_k$ (or $\rho \to \rho' \to \rho_k$) is called *the reduction of wave packets (reduction of quantum state)*. To treat the reduction of wave packets, let us consider a practical action of measurement or observation in more detail. That is, one needs an apparatus to perform a measurement. Unlike observable to be measured, an apparatus should not be a microscopic object but somehow possess a certain macroscopic scale. However, even when the apparatus has a macroscopic scale, the apparatus itself consists of microscopic parts like atoms or molecules to which quantum mechanics is applicable. Thus the microscopic interaction between an observable and an apparatus should be taken into consideration, so that we apply quantum mechanics to the composite (compound) system of a system to be measured and a system of the apparatus. Denote a system to be observed as $\mathcal{O}$ and the system of the apparatus as $\mathcal{M}$. Then we consider the change of a state by an interaction between two systems.

(i) Before a measurement of an observable B, suppose that the system $\mathcal{O}$ is in a state $\varphi^s = \sum c_k \varphi_k^s$ (prepared by an observable A such as $A\varphi_k^s = a_k \varphi_k^s$, where $\{\varphi_k^s\}$ is an ONS) and $\mathcal{M}$ is in a state φ^a. In the course of the measurement, an interaction described by a Hamiltonian H between the system and the apparatus takes place, and it induces the change of the compound system as

$$\varphi = \varphi^s \otimes \varphi^a \to U_t \varphi^s \otimes \varphi^a = \varphi(t),$$

where t is the length of a time interval needed to accomplish the measurement and $U_t = \exp(-itH)$. Note that to accomplish the measurement and to obtain the measured value, there must be somehow a one-to-one correspondence between the eigenvectors of A and the apparatus. Suppose there *exists* such a correspondence between the vector state $\varphi_k^s(t)$ of the system and the vector state $\varphi_k^a(t)$ of the apparatus with $\langle \varphi_k^a(t), \varphi_i^a(t) \rangle = \delta_{ki}$ at time t. Then the state vector after a measurement (its value is not read) should be

$$\varphi'(t) = \sum_k c_k \varphi_k^s(t) \otimes \varphi_k^a(t).$$

However, the state $\varphi'(t)$ does not coincide with $\varphi(t) = U_t \varphi_s \otimes \varphi_a$ in general unless $U_t = U_t^s \otimes U_t^a$, so that von Neumann claimed that an explanation of the change

$$\varphi \to \varphi'(t) \tag{5.12}$$

is a fundamental problem of the measurement.

(ii) Let us reconsider the measurement process discussed in (i) in terms of density operators. The initial states of $\mathcal{O}$ and $\mathcal{M}$ are given by

$$\rho^s = |\varphi^s\rangle\langle\varphi^s| \quad \text{and} \quad \rho^a = |\varphi^a\rangle\langle\varphi^a|,$$

and their compound state is represented as

$$\rho = \rho^s \otimes \rho^a = |\varphi\rangle\langle\varphi|, \quad \varphi = \varphi^s \otimes \varphi^a. \tag{5.13}$$

The interaction between two systems changes this state to

$$\rho(t) = |\varphi(t)\rangle\langle\varphi(t)| \tag{5.14}$$

by the unitary evolution U_t above.

However, as in the case when one does not take the interaction between a system and the apparatus into account, ρ should change into

$$\rho'(t) = \sum_k |c_k|^2 |\varphi_k(t)\rangle\langle\varphi_k(t)|, \quad \varphi_k(t) = \varphi_k^s(t) \otimes \varphi_k^a(t) \tag{5.15}$$

when the measurement induces the change described by (5.12). However, $\rho(t)$ in (5.14) is expressed as

$$\rho(t) = \rho'(t) + \sum_{k \neq i} c_k \overline{c_j} |\varphi_k(t)\rangle\langle\varphi_j(t)|.$$

Hence $\rho(t)$ and $\rho'(t)$ do not generally coincide. This difference is caused from the interference terms being a characteristic of microscopic world. We have taken the apparatus system into account so that these interference terms may vanish because the apparatus is not completely microscopic. Therefore, we have to check whether there exists an apparatus in which the interference terms will vanish. This problem is often called the *problem of observation* and it is related to the so-called *decoherence problem*, see Chap. 20.

Now let us discuss the relation between the reduction of wave packets and a mathematical operation of partial trace averaging-out the effect of the apparatus. If the correspondence of von Neumann (5.15) is realized, then the state at $t = 0$ is given by (5.13). Let us trace out the apparatus system from this state. Taking a trace by a CONS including the orthonormal vectors $\{\varphi_k^a(t)\}$ of the apparatus, we obtain the state of the system

$$\rho^{s\prime}(t) = \mathrm{tr}_a\,\rho(t) = \sum_k |c_k|^2 \left|\varphi_k^s(t)\right\rangle\!\left\langle\varphi_k^s(t)\right|.$$

Thus the desired state change of the system is realized as

$$\rho^s \to \rho^s \otimes \rho^a \to \rho(t) \to \rho^{s\prime}(t).$$

Note that to take a partial trace means to observe unit element I of $\mathcal{M}$ in the composite system $\mathcal{O} \otimes \mathcal{M}$. That is, for any observable Q in $\mathcal{O}$,

$$\mathrm{tr}\,\rho(t) Q \otimes I = \mathrm{tr}\,\rho'(t) Q \otimes I = \mathrm{tr}_a\,\rho^{s\prime}(t) Q.$$

Although the operation of taking partial trace is a common technique such as in the study of open systems, it is not obvious whether this operation can be applied to the problem of measurement asking the completeness of quantum theory. There have been several opinions and views on this problem, but we omit such discussions here.

To summarize, the essence of quantum measurement for a practical use consists of the following two points:

1. If we observe (measure) some observable with its spectrum measure $\{F_k\}$ in a state ρ which is prepared by means of a certain observable, the state ρ changes into the following state ρ':
$$\rho' = \sum_k F_k \rho F_k.$$
Here we ignored time dependence of the evolution.
2. The expectation value of observing $B \equiv \sum_j b_j E_j$ in the above state ρ' and the probability of getting the eigenvalue b_j of the observable are given by
$$\mathrm{tr}\,\rho' B \quad \text{and} \quad \mathrm{tr}\,\rho' E_j, \quad \text{respectively.}$$

Remark 5.9 In the case of observables with continuous spectrum, one has only to use POVM instead of F_k above, see Sect. 5.10.

5.9 Composite Systems and Operations

Let us consider a composite quantum system consisting of two quantum systems, described by a tensor product Hilbert space $\mathcal{H} \otimes \mathcal{K}$ of original Hilbert spaces $\mathcal{H}$ and $\mathcal{K}$. We will discuss in this section the reduction of a composed state.

The state of a quantum system may be changed due to a measurement (reduction of the wave function), or due to the action of an external field. Here we consider a particular case of the state changes, called operations and instrument. More general state change is due to the notion of lifting, which will be discussed in Chaps. 7 and 20.

Let us denote the set of all states ρ on some separable Hilbert space $\mathcal{H}$ by $\mathfrak{S}(\mathcal{H})$. An *operation* is a mapping (we call such a mapping a channel; see Chap. 7) Λ^* from $\mathfrak{S}(\mathcal{H})$ into itself. We will use the following model for an operation Λ^*. The considered quantum system, originally in state ρ, is coupled to another quantum system (called apparatus or environment) with separable Hilbert state space $\mathcal{H}'$ and density matrix ρ'. The systems interact during a finite time interval. By this interaction, the original state $\rho \otimes \rho'$ of the system plus apparatus changes into $U(\rho \otimes \rho')U^*$ with some unitary operator U on $\mathcal{H} \otimes \mathcal{H}'$. Then, a measurement of some property of the apparatus, corresponding to a projection operator E' on $\mathcal{H}'$, is performed. One gets the state

$$\widehat{\sigma} = (1 \otimes E')U(\rho \otimes \rho')U^*(1 \otimes E')$$

of the coupled system. Here the operator $\widehat{\sigma}$ is positively defined but it is not necessarily normalized, $\mathrm{tr}\,\widehat{\sigma} \neq 1$. Finally, the system is again considered as an isolated system, which is described by the reduced density matrix

$$\sigma = \mathrm{tr}'\,\widehat{\sigma}.$$

Here tr' means the partial trace with respect to the space $\mathcal{H}'$ only.

The mapping $\mathcal{O} : \rho \to \sigma$ is called an *operation*. Note that σ is not normalized. We may drop also the initial normalization $\mathrm{tr}\,\rho = 1$. One can prove that *any operation may be written as (the Kraus–Sudarshan representation)*

$$\mathcal{O}\rho = \sigma = \sum_{k \in K} A_k \rho A_k^*,$$

where A_k are operators on $\mathcal{H}$, and K is a finite or a countably-infinite set satisfying the condition

$$\sum_{k \in K} A_k^* A_k \leq 1.$$

Note that the series for σ and $\{A_k^* A_k\}$ are convergent with respect to the trace-norm and the ultraweak topology, respectively.

5.9.1 Operations and Completely Positive Mappings

In this subsection, we show that dual operations can be characterized as completely positive mappings (see Chap. 7). Denote $\mathbf{T}(\mathcal{H})$ the Banach space of trace-class operators and $\mathbf{B}(\mathcal{H})$ the C^*-algebra of all bounded operators on $\mathcal{H}$. Any operation $\mathcal{O}:\rho\to\sigma$, considered as a real-linear mapping of the set of non-normalized density matrices (i.e., nonnegative Hermitian trace-class operators), may be extended to a complex-linear mapping of $\mathbf{T}(\mathcal{H})$ into itself. The extension is defined by

$$\mathcal{O}X=\sum_{k\in K}A_kXA_k^*$$

for any $X\in\mathbf{T}(\mathcal{H})$. One can show that the sum over k converges, that $\mathcal{O}X\in\mathbf{T}(\mathcal{H})$, and moreover,

$$|\mathcal{O}X|\le|X|,$$

where the norm is defined by

$$|X|=\sup\left\{\left|\mathrm{tr}(BX)\right|;\ B\in\mathbf{B}(\mathcal{H}),\ \|B\|=1\right\}.$$

The space $\mathbf{B}(\mathcal{H})$ of all bounded operators, considered as a Banach space with the usual operator norm, is the conjugate space of $\mathbf{T}(\mathcal{H})$. Any $B\in\mathbf{B}(\mathcal{H})$ defines a continuous linear functional

$$F_B(X)=\mathrm{tr}(BX)$$

on $\mathbf{T}(\mathcal{H})$ and vice versa. The mapping $\widehat{\mathcal{O}}$ of $\mathbf{B}(\mathcal{H})$, conjugate to the mapping $\mathcal{O}$ of $\mathbf{T}(\mathcal{H})$, is defined for arbitrary $B\in\mathbf{B}(\mathcal{H})$ by

$$\mathrm{tr}\big((\widehat{\mathcal{O}}B)X\big)=\mathrm{tr}\big(B(OX)\big),$$

and the continuity of $\mathcal{O}$ implies

$$|\widehat{\mathcal{O}}B|\le|B|.$$

A representation of $\widehat{\mathcal{O}}$ is given by

$$\widehat{\mathcal{O}}B=\sum_{k\in K}A_k^*XA_k,$$

where the series is ultraweakly convergent.

The mapping $\widehat{\mathcal{O}}$ is positive, i.e., it maps the set of nonnegative Hermitian operators into itself. Actually, one can prove that the mapping $\widehat{\mathcal{O}}$ has a stronger positivity property. Consider an arbitrary $n\times n$ matrix $\mathbf{B}\equiv(B_{I,j})$ with matrix elements $B_{ij}\in\mathbf{B}(\mathcal{H})$. Such an operator matrix represents an operator on $\mathcal{H}\otimes\mathbb{C}^n$ with an n-dimensional Hilbert space $\mathbb{C}^n$. Denote by $\widehat{\mathcal{O}}\mathbf{B}$ the matrix with matrix elements $\widehat{\mathcal{O}}B_{ij}$. *The mapping* $\widehat{\mathcal{O}}$ *is called completely positive if* $\mathbf{B}\ge 0$ *implies* $\widehat{\mathcal{O}}\mathbf{B}\ge 0$ *for all* n.

Proposition 5.10 *The mapping $\widehat{\mathcal{O}}$, conjugated to the operation $\mathcal{O}$, is completely positive.*

The mapping Λ^* of a special type of the above $\mathcal{O}$ is called a *channel* if it sends a state to a state. The channel is a very important concept in studying quantum communication processes which will be discussed in Chap. 7.

5.9.2 Measurements as Unitary Transformations on Extended Space

We can represent any non-selective measurement as a unitary transformation on an extended Hilbert space.

Let $\mathcal{H}$ be a Hilbert space, ρ be a density operator, and E be the projection operator on $\mathcal{H}$. Then there exists a Hilbert space $\mathcal{K}$, a density operator w on $\mathcal{K}$, and a unitary operator U on $\mathcal{H}\otimes\mathcal{K}$ such that

$$E\rho E+(1-E)\rho(1-E)=\mathrm{tr}_{\mathcal{K}}\big(U(\rho\otimes w)U^*\big).$$

The state ρ as a result of measurement changes according to the rule (projection postulate for non-selective measurements):

$$\rho\to E\rho E+(1-E)\rho(1-E).$$

Therefore, this Kraus–Sudarshan theorem shows that any non-selective measurement can be represented as a unitary transformation on the Hilbert space which includes the measurement apparatus. More generally, we can represent in this form the action of any quantum operation:

Theorem 5.11 *For the above ρ and w, we have*

$$\sum_k A_k\rho A_k^*=\mathrm{tr}_{\mathcal{K}}\big(U(\rho\otimes w)U^*\big)$$

if $\sum_k A_k^*A_k=1$.

Proof Suppose that the indices k in A_k take the values $k=1,2,\dots$. Take for $\mathcal{K}$ a Hilbert space with orthogonal basis $\{g_i; i=0,1,2,\dots\}$. Then we decompose $\mathcal{H}\otimes\mathcal{K}$ into orthogonal subspaces $\mathcal{H}_i=\mathcal{H}\otimes g_i$ which we identify with $\mathcal{H}$. We write

$$\mathcal{H}\otimes\mathcal{K}=\mathcal{H}_0\oplus\widehat{\mathcal{H}},$$

where $\widehat{\mathcal{H}}\equiv\bigoplus_{k\ge1}\mathcal{H}_k$, $\mathcal{H}_k\equiv\mathcal{H}$, $k=0,1,2,\dots$. Vectors $f\in\mathcal{H}\otimes\mathcal{K}$ may be represented in matrix notation as

$$f=f_0\oplus\widehat{f}=\begin{pmatrix}f_0\\ \widehat{f}\end{pmatrix},$$

where $f_0 \in \mathcal{H}$ and $\widehat{f} \in \widehat{\mathcal{H}}$. Define, in terms of the operators $\{A_k\}$, the isometric operator $A : \mathcal{H} \to \widehat{\mathcal{H}}$ by

$$Af_0 = \bigoplus_{k \geq 1} A_k f_0.$$

In the matrix notation, the unitary operator U on $\mathcal{H} \otimes \mathcal{K}$ is defined as $U = \begin{pmatrix} 0 & A^* \\ A & 1-AA^* \end{pmatrix}$. Finally, the density operator w in $\mathcal{K}$ is taken to be the projection operator onto the vector g_0. With these definitions, one can check the validity of the formula in the formulation of the theorem. □

5.10 POVM and Instruments

An example of positive operator-valued measure (POVM) was considered in the previous section of this chapter. Here we discuss similar things in a more general setting suitable also for continuous measurements.

5.10.1 POVM

After von Neumann, a self-adjoint operator is often considered as an observable in quantum mechanics. In physics, however, one calls an observable any symmetric operator which is not necessary self-adjoint. Such an operator admits a spectral representation in terms of a *positive operator-valued measure* (*POVM*) which is, in general, not projection-valued.

Let Ω be a set with a σ-field $\mathcal{F}$. The pair $(\Omega, \mathcal{F})$ is called a *measurable space*. Let $\mathcal{H}$ be a Hilbert space and $\mathbf{B}(\mathcal{H})$ be the set of all bounded linear operators on $\mathcal{H}$.

Definition 5.12 A positive operator-valued measure on Ω is a map $M : \mathcal{F} \to \mathbf{B}(\mathcal{H})$ such that

(i) $M(B) \geq M(\varnothing) = 0$ for all $B \in \mathcal{F}$
(ii) If $\{B_n\}$ is a countable collection of disjoint sets in $\mathcal{F}$ then

$$M\left(\bigcup_{n=1}^{\infty} B_n\right) = \sum_{n=1}^{\infty} M(B_n),$$

where the series is weakly convergent.

A positive operator-valued measure is called an *observable* or *probability positive operator-valued measure* if also

$$M(\Omega) = 1.$$

Theorem 5.13 *For a symmetric operator A with a dense domain $\mathcal{D}(A) \subseteq \mathcal{H}$ there exists, in general, a non-unique POVM $M(d\lambda)$ such that*

$$\langle \psi, A\psi \rangle = \int \lambda \langle \psi, M(d\lambda)\psi \rangle, \quad \psi \in \mathcal{D}(A);$$

$$\|A\psi\|^2 = \int \lambda^2 \langle \psi, F(d\lambda)\psi \rangle, \quad \psi \in \mathcal{D}(A).$$

One can prove that any POVM $\{M(B)\}$ in $\mathcal{H}$ can be dilated to an orthogonal POVM $\{E(B)\}$ in a larger Hilbert space $\widetilde{\mathcal{H}}$, so that the following relation holds

$$M(B) = PE(B)P, \quad B \in \mathcal{F},$$

where P is the projection from $\widetilde{\mathcal{H}}$ onto $\mathcal{H}$.

Any *measurement with values in* Ω (it is called an $(\Omega, \mathcal{F})$-measurement) is described by an affine map $\rho \to W(\cdot\,; \rho)$ of the convex set of quantum states in $\mathcal{H}$ into the set of probability measures on $(\Omega, \mathcal{F})$. One can prove that any such map has the form

$$W(B; \rho) = \operatorname{tr}\left[M(B)\rho\right]$$

with some POVM $\{M(B); B \in \mathcal{F}\}$. Conversely, any POVM $\{M(B); B \in \mathcal{F}\}$ defines a probability measure $W(B; \rho)$ by this formula.

There is a one-to-one correspondence between POVM on a compact metrizable space Ω and the positive linear maps from the space of real continuous functions on Ω to $\mathbf{B}(\mathcal{H})$. It is given by the formula

$$\overline{M}(f) = \int_{\Omega} f(\omega) M(d\omega).$$

One has the normalization $\overline{M}(1) = 1$.

To describe the change of a quantum state after the measurement, the notion of an instrument was introduced.

5.10.2 Instrument

Let $\mathcal{T}$ be the Banach space of the self-adjoint trace-class operators on $\mathcal{H}$ with the trace norm and the cone $\mathcal{T}^+$ of non-negative trace-class operators. The states are defined as the non-negative trace-class operators of trace one, i.e., the density operators. A *positive linear map* $T : \mathcal{T} \to \mathcal{T}$ is a linear map such that if $x \in \mathcal{T}^+$ then $T(x) \in \mathcal{T}^+$. We denote $\mathcal{L}^+(\mathcal{T})$ the cone of all positive linear maps on $\mathcal{T}$.

Definition 5.14 A *positive map-valued* (PMV) *measure* on the measurable space $(\Omega, \mathcal{F})$ is a map $\Gamma : \mathcal{F} \to \mathcal{L}^+(\mathcal{T})$ such that

(i) $\Gamma(B) \geq \Gamma(\varnothing) = 0$, $B \in \mathcal{F}$.

(ii) If $\{B_n\}$ is a countable collection of disjoint sets in $\mathcal{F}$ then

$$\Gamma\left(\bigcup_{n=1}^{\infty} B_n\right) = \sum_{n=1}^{\infty} \Gamma(B_n),$$

where the sum is convergent in the strong topology.

The PMV measure Γ is called an *instrument* if it satisfies the further condition.

(iii) $\mathrm{tr}\,[\Gamma(\Omega)\rho] = \mathrm{tr}\,[\rho]$ for all $\rho \in \mathcal{T}$.

We will denote $\Gamma(B)\rho = \Gamma(B,\rho)$. The statistics of the measurement is given by the probability measure $W(B;\rho) = \mathrm{tr}\,[\Gamma(B)\rho]$. An instrument describes the state change $\rho \to \Gamma(B)\rho$ after the measurement, conditional on the value observed.

Theorem 5.15 *If $W(\cdot;\rho)$ is an $(\Omega,\mathcal{F})$-measurement on a Hilbert space $\mathcal{H}$ then there exist a unique POVM $\{M(B), B \in \mathcal{F}\}$ and an instrument Γ such that*

$$W(B;\rho) = \mathrm{tr}\left[M(B)\rho\right] = \mathrm{tr}\left[\Gamma(B)\rho\right], \quad B \in \mathcal{F}.$$

Proof The existence of the POVM and the instrument is not proved here (see [344] and [189]). We only show that there is an explicit formula for the instrument. Choose a countable partition $\{B_n\}$ of X into pairwise disjoint Borel sets and a sequence $\{\rho_n\}$ of density operators. Then the formula

$$\Gamma(B)\rho = \sum_n \mathrm{tr}\left[M(B \cap B_n)\rho\right]\rho_n$$

defines an instrument with the probability measure $W(B;\rho)$ and the POVM $F(B)$. □

To every instrument Γ on the value space $(\Omega,\mathcal{F})$ there is a unique observable $M(\cdot)$ such that

$$\mathrm{tr}\left[M(B)\rho\right] = \mathrm{tr}\left[\Gamma(B)\rho\right]$$

for all density operators ρ and $B \in \mathcal{F}$.

Let the value space $(\Omega,\mathcal{F})$ be a *standard Borel space*, i.e., a Borel space which is Borel isomorphic to a Borel subset of some complete separable metric space. It is known that every standard Borel space is Borel isomorphic to some zero-dimensional separable compact Hausdorff space.

If Γ_1 and Γ_2 are instruments on compact metrizable spaces Ω_1 and Ω_2, then it is possible to define their composition, an instrument Γ on $\Omega_2 \times \Omega_1$ which represents the measurement of Γ_1 first and then of Γ_2.

Theorem 5.16 *Let Γ_1 and Γ_2 be instruments on the standard Borel spaces Ω_1 and Ω_2. Then there exists a unique instrument Γ on $\Omega_2 \times \Omega_1$ such that $\Gamma(F \times E) = \Gamma_2(F)\Gamma_1(E)$ for all Borel subsets $F \subset \Omega_2$ and $E \subset \Omega_1$.*

The instrument Γ is called the *composition* of Γ_2 following Γ_1 and is denoted by $\Gamma_2 \circ \Gamma_1$. Let the instrument (Ω_1, Γ_1) determine the observable (Ω_1, M_1) and the instrument (Ω_2, Γ_2) determine the observable (Ω_2, M_2). The $\mathcal{T}^*$-valued measure $F \to \Gamma_1(\Omega_1)^* M_2(F)$ is an observable which is called the observable $M_2(\cdot)$, *conditioned* by the measurement of the observable $M_1(\cdot)$ with the instrument Γ_1.

The *joint distribution* of Γ_2 following Γ_1 in the state ρ is an observable $M(\cdot)$ on $\Omega_2 \times \Omega_1$ given by

$$M(G) = \mathrm{tr}\big[(\Gamma_2 \circ \Gamma_1)(G)\rho\big],$$

for G from the σ-field of the Borel sets in $\Omega_2 \times \Omega_1$. Its marginal distributions satisfy

$$M(F \times \Omega_1) = \Gamma_1(\Omega_1)^* M_2(F), \quad F \subset \Omega_2,$$

$$M(\Omega_2 \times E) = M_1(E), \quad E \subset \Omega_1.$$

5.10.3 Covariant POVM and Instrument

If we are interested in describing measurements which are symmetric under the action of a group such as the Galilean or the Poincaré groups, then we have to consider covariant POVM and instruments.

If G is a locally compact group, then a locally compact space Ω is called a G-space if there is a jointly continuous map $\Omega \times G \to \Omega$ such that $(xg_1)g_2 = x(g_1 g_2)$ for all $x \in \Omega$ and $g_1, g_2 \in G$. If B is a Borel set in Ω we define $B_g = \{y : y = xg, x \in B\}$.

Let U be a strongly continuous unitary representation of G in a Hilbert space $\mathcal{H}$. A POVM $M(\cdot)$ on Ω with values in $\mathbf{B}(\mathcal{H})$ is *covariant* if there exist U and G as above such that

$$U(g)^* M(B) U(g) = M(B_g)$$

for all $g \in G$ and all Borel sets $B \subseteq \Omega$.

If U is a unitary representation of a compact metrizable group G on the finite dimensional Hilbert space $\mathcal{H}$, there is the following representation for a covariant POVM $M(\cdot)$:

$$\overline{M}(f) = \int_G f(g) U(g)^* F U(g)\, dg,$$

where Ω is a G-space isomorphic to the set G/H of the right cosets of a stable subgroup H of a point in Ω, and f is a real continuous function on G which is constant on the right cosets, dg is the invariant measure on G, F is an operator on $\mathcal{H}$ such that $FU(h) = U(h)F$ for all $h \in H$.

A PMV measure Γ (in particular, an instrument) on Ω is called *covariant* if

$$U(g)^* \Gamma(B, \rho) U(g) = \Gamma(B_g, \rho_g)$$

for all $\rho \in \mathcal{T}$, $g \in G$ and $B \subseteq \Omega$. Here $\rho_g = U(g)^* \rho U(g)$.

If U is a unitary representation of a compact metrizable group G on the finite-dimensional Hilbert space $\mathcal{H}$, similarly to the covariant POVM, there is the following representation of a covariant instrument:

$$\overline{\Gamma}(f,\rho)=\int_G f(g)U(g)^*T\big(U(g)\rho U(g)^*\big)U(g)\,dg,$$

where T is a positive linear map $T:\mathcal{T}\to\mathcal{T}$ such that $U(h)T(\rho)U(h)^*=T(U(h)\rho U(h)^*)$ for all $h\in H$ and $\operatorname{tr}T(\rho)=\operatorname{tr}\rho$ for all $\rho\in\mathcal{T}$ such that $\rho U(g)=U(g)\rho$ for all $g\in G$.

Example 5.17 We give an example of a covariant instrument which represents an approximate position measurement [189]. Take the Hilbert space $\mathcal{H}=L^2(\mathbb{R}^3)$ and let U be the representation of the Euclidean group in $\mathcal{H}$. Take a bounded rotation-invariant function α on $\mathbb{R}^3$ such that $\int|\alpha(x)|^2\,d^3x=1$ and define the operator A_x on $\mathcal{H}$ by $(A_x\psi)(y)=\alpha(y-x)\psi(y)$ where $\psi\in\mathcal{H}$. Then the formula

$$\Gamma(B)\rho=\int_B A_x\rho A_x^*\,d^3x$$

defines an instrument on $\mathbb{R}^3$ which is covariant with respect to the Euclidean group. The associated POVM is the approximate position observable given by

$$M(B)\psi(x)=\big(1_B\circ\big(|\alpha|^2\big)\big)(x)\psi(x),$$

where 1_B is the characteristic function of the Borel set $B\subseteq\mathbb{R}^3$.

5.11 Seven Principles of Quantum Mechanics

We here summarize the principles of quantum mechanics in an axiomatic framework.

Most discussions on foundations and interpretations of quantum mechanics take place around the meaning of probability, measurements, reduction of the state, and entanglement. The list of basic axioms of quantum mechanics, as it was formulated by von Neumann, includes only general mathematical formalism of the Hilbert space and its statistical interpretation. From this point of view, any mathematical proposition on properties of operators in the Hilbert space can be considered as a quantum-mechanical result. From our point of view, such an approach is too general to be called foundations of quantum mechanics. We have to introduce more structures to treat a mathematical scheme as quantum mechanics.

These remarks are important for practical purposes. If we would agree about the basic axioms of quantum mechanics and if one proves a proposition in this framework then it could be considered as a quantum-mechanical result. Otherwise it can be a mathematical result without immediate relevance to quantum theory. An important example of such a case is related to Bell's inequalities. It is known that the correlation function of two spins computed in the 4-dimensional Hilbert space does not satisfy the Bell's inequalities. This result is often interpreted as the proof that quantum mechanics is inconsistent with Bell's inequalities. However, from the previous discussion it should be clear that such a claim is justified only if we agree to treat the 4-dimensional Hilbert space as describing a physical quantum-mechanical system. In some works on quantum information theory, a qubit, i.e., a 2-dimensional Hilbert space, is considered as a fundamental notion.

Let us note, however, that in fact the finite-dimensional Hilbert space should be considered only as a convenient approximation for a quantum-mechanical system, and if we want to investigate the fundamental properties of quantum mechanics then we have to work in an infinite-dimensional Hilbert space because only there the condition of locality in space and time can be formulated. There are certain problems where we cannot reduce the infinite-dimensional Hilbert space to a finite-dimensional subspace.

We shall present a list of seven axioms of quantum mechanics. The axioms are well known from various textbooks, but normally they are not combined together. Then, these axioms define an axiomatic quantum-mechanical framework. If some propositions are proved in this framework then it could be considered as an assertion in axiomatic quantum mechanics. If we fix the list, it can help to clarify some problems in the foundations of quantum mechanics.

For example, as we shall see, the seven axioms do not admit a nontrivial realization in the 4-dimensional Hilbert space. This axiomatic framework requires an infinite-dimensional Hilbert space. One can prove that Bell's inequalities might be consistent with the correlation function of the localized measurements of a spin computed in the infinite-dimensional Hilbert space (see Chap. 16). Therefore, in this sense, we can say that axiomatic quantum mechanics is consistent with Bell's inequalities and with local realism. It is well known that there are no Bell's type experiments without loopholes, so there is no contradiction between Bell's inequalities, axiomatic quantum mechanics and experiments.

There is a gap between an abstract approach to the foundation and the very successful pragmatic approach to quantum mechanics which is essentially reduced to the solution of the Schrödinger equation. If we are able to fill this gap then perhaps it will be possible to get some progress in the investigations of foundation because, in fact, the study of solutions of the Schrödinger equation led to the deepest and greatest achievements of quantum mechanics.

The key notion which can help build a bridge between the abstract formalism of the Hilbert space and the practically useful formalism of quantum mechanics is the notion of the ordinary 3-dimensional space. It is suggested that the *spatial properties of a quantum system* should be included in the list of basic axioms of quantum mechanics together with the standard notions of the Hilbert space, observables and states. Similar approach is well known in quantum field theory, but it is not very much used when we consider the foundations of quantum mechanics.

Quantum mechanics is essentially reduced to the solution of the Schrödinger equation. However, in many discussions on the foundations of quantum mechanics not only the Schrödinger equation is not considered, but even the space–time coordinates are not mentioned. Such views to the foundations of quantum mechanics are similar to the consideration of the foundations of electromagnetism but without mentioning the Maxwell equations.

Here a list of seven basic postulates (axioms) of quantum mechanics is presented. The axioms described below relate to:

1. Hilbert space
2. Measurements
3. Time
4. Space
5. Composed systems
6. Bose–Fermi alternative
7. Internal symmetries.

In particular, the list includes the axiom describing spatial properties of a quantum system which plays a crucial role in the standard formalism of quantum mechanics.

1. (Hilbert space) To a physical system one assigns a Hilbert space $\mathcal{H}$. The observables correspond to the self-adjoint operators in $\mathcal{H}$. The pure states correspond to the one-dimensional subspaces of $\mathcal{H}$. An arbitrary state is described by the density operator, i.e., a positive operator with the unit trace. For the expectation value $E_\rho(A)$ (or $\langle A\rangle_\rho$) of the observable A in the state described by the density operator ρ, we have the Born–von Neumann formula

$$E_\rho(A) = \mathrm{tr}(\rho A).$$

2. (Measurements) A measurement is an external intervention which changes the state of the system. These state changes are described by the concept of a state transformer, or an instrument. Let $\{\Omega, \mathcal{F}\}$ be a measurable space where Ω is a set and $\mathcal{F}$ is a σ-algebra its subsets. A state transformer Γ is a state transformation-valued measure $\Gamma = \{\Gamma_B;\ B \in \mathcal{F}\}$ on the measurable space. A state transformation Γ_B is a linear, positive, trace-norm contractive map on the set of trace-class operators in $\mathcal{H}$.

   An ideal state transformer $\Gamma = \{\Gamma_i;\ i = 1, 2, \dots\}$ associated with a discrete observable $A = \sum_{i=1}^{\infty} a_i E_i$ is given by the Dirac–von Neumann formula

$$\Gamma_i(\rho) = E_i \rho E_i,$$

   if it is known that the measurement outcome is a real number a_i. Here E_i is an orthogonal projection operator. Similar formulae hold for the positive operator-valued measure (POVM).
3. (Time evolution) The dynamics of the density operator ρ and of a state ψ in the Hilbert space which occurs with passage of time is given by

$$\rho(t) = U(t)\rho U(t)^*,$$
$$\psi(t) = U(t)\psi.$$

Here t is a real parameter (time), $U(t)$ is a unitary operator satisfying the abstract Schrödinger equation

$$i\hbar \frac{\partial}{\partial t} U(t) = H(t) U(t),$$

where $H(t)$ is a (possibly time dependent) self-adjoint energy operator (Hamiltonian) and $\hbar$ is the Planck constant.

4. (Space) There exists the 3-dimensional Euclidean space $\mathbb{R}^3$. Its group of motion is formed by the translation group T^3 and the rotation group $O(3)$. One supposes that in the Hilbert space $\mathcal{H}$ there is a unitary representation $U(a)$ of the translation by the vector $a \in \mathbb{R}^3$. If $(\Omega, \mathcal{F})$ is a measurable space and $\{F_B; B \in \mathcal{F}\}$ is the associated POVM then one has

$$U(a) F_B U(a)^* = F_{\alpha_a(B)},$$

where $\{\alpha_a : \mathcal{F} \to \mathcal{F}; a \in \mathbb{R}^3\}$ is the group of automorphisms.

One has also a projective representation of the rotation group $SO(3)$ which can be made into a unitary representation $U(R)$ of the covering group $SU(2)$, here $R \in SU(2)$. Hopefully, the distinction by the type of argument of U will be sufficient to avoid confusion. The irreducible representations of $SU(2)$ describe systems with integer and half-integer spins.

In solid state physics, one works with periodic potentials. In this case, instead of the group of motion of $\mathbb{R}^3$ one has a lattice L in $\mathbb{R}^3$ and its unitary representation.

5. (Composite systems) If there are two different systems with assigned Hilbert spaces $\mathcal{H}_1$ and $\mathcal{H}_2$ then the composite system is described by the tensor product

$$\mathcal{H} = \mathcal{H}_1 \otimes \mathcal{H}_2.$$

For composite systems with variable number of particles, the direct sum of the Hilbert spaces is used.

6. (Bose–Fermi alternative) The Hilbert space of an N-particle system is the N-fold tensor product of the single particle Hilbert spaces provided that the particles are not of the same species. For identical particles with integer spin (bosons), one uses the symmetrized N-fold tensor product $(\mathcal{H}^{\otimes N})_S$ of the Hilbert space $\mathcal{H}$. For identical particles with half-integer spin (fermions), one uses the anti-symmetrized N-fold tensor product $(\mathcal{H}^{\otimes N})_A$.
7. (Internal symmetries) There is a compact group G_{int} of internal symmetries and its unitary representation $U(\tau)$, $\tau \in G_{\text{int}}$ in the Hilbert space $\mathcal{H}$ which commutes with representations of the translation group $\mathbb{R}$ and the rotation group $SU(2)$. For instance, one could have the gauge group $G_{\text{int}} = U(1)$ which describes the electric charge. The group generates the superselection sectors.

Summary 5.18 Axiomatic quantum mechanics described by the presented seven axioms can be briefly formulated as follows. There is the space–time $\mathbb{R}^{\mathbf{1}} \times \mathbb{R}^{\mathbf{3}}$, the symmetry group $G = T^1 \times T^3 \times SU(2) \times G_{\text{int}}$, the Hilbert space $\mathcal{H}$ and the unitary representation $U(g)$ of G, here $g \in G$. Axiomatic quantum mechanics is given by the following data:

$$\{\mathcal{H}, U(g), \rho, (\Omega, \mathcal{F}, \alpha_g), \{F_B, \Gamma_B, B \in \mathcal{F}\}\}.$$

Here ρ is the density operator, $(\Omega, \mathcal{F})$ is the measurable space, α_g is the group of automorphisms of the σ-algebra $\mathcal{F}$, $\{F_B\}$ is POVM, and $\{\Gamma_B\}$ the state transformer.

Example 5.19 An example of a quantum system satisfying all seven axioms is given by the non-relativistic spin one-half particle with the Hilbert space $\mathcal{H} = \mathbb{C}^2 \otimes L^2(\mathbb{R}^3)$ and the Schrödinger–Pauli Hamiltonian and also by its multi-particle generalization.

We can add more axioms, of course. In particular, we did not postulate yet the covariance under the Poincaré or Galilei group (for the Galilei group one has a projective representation) but only invariance under spatial translations and rotations which we have in the non-relativistic theory as well as in the relativistic theory. We could also add the condition of the positivity of energy. Finally, we could postulate the standard non-relativistic Schrödinger equation for N bodies as a fundamental axiom of quantum mechanics.

Note also that in *relativistic quantum field theory* (see Chap. 16) all the axioms are valid except the second axiom on measurements which requires a special discussion. In fact, we have to add more axioms to get quantum field theory. One postulates that there exists an operator-valued generalized function $\Phi(f)$ in $\mathcal{H}$ which satisfies the axioms of locality, relativistic invariance, and spectrality.

Note that this axiomatic approach to quantum mechanics has been used for the investigation of quantum non-locality, see Chap. 8. It is shown that quantum non-locality in the sense of Bell exists only because the spatial properties of a quantum system are neglected. If we take the spatial degrees of freedom into account then local realism might be consistent with quantum mechanics and with performed experiments. If somebody wants to depart from local realism then he/she has to change quantum mechanics. The local realism representation in quantum mechanics was formulated as follows:

$$\operatorname{tr}(\rho A_m(x) B_n(y)) = E(\xi_m(x)\eta_n(y)).$$

Here A and B are observables depending on the space points x and y and on parameters m and n, while E is the expectation of two random fields $\xi_m(x)$ and $\eta_n(y)$. It would be important to prove the representation under more general assumptions. Such aspects will be discussed in Chap. 8.

5.12 Notes

There are a lot of textbooks on elementary quantum mechanics [203, 229, 237, 464, 519, 615, 672, 749]. The bra–ket notation is introduced in [203]. The monograph

[672] can be considered for its modern version and will be helpful for the students to study the quantum mechanics quickly. On the other hand, [749] has a precise explanation on the history of pre-quantum mechanics and will be suitable to study the quantum mechanics step by step. In [815], one can find many important papers in quantum mechanics, including the measurement problem. More recent works are treated in [644]. Regarding the measurement problem, [147] treats its mathematical aspect, and in [359, 615] one can find its philosophical aspect. On the conditional probability, [4] gives a concise explanation with an emphasis on the difference of quantum theory and classical theory. The Kraus theorem was first proved by Sudarshan [738] and Kraus in [452]. Holevo [346] presents the recent works in this area. The list of basic axioms of quantum mechanics was first formulated by von Neumann [806]; see also [34, 283, 404, 412, 494, 701, 793]. In [413], a consideration on the meaning of Bell's inequality was presented. The formulations of the seven axioms presented here are based on the material from [798]. For the mathematical treatment of the quantum field theory, see [65, 123, 310, 701, 736]. The definitions of a local realism in the sense of Bell and in the sense of Einstein and its relations with the contextual approach are considered in [413]. An application of POVM to the problem of localization of photons [736] is in [222, 346, 773]. The book of Davis [190] discusses POVM and instruments.

Many contributions to the quantum probability have been made by Accardi [4, 6, 7, 9, 34], Belavkin [91–93], Ohya [560], Petz [647], Hiai [331], and many others.

Chapter 6
Fundamentals of Classical Information Theory

In this chapter, we briefly review the basic facts of the classical information communication processes. The fundamental aspects of information theory according to Shannon [706] are composed of the following concepts: *message*; *entropy* describing the amount of information; *communication channel, mutual entropy, coding, and capacity* of the channel. We will discuss some coding theorems which are important results of the classical information theory.

6.1 Communication Processes and Channel

First of all, we discuss communication processes. Let A be a set (an *alphabet*) of certain *letters* (or *symbols*). A *message* is a sequence $m = a_i \cdots a_n$ of letters in A.

The set $]a_1, \dots, a_l[\equiv \{\Pi x_k \in A^{\mathbb{Z}} \equiv \Pi A; x_k = a_k, k = 1, \dots, l\}$, where ΠA is the infinite direct product of A, is called the *cylinder set* labeled by the message.

The cylinder set has the following properties:

1. $]a_1, \dots, a_l[=]b_1, \dots, b_l[$ iff $a_k = b_k$ for any k.
2. $]a_j, \dots, a_l[\subset]a_i, \dots, a_n[$ if $j \le i \le n \le l$.
3. $]a_j, \dots, a_l[= \cap\{]a_k[; j \le k \le l\}$.
4. $]a_j, \dots, a_l[= \cup\{]x_j, \dots, x_l[; x_k = a_k, j \le k \le l\}$.
5. $]a_j, \dots, a_l[^c = \cup\{]x_j, \dots, x_l[; x_k \ne a_k, j \le k \le l\}$.
6. $\cup\{]x_k[; x_k \in A, j \le k \le l\} = \cup\{]x_j, \dots, x_l[; x_k \in A, j \le k \le l\} = A^Z$.

Let $\mathcal{M}$ be the set of all cylinder sets, $\mathcal{M} \subseteq A^{\mathbb{Z}}$. Thus it is easily seen that the space $\mathcal{M}$ becomes a field; i.e., $E, F \subseteq \mathcal{M} \Rightarrow E \cup F,\ E \cap F,\ E^c \in \mathcal{M}$, so that $(A, \mathcal{F}_A)$ is called *the message space,* where $\mathcal{F}_A$ is the σ-field generated by $\mathcal{M}$.

Remark here that we call $\mathcal{M}$ itself the message space in the sequel.

In order to send information written by an element of this message space to a receiver, we first need to transfer the message into a proper form for a communication channel. This change of a message is called *a coding*. To be precise, a coding is a one-to-one map ξ from $\mathcal{M}$ to some space $\mathcal{X}$ which is called the *coded space*.

For instance, we have the following codings:

M. Ohya, I. Volovich, *Mathematical Foundations of Quantum Information and Computation and Its Applications to Nano- and Bio-systems*,
Theoretical and Mathematical Physics,
DOI 10.1007/978-94-007-0171-7_6, © Springer Science+Business Media B.V. 2011

1. When a message is expressed by binary a symbol 0 or 1, such a coding is a map from $\mathcal{M}$ to $\{0,1\}^{\mathbb{N}}$.

   Here 0, 1 can be represented, for example, by electric signals off and on.
2. In quantum coding, instead of electric signals we can use quantum states to represent 0, 1, for instance, vacuum and coherent states, respectively (see the next chapter for quantum codes).

Coding is a combination of several maps like the above codings 1 and 2, and one of the main goals of the coding theory is to find the most efficient coding and also efficient decoding for information transmission.

Generally, let $(\Omega, \mathcal{F}_\Omega, P(\Omega))$ be an input space and $\mathcal{X}$ be the coded input space, where $P(\Omega)$ is the set of all probability measures on Ω. This space $\mathcal{X}$ may be a classical object or a quantum object. For instance, if $\mathcal{X}$ is the set of density operators on a Hilbert space $\mathcal{H}$ for a quantum system, then the coded input system is described by the pair $(\mathbf{B}(\mathcal{H}), \mathfrak{S}(\mathcal{H}))$ discussed in Chap. 5.

In classical systems, Ω is $\mathcal{M}$ above. An output system is described similarly to the input system: The coded output space is denoted by $\overline{\mathcal{X}}$ and the decoded output space $\overline{\mathcal{M}}$ is made of another alphabet. A transmission (map) γ from $\mathcal{X}$ to $\overline{\mathcal{X}}$ is called a *channel*. It reflects all properties of a physical device. The mathematical description of a channel and some fundamental properties of it are discussed in Sect. 6.4. With a decoding $\bar{\xi}$, the whole information transmission process is written as

$$\mathcal{M} \xrightarrow{\xi} \mathcal{X} \xrightarrow{\gamma} \overline{\mathcal{X}} \xrightarrow{\bar{\xi}} \overline{\mathcal{M}}.$$

That is, a message $m \in \mathcal{M}$ is coded as $\xi(m)$ and it is sent to the output system through a channel γ, then the output coded message becomes $\gamma \circ \xi(m)$ and it is decoded as $\bar{\xi} \circ \gamma \circ \xi(m)$ at a receiver.

This transmission process is mathematically set as follows: N messages are sent to a receiver and the kth message $m^{(k)}$ occurs with the probability λ_k. Then the probability distribution (measure) of each message in the sequence $\{m^{(1)}, m^{(2)}, \dots, m^{(N)}\}$ of N messages is denoted by $p^{(N)} = \{\lambda_k\}$, which is a *state* in a classical system. If ξ is a classical coding, then $\xi(m)$ is a classical object such as an electric pulse. If ξ is a quantum coding, then $\xi(m)$ is a quantum object (state) such as a coherent state, which will be discussed in Chap. 7.

Example 6.1 Let $A \equiv \{0,1\}$, $\mathcal{M}_n \equiv \{]a_1, \dots, a_n[; a_k \in A\} = \{m^{(1)}, m^{(2)}, \dots, m^{(N)}\}$ (so $N = |A|^n = 2^n$) and $\mathcal{X} \equiv \{0,1\}^L$. In the case $n = 2$, $\mathcal{M}_n (= \mathcal{M}_2) = \{00, 01, 10, 11\}$, and any injection map ξ from $\mathcal{M}_2$ to $\mathcal{X} \equiv \{0,1\}^L$ is a coding. For instance, in the case $L = 4$, $\xi(00) = 0000$, $\xi(01) = 0101$, $\xi(10) = 1001$, $\xi(11) = 1011$.

Let $(\overline{\Omega}, \mathcal{F}_{\overline{\Omega}}, P(\overline{\Omega}))$ be an output probability space. Information, e.g., a message $\omega \in \mathcal{M}$, is coded by ξ and is sent through a channel γ to the output space $\overline{\Omega}$ after a certain decoding $\bar{\xi}$, which is expressed as

$$\omega \in \Omega \ (\text{e.g.}, = \mathcal{M}) \to \xi(\omega) \to \gamma \circ \xi(\omega) \to \bar{\xi} \circ \gamma \circ \xi(\omega) = \bar{\omega}.$$

The dual expression of one above which is called a channel and is mathematically written by

$$\mu \to \Xi^*(\mu) \longrightarrow \Gamma^* \circ \Xi^*(\mu) \longrightarrow \bar{\Xi}^* \circ \Gamma^* \circ \Xi^*(\mu) \equiv \overline{\mu},$$

where μ is a state (probability measure on $\mathcal{M}$) and Ξ^* denotes the coding, that is, the associated map of ξ in the sense that

$$\int f(\xi(\omega))\,d\mu = \int f(\omega)\,d(\Xi^*\mu), \quad \text{for any } f \in M(\mathcal{M}),$$

where Γ^* is the associated map of γ, $\bar{\Xi}^*$ is that of $\bar{\xi}$, and $M(\mathcal{M})$ is the set of all measurable functions on $\mathcal{M}$. The total channel (transformation) from $P(\mathcal{M})$, the set of all probability measures on $\mathcal{M}$, to $P(\overline{\mathcal{M}})$ is $\bar{\Xi}^* \circ \Gamma^* \circ \Xi^*$. In general, *a channel is a map from the set of input states to that of output states*, which is denoted by Λ^* in the sequel, $\Lambda^* = \bar{\Xi}^* \circ \Gamma^* \circ \Xi^*$.

The efficiency of information communication is measured by several quantities like mutual entropy, channel capacity, error probability, SNR (signal-to-noise ratio), some of which will be discussed in the subsequent sections.

6.2 Entropy

As is well-known, the notion of entropy was introduced by Clausius in order to discuss the thermal behavior of physical systems based on Carnot's work on the efficiency of a thermal engine. Boltzmann then gave a rigorous description of the entropy from the microscopic point of view, that is, the dynamical behavior of a large number of atoms. In quantum mechanics, the notion of entropy was considered by von Neumann. About 50 years after the works of Clausius, Boltzmann, Gibbs and others, Shannon gave new light on the notion of entropy and reformulated the entropy in terms of information [706].

The entropy of a state describing a physical system is a quantity expressing the uncertainty or randomness of the system. Shannon regarded this uncertainty attached to a physical system as the amount of information carried by the system, so that the entropy of a state can be read as the information carried by the state. His idea comes from the following consideration: *If a physical system has a large uncertainty and one gets the information of the system by some procedure, then the information so obtained is more valuable than that obtained from a system having less uncertainty.*

Let (X, p) be a complete event system. Namely, X is the set of all events, say $\{x_1, x_2, x_3, \ldots, x_n\}$, and p is a state (probability distribution) of X such that $p = \{p_1, p_2, \ldots, p_n\}$ with p_k $(= p(x_k) \equiv \text{Prob}(x = x_k))$, the occurrence probability of an event $x_k \in X$. Then the information (uncertainty) carried by (X, p), or simply by p, is given by

$$S(p) = -\lambda \sum_k p_k \log p_k,$$

where λ is a certain chosen constant (e.g., if λ is arranged so that the base of logarithm is 2, then $S(p)$ is often called a *bit*) and we take for simplicity $\lambda = 1$ in the sequel. This $S(p)$ is called the entropy of the state p or the system (X, p), which is sometimes denoted by $S(X)$, or $S(p_1, p_2, \dots, p_n)$. Put

$$\Delta_n = \left\{ p = \{p_k\}; \sum_k p_k = 1, p_k \geq 0 \ (k = 1, 2, \dots, n) \right\}.$$

For two complete event systems (X, p) $(p \in \Delta_n)$ and (Y, q) $(q \in \Delta_m)$, we denote a compound event system by (XY, r) $(r \in \Delta_{nm})$ with

$$XY \text{ (or } X \times Y) = \left\{ (x, y); x \in X, y \in Y \right\}$$

and

$$r = \{r_{ij}\} \equiv \left\{ p(x_i, y_j); x_i \in X, y_j \in Y \right\},$$

where $p(x, y)$ is the joint probability of the events $x \in X$ and $y \in Y$; $p(x, y) \equiv P(x_i \in X, y_j \in Y)$, satisfying the marginal conditions

$$\sum_{x \in X} p(x, y) = p(y) \quad \text{and} \quad \sum_{y \in Y} p(x, y) = p(x).$$

The entropy of the compound system (XY, r) is

$$S(r) \bigl(= S(XY)\bigr) = -\sum_{i,j} r_{ij} \log r_{ij}.$$

For the conditional probability $p(x \mid y) = p(x, y)/p(y)$ of $x \in X$ with respect to $y \in Y$, the conditional entropy $S(X \mid Y)$ is defined as

$$S(X \mid Y) = \sum_{y \in Y} S(X \mid y) p(y)$$

with

$$S(X \mid y) = -\sum_{x \in X} p(x \mid y) \log p(x \mid y),$$

which means the uncertainty still remaining in X after observing the system Y.

Let us consider the fundamental properties of the entropy in CS (Classical System).

Theorem 6.2 *For any $p, q \in \Delta_n$, we have the following properties*:

1. (*Positivity*) $S(p) \geq 0$.
2. (*Concavity*) $S(\lambda p + (1 - \lambda) q) \geq \lambda S(p) + (1 - \lambda) S(q)$ *for any* $\lambda \in [0, 1]$.
3. (*Symmetry*) *For any permutation π of indices of $p = \{p_k\}$*,

$$S(p_1, p_2, \dots, p_n) = S(p_{\pi(1)}, p_{\pi(2)}, \dots, p_{\pi(n)}).$$

4. (*Additivity*) *For any* $q \in \Delta_m$, *put* $r = p \otimes q \equiv \{p_i q_j\} \in \Delta_{nm}$. *Then*

$$S(p \otimes q) = S(p) + S(q).$$

5. (*Subadditivity*) *For any* $r = \{r_{ij}\} \in \Delta_{nm}$ *such that* $\sum_j r_{ij} = p_i$ *and* $\sum_i r_{ij} = q_j$, $S(r) \le S(p) + S(q)$.
6. (*Continuity*) $S(p_1, p_2, \ldots, p_n)$ *is a continuous function in each* p_k.
7. (*Expansibility*) $S(p_1, p_2, \ldots, p_n, 0) = S(p_1, p_2, \ldots, p_n)$.
8. (*Entropy increase*) *Let* A *be a doubly stochastic matrix. Then*

$$S(Ap) \ge S(p).$$

9. *For two complete event systems* (X, p) *and* (Y, q),

$$S(XY) = S(X) + S(Y \mid X) = S(Y) + S(X \mid Y) \le S(X) + S(Y).$$

10. (*Monotone increase*) $S(1/n, 1/n, \ldots, 1/n)$ *is monotone in* $n \in N$ *and* $\max\{S(p), S(q)\} \le S(r)$ *holds for all distributions* p, q *and* r *given in* (5).

Proof (1) Put $\eta(t) = -t \log t$ $(t \in [0, 1])$. Then, $S(p) = -\sum_{i=1}^n p_i \log p_i = \sum_{i=1}^n \eta(p_i) \ge 0$ because $\eta(t) \ge 0$. The equality holds iff $p_i = 1$ for some i and $p_j = 0$ $(i \ne j)$.

(2) This follows from the concavity of the function $\eta(t)$ for $t \in [0, 1]$.

(3) $S(p_{\pi(1)}, p_{\pi(2)}, \ldots, p_{\pi(n)}) = -\sum_{i=1}^n p_{\pi(i)} \log p_{\pi(i)} = -\sum_{i=1}^n p_i \log p_i = S(p)$ because π is just a permutation of indices.

(4) It follows from

$$\begin{aligned} S(p \otimes q) &= -\sum_{i,j} p_i q_j \log p_i q_j = -\sum_{i,j} p_i q_j \log p_i - \sum_{i,j} p_i q_j \log q_j \\ &= -\sum_i p_i \log p_i - \sum_j q_j \log q_j = S(p) + S(q). \end{aligned}$$

(5) The Klein's inequality ($\log \frac{1}{x} \ge 1 - x$ for $x \ge 0$) is applied as

$$\begin{aligned} S(p) + S(q) - S(r) &= -\sum_{i,j} r_{ij} \log p_i - \sum_{i,j} r_{ij} \log q_j + \sum_{i,j} r_{ij} \log r_{ij} \\ &= \sum_{i,j} r_{ij} \log \frac{r_{ij}}{p_i q_j} \ge \sum_{i,j} r_{ij} \left(1 - \frac{p_i q_j}{r_{ij}}\right) \\ &= \sum_{i,j} r_{ij} - \sum_{i,j} p_i q_j = 1 - 1 = 0. \end{aligned}$$

(6) This follows from the continuity of $\eta(t)(t \in [0, 1])$.

(7) It is clear from $0 \log 0 = 0$.

(8) Let A be $(a_{ij})_{i,j=1}^{n}$ and $q = Ap$. Then we can apply the concavity of $\eta(t)$ to get

$$S(q) = \sum_{i=1}^{n} \eta(q_i) = \sum_{i=1}^{n} \eta\left(\sum_{j=1}^{n} a_{ij} p_j\right)$$
$$\geq \sum_{i,j=1}^{n} a_{ij} \eta(p_j) = \sum_{j=1}^{n} \eta(p_j) = S(p).$$

(9) It follows from a simple computation:

$$S(XY) = -\sum_{x\in X, y\in Y} p(x,y)\log p(x,y) = -\sum_{x\in X, y\in Y} p(x,y)\log p(y \mid x)p(x)$$
$$= -\sum_{x\in X, y\in Y} p(x,y)\log p(y \mid x) - \sum_{x\in X, y\in Y} p(x,y)\log p(x)$$
$$= S(Y \mid X) - \sum_{x\in X} p(x)\log p(x) = S(X) + S(Y \mid X).$$

The inequality $S(XY) \leq S(X) + S(Y)$ is a re-expression of (5).

(10) The first statement is due to

$$S(1/n, 1/n, \ldots, 1/n) = \log n \uparrow (n \uparrow).$$

The second comes from (9). □

Now we discuss a mathematical formulation of the entropy in continuous systems. Let $(\Omega, \mathcal{F}, \mu)$ be a probability space and $F(\mathcal{G})$ be the set of all finite partitions of Ω for a σ-subfield $\mathcal{G}$ of $\mathcal{F}$ (i.e., $\tilde{A} = \{A_k; k = 1, \ldots, n < +\infty\} \in F(\mathcal{G})$ iff $A_k \in \mathcal{G}$, $A_k \cap A_j = \emptyset$ $(k \neq j)$ and $\bigcup_k A_k = \Omega$). We formulate the entropy of this state (measure) μ. Along the line of Shannon's philosophy for a discrete system, it is natural for us to define the entropy of μ as

$$S(\mu) = \sup\bigl\{S(\mu : \tilde{A}); \tilde{A} \in F(\mathcal{F})\bigr\}$$

with

$$S(\mu : \tilde{A}) = -\sum_{A_k \in \tilde{A}} \mu(A_k)\log \mu(A_k).$$

This definition is, of course, mathematically consistent, but the value $S(\mu)$ takes $+\infty$ for almost all states of continuous systems, for instance, for every Gaussian measure (see Chap. 20) μ in any real Hilbert space.

There exists another definition of the entropy for some continuous systems. Let f be a random variable and let $F(t)$ be the probability distribution associated to f, that is, $F(t)=\mu(\{\omega\in\Omega;\ f(\omega)\le t\})$. We only treat the case when there exists the density distribution $\rho_f(t)$ such that

$$F(t)=\int_{-\infty}^{t}\rho_f(x)\,dx.$$

This density distribution $\rho_f(t)$ corresponds to the distribution p for a discrete system. The probability that f takes the value between x and $x+\delta x$ is given by $\rho_f(a)\delta x$ according to the mean value theorem, where a is a proper value between x and $x+\delta x$. Therefore, dividing $\mathbb{R}$ as $\mathbb{R}=\bigcup_k I_k$, $I_k=[x_k,x_{k+1})$, $x_{k+1}=x_k+\delta x$, $I_k\cap I_j\neq\emptyset$ $(k\neq j)$, the probability p_k in I_k is given by $\rho_f(a_k)\delta x$ by a certain constant $a_k\in I_k$. The Shannon entropy of $\{p_k\}$ becomes

$$\begin{aligned}S&=-\sum_k\rho_f(a_k)\delta x\log\rho_f(a_k)\delta x\\&=-\sum_k\rho_f(a_k)\delta x\log\rho_f(a_k)-\sum_k\rho_f(a_k)\delta x\log\delta x,\end{aligned}$$

which becomes, as $\delta x\to 0$,

$$\begin{aligned}S&=-\int_{\mathbb{R}}\rho_f(x)\log\rho_f(x)\,dx-\int_{\mathbb{R}}\rho_f(x)\,dx\cdot\Big(\lim_{\delta x\to 0}\log\delta x\Big)\\&=+\infty.\end{aligned}$$

After subtracting the infinity from the above S (a kind of renormalization), the entropy with respect to (μ,f) is defined as

$$S(\mu:f)=-\int_{\mathbb{R}}\rho_f(x)\log\rho_f(x)\,dx.$$

More generally, the entropy for a density distribution ρ is given by

$$S(\rho)=-\int_{\mathbb{R}}\rho(x)\log\rho(x)\,dx,$$

which is usually called the differential entropy of ρ and was first introduced by Gibbs and Boltzmann independently.

As an example, for the Gaussian distribution ρ in the n-dimensional Hilbert space $\mathbb{R}^n$ with the covariance matrix A, we have

$$S(\rho)=\log(2\pi e)^{\frac{n}{2}}\left|\det A^{-1}\right|^{-\frac{1}{2}}.$$

However, this differential entropy has some inconvenient properties as a measure of information or uncertainty, for instance, it does not have positivity and scaling invariance, so that it cannot be regarded as the entropy of a state in continuous systems, but only as a certain measure indicating uncertainty of a state.

6.3 Relative Entropy

The relative entropy was studied by Kullback and Leibler [455] as an information measure representing the relative uncertainty of a probability distribution p with respect to that of q, which in other words indicates a sort of difference between p and q. In this subsection, we discuss some fundamental properties of the relative entropy. We first consider the relative entropy for two discrete probability distributions $p=\{p_i\}, q=\{q_j\}\in\Delta_n$, which is defined by

$$S(p,q)=\begin{cases}\sum_i p_i\log\frac{p_i}{q_i} & (p\ll q),\\ +\infty & (\text{otherwise}),\end{cases}$$

where $p\ll q$ means that $p_i=0$ whenever $q_i=0$ for each i.

Theorem 6.3 *For any $p,q\in\Delta_n$, we have:*

1. (*Positivity*) $S(p,q)\geq 0, =0$ *iff* $p=q$.
2. (*Joint convexity*) $S(\lambda p+(1-\lambda)q,\lambda r+(1-\lambda)t)\leq\lambda S(p,r)+(1-\lambda)S(q,t)$ *for any* $t,r\in\Delta_n$ *and* $\lambda\in[0,1]$.
3. (*Symmetry*) *For any permutation* π *of indices*
$$S(p_1,p_2,\dots,p_n,q_1,q_2,\dots,q_n)=S(p_{\pi(1)},\dots,p_{\pi(n)},q_{\pi(1)},\dots,q_{\pi(n)}).$$
4. (*Additivity*) $S(p\otimes r,q\otimes t)=S(p,q)+S(r,t)$ *for any* $t,r\in\Delta_m$.
5. (*Continuity*) $S(p_1,p_2,\dots,p_n,q_1,q_2,\dots,q_n)$ *is continuous in each variable* p_i *and* q_j.
6. (*Expansibility*) $S(p_1,p_2,\dots,p_n,0,q_1,q_2,\dots,q_n,0)=S(p,q)$.
7. (*Monotonicity*) $S(Ap,Aq)\leq S(p,q)$, *if* A *is a doubly stochastic matrix*.

Proof (1) $x-1\geq\log x$ implies $\log\frac{1}{x}\geq 1-x$. Therefore,

$$S(p,q)\geq\sum_{i=1}^n p_i\left(1-\frac{q_i}{p_i}\right)=\sum_i p_i-\sum_i q_i=1-1=0.$$

Moreover, $x-1=\log x$ iff $x=1$, which implies that $S(p,q)=0$ iff $p=q$.

(2) According to the definition of $S(p,q)$, for $p,q,r,t\in\Delta_n$, $\lambda\in[0,1]$, it is enough to show

$$\bigl(\lambda p_i+(1-\lambda)q_i\bigr)\log\frac{\lambda p_i+(1-\lambda)q_i}{\lambda r_i+(1-\lambda)t_i}\leq\lambda p_i\log\frac{p_i}{r_i}+(1-\lambda)q_i\log\frac{q_i}{t_i},$$

that is, we have only to prove the convexity of the following function $f(x)$ given by

$$f(x)=\bigl(xp_i+(1-x)q_i\bigr)\log\frac{xp_i+(1-x)q_i}{xr_i+(1-x)t_i},$$

which comes from

$$f''(x) = \frac{(p_i t_i - q_i t_i)^2}{\{xp_i + (1-x)q_i\}\{xr_i + (1-x)t_i\}^2} \geq 0.$$

(3) through (6) are easy to prove from the definition of $S(p, q)$. □

In the case of two probability measures μ and ν on a measurable space $(\Omega, \mathcal{F})$, the relative entropy is defined by

$$S(\mu, \nu) = \sup\left\{\sum_{j=1} \mu(A_j) \log \frac{\mu(A_j)}{\nu(A_j)};\ \{A_j\} \in F(\mathcal{F})\right\}.$$

In particular, the relative entropy restricted to a subalgebra $\mathcal{G} \subset \mathcal{F}$ is given by

$$S_{\mathcal{G}}(\mu, \nu) = \sup\left\{\sum_{j=1} \mu(A_j) \log \frac{\mu(A_j)}{\nu(A_j)};\ \{A_j\} \in F(\mathcal{G})\right\}.$$

This entropy is sometimes called the *coarse graining entropy*. The above relative entropy is re-expressed by means of the Radon–Nikodym theorem.

Using the Radon–Nikodym derivative f of μ w.r.t. ν, often denoted by $d\mu/d\nu$, one has

Theorem 6.4 (Gel'fand–Kolmogorov–Yaglom [136, 279, 764, 765]) *The relative entropy can be expressed as*

$$S_{\mathcal{G}}(\mu, \nu) = \begin{cases} \int_\Omega f \log f \, d\nu & (\mu \ll \nu), \\ +\infty & (\text{otherwise}), \end{cases}$$

where f is the Radon–Nikodym derivative $d\mu/d\nu$ and is a measurable function on $\mathcal{G}$.

Then we have the following theorem for a continuous version of relative entropy.

Theorem 6.5 *For any two probability measures μ and ν, we have*:

1. (*Positivity*) $S(\mu, \nu) \geq 0$, *and* $S(\mu, \nu) = 0$ *iff* $\mu = \nu$.
2. (*Joint convexity*) $S(\lambda\mu + (1-\lambda)\nu, \lambda\rho + (1-\lambda)\sigma) \leq \lambda S(\mu, \rho) + (1-\lambda) S(\nu, \sigma)$ *for any probability measures* ρ, σ *and* $\lambda \in [0, 1]$.
3. (*Symmetry*) *For an invertible mapping j from $\mathcal{F}$ to $\mathcal{F}$ such that $\mu \circ j$ is a probability measure and $j(\Omega) = \Omega$,*

$$S(\mu \circ j, \nu \circ j) = S(\mu, \nu).$$

4. (*Additivity*) *For any two probability measures ρ and σ,*

$$S(\mu \otimes \rho, \nu \otimes \sigma) = S(\mu, \nu) + S(\rho, \sigma);$$

where $\mu \otimes \rho \equiv \mu(A)\rho(B)$ for any $A < B \in \mathcal{F}$.

5. (*Lower semicontinuity*) *When* $\mu_n \to \mu$, $\nu_n \to \nu$ *in norm,*

$$S(\mu, \nu) \le \liminf S(\mu_n, \nu_n).$$

6. (*Monotonicity*) *For* σ*-subfields* $\mathcal{G}$ *and* $\mathcal{H}$ *with* $\mathcal{G} \subset \mathcal{H}$,

$$S_{\mathcal{G}}(\mu, \nu) \le S_{\mathcal{H}}(\mu, \nu).$$

Proof (1) Same as in the proof of Theorem 6.3.

(2) If $\mu \not\ll \rho$ or $\nu \not\ll \sigma$, then the inequality obviously holds because of $S(\mu, \rho) = \infty$ or $S(\nu, \sigma) = \infty$. Therefore, we prove the above inequality in the case when $\mu \ll \rho$ and $\nu \ll \sigma$. For any $\{A_i\} \in F(\mathcal{F})$, we put $\mu_i = \mu(A_i)$, $\nu_i = \nu(A_i)$, $\rho_i = \rho(A_i)$ and $\sigma_i = \sigma(A_i)$. Then the inequality is proved similarly as in the last theorem.

(3) Clear from the definition of $S(\mu, \nu)$.

(4) If $\mu \not\ll \nu$ or $\rho \not\ll \sigma$, then the inequality obviously holds because of $S(\mu, \nu) = \infty$ or $S(\rho, \sigma) = \infty$. Therefore, it is enough to prove the above inequality in the case when $\mu \ll \nu$ and $\rho \ll \sigma$. For any finite partition $\{A_i \otimes A_j\} \in F(\mathcal{F} \otimes \mathcal{F})$, the statement is due to the following equality:

$$\begin{aligned}
&\sum_{i,j} \mu \otimes \rho(A_i \otimes A_j) \log \frac{\mu \otimes \rho(A_i \otimes A_j)}{\nu \otimes \sigma(A_i \otimes A_j)} \\
&\quad = \sum_{i,j} \mu(A_i)\rho(A_j) \log \frac{\mu(A_i)\rho(A_j)}{\nu(A_i)\sigma(A_j)} \\
&\quad = \sum_{i} \mu(A_i) \log \frac{\mu(A_i)}{\nu(A_i)} + \sum_{j} \rho(A_j) \log \frac{\rho(A_j)}{\sigma(A_j)}.
\end{aligned}$$

Taking the supremum, we obtain the required equality.

(5) Since $\mu_n(A) \to \mu(A)$, $\nu_n(A) \to \nu(A)$, for any $A \in \mathcal{F}$, we have

$$\begin{aligned}
\nu(A) > 0 \quad &\Longrightarrow \quad \lim_{n\to\infty} \mu_n(A) \log \frac{\mu_n(A)}{\nu_n(A)} = \mu(A) \log \frac{\mu(A)}{\nu(A)}, \\
\nu(A) = 0, \mu(A) > 0 \quad &\Longrightarrow \quad \lim_{n\to\infty} \mu_n(A) \log \frac{\mu_n(A)}{\nu_n(A)} = \infty = \mu(A) \log \frac{\mu(A)}{\nu(A)}, \\
\nu(A) = \mu(A) = 0 \quad &\Longrightarrow \quad \mu_n(A) \log \frac{\mu_n(A)}{\nu_n(A)} \ge \mu_n(A) - \nu_n(A) \to 0 \\
&\Longrightarrow \quad \liminf_{n\to\infty} \mu_n(A) \log \frac{\mu_n(A)}{\nu_n(A)} \ge 0 = \mu(A) \log \frac{\mu(A)}{\nu(A)}.
\end{aligned}$$

Therefore, the definition of $S(\mu, \nu)$ implies $\liminf_{n\to\infty} S(\mu_n(A), \nu_n(A)) \ge S(\mu, \nu)$.

(6) Obvious from the definition of $S(\mu, \nu)$. □

6.4 Mutual Entropy

6.4.1 Discrete Case

The mutual entropy (information) using the relative entropy generalizes the original definition of the mutual information by Shannon, but they are essentially in the same context for finite event systems, so that we call both the mutual entropy.

This entropy is a very important quantity in communication theory because it has an essential physical meaning, that is, when we send information through a certain channel, we hope to know how much information can be correctly transmitted from an input system to an output system. It is the mutual entropy that represents this amount of information. Therefore, mutual entropy does depend on an input (initial) state and a transmission channel.

As we mentioned in the beginning of this chapter, the total channel (transformation) from $P(\mathcal{M})$, the set of all probability measures on $\mathcal{M}$, to $P(\overline{\mathcal{M}})$ is $\bar{\varXi}^* \circ \varGamma^* \circ \varXi^*$, which is denoted by $\varLambda^*$ in the sequel.

In the case when an input state is a discrete probability distribution and a channel is expressed by a transition probability, the mutual entropy of Shannon is expressed as

$$I(p; \varLambda^*) \equiv S(r, p \otimes q) = \sum_{i,j} r_{ij} \log \frac{r_{ij}}{p_i q_j}$$

where the joint probability distribution $r = \{r_{ij}\}$ and the output state (probability distribution) $q = \{q_j\}$ are determined by an input state $p = \{p_i\}$ and a transition probability (channel) $\varLambda^* = (p(j \mid i))$, i.e., $p(j \mid i) \equiv P(\overline{X} = \overline{x}_j \mid X = x_i)$ such that

$$r_{ij} = p(j \mid i) p_i, \qquad q_j = \sum_i r_{ij}.$$

The mutual entropy is often expressed by $I(X, \overline{X})$ where X and $\overline{X}$ describe the input and output event systems, respectively. In such cases, for two event systems X and Y, the mutual entropy can be written as

$$I(X \wedge Y) = \sum_{x \in X, y \in Y} p(x, y) \log \frac{p(x, y)}{p(x) p(y)}$$

where $p(x)$ is the probability of $X = x$; $P(X = x)$, and $p(x, y)$ is the joint probability $P(X = x, Y = y)$.

The fundamental properties of the mutual entropy of Shannon are as follows:

Theorem 6.6 *For an initial state (probability) p, a channel $\varLambda^*$, and the final state $q = \varLambda^* p$ as above, we have*

1. $I(p; \varLambda^*) = S(p) + S(q) - S(r)$ *if* $S(r)$ *is finite, where* r *is a joint distribution in* $P(X \otimes \overline{X})$.
2. $0 \le I(p; \varLambda^*) \le \min\{S(p), S(q)\}$.

Proof (1) When $S(r)$ is finite,

$$\begin{aligned} I(p;\Lambda^*) &= \sum_{i,j} r_{i,j}\log\frac{r_{i,j}}{p_i q_j} = \sum_{i,j} r_{i,j}\log r_{i,j} - \sum_{i,j} r_{i,j}\log p_i q_j \\ &= S(p)+S(q)-S(r). \end{aligned}$$

(2) It holds that

$$\begin{aligned} 0 \le I(p;\Lambda^*) &= \sum_{i,j} r_{i,j}\log\frac{p(i\mid j)}{p_i} = \sum_{i,j} r_{i,j}\bigl(\log p(i\mid j) - \log p_i\bigr) \\ &\le -\sum_{i,j} r_{i,j}\log p_i = -\sum_{i,j} p_i\log p_i = S(p). \end{aligned}$$

The other inequality is obtained in the sequel. □

Exercise 6.7 Prove the following equalities: $I(X,Y) = S(X) - S(X \mid Y) = S(Y) - S(Y \mid X)$.

We will discuss more about the mutual entropy. Let X, Y, Z be three event systems. Then the conditional mutual entropy and the mutual entropy between XY and Z are defined as follows:

$$I(X,Y\mid Z) = \sum_{x\in X, y\in Y, z\in Z} p(x,y,z)\log\frac{p(x,y\mid z)}{p(x\mid z)p(y\mid z)}$$

and

$$I(XY,Z) = \sum_{x\in X, y\in Y, z\in Z} p(x,y,z)\log\frac{p(x,y,z)}{p(x,y)p(z)}.$$

Then it is easy to prove the following equalities:

1. $I(X,Y\mid Z) = S(X\mid Z) - S(X\mid YZ) = S(Y\mid Z) - S(Y\mid XZ)$.
2. $I(XY,Z) = S(XY) - S(XY\mid Z) = S(Z) - S(Z\mid XY) = I(X,Z) + I(Y,Z\mid X) = I(Y,Z) + I(X,Z\mid Y)$.
3. The common information (entropy) contained in all three X, Y, Z is computed as

$$\begin{aligned} I(X,Y,Z) &= I(X,Y) - I(X,Y\mid Z) \\ &= I(Y,Z) - I(Y,Z\mid X) \\ &= I(X,Z) - I(X,Z\mid Y). \end{aligned}$$

6.4.2 Continuous Case

In the continuous case, i.e., when input and output states are given by probability measures and a channel is a map between their sets of all probability measures, the mutual entropy was introduced by Gelfand and Yaglom with an aim of studying Gaussian channels. The entropy, relative entropy and mutual entropy are defined as follows:

$$\begin{aligned} S(\mu) &= \sup\left\{-\sum_{k=1}^{n}\mu(A_k)\log\mu(A_k);\ \{A_k\}\in F(\mathcal{F})\right\}, \\ S(\mu,\nu) &= \sup\left\{\sum_{k=1}^{n}\mu(A_k)\log\frac{\mu(A_k)}{\nu(A_k)};\{A_k\}\in F(\mathcal{F})\right\}, \\ I(\mu;\Lambda^*) &= S(\Phi,\mu\otimes\Lambda^*\mu), \end{aligned} \tag{6.1}$$

where $F(\mathcal{F})$ is the set of all finite partitions of Ω, that is, $\{A_k\}\in F(\mathcal{F})$ iff $A_k\in\mathcal{F}$ with $A_k\cap A_j=\emptyset$ $(k\neq j)$ and $\bigcup_{k=1}^{n}A_k=\Omega$.

Let $(\Omega,\mathcal{F})$ be an input system and $(\overline{\Omega},\overline{\mathcal{F}})$ be an output system, where more rigorously, Ω and $\bar{\Omega}$ are Hausdorff spaces. For an input state $\mu\in P(\Omega)$, the output state is $\overline{\mu}=\Lambda^*\mu$, and the compound state (joint probability measure) Φ is given by

$$\Phi(Q\times\overline{Q})=\int_Q\lambda\left(1_{\overline{Q}}\right)d\mu,\quad Q\in\mathcal{F},\ \overline{Q}\in\overline{\mathcal{F}},$$

where λ is the dual map of Λ^* sending an input random variable on $(\Omega,\mathcal{F})$ to an output random variable on $(\overline{\Omega},\overline{\mathcal{F}})$, which is a generalized expression of $\overline{\xi}\circ\gamma\circ\xi$ in Sect. 6.1, and $1_{\overline{Q}}$ is the characteristic function on $\overline{\Omega}$, namely,

$$1_{\overline{Q}}(\omega)=\begin{cases}1, & \text{if } \omega\in\overline{Q},\\ 0, & \text{if } \omega\notin\overline{Q}.\end{cases}$$

The above compound state Φ satisfies the following marginal conditions

$$\Phi(Q\times\overline{\Omega})=\mu(Q),\qquad \Phi(\Omega\times\overline{Q})=\Lambda^*\mu(\overline{Q}).$$

The channel above $\Lambda^*:P(\Omega)\to P(\bar{\Omega})$ can be written as

$$\Lambda^*\mu(\overline{Q})=\int_\Omega\lambda\left(1_{\overline{Q}}\right)(x)\mu(dx),\quad \mu\in P(\Omega),\ \overline{Q}\in\overline{\mathcal{F}}. \tag{6.2}$$

The functional $\lambda(1_{\overline{Q}})(\cdot)$ can be represented by the function $\lambda:\Omega\times\overline{\mathcal{F}}\to\mathbb{R}^+$ so that the following conditions are satisfied:

1. $\lambda(x,\cdot)\in P(\bar{\Omega})$.
2. $\lambda(\cdot,\overline{Q})(=\lambda(1_{\overline{Q}})(\cdot))\in M(\Omega)$, the set of all measurable functions on $(\Omega,\mathcal{F})$.

Remark 6.8 For the message spaces $\Omega=\mathcal{M}\equiv A^{\mathbf{Z}}$ and $\overline{\Omega}=\bar{\mathcal{M}}\equiv\overline{A}^{\mathbf{Z}}$, the conditional probability is just one example of the above channel λ.

Definition 6.9 For a given channel Λ^*, the mutual entropy is defined by

$$I(\mu; \Lambda^*) \equiv S(\Phi, \Phi_0),$$

where $S(\cdot, \cdot)$ is the relative entropy and Φ_0 is the direct product measure of μ and ν given as

$$\Phi_0 \equiv \mu \otimes \nu = \mu \otimes \Lambda^* \mu.$$

The case when μ is a Gaussian measure and Λ^* is a Gaussian channel has been studied by Gelfand, Yaglom and others. As discussed in the previous section, the entropy for a continuous system is essentially infinite, but the mutual entropy can be finite. In any case, we manage to have the same fundamental inequality as that in Theorem 6.6:

$$0 \le I(\mu; \Lambda^*) \le S(\mu),$$

where $S(\mu)$ is defined by (6.1).

Finally, we mention the mutual (differential) entropy when the measures $\mu, \Lambda^* \mu, \Phi$ have the density distributions $p(x), q(y)$ on $\mathbb{R}^n$, $r(x, y)$ on $\mathbb{R}^{2n}$. In this case, the mutual entropy is often expressed by

$$I = \int r(x, y) \log \frac{r(x, y)}{p(x)p(y)} \, dx \, dy.$$

Though this expression is similar to the differential entropy, it will not diverge so often because the divergent parts in r and $p \otimes q$ cancel each other.

6.5 Entropy of Information Source

Here we consider the entropy per one letter (unit) of an input space (alphabet space) A. Let $\mathcal{M}_l$ be the set of all messages m with length l, that is, $m =]a_1, \ldots, a_l[$, and let μ be a state (probability measure) defined on the σ-field $\mathcal{F}_A$ generated by $\mathcal{M}_l$. The entropy of $(\mathcal{M}_l, \mu)$ is given by

$$S(\mathcal{M}_l) = - \sum_{m \in \mathcal{M}_l} \mu(m) \log \mu(m),$$

so that the entropy per letter is given by

$$\frac{S(\mathcal{M}_l)}{l}.$$

Thus, if the limit of this quantity as $l \to \infty$ exists, the limit is called the *entropy rate* of the *input source* (A, μ), and we denote it by $\widetilde{S}(\mu)$. Now we consider a shift transformation T defined as

$$T : \Pi a_k \in A^Z \longmapsto \Pi a'_k \in A^Z, \quad a'_k \equiv a_{k+1}.$$

Definition 6.10

1. A state μ is said to be T-stationary if $\mu(T^{-1}E) = \mu(E)$ for any $E \in \mathcal{F}_A$. The set of all stationary states is denoted by $\mathcal{S}_T(A) \subset \mathcal{S}(A)$ (the set of all states on $\mathcal{F}_A$).
2. A T-stationary state μ is ergodic if $E \in \mathcal{F}_A$ is μ-a.e. T-invariant (i.e., $\mu((T^{-1}E \cup E) \cap (T^{-1}E \cap E)^c) = 0$) implies $\mu(E) = 0$ or 1. The set of all ergodic states is denoted by $\mathcal{S}_E(A) \subset \mathcal{S}_T(A) \subset \mathcal{S}(A)$.
3. A state μ is of Bernoulli type if $\mu(]a_1, \dots, a_l[) = \prod_{k=1}^{l} \mu(]a_k[)(\equiv \prod_{k=1}^{l} P(a_k))$ is satisfied.

Remark 6.11 (i) $\mathcal{S}_E(A)$ is the set of all extreme elements of $\mathcal{S}_T(A)$, that is, $\mathcal{S}_E(A) = \operatorname{ex}\mathcal{S}_T(A)$ due to Krein–Milman theorem. (ii) The Bernoulli input source (A, μ) is sometimes called memoryless.

Theorem 6.12 *If μ is T-stationary, then the limit $l \to \infty$ of the average entropy $\frac{S(\mathcal{M}_l)}{l}$ exists.*

Proof Let us denote an l-length message by $m^l =]a_1, \dots, a_l[$ and $S(\mathcal{M}_l)$ simply by S^l. Then by stationarity, it holds

$$S^{l+h} \le S^l + S^h,$$

which implies $S^{nl} \le l S^n$ for any natural numbers n, l, so that we have

$$\frac{S^l}{l} \le S^1.$$

Here S^1 is the entropy of a message of length 1, thus $S^1 \le \log|A| < +\infty$. Putting

$$\widetilde{S} = \inf_l \frac{S^l}{l}, \tag{6.3}$$

for any $\epsilon > 0$ there exists a natural number k such that $\frac{S^k}{k} < \widetilde{S} + \epsilon$. Take another natural number $n > k$ and put $r_n = [\frac{n}{k}] + 1$ (Gauss Symbol). Then it holds

$$S^n \le S^{r_n k} \le r_n S^k,$$

which implies

$$\limsup_{n\to\infty} \frac{S^n}{n} \le \limsup_{n\to\infty} \frac{r_n}{r_n - 1}(\widetilde{S} + \epsilon) = \widetilde{S} + \epsilon.$$

By (6.3), for any l it holds $\frac{S^n}{n} \ge \widetilde{S}$, hence $\widetilde{S} = \lim_{l\to\infty} \frac{S^l}{l}$. □

Remark 6.13 This theorem is true without the condition of T-stationary if the number of elements of A is finite (i.e., $|A| < +\infty$).

Remark 6.14 The limit $\limsup_{n\to\infty} \frac{S^n}{n}$ always exists and it relates to the "transmission rate", as is discussed below.

Exercise 6.15 Prove that the entropy rate for a Bernoulli state μ is given by

$$\widetilde{S}(\mu) = -\sum_{a\in A} \mu(]a[) \log \mu(]a[).$$

6.6 Capacity

Here we discuss the capacity of a channel in the process $\mathcal{X} \xrightarrow{\gamma} \mathcal{Y}$, or equivalently, $\mu \to \Lambda^*\mu$, where $\mathcal{X}$ and $\mathcal{Y}$ are arbitrary probability spaces (or random variables associated with those spaces) and Λ^* is a channel sending a state in $\mathcal{X}$ to that in $\mathcal{Y}$. In actual communication processes, the capacity is more subtle because in that case one asks for the information (entropy) and the mutual entropy per unit time (a letter), that is, we have to think about the rate of such entropies, which we will discuss in the forthcoming sections.

The channel capacity is the supremum of the mutual entropy over all μ in a suitable set of input states $\mathfrak{S}$:

$$C = \sup\{I(\mu; \Lambda^*); \mu \in \mathfrak{S}\}.$$

In particular, for the discrete case, let X, Y be the event systems of input and output, and let an input $\{x_i\}_{i=1}^n \in X$ occur with the probability $p \equiv p(x_i)$ and the output be $\{y_j\} \in Y$ associated with a channel Λ^* determined by an $m \times n$ matrix of the conditional probability, $\Lambda^* = (p(y_j|x_i))$. The mutual entropy between X and Y is written as

$$I(p; \Lambda^*) = \sum_{i=1}^{n}\sum_{j=1}^{m} p(x_i)p(y_j|x_i) \log \frac{p(y_j|x_i)}{p(y_j)},$$

so that the channel capacity is the supremum of the mutual entropy over all p in a suitable set of input states $\mathfrak{S}$:

$$C = \{I(p; \Lambda^*); p \in \mathfrak{S}\}.$$

Example 6.16 Let $X = Y = \{0, 1\}$ and the error probability (defining a channel) be given by $p(0|1) = p(1|0) = e$ (this channel is called of X-type). Then the mutual entropy is expressed by means of the input probability $p \equiv p(0)$ such that

$$I(X, Y) = -t \log t - (1-t) \log(1-t) + e \log e + (1-e) \log(1-e),$$

where $t = p + e - 2ep$. By maximizing this mutual entropy w.r.t. p, we obtain the capacity as

$$C = 1 + e \log e + (1-e) \log(1-e).$$

Following a conventional argument, let us consider a continuous input source X with the probability distribution f, and an independent additive Gaussian white noise Z; $g(z) \in N(0, \sigma_Z^2)$. The output is $Y = X + Z$, and its distribution is

$$h(y) = f(x)g(z),$$

so that the mean of Y^2 is calculated as $\langle Y^2 \rangle = \sigma_Z^2 + \langle X^2 \rangle = N + P$, where P is a constant (signal power) satisfying $\int x^2 f(x)\,dx \le P$. The channel capacity C is the maximum of the mutual entropy $I(X;Y)$ subject to the conditions $\int x^2 f(x)\,dx \le P$. Thus the capacity is

$$C = \frac{1}{2}\log\left(1 + \frac{P}{N}\right),$$

the maximum being attained for $X \in N(0, P)$. For a white noise of limited bandwidth W, there are $2W$ independent sample values, so the capacity becomes

$$C = W \log\left(1 + \frac{P}{N}\right).$$

6.7 McMillan's Theorem

Let $\tilde{C}$ be a finite (measurable) partition of $\mathcal{M}$, that is, $C_k \in \tilde{C} \subset \mathcal{F}_A$ $(k = 1, \ldots, n)$, $C_k \cap C_j = \phi$ $(k \neq j)$ and $\bigcup_{k=1}^{n} C_k = \mathcal{M}$. We denote the σ-field generated by all elements of $\tilde{C}$ by the same notation $\tilde{C}$. Then the entropy function and the conditional entropy function with respect to the two σ-fields $\tilde{C}$ and $\tilde{D}$ are denoted as

$$\hat{S}(\tilde{C}) = -\sum_{C \in \tilde{C}} \log \mu(C) 1_C,$$

$$\hat{S}(\tilde{C} \mid \tilde{D}) = -\sum_{C \in \tilde{C}, D \in \tilde{D}} \log \mu(C \mid D) 1_{C \cap D},$$

where 1_C is the characteristic function of C. Moreover, a σ-field generated by $T^{-k+1}(\tilde{C})$ $(k = 1, 2, \ldots, n)$ is denoted as

$$\bigvee_{k=1}^{n-1} T^{-k+1}(\tilde{C}).$$

Then *McMillan's theorem* [118, 232, 233, 456] is the following.

Theorem 6.17 (McMillan's theorem)

1. $\frac{1}{n}\hat{S}(\bigvee_{k=1}^{n-1} T^{-k+1}(\tilde{C}))$ *converges a T-invariant function h μ-a.e. and in L^1.*
2. *If μ is ergodic, then $h = \int_\Omega \hat{S}(\tilde{C} \mid \bigvee_{k=1}^{\infty} T^{-k}\tilde{C})\,d\mu$.*

Moreover, for $\mathcal{M}_l \equiv \{m^l =]a_1, \ldots, a_l[; a_k \in A, 1 \le k \le l\}$, it holds that $\mathcal{M}_n = \mathcal{M}_1 \bigvee T^{-1}\mathcal{M}_{n-1} = \bigvee_{k=0}^{n-1} T^{-k}\mathcal{M}_1$. In McMillan's theorem, we can take any partition $\mathcal{M}_n$ instead of $\tilde{C}$, so that we have:

Theorem 6.18

1. *For a stationary information source* (A, μ), *there exists a* T*-invariant function* h *such that* $-\frac{1}{n}\sum_{m\in\mathcal{M}_n} \log \mu(m)1_m$ *converges to* h, *as* $n \to \infty$, *a.e. and in* L^1.
2. *If* μ *is ergodic, then* $h = \widetilde{S}(\mu)$.

McMillan's theorem tells that if an information source (A, μ) is ergodic, then for any positive ε and δ we have

$$\mu\left(\left\{x \in A^Z; \left|\frac{1}{n}\hat{S}\left(\bigvee_{k=0}^{n-1} T^{-k}(\tilde{C})\right)(x) - \widetilde{S}(\mu)\right| > \varepsilon\right\}\right) < \delta$$

for sufficiently large n. From this fact, we can divide $\mathcal{M}_n$ into two parts $\mathcal{M}_n^{(1)}$ and $\mathcal{M}_n^{(2)}$ such that

$$\left|\frac{1}{n}\log \mu\left(M^{(1)}\right) + \widetilde{S}(\mu)\right| < \varepsilon \quad \text{for any } M^{(1)} \in \mathcal{M}_n^{(1)}$$

and

$$\mu\left(\bigcup_{M^{(2)}\in\mathcal{M}_n^{(2)}} M^{(2)}\right) < \delta.$$

That is, for any $M^{(1)} \in \mathcal{M}_n^{(1)}$ the occurrence probability $\mu(M^{(1)})$ is almost equal to $e^{-n\widetilde{S}(\mu)}$, which means the number of elements in $\mathcal{M}_n^{(1)}$ is almost equal to $e^{n\widetilde{S}(\mu)}$, so that when we take a message large enough (i.e., $n \gg 1$), the information per a letter of that message can be equal to the information of the input source. Remark that $|\mathcal{M}_n| = |A|^n = e^{n\log|A|}$ implies $\mathcal{M}_n^{(1)} = \mathcal{M}_n$ iff $\widetilde{S}(\mu) = \log|A|$.

6.8 Shannon's Coding Theorem

As discussed in Sect. 6.1, the whole information transmission process is written as

$$\mathcal{M} \xrightarrow{\xi} \mathcal{X} \xrightarrow{\gamma} \overline{\mathcal{X}} \xrightarrow{\overline{\xi}} \overline{\mathcal{M}}.$$

The coded message x is sent to an output alphabet message space $\overline{\mathcal{M}}$ through a channel γ and a decoding map $\overline{\xi}$. For notational simplicity, we neglect the coding and decoding for a while, that is, we consider a process $\mathcal{M} \xrightarrow{\overline{\xi}\circ\gamma\circ\xi} \overline{\mathcal{M}}$. Note that the map $\overline{\xi}\circ\gamma\circ\xi$ can be expressed by a map (kernel) $\lambda : \Omega \times \bar{\mathcal{F}} \to \mathbb{R}^+$, so that we call λ itself a channel.

6.8.1 Channel, Transmission Rate and Capacity Rate

Before going to discuss coding theorems, we state some fundamental properties of a channel $\lambda : \Omega \times \bar{\mathcal{F}} \to \mathbb{R}^+$ with $\Omega = \mathcal{M}$ and $\overline{\Omega} = \overline{\mathcal{M}}$. Let us use the same notation T for a shift on $\Omega, \overline{\Omega}$ and $\Omega \times \overline{\Omega}$ (i.e., $T(m, \overline{m}) \equiv (Tm, T\overline{m})$ for any $(m, \overline{m}) \in \Omega \times \overline{\Omega}$), and denote the set of all channels by $\mathcal{C}(\Omega, \overline{\Omega})$.

Remark 6.19 In order to take the coding and decoding into consideration, we need to replace Ω with the coded input space $\mathcal{X}$ and $\overline{\Omega}$ with the coded output space $\overline{\mathcal{X}}$, and we consider two one-to-one mappings coding ξ (between $\mathcal{M}$ and $\mathcal{X}$) and decoding $\overline{\xi}$ (between $\overline{\mathcal{X}}$ and $\overline{\mathcal{M}}$). For instance, a message $m \in \mathcal{M} = A^{\mathbb{Z}}$ is coded by ξ so that a message of the length l, i.e., $m^l =]a_1, \dots, a_l[\in \mathcal{M}_l \subset \mathcal{M}$, is coded to $x^l \equiv]x_1, \dots, x_l[\in \mathcal{X}_l \subset \mathcal{X}$.

Definition 6.20

1. A channel $\lambda : \Omega \times \bar{\mathcal{F}} \to \mathbb{R}^+$ is stationary if $\lambda(Tm, \overline{Q}) = \lambda(m, T^{-1}\overline{Q})$ for any $m \in \Omega$ and any $\overline{Q} \in \bar{\mathcal{F}}_{\overline{\Omega}}$. The set of all stationary channels is denoted by $\mathcal{C}_T(\Omega, \overline{\Omega})$.
2. A channel λ has nth memory if $\lambda(m,]b_i, \dots, b_j[) = \lambda(m',]b_i, \dots, b_j[)$ for any $]b_i, \dots, b_j[\in \overline{\Omega}$ and any $m = \Pi_k a_k, m' = \Pi_k a'_k$ with $a_k = a'_k$ $(i - n \le k \le j)$.
3. A stationary channel λ is ergodic if the compound measure Φ of μ and λ is ergodic on $\Omega \times \overline{\Omega}$ for any ergodic measure μ on Ω.The set of all ergodic channels is denoted by $\mathcal{C}_E(\Omega, \overline{\Omega}) \subset \mathcal{C}_T(\Omega, \overline{\Omega})$. Note that $\mathcal{C}_E(\Omega, \overline{\Omega}) = \mathrm{ex}\mathcal{C}_T(\Omega, \overline{\Omega})$, the set of all extreme points of $\mathcal{C}_T(\Omega, \overline{\Omega})$.
4. A stationary channel λ has finite (n-step) dependence if the following condition is met: There exists a natural number n such that for any integers m, r, t, s $(m \le r \le t \le s)$ if $s - r > n$, then the following equality holds for any $\overline{Q}_{m,r} \equiv]b_m, \dots, b_r[, \overline{Q}_{s,t} \equiv]b_s, \dots, b_t[\in \overline{\Omega}$ and any $x \in \Omega$
$$\lambda(x, \overline{Q}_{m,r} \cap \overline{Q}_{s,t}) = \lambda(x, \overline{Q}_{m,r})\lambda(x, \overline{Q}_{s,t}).$$
5. A channel λ is unpredictable if for a pair of messages $m = \Pi_k a_k, m' = \Pi_k a'_k$ with $a_k = a'_k$ $(k \le t)$ it holds that
$$\lambda(m,]b_t[) = \lambda(m',]b_t[).$$

Let us consider all messages of length n in the input and output spaces ($\Omega = \mathcal{M} \equiv A^{\mathbf{Z}}$ and $\overline{\Omega} = \overline{\mathcal{M}} \equiv B^{\mathbf{Z}}$ (B is the set of letters in the output), respectively), and we denote them by $\mathcal{M}_n$ and $\overline{\mathcal{M}}_n$, respectively. We called $\frac{S(\mathcal{M}_n)}{n}$ the entropy per letter in the previous section, and we will here denote it by $\frac{S(\mu;\mathcal{M}_n)}{n}$ to show the state-dependence precisely. Then the mutual entropy $I(\mu; \mathcal{M}_n, \overline{\mathcal{M}}_n)$ w.r.t. $\mathcal{M}_n$ and $\overline{\mathcal{M}}_n$ is defined by
$$I(\mu; \mathcal{M}_n, \overline{\mathcal{M}}_n) = S(\Lambda^* \mu; \overline{\mathcal{M}}_n) + \sum_{C \in \mathcal{M}_n} \sum_{D \in \overline{\mathcal{M}}_n} \Phi(C \times D) \log \frac{\Phi(C \times D)}{\mu(C)}$$

on $\mathcal{M}_n \times \overline{\mathcal{M}}_n$ with the joint measure Φ, where $\Lambda^*\mu$ is given by (6.2). One can easily show the above is equal to

$$I(\mu; \mathcal{M}_n, \overline{\mathcal{M}}_n) = S(\mu; \mathcal{M}_n) + S(\Lambda^*\mu; \overline{\mathcal{M}}_n) - S(\Phi; \mathcal{M}_n \times \overline{\mathcal{M}}_n)$$

so that the mutual entropy rate is

$$\frac{I(\mu; \mathcal{M}_n, \overline{\mathcal{M}}_n)}{n}.$$

Then we can prove the following facts for a stationary channel Λ^* (i.e., the above λ).

Theorem 6.21

1. *The limit of* $\frac{I(\mu; \mathcal{M}_n, \overline{\mathcal{M}}_n)}{n}$ *as* $n \to \infty$ *exists and is*

$$\widetilde{I}(\mu; \Lambda^*) \equiv \lim_{n\to\infty} \frac{I(\mu; \mathcal{M}_n, \overline{\mathcal{M}}_n)}{n} = \widetilde{S}(\mu) + \widetilde{S}(\Lambda^*\mu) - \widetilde{S}(\Phi).$$

2. $\widetilde{I}(\cdot; \Lambda^*)$ *is affine on the set of stationary states* $\mathcal{S}_T(\mathcal{M})$:

$$\widetilde{I}(\alpha\mu + \beta\nu; \Lambda^*) = \alpha\widetilde{I}(\mu; \Lambda^*) + \beta\widetilde{I}(\nu; \Lambda^*) \quad \text{for any } \mu, \nu \in \mathcal{S}_T(\mathcal{M}).$$

3. *If the channel* Λ^* *(i.e., the above* λ*) has finite memory and finite dependence, then* $\widetilde{I}(\cdot; \Lambda^*)$ *is upper semi-continuous on* $\mathcal{S}_T(\mathcal{M})$.

The above limit $\widetilde{I}(\mu; \Lambda^*)$ is called the *transmission rate* (*or velocity*). Now let us define two capacities of a channel.

Definition 6.22

1. The stationary channel capacity (rate) is the supremum of the transmission rate:

$$C_s \equiv \sup\{\widetilde{I}(\mu; \Lambda^*); \mu \in \mathcal{S}_T(\mathcal{M})\}.$$

2. The ergodic channel capacity (rate) is

$$C_e \equiv \sup\{\widetilde{I}(\mu; \Lambda^*); \mu \in \mathcal{S}_E(\mathcal{M}), \Phi \in \mathcal{S}_E(\mathcal{M}\times\overline{\mathcal{M}})\}.$$

An ergodic λ is particularly important, as seen in the next theorem.

Theorem 6.23

1. *For a stationary channel* λ, *we have*

$$0 \le C_e \le C_s \le \log|A||B|.$$

2. *For an ergodic channel* λ, *we have*

$$C_e = C_s.$$

Theorem 6.24 *If a stationary channel λ has finite memory and finite dependence, there exists an ergodic measure μ such that*

$$C_e = \widetilde{I}(\mu; \Lambda^*).$$

6.8.2 Coding Theorem

Now we are going to discuss a coding theorem. First, we state a hypothesis of a channel needed for all theorems below:

⟨Hypothesis C⟩ Channel λ is ergodic, of finite memory and unpredictable.

The following theorem is useful for coding theorems.

Theorem 6.25 (Feinstein's Theorem) *Under* ⟨Hypothesis C⟩, *for any $\varepsilon > 0$ and any $\delta \in (0, C_e)$, by taking natural numbers $n = n(\varepsilon, \delta)$ and N properly, there exist N messages $u_1, \dots, u_N \in \mathcal{M}_{n+t}$ (t is the memory length) and $v_1, \dots, v_N \subset \overline{\Omega}$ such that*

(i) $v_i \cap v_j = \varnothing$ $(i \neq j)$
(ii) $x \in u_j \Rightarrow \lambda(x, v_j) > 1 - \varepsilon$ *for any* j
(iii) $N > e^{n(C_e - \delta)}$.

From statement (i), if the output message is v_j, we can know the input message with the probability nearly equal to 1.

Now we have to consider the whole process of communication:

$$\mathcal{M} \xrightarrow{\xi} \mathcal{X} \xrightarrow{\gamma} \overline{\mathcal{X}} \xrightarrow{\bar{\xi}} \overline{\mathcal{M}}.$$

Practically, recall that $m^l =]a_1, \dots, a_l[\in \mathcal{M}_l \subset \mathcal{M} = A^{\mathbb{Z}} \xrightarrow{\xi} x^{l'} \equiv]x_1, \dots, x_{l'}[\in \mathcal{X}_{l'} \subset \mathcal{X}$.

Let (A, μ_0) be a given input source. One of the goals of communication theory is the following Shannon's Coding theorem.

Theorem 6.26 (First Coding Theorem) *For an ergodic input source (A, μ_0), if $\widetilde{S}(\mu_0) < C_e$ (the capacity of the channel λ) and* ⟨Hypothesis C⟩ *is satisfied, then for any $\varepsilon > 0$ there exists a natural number n such that in the communication process,*

$$m \in \mathcal{M}_n \to \xi(m) \in \mathcal{X}_{n+t} \to \gamma \circ \xi(m) \in \overline{\mathcal{X}} \to \bar{\xi} \circ \gamma \circ \xi(m) = \overline{m} \in \overline{\mathcal{M}},$$

we can recognize a message ω sent from the input source with the probability higher than $1 - \varepsilon$ by the associated output message $\bar{\omega}$.

This theorem means that $\widetilde{S}(\mu_0) < C_e$ with ⟨Hypothesis C⟩ implies $\lambda(\xi(m), \overline{m}) > 1 - \varepsilon$. That is, we can estimate the input message from the output message with probability of nearly 1. In other words, the error probability can be nearly zero in the situation of this theorem.

Theorem 6.27 (Second Coding Theorem) *Under the same conditions of Theorem* 6.26, *there exist a coding and a decoding such that the transmission rate is arbitrarily close to* $\widetilde{S}(\mu_0)$.

This theorem guarantees that for a channel satisfying $\widetilde{S}(\mu_0) < C_e$, there exist a coding and a decoding such that they code a message of the input source (A, μ_0) without error.

6.8.3 Interpretation of Coding Theorem

We will see the intuitive meaning of the coding theorem by taking a simple situation. As we mentioned before, the conditional probability $p(\cdot|\cdot)$ is just one example of a channel λ. In the sequel, we take this conditional probability as the channel.

Let us assume that the output alphabet message space $\overline{\mathcal{M}}$ is embedded in $\mathcal{M}$. When the channel γ is expressed by a transition matrix $\Gamma = (p(y|x))_{x\in\mathcal{X}, y\in\overline{\mathcal{X}}}$, the whole transmission channel λ can be expressed by $\Lambda^* = (p(\overline{m}|m))$ because the maps ξ and $\overline{\xi}$ are injections so that they preserve the conditional probability. In the above general discussion, $p(\overline{m}|m)$ is expressed as $\lambda(\xi(m), \overline{m})$.

Thus in the case when $\mathcal{M}$ is properly imbedded in $\overline{\mathcal{M}}$, the error probability for a sent message m is

$$p_e^m \equiv 1 - p(m|m).$$

The maximum error probability and the average error probability are respectively defined by

$$p_e = \max\left\{p_e^m; m \in \mathcal{M}\right\} \quad \text{and} \quad \widetilde{p}_e = \frac{1}{|\mathcal{M}|}\sum_{m\in\mathcal{M}} p_e^m.$$

Remark that these error probabilities do depend on a coding ξ and a decoding $\overline{\xi}$ taken, so that we write $p_e = p_e(\xi, \overline{\xi})$ and $\widetilde{p}_e = \widetilde{p}_e(\xi, \overline{\xi})$ to exhibit this dependence.

If an information source produces R letters per unit time, then it produces tR letters in t time units. This means that the information source produces $n \equiv [2^{tR}]$ different sequences in t time units, so that the set of all sequences is $\mathcal{M}_n$. When t is an integer, one has $n = 2^{tR} = |\mathcal{M}_n|$, thus $R = \log|\mathcal{M}_n|/t$, so that R is equal to the entropy rate in the case when every message occurs equally. In the case when every message does not occur equally but the information source has a probability μ_0 describing the occurrence of messages, the McMillan's theorem tells us that the above R is nearly equal to the entropy rate $\widetilde{S}(\mu_0)$.

Let us remind that the capacity (rate) C is the maximum (supremum) of the transmission rate $\widetilde{I}(\mu;\Lambda^*)$; $C=\sup\{\widetilde{I}(\mu;\Lambda^*);\mu\}$.

Very intuitive expression of the coding theorem is: *Let us consider an information transmission sending R letters per unit time. If $R<C$, then there exist a coding ξ and a decoding $\overline{\xi}$ such that the mean error probability $\widetilde{p_e}(\xi,\overline{\xi})$ becomes arbitrarily small.*

A bit more mathematical expression is the following: *If the entropy rate $\widetilde{S}(\mu_0)$ of an information source (A,μ_0) is less than the capacity C of a channel, then for any $\varepsilon>0$, there exist a coding ξ and a decoding $\overline{\xi}$ such that $\widetilde{p_e}(\xi,\overline{\xi})<\varepsilon$.*

The most rigorous expression is the one given in the previous subsection.

The proof of the coding theorem is elementary when μ_0 is Bernoulli and the channel Λ^* is memoryless (i.e., $p(]b_1,\ldots,b_l[|]a_1,\ldots,a_l[)=\prod_{k=1}^{l}p(b_k\mid a_k)$). However, the proof of the general case is rather tedious.

6.9 Notes

The notion of entropy was introduced by Clausius, and Boltzman defined the entropy from a microscopic point of view. The differential entropy was first introduced by Gibbs and Boltzman. The information communication processes can be found in the books [74, 180, 359, 388, 456, 516, 742]. The original work of Shannon on the notion of entropy is due to [706] and it is reformulated within the probabilistic frameworks in [69, 74, 308, 355, 357, 359, 403, 456, 458, 517, 766]. The difficulty of the differential entropy can be found in [566]. Kullback and Leibler information was introduced in [455]. The measure-theoretic extension of the Kullback–Leibler entropy was studied by Gelfand, Kolmogorov, Yaglom, Umegaki, and many others [136, 279, 764, 765]. The capacity of an ergodic channel was discussed by [136, 403, 630]. The McMillan's theorem and the coding theorem were presented in [118, 232, 233, 456]. The physical meaning of McMillan's theorem is discussed in [118, 357].

Chapter 7
Fundamentals of Quantum Information

In this chapter basic notions of quantum information will be considered. A channel is a mapping of an input state to an output state. The amount of information transmitted from input to output is described by the mutual entropy.

As we discussed in Chap. 1, the computational process in a computer consists of input and its transformation (computation) and output, which is exactly the same as the communication process of information as seen in Chap. 6 and this chapter. In this chapter, we discuss mathematical properties of von Neumann quantum entropy $S(\rho)$, quantum channel Λ^*, quantum relative entropy $S(\rho,\sigma)$, quantum mutual entropy $I(\rho;\Lambda^*)$, coherent entropy and their generalizations to C^*-systems.

We first summarize the mathematical description of both classical and quantum systems which were explained in the previous chapters.

Passing to quantum systems, we denote the set of all bounded linear operators on a Hilbert space $\mathcal{H}$ by $\mathbf{B}(\mathcal{H})$, and the set of all density operators on $\mathcal{H}$ by $\mathfrak{S}(\mathcal{H})$.

More generally, let $\mathcal{A}$ be a C^*-algebra (i.e., complex normed algebra with involution $*$ such that $\|A\|=\|A^*\|$, $\|A^*A\|=\|A\|^2$ and complete w.r.t. $\|\cdot\|$) and $\mathfrak{S}(\mathcal{A})$ be the set of all states on $\mathcal{A}$ (i.e., positive continuous linear functionals φ on $\mathcal{A}$ such that $\varphi(I)=1$ if $I\in\mathcal{A}$).

If $\Lambda:\mathcal{A}\to\mathcal{B}$ is a map from an algebra $\mathcal{A}$ to an algebra $\mathcal{B}$ then its dual map $\Lambda^*:\mathfrak{S}(\mathcal{B})\to\mathfrak{S}(\mathcal{A})$ is called a *channel*. That is, $\operatorname{tr}\rho\Lambda(A)=\operatorname{tr}\Lambda^*(\rho)A$. Remark that the algebra $\mathcal{B}$ sometimes will be denoted $\overline{\mathcal{A}}$.

Definition 7.1 Let $(\mathcal{A},\mathfrak{S}(\mathcal{A}))$ be an input system and $(\overline{\mathcal{A}},\mathfrak{S}(\overline{\mathcal{A}}))$ be an output system. Take any $\varphi,\psi\in\mathfrak{S}(\mathcal{A})$.

1. Λ^* is linear if $\Lambda^*(\lambda\varphi+(1-\lambda)\psi)=\lambda\Lambda^*\varphi+(1-\lambda)\Lambda^*\psi$ holds for any $\lambda\in[0,1]$.
2. Λ^* is completely positive (CP) if Λ^* is linear and its dual $\Lambda:\overline{\mathcal{A}}\to\mathcal{A}$ satisfies

$$\sum_{i,j=1}^{n} A_i^*\Lambda(\overline{A}_i^*\overline{A}_j)A_j\ge 0$$

for any $n\in\mathbb{N}$ and any $\{\overline{A}_i\}\subset\overline{\mathcal{A}}$, $\{A_i\}\subset\mathcal{A}$.

M. Ohya, I. Volovich, *Mathematical Foundations of Quantum Information and Computation and Its Applications to Nano- and Bio-systems*,
Theoretical and Mathematical Physics,
DOI 10.1007/978-94-007-0171-7_7,

Table 7.1 Descriptions of CDS, QDS and GQDS

	CDS	QDS	GQDS
Observable	real r.v. in $M(\Omega)$	Hermitian operator A on $\mathcal{H}$ (self adjoint operator in $\mathbf{B}(\mathcal{H})$)	Self-adjoint element A in C^*-algebra $\mathcal{A}$
State	Probability measure $\mu \in P(\Omega)$	Density operator ρ on $\mathcal{H}$	Functional (state) $\varphi \in \mathfrak{S}$ with $\varphi(I) = 1$
Expectation	$\int_\Omega f\, d\mu$	$\operatorname{tr} \rho A$	$\varphi(A)$

Such an algebraic approach contains both the classical and quantum theories. The description of a classical probabilistic system (CDS), a quantum dynamical system(QDS), and a general quantum dynamical system (GQDS) are given in Table 7.1.

7.1 Communication Processes

We discussed the communication process in classical systems in Chap. 6. Here we discuss the quantum communication processes.

Let $\mathcal{M}$ be the infinite direct product of the alphabet A, that is, $\mathcal{M} = A^Z \equiv \prod_{-\infty}^{\infty} A$, called a message spaceas in the previous chapter. A coding is a one-to-one map ξ from $\mathcal{M}$ to some space X which is called the coded space. This space X may be a classical object or a quantum object. The classical case was discussed in Chap. 6. For a quantum system, X may be the space of quantum observables on a Hilbert space $\mathcal{H}$, then the coded input system is described by $(\mathbf{B}(\mathcal{H}), \mathfrak{S}(\mathcal{H}))$. The coded output space is denoted by $\tilde{X}$, and the decoded output space $\tilde{\mathcal{M}}$ is made of another alphabet. A transmission (map) γ from X to $\tilde{X}$ (actually, its dual map as discussed below) is called a channel, which reflects the property of a physical device. With a decoding $\tilde{\xi}$, the whole information transmission process is written as

$$\mathcal{M} \xrightarrow{\xi} X \xrightarrow{\gamma} \tilde{X} \xrightarrow{\tilde{\xi}} \tilde{\mathcal{M}}.$$

That is, a message $m \in \mathcal{M}$ is coded to $\xi(m)$ and it is sent to the output system through a channel γ, then the output coded message becomes $\gamma \circ \xi(m)$ and it is decoded to $\tilde{\xi} \circ \gamma \circ \xi(m)$ at a receiver.

Then the occurrence probability of each message in the sequence $(m^{(1)}, m^{(2)}, \ldots, m^{(N)})$ of N messages is denoted by $\rho = \{p_k\}$, which is a state in a classical system. If ξ is a quantum coding, then $\xi(m)$ is a quantum object (state) such as a coherent state. Here we consider such a quantum coding, that is, $\xi(m^{(k)})$ is a quantum state, and we denote $\xi(m^{(k)})$ by σ_k. Thus the coded state for the sequence $(m^{(1)}, m^{(2)}, \ldots, m^{(N)})$ is written as

$$\sigma = \sum_k p_k \sigma_k.$$

This state is transmitted through the dual map of γ which is called a channel in the sequel. This channel (the dual of γ) is expressed by a completely positive mapping Γ^*, in the sense of Chap. 5, from the state space of X to that of $\tilde{X}$, hence the output coded quantum state $\tilde{\sigma}$ is $\Gamma^*\sigma$. Since the information transmission process can be understood as a process of state (probability) change, when Ω and $\tilde{\Omega}$ are classical and X and $\tilde{X}$ are quantum, the process is written as

$$P(\Omega) \xrightarrow{\Xi^*} \mathfrak{S}(\mathcal{H}) \xrightarrow{\Gamma^*} \mathfrak{S}(\tilde{\mathcal{H}}) \xrightarrow{\tilde{\Xi}^*} P(\tilde{\Omega}),$$

where Ξ^* (resp., $\tilde{\Xi}^*$) is the channel corresponding to the coding ξ (resp., $\tilde{\xi}$) and $\mathfrak{S}(\mathcal{H})$ (resp., $\mathfrak{S}(\tilde{\mathcal{H}})$) is the set of all density operators (states) on $\mathcal{H}$ (resp., $\tilde{\mathcal{H}}$).

We have to be careful when studying the objects in the above transmission process. Namely, we have to make clear which object is going to be studied. For instance, if we want to know the information of a quantum state through a quantum channel γ (or Γ^*), then we have to take X so as to describe a quantum system like a Hilbert space, and we need to start the study from a quantum state in the quantum space X not from a classical state associated to messages. We have a similar situation when we treat a state change (computation) in a quantum computer.

7.2 Quantum Entropy for Density Operators

The entropy of a quantum state was introduced by von Neumann. This entropy of a state ρ is defined by

$$S(\rho) = -\operatorname{tr} \rho \log \rho.$$

For a state ρ, there exists a unique spectral decomposition

$$\rho = \sum_k \mu_k P_k,$$

where μ_k is an eigenvalue of ρ and P_k is the associated projection for each μ_k. The projection P_k is not one-dimensional when μ_k is degenerated, so that the spectral decomposition can be further decomposed into one-dimensional projections. Such a decomposition is called a Schatten decomposition, namely,

$$\rho = \sum_{kj} \mu_{kj} E_{kj},$$

where E_{kj} is the one-dimensional projection associated with μ_k and the degenerated eigenvalue μ_k repeats $\dim P_k$ times; for instance, if the eigenvalue μ_1 has degeneracy 3, then $\mu_{11} = \mu_{12} = \mu_{13} < \mu_2$. To simplify notations, we shall write the Schatten decomposition as

$$\rho = \sum_k \mu_k E_k,$$

where the numbers $\{\mu_k\}$ form a probability distribution $\{\mu_k\}$:

$$\sum_k \mu_k = 1, \quad \mu_k \geq 0.$$

This Schatten decomposition is not unique unless every eigenvalue is non-degenerated. Then the entropy (von Neumann entropy) $S(\rho)$ of a state ρ equals to the Shannon entropy of the probability distribution $\{\mu_k\}$:

$$S(\rho) = -\sum_k \mu_k \log \mu_k.$$

Therefore, the von Neumann entropy contains the Shannon entropy as a special case.

Let us summarize the fundamental properties of the entropy $S(\rho)$.

Theorem 7.2 *For any density operator $\rho \in \mathfrak{S}(\mathcal{H})$, the following statements hold*:

1. *(Positivity)* $S(\rho) \geq 0$.
2. *(Symmetry) Let* $\rho' = U\rho U^*$ *for an unitary operator* U. *Then*
$$S(\rho') = S(\rho).$$
3. *(Concavity)* $S(\lambda\rho_1 + (1-\lambda)\rho_2) \geq \lambda S(\rho_1) + (1-\lambda)S(\rho_2)$ *for any* $\rho_1, \rho_2 \in \mathfrak{S}(\mathcal{H})$ *and* $\lambda \in [0, 1]$.
4. *(Additivity)* $S(\rho_1 \otimes \rho_2) = S(\rho_1) + S(\rho_2)$ *for any* $\rho_k \in \mathfrak{S}(\mathcal{H}_k)$.
5. *(Subadditivity) For the reduced states* ρ_1, ρ_2 *of* $\rho \in \mathfrak{S}(\mathcal{H}_1 \otimes \mathcal{H}_2)$,
$$S(\rho) \leq S(\rho_1) + S(\rho_2).$$
6. *(Lower semicontinuity) If* $\|\rho_n - \rho\|_1 \to 0$ $(\equiv \mathrm{tr}\,|\rho_n - \rho| \to 0)$ *as* $n \to \infty$, *then*
$$S(\rho) \leq \lim_{n\to\infty} \inf S(\rho_n).$$
7. *(Continuity) Let* ρ_n, ρ *be elements in* $\mathfrak{S}(\mathcal{H})$ *satisfying the following conditions*: (i) $\rho_n \to \rho$ *weakly as* $n \to \infty$, (ii) $\rho_n \leq A$ $(\forall n)$ *for some compact operator* A, *and* (iii) $-\sum_k a_k \log a_k < +\infty$ *for the eigenvalues* $\{a_k\}$ *of* A. *Then* $S(\rho_n) \to S(\rho)$.
8. *(Strong subadditivity) Let* $\mathcal{H} = \mathcal{H}_1 \otimes \mathcal{H}_2 \otimes \mathcal{H}_3$ *and denote the reduced states* $\mathrm{tr}_{\mathcal{H}_i \otimes \mathcal{H}_j} \rho$ *by* ρ_k *and* $\mathrm{tr}_{\mathcal{H}_k} \rho$ *by* ρ_{ij}. *Then* $S(\rho) + S(\rho_2) \leq S(\rho_{12}) + S(\rho_{23})$ and $S(\rho_1) + S(\rho_2) \leq S(\rho_{13}) + S(\rho_{23})$.
9. *(Entropy increase)* (i) *Let* $\mathcal{H}$ *be a finite dimensional space. If the channel* Λ^* *is completely positive and unital, that is, the dual map* Λ *of the channel* Λ^* *satisfies* $\Lambda I = I$, *then* $S(\Lambda^*\rho) \geq S(\rho)$. (ii) *For an arbitrary Hilbert space* $\mathcal{H}$, *if the dual map* Λ *of the channel* Λ^* *satisfies* $\Lambda(\rho) \in \mathfrak{S}(\mathcal{H})$, *then* $S(\Lambda^*\rho) \geq S(\rho)$.
10. *(Araki–Lieb inequality) For the reduced states* ρ_1, ρ_2 *of* $\rho \in \mathfrak{S}(\mathcal{H}_1 \otimes \mathcal{H}_2)$,
$$\left|S(\rho_1) - S(\rho_2)\right| \leq S(\rho) \leq S(\rho_1) + S(\rho_2).$$

In order to prove the theorem, we prepare a lemma.

Lemma 7.3 *Let f be a convex C^1 function on a proper domain and $\rho, \sigma \in \mathfrak{S}(\mathcal{H})$. Then*

1. (*Klein's inequality*) $\mathrm{tr}\{f(\rho) - f(\sigma) - (\rho - \sigma) f'(\sigma)\} \geq 0$.
2. (*Peierls inequality*) $\sum_k f(\langle x_k, \rho x_k \rangle) \leq \mathrm{tr}\, f(\rho)$ *for any CONS* $\{x_k\}$ *in* $\mathcal{H}$.
   (*Remark*: $\rho = \sum_n \mu_n E_n \to f(\rho) = \sum_n f(\mu_n) E_n$.)

Proof (1) Let $\{x_n\}$ and $\{y_m\}$ be two CONS containing all eigenvalues x_n and y_m of ρ and σ, respectively, such that $\rho x_n = \lambda_n x_n, \sigma y_m = \mu_m y_m$. It is easy to see

$$\begin{aligned}
&\mathrm{tr}\bigl\{f(\rho) - f(\sigma) - (\rho - \sigma) f'(\sigma)\bigr\} \\
&\quad = \sum_n \Bigl\{ f(\lambda_n) - \sum_m \bigl|\langle x_n, y_m \rangle\bigr|^2 f(\mu_m) - \sum_m \bigl|\langle x_n, y_m \rangle\bigr|^2 (\lambda_n - \mu_m) f'(\mu_m) \Bigr\} \\
&\quad = \sum_{n,m} \bigl|\langle x_n, y_m \rangle\bigr|^2 \bigl\{ f(\lambda_n) - f(\mu_m) - (\lambda_n - \mu_m) f'(\mu_m) \bigr\},
\end{aligned}$$

which is positive because f is convex.

(2) Let us take σ in (1) such that

$$\langle x_k, \sigma x_j \rangle = \langle x_k, \rho x_j \rangle \delta_{kj}.$$

Then

$$\mathrm{tr}\, f(\sigma) = \sum_k f\bigl(\langle x_k, \rho x_k \rangle\bigr), \qquad \mathrm{tr}(\rho - \sigma) f'(\sigma) = 0,$$

which imply

$$0 \leq \mathrm{tr}\bigl\{f(\rho) - f(\sigma) - (\rho - \sigma) f'(\sigma)\bigr\} = \mathrm{tr}\, f(\rho) - \sum_k f\bigl(\langle x_k, \rho x_k \rangle\bigr).$$

□

Proof of Theorem 7.2 (1) For the Schatten decomposition of $\rho = \sum_n \mu_n E_n$, it is easy to get

$$S(\rho) = -\sum_n \mu_n \log \mu_n,$$

which is non-negative.

(2)

$$\begin{aligned}
S(U\rho U^*) &= -\mathrm{tr}\, U\rho U^* \log U\rho U^* = -\mathrm{tr}\, U\rho U^* U (\log \rho) U^* \\
&= -\mathrm{tr}\, U\rho \log \rho U^* = -\mathrm{tr}\, U^* U\rho \log \rho = S(\rho).
\end{aligned}$$

(3) Put $\eta(t) = -t \log t$. Since this function $\eta(t)$ is concave, for the CONS $\{x_n\}$ containing all eigenvectors of $\lambda\rho_1 + (1-\lambda)\rho_2$, we obtain

$$\begin{aligned} S\big(\lambda\rho_1 + (1-\lambda)\rho_2\big) &= \sum_n \eta\big(\lambda\langle x_n, \rho_1 x_n\rangle + (1-\lambda)\langle x_n, \rho_2 x_n\rangle\big) \\ &\geq \lambda \sum_n \eta\big(\langle x_n, \rho_1 x_n\rangle\big) + (1-\lambda)\sum_n \eta\big(\langle x_n, \rho_2 x_n\rangle\big) \\ &\geq \lambda \sum_n \big\langle x_n, \eta(\rho_1) x_n\big\rangle + (1-\lambda)\sum_n \big\langle x_n, \eta(\rho_2) x_n\big\rangle \\ &= \lambda S(\rho_1) + (1-\lambda) S(\rho_2), \end{aligned}$$

where the last inequality comes from Peierls' inequality and concavity of η.

(4) Let $\rho_1 = \sum_n \lambda_n E_n$ and $\rho_2 = \sum_m \mu_m F_m$ be the Schatten decompositions. Then $\rho_1 \otimes \rho_2 = \sum_{n,m} \lambda_n \mu_m E_n \otimes F_m$ is the Schatten decomposition of $\rho_1 \otimes \rho_2$. Hence

$$S(\rho_1 \otimes \rho_2) = -\sum_{n,m} \lambda_n \mu_m \log \lambda_n \mu_m = S(\rho_1) + S(\rho_2).$$

(5) Applying Klein's inequality to the function $f(t) = -\eta(t) = t \log t$, we have

$$\operatorname{tr} \rho \log \rho - \operatorname{tr} \rho_1 \otimes \rho_2 \log \rho_1 \otimes \rho_2 \geq \operatorname{tr} \rho - \operatorname{tr} \rho_1 \otimes \rho_2 = 0,$$

but the left-hand side is equal to $S(\rho_1) + S(\rho_2) - S(\rho)$, so we get the desired inequality.

(6) Define

$$S_\alpha(\rho) = \frac{1}{1-\alpha} \log\big(\operatorname{tr} \rho^\alpha\big)$$

for any $\alpha \in \mathbb{R}^+$ with $\alpha \neq 1$ ($S_\alpha(\rho)$ is often called the αth Renyi entropy in quantum systems). Substituting $\rho = \sum_n \mu_n E_n$,

$$S_\alpha(\rho) = \frac{1}{1-\alpha} \log\left(\sum_n \mu_n^\alpha\right) \to -\sum_n \mu_n \log \mu_n = S(\rho) \quad (\alpha \to 1).$$

Since it is easily shown that $S_{\alpha'}(\rho) \leq S_\alpha(\rho)$ $(1 < \alpha \leq \alpha')$,

$$S(\rho) = \lim_{\alpha \to 1} S_\alpha(\rho) \leq \sup\big\{S_\alpha(\rho); \alpha > 1\big\}.$$

We may consider that every eigenvalue of $|\rho_n - \rho|$ is less than 1 for a sufficiently large n because of $\|\rho_n - \rho\|_1 \to 0$ as $n \to \infty$. Therefore,

$$\big|\operatorname{tr} \rho_n^\alpha - \operatorname{tr} \rho^\alpha\big| \leq \operatorname{tr} |\rho_n - \rho|^\alpha \leq \operatorname{tr} |\rho_n - \rho| \to 0 \quad (n \to \infty),$$

which means that $S_\alpha(\rho)$ is a continuous function w.r.t. the trace norm $\|\cdot\|_1$ for any $\alpha > 1$. $S(\rho)$ is the supremum of a continuous function, hence it is lower semicontinuous.

We omit the proofs of (7) and (8); see [578].

(9) We here prove (i) only. It follows from monotonicity of the relative entropy which is proved in the next section. We have $S(\Lambda^*\rho, \Lambda^*\sigma) \le S(\rho,\sigma)$. If $\dim \mathcal{H} = n$ one has

$$S(\rho, I/n) = -S(\rho) + \log n.$$

Since

$$S(\Lambda^*\rho, \Lambda^* I/n) = S(\Lambda^*\rho, I/n) = -S(\Lambda^*\rho) + \log n,$$

the result follows. The proof of (ii) comes from the convex analysis of operators, which we omit here (see [570]).

(10) This follows from the proof of (5). □

7.3 Relative Entropy for Density Operators

For two states $\rho, \sigma \in \mathfrak{S}(\mathcal{H})$, the *relative entropy* is first defined by Umegaki

$$S(\rho,\sigma) = \begin{cases} \operatorname{tr}\rho(\log\rho - \log\sigma) & (\rho \ll \sigma), \\ \infty & \text{otherwise,} \end{cases}$$

where $\rho \ll \sigma$ means that $\operatorname{tr}\sigma A = 0 \Rightarrow \operatorname{tr}\rho A = 0$ for $A \ge 0 \Leftrightarrow \overline{\operatorname{ran}\rho} \subset \overline{\operatorname{ran}\sigma}$. The main properties of relative entropy are summarized as:

Theorem 7.4 *The relative entropy has the following properties*:

1. *(Positivity)* $S(\rho,\sigma) \ge 0$, $= 0$ *iff* $\rho = \sigma$.
2. *(Joint convexity)* $S(\lambda\rho_1 + (1-\lambda)\rho_2, \lambda\sigma_1 + (1-\lambda)\sigma_2) \le \lambda S(\rho_1,\sigma_1) + (1-\lambda)S(\rho_2,\sigma_2)$ *for any* $\lambda \in [0,1]$.
3. *(Additivity)* $S(\rho_1 \otimes \rho_2, \sigma_1 \otimes \sigma_2) = S(\rho_1,\sigma_1) + S(\rho_2,\sigma_2)$.
4. *(Lower semicontinuity) If* $\lim_{n\to\infty} \|\rho_n - \rho\|_1 = 0$ *and* $\lim_{n\to\infty} \|\sigma_n - \sigma\|_1 = 0$, *then* $S(\rho,\sigma) \le \liminf_{n\to\infty} S(\rho_n,\sigma_n)$. *Moreover, if there exists a positive number* λ *satisfying* $\rho_n \le \lambda\sigma_n$, *then* $\lim_{n\to\infty} S(\rho_n,\sigma_n) = S(\rho,\sigma)$.
5. *(Monotonicity) For a channel* Λ^* *from* $\mathfrak{S}$ *to* $\overline{\mathfrak{S}}$, $S(\Lambda^*\rho, \Lambda^*\sigma) \le S(\rho,\sigma)$.
6. *(Lower bound)* $\|\rho - \sigma\|^2/2 \le S(\rho,\sigma)$.
7. *(Invariance under the unitary mapping)* $S(U\rho U^*; U\sigma U^*) = S(\rho;\sigma)$ *where* U *is a unitary operator.*

Proof (1) $f(t) = t\log t$ is a convex function and $f'(t) = \log t + 1$ holds. By Klein's inequality, we have

$$\begin{aligned} 0 &\le \operatorname{tr}\bigl\{f(\rho) - f(\sigma) - (\rho-\sigma)f'(\sigma)\bigr\} \\ &= \operatorname{tr}\bigl\{\rho\log\rho - \sigma\log\sigma - (\rho-\sigma)(\log\sigma + I)\bigr\} \\ &= \operatorname{tr}\bigl\{\rho\log\rho - \rho\log\sigma - (\rho-\sigma)\bigr\} \\ &= S(\rho;\sigma) - \operatorname{tr}\rho + \operatorname{tr}\sigma = S(\rho;\sigma). \end{aligned}$$

(3) For any $\rho_1, \sigma_1 \in \mathfrak{S}(\mathcal{H}_1)$ and $\rho_2, \sigma_2 \in \mathfrak{S}(\mathcal{H}_2)$,

$$S(\rho_1 \otimes \rho_2; \sigma_1 \otimes \sigma_2) = \operatorname{tr} \rho_1 \otimes \rho_2 (\log \rho_1 \otimes \rho_2 + \log \sigma_1 \otimes \sigma_2).$$

Using the equality $\rho_1 \otimes \rho_2 = (\rho_1 \otimes I)(I \otimes \rho_2)$, we get

$$\begin{aligned} &S(\rho_1 \otimes \rho_2; \sigma_1 \otimes \sigma_2) \\ &\quad = \operatorname{tr} \rho_1 \otimes \rho_2 (\log \rho_1 \otimes I + \log I \otimes \rho_2) - \operatorname{tr} \rho_1 \otimes \rho_2 (\log \sigma_1 \otimes I + \log I \otimes \sigma_2) \\ &\quad = \operatorname{tr} \rho_1 \log \rho_1 \otimes \rho_2 + \operatorname{tr} \rho_1 \otimes \rho_2 \log \rho_2 - \operatorname{tr} \rho_1 \log \sigma_1 \otimes \rho_2 - \operatorname{tr} \rho_1 \otimes \rho_2 \log \sigma_2 \\ &\quad = \operatorname{tr} \rho_1 (\log \rho_1 - \log \sigma_1) + \operatorname{tr} \rho_2 (\log \rho_2 - \log \sigma_2) = S(\rho_1; \sigma_1) + S(\rho_2; \sigma_2). \end{aligned}$$

(4) We define $S_\lambda(\rho; \sigma)$ by

$$S_\lambda(\rho; \sigma) \equiv \frac{1}{\lambda} \bigl\{ S\bigl(\lambda\rho + (1-\lambda)\sigma\bigr) - \lambda S(\rho) - (1-\lambda) S(\sigma) \bigr\}.$$

From the concavity of $S(\cdot)$, $S_\lambda(\rho; \sigma) \geq 0$ holds. Put $g(\lambda) \equiv \lambda S_\lambda(\rho; \sigma)$. Then $g : [0,1] \to \mathbb{R}^+$ and $g''(\lambda) \leq 0$, so that $g(\lambda)$ is a concave function with respect to λ. Moreover, we get $[\frac{d}{d\lambda} g(\lambda)]_{\lambda=0} = S(\rho; \sigma)$ and the following equality

$$\begin{aligned} \left[\frac{d}{d\lambda} g(\lambda) \right]_{\lambda=0} &= g'(0) = \lim_{\lambda \to 0} \frac{g(\lambda) - g(0)}{\lambda} \\ &= \lim_{\lambda \to 0} \frac{g(\lambda)}{\lambda} = \lim_{\lambda \to 0} S_\lambda(\rho; \sigma). \end{aligned}$$

Therefore, we obtain

$$S(\rho; \sigma) = \lim_{\lambda \to 0} S_\lambda(\rho; \sigma).$$

Now we put $\hat{S}(\rho) = -\rho \log \rho$ and show that $\hat{S}(\rho)$ is an operator-concave function. For any $t \in \mathbb{R}^+$,

$$-t \log t = \int_0^\infty \left\{ 1 - \frac{u}{t+u} - \frac{t}{1+u} \right\} du$$

holds, so that the concavity of $\hat{S}(\rho)$ is equivalent to the convexity of $\frac{I}{\rho + I}$. Hence, we have only to show the following inequality:

$$\frac{I}{\frac{(\rho_1+\rho_2)}{2} + I} \leq \frac{1}{2} \left(\frac{I}{\rho_1 + I} \right) + \frac{1}{2} \left(\frac{I}{\rho_2 + I} \right).$$

The above inequality is satisfied iff the following inequality holds:

$$\frac{4I}{(\rho_2+I)^{-1/2}(\rho_1+I)(\rho_2+I)^{-1/2} + I} \leq \frac{I}{(\rho_2+I)^{-1/2}(\rho_1+I)(\rho_2+I)^{-1/2}} + I.$$

Now we put $B \equiv (\rho_2 + I)^{-1/2}$ and $C \equiv \rho_1 + I$, $A \equiv BCB$. Then $A > 0$. Let a spectral decomposition of A be

$$A = \int_0^\infty t\, dE(t),$$

then we have

$$\frac{4I}{A+I} = \int_0^\infty \frac{4}{t+1}\, dE(t), \qquad \frac{I}{A} + I = \int_0^\infty \left(\frac{1}{t} + 1\right) dE(t).$$

Therefore,

$$\frac{4I}{A+I} \le \frac{I}{A} + I$$

is satisfied.

Here, let P_n be a projection from $\mathcal{H}$ to an n-dimensional closed subspace $\mathcal{H}_n$ of $\mathcal{H}$ with $\mathcal{H}_n \subset \mathcal{H}_{n+1}$ and satisfying $P_n \uparrow I$, $P_n \le P_{n+1} \le \cdots \le I$. For any $Q \ge 0$, $\operatorname{tr} QP_n \uparrow \operatorname{tr} Q$ holds. If we take

$$\hat{S}_\lambda(\rho;\sigma) \equiv \frac{1}{\lambda}\left\{\hat{S}\big(\lambda\rho + (1-\lambda)\sigma\big) - \lambda\hat{S}(\rho) - (1-\lambda)\hat{S}(\sigma)\right\}$$

as Q, then we obtain

$$\operatorname{tr} \hat{S}_\lambda(\rho;\sigma) P_n \uparrow \operatorname{tr} \hat{S}_\lambda(\rho;\sigma) = S_\lambda(\rho;\sigma).$$

Hence,

$$S(\rho;\sigma) = \lim_{\lambda\to 0} S_\lambda(\rho;\sigma) = \lim_{\lambda\to 0} \sup_n \operatorname{tr} \hat{S}(\rho;\sigma) P_n = \sup_n \lim_{\lambda\to 0} \operatorname{tr} \hat{S}(\rho;\sigma) P_n.$$

Due to $\dim \mathcal{H}_n = \dim P_n\mathcal{H} = n < +\infty$, by putting $F(\rho_n, \sigma_n) \equiv \lim_{\lambda\to 0} \operatorname{tr} \hat{S}_\lambda(\rho;\sigma) \times P_n$, $F(\rho_n, \sigma_n)$ is expressed by finitely many of eigenvalues of ρ and σ. When $\rho_n \to \rho, \sigma_n \to \sigma$, each eigenvalue $\lambda_n^{(j)}, \mu_n^{(j)}$ of ρ_n and σ_n, respectively, satisfies $\lambda_n^{(j)} \to \lambda^{(j)}, \mu_n^{(j)} \to \mu^{(j)}$, where $\lambda^{(j)}$, $\mu^{(j)}$ are the eigenvalues of ρ and σ, respectively, so that $F(\rho_n, \sigma_n) \to F(\rho, \sigma)$ is obtained. Therefore, $S(\rho;\sigma)$ is lower semicontinuous because it is the supremum of a continuous function: $S(\rho;\sigma) \le \liminf_{n\to\infty} S(\rho_n;\sigma_n)$.

(5) By the Stinespring's dilation theorem, every completely positive map can be decomposed into a product of the following three operations: (1) tensoring with a second system in a specified state, (2) unitary transformation, and (3) reduction to a subsystem. To write it in an explicit form, we use the Kraus representation

$$\Lambda^*\rho = \sum_{i=1}^n A_i \rho A_i^*, \qquad \sum_{i=1}^n A_i^* A_i = 1.$$

There is a unitary operator U in $H \otimes \mathbb{C}^n$ and a one-dimensional projector operator P in $\mathbb{C}^n$ such that one has

$$U(\rho \otimes P)U^* = \sum_{i,j=1}^{n} A_i \rho A_j^* \otimes |i\rangle\langle j|$$

where $\{|i\rangle\}$ is an orthonormal basis in $\mathbb{C}^n$. Then we get

$$\Lambda^* \rho = \operatorname{tr}_2\bigl(U(\rho \otimes P)U^*\bigr).$$

Now we use the subadditivity of the relative entropy

$$S(\operatorname{tr}_2 A, \operatorname{tr}_2 B) \le S(A, B)$$

where tr_2 is a partial trace to obtain

$$\begin{aligned} S(\Lambda^*\rho, \Lambda^*\sigma) &= S\bigl(\operatorname{tr}_2\bigl(U(\rho \otimes P)U^*\bigr), \operatorname{tr}_2\bigl(U(\sigma \otimes P)U^*\bigr)\bigr) \\ &\le S\bigl(U(\rho \otimes P)U^*, U(\sigma \otimes P)U^*\bigr) = S(\rho \otimes P, \sigma \otimes P) \\ &= S(\rho, \sigma) + S(P, P) = S(\rho, \sigma). \end{aligned}$$

The monotonicity of the relative entropy is proved.

(7) It follows that

$$\begin{aligned} S(U\rho U^*; U\sigma U^*) &= -\operatorname{tr} U\rho U^*(\log U\rho U^* - \log U\sigma U^*) \\ &= \operatorname{tr} U\rho U^* U(\log \rho - \log \sigma)U^* \\ &= \operatorname{tr} U^* U\rho(\log \rho - \log \sigma) \\ &= \operatorname{tr} \rho(\log \rho - \log \sigma) = S(\rho; \sigma). \end{aligned}$$ □

Let us extend the relative entropy to two positive operators instead of two states. If A and B are two positive Hermitian operators (not necessarily states, i.e., not necessarily with unit traces) then we set

$$S(A, B) = \operatorname{tr} A(\log A - \log B).$$

The following *Bogoliubov inequality* holds.

Theorem 7.5 (Bogoliubov Inequality) *One has*

$$S(A, B) \ge \operatorname{tr} A(\log \operatorname{tr} A - \log \operatorname{tr} B).$$

Proof Let H_0 and V be two bounded Hermitian operators such that e^{-H_0} and $e^{-(H_0+V)}$ are positive-definite. Let us consider the function (free energy)

$$f(\lambda) = -\log \operatorname{tr} e^{-H_\lambda}$$

where

$$H_\lambda = H_0 + \lambda V$$

and $\lambda \geq 0$. Let us prove that the function $f(\lambda)$ is convex, $f''(\lambda) \leq 0$. This will lead to the Bogoliubov inequality. By using the relation

$$\frac{\partial}{\partial \lambda} e^{-H_\lambda} = -\int_0^1 dt e^{-tH_\lambda} V e^{-(1-t)H_\lambda}$$

one obtains

$$\frac{\partial^2}{\partial \lambda^2} f(\lambda) = -\left(\operatorname{tr} e^{-H_\lambda}\right)^{-1} \int_0^1 dt \operatorname{tr}\left(\left(V - \langle V \rangle\right) e^{-tH_\lambda} \left(V - \langle V \rangle\right) e^{-(1-t)H_\lambda}\right),$$

which is nonpositive. Here

$$\langle V \rangle = \left(\operatorname{tr} e^{-H_\lambda}\right)^{-1} \operatorname{tr} V e^{-H_\lambda}.$$

Therefore,

$$f'(\lambda_1) \geq f'(\lambda_2) \quad \text{if } \lambda_1 \leq \lambda_2.$$

In particular, one has

$$f(1) - f(0) = \int_0^1 d\lambda\, f'(\lambda) \leq f'(0).$$

Since

$$f'(0) = \left(\operatorname{tr} e^{-H_0}\right)^{-1} \operatorname{tr} V e^{-H_0},$$

from the last inequality we get

$$-\log \operatorname{tr} e^{-(H_0+V)} + \log \operatorname{tr} e^{-H_0} \leq \left(\operatorname{tr} e^{-H_0}\right)^{-1} \operatorname{tr} V e^{-H_0}.$$

In this form, the inequality was obtained by Bogoliubov. Now if we set

$$e^{-H_0} = A, \qquad e^{-(H_0+V)} = B, \quad V = \log A - \log B,$$

we obtain

$$\log \operatorname{tr} A - \log \operatorname{tr} B \leq (\operatorname{tr} A)^{-1} \operatorname{tr} A(\log A - \log B).$$

□

We remark here that in some cases the above quantum relative entropy can be approximated by the classical relative entropy [331].

7.4 More About Channels

A general quantum system containing all systems such as discrete and continuous in both classical and quantum setting is described by a C^*-algebra or a von Neumann algebra as we mentioned before, so that we discuss the channeling transformation in C^*-algebraic contexts. However, it is enough for the readers who are not familiar with C^*-algebras to imagine a usual quantum system, for instance, regard $\mathcal{A}$ and $\mathfrak{S}(\mathcal{A})$ below as $\mathbf{B}(\mathcal{H})$ and $\mathfrak{S}(\mathcal{H})$, respectively. Let $\mathcal{A}$ and $\overline{\mathcal{A}}$ be C^*-algebras, and $\mathfrak{S}(\mathcal{A})$ and $\mathfrak{S}(\overline{\mathcal{A}})$ be the sets of all states on $\mathcal{A}$ and $\overline{\mathcal{A}}$, respectively.

A *channel* is a mapping from $\mathfrak{S}(\mathcal{A})$ to $\mathfrak{S}(\overline{\mathcal{A}})$. There exist channels with various properties.

Definition 7.6 Let $(\mathcal{A}, \mathfrak{S}(\mathcal{A}))$ be an input system and $(\overline{\mathcal{A}}, \mathfrak{S}(\overline{\mathcal{A}}))$ be an output system. Take any $\varphi, \psi \in \mathfrak{S}(\mathcal{A})$.

1. Λ^* is of Schwarz type if $\Lambda(\overline{A}^*) = \Lambda(\overline{A})^*$ and $\Lambda(\overline{A})^* \Lambda(\overline{A}) \le \Lambda(\overline{A}^* \overline{A})$.
2. Λ^* is stationary if $\Lambda \circ \alpha_t = \overline{\alpha}_t \circ \Lambda$ for any $t \in \mathbb{R}$.
   (Here α_t and $\overline{\alpha}_t$ are groups of automorphisms of the algebras $\mathcal{A}$ and $\overline{\mathcal{A}}$, respectively.)
3. Λ^* is ergodic if Λ^* is stationary and $\Lambda^*(\mathrm{ex} I(\alpha)) \subset \mathrm{ex} I(\overline{\alpha})$.
   (Here $\mathrm{ex}\, I(\alpha)$ is the set of extreme points of the set of all stationary states $I(\alpha)$.)
4. Λ^* is orthogonal if for any two orthogonal states $\varphi_1, \varphi_2 \in \mathfrak{S}(\mathcal{A})$ (denoted by $\varphi_1 \perp \varphi_2$) one has $\Lambda^* \varphi_1 \perp \Lambda^* \varphi_2$.
5. Λ^* is deterministic if Λ^* is orthogonal and bijective.
6. For a subset $\mathfrak{S}_0$ of $\mathfrak{S}(\mathcal{A})$, Λ^* is chaotic for $\mathfrak{S}_0$ if $\Lambda^* \varphi_1 = \Lambda^* \varphi_2$ for any $\varphi_1, \varphi_2 \in \mathfrak{S}_0$.
7. Λ^* is chaotic if Λ^* is chaotic for $\mathfrak{S}(\mathcal{A})$.
8. (Kraus–Sudarshan representation) A completely positive channel Λ^* can be represented as

$$\Lambda^* \rho = \sum_i A_i \rho A_i^*, \qquad \sum_i A_i^* A_i \le 1.$$

   Here A_i are bounded operators on H.

Most of channels appearing in physical processes are CP channels. Examples of such channels are provided below. Take a density operator $\rho \in \mathfrak{S}(\mathcal{H})$ as an input state.

Example 7.7 (Unitary evolution) Let H be the Hamiltonian of a system. Then

$$\rho \to \Lambda^* \rho = U_t \rho U_t^*,$$

where $t \in \mathbb{R}$, $U_t = \exp(-itH)$, is a CP channel.

Example 7.8 (Semigroup evolution) Let V_t $(t \in \mathbb{R}^+)$ be a one parameter semigroup on $\mathcal{H}$. Then

$$\rho \to \Lambda^* \rho = V_t \rho V_t^*, \quad t \in \mathbb{R}^+,$$

is a CP channel.

Example 7.9 (Quantum measurement) If a measuring apparatus is prepared by a positive operator-valued measure $\{Q_n\}$ then the state ρ changes to a state $\Lambda^*\rho$ after this measurement,

$$\rho \to \Lambda^*\rho = \sum_n Q_n \rho Q_n.$$

Example 7.10 (Reduction) If a system Σ_1 interacts with an external system Σ_2 described by another Hilbert space $\mathcal{K}$ and the initial states of Σ_1 and Σ_2 are ρ and σ, respectively, then the combined state θ_t of Σ_1 and Σ_2 at time t after the interaction between two systems is given by

$$\theta_t \equiv U_t(\rho \otimes \sigma)U_t^*,$$

where $U_t = \exp(-itH)$ with the total Hamiltonian H of Σ_1 and Σ_2. A channel is obtained by taking the partial trace w.r.t. $\mathcal{K}$ such as

$$\rho \to \Lambda^*\rho \equiv \mathrm{tr}_{\mathcal{K}}\, \theta_t.$$

Example 7.11 (Optical communication processes) A quantum communication process is described by the following scheme.

$$\begin{array}{ccccc} & \nu \in \mathfrak{S}(\mathcal{K}) & & & \\ & \downarrow & & & \\ \mathfrak{S}(\mathcal{H}) \ni \rho & \longrightarrow & & & \bar{\rho} = \Lambda^*\rho \in \mathfrak{S}(\mathcal{H}) \\ & & & \downarrow & \\ & & & \text{Loss} & \end{array}$$

$$\begin{array}{ccc} \mathfrak{S}(\mathcal{H}) & \xrightarrow{\Lambda^*} & \mathfrak{S}(\mathcal{H}) \\ \gamma^* \downarrow & & \uparrow a^* \\ \mathfrak{S}(\mathcal{H} \otimes \mathcal{K}) & \xrightarrow[\pi^*]{} & \mathfrak{S}(\mathcal{H} \otimes \mathcal{K}). \end{array}$$

The above maps γ^*, a^* are given by

$$\gamma^*(\rho) = \rho \otimes \nu, \quad \rho \in \mathfrak{S}(\mathcal{H}),$$
$$a^*(\theta) = \mathrm{tr}_{\mathcal{K}}\, \theta, \quad \theta \in \mathfrak{S}(\mathcal{H} \otimes \mathcal{K}),$$

where ν is noise coming from the outside of the system. The map π^* is a certain channel determined by physical properties of the device transmitting information. Hence the channel for the above process is given as $\Lambda^*\rho \equiv \mathrm{tr}_{\mathcal{K}}\, \pi^*(\rho \otimes \nu) = (a^* \circ \pi^* \circ \gamma^*)(\rho)$.

Example 7.12 (Attenuation process) An example of the channel from Example 7.11 is the attenuation channel defined as follows: Take $\rho = |\theta\rangle\langle\theta|$, a coherent state, and $\nu = |0\rangle\langle 0|$, a vacuum state. Then

$$\Lambda^*\rho = |\alpha\theta\rangle\langle\alpha\theta|, \quad 0 \le |\alpha|^2 \le 1.$$

Example 7.13 (Noisy optical channel) When there exists noise ξ, the channel is

$$\Lambda^* \rho \equiv \mathrm{tr}_{\mathcal{K}_2} \Pi^* (\rho \otimes \xi).$$

Further, when ξ is the m_1 photon number state $|m_1\rangle\langle m_1|$ of $\mathcal{K}_1$,

$$\Lambda^* \rho = \mathrm{tr}_{\mathcal{K}_2} V (\rho \otimes |m_1\rangle\langle m_1|o) V^*.$$

Take

$$V : \mathcal{H}_1 \otimes \mathcal{K}_1 \to \mathcal{H}_2 \otimes \mathcal{K}_2$$

as

$$V\big(|n_1\rangle \otimes |m_1\rangle\big) \equiv \sum_{j=0}^{n_1+m_1} c_j^{n_1,m_1} |j\rangle \otimes |n_1 + m_1 - j\rangle$$

with

$$c_j^{n_1,m_1} \equiv \sum_{r=L}^{K} (-1)^{n_1+j-r} \frac{\sqrt{n_1! m_1! j! (n_1 + m_1 - j)!}}{r!(n_1 - j)!(j - r)!(m_1 - j + r)!} \times \sqrt{\eta^{m_1-j+2r} (1 - \eta)^{n_1+j-2r}},$$

where

$$K = \min\{n_1, j\}, \qquad L \equiv \max\{m_1 - j, 0\}.$$

Example 7.14 (Single qubit quantum channels) Any density operator in $\mathbb{C}^2$ can be written in the form

$$\rho = \frac{1}{2}(I + a \cdot \sigma)$$

where $\sigma = (\sigma_1, \sigma_2, \sigma_3)$ are the Pauli matrices, $a = (a_1, a_2, a_3)$ are real parameters,

$$a_1^2 + a_2^2 + a_3^2 \leq 1,$$

and

$$a \cdot \sigma = \sum_{i=1}^{3} a_i \sigma_i.$$

One can prove that a completely positive, trace-preserving map is unitary equivalent to the following simple map

$$\Lambda_\lambda^* \left(\frac{1}{2}(I + a \cdot \sigma) \right) = \frac{1}{2}(I + b \cdot \sigma).$$

Here

$$b_i = \lambda_i a_i, \quad i = 1, 2, 3,$$

where λ_i are real numbers which obey certain conditions, see Theorem 7.15 below.

Let us prove complete positivity for the above channels.

Proof Examples 7.7 and 7.8 are trivial, and Example 7.12 is a special case of Example 7.11, so we only will give proofs for the remaining examples.

(Example 7.9) For any density operator ρ and any POV $\{Q_k\}$,

$$\Lambda^* \rho = \sum_k Q_k^{1/2} \rho Q_k^{1/2}.$$

Since ρ is positive, $Q_k^{1/2} \rho Q_k^{1/2}$ is positive. Hence $\Lambda^* \rho$ is positive. Moreover, we get

$$\begin{aligned}
\operatorname{tr} \Lambda^* \rho &= \operatorname{tr}\Big(\sum_k Q_k^{1/2} \rho Q_k^{1/2}\Big) \\
&= \sum_k \operatorname{tr}(Q_k \rho) \\
&= \operatorname{tr}\Big(\sum_k Q_k \rho\Big) \\
&= \operatorname{tr} \rho = 1.
\end{aligned}$$

Therefore, we obtain

$$\Lambda^* \rho \in \mathfrak{S}(\mathcal{H}).$$

From the definition of Λ^*, the dual map $\Lambda : \mathbf{B}(\mathcal{H}) \to \mathbf{B}(\mathcal{H})$ of Λ^* is given by

$$\Lambda(A) = \sum_k Q_k^{1/2} A Q_k^{1/2}, \quad A \in \mathbf{B}(\mathcal{H}).$$

Thus, for any vector $x \in \mathcal{H}$ and any $n \in \mathbb{N}$, we have

$$\begin{aligned}
\Big\langle x, \sum_{i,j=1}^{n} B_i^* \Lambda(A_i^* A_j) B_j x \Big\rangle &= \Big\langle x, \sum_{i,j} B_i^* \sum_k Q_k^{1/2} (A_i^* A_j) Q_k^{1/2} B_j x \Big\rangle \\
&= \sum_{i,j} \sum_k \langle x, B_i^* Q_k^{1/2} (A_i^* A_j) Q_k^{1/2} B_j x \rangle \\
&= \sum_{i,j} \sum_k \langle A_i Q_k^{1/2} B_i x, A_j Q_k^{1/2} B_j x \rangle
\end{aligned}$$

$$= \sum_k \left\langle \sum_i A_i Q_k^{1/2} B_i x, \sum_j A_j Q_k^{1/2} B_j x \right\rangle \geq 0$$

for any $n \in \mathbb{N}$ and any $A_i, B_i \in \mathbf{B}(\mathcal{H})$. Therefore, Λ^* is a channel.

(Example 7.10) For any $x \in \mathcal{H}$, any CONS $\{y_k\} \in \mathcal{K}$, we get

$$\begin{aligned}
\langle x, \Lambda^* \rho x \rangle &= \langle x, \operatorname{tr}_{\mathcal{K}} U_t(\rho \otimes \sigma) U_t^* x \rangle \\
&= \sum_k \langle x \otimes y_k, U_t(\rho \otimes \sigma) U_t^* x \otimes y_k \rangle \\
&= \sum_k \langle x \otimes y_k, U_t(\rho \otimes \sigma) U_t^* x \otimes y_k \rangle \\
&= \sum_k \langle U_t^* x \otimes y_k, (\rho \otimes \sigma) U_t^* x \otimes y_k \rangle \geq 0,
\end{aligned}$$

and

$$\begin{aligned}
\operatorname{tr} \Lambda_t^* \rho &= \operatorname{tr}_{\mathcal{H}} \operatorname{tr}_{\mathcal{K}} U_t(\rho \otimes \sigma) U_t^* = \operatorname{tr} U_t(\rho \otimes \sigma) U_t^* \\
&= \operatorname{tr} U_t^* U_t(\rho \otimes \sigma) = \operatorname{tr}(\rho \otimes \sigma) = 1.
\end{aligned}$$

Thus $\Lambda^* \rho$ is a state. The dual map $\Lambda_t : \mathbf{B}(\mathcal{H}) \to \mathbf{B}(\mathcal{H})$ of the channel Λ^* is defined by

$$\Lambda_t(Q) = \operatorname{tr}_{\mathcal{K}} \sigma U_t^* (Q \otimes I) U_t$$

for any $Q \in \mathbf{B}(\mathcal{H})$. The completely positivity of Λ_t^* can be proved as follows:

$$\begin{aligned}
&\left\langle x, \sum_{i,j=1}^n B_i^* \Lambda_t(A_i^* A_j) B_j x \right\rangle \\
&\quad = \left\langle x, \sum_{i,j=1}^n B_i^* \operatorname{tr}_{\mathcal{K}} (I \otimes \sigma) U_t^* (A_i^* A_j \otimes I) U_t B_j x \right\rangle \\
&\quad = \sum_{i,j=1}^n \sum_k \langle x \otimes y_k, (B_i^* \otimes I)\big(I \otimes \sigma^{1/2}\big) U_t^* (A_i^* A_j \otimes I)\big(I \otimes \sigma^{1/2}\big) \\
&\qquad \times U_t(B_j \otimes I) x \otimes y_k \rangle \\
&\quad = \sum_k \left\langle \sum_i^n A_i U_t B_i x \otimes \sigma^{1/2} y_k, \sum_j^n A_j U_t B_j x \otimes \sigma^{1/2} y_k \right\rangle \geq 0,
\end{aligned}$$

for any $n \in \mathbb{N}$ and any $A_i, B_i \in \mathbf{B}(\mathcal{H})$.

(Example 7.11) We have only to show the completely positivity of γ and a because π is completely positive. For any $A_i \in \mathbf{B}(\mathcal{H} \otimes \mathcal{K})$, $B_j \in \mathbf{B}(\mathcal{H})$, CONS $\{x_k\}$

of $\mathcal{H}$, CONS $\{y_l\}$ of $\mathcal{K}$, $\xi \in \mathfrak{S}(\mathcal{K})$, $x \in \mathcal{H}$, and any $n \in \mathbb{N}$, we have

$$\begin{aligned}
&\left\langle x, \sum_{i,j=1}^{n} B_i^* \gamma(A_i^* A_j) B_j x \right\rangle \\
&= \sum_{i,j=1}^{n} \langle B_i x, \operatorname{tr}_{\mathcal{K}} \xi A_i^* A_j B_j x \rangle \\
&= \sum_{i,j=1}^{n} \sum_{m} \big\langle B_i x \otimes y_m, (I \otimes \xi) A_i^* A_j B_j x \otimes y_m \big\rangle \\
&= \sum_{i,j=1}^{n} \sum_{m} \sum_{k,l} \big\langle B_i x \otimes y_m, (I \otimes \xi) A_i^* x_k \otimes y_l \; x_k \otimes y_l, A_j B_j x \otimes y_m \big\rangle \\
&= \sum_{i,j=1}^{n} \sum_{k,l} \big\langle x_k \otimes y_l, A_j \big(|B_j x \; B_i x| \otimes I\big)(I \otimes \xi) A_i^* x_k \otimes y_l \big\rangle \\
&= \sum_{k,l} \sum_{i,j=1}^{n} \big\langle x_k \otimes y_l, A_j (B_j \otimes I)\big(|x\rangle\langle x| \otimes I\big)(B_i^* \otimes I)\big(I \otimes \xi^{\frac{1}{2}}\big)\big(I \otimes \xi^{\frac{1}{2}}\big) \\
&\quad \times A_i^* x_k \otimes y_l \big\rangle \\
&= \sum_{m} \sum_{k,l} \sum_{j=1}^{n} \big\langle x_k \otimes y_l, A_j \big(I \otimes \xi^{\frac{1}{2}}\big)(B_j \otimes I) x \otimes y_m \big\rangle \\
&\quad \times \overline{\sum_{i=1}^{n} \big\langle x_k \otimes y_l, A_i \big(I \otimes \xi^{\frac{1}{2}}\big)(B_i \otimes I) x \otimes y_m \big\rangle} \\
&= \sum_{m} \sum_{k,l} \left| \sum_{j=1}^{n} \big\langle x_k \otimes y_l, A_j \big(I \otimes \xi^{\frac{1}{2}}\big)(B_j \otimes I) x \otimes y_m \big\rangle \right|^2 \geq 0.
\end{aligned}$$

Thus γ is a channel. For any $A_i \in \mathbf{B}(\mathcal{H})$, $B_j \in B(\mathcal{H} \otimes \mathcal{K})$, CONS $\{x_k\}$ of $\mathcal{H}$, CONS $\{y_l\}$ of $\mathcal{K}$, $x \otimes y \in \mathcal{H} \otimes \mathcal{K}$, and any $n \in \mathbb{N}$, we obtain

$$\begin{aligned}
&\left\langle x \otimes y, \sum_{i,j=1}^{n} B_i^* a(A_i^* A_j) B_j x \otimes y \right\rangle \\
&= \sum_{i,j=1}^{n} \big\langle B_i x \otimes y, (A_i^* A_j \otimes I) B_j x \otimes y \big\rangle \\
&= \sum_{i,j=1}^{n} \big\langle B_i x \otimes y, (A_i^* \otimes I)(A_j \otimes I) B_j x \otimes y \big\rangle
\end{aligned}$$

$$
\begin{aligned}
&= \left\langle \sum_{i=1}^{n} (A_i \otimes I) B_i x \otimes y, \sum_{j=1}^{n} (A_j \otimes I) B_j x \otimes y \right\rangle \\
&= \sum_{k,l} \left\langle \sum_{i=1}^{n} (A_i \otimes I) B_i x \otimes y, x_k \otimes y_l \right\rangle \\
&\quad \times \left\langle x_k \otimes y_l, \sum_{j=1}^{n} (A_j \otimes I) B_j x \otimes y \right\rangle \\
&= \sum_{k,l} \left| \left\langle \sum_{i=1}^{n} (A_i \otimes I) B_i x \otimes y, x_k \otimes y_l \right\rangle \right|^2 \geq 0.
\end{aligned}
$$

Therefore, a^* is a channel. Since the composition of completely positive maps is completely positive, the mapping Λ is completely positive, that is, Λ^* is a channel.

(Example 7.14) This comes from Theorem 7.15. □

Theorem 7.15 *A mapping Γ_λ in $\mathbf{B}(\mathbb{C}^2)$ is completely positive and trace-preserving if and only if it has the form*

$$
\Gamma_\lambda^* \rho = U\big[\Lambda_\lambda^*(V \rho V^*)\big] U^*
$$

where U and V are unitary operators in $\mathbb{C}^2$ and λ_i are real numbers, $|\lambda_i| \leq 1, i = 1, 2, 3,$ *which satisfy the following inequalities*:

$$
(\lambda_1 + \lambda_2)^2 \leq (1 + \lambda_3)^2, \qquad (\lambda_1 + \lambda_2)^2 \leq (1 + \lambda_3)^2,
$$

$$
\big(1 - \lambda_1^2 - \lambda_2^2 - \lambda_3^2\big)^2 \geq 4\big(\lambda_1^2 \lambda_2^2 + \lambda_2^2 \lambda_3^2 + \lambda_3^2 \lambda_1^2 - 2\lambda_1 \lambda_2 \lambda_3\big).
$$

The theorem is proved in [669].

7.4.1 von Neumann Entropy for the Qubit State

Any density operator in $\mathbb{C}^2$ has the form

$$
\rho = \frac{1}{2}(I + a \cdot \sigma).
$$

Its eigenvalues are $1 \pm a$, where $a = (a_1^2 + a_2^2 + a_3^2)^{1/2}$, $0 \leq a \leq 1$, and the von Neumann entropy is

$$
S(\rho) = -\operatorname{tr} \rho \log \rho = F(a),
$$

where

$$
F(a) = -\frac{1}{2}(1 + a) \log \frac{1}{2}(1 + a) - \frac{1}{2}(1 - a) \log \frac{1}{2}(1 - a).
$$

Note that $F(a)$ is a decreasing function for $0 < a < 1$ since

$$F'(a) = \frac{1}{2}\log\left(1 - \frac{2a}{1+a}\right) < 0.$$

For the channel

$$\Lambda_\lambda^*(\rho) = \Lambda_\lambda^*\left(\frac{1}{2}(I + a \cdot \sigma)\right) = \frac{1}{2}(I + b \cdot \sigma),$$

$b_i = \lambda_i a_i$, $|\lambda_i| \le 1$, $i = 1, 2, 3$, we get

$$S\big(\Lambda_\lambda^*(\rho)\big) = F(b).$$

We obtain the entropy increase

$$S(\rho) \le S\big(\Lambda_\lambda^*(\rho)\big)$$

because $F(a)$ is a decreasing function, $F(a) \le F(b)$ if $b \le a$.The last inequality is valid since $|\lambda_i| \le 1, i = 1, 2, 3$.

7.5 Quantum Mutual Entropy

Quantum relative entropy was introduced by Umegaki and generalized by Araki and Uhlmann. Then a quantum analogue of Shannon‘s mutual entropy was considered by Levitin, Holevo and Ingarden for the classical input and output passing through a possibly quantum channel, in which case, as discussed below, the Shannon theory is essentially applied. Thus we call such quantum mutual entropy semi-quantum mutual entropy in the sequel. The fully quantum mutual entropy, namely, for the quantum input and output with a quantum channel, was introduced by Ohya, which is called the *quantum mutual entropy*. It was generalized to a general quantum system described by a C^*-algebra.

The quantum mutual entropy clearly contains the semi-quantum mutual entropy as shown below. We mainly discuss the quantum mutual entropy in a usual quantum system described by a Hilbert space, and its generalization to C^*-systems will be explained briefly for future use (e.g., relativistic quantum information) in the last section of this chapter. Note that the general mutual entropy contains all other cases including the measure theoretic definition of Gelfand and Yaglom.

Let $\mathcal{H}$ be a Hilbert space for the input space, andthe output space is described by another Hilbert space $\tilde{\mathcal{H}}$, one often takes $\mathcal{H} = \tilde{\mathcal{H}}$. A channel from the input system to the output system is a mapping Λ^* from $\mathfrak{S}(\mathcal{H})$ to $\mathfrak{S}(\tilde{\mathcal{H}})$.

An input state $\rho \in \mathfrak{S}(\mathcal{H})$ is sent to the output system through a channel Λ^*, so that the output state is written as $\tilde{\rho} \equiv \Lambda^*\rho$. Then it is important to investigate how much information of ρ is correctly sent to the output state $\Lambda^*\rho$. This amount of information transmitted from input to output is expressed by the mutual entropy (or mutual information).

The quantum mutual entropy was introduced on the basis of von Neumann entropy for purely quantum communication processes. The mutual entropy depends on an input state ρ and a channel Λ^*, so it is denoted by $I(\rho; \Lambda^*)$, which should satisfy the following conditions:

1. The quantum mutual entropy is well-matched to the von Neumann entropy, that is, if a channel is trivial, i.e., $\Lambda^* =$ identity map, then the mutual entropy equals to the von Neumann entropy: $I(\rho; \mathrm{id}) = S(\rho)$.
2. When the system is classical, the quantum mutual entropy reduces to the classical one.
3. Shannon's type fundamental inequality $0 \le I(\rho; \Lambda^*) \le S(\rho)$ holds.

In order to define the quantum mutual entropy, we need the quantum relative entropy and the joint state (it is called a "compound state" in the sequel) describing the correlation between an input state ρ and the output state $\Lambda^*\rho$ through a channel Λ^*. A finite partition of Ω in the classical case corresponds to an orthogonal decomposition $\{E_k\}$ of the identity operator I of $\mathcal{H}$ in the quantum case because the set of all orthogonal projections is considered to have an event system in a quantum system as discussed in Chap. 5. It is known that the following equality holds

$$\sup\left\{-\sum_k \operatorname{tr}\rho E_k \log \operatorname{tr}\rho E_k; \{E_k\}\right\} = -\operatorname{tr}\rho\log\rho,$$

and the supremum is attained when $\{E_k\}$ is a Schatten decomposition of $\rho = \sum_k \mu_k E_k$. Therefore, the Schatten decomposition is used to define the compound state and the quantum mutual entropy.

The compound state Φ_E (corresponding to a joint state in classical systems) of ρ and $\Lambda^*\rho$ was introduced by Ohya in 1983. It is given by

$$\Phi_E = \sum_k \mu_k E_k \otimes \Lambda^* E_k,$$

where E stands for a Schatten decomposition $\{E_k\}$ of ρ so that the compound state depends on how we decompose the state ρ into basic states (elementary events), in other words, how we see the input state. It is easy to see that $\operatorname{tr}\Phi_E = 1$, $\Phi_E > 0$.

Applying the relative entropy $S(\cdot,\cdot)$ to two compound states Φ_E and $\Phi_0 \equiv \rho \otimes \Lambda^*\rho$ (the former includes a certain correlation of input and output and the later does not), we can define the *quantum mutual entropy* (*information*) as

$$I(\rho; \Lambda^*) = \sup\bigl\{S(\Phi_E, \Phi_0); E = \{E_k\}\bigr\},$$

where the supremum is taken over all Schatten decompositions of ρ because this decomposition is not always unique unless every eigenvalue of ρ is not degenerated. Some computations reduce it to the following form for a linear channel.

Theorem 7.16 *We have*

$$I(\rho; \Lambda^*) = \sup\left\{\sum_k \mu_k S(\Lambda^* E_k, \Lambda^*\rho); E = \{E_k\}\right\}.$$

Proof Let $\{x_n\}$ be a CONS of $\mathcal{H}_1$ containing all eigenvectors of ρ. Namely, we put $\rho = \sum_n \mu_k E_n$ and $E_n = |x_n\rangle\langle x_n|$ for every n. Then $\rho E_n = \mu_n E_n$ if x_n is an eigenvector of ρ, and $\rho E_n = 0$ otherwise. We, in addition, put $\{y_k^{(n)}\}$ be a CONS of $\mathcal{H}_2$ which diagonalizes $\Lambda^* E_n$, that is, $\Lambda^* E_n = \sum_k \nu_k^{(n)} |y_k^{(n)}\rangle\langle y_k^{(n)}|$. Then $\{x_n \otimes y_k^{(n)}\}$ becomes a CONS of the tensor product Hilbert space $\mathcal{H}_1 \otimes \mathcal{H}_2$.

By use of this CONS, we can expand Φ_E as $\Phi_E = \sum_n \sum_k \mu_n \nu_k^{(n)} |x_n \otimes y_k^{(n)}\rangle\langle x_n \otimes y_k^{(n)}|$.

Thus we obtain

$$\begin{aligned}
\operatorname{tr} \Phi_E \log \Phi_E &= \sum_n \sum_k \mu_n \nu_k^{(n)} \log\bigl(\mu_n \nu_k^{(n)}\bigr) \\
&= \sum_n \sum_k \mu_n \nu_k^{(n)} \bigl(\log \mu_n + \log \nu_k^{(n)}\bigr) \\
&= \sum_n \mu_n \log \mu_n + \sum_n \mu_n \sum_k \nu_k^{(n)} \log \nu_k^{(n)} \\
&= -S(\rho) - \sum_n \mu_n S(\Lambda^* E_n)
\end{aligned}$$

and

$$\begin{aligned}
\operatorname{tr} \Phi_E \log \Phi_0 &= \operatorname{tr}\Bigl(\sum_n (\mu_n E_n \otimes \Lambda^* E_n) \log(\rho \otimes \Lambda^* \rho)\Bigr) \\
&= \sum_n \mu_n \operatorname{tr}\bigl((E_n \otimes \Lambda^* E_n)(\log \rho \otimes \mathbf{I} + \log \mathbf{I} \otimes \Lambda^* \rho)\bigr) \\
&= -S(\rho) - S(\Lambda^* \rho),
\end{aligned}$$

where we have used the affinity of the channel. Finally, we obtain the relation

$$\begin{aligned}
S(\Phi_E; \Phi_0) &= S(\Lambda^* \rho) - \sum_n \mu_n S(\Lambda^* E_n) \\
&= \sum_n \mu_n S(\Lambda^* E_n; \Lambda^* \rho).
\end{aligned}$$

□

It is easy to see that the quantum mutual entropy satisfies conditions (1)–(3) mentioned above.

When the input system is classical, an input state ρ is given by a probability distribution, or a probability measure. In either case, the Schatten decomposition of ρ is unique, namely, for the case of a probability distribution $\rho = \{\mu_k\}$,

$$\rho = \sum_k \mu_k \delta_k,$$

where δ_k is the delta measure, that is,

$$\delta_k(j) = \delta_{k,j} = \begin{cases} 1 & (k = j), \\ 0 & (k \neq j), \end{cases} \quad \forall j.$$

Therefore, for any channel Λ^*, the mutual entropy becomes

$$I(\rho; \Lambda^*) = \sum_k \mu_k S(\Lambda^* \delta_k, \Lambda^* \rho),$$

which equals to the following usual expression when one of the two terms is finite for an infinite-dimensional Hilbert space:

$$I(\rho; \Lambda^*) = S(\Lambda^* \rho) - \sum_k \mu_k S(\Lambda^* \delta_k).$$

The above equality has been taken by Holevo and Levitin (HL for short in the sequel) as the one associated with a classical–quantum channel. Thus the Ohya's quantum mutual entropy (we call it the quantum mutual entropy in the sequel) contains the HL quantum mutual entropy (we call it the semi-quantum mutual entropy in the sequel) as a special one. We will elaborate more on this in the next chapter.

Note that the definition of the quantum mutual entropy might be written as

$$I_F(\rho; \Lambda^*) = \sup\left\{ \sum_k \mu_k S(\Lambda^* \rho_k, \Lambda^* \rho); \rho = \sum_k \mu_k \rho_k \in F(\rho) \right\},$$

where $F(\rho)$ is the set of all orthogonal finite decompositions of ρ. Here ρ_k being orthogonal to ρ_j (denoted by $\rho_k \perp \rho_j$) means that the range of ρ_k is orthogonal to that of ρ_j. We state this equality in the next theorem.

Theorem 7.17 *One has* $I(\rho; \Lambda^*) = I_F(\rho; \Lambda^*)$.

Proof The inequality $I(\rho; \Lambda^*) \le I_F(\rho; \Lambda^*)$ is obvious. Let us prove the converse. Each ρ_k in an orthogonal decomposition of ρ is further decomposed into one-dimensional projections $\rho_k = \sum_j \mu_j^{(k)} E_j^{(k)}$, a Schatten decomposition of ρ_k. From the following equalities of the relative entropy proved by Araki: (1) $S(a\rho, b\sigma) = aS(\rho, \sigma) - a \log \frac{b}{a}$, for any positive number a, b; (2) $\rho_1 \perp \rho_2 \to S(\rho_1 + \rho_2, \sigma) = S(\rho_1, \sigma) + S(\rho_2, \sigma)$, we have

$$\begin{aligned} \sum_k \mu_k S(\Lambda^* \rho_k, \Lambda^* \rho) &= \sum_{k,j} \mu_k \mu_j^{(k)} S\big(\Lambda^* E_j^{(k)}, \Lambda^* \rho\big) + \sum_{k,j} \mu_k \mu_j^{(k)} \log \mu_j^{(k)} \\ &\le \sum_{k,j} \mu_k \mu_j^{(k)} S\big(\Lambda^* E_j^{(k)}, \Lambda^* \rho\big), \end{aligned}$$

which implies the converse inequality $I(\rho; \Lambda^*) \ge I_F(\rho; \Lambda^*)$ because $\sum_{k,j} \mu_k \mu_j^{(k)} \times E_j^{(k)}$ is a Schatten decomposition of ρ. Thus $I(\rho; \Lambda^*) = I_F(\rho; \Lambda^*)$. □

Moreover, the following fundamental inequality follows from the monotonicity of relative entropy:

Theorem 7.18 (Shannon's Inequality)

$$0 \le I(\rho; \Lambda^*) \le \min\{S(\rho), S(\Lambda^*\rho)\}.$$

Proof $S(\Phi_E; \Phi_0) \ge 0$ implies $I(\rho; \Lambda^*) \ge 0$. When $S(\Lambda^*\rho)$ is finite,

$$\begin{aligned} S(\Phi_E; \Phi_0) &= \sum_n \mu_n S(\Lambda^* E_n; \Lambda^*\rho) \\ &= S(\Lambda^*\rho) - \sum_n \mu_n S(\Lambda^* E_n) \le S(\Lambda^*\rho), \end{aligned}$$

which implies $I(\rho; \Lambda^*) \le S(\Lambda^*\rho)$. When $S(\Lambda^*\rho)$ is infinite, the inequality $I(\rho; \Lambda^*) \le S(\Lambda^*\rho) = +\infty$ is obvious. Since Λ^* is of a Schwarz type, we have

$$\begin{aligned} S(\Phi_E; \Phi_0) &= \sum_n \mu_n S(\Lambda^* E_n; \Lambda^*\rho) \le \sum_n \mu_n S(E_n; \rho) \\ &= \sum_n \mu_n (\operatorname{tr} E_n \log E_n - \operatorname{tr} E_n \log \rho) = -\operatorname{tr} \rho \log \rho = S(\rho). \end{aligned}$$

Taking the supremum over E, we get $I(\rho; \Lambda^*) \le S(\rho)$. □

For given two channels Λ_1^* and Λ_2^*, one has the *quantum data processing inequality* which is

$$S(\rho) \ge I(\rho, \Lambda_1^*) \ge I(\rho, \Lambda_2^* \circ \Lambda_1^*).$$

The second inequality follows from monotonicity of the relative entropy.

This is analogous to the classical data processing inequality for a Markov process $X \to Y \to Z$:

$$S(X) \ge I(X, Y) \ge I(X, Z)$$

where $I(X, Y)$ is the mutual information between random variables X and Y.

The mutual entropy is a measure for not only information transmission but also description of state change, so that this quantity can be applied to several topics in quantum dynamics. It can be also applied to some topics on a quantum computer or in computation to see the ability of information transmission. These applications are discussed in Chaps. 8 and 9.

7.5.1 Generalized (or Quasi) Quantum Mutual Entropy

Let us discuss a generalized (or quasi) quantum mutual entropy here. This mutual entropy is defined for given two states ρ and σ without a channel. For given states

ρ and σ defined on Hilbert spaces $\mathcal{H}$ and $\mathcal{K}$, respectively, a compound state Φ of these ρ and σ is a state satisfying the following marginal conditions: $\mathrm{tr}_{\mathcal{K}}\,\Phi = \rho$ and $\mathrm{tr}_{\mathcal{H}}\,\Phi = \sigma$. Such a (quasi) compound state is not unique, we denote the set of all compound states by $\mathcal{C}(\rho, \sigma)$.

Definition 7.19

1. For given states ρ and σ, the generalized quantum mutual entropy for a fixed a compound state Φ of ρ and σ is

$$I(\rho, \sigma; \Phi) \equiv S(\Phi, \Phi_0),$$

   where $\Phi_0 = \rho \otimes \sigma$.
2. For given states ρ and σ, the generalized quantum mutual entropy is $I(\rho, \sigma) \equiv \sup\{I(\rho, \sigma; \Phi); \Phi \in \mathcal{C}(\rho, \sigma)\}$.

The generalized quantum mutual entropy can be used to define a measure to characterize the entangled states, which will be discussed in Chap. 8.

7.5.2 Ergodic Type Theorem

We prove one ergodic type theorem for quantum mutual entropy.

Theorem 7.20 *Let a state φ be given by $\varphi(\cdot) = \mathrm{tr}\,\rho\cdot$. Then*

1. *If a channel Λ^* is deterministic, then $I(\rho; \Lambda^*) = S(\rho)$.*
2. *If a channel Λ^* is chaotic, then $I(\rho; \Lambda^*) = 0$.*
3. *If ρ is a faithful state and the every eigenvalue of ρ is non-degenerate, then $I(\rho; \Lambda^*) = S(\Lambda^*\rho)$.*

   (Here ρ is said to be faithful if $\mathrm{tr}\,\rho A^*A = 0$ *implies* $A = 0$.*)*

Proof (1) Since Λ^* is deterministic, for each elementary event E_k of ρ, $\Lambda^* E_k$ is a pure state and $\Lambda^*\rho = \sum_k \mu_k \Lambda^* E_k$ is an extremal decomposition of $\Lambda^*\rho$ into pure states. Hence, $S(\Lambda^*\rho) = S(\rho) = -\sum_k \mu_k \log \mu_k$. We have $S(\sigma_E, \sigma_0) = S(\Lambda^*\rho)$ because $\Lambda^* E_k$ is pure (so $S(\Lambda^* E_k) = 0$).

Therefore, we get $I(\rho; \Lambda^*) = S(\rho)$.

(2) Since Λ^* is chaotic, the compound state σ_E is equivalent to σ_0. Indeed

$$\begin{aligned}\sigma_E &= \sum_k \mu_k E_k \otimes \Lambda^* E_k \\ &= \sum_k \mu_k E_k \otimes \Lambda^* \rho \\ &= \sigma_0.\end{aligned}$$

Thus we get $I(\rho; \Lambda^*) = \sup_E S(\sigma_E; \sigma_0) = S(\sigma_0; \sigma_0) = 0$.

(3) When ρ is a faithful state and every eigenvalue of ρ is non-degenerate, then $\Lambda^* E_k$ is a pure state. From this fact, we have

$$I(\rho; \Lambda^*) = \sup_E S(\sigma_E; \sigma_0) = S(\Lambda^* \rho)$$

because of $S(\Lambda^* E_k) = 0$. □

7.6 Entropy Exchange and Coherent Information

First, we define the entropy exchange. If a quantum operation Λ^* is represented as

$$\Lambda^*(\rho) = \sum_i A_i \rho A_i^*, \quad \sum_i A_i^* A_i \leq 1$$

then the *entropy exchange* of the quantum operation Λ^* with input state ρ is defined to be

$$S_e(\rho, \Lambda^*) = S(W) = -\operatorname{tr}(W \log W),$$

where the entries of the matrix W are

$$W_{ij} = \frac{\operatorname{tr}(A_i \rho A_j^*)}{\operatorname{tr}(\Lambda^*(\rho))}.$$

Remark that if $\sum_i A_i^* A_i = 1$ holds, then the quantum operation Λ^* is a channel.

Definition 7.21 The *coherent information* is defined by

$$I_c(\rho, \Lambda^*) = S\left(\frac{\Lambda^*(\rho)}{\operatorname{tr}(\Lambda^*(\rho))}\right) - S_e(\rho, \Lambda^*).$$

Let ρ be a quantum state and Λ_1^*, Λ_2^* be trace-preserving quantum operations. Then

$$S(\rho) \geq I_c(\rho, \Lambda_1^*) \geq I_c(\rho, \Lambda_2^* \circ \Lambda_1^*),$$

which is a similar property to that of quantum mutual entropy.

Another entropy is defined by this coherent information with the von Neumann entropy $S(\rho)$, namely,

$$I_{CM}(\sigma, \Lambda^*) \equiv S(\sigma) + S(\Lambda^* \sigma) - S_e(\rho, \Lambda^*).$$

We call this mutual type information the *coherent mutual entropy* here.

However, this coherent information cannot be considered as a candidate of the mutual entropy for communication due to a theorem of the next section.

7.7 Comparison of Various Quantum Mutual Type Entropies

There exist several different kinds of information similar to mutual entropy. We compare these mutual type entropies.

Let $\{x_n\}$ be a CONS in the input Hilbert space $\mathcal{H}_1$, and let a quantum channel Λ^* be given by

$$\Lambda^*(\bullet) \equiv \sum_n A_n \bullet A_n,$$

where $A_n \equiv |x_n\rangle\langle x_n|$ is a one-dimensional projection satisfying

$$\sum_n A_n = I.$$

Theorem 7.22 *When* $\{A_j\}$ *is a projection-valued measure and* $\dim(\operatorname{ran} A_j) = 1$, *for an arbitrary state* ρ *we have*

1. $I(\rho, \Lambda^*) \le \min\{S(\rho), S(\Lambda^*\rho)\}$
2. $I_C(\rho, \Lambda^*) = 0$
3. $I_{CM}(\rho, \Lambda^*) = S(\rho)$.

Proof For any density operator $\rho \in \mathfrak{S}(\mathcal{H}_1)$ and the channel Λ^* given above, we have $W_{ij} = \operatorname{tr} A_i \rho A_j = \delta_{ij}\langle x_i, \rho x_j\rangle = \langle x_i, \rho x_i\rangle = W_{ii}$ so that one has

$$W = \begin{pmatrix} \langle x_1, \rho x_1\rangle & & 0 \\ & \ddots & \\ 0 & & \langle x_N, \rho x_N\rangle \end{pmatrix} = \sum_n \langle x_n, \rho x_n\rangle A_n.$$

Then the entropy exchange of ρ with respect to the quantum channel Λ^* is

$$S_e(\rho, \Lambda^*) \equiv S(W) = -\sum_n \langle x_n, \rho x_n\rangle \log\langle x_n, \rho x_n\rangle.$$

Since

$$\Lambda^*\rho = \sum_n A_n \rho A_n = \sum_n \langle x_n, \rho x_n\rangle A_n = W,$$

the coherent information of ρ with respect to the quantum channel Λ^* is given as

$$I_C(\rho, \Lambda^*) \equiv S(\Lambda^*\sigma) - S_e(\rho, \Lambda^*) = S(W) - S(W) = 0$$

for any $\rho \in \mathfrak{S}(\mathcal{H}_1)$. The coherent mutual entropy is computed as

$$\begin{aligned} I_{CM}(\rho, \Lambda^*) &\equiv S(\rho) + S(\Lambda^*\sigma) - S_e(\rho, \Lambda^*) \\ &= S(\rho) + I_C(\rho, \Lambda^*) = S(\rho) \end{aligned}$$

for any $\rho \in \mathfrak{S}(\mathcal{H}_1)$.

The quantum mutual entropy becomes

$$I(\rho, \Lambda^*) = \sup\left\{S(\Lambda^*\rho) - \sum_n \mu_n S(\Lambda^* E_n)\right\},$$

where the sup is taken over all Schatten decompositions $\rho = \sum_m \mu_m E_m$, $E_m = |y_m\rangle\langle y_m|$, $\langle y_n, y_m\rangle = \delta_{nm}$.

So we obtain

$$\begin{aligned} I(\rho, \Lambda^*) &= S(\Lambda^*\rho) - \sum_m \mu_m S(\Lambda^* E_m) \\ &= S(\Lambda^*\rho) - \sum_m \mu_m \sum_k \eta(\tau_k^m) \le \min\{S(\rho), S(\Lambda^*\rho)\}, \end{aligned}$$

where $\eta(t) \equiv -t \log t$ and $\tau_k^m \equiv |\langle x_k, y_m\rangle|^2$. This means that $I_o(\rho, \Lambda^*)$ takes various values depending on the input state ρ, for instance,

1. $\sum_m \mu_m \sum_k \eta(\tau_k^m) = 0 \quad \Longrightarrow \quad I(\rho, \Lambda^*) = S(\Lambda^*\rho),$
2. $\sum_m \mu_m \sum_k \eta(\tau_k^m) > 0 \quad \Longrightarrow \quad I(\rho, \Lambda^*) < S(\Lambda^*\rho).$

□

We can further prove that the coherent information vanishes for a general class of channels.

Proposition 7.23 *Let a CONS $\{x_n\}$ be given in the input Hilbert space, and let a sequence of the density operators $\{\rho_n\}$ be given in the output Hilbert space. Consider a channel Λ^* given by*

$$\Lambda^*(\rho) = \sum_n \langle x_n|\rho|x_n\rangle \rho_n$$

where ρ is any state in the input Hilbert space. (One can check that it is a trace-preserving CP map.) Then the coherent information vanishes, that is, $I_C(\rho, \Lambda^*) = 0$ for any state ρ.

Proof Take a spectral representation for ρ_n:

$$\rho_n = \sum_\alpha \lambda_{\alpha n} |\psi_{\alpha n}\rangle\langle\psi_{\alpha n}|.$$

Then one can write a Kraus–Sudarshan representation for the channel Λ^*:

$$\Lambda^*(\rho) = \sum_{n,\alpha} A_{\alpha n} \rho A_{\alpha n}^*$$

where

$$A_{\alpha n} = \lambda_{\alpha n}^{1/2} |\psi_{\alpha n}\rangle\langle x_n|.$$

Now, the coherent information is

$$I_C(\rho, \Lambda^*) \equiv S(\Lambda^*\rho) - S_e(\rho, \Lambda^*),$$

where the entropy exchange is given by

$$S_e(\rho, \Lambda^*) \equiv S(W).$$

Here the operator W is defined as

$$W = \sum_{n,\alpha} |\psi_{\alpha n}\rangle \operatorname{tr}(A_{\alpha n}\rho A_{\alpha n}^*)\langle\psi_{\alpha n}|.$$

But it is equal to $\Lambda^*(\rho)$, i.e.,

$$\Lambda^*(\rho) = W,$$

since

$$\operatorname{tr}(A_{\alpha n}\rho A_{\alpha n}^*) = \lambda_{\alpha n}\langle x_n|\rho|x_n\rangle.$$

Therefore, we obtain

$$I_C(\rho, \Lambda^*) = S(\Lambda^*\rho) - S(W) = 0. \qquad \square$$

Remark 7.24 The channel of the form $\Lambda^*(\rho) = \sum_n \langle x_n|\rho|x_n\rangle\rho_n$ can be considered as the *classical–quantum channel* iff the classical probability distribution $\{p_n = \langle x_n|\rho|x_n\rangle\}$ is given a priori.

7.8 Lifting and Beam Splitting

There exists a special channel named "lifting", and it is a useful concept to characterize quantum communication or stochastic processes. It can be a mathematical tool to describe a process in a quantum algorithm, so that we will explain its foundations here.

Definition 7.25 Let $\mathcal{A}_1, \mathcal{A}_2$ be C^*-algebras and let $\mathcal{A}_1 \otimes \mathcal{A}_2$ be a fixed C^*-tensor product of $\mathcal{A}_1$ and $\mathcal{A}_2$. A lifting from $\mathcal{A}_1$ to $\mathcal{A}_1 \otimes \mathcal{A}_2$ is a weak $*$-continuous map

$$\mathcal{E}^* : \mathfrak{S}(\mathcal{A}_1) \to \mathfrak{S}(\mathcal{A}_1 \otimes \mathcal{A}_2).$$

If $\mathcal{E}^*$ is affine and its dual is a completely positive map, we call it a linear lifting; if it maps pure states into pure states, we call it pure.

The algebra $\mathcal{A}_2$ can be that of the output, namely, $\overline{\mathcal{A}}$ above. Note that to every lifting from $\mathcal{A}_1$ to $\mathcal{A}_1 \otimes \mathcal{A}_2$ we can associate two channels: one from $\mathcal{A}_1$ to $\mathcal{A}_1$, defined by

$$\Lambda^*\rho_1(a_1) \equiv (\mathcal{E}^*\rho_1)(a_1 \otimes 1) \quad \forall a_1 \in \mathcal{A}_1,$$

the other from $\mathcal{A}_1$ to $\mathcal{A}_2$, defined by

$$\Lambda^*\rho_1(a_2) \equiv (\mathcal{E}^*\rho_1)(1 \otimes a_2) \quad \forall a_2 \in \mathcal{A}_2.$$

In general, a state $\varphi \in \mathfrak{S}(\mathcal{A}_1 \otimes \mathcal{A}_2)$ such that

$$\varphi\,|_{\mathcal{A}_1\otimes 1} = \rho_1 \quad \text{and} \quad \varphi\,|_{1\otimes\mathcal{A}_2} = \rho_2$$

is called a compound state of the states $\rho_1 \in \mathfrak{S}(\mathcal{A}_1)$ and $\rho_2 \in \mathfrak{S}(\mathcal{A}_2)$. In the classical probability theory, also the term *coupling between* ρ_1 *and* ρ_2 is used.

The following problem is important in several applications: Given a state $\rho_1 \in \mathfrak{S}(\mathcal{A}_1)$ and a channel $\Lambda^* : \mathfrak{S}(\mathcal{A}_1) \to \mathfrak{S}(\mathcal{A}_2)$, find a standard lifting $\mathcal{E}^* : \mathfrak{S}(\mathcal{A}_1) \to \mathfrak{S}(\mathcal{A}_1 \otimes \mathcal{A}_2)$ such that $\mathcal{E}^*\rho_1$ is a compound state of ρ_1 and $\Lambda^*\rho_1$. Several particular solutions of this problem have been proposed by Ohya, Ceccini and Petz, however, an explicit description of all the possible solutions to this problem is still missing.

Definition 7.26 A lifting from $\mathcal{A}_1$ to $\mathcal{A}_1 \otimes \mathcal{A}_2$ is called non-demolition for a state $\rho_1 \in \mathfrak{S}(\mathcal{A}_1)$ if ρ_1 is invariant for Λ^* i.e., if for all $a_1 \in \mathcal{A}_1$

$$(\mathcal{E}^*\rho_1)(a_1 \otimes 1) = \rho_1(a_1).$$

The idea of this definition being that the interaction with system 2 does not alter the state of system 1.

Definition 7.27 Let $\mathcal{A}_1, \mathcal{A}_2$ be C^*-algebras and let $\mathcal{A}_1 \otimes \mathcal{A}_2$ be a fixed C^*-tensor product of $\mathcal{A}_1$ and $\mathcal{A}_2$. A transition expectation from $\mathcal{A}_1 \otimes \mathcal{A}_2$ to $\mathcal{A}_1$ is a completely positive linear map $\mathcal{E}^* : \mathcal{A}_1 \otimes \mathcal{A}_2 \to \mathcal{A}_1$ satisfying

$$\mathcal{E}^*(1_{\mathcal{A}_1} \otimes 1_{\mathcal{A}_2}) = 1_{\mathcal{A}_1}.$$

An input signal is transmitted and received by an apparatus which produces an output signal. Here $\mathcal{A}_1$ (resp., $\mathcal{A}_2$) is interpreted as the algebra of observables of the input (resp., output) signal and $\mathcal{E}^*$ describes the interaction between the input signal and the receiver as well as the preparation of the receiver. If $\rho_1 \in \mathfrak{S}(\mathcal{A}_1)$ is the input signal, then the state $\Lambda^*\rho_1 \in \mathfrak{S}(\mathcal{A}_2)$ is the state of the (observed) output signal. Therefore, in the reduction dynamics discussed before, the correspondence from a state ρ to the interacting state $\theta_t \equiv U_t(\rho \otimes \rho)U_t^*$ gives us a time-dependent lifting.

Further, another important lifting related to this signal transmission is due to a quantum communication process discussed above. In several important applications, the state ρ_1 of the system before the interaction (preparation, input signal) is not known and one would like to know this state knowing only $\Lambda^*\rho_1 \in \mathfrak{S}(\mathcal{A}_2)$, i.e., the state of the apparatus after the interaction (output signal). From a mathematical point of view, this problem is not well-posed since the map Λ^* is usually not invertible. The best one can do in such cases is to acquire a control on the description of those input states which have the same image under Λ^* and then choose among them according to some statistical criterion.

In the following, we rewrite some communication processes by using liftings.

Example 7.28 (Isometric lifting) Let $V : \mathcal{H}_1 \to \mathcal{H}_1 \otimes \mathcal{H}_2$ be an isometry

$$V^* V = 1_{\mathcal{H}_1}.$$

Then the map

$$\mathcal{E} : x \in \mathbf{B}(\mathcal{H}_1) \otimes \mathbf{B}(\mathcal{H}_2) \to V^* x V \in \mathbf{B}(\mathcal{H}_1)$$

is a transition expectation in the sense of Accardi, and the associated lifting maps a density matrix w_1 in $\mathcal{H}_1$ into $\mathcal{E}^* w_1 = V w_1 V^*$ in $\mathcal{H}_1 \otimes \mathcal{H}_2$. Liftings of this type are called isometric. Every isometric lifting is a pure lifting, which is applied to some quantum algorithms such as Shor's.

Example 7.29 (The attenuation (or beam splitting) lifting) It is a particular isometric lifting characterized by the properties:

$$\mathcal{H}_1 = \mathcal{H}_2 =: \Gamma(\mathbb{C}) \text{ (Fock space over } \mathbb{C}) = L^2(\mathbb{R}),$$
$$V : \Gamma(\mathbb{C}) \to \Gamma(\mathbb{C}) \otimes \Gamma(\mathbb{C})$$

is a map defined by the expression

$$V|\theta\rangle = |\alpha\theta\rangle \otimes |\beta\theta\rangle,$$

where $|\theta\rangle$ is the normalized coherent vector parameterized by $\theta \in \mathbb{C}$ and $\alpha, \beta \in \mathbb{C}$ are such that

$$|\alpha|^2 + |\beta|^2 = 1.$$

Notice that this lifting maps coherent states into products of coherent states. So it maps the simplex of the so-called classical states (i.e., the convex combinations of coherent vectors) into itself. Restricted to these states it is of convex product type explained below, but it is not of convex product type on the set of all states.

Denoting, for $\theta \in \mathbb{C}$, ω_θ the coherent state on $\mathbf{B}(\Gamma(\mathbb{C}))$, namely,

$$\omega_\theta(b) = \langle \theta, b\theta \rangle \quad b \in \mathbf{B}\big(\Gamma(\mathbb{C})\big),$$

for any $b \in \mathbf{B}(\Gamma(\mathbb{C}))$

$$(\mathcal{E}^* \omega_\theta)(b \otimes 1) = \omega_{\alpha\theta}(b),$$

so that this lifting is not non-demolition. These equations mean that, by the effect of the interaction, a coherent signal (beam) $|\theta\rangle$ splits into two signals (beams) still coherent, but of lower intensity, and the total intensity (energy) is preserved by the transformation.

Finally, we mention two important beam splittings which are used to discuss quantum gates and quantum teleportation later.

1. (Superposed beam splitting)

$$V_s|\theta\rangle \equiv V_0\big(|\theta\rangle \otimes |0\rangle - i|0\rangle \otimes |\theta\rangle\big)$$
$$= \frac{1}{\sqrt{2}}\big(|\alpha\theta\rangle \otimes |\beta\theta\rangle - i|\beta\theta\rangle \otimes |\alpha\theta\rangle\big).$$

2. (Beam splitting with two inputs and two outputs) Let $|\theta\rangle$ and $|\gamma\rangle$ be two input coherent vectors. Then

$$V_d\big(|\theta\rangle \otimes |\gamma\rangle\big) \equiv V\big(|\theta\rangle \otimes |\gamma\rangle\big)$$
$$= |\alpha\theta + \beta\gamma\rangle \otimes |-\bar{\beta}\theta + \bar{\alpha}\gamma\rangle,$$

where V_0 and V are given in Sect. 7.4. These extend linearly to an isometry, and their isometric liftings are neither of convex product type nor of non-demolition type.

Example 7.30 (Compound lifting) Let $\Lambda^* : \mathfrak{S}(\mathcal{A}_1) \to \mathfrak{S}(\mathcal{A}_2)$ be a channel. For any $\rho_1 \in \mathfrak{S}(\mathcal{A}_1)$ in the closed convex hull of the external states, fix a decomposition of ρ_1 as a convex combination of extremal states in $\mathfrak{S}(\mathcal{A}_1)$

$$\rho_1 = \int_{\mathfrak{S}(\mathcal{A}_1)} \omega_1 \, d\mu$$

where μ is a Borel measure on $\mathfrak{S}(\mathcal{A}_1)$ with support in the extremal states, and define

$$\mathcal{E}^*\rho_1 \equiv \int_{\mathfrak{S}(\mathcal{A}_1)} \omega_1 \otimes \Lambda^*\omega_1 \, d\mu.$$

Then $\mathcal{E}^* : \mathfrak{S}(\mathcal{A}_1) \to \mathfrak{S}(\mathcal{A}_1 \otimes \mathcal{A}_2)$ is a lifting, nonlinear even if Λ^* is linear, and it is of a non-demolition type. The most general lifting, the mapping $\mathfrak{S}(\mathcal{A}_1)$ into the closed convex hull of the extremal product states on $\mathcal{A}_1 \otimes \mathcal{A}_2$ is essentially of this type. This nonlinear non-demolition lifting was first discussed by Ohya to define the compound state and the mutual entropy as explained before. However, the above is a bit general because we shall weaken the condition that μ is concentrated on the extremal states.

Therefore, once a channel is given, by which a lifting of convex product type can be constructed. For example, the von Neumann quantum measurement process is written, in the terminology of lifting, as follows: Having measured a compact observable $A = \sum_n a_n P_n$ (spectral decomposition with $\sum_n P_n = I$) in a state ρ, the state after this measurement will be

$$\Lambda^*\rho = \sum_n P_n \rho P_n$$

and a lifting $\mathcal{E}^*$ of convex product type associated to this channel Λ^* and to a fixed decomposition of ρ as $\rho = \sum_n \mu_n \rho_n$ ($\rho_n \in \mathfrak{S}(\mathcal{A}_1)$) is given by

$$\mathcal{E}^* \rho = \sum_n \mu_n \rho_n \otimes \Lambda^* \rho_n.$$

Before closing this section, we reconsider a noisy channel, attenuation channel, and amplifier process (lifting) in optical communication.

Attenuation channel. In particular, when $m_1 = 0$ for the noisy channel, Λ^* is called the attenuation channel. We discuss this attenuation channel in the context of the Weyl algebra given in Chap. 4.

Let T be a symplectic transformation of $\mathcal{H}$ to $\mathcal{H} \oplus \mathcal{K}$, i.e., $\sigma(f, g) = \sigma(Tf, Tg)$. Then there is a homomorphism $\alpha_T : \mathrm{CCR}(\mathcal{H}) \to \mathrm{CCR}(\mathcal{H} \oplus \mathcal{K})$ such that

$$\alpha_T\bigl(W(f)\bigr) = W(Tf). \tag{7.1}$$

We may regard the Weyl algebra $\mathrm{CCR}(\mathcal{H} \oplus \mathcal{K})$ as $\mathrm{CCR}(\mathcal{H}) \otimes \mathrm{CCR}(\mathcal{K})$, and given a state ψ on $\mathrm{CCR}(\mathcal{H})$, a channeling transformation arises as

$$(\Lambda^* \omega)(A) = (\omega \otimes \psi)\bigl(\alpha_T(A)\bigr), \tag{7.2}$$

where the input state ω is an arbitrary state of $\mathrm{CCR}(\mathcal{H})$ and $A \in \mathrm{CCR}(\mathcal{H})$ (this ψ is a noise state above). To see a concrete example discussed in [559], we choose $\mathcal{H} = \mathcal{K}$, $\psi = \varphi$ and

$$F(\xi) = a\xi \oplus b\xi. \tag{7.3}$$

If $|a|^2 + |b|^2 = 1$ holds for the numbers a and b, this F is an isometry, and a symplectic transformation, and we arrive at the channeling transformation

$$(\Lambda^* \omega) W(g) = \omega\bigl(W(ag)\bigr) e^{-\frac{1}{2}\|bg\|^2} \quad (g \in \mathcal{H}). \tag{7.4}$$

In order to have an alternative description of Λ^* in terms of density operators acting of $\Gamma(\mathcal{H})$, we introduce the linear operator $V : \Gamma(\mathcal{H}) \to \Gamma(\mathcal{H}) \otimes \Gamma(\mathcal{H})$ defined by

$$V\bigl(\pi_F(A)\bigr)\Phi = \pi_F\bigl(\alpha_T(A)\bigr)\Phi \otimes \Phi.$$

Then we have

$$V\bigl(\pi_F\bigl(W(f)\bigr)\bigr)\Phi = \bigl(\pi_F\bigl(W(af)\bigr) \otimes \pi_F\bigl(W(bf)\bigr)\bigr)\Phi \otimes \Phi,$$

hence

$$V\Phi_f = \Phi_{af} \otimes \Phi_{bf}.$$

Lemma 7.31 *Let ω be a state of* $\mathrm{CCR}(\mathcal{H})$ *which has a density D in the Fock representation. Then the output state $\Lambda^*\omega$ of the attenuation channel has the density* $\mathrm{tr}_2\, V D V^*$ *in the Fock representation.*

Proof Since we work only in the Fock representation, we skip π_F in the formulas. First, we show that

$$V^*\big(W(f)\otimes I\big)V = W(af)e^{-\frac{1}{2}\|bf\|^2} \tag{7.5}$$

for every $f \in \mathcal{H}$. (This can be done by computing the quadratic form of both operators on coherent vectors.) Now we proceed as follows:

$$\begin{aligned}\operatorname{tr}(\operatorname{tr}_2 VDV^*)W(f) &= \operatorname{tr} VDV^*\big(W(f)\otimes I\big)\\ &= \operatorname{tr} DV^*\big(W(f)\otimes I\big)V\\ &= \operatorname{tr} DW(af)e^{-\frac{1}{2}\|bf\|^2}\\ &= \omega\big(W(af)\big)e^{-\frac{1}{2}\|bf\|^2},\end{aligned}$$

which is nothing but $(\Lambda^*\omega)(W(f))$ due to (7.4). □

The lemma says that Λ^* is really the same to the noisy channel with $m = 0$.

We note that Λ, the dual of Λ^*, is a so-called quasi-free completely positive mapping of $\mathrm{CCR}(\mathcal{H})$ given as

$$\Lambda\big(W(f)\big) = W(af)e^{-\frac{1}{2}\|bf\|^2} \tag{7.6}$$

(cf. [194], or Chap. 8 of [647]).

Proposition 7.32 *If ψ is a regular state of* $\mathrm{CCR}(\mathcal{H})$, *that is, $t \mapsto \psi(W(tf))$ is a continuous function on $\mathbb{R}$ for every $f \in \mathcal{H}$, then*

$$(\Lambda^*)^n(\psi) \to \varphi$$

pointwise, where φ is a Fock state.

Proof It is enough to look at the formula

$$\big((\Lambda^*)^n\psi\big)\big(W(f)\big) = \varphi\big(W\big(a^n f\big)\big)\exp -\frac{1}{2}\left(\sum_{k=0}^{n-1}\big\|a^k bf\big\|^2\right),$$

and the statement is proved. □

It is worth noting that the singular state

$$\tau\big(W(f)\big) = \begin{cases} 0, & \text{if } f \neq 0,\\ 1, & \text{if } f = 0\end{cases} \tag{7.7}$$

is an invariant state of $\mathrm{CCR}(\mathcal{H})$. On the other hand, the proposition above applies to states possessing density operator in the Fock representation. Therefore, we have

Corollary 7.33 *Λ^* regarded as a channel of $\mathbf{B}(\Gamma(\mathcal{H}))$ has a unique invariant state, the Fock state, and correspondingly Λ^* is ergodic.*

Λ^* is not only ergodic but it is completely dissipative in the sense that $\Lambda(A^*A) = \Lambda(A^*)\Lambda(A)$ may happen only in the trivial case when A is a multiple of the identity, which was discussed by M. Fannes and A. Verbeure. In fact,

$$\Lambda^* = (\mathrm{id} \otimes \omega)\alpha_T \tag{7.8}$$

where α_T is given by (7.1) and (7.3), and $\omega(W(f)) = \exp(-\|bf\|^2)$ is a quasi-free state.

Amplifier channel. To recover the loss, we need to amplify the signal (photon). In quantum optics, a linear amplifier is usually expressed by means of annihilation operators a and b on $\mathcal{H}$ and $\mathcal{K}$, respectively:

$$c = \sqrt{G}a \otimes I + \sqrt{G-1}I \otimes b^*$$

where G (≥ 1) is a constant and c satisfies CCR (i.e., $[c, c^*] = I$) on $\mathcal{H} \otimes \mathcal{K}$. This expression is not convenient to compute several pieces of information like entropy. The lifting expression of the amplifier is good for such use, and it is given as follows:

Let $c = \mu a \otimes I + \nu I \otimes b^*$ with $|\mu|^2 - |\nu|^2 = 1$ and $|\gamma\rangle$ be the eigenvector of c, i.e., $c|\gamma\rangle = \gamma|\gamma\rangle$. For two coherent vectors $|\theta\rangle$ on $\mathcal{H}$ and $|\theta'\rangle$ on $\mathcal{K}$, $|\gamma\rangle$ can be written by the squeezing expression, $|\gamma\rangle = |\theta \otimes \theta'; \mu, \nu\rangle$, and the lifting is defined by an isometry

$$V_{\theta'}|\theta\rangle = |\theta \otimes \theta'; \mu, \nu\rangle$$

such that

$$\mathcal{E}^*\rho = V_{\theta'}\rho V_{\theta'}^*, \quad \rho \in \mathfrak{S}(\mathcal{H}).$$

The channel of the amplifier is

$$\Lambda^*\rho = \mathrm{tr}_{\mathcal{K}}\,\mathcal{E}^*\rho.$$

7.9 Entropies for General Quantum States

Let us briefly discuss some basic facts of the entropy theory for general quantum systems, which might be needed to treat communication (computation) process from a general point of view, that is, independently from classical or quantum.

Let $(\mathcal{A}, \mathfrak{S}(\mathcal{A}))$ be a C^*-system. The entropy (uncertainty) of a state $\varphi \in \mathcal{S}$ seen from the reference system, a weak $*$-compact convex subset of the whole state space $\mathfrak{S}(\mathcal{A})$ on the C^*-algebra $\mathcal{A}$, was introduced by Ohya. This entropy contains the von Neumann's and classical entropies as special cases.

Every state $\varphi \in \mathcal{S}$ has a maximal measure μ pseudo-supported on $\operatorname{ex}\mathcal{S}$ (extreme points in $\mathcal{S}$) such that

$$\varphi = \int_{\mathcal{S}} \omega \, d\mu.$$

The measure μ giving the above decomposition is not unique unless $\mathcal{S}$ is a Choquet simplex (i.e., for the set $\hat{\mathcal{S}} \equiv \{\lambda\omega; \omega \in \mathcal{S}\}$, define an order such that $\varphi_1 \succeq \varphi_2$ iff $\varphi_1 - \varphi_2 \in \hat{\mathcal{S}}$, $\mathcal{S}$ is a Choquet simplex if $\hat{\mathcal{S}}$ is a lattice for this order), so that we denote the set of all such measures by $M_\varphi(\mathcal{S})$. Take

$$D_\varphi(\mathcal{S}) \equiv \left\{ \mu \in M_\varphi(\mathcal{S}); \exists\{\mu_k\} \subset \mathbb{R}^+ \text{ and } \{\varphi_k\} \subset \operatorname{ex}\mathcal{S} \right.$$
$$\left. \text{s.t. } \sum_k \mu_k = 1, \mu = \sum_k \mu_k \delta(\varphi_k) \right\},$$

where $\delta(\varphi)$ is the delta measure concentrated on $\{\varphi\}$. Put

$$H(\mu) = -\sum_k \mu_k \log \mu_k$$

for a measure $\mu \in D_\varphi(\mathcal{S})$.

Definition 7.34 The entropy of a general state $\varphi \in \mathcal{S}$ w.r.t. $\mathcal{S}$ is defined by

$$S^{\mathcal{S}}(\varphi) = \begin{cases} \inf\{H(\mu); \mu \in D_\varphi(\mathcal{S})\} & (D_\varphi(\mathcal{S}) \neq \emptyset), \\ \infty & (D_\varphi(\mathcal{S}) = \emptyset). \end{cases}$$

When $\mathcal{S}$ is the total space $\mathfrak{S}(\mathcal{A})$, we simply denote $S^{\mathcal{S}}(\varphi)$ by $S(\varphi)$.

This entropy (called the *mixing $\mathcal{S}$-entropy*, or the *C^*-entropy*) of a general state φ satisfies the following properties.

Theorem 7.35 *When $\mathcal{A} = \mathbf{B}(\mathcal{H})$ and $\alpha_t = Ad(U_t)$ (i.e., $\alpha_t(A) = U_t^* A U_t$ for any $A \in \mathcal{A}$) with a unitary operator U_t, for any state φ given by $\varphi(\cdot) = \operatorname{tr}\rho\,\cdot$ with a density operator ρ, the following facts hold*:

1. $S(\varphi) = -\operatorname{tr}\rho\log\rho$.
2. *If φ is an α-invariant faithful state and every eigenvalue of ρ is non-degenerate, then $S^{I(\alpha)}(\varphi) = S(\varphi)$, where $I(\alpha)$ is the set of all α-invariant faithful states.*
3. *If $\varphi \in K(\alpha)$, then $S^{K(\alpha)}(\varphi) = 0$, where $K(\alpha)$is the set of all KMS states.*

Theorem 7.36 *For any $\varphi \in K(\alpha)$, we have*:

1. $S^{K(\alpha)}(\varphi) \le S^{I(\alpha)}(\varphi)$
2. $S^{K(\alpha)}(\varphi) \le S(\varphi)$.

This $\mathcal{S}$ (or mixing) entropy gives a measure of the uncertainty observed from the reference system $\mathcal{S}$ so that it has the following merits: Even if the total entropy

$S(\varphi)$ is infinite, $S^{\mathcal{S}}(\varphi)$ is finite for some $\mathcal{S}$, hence it explains a sort of symmetry breaking in $\mathcal{S}$. Other similar properties of $S(\rho)$ hold for $S^{\mathcal{S}}(\varphi)$. This entropy can be applied to characterize normal states and quantum Markov chains in von Neumann algebras.

The relative entropy for two general states φ and ψ was introduced by Araki and Uhlmann, and their relation is considered by Donald and Hiai et al.

Araki's Definition Let $\mathcal{N}$ be a σ-finite von Neumann algebra acting on a Hilbert space $\mathcal{H}$ and φ, ψ be normal states on $\mathcal{N}$ given by $\varphi(\cdot) = \langle x, \cdot x\rangle$ and $\psi(\cdot) = \langle y, \cdot y\rangle$ with $x, y \in \mathcal{K}$ (a positive natural cone) $\subset \mathcal{H}$. The operator $S_{x,y}$ is defined by

$$S_{x,y}(Ay + z) = s^{\mathcal{N}}(y)A^*x, \quad A \in \mathcal{N}, s^{\mathcal{N}'}(y)z = 0,$$

on the domain $\mathcal{N}y + (I - s^{\mathcal{N}'}(y))\mathcal{H}$, where $s^{\mathcal{N}}(y)$ is the projection from $\mathcal{H}$ to $\{\mathcal{N}'y\}^-$, the $\mathcal{N}$-support of y. Using this $S_{x,y}$, the relative modular operator $\Delta_{x,y}$ is defined as $\Delta_{x,y} = (S_{x,y})^*\overline{S_{x,y}}$, whose spectral decomposition is denoted by $\int_0^\infty \lambda\, de_{x,y}(\lambda)$ ($\overline{S_{x,y}}$ is the closure of $S_{x,y}$). Then the Araki relative entropy is given by

Definition 7.37

$$S(\psi, \varphi) = \begin{cases} \int_0^\infty \log\lambda\, d\langle y, e_{x,y}(\lambda)y\rangle & (\psi \ll \varphi), \\ \infty & \text{otherwise}, \end{cases}$$

where $\psi \ll \varphi$ means that $\varphi(A^*A) = 0$ implies $\psi(A^*A) = 0$ for $A \in \mathcal{N}$.

Uhlmann's Definition Let $\mathcal{L}$ be a complex linear space and p, q be two seminorms on $\mathcal{L}$. Moreover, let $H(\mathcal{L})$ be the set of all positive Hermitian forms α on $\mathcal{L}$ satisfying $|\alpha(x, y)| \le p(x)q(y)$ for all $x, y \in \mathcal{L}$. Then the quadratic mean $QM(p, q)$ of p and q is defined by

$$QM(p, q)(x) = \sup\{\alpha(x, x)^{1/2}; \alpha \in H(\mathcal{L})\}, \quad x \in \mathcal{L}.$$

There exists a family of seminorms $p_t(x)$ of $t \in [0, 1]$ for each $x \in \mathcal{L}$ satisfying the following conditions:

1. For any $x \in \mathcal{L}$, $p_t(x)$ is continuous in t
2. $p_{1/2} = QM(p, q)$
3. $p_{t/2} = QM(p, p_t)$
4. $p_{(t+1)/2} = QM(p_t, q)$.

This seminorm p_t is denoted by $QI_t(p, q)$ and is called the quadratic interpolation from p to q. It is shown that for any positive Hermitian forms α, β, there exists a unique function $QF_t(\alpha, \beta)$ of $t \in [0, 1]$ with values in the set $H(\mathcal{L})$ such that $QF_t(\alpha, \beta)(x, x)^{1/2}$ is the quadratic interpolation from $\alpha(x, x)^{1/2}$ to $\beta(x, x)^{1/2}$. The relative entropy functional $S(\alpha, \beta)(x)$ of α and β is defined as

$$S(\alpha, \beta)(x) = -\liminf_{t\to 0} \frac{1}{t}\{QF_t(\alpha, \beta)(x, x) - \alpha(x, x)\}$$

for $x \in \mathcal{L}$. Let $\mathcal{L}$ be a *-algebra $\mathcal{A}$ and φ, ψ be positive linear functionals on $\mathcal{A}$ defining two Hermitian forms φ^L, ψ^R such as $\varphi^L(A,B) = \varphi(A^*B)$ and $\psi^R(A,B) = \psi(BA^*)$.

Definition 7.38 The Uhlmann relative entropy of φ and ψ is defined by

$$S(\psi, \varphi) = S\bigl(\psi^R, \varphi^L\bigr)(I).$$

Ohya's Definition Next we discuss the mutual entropy in C^*-systems. For any $\varphi \in \mathcal{S} \subset \mathfrak{S}(\mathcal{A})$ and a channel $\Lambda^* : \mathfrak{S}(\mathcal{A}) \to \mathfrak{S}(\overline{\mathcal{A}})$, define the compound states by

$$\Phi_\mu^{\mathcal{S}} = \int_{\mathcal{S}} \omega \otimes \Lambda^* \omega \, d\mu$$

and

$$\Phi_0 = \varphi \otimes \Lambda^* \varphi.$$

The first compound state generalizes the joint probability in classical systems, and it exhibits the correlation between the initial state φ and the final state $\Lambda^*\varphi$.

Definition 7.39 The mutual entropy w.r.t. $\mathcal{S}$ and μ is

$$I_\mu^{\mathcal{S}}(\varphi; \Lambda) = S\bigl(\Phi_\mu^{\mathcal{S}}, \Phi_0\bigr),$$

and the mutual entropy w.r.t. $\mathcal{S}$ is defined as

$$I^{\mathcal{S}}(\varphi; \Lambda^*) = \lim_{\varepsilon \to 0} \sup\bigl\{ I_\mu^{\mathcal{S}}(\varphi; \Lambda^*); \mu \in F_\varphi^\varepsilon(\mathcal{S})\bigr\},$$

where

$$F_\varphi^\varepsilon(\mathcal{S}) = \begin{cases} \{\mu \in D_\varphi(\mathcal{S}); S^{\mathcal{S}}(\varphi) \le H(\mu) \le S^{\mathcal{S}}(\varphi) + \varepsilon < +\infty\}, \\ M_\varphi(\mathcal{S}) \text{ if } S^{\mathcal{S}}(\varphi) = +\infty. \end{cases}$$

The following fundamental inequality is satisfied in almost all physical cases:

$$0 \le I^{\mathcal{S}}(\varphi; \Lambda^*) \le S^{\mathcal{S}}(\varphi).$$

The main properties of the relative entropy and the mutual entropy are shown in the following theorem.

Theorem 7.40

1. (*Positivity*) $S(\varphi, \psi) \ge 0$ and $S(\varphi, \psi) = 0$ iff $\varphi = \psi$.
2. (*Joint convexity*) $S(\lambda\psi_1 + (1-\lambda)\psi_2, \lambda\varphi_1 + (1-\lambda)\varphi_2) \le \lambda S(\psi_1, \varphi_1) + (1-\lambda) S(\psi_2, \varphi_2)$ *for any* $\lambda \in [0, 1]$.
3. (*Additivity*) $S(\psi_1 \otimes \psi_2, \varphi_1 \otimes \varphi_2) = S(\psi_1, \varphi_1) + S(\psi_2, \varphi_2)$.

4. (*Lower semicontinuity*) *If* $\lim_{n\to\infty} \|\psi_n - \psi\| = 0$ *and* $\lim_{n\to\infty} \|\varphi_n \to \varphi\| = 0$, *then* $S(\psi, \varphi) \le \lim_{n\to\infty} \inf S(\psi_n, \varphi_n)$. *Moreover, if there exists a positive number* λ *satisfying* $\psi_n \le \lambda\varphi_n$, *then* $\lim_{n\to\infty} S(\psi_n, \varphi_n) = S(\psi, \varphi)$.
5. (*Monotonicity*) *For a channel* Λ^* *from* $\mathfrak{S}$ *to* $\overline{\mathfrak{S}}$,

$$S(\Lambda^*\psi, \Lambda^*\varphi) \le S(\psi, \varphi).$$

6. (*Lower bound*) $\|\psi - \varphi\|^2/4 \le S(\psi, \varphi)$.

Remark 7.41 This theorem is a generalization of that for the density operators.

Before closing this section, we mention the dynamical entropy introduced by Connes, Narnhofer and Thirring.

Connes–Narnhofer–Thirring Entropy The CNT entropy $H_\varphi(\mathcal{M})$ of a C^*-subalgebra $\mathcal{M} \subset \mathcal{A}$ is defined by

$$H_\varphi(\mathcal{M}) \equiv \sup\left\{\sum_j \mu_j S(\varphi_j \lceil_{\mathcal{M}}, \varphi \lceil_{\mathcal{M}}); \varphi = \sum_j \mu_j \varphi_j\right\},$$

where the supremum is taken over all finite decompositions $\varphi = \sum_j \mu_j \varphi_j$ of φ, and $\varphi \lceil_{\mathcal{M}}$ is the restriction of φ to $\mathcal{M}$. This entropy is the mutual entropy when a channel is the restriction to subalgebra and the decomposition is orthogonal. There are some relations between the mixing entropy $S^{\mathcal{S}}(\varphi)$ and the CNT entropy.

1. For any state φ on a unital C^*-algebra $\mathcal{A}$,

$$S(\varphi) = H_\varphi(\mathcal{A}).$$

2. Let $(\mathcal{A}, G, \alpha)$ with a certain group G be a W^*-dynamical system and let φ be a G-invariant normal state of $\mathcal{A}$, then

$$S^{I(\alpha)}(\varphi) = H_\varphi\big(\mathcal{A}^\alpha\big),$$

where $\mathcal{A}^\alpha$ is the fixed points algebra of $\mathcal{A}$ w.r.t. α.
3. Let $\mathcal{A}$ be the C^*-algebra $\mathbf{C}(\mathcal{H})$ of all compact operators on a Hilbert space $\mathcal{H}$, and G be a group, α be a $*$-automorphic action of G-invariant density operator. Then

$$S^{I(\alpha)}(\rho) = H_\rho\big(\mathcal{A}^\alpha\big).$$

4. There exists a model such that

$$S^{I(\alpha)}(\varphi) > H_\varphi\big(\mathcal{A}^\alpha\big) = 0.$$

7.10 Sufficiency and Relative Entropy

The notion of sufficiency plays an important role in statistics. In classical statistics, the sufficiency is defined as follows:

Let $(\Omega, \mathcal{F}, P(\Omega))$ be a probability space, P be a subset of $P(\Omega)$ and $\mathcal{G}$ be a subalgebra of $\mathcal{F}$. $\mathcal{G}$ is said to be sufficient for P if for any $A \in \mathcal{F}$, there exists a $\mathcal{G}$-measurable function h satisfying the equality

$$h = E_\mu(1_A \mid \mathcal{G}) \quad \mu\text{-a.e.}, \forall \mu \in P.$$

Moreover, $\mathcal{G}$ is said to be pairwise sufficiency for P if $\mathcal{G}$ is sufficient for any pair $\{\mu, \nu\}$ in P.

We have the following important theorem.

Theorem 7.42

1. *If the set P is uniform (i.e., any μ and ν in P are absolutely continuous with respect to each other), $\mathcal{G}$ is sufficient for P iff the Radon–Nikodym derivative $\frac{d\nu}{d\mu}$ for any $\mu, \nu \in P$ is in the set $\mathfrak{M}_{\mathcal{G}}$ of all $\mathcal{G}$-measurable functions in the sense of μ-a.e.*
2. *$\mathcal{G}$ is pairwise sufficiency for P iff $\frac{d\mu}{d(\mu+\nu)} \in \mathfrak{M}_{\mathcal{G}}$ $(\mu + \nu)$-a.e.*

Let D be the set of all measures such that there exists a measure λ satisfying $\mu \ll \lambda$ for any μ in D.

Theorem 7.43

1. *D is a countable subset of $P(\Omega)$.*
2. *G is sufficient for P iff $\mathcal{G}$ is pairwise sufficient for P.*

Kullback–Leibler obtained the following important characterization of sufficiency by means of the classical relative entropy.

Theorem 7.44 *Take μ and ν in $P(\Omega)$.*

1. *If $\mathcal{G}$ is sufficient for $\{\mu, \nu\}$, then $S(\mu, \nu) = S_{\mathcal{G}}(\mu, \nu)$.*
2. *If $S(\mu, \nu) = S_{\mathcal{G}}(\mu, \nu) < +\infty$, then $\mathcal{G}$ is sufficient for $\{\mu, \nu\}$.*

Csiszar proved a theorem concerning the distance between two measures.

Theorem 7.45 *For any μ and ν in $P(\Omega)$, we have $\|\mu - \nu\| \le \sqrt{2S(\mu, \nu)}$.*

The quantum version of the sufficiency was first studied by Umegaki [762, 763] and it is developed by Gudder–Marchard [302], Hiai–Ohya–Tsukada [328, 329] and Petz [646].

Let $\mathfrak{N}$ be a σ-finite von Neumann algebra, $\mathfrak{S}$ be the set of all normal states on $\mathfrak{N}$, $\alpha(R)$ be the set of all strongly continuous one-parameter groups of automorphisms of $\mathfrak{N}$, and let $E_\varphi(\cdot \mid \mathfrak{M})$ be the conditional expectation from $\mathfrak{N}$ to its subalgebra $\mathfrak{M}$.

Definition 7.46

1. $\mathfrak{M}$ is said to be sufficient for a set of $\mathcal{S}$ of $\mathfrak{S}$ if for any $\varphi \in \mathcal{S}$ and any $A \in \mathfrak{N}$, there exists the conditional expectation $E_\varphi(\cdot \mid \mathfrak{M})$ satisfying $A_0 = E_\varphi(A \mid \mathfrak{M})$ φ-a.e., where $B = C$ φ-a.e. means that $\varphi(|B - C|) = 0$.
2. $\mathfrak{M}$ is said to be minimal sufficient for a set of $\mathcal{S}$ if $\mathfrak{M}$ is the smallest subalgebra to be sufficient for $\mathcal{S}$.

One can easily prove following facts:

1. For two states φ, ψ with $\psi \ll \varphi$, $\mathfrak{M}$ is sufficient for $\{\varphi, \psi\}$ iff there exists the conditional expectation $E_\varphi(\cdot \mid \mathfrak{M})$ such that $\psi(A) = \psi(E_\varphi(A \mid \mathfrak{M}))$, $A \in \mathfrak{N}$.
2. If $\mathfrak{M}$ is sufficient for $\{\varphi, \psi\}$, then $\psi = \varphi$ on $\mathfrak{N}$ iff $\psi = \varphi$ on $\mathfrak{M}$.
3. If $\mathcal{S}$ contains a faithful state φ, then $\mathfrak{M}$ is sufficient for $\mathcal{S}$ iff $\mathfrak{M}$ is sufficient for all pairs $\{\varphi, \psi\}$, $\forall \psi \in \mathcal{S}$.
4. If $\mathfrak{M}$ is sufficient for $\mathcal{S}$, then any subalgebra $\mathfrak{M}_0$ containing $\mathfrak{M}$ is sufficient for $\mathcal{S}$ if there exists the conditional expectation $E_\varphi(\cdot \mid \mathfrak{M}_0)$.

Let us define two sets as

$$\mathfrak{N}_\alpha \equiv \left\{A \in \mathfrak{N};\, \alpha_t(A) = A, t \in \mathbb{R}\right\},$$
$$\mathfrak{N}_\varphi \equiv \left\{A \in \mathfrak{N};\, \sigma_t^\varphi(A) = A, t \in \mathbb{R}\right\},$$

where $\{\sigma_t^\varphi; t \in \mathbb{R}\}$ is the modular automorphism group w.r.t. φ.

Then we have important theorems as follows:

Theorem 7.47 *Let φ be a faithful state.*

1. *$\mathfrak{N}_\varphi$ is sufficient for a pair $\{\varphi, \psi\}$ iff $\psi \in I(\sigma^\varphi)$, the set of all invariant states w.r.t. the modular automorphism σ_t^φ $(t \in \mathbb{R})$.*
2. *$\mathfrak{N}_\varphi$ is minimal sufficient for $I(\sigma^\varphi)$.*
3. *If $\mathfrak{N}$ is $\mathbb{R}$-finite, then $\mathfrak{N}_\alpha$ is sufficient for $I(\alpha)$.*
4. *The center $\mathfrak{Z} \equiv \mathfrak{N} \cap \mathfrak{N}'$ is sufficient for $\{\varphi, \psi\}$ iff $\psi \in K(\sigma^\varphi)$, the set of all KMS states w.r.t. σ^φ.*
5. *$\mathfrak{Z}$ is minimal sufficient for $K(\sigma^\varphi)$.*
6. *If $\mathfrak{M}$ is sufficient for $\{\varphi, \psi\}$, then $S(\varphi, \psi) = S_{\mathfrak{M}}(\varphi, \psi)$.*

The Csiszar's inequality is generalized to quantum case.

Theorem 7.48 *For two states φ, ψ in $\mathfrak{S}$, $\|\varphi - \psi\| \le \sqrt{2S(\varphi, \psi)}$.*

Using this theorem and the modular operator, one has

Theorem 7.49 *Let two states φ, ψ be faithful and $\mathfrak{M} \subset \mathfrak{N}_\varphi$. Put $\psi' \equiv \psi \circ E_\varphi(\cdot \mid \mathfrak{M})$. When $S_{\mathfrak{M}}(\psi, \varphi) < +\infty$, one has*

1. *$S(\psi, \psi') = S(\psi, \varphi) - S_{\mathfrak{M}}(\psi, \varphi)$, and*
2. *$\|\psi' - \psi\| \le \sqrt{2S(\psi, \varphi) - S_{\mathfrak{M}}(\psi, \varphi)}$.*

The theorem stated above can be applied to classify the states.

Theorem 7.50

1. *If* $\varphi, \psi \in I(\alpha)$, *then* $S(\varphi, \psi) = S_{\mathfrak{N}_\alpha}(\varphi, \psi)$.
2. *If* $\varphi, \psi \in I(\alpha)$, *then* $S(\varphi, \psi) = S_{\mathfrak{Z}}(\varphi, \psi)$.
3. *For any* $\varphi \in K(\alpha)$ *and* $\psi \in \mathfrak{S}$, $\psi \in I(\alpha)$ *if* $S(\psi, \varphi) = S_{\mathfrak{N}_\alpha}(\psi, \varphi) < +\infty$.
4. *For any* $\varphi \in K(\alpha)$ *and* $\psi \in \mathfrak{S}, \psi \in I(\alpha)$ *if* $S(\psi, \varphi) = S_{\mathfrak{Z}}(\psi, \varphi) < +\infty$.

The concept of sufficiency can be used to find the smallest algebra that is enough to characterize a certain set of states of interest for a particular study. The proofs of theorems of this section can be found in the book [330] and the papers [328, 329].

7.11 Notes

The quantum entropy for a density operator was defined by von Neumann [806] about 20 years before the Shannon entropy appeared. The Schatten decomposition is discussed in [685]. The properties of entropy are summarized in [578, 814]. The quantum relative entropy was first defined by Umegaki [761] and its property was studied by Lindblad [481]. Its generalization to von Neumann algebra was done by Araki [63, 64] and further generalization to $*$-algebra was done by Uhlmann, where the monotonicity was proved. Main properties of the relative entropy are summarized from the articles [63, 64, 206, 328, 329, 481, 578, 646, 760]. Bogoliubov inequality was introduced in [122]. The quantum mutual entropy was introduced by Holevo, Livitin, Ingarden [341, 342, 358, 477] for the classical input and output passing through a possibly quantum channel. Belavkin and Stratonovich studied quantum signals processing by similar classical quantum mutal type of entropy [90]. The fully quantum-mechanical mutual entropy was defined by Ohya [559], and its generalization to C^*-algebra was done in [561]. The proof of the equality $\sup\{-\sum_k \operatorname{tr} \rho E_k \log \operatorname{tr} \rho E_k; \{E_k\}\} = -\operatorname{tr} \rho \log \rho$ can be found in [578]. Theorems 7.17 and 7.18 are proved in [559]. Applications of the mutual entropy can be found in various fields [11, 51, 535, 536, 562, 563, 570, 585]. The mathematical discussion of a channel was given in [559, 570, 587], and the concept of lifting was introduced in [19]. The ergodic type Theorem 7.20 for the quantum mutual entropy was proved by Ohya [570]. The characterization of quantum communication or stochastic processes is discussed in [12, 244], and the beam splitting was rigorously studied by Fichtner, Freutenberg and Liebscher [242–244]. Several particular solutions of the compound states have been proposed in [559, 561, 562]. The transition expectation was introduced by Accardi [6] to study quantum Markov process [8]. Details of the noisy channel were given in [585]. In quantum optics, a linear amplifier has been discussed by several authors [523, 828], and its rigorous expression given here is in [587].

The comparison of several mutual type entropies is studied in [601].

The entropy exchange of a quantum operation and the coherent information were discussed in [110, 716]. Further discussion on these two entropies is given in [84].

The entropy (uncertainty) of a state in a C^*-algebra A was introduced in [561] and its properties are discussed in [51, 359, 535]. For Choquet simplex, we refer to [162]. The relative entropy for two general states was introduced by Araki [63, 64] and Uhlmann [760] and their relation is considered in [206, 328, 329]. The mutual entropy in a C^*-algebra was introduced by Ohya [570]. Theorem 7.40 is a summary of the main results in the papers [63, 64, 206, 328, 329, 646, 760]. Connes–Narnhofer–Thirring entropy was introduced in [176], and the relation between this entropy and the mutual entropy was studied in [535, 536]. Other references of quantum entropy are extensively discussed in the book [578]. Bogoliubov inequality was proved in [120].

Chapter 8
Locality and Entanglement

We will discuss in this chapter various notions of locality and quantum entanglement. Bell's approach to the problem of quantum nonlocality does not include the spatial dependence of entangled states which is crucial for this problem. We will present a new approach, suggested by Volovich, to the problem of quantum nonlocality which is based on the consideration of the spatially depending entangled states and which restores locality.

We pay a special attention to the spatial dependence of the correlation functions and show that the quantum mechanical correlations are asymptotically consistent with the local realistic hidden-variables representation. The dependence on the spatial variables is crucial for the consideration of the problem of locality and entangled states. In non-relativistic quantum mechanics, the wave function of two particles is $\psi = (\psi_{ij}(\mathbf{r}_1, \mathbf{r}_2, t))$ where i and j are spinor indices, t is time and $\mathbf{r}_1$ and $\mathbf{r}_2$ are vectors in a three-dimensional space. Information is physical and, moreover, it is localized in space–time. If one makes local measurements of spin $\sigma \cdot a$ in the spatial region $\mathcal{O}_A$ and spin $\sigma \cdot b$ (a, b are unit vectors in $\mathbb{R}^3$, σ are Pauli matrices) in the region $\mathcal{O}_B$ then the appropriate quantum correlation function has the form

$$\langle\psi|(\sigma \cdot a)P_{\mathcal{O}_A} \otimes (\sigma \cdot b)P_{\mathcal{O}_B}|\psi\rangle.$$

Here $P_{\mathcal{O}_A}$ (resp., $P_{\mathcal{O}_B}$) is the projection operator to the region $\mathcal{O}_A$ (resp., $\mathcal{O}_B$).

The spatial dependence in this quantum mechanical correlation function leads to phenomena of disentanglement when particles separate and to a modification of Bell's theorem about the local realistic representation of quantum correlations.

Separability and entanglement of quantum states are discussed, and their characterization based on various recent studies are presented. It is shown also that the Einstein–Podolsky–Rosen (EPR) model and the Bohm model of entangled states are not equivalent.

Locality, entanglement and quantum field theory are discussed also in Chap. 16.

M. Ohya, I. Volovich, *Mathematical Foundations of Quantum Information and Computation and Its Applications to Nano- and Bio-systems*,
Theoretical and Mathematical Physics,
DOI 10.1007/978-94-007-0171-7_8, © Springer Science+Business Media B.V. 2011

8.1 EPR Model and Bohm Model

Remarkable experimental and theoretical results achieved in quantum optics, quantum computing, teleportation, and cryptography in recent years are based on the previous investigations of fundamental properties of quantum mechanics. Especially important are the properties of non-factorizable entangled states introduced by EPR in 1935 which were named by Schrödinger as the most characteristic feature of quantum mechanics. After the works of Bohm (1951) and Bell (1965), these investigations led to the recent important experimental research in quantum optics.

Though the EPR work dealt with continuous variables, most of the further activity have concentrated almost exclusively on systems of discrete spin variables following Bohm's [124] and Bell's [95] works. It was shown by Khrennikov and Volovich [413, 414] that, in fact, Bohm's formulation is not equivalent to the original EPR model, and there exists a local hidden-variable representation for the EPR correlation function of positions or momenta of two entangled particles.

The discovery of Bell's inequality is one of the important achievements in studying of fundamental problems of quantum theory. Results of EPR and Bell are often interpreted as showing the impossibility of a local realistic representation of quantum mechanics. After Bell discovered the inequality, its variants have appeared. One of the simplest versions is the Clauser–Horn–Shimony–Holt (CHSH) inequality. These variants are not equivalent with each other, however, the important common point is that the violation of any of them implies a nonclassical non-local correlation, one of entanglements.

The precise meaning of the local or non-local correlation we shall discuss in this chapter.

Bell's inequalities do not include an explicit dependence upon the space–time variables though such dependence is crucial for the whole discussion of the problem of locality or non-locality in quantum mechanics. A modification of the Bell framework was proposed by Volovich [790, 793, 797] which includes the dependence on spatial variables and which restores locality.

In this chapter, we discuss the EPR paradox, Bell's inequalities and its variants (e.g., CHSH inequality), and we will discuss them from the point of view of the quantum field theory in Chap. 16, where we stress that the fundamental notion in physics is not the notion of a particle but of a quantum field.

It will be shown in Sect. 8.5 (Local Observations) of this chapter that Bell's inequalities are consistent with the quantum theory if we take into account the spatial dependence of the wave function and locality of detectors separated at a large enough distance.

8.1.1 Various Localities

There are various notions of locality and causality. In classical physics, a cause should always precede its effect. In the special relativity theory, this requirement is strengthened so as to limit causes to the past light cone of the event to be explained;

also an event cannot be a cause of any events outside the former event's future light cone. These restrictions are consistent with the requirement that causal influences cannot travel faster than the speed of light.

Einstein formulated locality in the following way:

"*But on one supposition we should, in my opinion, absolutely hold fast: the real factual situation of the system* S_2 *is independent of what is done with the system* S_1, *which is spatially separated from the former.*"

There are two well known mathematical formulations of local causality in the quantum field theory. They are conditions of local commutativity [698, 736, 811] and Bogolyubov's local causality [122]. Local commutativity means that the commutators (or anti-commutators) of the field operators vanish for space-like distances. There is a similar formulation of local commutativity in the algebraic approach to the quantum field theory.

The Bogolyubov causality condition is formulated in terms of the evolution operator (the S-matrix). Here we formulate it more generally in terms of channel transformations. Our physical system is under action of various external conditions, i.e., under the influence of various classical fields (electromagnetic, gravitational, etc.). Let $G = \{g(x)\}$ be a set of classical fields defined on the Minkowski space $M = \{x = (\mathbf{r}, t) | \mathbf{r} \in \mathbb{R}^3, t \in \mathbb{R}\}$ with the inner product $\langle x, x \rangle = (ct)^2 - |\mathbf{r}|^2$, where c is the speed of light. We suppose that the sets of the classical fields G_1 and G_2 are such that the union of the supports of the fields G_1 is earlier or space-like to the union of the supports of the fields G_2, i.e., $\operatorname{supp} G_1 \lesssim \operatorname{supp} G_2$. Let the channel transformation Λ^* be dependent on the fields G. Then the generalized *Bogolyubov local causality* condition reads:

$$\Lambda^*_{g_1+g_2} = \Lambda^*_{g_2} \Lambda^*_{g_1}, \quad g_1 \in G_1, g_2 \in G_2.$$

If we try to relate these mathematical formulations of local causality with physical description of observations then some questions occur. For instance, there is the Landau and Peierls problem that one cannot speak about the localization of relativistic quantum particles in space but only about their momenta and energies. There is a known question whether we can prepare a relativistic quantum particle in a Newton–Wigner state with a good localization in a bounded region in space. And if so, whether it instantly develops a non-causal behavior, i.e., the probability of finding the particle at any later time arbitrary far away is not zero.

We shall show that, in fact, the notions of chance and probability are not well suited to describe the space–time locally causal behavior of classical and quantum systems.

We consider various mathematical formulations of the notion of locality and local realism, in particular what can be called Bell's locality and local realism, Einstein's local realism, and local realistic representation for quantum correlations.

Many of these discussions are very much linked to the deep examination of the space–time dependence of measurements, namely, the locality of observation. The space–time dependence is not explicitly indicated in many important achievements of the modern quantum information theory. We emphasize the importance of the investigation of quantum information effects in space and time. Transmission and processing of (quantum) information is a physical process in space–time. *Information transmission is physical and, moreover, it is localized.* Therefore, a formulation of the basic notions in the quantum information theory, such as the notions of composite systems, entangled states, and the channel, should include space–time variables [794, 797].

In 1935, Einstein, Podolsky and Rosen (EPR) advanced an argument about incompleteness of quantum mechanics [218]. They proposed a *gedanken* experiment involving a system of two particles spatially separated but correlated in position and momentum and argued that two non-commuting variables (position and momentum of a particle) can have simultaneous physical reality. They concluded that the description of physical reality given by quantum mechanics, which does not permit such a simultaneous reality, is incomplete.

Though the EPR work dealt with continuous variables, most of the further activity have concentrated almost exclusively on systems of discrete spin variables following to Bohm's [124] and Bell's [95] works.

Entangled states, i.e., the states of two particles with the wave function which is not a product of the wave functions of a single particle, have been studied in many theoretical and experimental works starting from the works of Einstein, Podolsky and Rosen, Bohm and Bell.

Bell's theorem [95] states that there are quantum spin correlation functions that cannot be represented as classical correlation functions of separated random variables. It has been interpreted as incompatibility of the requirement of locality with the statistical predictions of quantum mechanics. For a recent discussion of Bell's theorem, see, for example, [169] (see also the references in [791]). It is now widely accepted, as a result of Bell's theorem and related experiments, that "Einstein's local realism" must be rejected.

Let us stress here that there are various interpretations of the notion of locality. Locality in the sense of Bell's representation does not mean locality in the sense of some dependence on spatial variables. There is no explicit dependence on the spatial variables at all in the Bell's representation. "Bell's locality" in his hidden-variable representation means just the factorization of random variables which do not depend on the spatial variables. The spatial dependence of the wave function in quantum mechanics is also neglected in Bell's discussion. The role of locality in the three dimensional space in quantum mechanics and in the hidden variables approach is considered in [790, 791] and will be discussed in this chapter.

8.1.2 Probability and Local Causality

In the quantum field theory, there are well known conditions of local causality (local commutativity and Bogolyubov's causality). However, even for classical systems, the notions of chance and probability are not quite consistent with the notion of causality which is formulated in space–time.

Consider the following approach. Let A be an event localized in a space–time region $\mathcal{O}_A$ (for instance, detection of a particle) and B an event localized in a space–time region $\mathcal{O}_B$. Suppose that the regions $\mathcal{O}_A$ and $\mathcal{O}_B$ are space-like separated.

Namely, for $(t_1, \mathbf{x}_1) \in \mathcal{O}_A$, $(t_2, \mathbf{x}_2) \in \mathcal{O}_B$, one has $|\mathbf{x}_1 - \mathbf{x}_2|/|t_1 - t_2| > c$, where c is the speed of light. Local causality demands that events in $\mathcal{O}_A$ should not be causes of events in $\mathcal{O}_B$, and vice versa. Therefore, one could expect that the conditional probability $P(A|B)$ of the event A in $\mathcal{O}_A$ given the occurrence of event B in $\mathcal{O}_B$ should not depend on the event B, i.e.,

$$P(A|B) = P(A).$$

One could try stating the last formula as the local causality condition. However, it is not true. Indeed, suppose there is only one classical or quantum particle in the space with some distribution, i.e., at every instant of time there is only one particle. We can assume that the probability $P(A)$ of detecting the particle in the region $\mathcal{O}_A$ is $1/2$, $P(A) = 1/2$, and also the probability $P(B) = 1/2$ of detecting the particle in the region $\mathcal{O}_B$. Now, if we detect the particle in the region $\mathcal{O}_B$ then the conditional probability $P(A|B)$ that the particle can be detected in the region $\mathcal{O}_A$ will be just zero, and we obtain $0 = P(A|B) \neq P(A) = 1/2$. That is, the independence in the sense of the space–like separation will not be a correct interpretation of the independency in the probabilistic sense.

It is important to remark that from the above example, the definition of the conditional probability (independency, locality) is very subtle, so that we have to reconsider it carefully. It is interesting to note that we deal here with *the reduction (or collapse) of the distribution function in the classical theory* similar to the infamous collapse of the wave function in quantum mechanics.

Bell's Local Causality

Slightly more elaborated condition was suggested by Bell. We shall see, however, that his condition is also not valid either in the classical or in the quantum theory. First, Bell introduced the notion of "beables" which can be described in classical terms. The beables must include the settings of switches and knobs on experimental equipment, the currents in coils, and the readings of instruments. These beables play a role analogous to the classical fields G in the Bogolyubov approach.

Let N denote a specification of all the beables, of some theory, belonging to the overlap of the backward light cones of space-like separated regions $\mathcal{O}_1$ and $\mathcal{O}_2$. Let K be a specification of some beables from the remainder of the backward light cone of $\mathcal{O}_1$ to N and B of some beables in the region $\mathcal{O}_2$, and let M be a specification of some beables from the remainder of the backward light cone of $\mathcal{O}_2$ to N. *Bell postulates* (J.S. Bell, The theory of local beables. TH-2053-CERN, 28 July 1975) *the following condition of local causality*:

$$P(A|K, M, N, B) = P(A|K, N).$$

The Bell condition means that supplementary information from the region $\mathcal{O}_2$ is redundant.

Classical and quantum theory is not locally causal in the sense of Bell. Indeed, Bell himself noticed that ordinary quantum mechanics, even the relativistic quantum field theory, is not locally causal in the sense of his local causality condition. He considered the following example. Suppose we have a radioactive nucleus which can emit a single α-particle, surrounded by α-particle counters at a considerable distance. So long as it is not specified that some other counter registers, there is a chance for a particular counter that it registers. But if it is specified that some other counter does register, even in a region of space–time outside the relevant backward light cone, the chance that the given counter registers is zero. We simply do not have the above local causality condition.

It is interesting that then, based on this wrong local causality condition, Bell proceeded to derive his inequalities for the correlation functions. The main point in the derivation is the representation of the joint probability distribution in the product form:

$$P(A,B|K,M,N) = P(A|K,N)P(B|M,N).$$

It follows from the Bayes formula

$$P(A,B|K,M,N) = P(A|K,M,N,B)P(B|K,M,N)$$

and the local causality condition above.

8.1.3 EPR Model vs. Bohm and Bell Model

The original EPR system involving continuous variables has been considered by Bell in [96]. He has mentioned that if one admits "measurement" of arbitrary "observables" on arbitrary states than it is easy to mimic his work on spin variables (just take a two-dimensional subspace and define an analogue of spin operators). The problem which he was discussing in [96] is narrower, restricted to measurement of positions only, on two non-interacting spinless particles in free space. Bell used the Wigner distribution [96] approach to quantum mechanics. The original EPR state has a non-negative Wigner distribution. Bell argues that it gives a local, classical model of hidden variables, and therefore the EPR state should not violate local realism. He then considers a state with non-positive Wigner distribution and demonstrates that this state violates local realism.

Bell's proof of violation of local realism in the phase space has been criticized in [390] because of the use of an unnormalizable Wigner distribution. Then in [80] it was demonstrated that the Wigner function of the EPR state, though positive definite, provides evidence of the non-local character of this state if one measures a displaced parity operator.

To the original EPR problem here we apply the method which was used by Bell in his well known paper [95]. He has shown that the correlation function of two spins cannot be represented by classical correlations of separated bounded random variables. This Bell's theorem has been interpreted as incompatibility of local realism with quantum mechanics. We shall show that, in contrast to Bell's theorem for spin correlation functions, the correlation function of positions (or momenta) of two particles always admits a representation in the form of classical correlation of separated random variables. This result looks rather surprising since one often thinks that the Bohm–Bell reformulation of the EPR paradox is equivalent to the original one.

8.1.4 Bell's Locality

Bell has suggested some form of locality which we shall formulate as follows. Suppose that in a Hilbert space $\mathcal{H}$ we have two families of quantum-mechanical observables, say $\{A_i\}$ and $\{B_j\}$, such that $[A_i, B_j] = 0$ for all indices i and j, though operators A_i and B_j not necessary commute among themselves. Then for any density operator ρ in $\mathcal{H}$ there should exist a classical probability space $(\Omega, \mathcal{F}, dP)$ and two families of classical random variables $a_i(\lambda)$ and $b_j(\lambda)$, $\lambda \in \Omega$ such that the quantum correlation functions are equal to the classical correlation functions:

$$\mathrm{tr}(\rho A_i B_j) = \int_\Omega a_i(\lambda) b_j(\lambda)\, dP(\lambda)$$

for all indices i and j. Here $dP(\lambda)$ is the probability measure. It is supposed that the function $a_i(\lambda)$ takes values in the spectrum of A_i, and the function $b_j(\lambda)$ takes values in the spectrum of B_j. Bell's theorem says that there are such ρ, A_i and B_j which cannot be represented in this form.

We call this relation the *Bell locality* because of the following interpretation. Consider an experiment when at apparatus 1 one makes the measurements of the observables A_i, and at apparatus 2 one makes the measurements of the observables B_j, and the two apparatus are (space-like) separated. The parameter λ, Bell argued, represents a state of the system and is called the *hidden variable*. The random variables $a_i(\lambda)$ and $b_j(\lambda)$ describe the outcomes of the measurements at apparatus 1 and 2, respectively. One can call this formula a *local realistic representation in the sense of Bell. The locality here means that the measure $dP(\lambda)$ does not depend on indices i and j*. One can always find a representation

$$\mathrm{tr}(\rho A_i B_j) = \int_\Omega a_i(\lambda) b_j(\lambda)\, dP_{ij}(\lambda),$$

but it is not considered local in the sense of Bell. Remark that adaptive expression of this equality will be discussed later in the chapter.

8.1.5 Discussion of Bell's Locality

In the Bohm's formulation, one considers a pair of spin one-half particles formed in the singlet spin state and moving freely towards two detectors. If one neglects the space part of the wave function then one has the Hilbert space $\mathbb{C}^2 \otimes \mathbb{C}^2$, and the quantum mechanical correlation of two spins in the singlet state $\psi_{\text{spin}} \in \mathbb{C}^2 \otimes \mathbb{C}^2$ is

$$D_{\text{spin}}(a,b) = \langle \psi_{\text{spin}} | \sigma \cdot a \otimes \sigma \cdot b | \psi_{\text{spin}} \rangle = -a \cdot b. \tag{8.1}$$

Here $a = (a_1, a_2, a_3)$ and $b = (b_1, b_2, b_3)$ are two unit vectors in the three-dimensional space $\mathbb{R}^3$, $\sigma = (\sigma_1, \sigma_2, \sigma_3)$ are the Pauli matrices, $\sigma \cdot a = \sum_{i=1}^{3} \sigma_i a_i$, and $\psi_{\text{spin}} = (|01\rangle - |10\rangle)/\sqrt{2}$. Note that $|01\rangle = |0\rangle \otimes |1\rangle$.

Bell's theorem (see Sect. 8.1.8 below) states that the function $D_{\text{spin}}(a,b)$ in (8.1) cannot be represented in the form

$$\int \xi_1(a,\lambda)\xi_2(b,\lambda)\, d\rho(\lambda),$$

i.e.,

$$D_{\text{spin}}(a,b) \neq \int \xi_1(a,\lambda)\xi_2(b,\lambda)\, d\rho(\lambda). \tag{8.2}$$

Here $\xi_1(a,\lambda)$ and $\xi_2(b,\lambda)$ are random fields on the sphere, which satisfy the bound

$$\left|\xi_n(a,\lambda)\right| \leq 1, \quad n = 1, 2, \tag{8.3}$$

and $d\rho(\lambda)$ is a positive probability measure, $\int d\rho(\lambda) = 1$. The parameters λ are interpreted as hidden variables in a realist theory.

The proof of the theorem is based on Bell's (or CHSH) inequalities. We would like to stress that the main point in the proof is actually not the discreteness of classical or quantum spin variables, but the bound (8.3) for classical random fields.

8.1.6 Example of Local Realist Representation for Spins

If we relax the bound, then one can exhibit a locally realistic model which reproduces quantum correlation of two spins. Indeed, let us take the probability space Ω with just 3 points, $\Omega = \{1, 2, 3\}$, and the expectation

$$Ef = \frac{1}{3} \sum_{\lambda=1}^{3} f(\lambda).$$

Let the random fields be

$$\xi_1(a,\lambda) = -\xi_2(a,\lambda) = \sqrt{3}a_\lambda, \quad \lambda = 1, 2, 3.$$

Then one has the relation:

$$-(a,b)=\frac{1}{3}\sum_{\lambda=1}^{3}\sqrt{3}a_\lambda\big(-\sqrt{3}b_\lambda\big)=E\xi_1(a)\xi_2(b).$$

The Bell's theorem (8.1) is not valid in this case because we do not have the bound (8.3). Instead, we have

$$\big|\xi_n(a,\lambda)\big|\le\sqrt{3}.$$

8.1.7 Local Realist Representation for EPR Correlations

Now let us apply Bell's approach employing correlation functions to the original EPR case. The Hilbert space of two 1-dimensional particles is $L^2(\mathbb{R})\otimes L^2(\mathbb{R})$, and the canonical coordinates and momenta are q_1, q_2, p_1, p_2 which obey the commutation relations

$$[q_m,p_n]=\mathrm{i}\delta_{mn},\qquad [q_m,q_n]=0,\qquad [p_m,p_n]=0,\quad m,n=1,2.$$

The EPR paradox can be described as follows. There is such a state ψ of two particles that by measuring p_1 or q_1 of the first particle, we can predict with certainty, and without interacting with the second particle, either the value of p_2 or the value of q_2 of the second particle. In the first case, p_2 is an element of physical reality, in the second such is q_2. Then, these realities must exist in the second particle before any measurement on the first particle since it is assumed that the particles are separated by a space-like interval. However, the realities cannot be described by quantum mechanics because they are incompatible; coordinates and momenta do not commute. So the EPR concludes that quantum mechanics is not complete. Note that the EPR state actually is not a normalized state since it is represented by the delta-function, $\psi=\delta(x_1-x_2-a)$.

An important point in the EPR consideration is that one can choose what we measure: either the value of p_1 or the value of q_1. For a mathematical formulation of a free choice, we introduce canonical transformations of our variables with parameter α:

$$q_n(\alpha)=q_n\cos\alpha-p_n\sin\alpha,$$

$$p_n(\alpha)=q_n\sin\alpha+p_n\cos\alpha;\quad n=1,2.$$

Then one gets

$$\big[q_m(\alpha),p_n(\alpha)\big]=\mathrm{i}\delta_{mn};\quad m,n=1,2.$$

In particular, one has $q_n(0)=q_n$, $q_n(3\pi/2)=p_n$, $n=1,2$.

Now let us consider the correlation function

$$D(\alpha_1,\alpha_2)=\langle\psi|q_1(\alpha_1)\otimes q_2(\alpha_2)|\psi\rangle. \tag{8.4}$$

The correlation function $D(\alpha_1, \alpha_2)$ in (8.4) is an analogue of the Bell's correlation function $D_{\text{spin}}(a, b)$ in (8.1). Bell in [96] has suggested to consider the correlation function of just the free evolutions of the particles at different times (see below).

We shall prove the following local realist representation for the correlation function.

Theorem 8.1 *Quantum-mechanical correlation function in* (8.4) *can be represented in the form*

$$\langle\psi|q_1(\alpha_1)\otimes q_2(\alpha_2)|\psi\rangle = \int \xi_1(\alpha_1,\lambda)\xi_2(\alpha_2,\lambda)\,d\rho(\lambda). \tag{8.5}$$

Here $\xi_n(\alpha_n,\lambda), n = 1, 2$ *are two real functions (random processes), possibly unbounded, and* $d\rho(\lambda)$ *is a probability measure,* $\int d\rho(\lambda) = 1$.

Proof Let us prove that there are required functions $\xi_n(\alpha_n,\lambda)$ for an arbitrary state ψ. We rewrite the correlation function $D(\alpha_1,\alpha_2)$ of (8.4) in the form

$$\begin{aligned}\langle\psi|q_1(\alpha_1)\otimes q_2(\alpha_2)|\psi\rangle &= \langle q_1q_2\rangle\cos\alpha_1\cos\alpha_2 - \langle p_1q_2\rangle\sin\alpha_1\cos\alpha_2\\ &\quad - \langle q_1p_2\rangle\cos\alpha_1\sin\alpha_2 + \langle p_1p_2\rangle\sin\alpha_1\sin\alpha_2.\end{aligned}$$

Here we used the notation

$$\langle q_1q_2\rangle = \langle\psi|q_1q_2|\psi\rangle.$$

Now let us set

$$\begin{aligned}\xi_1(\alpha_1,\lambda) &= f_1(\lambda)\cos\alpha_1 - g_1(\lambda)\sin\alpha_1,\\ \xi_2(\alpha_2,\lambda) &= f_2(\lambda)\cos\alpha_2 - g_2(\lambda)\sin\alpha_2.\end{aligned}$$

Here the real functions $f_n(\lambda), g_n(\lambda),\ n = 1, 2$, are such that

$$\begin{aligned}E(f_1f_2) &= \langle q_1q_2\rangle, \qquad & E(g_1f_2) &= \langle p_1q_2\rangle,\\ E(f_1g_2) &= \langle q_1p_2\rangle, \qquad & E(g_1g_2) &= \langle p_1p_2\rangle.\end{aligned} \tag{8.6}$$

We use for the expectation the notation $E(f_1f_2) = \int f_1(\lambda)f_2(\lambda)\,d\rho(\lambda)$. To solve the system of equations (8.6), we take

$$f_n(\lambda) = \sum_{\mu=1}^{2} F_{n\mu}\eta_\mu(\lambda), \qquad g_n(\lambda) = \sum_{\mu=1}^{2} G_{n\mu}\eta_\mu(\lambda),$$

where $F_{n\mu}$, $G_{n\mu}$ are constants and $E(\eta_\mu\eta_\nu) = \delta_{\mu\nu}$. We denote

$$\langle q_1q_2\rangle = A, \qquad \langle p_1q_2\rangle = B, \qquad \langle q_1p_2\rangle = C, \qquad \langle p_1p_2\rangle = D.$$

A solution of (8.6) may be given, for example, by

$$f_1 = A\eta_1, \qquad f_2 = \eta_1,$$
$$g_1 = B\eta_1 + \left(D - \frac{BC}{A}\right)\eta_2, \qquad g_2 = \frac{C}{A}\eta_1 + \eta_2.$$

Hence the representation of the quantum correlation function in terms of the separated classical random processes (8.5) is proved. □

The condition of reality of the functions $\xi_n(\alpha_n, \lambda)$ is important. It means that the range of $\xi_n(\alpha_n, \lambda)$ is the set of the eigenvalues of the operator $q_n(\alpha_n)$. If we relax this condition, then one can get a local hidden-variable representation just by using an expansion of unity:

$$\begin{aligned} \langle\psi|q_1(\alpha_1)q_2(\alpha_2)|\psi\rangle &= \sum_\lambda \langle\psi|q_1(\alpha_1)|\lambda\rangle\langle\lambda|q_2(\alpha_2)|\psi\rangle \\ &= \sum_\lambda \xi_1(\alpha_1, \lambda)\xi_2(\alpha_2, \lambda), \end{aligned}$$

where

$$\xi_1(\alpha_1, \lambda) = \langle\psi|q_1(\alpha_1)|\lambda\rangle, \qquad \xi_2(\alpha_2, \lambda) = \langle\lambda|q_2(\alpha_2)|\psi\rangle.$$

For a discussion of this point in the context of a noncommutative spectral theory, see [791].

Similarly, one can prove a representation

$$\langle\psi|q_1(t_1) \otimes q_2(t_2)|\psi\rangle = \int \xi_1(t_1, \lambda)\xi_2(t_2, \lambda)\, d\rho(\lambda)$$

where $q_n(t) = q_n + p_n t$, $n = 1, 2$, is a free quantum evolution of the particles. It is enough to take

$$\xi_1(t_1, \lambda) = f_1(\lambda) + g_1(\lambda)t_1, \qquad \xi_2(t_2, \lambda) = f_2(\lambda) + g_2(\lambda)t_2.$$

To summarize, it is shown that, in contrast to the Bell's theorem for the spin or polarization variables, for the original EPR correlation functions which deal with positions and momenta one can get a local (in the sense of Bell) realistic representation in terms of separated random processes. The representation is obtained for any state including entangled states. Therefore, the original EPR model does not lead to quantum non-locality in the sense of Bell even for entangled states. One can get quantum non-locality in the EPR situation only if we (rather artificially) restrict ourself in the measurements with a two-dimensional subspace of the infinite dimensional Hilbert space corresponding to the position or momentum observables.

If we adopt Bell's approach to the local realism (i.e., of using classical separated stochastic processes but without explicit dependence on spatial variables) then one can say that the original EPR model admits a locally realistic description, in contrast to what was expected for the model. It follows also that the phenomena of quantum non-locality in the sense of Bell depends not only on the properties of entangled states but also on particular observables which we want to measure (bounded spin-like or unbounded momentum and position observables). An interrelation of the roles of entangled states and the bounded observables in considerations of local realism and quantum non-locality deserves a further theoretical and experimental study.

8.1.8 On Bell's Theorem

Bell proved [95] that there are quantum spin correlation functions in entangled states that cannot be represented as classical correlation functions of separated random variables. *Bell's theorem* can be formulated as the following inequality, see [790]:

$$\langle\psi_{\text{spin}}|\sigma\cdot a\otimes\sigma\cdot b|\psi_{\text{spin}}\rangle \neq \int_\Omega \xi_a(\lambda)\eta_b(\lambda)\,dP(\lambda),$$

where $\xi_a(\lambda)$ and $\eta_b(\lambda)$ are the functions depending on unit vectors a and b with the bounds $|\xi_a(\lambda)|\leq 1$, $|\eta_b(\lambda)|\leq 1$, and $(\Omega,\mathcal{F},P)$ is a probability space. The inequality here means that there cannot be an equality for all a and b. Another, shorter formulation of the theorem is:

$$\cos(\alpha-\beta)\neq E(\xi_\alpha\eta_\beta),$$

where $\xi_\alpha=\xi_\alpha(\lambda)$ and $\eta_\beta=\eta_\beta(\lambda)$ are two random processes [343] such that $|\xi_\alpha(\lambda)|\leq 1$, $|\eta_\beta(\lambda)|\leq 1$, and E is the expectation,

$$E(\xi_\alpha\eta_\beta)=\int_\Omega \xi_\alpha(\lambda)\eta_\beta(\lambda)\,dP(\lambda).$$

Here the function $\cos(\alpha-\beta)$ describes a quantum-mechanical correlation of spins of two entangled particles with α and β representing the angles of the measured spins since we have

$$\langle\psi_{\text{spin}}|\sigma\cdot a\otimes\sigma\cdot b|\psi_{\text{spin}}\rangle=-a\cdot b=\cos(\alpha-\beta).$$

The proof of the inequality is based on the Bell–CSHS inequality which reads (see the next section)

$$\left|P(\alpha,\beta)-P(\alpha,\beta')\right|+\left|P(\alpha',\beta)+P(\alpha',\beta')\right|\leq 2$$

for arbitrary angles $\alpha,\beta,\alpha',\beta'$. Here $P(\alpha,\beta)=E(\xi_\alpha\eta_\beta)$ is the classical correlation function.

Bell's theorem has been interpreted as incompatibility of the requirement of locality with the statistical predictions of quantum mechanics [95]. For a recent discussion of Bell's theorem and Bell's inequalities, see, for example, [48, 813] and references therein. It is now widely accepted, as a result of Bell's theorem and related experiments, that "local realism" must be rejected.

Let us discuss first the quantum-mechanical correlation function. The correlation function

$$\langle\psi_{\text{spin}}|\sigma\cdot a\otimes\sigma\cdot b|\psi_{\text{spin}}\rangle$$

describes quantum correlations of two spins in the two qubit Hilbert space when the space–time dependence of the wave functions of the particles is neglected. Let us note, however, that the very formulation of the problem of locality in quantum mechanics prescribes a special role to the position in the ordinary three-dimensional space. Therefore, the problem of local in space observations should not be neglected in discussions of the problem of locality in relation to Bell's inequalities.

If we want to speak about locality in the quantum theory then we have to somehow localize our particles. For example, we could approximately measure the density of the energy or the position of the particles simultaneously with the spin. Only then we could come to some conclusions about the relevance of the spin correlation function to the problem of locality. Therefore, one has to consider a *quantum-mechanical correlation function which describes localized observations and one cannot neglect the spatial dependence of the wave function.*

Let us stress that we discuss here not a problem of interpretation of the quantum theory, but a problem of how to make correct quantum-mechanical computations describing an experiment with two detectors localized in space. It was pointed out [790] that if we make *local* observations of spins then the space–time part of the wave function leads to an extra factor in quantum correlations, and as a result the ordinary conclusion from the Bell theorem about the non-locality of the quantum theory fails.

We present a modification of Bell's equation which includes space and time variables. The function $\cos(\alpha-\beta)$ describes the quantum mechanical correlation of spins of two entangled particles if we neglect the space–time dependence of the wave function. It was shown in [790] that if one takes into account the space part of the wave function then the quantum correlation describing local observations of spins in the simplest case will take the form of $g\cos(\alpha-\beta)$ instead of just $\cos(\alpha-\beta)$. Here the parameter g describes the location of the system in space and time. In this case, one gets *a modified Bell's equation*

$$g\cos(\alpha-\beta)=E(\xi_\alpha\eta_\beta).$$

One can prove that if the distance between detectors is large enough then the factor g becomes small and there exists a solution of the modified equation. This result has applications to the security of certain quantum-cryptographic protocols [792, 793]. We will show that, in fact, at large separation between particles all reasonable quantum states become disentangled (factorized).

It is important to study also a more general question of which class of functions $f(s,t)$ admits a representation of the form

$$f(s,t)=E(x_s y_t),$$

where x_s and y_t are bounded stochastic processes, and also an analogous question for the functions of several variables $f(t_1, \dots, t_n)$. Such considerations lead to a *noncommutative* generalization of von Neumann's spectral theorem which will be considered below.

In the previous section, it was mentioned that the vacuum state ω_0 in the free quantum field theory is a nonfactorized (entangled) state for observables belonging to space-like separated regions:

$$\omega_0\big(\varphi(x)\varphi(y)\big) - \omega_0\big(\varphi(x)\big)\omega_0\big(\varphi(y)\big) \neq 0.$$

Here $\varphi(x)$ is a free scalar field in the Minkowski space–time and $(x-y)^2 > 0$. Hence there is a statistical dependence between causally disconnected regions. We will discuss this aspect more in Chap. 16.

However, one has an asymptotic factorization of the vacuum state for large separations of the space-like regions. Moreover, one proves that in the quantum field theory there is an asymptotic factorization for any reasonable state and any local observables. Therefore, at large distances any reasonable state becomes disentangled. We have the relation

$$\lim_{|l|\to\infty}\big[\omega\big(\alpha_l(A)B\big) - \omega\big(\alpha_l(A)\big)\omega(B)\big] = 0.$$

Here ω is a state from a rather wide class of the states which includes entangled states, A and B are two local observables, and $\alpha_l(A)$ is the translation of the observable A along the three-dimensional vector l. As a result, a violation of Bell's inequalities (see below) can be observed without inconsistency with principles of relativistic quantum theory only if the distance between detectors is rather small. We suggest a further experimental study of entangled states in space–time by studying the dependence of the correlation functions on the distance between detectors.

It would be useful if the local algebraic approach to the quantum theory [65, 123, 310] could be developed in this direction.

This fact leads also to important consequences for quantum teleportation and quantum cryptography, see [792].

Bell's theorem constitutes an important part in quantum cryptography [219]. It is now generally accepted that techniques of quantum cryptography, discussed in Chap. 17, can allow secure communications between distant parties [818], see [792] and the references therein. The promise of secure cryptographic quantum key distribution schemes is based on the use of quantum entanglement in the spin space and on quantum no-cloning theorem. An important contribution of quantum cryptography is a mechanism for detecting eavesdropping.

However, in certain current quantum cryptography protocols the space part of the wave function is neglected. But just the space part of the wave function describes the behavior of particles in the ordinary real three-dimensional space. As a result, such schemes can be secure against eavesdropping attacks in the abstract spin space, but could be insecure in the real three-dimensional space. We will discuss how one can try to improve the security of quantum cryptography schemes in space by using a special preparation of the space part of the wave function, see [791].

A apace–time description is important for quantum computation [487]. Some problems of quantum teleportation in space have been discussed in [248].

8.2 Bell's Theorem

8.2.1 Bell's Theorem and Stochastic Processes

In the presentation of Bell's theorem, we will follow [794] where one can also find more references. Bell's theorem reads:

$$\cos(\alpha-\beta)\neq E(\xi_\alpha\eta_\beta), \tag{8.7}$$

where ξ_α and η_β are two random processes such that $|\xi_\alpha|\leq 1$, $|\eta_\beta|\leq 1$, and E is the expectation. In more details,

Theorem 8.2 *There exist no probability space* $(\Omega,\mathcal{F},d\rho(\lambda))$ *and no pair of stochastic processes* $\xi_\alpha=\xi_\alpha(\lambda)$, $\eta_\beta=\eta_\beta(\lambda)$, $0\leq\alpha,\beta\leq 2\pi$ *which obey* $|\xi_\alpha(\lambda)|\leq 1$, $|\eta_\beta(\lambda)|\leq 1$ *such that the following equation is valid*

$$\cos(\alpha-\beta)=E(\xi_\alpha\eta_\beta) \tag{8.8}$$

for all α *and* β.

Before we prove this theorem, we need mathematical preparation. Here Ω is a set, $\mathcal{F}$ is a σ-field of subsets and $d\rho(\lambda)$ is a probability measure, i.e., $d\rho(\lambda)\geq 0$, $\int d\rho(\lambda)=1$. The expectation is

$$E(\xi_\alpha\eta_\beta)=\int_\Omega \xi_\alpha(\lambda)\eta_\beta(\lambda)\,d\rho(\lambda).$$

One can write (8.8) as an integral equation

$$\cos(\alpha-\beta)=\int_\Omega \xi_\alpha(\lambda)\eta_\beta(\lambda)\,d\rho(\lambda). \tag{8.9}$$

We say that the integral equation (8.9) has no solutions $(\Omega,\mathcal{F},d\rho(\lambda),\xi_\alpha,\eta_\beta)$ with the bound $|\xi_\alpha|\leq 1$, $|\eta_\beta|\leq 1$.

We will prove the theorem below. Let us discuss now the physical interpretation of this result.

Consider a pair of spin one-half particles formed in the singlet spin state and moving freely towards two detectors. If one neglects the space part of the wave function then one has the Hilbert space $\mathbb{C}^2 \otimes \mathbb{C}^2$ and the quantum-mechanical correlation of two spins in the singlet state $\psi_{\text{spin}} \in \mathbb{C}^2 \otimes \mathbb{C}^2$ is

$$D_{\text{spin}}(a,b) = \langle \psi_{\text{spin}} | \sigma \cdot a \otimes \sigma \cdot b | \psi_{\text{spin}} \rangle = -a \cdot b. \tag{8.10}$$

Here $a = (a_1, a_2, a_3)$ and $b = (b_1, b_2, b_3)$ are two unit vectors in the three-dimensional space $\mathbb{R}^3$, $\sigma = (\sigma_1, \sigma_2, \sigma_3)$ are the Pauli matrices,

$$\sigma_1 = \begin{pmatrix} 0 & 1 \\ 1 & 0 \end{pmatrix}, \qquad \sigma_2 = \begin{pmatrix} 0 & -\mathrm{i} \\ \mathrm{i} & 0 \end{pmatrix}, \qquad \sigma_3 = \begin{pmatrix} 1 & 0 \\ 0 & -1 \end{pmatrix},$$

$$\sigma \cdot a \equiv \sum_{i=1}^{3} \sigma_i a_i,$$

and

$$\psi_{\text{spin}} = \frac{1}{\sqrt{2}} \left(\begin{pmatrix} 0 \\ 1 \end{pmatrix} \otimes \begin{pmatrix} 1 \\ 0 \end{pmatrix} - \begin{pmatrix} 1 \\ 0 \end{pmatrix} \otimes \begin{pmatrix} 0 \\ 1 \end{pmatrix} \right).$$

If the vectors a and b belong to the same plane then one can write $-a \cdot b = \cos(\alpha - \beta)$, and hence Bell's theorem states that the function $D_{\text{spin}}(a,b)$ in (8.10) cannot be represented in the form

$$P(a,b) = \int \xi(a,\lambda)\eta(b,\lambda)\, d\rho(\lambda), \tag{8.11}$$

i.e.,

$$D_{\text{spin}}(a,b) \neq P(a,b). \tag{8.12}$$

Here $\xi(a,\lambda)$ and $\eta(b,\lambda)$ are random fields on the sphere, $|\xi(a,\lambda)| \leq 1$, $|\eta(b,\lambda)| \leq 1$, and $d\rho(\lambda)$ is a positive probability measure, $\int d\rho(\lambda) = 1$. The parameters λ are interpreted as hidden variables in a realist theory. It is clear that (8.12) can be reduced to (8.7).

8.2.2 CHSH Inequality

To prove Theorem 8.2, we will use the following theorem which is a slightly generalized CHSH result.

Theorem 8.3 *Let f_1, f_2, g_1, and g_2 be random variables (i.e., measurable functions) on the probability space $(\Omega, \mathcal{F}, d\rho(\lambda))$ such that*

$$\left| f_i(\lambda) g_j(\lambda) \right| \leq 1, \quad i,j = 1,2.$$

Let

$$P_{ij} \equiv E(f_i g_j), \quad i,j = 1,2.$$

Then

$$|P_{11} - P_{12}| + |P_{21} + P_{22}| \le 2.$$

Proof One has

$$P_{11} - P_{12} = E(f_1 g_1) - E(f_1 g_2)$$
$$= E\big(f_1 g_1(1 \pm f_2 g_2)\big) - E\big(f_1 g_2(1 \pm f_2 g_1)\big).$$

Hence

$$|P_{11} - P_{12}| \le E(1 \pm f_2 g_2) + E(1 \pm f_2 g_1) = 2 \pm (P_{22} + P_{21}).$$

Now let us note that if x and y are two real numbers then

$$|x| \le 2 \pm y \to |x| + |y| \le 2.$$

Therefore, taking $x = P_{11} - P_{12}$ and $y = P_{22} + P_{21}$, one gets the bound

$$|P_{11} - P_{12}| + |P_{21} + P_{22}| \le 2.$$

The theorem is proved. □

The last inequality is called the *CHSH inequality*. By using notations of (8.11), one has

$$\big|P(a,b) - P(a,b')\big| + \big|P(a',b) + P(a',b')\big| \le 2$$

for any four unit vectors a, b, a', b'. So then, we have to prove Theorem 8.2.

Proof Let us denote

$$f_i(\lambda) = \xi_{\alpha_i}(\lambda), \qquad g_j(\lambda) = \eta_{\beta_j}(\lambda), \quad i, j = 1, 2$$

for some α_i, β_j. If one were to have

$$\cos(\alpha_i - \beta_j) = E(f_i g_j),$$

then, due to Theorem 8.3, one should have

$$\big|\cos(\alpha_1 - \beta_1) - \cos(\alpha_1 - \beta_2)\big| + \big|\cos(\alpha_2 - \beta_1) + \cos(\alpha_2 - \beta_2)\big| \le 2.$$

However, for $\alpha_1 = \pi/2$, $\alpha_2 = 0$, $\beta_1 = \pi/4$, $\beta_2 = -\pi/4$ we obtain

$$\big|\cos(\alpha_1 - \beta_1) - \cos(\alpha_1 - \beta_2)\big| + \big|\cos(\alpha_2 - \beta_1) + \cos(\alpha_2 - \beta_2)\big| = 2\sqrt{2}$$

which is greater than 2. This contradiction proves Theorem 8.2. □

It will be shown below that if one takes into account the space part of the wave function then the quantum correlation in the simplest case will take the form

$g\cos(\alpha-\beta)$ instead of just $\cos(\alpha-\beta)$ where the parameter g describes the location of the system in space and time. In this case, one can get a representation

$$g\cos(\alpha-\beta)=E(\xi_\alpha\eta_\beta)$$

if g is small enough. The factor g gives a contribution to visibility or efficiency of detectors that are used in the phenomenological description of detectors.

8.3 Various Local Realisms

Einstein, Podolsky and Rosen presented an argument to show that there are situations in which the scheme of the quantum theory seems to be incomplete [218]. They proposed a *gedanken* experiment involving a system of two particles spatially separated but correlated in position and momentum, and argued that two non-commuting variables (position and momentum of a particle) can have simultaneous physical reality. They concluded that the description of physical reality given by quantum mechanics is incomplete because it does not permit such a simultaneous reality, due to the uncertainty principle.

Though the EPR work dealt with continuous position and momentum variables, most of the further activity has concentrated almost exclusively on systems of discrete spin variables following Bohm's [124] and Bell's [95] works.

Bell's theorem [95] discussed above says that there are quantum spin correlation functions that cannot be represented as classical correlation functions of separated random variables. It has been interpreted as incompatibility of the requirement of locality with the statistical predictions of quantum mechanics [95].

However, it was shown in [790, 791, 797] that in the derivation of such a conclusion the fundamental fact that space–time exists was neglected. Moreover, if we take into account the spatial dependence of the wave function then the standard formalism of quantum mechanics might be consistent with local realism.

We will give two different definitions of the notions of local realism which we call Bell's and Einstein's local realism. We demonstrate that if we do not neglect the space–time structure in the standard quantum-mechanical formalism then quantum mechanics actually is consistent with local realism. Since detectors of particles are obviously located somewhere in space it shows that loopholes are unavoidable in experiments aimed to establish a violation of Bell's inequalities [82].

8.3.1 Bell's Local Realism

A mathematical formulation of Bell's local realism may be given by the relation (in more details it is discussed in the next section)

$$\langle\psi|A(a)B(b)|\psi\rangle=E\big(\xi(a)\eta(b)\big). \tag{8.13}$$

Here $A(a)$ and $B(b)$ are self-adjoint operators which commute on a natural domain, and a and b are certain indices. Here E is a mathematical expectation, $\xi(a)$ and $\eta(b)$ are two stochastic processes, and ψ is a vector from a Hilbert space. Then we say that the triplet

$$\{A(a), B(b), \psi\}$$

satisfies the *Bell's local realism* (BLR) condition.

Bell proved that a two spin quantum correlation function which is equal to just $-a \cdot b$, where a and b are two three-dimensional vectors, cannot be represented in the form (8.13), i.e.,

$$\langle \psi_{\text{spin}} | \sigma \cdot a \otimes \sigma \cdot b | \psi_{\text{spin}} \rangle \neq E\big(\xi(a)\eta(b)\big), \tag{8.14}$$

if one has the bound $|\xi(a)| \leq 1$, $|\eta(b)| \leq 1$. Here $a = (a_1, a_2, a_3)$ and $b = (b_1, b_2, b_3)$ are two unit vectors in the three-dimensional space $\mathbb{R}^3$, and $\sigma = (\sigma_1, \sigma_2, \sigma_3)$ are the Pauli matrices.

Therefore, the correlation function of two spins does not satisfy the BLR condition (8.13). In this sense, sometimes one speaks about quantum non-locality.

8.3.2 Space and Time in Axioms of Quantum Mechanics

Note, however, that in the previous discussion the space–time parameters were not explicitly involved though one speaks about non-locality. Actually, the "local realism" in the Bell's sense as it was formulated above in (8.13) is a notion which has nothing to do with the notion of locality in the ordinary three-dimensional space. Therefore, we also define another notion which we will call the condition of local realism in the sense of Einstein.

To explain the notion, let us first remind that the usual axiomatic approach to quantum theory involves only the Hilbert space, observable, the density operator ρ, and the von Neumann formula for the probability $P(B)$ of the outcome B, $P(B) = \operatorname{tr} \rho E_B$ where $\{E_B\}$ is POVM associated with a measurable space $(\Omega, \mathcal{F})$, here B belongs to the σ-algebra $\mathcal{F}$. It was stressed in [793] that in a more realistic axiomatic approach to quantum mechanics one has to include an axiom on the existence of space and time. It can be formulated as follows

$$U(d) E_B U(d)^* = E_{\alpha_d(B)}.$$

Here $U(d)$ is the unitary representation of the group of translations in time and in the three-dimensional space, and $\alpha_d : \mathcal{F} \to \mathcal{F}$ is the group of automorphisms.

8.3.3 Einstein's Local Realism

Let two families of self-adjoint operators $\{A(a, \mathcal{O})\}$ and $\{B(b, \mathcal{O})\}$ be given in a Hilbert space $\mathcal{H}$, parameterized by the regions $\mathcal{O}$ in the Minkowski space–time.

Suppose that one has a representation

$$\langle\psi|A(a,\mathcal{O}_1)B(b,\mathcal{O}_2)|\psi\rangle = E\big(\xi(a,\mathcal{O}_1)\eta(b,\mathcal{O}_2)\big)$$

for $a, b, \mathcal{O}_1, \mathcal{O}_2$ for which the operators commute. Then we say that the quadruplet

$$\big\{A(a,\mathcal{O}_1), B(b,\mathcal{O}_2), U(d), \psi\big\}$$

satisfies the *Einstein's local realism* (ELR) condition.

8.3.4 Local Realistic Representation for Quantum Spin Correlations

Quantum correlation describing the localized measurements of spins in the regions $\mathcal{O}_1$ and $\mathcal{O}_2$ includes the projection operators $P_{\mathcal{O}_1}$ and $P_{\mathcal{O}_2}$. In contrast to Bell's theorem (8.14), now there exists a local realist representation [791]

$$\langle\psi|\sigma\cdot aP_{\mathcal{O}_1}\otimes\sigma\cdot bP_{\mathcal{O}_2}|\psi\rangle = E\big(\xi(\mathcal{O}_1,a)\eta(\mathcal{O}_2,b)\big) \tag{8.15}$$

if the distance between the regions $\mathcal{O}_1$ and $\mathcal{O}_2$ is large enough. Here all classical random variables are bounded by 1.

Since detectors of particles are obviously located somewhere in space, it shows that loopholes are unavoidable in experiments aimed to establish a violation of Bell's inequalities. Though there were some reports on experimental derivation of violation of Bell's inequalities, in fact, such violations were always based on additional assumptions besides local realism. No genuine Bell's inequalities have been violated since some loopholes were always in the experiments; for a review, see, for example, [169, 673]. There were many discussions of proposals for experiments which could avoid the loopholes; however, up to now a convincing proposal has still not been suggested.

One can compare the situation with attempts to measure the position and momentum of a particle in a single experiment. Also one could speak about some technical difficulties (similar to the efficiency of detectors loophole) and hope that someone could come with a proposal to make an experiment without loopholes. However, we know from the uncertainty relation for the measurement of momentum and position that it is not possible. Similarly, formula (8.15) shows that a loophole-free experiment in which a violation of Bell's inequalities could be observed is impossible if the distance between detectors is large enough. Therefore, loopholes in Bell's experiments are irreducible.

8.3.5 Correlation Functions and Local Realism

A mathematical formulation of Bell's local realism may be given as follows. Let two families of self-adjoint operators $\{A(a)\}$ and $\{B(b)\}$ be given in a Hilbert space $\mathcal{H}$, which commute (i.e., $[A(a), B(b)] = 0$) on a natural domain. Here a and b are elements of two arbitrary sets of indices. Suppose that one has a representation

$$\langle\psi|A(a)B(b)|\psi\rangle = E\big(\xi(a)\eta(b)\big) \tag{8.16}$$

for any a, b, where E is the mathematical expectation, and $\xi(a)$ and $\eta(b)$ are two stochastic processes such that the range of $\xi(a)$ is the spectrum of $A(a)$ and the range of $\eta(b)$ is the spectrum of $B(b)$. Here ψ is a vector from $\mathcal{H}$. Then we say that *the triplet*

$$\big(\{A(a)\}, \{B(b)\}, \psi\big)$$

satisfies the BLR (*Bell's local realism*) *condition.*

Bell proved that a two spin quantum correlation function which is equal to just $-a \cdot b$, where a and b are two three-dimensional vectors, cannot be represented in the form (8.13) if one has the bound $|\xi(a)| \leq 1$, $|\eta(b)| \leq 1$. Therefore, the correlation function of two spins does not satisfy to the BLR condition (8.16). In this sense, sometimes one speaks about quantum non-locality.

Note, however, that in the previous discussion the space–time parameters were not explicitly involved though one speaks about non-locality. Actually, the "local realism" in the Bell's sense as it was formulated above in (8.13) is a very general notion which has nothing to do with notion of locality in the ordinary three-dimensional space. We will define now another notion which we will call the condition of local realism in the sense of Einstein. First, let us recall that in the quantum field theory the condition of locality (local commutativity) reads:

$$\big[F(x), G(y)\big] = 0$$

if the space–time points x and y are space-like separated. Here $F(x)$ and $G(y)$ are two Bose field operators (for Fermi fields we have anti-commutators).

Suppose that in the Hilbert space $\mathcal{H}$ there is a unitary representation U of the inhomogeneous Lorentz group, and let a family of self-adjoint operators $\{A(a, \mathcal{O})\}$ parameterized by the regions $\mathcal{O}$ be given in Minkowski space–time where a is an arbitrary index. Let us suppose that the unitary operator translations act as

$$U(d)A(a, \mathcal{O})U(d)^* = A\big(a, \mathcal{O}(d)\big)$$

where d is a four-dimensional vector and $\mathcal{O}(d)$ is a shift of $\mathcal{O}$ at d. Let also a family of operators $\{B(b, \mathcal{O})\}$ be given with similar properties. Suppose that one has a representation

$$\langle\psi|A(a, \mathcal{O}_1)B(b, \mathcal{O}_2)|\psi\rangle = E\big(\xi(a, \mathcal{O}_1)\eta(b, \mathcal{O}_2)\big) \tag{8.17}$$

for a, b, $\mathcal{O}_1$, $\mathcal{O}_2$ for which the operators commute

$$\big[A(a, \mathcal{O}_1), B(b, \mathcal{O}_2)\big] = 0.$$

The correlation function (8.17) describes the results of a simultaneous measurement. Moreover, we suppose that the range of $\xi(a, \mathcal{O}_1)$ is the spectrum of $A(a, \mathcal{O}_1)$ and the range of $\eta(b, \mathcal{O}_2)$ is the spectrum of $B(b, \mathcal{O}_2)$. Then we say that *the quadruplet*

$$\left(\left\{A(a, \mathcal{O}_1)\right\}, \left\{B(b, \mathcal{O}_2)\right\}, U, \psi\right)$$

satisfies the ELR (Einstein's local realism) condition.

For Fermi fields which anti-commute we assume the same relation (8.17) but the random fields ξ and η should now be anti-commutative random fields (superanalysis and probability of anti-commutative variables are considered in [406, 789]).

One can write an analogue of the presented notions in the case when the region $\mathcal{O}$ shrinks to one point (in such a case, we have an operator $A(a, x)$) and also for n-point correlation functions

$$\langle\psi|A_1(a_1, x_1)\cdots A_n(a_n, x_n)|\psi\rangle = E\left(\xi_1(a_1, x_1)\cdots\xi_n(a_n, x_n)\right).$$

A non-commutative spectral theory related with such representations is considered in [797].

8.4 Entangled States in Space–Time. Disentanglement

Entangled states, i.e., the states of two particles with the wave function which is not a product of the wave functions of single particles, have been studied in many theoretical and experimental works starting from the paper of Einstein, Podolsky and Rosen, see, e.g., [48].

Let us consider two particles with spin $1/2$. As we know, a particle with spin $1/2$ is described by the Dirac equation. We start with the discussion of non-relativistic approximation. Then one has the Pauli equation. For a consideration of the relativistic particles see Sect. 16.5.

The Hilbert space assigned to one particle with spin $1/2$ in non-relativistic approximation is $\mathbb{C}^2 \otimes L^2(\mathbb{R}^3)$, and the Hilbert space of two particles is $\mathbb{C}^2 \otimes L^2(\mathbb{R}^3) \otimes \mathbb{C}^2 \otimes L^2(\mathbb{R}^3)$. The two-particle wave function is

$$\psi = \left(\psi_{\alpha\beta}(\mathbf{r}_1, \mathbf{r}_2, t)\right),$$

where α and β are spinor indices, t is time, and $\mathbf{r}_1$ and $\mathbf{r}_2$ are vectors in the three-dimensional space.

We suppose that there are two detectors (A and B) which are located in space $\mathbb{R}^3$ within two bounded regions $\mathcal{O}_A$ and $\mathcal{O}_B$, respectively, well separated from one another. If one makes a local observation of the projection of spin to the direction along a unit vector a in the region $\mathcal{O}_A$ then this means that one measures not only the spin observable $\sigma \cdot a$ where σ are Pauli matrices but also some another observable which describes the localization of the particle like the energy density or the projection operator $P_{\mathcal{O}}$ to the region $\mathcal{O}$. We will consider here correlation functions of the projection operators $P_{\mathcal{O}}$.

A quantum correlation function describing the localized measurements of spin $\sigma \cdot a$ in the region $\mathcal{O}_A$ and spin $\sigma \cdot b$ (b is a unit vector in $\mathbb{R}^3$) in the region $\mathcal{O}_B$ is

$$\langle\psi|(\sigma \cdot a) \otimes P_{\mathcal{O}_A} \otimes (\sigma \cdot b) \otimes P_{\mathcal{O}_B}|\psi\rangle.$$

Let us consider the simplest case when the wave function ($\psi_{\alpha\beta}(\mathbf{r}_1, \mathbf{r}_2, t)$) does not depend on time and, moreover, it has the form of a product of the spin function ψ_{spin} and a complex-valued function $\phi(\mathbf{r}_1, \mathbf{r}_2)$ depending on the spatial variables $\mathbf{r}_1$ and $\mathbf{r}_2$,

$$\psi = \psi_{\text{spin}}\phi(\mathbf{r}_1, \mathbf{r}_2).$$

Then one has

$$\begin{aligned}
&\langle\psi|(\sigma \cdot a) \otimes P_{\mathcal{O}_A} \otimes (\sigma \cdot b) \otimes P_{\mathcal{O}_B}|\psi\rangle \\
&\quad = \langle\psi_{\text{spin}}|(\sigma \cdot a) \otimes (\sigma \cdot b)|\psi_{\text{spin}}\rangle\langle\phi|P_{\mathcal{O}_A} \otimes P_{\mathcal{O}_B}|\phi\rangle \\
&\quad = g(\mathcal{O}_A, \mathcal{O}_B)D_{\text{spin}}(a, b)
\end{aligned}$$

where

$$D_{\text{spin}}(a, b) = \langle\psi_{\text{spin}}|(\sigma \cdot a) \otimes (\sigma \cdot b)|\psi_{\text{spin}}\rangle$$

and where the function

$$g(\mathcal{O}_A, \mathcal{O}_B) = \langle\phi|P_{\mathcal{O}_A} \otimes P_{\mathcal{O}_B}|\phi\rangle = \int_{\mathcal{O}_A \times \mathcal{O}_B} \left|\phi(\mathbf{r}_1, \mathbf{r}_2)\right|^2 d\mathbf{r}_1\, d\mathbf{r}_2$$

describes the correlation of particles in space. It is the probability to find one particle in the region $\mathcal{O}_A$ and another particle in the region $\mathcal{O}_B$.

One has

$$0 \le g(\mathcal{O}_A, \mathcal{O}_B) \le 1.$$

If $\mathcal{O}_A$ and $\mathcal{O}_B$ are bounded regions, and $\mathcal{O}_A(l)$ is a translation of $\mathcal{O}_A$ to the three-dimensional vector l then one can prove

$$\lim_{|l| \to \infty} g\big(\mathcal{O}_A(l), \mathcal{O}_B\big) = 0.$$

We denote

$$\omega(Q) = \langle\psi|Q|\psi\rangle$$

for an observable Q. Then

$$\omega(\sigma \cdot a P_{\mathcal{O}_A} \otimes \sigma \cdot b P_{\mathcal{O}_B}) = g(\mathcal{O}_A, \mathcal{O}_B)D_{\text{spin}}(a, b).$$

We take ψ_{spin} such that

$$\langle\psi_{\text{spin}}|\sigma \cdot a \otimes I|\psi_{\text{spin}}\rangle = \langle\psi_{\text{spin}}|I \otimes \sigma \cdot b|\psi_{\text{spin}}\rangle = 0,$$

so we have

$$\omega(\sigma \cdot a P_{\mathcal{O}_A} \otimes I) = \omega(I \otimes \sigma \cdot b P_{\mathcal{O}_B}) = 0.$$

Therefore, we have proved the following proposition which says that the state $\psi = \psi_{\text{spin}}\phi(\mathbf{r}_1, \mathbf{r}_2)$ becomes disentangled (factorized) when the distance between $\mathcal{O}_A$ and $\mathcal{O}_B$ becomes large.

Proposition 8.4 *One has the following property of the asymptotic factorization (disentanglement) at large distances:*

$$\lim_{|l|\to\infty} \big[\omega(\sigma \cdot a P_{\mathcal{O}_A(l)} \otimes \sigma \cdot b P_{\mathcal{O}_B}) - \omega(\sigma \cdot a P_{\mathcal{O}_A(l)} \otimes I)\omega(I \otimes \sigma \cdot b P_{\mathcal{O}_B})\big] = 0,$$

or

$$\lim_{|l|\to\infty} \omega(\sigma \cdot a P_{\mathcal{O}_A(l)} \otimes \sigma \cdot b P_{\mathcal{O}_B}) = 0.$$

8.5 Local Observations

8.5.1 Modified Bell's Equation

In the previous section, the space part of the wave function of the particles was neglected. However, exactly the space part is relevant to the discussion of locality. The Hilbert space assigned to one particle with spin 1/2 is $\mathbb{C}^2 \otimes L^2(\mathbb{R}^3)$, and the Hilbert space of two particles is $\mathbb{C}^2 \otimes L^2(\mathbb{R}^3) \otimes \mathbb{C}^2 \otimes L^2(\mathbb{R}^3)$. The complete wave function is $\psi = (\psi_{\alpha\beta}(\mathbf{r}_1, \mathbf{r}_2, t))$ where α and β are spinor indices, t is time, and $\mathbf{r}_1$ and $\mathbf{r}_2$ are vectors in the three-dimensional space.

We suppose that there are two detectors (A and B) which are located in space $\mathbb{R}^3$ within the two localized regions $\mathcal{O}_A$ and $\mathcal{O}_B$, respectively, well separated from one another. If one makes a local observation in the region $\mathcal{O}_A$ then this means that one measures not only the spin observable σ_i but also some another observable which describes the localization of the particle like the energy density or the projection operator $P_{\mathcal{O}}$ to the region $\mathcal{O}$. We will consider here correlation functions of the projection operators $P_{\mathcal{O}}$.

A quantum correlation describing the localized measurements of spins in the regions $\mathcal{O}_A$ and $\mathcal{O}_B$ is

$$\langle\psi|\sigma \cdot a P_{\mathcal{O}_A} \otimes \sigma \cdot b P_{\mathcal{O}_B}|\psi\rangle.$$

Now one inquires whether one can write a representation

$$\langle\psi|\sigma \cdot a P_{\mathcal{O}_A} \otimes \sigma \cdot b P_{\mathcal{O}_B}|\psi\rangle = \int \xi(a, \mathcal{O}_A, \lambda)\eta(b, \mathcal{O}_B, \lambda)\, d\rho(\lambda), \tag{8.18}$$

where $|\xi(a, \mathcal{O}_A(l), \lambda)| \leq 1$, $|\eta(b, \mathcal{O}_B, \lambda)| \leq 1$.

Remark 8.5 A local modified equation reads

$$\left|\phi(\mathbf{r}_1,\mathbf{r}_2,t)\right|^2\cos(\alpha-\beta)=E\big(\xi(\alpha,\mathbf{r}_1,t)\eta(\beta,\mathbf{r}_2,t)\big).$$

Let us consider the simplest case when the wave function has the form of the product of the spin function and the space function $\psi=\psi_{\mathrm{spin}}\phi(\mathbf{r}_1,\mathbf{r}_2)$. Then one has (see the previous section)

$$\langle\psi|\sigma\cdot aP_{\mathcal{O}_A}\otimes\sigma\cdot bP_{\mathcal{O}_B}|\psi\rangle=g(\mathcal{O}_A,\mathcal{O}_B)D_{\mathrm{spin}}(a,b),$$

where the function

$$g(\mathcal{O}_A,\mathcal{O}_B)=\int_{\mathcal{O}_A\times\mathcal{O}_B}\left|\phi(\mathbf{r}_1,\mathbf{r}_2)\right|^2 d\mathbf{r}_1\,d\mathbf{r}_2$$

describes the correlation of particles in space.

If $\mathcal{O}_A$ is a bounded region and $\mathcal{O}_A(l)$ is a translation of $\mathcal{O}_A$ to the three-dimensional vector l then one has

$$\lim_{|l|\to\infty}g\big(\mathcal{O}_A(l),\mathcal{O}_B\big)=0.$$

If we are interested in the conditional probability of finding the projection of spin along vector a for the particle 1 in the region $\mathcal{O}_A(l)$ and the projection of spin along the vector b for the particle 2 in the region $\mathcal{O}_B$ then we have to divide both sides of (8.18) by $g(\mathcal{O}_A(l),\mathcal{O}_B)$.

Note that here the classical random variable $\xi=\xi(a,\mathcal{O}_A(l),\lambda)$ is not only separated in the sense of Bell (i.e., it depends only on a) but it is also local in the three-dimensional space since it depends only on the region $\mathcal{O}_A(l)$. The classical random variable η is also local in three-dimensional space since it depends only on $\mathcal{O}_B$. Note also that since the eigenvalues of the projector $P_{\mathcal{O}}$ are 0 or 1 then one should have $|\xi(a,\mathcal{O}_A,\lambda)|\le 1$.

Due to the property of the asymptotic factorization and the vanishing of the quantum correlation for large $|l|$, there exists a trivial asymptotic classical representation of the form (8.18) with $\xi=\eta=0$.

We can do even better and find a classical representation which will be valid uniformly for large $|l|$.

If g does not depend on $\mathcal{O}_A$ and $\mathcal{O}_B$ then instead of (8.8) in Theorem 8.2 we could have a modified equation

$$g\cos(\alpha-\beta)=E(\xi_\alpha\eta_\beta). \tag{8.19}$$

The factor g is important. In particular, one can write the following representation [789] for $0\le g\le 1/2$:

$$g\cos(\alpha-\beta)=\int_0^{2\pi}\sqrt{2g}\cos(\alpha-\lambda)\sqrt{2g}\cos(\beta-\lambda)\,\frac{d\lambda}{2\pi}.$$

Therefore, if $0 \le g \le 1/2$ then there exists a solution of (8.19) where

$$\xi_\alpha(\lambda) = \sqrt{2g}\cos(\alpha - \lambda), \qquad \eta_\beta(\lambda) = \sqrt{2g}\cos(\beta - \lambda),$$

and $|\xi_\alpha| \le 1$, $|\eta_\beta| \le 1$. If $g > 1/\sqrt{2}$ then it follows from Theorem 8.3 that there is no solution to (8.19). We have obtained

Theorem 8.6 *If $g > 1/\sqrt{2}$ then there is no solution $(\Omega, \mathcal{F}, d\rho(\lambda), \xi_\alpha, \eta_\beta)$ to* (8.19) *with the bounds $|\xi_\alpha| \le 1$, $|\eta_\beta| \le 1$. If $0 \le g \le 1/2$ then there exists a solution to* (8.19) *with the bounds $|\xi_\alpha| \le 1$, $|\eta_\beta| \le 1$.*

Remark 8.7 Further results on solutions of the modified equation have been obtained by Guschchin, Bochkarev and Prokhorenko. Local variable models for inefficient detectors are presented in [472, 673].

Now let us construct a hidden-variable representation for the quantum-mechanical correlation function. We have

$$g\big(\mathcal{O}_A(l), \mathcal{O}_B\big) = \int_{\mathcal{O}_A(l)\times\mathcal{O}_B} \big|\phi(\mathbf{r}_1, \mathbf{r}_2)\big|^2 d\mathbf{r}_1\, d\mathbf{r}_2.$$

There exists such an $L > 0$ that

$$\int_{B_L \times \mathbb{R}^3} \big|\phi(\mathbf{r}_1, \mathbf{r}_2)\big|^2 d\mathbf{r}_1\, d\mathbf{r}_2 = \varepsilon < 1/2,$$

where $B_L = \{\mathbf{r} \in \mathbb{R}^3 : |\mathbf{r}| \ge L\}$. We have the following

Theorem 8.8 *Suppose that the wave function of two particles has the form $\psi = \psi_{\mathrm{spin}}\phi(\mathbf{r}_1, \mathbf{r}_2)$. Then for a large enough $|l|$ there exists the following representation of the quantum correlation function*:

$$\begin{aligned}\langle\psi|\sigma\cdot a P_{\mathcal{O}_A(l)} \otimes \sigma\cdot b P_{\mathcal{O}_B}|\psi\rangle &= g\big(\mathcal{O}_A(l), \mathcal{O}_B\big)\cos(\alpha - \beta)\\ &= E\big(\xi\big(\alpha, \mathcal{O}_A(l)\big)\eta(\beta, \mathcal{O}_B)\big),\end{aligned}$$

where the classical random variables ξ and η are bounded by 1.

Proof To prove the theorem let us make an additional assumption that the classical random variable has the form of a product of two independent classical random variables $\xi(\alpha, \mathcal{O}_A) = \xi_{\mathrm{space}}(\mathcal{O}_A)\xi_{\mathrm{spin}}(\alpha)$ and similarly for η. We write

$$\begin{aligned}&g\big(\mathcal{O}_A(l), \mathcal{O}_B\big)\cos(\alpha - \beta)\\ &\quad = \int_{\mathcal{O}_A(l), \mathcal{O}_B} \frac{1}{\varepsilon}\big|\phi(\mathbf{r}_1, \mathbf{r}_2)\big|^2 d\mathbf{r}_1\, d\mathbf{r}_2 \cdot \varepsilon\cos(\alpha - \beta)\\ &\quad = E\big(\xi_{\mathrm{space}}\big(\mathcal{O}_A(l)\big)\eta_{\mathrm{space}}(\mathcal{O}_B)\big)E\big(\xi_{\mathrm{spin}}(\alpha)\xi_{\mathrm{spin}}(\beta)\big).\end{aligned}$$

Here $\xi_{\text{space}}(\mathcal{O}_A(l))$ and $\eta_{\text{space}}(\mathcal{O}_B)$ are random variables on the probability space $\Omega = B_L \times \mathbb{R}^3$ with the probability measure

$$dP(\mathbf{r}_1, \mathbf{r}_2) = \frac{1}{\varepsilon}\left|\phi(\mathbf{r}_1, \mathbf{r}_2)\right|^2 d\mathbf{r}_1\, d\mathbf{r}_2$$

of the form

$$\xi_{\text{space}}\bigl(\mathcal{O}_A(l), \mathbf{r}_1, \mathbf{r}_2\bigr) = \chi_{\mathcal{O}_A(l)}(\mathbf{r}_1), \qquad \eta_{\text{space}}(\mathcal{O}_B, \mathbf{r}_1, \mathbf{r}_2) = \chi_{\mathcal{O}_B}(\mathbf{r}_2),$$

where $\chi_{\mathcal{O}}(\mathbf{r})$ is the characteristic function of the region $\mathcal{O}$. We assume that $\mathcal{O}_A(l)$ belongs to B_L. Then we have

$$E\bigl(\xi_{\text{space}}\bigl(\mathcal{O}_A(l)\bigr)\eta_{\text{space}}(\mathcal{O}_B)\bigr) = \frac{1}{\varepsilon} g\bigl(\mathcal{O}_A(l), \mathcal{O}_B\bigr).$$

Further, let $\xi_{\text{spin}}(\alpha)$ be a random process on the circle $0 \le \varphi \le 2\pi$ with the probability measure $d\varphi/2\pi$ of the form

$$\xi_{\text{spin}}(\alpha, \varphi) = \sqrt{2\varepsilon}\cos(\alpha - \varphi).$$

Then we have

$$E\bigl(\xi_{\text{spin}}(\alpha)\xi_{\text{spin}}(\beta)\bigr) = \varepsilon\cos(\alpha - \beta).$$

The theorem is proved. □

8.6 Separability and Entanglement

In Sect. 8.4, we discussed the EPR entangled states. The name of entangled state is given by Schrödinger as the vector state of two particles which is not a product of the wave functions of single particles. He said "Entanglement is the characteristic trait of quantum mechanics, the one that enforces its entire departure from classical lines of thought" [693]. During the last two decades, it was observed that these fundamental nonclassical states (i.e., entangled states) can be used in several fields. Now it was found that the effective characterization of entanglement or of separability, the complementary to the entanglement, of general mixed quantum states is a hard problem.

Definition 8.9 A state θ (i.e., a density operator) on the tensor product $\mathcal{H} \otimes \mathcal{K}$ of two Hilbert spaces $\mathcal{H}$ and $\mathcal{K}$ is called a compound state in general, and it is *separable* if it can be represented in the form

$$\theta = \sum_k \lambda_k \rho_k \otimes \sigma_k,$$

where $\lambda_k \ge 0$ with $\sum_k \lambda_k = 1$, ρ_k and σ_k are states in $\mathcal{H}$ and $\mathcal{K}$. An *entangled state* is a state not belonging to the set of all separable states, so an entangled state is simply considered as an inseparable state.

In 1996, Peres introduced the concept of the PPT (Partial Positive Transpose) criterion to study the separability [645]. Horodeckis [349] gave a necessary and sufficient condition of separability in terms of positive maps via the so-called Jamiołkowski isomorphism [379]. In the case of two- or three-dimensional Hilbert space, it is shown by Størmer [737] that every positive map can be decomposed into completely positive maps and the transpose operation. Using such a decomposition, it was shown that the positive map criterion can be equivalent to the PPT condition. Most of such studies of entanglement have been done in finite-dimensional Hilbert spaces. In the infinite-dimensional case, Belavkin and Ohya [93, 94] rigorously studied mathematical structure of a compound state in terms of Hilbert–Schmidt operators, called the entangling operators, and gave a finer classification of compound states. It is shown that an entangling operator gives a classification which is a generalization of the PPT condition. In the finite-dimensional case, the Belavkin and Ohya criterion is equivalent to the PPT criterion.

In this section, we discuss these criteria of separable and entangled states.

Let us give a typical example of an entangled state. Two systems are described by separable Hilbert spaces $\mathcal{H}$ and $\mathcal{K}$, respectively. For any orthonormal vectors $x_1, x_2 \in \mathcal{H}$ and $y_1, y_2 \in \mathcal{K}$, define

$$z = \frac{1}{\sqrt{2}} x_1 \otimes y_1 + \frac{1}{\sqrt{2}} x_2 \otimes y_2.$$

Then the state $|z\rangle\langle z|$ is a pure state on $\mathcal{H} \otimes \mathcal{K}$, which satisfies the marginal conditions

$$\operatorname{tr}_{\mathcal{K}} |z\rangle\langle z| = \frac{1}{2}\big(|x_1\rangle\langle x_1| + |x_2\rangle\langle x_2|\big) \equiv \rho,$$
$$\operatorname{tr}_{\mathcal{H}} |z\rangle\langle z| = \frac{1}{2}\big(|y_1\rangle\langle y_1| + |y_2\rangle\langle y_2|\big) \equiv \sigma.$$

Thus the state $|z\rangle\langle z|$ is usually called the entangled state with two marginal states ρ and σ.

The notions of separability and entanglement are currently used as above in order to describe the correlation of two states. In some cases, these notions do not properly represent the correlation in both classical and quantum systems. For instance, the correlation in a separable state can be stronger than that of an entangled state. Although we follow conventional terminology in this book, we will give comments related to the above facts in a few other places, too.

8.6.1 Entangling Operator

Given a density operator θ on $\mathcal{H} \otimes \mathcal{K}$, define the maps $\phi : \mathbf{B}(\mathcal{K}) \to \mathbf{B}(\mathcal{H})_* \equiv \{A \in \mathbf{B}(\mathcal{H}); \operatorname{tr}\sqrt{A^* A} < \infty\}$ and $\phi^* : \mathbf{B}(\mathcal{H}) \to \mathbf{B}(\mathcal{K})_*$ as

$$\begin{aligned} &\phi(B) = \operatorname{tr}_{\mathcal{K}}(I \otimes B)\theta, \qquad \phi^*(A) = \operatorname{tr}_{\mathcal{H}}(A \otimes I)\theta, \\ &\theta(A \otimes B) = \operatorname{tr}_{\mathcal{H}} \phi(B) A = \operatorname{tr}_{\mathcal{K}} \phi(B) A, \end{aligned} \tag{8.20}$$

for any $A \in \mathbf{B}(\mathcal{H})$ and $B \in \mathbf{B}(\mathcal{K})$. We have the following theorem for the separability.

Theorem 8.10 *If the density matrix θ on $\mathcal{H} \otimes \mathcal{K}$ is separable then the maps ϕ and ϕ^* are CP maps.*

To prove this theorem, we need some preparations, which will be given in the following discussion.

For a state σ on the Hilbert space $\mathcal{K}$, by a Schatten decomposition, this state can be written

$$\sigma = \sum_n p_n |e_n\rangle\langle e_n|,$$

where $\{e_n\}$ is a complete orthonormal system (CONS) of $\mathcal{K}$ and p_n are non-negative numbers. To express the mixed state σ, one can employ a redundant Hilbert space and a pure state over the composite system. That is, let us prepare another Hilbert space $\mathcal{H}$ whose dimension should be larger than the rank of the density operator σ. Then, a choice $\mathcal{H} \equiv l^2(\mathbb{Z})$ suffices, for instance. Taking a CONS $\{|n\rangle\}$ of $\mathcal{H}$, we define a normalized vector $\Omega \in \mathcal{H} \otimes \mathcal{K}$ as

$$\Omega \equiv \sum_n \mu_n |n\rangle \otimes |e_n\rangle, \tag{8.21}$$

where μ_n is a complex number satisfying $|\mu_n|^2 = p_n$. The vector Ω is called the *purification* of the mixed state σ. Then one can easily show that one of the marginal states of $|\Omega\rangle\langle\Omega|$ is nothing but the original state $\sigma = \mathrm{tr}_{\mathcal{H}} |\Omega\rangle\langle\Omega|$.

More generally, any normal state φ on $\mathbf{B}(\mathcal{K})$ is written as

$$\varphi(\cdot) \equiv \mathrm{tr}\,\sigma\,\cdot, \qquad \sigma \in \mathfrak{S}(\mathcal{K}),$$

and there exists a Hilbert–Schmidt operator H from an another Hilbert space $\mathcal{H}$ to $\mathcal{K}$ (i.e., $\sum_k \|Hx_k\|^2 < +\infty$ for any CONS $\{x_k\}$ in $\mathcal{H}$) such that

$$\varphi(B) = \mathrm{tr}_{\mathcal{H}}\, H^* B H = \mathrm{tr}_{\mathcal{K}}\, \sigma B, \quad B \in \mathbf{B}(\mathcal{K}), \tag{8.22}$$

where the dimension of $\mathcal{H}$ should be larger than the rank of σ as above. The above operator H is defined by a map from $\mathcal{H}$ to $\mathcal{K}$ as

$$H \equiv \sum_n \mu_n |e_n\rangle\langle n|, \tag{8.23}$$

where μ_n is a complex number satisfying $|\mu_n|^2 = p_n$, $\{e_n\}$ and $\{|n\rangle\}$ are CONSs in the purification vector Ω above. We have

$$\sigma = HH^*.$$

The Hilbert–Schmidt operator H is called the *entangling operator*.

We now equip H with a complex conjugation $J_{\mathcal{H}}$ on the standard base $\{|n\rangle\}$ in H,

$$J_{\mathcal{H}}x \equiv J_{\mathcal{H}} \sum \langle n|x\rangle |n\rangle = \sum \overline{\langle n|x\rangle} |n\rangle, \quad x \in \mathcal{H},$$

defining an isometric transposition "$\sim$"

$$\tilde{A} \equiv J_{\mathcal{H}} A^* J_{\mathcal{H}}$$

on $\mathbf{B}(\mathcal{H})$, which we call the "H-transposition" operation, i.e., the equation $\langle m|\tilde{A}|n\rangle = \langle n|A|m\rangle$ holds on the standard base $\{|n\rangle\}$ in $\mathcal{H}$.

Using the entangling operator H, under this preparation, a pure entangled state ω on $\mathcal{H} \otimes \mathcal{K}$ (i.e., a normalized vector Ω in $\mathcal{H} \otimes \mathcal{K}$) can be achieved as

$$\begin{aligned} \omega(A \otimes B) &= \langle \Omega|A \otimes B|\Omega\rangle \\ &= \mathrm{tr}_{\mathcal{H}}\, \tilde{A} H^* B H = \mathrm{tr}_{\mathcal{K}}\, B H \tilde{A} H^* \end{aligned} \tag{8.24}$$

for all $A \in \mathbf{B}(\mathcal{H})$ and $B \in \mathbf{B}(\mathcal{K})$ with marginals

$$\begin{aligned} \psi(A) &= \omega(A \otimes I) = \mathrm{tr}_{\mathcal{H}}\, \tilde{A} H^* H = \mathrm{tr}_{\mathcal{H}}\, A \widetilde{H^* H} \equiv \mathrm{tr}_{\mathcal{H}}\, \rho A, \\ \varphi(B) &= \omega(I \otimes B) = \mathrm{tr}_{\mathcal{K}}\, B H H^* \equiv \mathrm{tr}_{\mathcal{K}}\, \sigma B, \end{aligned}$$

where Ω in $\mathcal{H} \otimes \mathcal{K}$ is given by

$$\langle x \otimes y|\Omega\rangle \equiv \langle y|H J_{\mathcal{H}} x\rangle \tag{8.25}$$

for all $x \in \mathcal{H}$ and $y \in \mathcal{K}$. We know that the normalized vector Ω is represented in the decomposed form (8.21) due to the following simple computation:

$$\begin{aligned} \langle y|H J_{\mathcal{H}} x\rangle &= \sum_n \mu_n \langle y|e_n\rangle \langle n|J_{\mathcal{H}} x\rangle \\ &= \sum_n \mu_n \langle x|n\rangle \langle y|e_n\rangle = \langle x \otimes y|\Omega\rangle, \end{aligned}$$

i.e., Ω is the purification of σ given above. So we have

$$\begin{aligned} \mathrm{tr}_{\mathcal{H}}\, \tilde{A} H^* B H &= \sum_{m,n,k} \overline{\mu_n} \mu_k \langle m|\tilde{A}|n\rangle \langle e_n|B|e_k\rangle \langle k|m\rangle \\ &= \sum_{m,n} \overline{\mu_n} \mu_m \langle n|A|m\rangle \langle e_n|B|e_m\rangle \\ &= \langle \Omega|A \otimes B|\Omega\rangle. \end{aligned}$$

Note that the marginal density ρ has the symmetry property with respect to the complex conjugation $J_{\mathcal{H}}$:

$$\rho = \widetilde{H^* H} = H^* H = \widetilde{\rho} = \overline{\rho},$$

where $\overline{A} \equiv J_{\mathcal{H}} A J_{\mathcal{H}}$, that is a "$\mathcal{H}$-complex conjugation" operation. Since ρ is the marginal state of $|\Omega\rangle\langle\Omega|$ having a diagonal representation, $\rho = \sum p_n |n\rangle\langle n|$ in the standard base in $\mathcal{H}$.

Let us extend this operator H so as to discuss several types of compound states.

For a given state (density operator) θ in a compound system $\mathcal{H} \otimes \mathcal{K}$, there is a Schatten decomposition such that

$$\theta = \sum_k p_k |e_k\rangle\langle e_k|,$$

where $e_k \in \mathcal{H} \otimes \mathcal{K}$ and $\langle e_k|e_l\rangle = \delta_{kl}$, $p_k \geq 0$, $\sum_k p_k = 1$. This mixed state can be represented as an entangling operator $H_{\mathcal{F}}$ from a Hilbert space $\mathcal{F}$ to $\mathcal{H} \otimes \mathcal{K}$, where $\mathcal{F}$ can be taken as the subspace of $l^2(\mathbb{Z})$ with a complex conjugation $J_{\mathcal{F}}$ on a standard base $\{|k\rangle\}$. This $H_{\mathcal{F}}$ is given as

$$H_{\mathcal{F}} \equiv \sum_k \mu_k |e_k\rangle\langle k|,$$

where μ_k is a complex number satisfying $|\mu_k|^2 = p_k$. Thus we obtain the following lemma.

Lemma 8.11 *Using the entangling operator $H_{\mathcal{F}}$, the normal compound state $\omega(\cdot) \equiv \operatorname{tr} \theta \cdot$ is expressed as*

$$\omega(A \otimes B) = \operatorname{tr}_{\mathcal{F}} H_{\mathcal{F}}^* (A \otimes B) H_{\mathcal{F}}. \tag{8.26}$$

Here $\mathcal{H}$ is also equipped with a complex conjugation $J_{\mathcal{H}}$ on the standard base $\{|n\rangle\}$ in $\mathcal{H}$. Without loss of generality, the marginal density ρ on $\mathcal{H}$ can be diagonalized as

$$\rho = \operatorname{tr}_{\mathcal{K}} \theta = \sum \lambda_n |n\rangle\langle n|, \quad \lambda_n \geq 0, \ \sum \lambda_n = 1. \tag{8.27}$$

Using the purification Ω of the density operator θ, we can define another entangling operator $H_{\mathcal{H}} : \mathcal{H} \mapsto \mathcal{F} \otimes \mathcal{K}$ by

$$\begin{aligned}\langle z \otimes y | H_{\mathcal{H}} J_{\mathcal{H}} x\rangle &\equiv \langle x \otimes y \otimes z | \Omega\rangle \\ &= \langle x \otimes y | H_{\mathcal{F}} J_{\mathcal{F}} z\rangle, \quad \forall x \in \mathcal{H},\ y \in \mathcal{K}, z \in \mathcal{F}.\end{aligned} \tag{8.28}$$

For any CONS $\{y_m\}$ in $\mathcal{K}$, the entangling operator $H_{\mathcal{F}}$ can be written in a matrix representation form:

$$\begin{aligned}H_{\mathcal{F}} &= \sum_{n,m,k} |n \otimes y_m\rangle\langle n \otimes y_m | e_k\rangle \mu_k \langle k| \\ &= \sum_{n,k} |n\rangle \otimes h_k(n) \langle k|,\end{aligned} \tag{8.29}$$

where $h_k(n) = \sum_m y_m \langle n \otimes d_m | e_k \rangle \mu_k \in \mathcal{K}$, and it gives the purification of θ as $\Omega = \sum_{n,k} |n\rangle \otimes h_k(n) \otimes |k\rangle$. The definition (8.28) means that the entangling operator $H_{\mathcal{H}}$ can be given as a transposition of $H_{\mathcal{F}}$ in the sense of its matrix representation such as

$$H_{\mathcal{H}} = \sum_{n,k} |k\rangle \otimes h_k(n)\langle n| = \sum_n |H_n\rangle\langle n|, \tag{8.30}$$

where $H_n = \sum_k |k\rangle \otimes h_k(n)$. Using this decomposition, it is easy to see that the operator $H_{\mathcal{H}}$ satisfies the following theorem [93, 94]:

Theorem 8.12 *The compound state* ω *in* (8.26) *can be achieved as an entanglement*

$$\omega(A \otimes B) = \langle \Omega | I \otimes A \otimes B | \Omega \rangle = \mathrm{tr}_{\mathcal{H}}\, \tilde{A} H_{\mathcal{H}}^* (I \otimes B) H_{\mathcal{H}} = \mathrm{tr}_{\mathcal{H}}\, H_{\mathcal{H}} \tilde{A} H_{\mathcal{H}}^* (I \otimes B)$$

with its marginals $\rho = \widetilde{H_{\mathcal{H}}^* H_{\mathcal{H}}} = H_{\mathcal{H}}^* H_{\mathcal{H}} \in \mathfrak{S}(\mathcal{H})$ *and* $\sigma = \mathrm{tr}_{\mathcal{H}}\, H_{\mathcal{H}} H_{\mathcal{H}}^* \in \mathfrak{S}(\mathcal{K})$.

Note that if a given state ω is pure, then a Hilbert space $\mathcal{F}$ becomes $\mathbb{C}$, i.e., $H_{\mathcal{H}} = H$ in (8.23). In the following, we will denote the entangling operator defined in (8.28) by the symbol H indifferently.

8.6.2 True Quantum Entanglement, d- and c-Entanglements

Now let us see how the entangling operator can be used to classify the compound states.

Put

$$\phi(B) \equiv H^* \widetilde{(I \otimes B)} H \; \big(= J H^* (I \otimes B^*) H J \big), \quad B \in \mathbf{B}(\mathcal{K}) \tag{8.31}$$

and

$$\phi^*(A) \equiv \mathrm{tr}_{\mathcal{F}}\, H \tilde{A} H^*, \quad A \in \mathbf{B}(\mathcal{H}), \tag{8.32}$$

then one can expresses ω as

$$\omega(A \otimes B) = \mathrm{tr}_{\mathcal{H} \otimes \mathcal{K}}\, \theta(A \otimes B) = \mathrm{tr}_{\mathcal{H}}\, A\phi(B) = \mathrm{tr}_{\mathcal{K}}\, B\phi^*(A). \tag{8.33}$$

This ϕ maps from $\mathbf{B}(\mathcal{K})$ to the predual space $\mathbf{B}(\mathcal{H})_*$, and ϕ^* maps from $\mathbf{B}(\mathcal{H})$ to $\mathbf{B}(\mathcal{K})_*$.

Note that the above definitions of (8.31) and (8.32) can be written simply as (8.20) in the beginning of the previous subsection.

From the discussion in the previous subsection, we obtain that any density operator θ can be represented in the form

$$\theta = \sum_{n,m} |m\rangle\langle n| \otimes \mathrm{tr}_{\mathcal{F}} |H_m\rangle\langle H_n| \tag{8.34}$$

with the marginal expectations

$$\phi^*(A) = \sum_{n,m} \langle n|A|m\rangle \,\mathrm{tr}_{\mathcal{F}} |H_m\rangle\langle H_n|, \tag{8.35}$$

$$\phi(B) = \sum_{n,m} |m\rangle\langle n| \langle H_n|(I \otimes B)|H_m\rangle \tag{8.36}$$

with the corresponding orthogonality

$$\mathrm{tr}_{\mathcal{K}\otimes\mathcal{F}} |H_m\rangle\langle H_n| = \lambda_n \delta_{n,m} = \langle H_n|H_m\rangle. \tag{8.37}$$

Note that the diagonal representation of ρ in (8.27) is characterized by the above weak orthogonality property.

When the weak orthogonality property becomes stronger such as

$$\mathrm{tr}_{\mathcal{F}} |H_m\rangle\langle H_n| = \lambda_n \sigma_n \delta_{nm},$$

where $\sigma_n \in \mathfrak{S}(\mathcal{K})$, $\phi^*(A)$ is

$$\phi^*(A) = \sum_{n} \langle n|A|n\rangle \otimes \lambda_n \sigma_n. \tag{8.38}$$

Note that the map $B(\in \mathbf{B}(\mathcal{K})) \to \widetilde{\phi(B)}$ given by

$$\begin{aligned}\widetilde{\phi(B)} &= \sum_{n,m} |m\rangle\langle n| \langle H_m|(I \otimes B)|H_n\rangle \\ &= H^*(I \otimes B)H \; \big(\in \mathbf{B}(\mathcal{H})_*\big)\end{aligned}$$

is the complete positive map (CP for short; that is, $\sum_{i,j} B_i^* \phi^*(A_i^* A_j) B_j \geq 0$ for any $\{A_i\} \subset \mathbf{B}(\mathcal{H})$ and $\{B_i\} \subset \mathbf{B}(\mathcal{K})$) in the Steinspring form, and the map $A(\in \mathbf{B}(\mathcal{H})) \to \phi^*(\tilde{A})$ given by

$$\begin{aligned}\phi^*(\tilde{A}) &= \sum_{n,m} \langle m|A|n\rangle \,\mathrm{tr}_{\mathcal{F}} |H_m\rangle\langle H_n| \\ &= \sum_{k} \big(\langle k| \otimes I\big) H A H^* \big(|k\rangle \otimes I\big) \; \big(\in \mathbf{B}(\mathcal{K})_*\big)\end{aligned}$$

is also CP in the Kraus form. However, ϕ and ϕ^* are positive, but they are not necessarily CP, unless $\widetilde{\mathbf{B}(\mathcal{K})} = \mathbf{B}(\mathcal{K})$ or $\widetilde{\mathbf{B}(\mathcal{H})} = \mathbf{B}(\mathcal{H})$ (i.e., $\mathbf{B}(\mathcal{K})$ or $\mathbf{B}(\mathcal{H})$) is abelian.

We introduce the concept of co-CP maps. A map ϕ is co-CP if its composition with the transposition is a CP map, that is,

$$\sum_{i,j} B_i^* \phi^*(A_j^* A_i) B_j \geq 0 \quad \text{for any } \{A_i\} \subset \mathbf{B}(\mathcal{H}) \text{ and } \{B_i\} \subset \mathbf{B}(\mathcal{K}).$$

So our maps given by the entangling operator H are always co-CP maps.

We shall prove Theorem 8.10.

Proof If the density operator θ of a normal compound state ω on $\mathbf{B}(\mathcal{H} \otimes \mathcal{K})$ is given as $\theta = \sum p_n \rho_n \otimes \sigma_n$, $\rho_n \in \mathfrak{S}(\mathcal{H})$, $\sigma_n \in \mathfrak{S}(\mathcal{K})$, then, using (8.20), the map ϕ^* can be represented as

$$\phi^*(A) = \sum p_n \operatorname{tr} A \widetilde{\rho_n} \cdot \sigma_n.$$

Now for all $A_i \in \mathbf{B}(\mathcal{H})$ we have

$$\operatorname{tr}_{\mathcal{H}} A_i^* A_j \widetilde{\rho_n} = \operatorname{tr}_{\mathcal{H}} \widetilde{A_i^* A_j \rho_n} = \operatorname{tr}_{\mathcal{H}} A_j^* A_i \widetilde{\rho_n}.$$

This equation means that ϕ^* of the separable state ω is always co-CP and also CP. The complete positivity of ϕ can be shown easily by the symmetry of ρ_n in its eigen-representation. □

We have the following definition of the entanglement.

Definition 8.13 The dual map $\phi^* : \mathbf{B}(\mathcal{H}) \to \mathbf{B}(\mathcal{K})_*$ to co-CP map $\phi : \mathbf{B}(\mathcal{K}) \to \mathbf{B}(\mathcal{H})_*$, normalized as $\operatorname{tr}_{\mathcal{H}} \phi(I) = 1$, is called the (generalized) entanglement of the state $\rho = \phi(I)$ on $\mathbf{B}(\mathcal{H})$ to the state $\sigma = \phi^*(I)$ on $\mathbf{B}(\mathcal{K})$. The entanglement ϕ^* is called true quantum if it is not completely positive. All other entanglements are not true quantum.

This definition can be easily applied to the cases of the infinite-dimensional Hilbert spaces.

We summarize some notations for the subsequent use. A normal compound state ω with its marginals ρ and σ is expressed by a density operator θ in $\mathcal{H} \otimes \mathcal{K}$; that is, $\omega(\cdot) = \operatorname{tr} \theta \cdot$, and θ is written in the following forms due to the strength of the correlation between two marginal states. For any θ, we have the representations (8.34)–(8.37). We classify the compound states by using the properties of the maps ϕ and ϕ^*.

1. (q-entanglement) We call a compound state θ q-entangled if ϕ^* is not CP. We denote a true quantum entanglement by ϕ_q^* (i.e., ϕ_q is not CP) and a q-compound state by θ_q so that

$$\phi_q^*(A) = \sum_{n,m} \langle n|A|m\rangle \operatorname{tr}_{\mathcal{F}} |H_m\rangle\langle H_n|,$$

$$\theta_q = \sum_{n,m} |m\rangle\langle n| \otimes \mathrm{tr}_{\mathcal{F}} |H_m\rangle\langle H_n|,$$

with the weak orthogonality condition

$$\mathrm{tr}_{\mathcal{K}\otimes\mathcal{F}} |H_m\rangle\langle H_n| = \lambda_n \delta_{n,m} = \langle H_n|H_m\rangle.$$

The set of all true quantum entanglements is denoted by $\mathcal{E}_q$.

2. (d-entanglement) If an entangling map ϕ^* satisfies the following conditions (i)–(iii) then ϕ^* is called a d-entanglement:

$$\text{(i)} \quad \phi^*(A) = \sum_n \langle n|A|n\rangle \, \mathrm{tr}_{\mathcal{F}} |H_m\rangle\langle H_n| = \sum_n \lambda_n \langle n|A|n\rangle \sigma_n,$$
$$\text{(ii)} \quad \theta = \sum_n |n\rangle\langle n| \otimes \mathrm{tr} \, |H_m\rangle\langle H_n| = \sum_n |n\rangle\langle n| \otimes \lambda_n \sigma_n,$$

with the strong orthogonality condition

$$\text{(iii)} \quad \mathrm{tr}_{\mathcal{F}} |H_m\rangle\langle H_n| = \lambda_n \sigma_n \delta_{nm},$$

where $\sigma_n \in \mathfrak{S}(\mathcal{K})$. Let us denote a d-entanglement by ϕ_d^* and its compound state by θ_d. The set of all d-entanglements is denoted by $\mathcal{E}_d$.

3. (c-entanglement) An entanglement ϕ^* is called a c-entanglement if it has the same form as a d-entanglement, but $\{\sigma_n\}$ are commutative. We denote a c-entanglement by ϕ_c^* and its compound state by θ_c. $\mathcal{E}_c$ is the set of all c-entanglements.

It is clear that $\mathcal{E}_d$ and $\mathcal{E}_c$ belong to the set of all not true quantum states.

Remark 8.14 In conventional discussions, the above d- and c-entanglements are classified as the separable states. However, there exist several important applications with a quantum correlated state written as a d-entanglement, such as quantum measurement and filtering, quantum compound state, and lifting. So, we named such states as above. This observation is one of the reasons why we mentioned in the beginning of this section that conventional notions of separability and entanglement are not always proper.

In the case of pure states, we know that the condition of a true quantum entanglement characterizes the pure entanglement state, i.e., non-product pure state [504]. Before discussing the characterization of mixture entangled states, we will show how the true quantum entanglement characterizes the pure entangled states.

Let ω be a pure state on $\mathbf{B}(\mathcal{H} \otimes \mathcal{K})$. For any pure state, there exists a vector $\Omega \in \mathcal{H} \otimes \mathcal{K}$ such that

$$\omega(A \otimes B) = \langle \Omega|(A \otimes B)|\Omega\rangle.$$

Using (8.20), one has

$$\phi^*(A) = \mathrm{tr}_{\mathcal{H}}(A \otimes I_{\mathcal{K}})|\Omega\rangle\langle\Omega|.$$

Now, consider the CP condition of ϕ^*. Let $\{x_k\}$ be a CONS in $\mathcal{H}$ and $\{y_l\}$ be a CONS in $\mathcal{K}$. Assume that Ω is given by

$$\Omega = \sum_k \lambda_k |x_k\rangle \otimes |y_k\rangle \quad \left(\lambda_k \in \mathbb{C}, \sum_k |\lambda_k|^2 = 1\right)$$

where at least two elements of $\{\lambda_k\}$ are non-zero. In order to show that ϕ^* is non-CP, some preliminaries are necessary. The set $M_n(\mathcal{A})$ denotes the C^*-algebra of $n \times n$ matrices with entries in $\mathcal{A}$. Let $\{e_{ij}\}$ be the canonical basis for $M_n(\mathbb{C}) \equiv M_n$, i.e., the $n \times n$matrices with a 1 in row i, column j, and zeros elsewhere. It is well known that every element in $\mathcal{A} \odot M_n$ can be written as $\sum A_{ij} \otimes e_{ij}$, where the A_{ij}'s (being in $\mathcal{A}$) are unique. The map

$$\Theta : \mathcal{A} \odot M_n \to M_n(\mathcal{A}), \qquad \sum A_{ij} \otimes e_{ij} \to (A_{ij})$$

is linear, multiplicative, *-preserving, and bijective. Therefore, it should be clear that CP of ϕ^* is equivalent to positivity of the operator $\sum_{i,j=1}^n e_{ij} \otimes \phi^*(A_i^* A_j)$ for any n.

Let $\{e_i\}$ be a CONS in $\mathbb{C}^n$. One has

$$\begin{aligned}
\phi^*(A_i^* A_j) &= \operatorname{tr}_{\mathcal{H}}(A_i^* A_j \otimes I)|\Omega\rangle\langle\Omega| \\
&= \sum_{k,l} \operatorname{tr}_{\mathcal{H}}(A_i^* A_j \otimes I)|\lambda_k x_k \otimes y_k\rangle\langle\lambda_l x_l \otimes y_l| \\
&= \sum_{k,l} \lambda_k \overline{\lambda_l} \operatorname{tr}_{\mathcal{H}}\big(A_i^* A_j |x_k\rangle\langle y_l|\big)|z_k\rangle\langle z_l| \\
&= \sum_{k,l} \lambda_k \overline{\lambda_l} \langle x_l|A_i^* A_j|x_k\rangle |y_k\rangle\langle y_l|.
\end{aligned}$$

Thus

$$\sum_{i,j=1}^n e_{ij} \otimes \phi^*(A_i^* A_j) = \sum_{i,j=1}^n \sum_{k,l} \lambda_k \overline{\lambda_l} \langle x_l|A_i^* A_j|x_k\rangle |e_i\rangle\langle e_j| \otimes |y_k\rangle\langle y_l|.$$

Put $A_i \equiv |x\rangle\langle x_i|$ $(x \in \mathcal{H},\ \|x\| = 1)$, then the above LHS is

$$\big(\phi^*\big(|x_i\rangle\langle x_j|\big)\big) = \sum_{i,j=1}^n \lambda_j \overline{\lambda_i} |e_i\rangle\langle e_j| \otimes |y_j\rangle\langle y_i|.$$

The positivity of $\{\phi^*(|x_i\rangle\langle x_j|)\}$ entails that of $\langle\Psi|(\sum \phi^*(|v_i\rangle\langle v_j|) \otimes e_{ij})|\Psi\rangle$ for any $\Psi \in \mathbb{C}^n \otimes \mathcal{K}$. Let us take $\Psi_\pm$ of the form

$$\Psi_\pm = e_k \otimes y_l \pm e_l \otimes y_k \quad (k \neq l).$$

Then

$$\langle \Psi_\pm | \left(\sum \phi^* \left(|v_i\rangle\langle v_j| \right) \otimes e_{ij} \right) | \Psi_\pm \rangle = \pm 2 \operatorname{Re} \lambda_k \overline{\lambda_l} \in \mathbb{R}.$$

If $2 \operatorname{Re} \lambda_k \overline{\lambda_l}$ is positive then

$$\langle \Psi_- | \left(\sum \phi^* \left(|v_i\rangle\langle v_j| \right) \otimes e_{ij} \right) | \Psi_- \rangle = -2 \operatorname{Re} \lambda_k \overline{\lambda_l} < 0.$$

Also if $2 \operatorname{Re} \lambda_k \overline{\lambda_l}$ is negative then

$$\langle \Psi_+ | \left(\sum \phi^* \left(|v_i\rangle\langle v_j| \right) \otimes e_{ij} \right) | \Psi_+ \rangle = 2 \operatorname{Re} \lambda_k \overline{\lambda_l} < 0.$$

This means that ϕ^* is non-CP.

8.6.3 Criteria of Entangled States

As we have mentioned at the beginning of this section, the characterization of entangled states is not easy. There are several operational criteria which enable one to detect quantum entangled states. One of them is the so-called PPT (positive partial transpose) criterion introduced by Peres and Horodeckis [349, 645]. In the low-dimensional case, this criterion characterizes the entangled states. In this subsection, we discuss the PPT criterion and show that the criteria using entanglement maps generalize the PPT criterion.

PPT Criterion

The $\mathcal{K}$-partial transpose operation of a compound state $\theta \in \mathfrak{S}(\mathcal{H} \otimes \mathcal{K})$ is denoted by $\theta^{T_\mathcal{K}}$ so that

$$\theta \in \mathfrak{S}(\mathcal{H} \otimes \mathcal{K}) \longrightarrow \theta^{T_\mathcal{K}} = (I \otimes T)\theta,$$

where T is the transpose operation on $\mathbf{B}(\mathcal{K})$. For example, θ is decomposable as $\theta = \sum_{n,m} |m\rangle\langle n| \otimes B_{mn}$, where $\{|n\rangle\}$ is the standard basis in $\mathcal{H}$ and $B_{mn} \in \mathbf{B}(\mathcal{K})$. Then

$$\theta \longrightarrow \theta^{T_\mathcal{K}} = \sum_{m,n} |m\rangle\langle n| \otimes B_{mn}^T.$$

Definition 8.15 A compound state θ is called a PPT state if $\theta^{T_\mathcal{K}}$ is positive. All other states are called NPT states.

It is easy to show that PPT condition is a necessary condition of a separable state. If a state θ is separable and written as $\sum_k \lambda_k \rho_k \otimes \sigma_k$, then $\theta^{T_{\mathcal{K}}} = \sum_k \lambda_k \rho_k \otimes \sigma_k^T \geq 0$ because σ_k^T is positive. However, it is known that the converse statement does not hold. Let us review here the argument of Horodeckis' works.

Horodeckis gave a non-operational characterization of a separable state:

Theorem 8.16 *A state $\theta \in \mathfrak{S}(\mathcal{H} \otimes \mathcal{K})$ is separable iff*

$$\operatorname{tr} \theta W \geq 0$$

for any Hermitian operator W satisfying $\operatorname{tr}(\rho \otimes \sigma) W \geq 0$ *for all pure states $\rho \in \mathfrak{S}(\mathcal{H})$ and $\sigma \in \mathfrak{S}(\mathcal{K})$.*

This theorem is a direct application of Hahn–Banach theorem. The Hermitian operator W which detects an entanglement of θ is called the "entanglement witness", a term introduced by Terhal [745].

The witness operator can be translated into the language of positive maps via the Jamiołkowski isomorphism in the finite-dimensional case. Each entanglement witness W on $\mathcal{H} \otimes \mathcal{K}$ corresponds to a positive map Λ from $\mathbf{B}(\mathcal{H})$ to $\mathbf{B}(\mathcal{K})$. This isomorphism is defined by

$$\Lambda \longrightarrow W_{\mathbb{J}}(\Lambda) = (I \otimes \Lambda)\mathbb{J}, \tag{8.39}$$

where $\mathbb{J} \equiv \sum_{n,m}^{\dim \mathcal{H}} |n\rangle\langle m| \otimes |m\rangle\langle n|$ on $\mathcal{H} \otimes \mathcal{H}$. It is very important that the maps corresponding to entanglement witnesses are positive, but not CP. This fact can be represented by the following theorem which is a translation of the previous theorem via the Jamiołkowski isomorphism:

Proposition 8.17 *A state $\theta \in \mathfrak{S}(\mathcal{H} \otimes \mathcal{K})$ is separable iff*

$$(I \otimes \Lambda)\theta \geq 0 \tag{8.40}$$

for any positive map Λ from $\mathbf{B}(\mathcal{H})$ *to* $\mathbf{B}(\mathcal{K})$.

For the low-dimensional case (i.e., the cases such as $\mathbb{C}^2 \to \mathbb{C}^2$ or $\mathbb{C}^3 \to \mathbb{C}^2$) any positive map Λ can be decomposed as

$$\Lambda = \Lambda_1^{\mathrm{CP}} + \Lambda_2^{\mathrm{CP}} \circ T, \tag{8.41}$$

where Λ_i^{CP} are CP maps.

The proof of this decomposition was given by Størmer [737].

Using the decomposability property of positive maps, the condition $(I \otimes \Lambda)\theta \geq 0$ reads

$$\big(I \otimes \Lambda_1^{\mathrm{CP}}\big)\theta + \big(I \otimes \Lambda_2^{\mathrm{CP}}\big)\theta^{T_{\mathcal{K}}} \geq 0. \tag{8.42}$$

This means that the condition $\theta^{T_{\mathcal{K}}} \geq 0$ becomes a necessary and sufficient condition of separability in the low-dimensional case, for dimensions two or three only.

Majewski gave a characterization of PPT states by using Tomita–Takesaki scheme (see Chap. 4) to describe the transposition. Here we review his approach (see [460, 501, 502]).

Let $\mathcal{H}$ be a finite dimensional Hilbert space. Using an invertible density matrix ρ we can define a faithful state ω on $\mathbf{B}(\mathcal{H})$ as $\omega(A) = \operatorname{tr}\rho A$ for $A \in \mathbf{B}(\mathcal{H})$. Let us consider the GNS triple $(\mathcal{H}_\omega, \pi_\omega, \Omega)$ associated with $(\mathbf{B}(\mathcal{H}), \omega)$. Such triples are given by the following:

- GNS Hilbert space $\mathcal{H}_\omega = \overline{\{\pi_\omega(A)\Omega;\, A \in \mathbf{B}(\mathcal{H})\}}^{(\cdot,\cdot)}$ with $(A\Omega, B\Omega) \equiv \operatorname{tr} A^* B\rho$ for $A, B \in \mathbf{B}(\mathcal{H})$
- Cyclic vector $\Omega = \rho^{1/2}$
- Representation $\pi_\omega(A)\Omega = A\Omega$.

In the GNS representation, the modular conjugation J_m is just the Hermitian involution, i.e., $J_m A\rho^{1/2} = \rho^{1/2} A^*$, and the modular operator Δ is equal to the map $\rho J_m \rho^{-1} J_m$. We remark that ρ^{-1} is, in general, an unbounded operator. Hence, the domain of Δ should be considered. Since, (i) $\{A\rho^{\frac{1}{2}};\, A \in \mathbf{B}(\mathcal{H})\}$ is a dense subset in the set $\mathbf{S}(\mathcal{H})$ of all Hilbert–Schmidt operators, (ii) $\alpha_t(\sigma) = \rho^{\mathrm{i}t}\sigma\rho^{-\mathrm{i}t}$ is a one-parameter group of automorphisms on $\mathbf{S}(\mathcal{H})$, there exists [130] a set of α-analytic elements $\mathbf{B}(\mathcal{H})_\alpha$. Thus $\Delta\sigma = \alpha_t(\sigma)|_{t=-\mathrm{i}} = \rho\sigma\rho^{-1}$ is well defined for $\sigma \in \mathbf{B}(\mathcal{H})_\alpha$. In particular, the polar decomposition of Tomita's operator (see Chap. 4) is written as

$$SA\Omega = A^*\Omega = J_m \Delta^{\frac{1}{2}} A\Omega.$$

Note that $\{\pi_\omega(A)\Omega;\, A \in \mathbf{B}(\mathcal{H})\} \subseteq D(\Delta^{\frac{1}{2}})$, where $D(\cdot)$ stands for the domain. In order to discuss the transposition on $\pi_\omega(\mathbf{B}(\mathcal{H}))$, we introduce the following two conjugations: J_c on $\mathcal{H}$ and J on $\mathcal{H}_\omega$. Due to the faithfulness of ω, the eigenvectors $\{e_i\}$ of ρ form an orthogonal basis in $\mathcal{H}$. Hence we can define

$$J_c x = \sum_i \overline{\langle e_i, x\rangle} e_i$$

for every $x \in \mathcal{H}$. Due to the fact that $\{e_{ij} = |e_i\rangle\langle e_j|\}$ forms an orthogonal basis in $\mathcal{H}_\omega$, we can also define a conjugation J on $\mathcal{H}_\omega$

$$JA\Omega = \sum_i \overline{(e_{ij}, A\Omega)} e_{ij}$$

with $J\Omega = \Omega$.

Following the construction presented in [460, 501], we can define a transposition on $\mathbf{B}(\mathcal{H})$ as the map $A \in \mathbf{B}(\mathcal{H}) \mapsto A^t \equiv J_c A^* J_c$. By τ_0 we will denote the map induced on $\mathcal{H}_\omega$ by transposition, i.e.,

$$\tau_0 A\Omega = A^t\Omega.$$

The main properties of τ_0 are the following:

Proposition 8.18 [460]

1. *Let $A \in \mathbf{B}(\mathcal{H})$ and $\xi \in \mathcal{H}_\omega$. Then*

$$A^t \xi = J A^* J \xi.$$

2. *The map τ_0 has its polar decomposition, i.e.,*

$$\tau_0 = U \Delta^{1/2},$$

where U is the unitary operator on $\mathcal{H}_\omega$ defined by

$$U = \sum_{ij} |e_{ij})(e_{ij}|, \tag{8.43}$$

and the sum defining the operator U is understood in the weak operator topology.

In the above setting, we can introduce the natural cone $\mathcal{P}$ [62, 174] associated with $(\pi_\omega(\mathbf{B}(\mathcal{H})), \Omega)$,

$$\mathcal{P} = \overline{\{\Delta^{1/4} A \Omega : A \geq 0, A \in \pi_\omega(\mathbf{B}(\mathcal{H}))\}}^{(\cdot,\cdot)}.$$

The relationship between the Tomita–Takesaki scheme and transposition is given in the following:

Proposition 8.19 [460] *Let $\xi \mapsto \omega_\xi$ be a homeomorphism between the natural cone $\mathcal{P}$ and the set of normal states on $\pi_\omega(\mathbf{B}(\mathcal{H}))$such that*

$$\omega_\xi(A) = (\xi, A\xi), \quad A \in \mathbf{B}(\mathcal{H}).$$

For every state ω define $\omega^\tau(A) = \omega(A^t)$. If $\xi \in \mathcal{P}$ then the unique vector in $\mathcal{P}$ mapped into the state ω_ξ^τ by the homeomorphism described above is equal to $U\xi$, i.e.,

$$\omega_\xi^\tau(A) = (U\xi, AU\xi), \quad A \in \mathbf{B}(\mathcal{H}).$$

Let $\mathcal{H}_1$ and $\mathcal{H}_2$ be finite dimensional Hilbert spaces. We want to emphasize that finite-dimensionality of Hilbert spaces is assumed only in the proof of Theorem 8.20. More precisely, due to technical questions concerning the domain of the modular operator Δ, we could prove this theorem only for the finite-dimensional case (see [460]). On the other hand, we emphasize that the description of a composite system based on Tomita's approach is very general. It relies on the construction of the tensor product of standard forms of von Neumann algebras, and this description can be done in a very general way (so the infinite-dimensional case is included, cf. [186]).

Again we will consider a composite system $1 + 2$. Suppose that a subsystem 1 is described by $\mathcal{A}_1 = \mathbf{B}(\mathcal{H}_1)$ equipped with a faithful state ω_1 given by an invertible density matrix ρ_1 as $\omega_1(A) \equiv \operatorname{tr} \rho_1 A$. Similarly, let $\mathcal{A}_2 = \mathbf{B}(\mathcal{H}_2)$ as another subsystem $\mathcal{A}_2$, ρ_2 be an invertible density matrix in $\mathbf{B}(\mathcal{H}_2)$, and ω_2 be a state on $\mathcal{A}_2$ such that $\omega_2(B) \equiv \operatorname{tr} \rho_2 B$ for $B \in \mathcal{A}_2$. By $(\mathcal{K}, \pi, \Omega)$, $(\mathcal{K}_1, \pi_1, \Omega_1)$, and $(\mathcal{K}_2, \pi_2, \Omega_2)$ we denote the GNS representations of $(\mathcal{A}_1 \otimes \mathcal{A}_2, \omega_1 \otimes \omega_2)$, $(\mathcal{A}_1, \omega_1)$, and $(\mathcal{A}_2, \omega_2)$, respectively. Then the triple $(\mathcal{K}, \pi, \Omega)$ can be given by following identifications (cf. [186, 502]):

$$\mathcal{K} = \mathcal{K}_1 \otimes \mathcal{K}_2, \qquad \pi = \pi_1 \otimes \pi_2, \qquad \Omega = \Omega_1 \otimes \Omega_2.$$

With these identification we have

$$J_m = J_1 \otimes J_2, \qquad \Delta = \Delta_1 \otimes \Delta_2$$

where J_m, J_1, and J_2 are modular conjugations and Δ, Δ_1, Δ_2 are modular operators for $(\pi(\mathcal{A}_1 \otimes \mathcal{A}_2)'', \Omega)$, $(\pi_1(\mathcal{A}_1)'', \Omega_1)$, and $(\pi_2(\mathcal{A}_2)'', \Omega_2)$, respectively. Due to the finite-dimensionality of the corresponding Hilbert spaces, we will identify $\pi_1(\mathcal{A}_1)'' = \pi_1(\mathcal{A}_1)$, etc. Moreover, we will also write $A\Omega_1$ and $B\Omega_2$ instead of $\pi_1(A)\Omega_1$ and $\pi_2(B)\Omega_2$ for $A \in \mathcal{A}_1$, $B \in \mathcal{A}_2$ without confusion. Furthermore, we denote the finite dimension of $\mathcal{H}_2$ by n. Thus $\mathbf{B}(\mathcal{H}_2) \equiv \mathbf{B}(\mathbb{C}^n) \equiv M_n(\mathbb{C})$. To put some emphasis on the dimensionality of the "reference" subsystem $\mathcal{A}_2$, we denote by $\mathcal{P}_n$ the natural cone of $(M_n^\pi(\mathcal{A}_1), \omega_1 \otimes \omega_0)$, where $\pi(\mathcal{A}_1 \otimes M_n(\mathbb{C}))$ is denoted by $M_n^\pi(\mathcal{A}_1)$ and ω_0 is a faithful state on $M_n(\mathbb{C})$.

In order to characterize the set of PPT states, Majewski introduced the notion of the "transposed cone" $\mathcal{P}_n^\tau = (I \otimes U)\mathcal{P}_n$, where τ is transposition on $M_n(\mathbb{C})$ and U is the unitary operator given in (8.43) with the eigenvectors of the density matrix ρ_0 corresponding to ω_0.

Then the construction $\mathcal{P}_n$ and $\mathcal{P}_n^\tau$ may be realized as follows:

$$\mathcal{P}_n = \overline{\left\{\Delta^{1/4}[A_{ij}]\Omega : [A_{ij}] \in M_n^\pi(\mathcal{A}_1)^+\right\}},$$
$$\mathcal{P}_n^\tau = \overline{\left\{\Delta^{1/4}[A_{ji}]\Omega : [A_{ij}] \in M_n^\pi(\mathcal{A}_1)^+\right\}}.$$

Consequently, we arrived at the following.

Theorem 8.20 [460] *In the finite-dimensional case,*

$$\mathcal{P}_n^\tau \cap \mathcal{P}_n = \left\{\Delta^{1/4}[A_{ij}]\Omega : [A_{ij}] \geq 0, [A_{ji}] \geq 0\right\}.$$

Corollary 8.21

1. *There is a one-to-one correspondence between the set of PPT states and* $\mathcal{P}_n^\tau \cap \mathcal{P}_n$.
2. *There is a one-to-one correspondence between the set of separable states and* $\mathcal{P}_1 \otimes \mathcal{P}_2$ *(cf.* [502]).

Now we will discuss the relation between the Hilbert space description of PPT states and BO characterization. First, we note that Tomita's approach leads to the following representation of a compound state ω:

$$\omega\Bigl(\sum_i A_i \otimes B_i\Bigr) = \sum_i (\xi, A_i \otimes B_i \xi) = \sum_i \varphi_{\xi, A_i}(B_i), \tag{8.44}$$

where $\varphi_{\xi,A_i}(B_i) \equiv (\xi, A_i \otimes B_i \xi)$ and $\xi \in \mathcal{P}_n$. Here we have used the well known result from Tomita–Takesaki theory saying that for any normal state ω on a von Neumann algebra with cycling and separating vector Ω there is a unique vector ξ in the natural cone $\mathcal{P}_n$ such that $\omega(A) = (\xi, A\xi)$.

Let us observe that for $A \in \mathcal{A}_1$, $A \geq 0$, and any $B \in \mathcal{A}_2$,

$$\begin{aligned}\omega\bigl(A \otimes B^t\bigr) &= \bigl(\xi, A \otimes B^t \xi\bigr) = \bigl(\chi_{\xi_A}, B^t \chi_{\xi_A}\bigr)\\ &= (U\chi_{\xi_A}, BU\chi_{\xi_A})\\ &= (\chi_{\xi_A}, UBU\chi_{\xi_A}) = \omega(A \otimes UBU),\end{aligned}$$

where χ_{ξ_A} is the Tomita representation of $\mathrm{tr}_{\mathcal{K}_2}(\varphi(A)\cdot)$ with $\varphi(A) = \mathrm{tr}_{\mathcal{K}_1} |\xi)(\xi| A \otimes I$, and we have used the notation given in Proposition 8.19 and the fact that the partial trace $\mathrm{tr}_{\mathcal{K}_1}(\cdot)$ is a well defined conditional expectation.

As $\omega(A \otimes B)$ is linear in $\mathcal{A}$, and any A can be written as a sum of four positive elements (Jordan decomposition) the previous results can be extended to

$$\omega\bigl(A \otimes B^t\bigr) = \omega(A \otimes UBU) \tag{8.45}$$

for any $A \in \mathcal{A}_1$ and $B \in \mathcal{A}_2$.

Now we are in a position to compare the strategy by the BO approach with the Majewski approach. First, we note that the maps $\varphi_{\xi,A_i}(B_i)$ in (8.44) can be considered as

$$\mathbf{B}(\mathcal{K}_1) \ni A \mapsto \varphi_{\xi,A}(\cdot) \in \mathbf{B}(\mathcal{K}_2)_*.$$

Second, note that the positivity of a compound state ω implies

$$0 \leq \omega\Bigl(\sum_{i,j} A_i^* A_j \otimes B_i^* B_j\Bigr) = \sum\nolimits_{i,j} \varphi_{\xi, A_i^* A_j}(B_i^* B_j).$$

By using the same vector $\xi \in P_n$, let us define $\omega^\tau \in (\mathcal{A}_1 \otimes \mathcal{A}_2)_*$ by

$$\omega^\tau\Bigl(\sum_i A_i \otimes B_i\Bigr) = \sum\nolimits_i \varphi^\tau_{\xi,A_i}(B_i),$$

where $\varphi^\tau_{\xi,A_i}(B_i) \equiv (\xi, (A_i \otimes B_i^t)\xi)$. The positivity of ω^τ implies

$$\omega^\tau\Bigl(\sum_{i,j} A_i^* A_j \otimes B_i^* B_j\Bigr) = \sum_{i,j} \varphi^\tau_{\xi, A_i^* A_j}(B_i^* B_j)$$

$$= \sum_{i,j} \big(\xi, \big(A_i^* A_j \otimes (B_i^* B_j)^t\big)\xi\big)$$

$$= \sum_{i,j} \big((I \otimes U)\xi, (A_i^* A_j \otimes B_i^* B_j)\xi(I \otimes U)\big) \geq 0,$$

where in the last equality we have used (8.45). Hence, CP and co-CP of an entangling mapping are equivalent to $\xi \in \mathcal{P}_n^\tau \cap \mathcal{P}_n$. Consequently, we conclude the following.

Theorem 8.22 *The description of PPT states by $\xi \in \mathcal{P}_n^\tau \cap \mathcal{P}_n$ can be recognized as the dual description of PPT states by $\mathcal{E}/\mathcal{E}_q$.*

8.6.4 Degrees of Entanglement

It is important to measure the degree of entanglement of the entangled states as discussed above. There exist several such measures, we discuss two of them in this section. The entanglement degree for mixed states has been studied by some entropic measures such as quantum relative entropy and quantum mutual entropy. Vedral et al. [537] defined the degree of an entangled state θ as the minimum distance between θ and all disentangled states $\theta_0 \in \mathcal{D}$.

Definition 8.23 $D(\theta) \equiv \min\{d(\theta \| \theta_0); \theta_0 \in \mathcal{D}\}$, where d is any measure of a sort of distance between the two states θ and θ_0, properly defined.

As an example of d, one can use the relative entropy $S(\theta, \theta_0) \equiv \operatorname{tr}\theta(\log\theta - \log\theta_0)$, then

$$D(\theta) = \min\big\{S(\theta, \theta_0); \theta_0 \in \mathcal{D}\big\}.$$

Since this measure has to take the minimum over all disentangled states, it is difficult to compute, in particular, for an infinite-dimensional Hilbert space.

As another example of d which gives just a distance between the two states, Majewski employed the geometry of the Hilbert space [503].

Definition 8.24 Let ξ be a vector in the natural cone $\mathcal{P}$ corresponding to a normal state of a composite system $1+2$.

1. The degree of entanglement is given by

$$D_e(\xi) = \inf_\eta\big\{\|\xi - \eta\|; \eta \in \mathcal{P}_1 \otimes \mathcal{P}_2\big\}.$$

2. The degree of genuine entanglement is defined by

$$D_{ge}(\xi) = \inf_\eta\big\{\|\xi - \eta\|; \eta \in \mathcal{P}_n \cap \mathcal{P}_n^\tau\big\}.$$

We will briefly discuss the geometric idea behind these definitions. The key to the argument is the concept of convexity (in Hilbert spaces). Namely, we observe the following:

1. $\mathcal{P} \supset \mathcal{P}_1 \otimes \mathcal{P}_2$ is a convex subset.
2. $\mathcal{P} \supset \mathcal{P}_n \cap \mathcal{P}_n^\tau$ is a convex subset.
3. The theory of Hilbert spaces says that $\exists! \xi_0 \in \mathcal{P}_1 \otimes \mathcal{P}_2$ such that $D_e(\xi) = \|\xi - \xi_0\|$.
4. Analogously, $\exists! \eta_0 \in \mathcal{P}_n \cap \mathcal{P}_n^\tau$, such that $D_{ge}(\xi) = \|\xi - \eta_0\|$.

The point to note here is that we used the well known property of convex subsets in a Hilbert space: a closed convex subset W in a Hilbert space $\mathcal{H}$ contains a unique vector having the smallest norm. This ensures the existence of vectors ξ_0 and η_0 introduced respectively in (3) and (4) above.

To illustrate the above measures of entanglement, we present an example. It is based on a modification of Kadison–Ringrose argument (cf. [397] and Tomita–Takesaki Theory (see Chap. 4)).

Example 8.25 Let $\{e_1, e_2, e_3\}$ (resp., $\{f_1, f_2, f_3\}$) be an orthonormal basis in the three-dimensional Hilbert space $\mathcal{H}$ (resp., $\mathcal{K}$). By P we denote the following rank one orthogonal projector:

$$\begin{aligned} P &= \frac{1}{3} |e_1 \otimes f_1 + e_2 \otimes f_2 + e_3 \otimes f_3\rangle\langle e_1 \otimes f_1 + e_2 \otimes f_2 + e_3 \otimes f_3| \\ &\equiv |x\rangle\langle x| \in \mathbf{B}(\mathcal{H} \otimes \mathcal{K})^+. \end{aligned}$$

Let S be an operator of the form

$$S = \sum_{i=1}^{k} A_i \otimes B_i,$$

where $k < +\infty$ and $A_i \in \mathbf{B}(\mathcal{H})^+$, $B_i \in \mathbf{B}(\mathcal{K})^+$. It can be shown (see [397]) that

$$\|P - S\| \geq \frac{1}{6},$$

where $\|\cdot\|$ stands for the operator norm. Any separable state on $\mathbf{B}(\mathcal{H}) \otimes \mathbf{B}(\mathcal{K})$ can be expressed in the form

$$\varrho_0 = \sum_{i=1}^{l} \omega_{z_i} \otimes \omega_{y_i},$$

where $l < +\infty$, $z_1, \ldots, z_l \in \mathcal{H}$, $y_1, \ldots, y_l \in \mathcal{K}$, and the vector state ω_z is defined as $\omega_z \equiv (z, Az)$. Then, again, following Kadison–Ringrose exercise one can show that

$$\|\omega_x - \varrho_0\| \geq \frac{1}{6}.$$

On the other hand (see [130]), if a vector ξ (resp., vector η) $\in \mathcal{P}$ defines a normal positive form ω_ξ (resp., ω_η), then one has

$$\|\xi-\eta\|^2 \le \|\omega_\xi-\omega_\eta\| \le \|\xi-\eta\|\|\xi+\eta\|.$$

Let, in the composite system $1+2$ described in the previous subsection, $\mathcal{H}_1$ and $\mathcal{H}_2$ be three-dimensional Hilbert spaces. Recall that $\omega=\omega_1\otimes\omega_2$, where $\omega_1(\cdot)=\mathrm{tr}_{\mathcal{H}_1}(\rho_1\cdot)$ and $\omega_2(\cdot)=\mathrm{tr}_{\mathcal{H}_2}(\rho_2\cdot)$ are faithful states. Put $\omega(\cdot)=\mathrm{tr}_{\mathcal{H}_1\otimes\mathcal{H}_2}(\varrho\cdot)$. $\mathcal{A}_1\otimes\mathcal{A}_2$ is isomorphic to $\pi_{\omega_1}(\mathcal{A}_1)\otimes\pi_{\omega_2}(\mathcal{A}_2)$. Moreover, states on $\pi_{\omega_1}(\mathcal{A}_1)\otimes\pi_{\omega_2}(\mathcal{A}_2)$ are described by vectors in the natural cone $\varrho^{\frac{1}{4}}(\mathcal{A}_1\otimes\mathcal{A}_2)^+\varrho^{\frac{1}{4}}$. The inequalities discussed in the first part of the example lead to the following estimation of the degree of entanglement for the state given by the vector $\varrho^{\frac{1}{4}}\mathcal{P}\varrho^{\frac{1}{4}}$:

$$D_e\big(\varrho^{\frac{1}{4}}\mathcal{P}\varrho^{\frac{1}{4}}\big)\ge\frac{1}{12}.$$

We end this example remarking that the same arguments applied to $x'=\frac{1}{\sqrt{2}}(e_1\otimes f_1+e_2\otimes f_2)$ in the two-dimensional case lead to

$$D_e\big(\varrho^{\frac{1}{4}}\mathcal{P}\varrho^{\frac{1}{4}}\big)\ge\frac{1}{8}.$$

A computable degree of entanglement was introduced by Belavkin, Matsuoka and Ohya [39, 93, 94, 513].

We propose a new degree to measure the difference (or correlation) between the state θ and the direct tensor product $\rho\otimes\sigma$; the marginal state of θ.

Let θ be a state with the marginal density operators ρ and σ.

Definition 8.26 Let $\mathcal{H}$, $\mathcal{K}$ be separable Hilbert spaces, and let θ be a density operator in $\mathbf{B}(\mathcal{H}\otimes\mathcal{K})$ with its marginal densities denoted by ρ and σ in $\mathbf{B}(\mathcal{H})$, $\mathbf{B}(\mathcal{K})$, respectively.

The quasi-mutual entropy of ρ and σ w.r.t. θ is defined by

$$I_\theta(\rho,\sigma)\equiv\mathrm{tr}\,\theta(\log\theta-\log\rho\otimes\sigma). \tag{8.46}$$

The degree of entanglement of θ, denoted by $D_{\mathrm{EN}}(\theta)$, is defined by

$$D_{\mathrm{EN}}(\theta;\rho,\sigma)\equiv\frac{1}{2}\big\{S(\rho)+S(\sigma)\big\}-I_\theta(\rho,\sigma), \tag{8.47}$$

where $S(\cdot)$ is the von Neumann entropy.

Recalling that, for density operators θ, γ in $\mathbf{B}(\mathcal{H})$, the relative entropy θ and γ is defined by

$$S(\theta,\gamma)\equiv\mathrm{tr}\,\theta(\log\theta-\log\gamma),$$

we see that $I_\theta(\rho,\sigma)$ is the relative entropy of the tensor product of its marginal states ρ and σ of θ. Since it is known that the relative entropy is a kind of difference between states, it is clear why the degree of entanglement of θ given by (8.47) is a measure of how far θ is from the product state $\rho\otimes\sigma$. Moreover, we can see that D_{EN} is a kind of symmetrized quantum conditional entropy. In the classical case, the conditional entropy always takes non-negative values; however, our new criterion D_{EN} can be negative according to the strength of correlation between ρ and σ [40].

If $\theta\in\mathfrak{S}_1\otimes\mathfrak{S}_2$ is an entangled pure state with the marginal states ρ,σ, then von Neumann entropy $S(\theta)=0$. Moreover, from the Araki–Lieb inequality [61]:

$$\left|S(\rho)-S(\sigma)\right|\le S(\theta)\le S(\rho)+S(\sigma),$$

we have $S(\rho)=S(\sigma)$. It follows that

$$\begin{aligned} I_\theta(\rho,\sigma) &= \operatorname{tr}\theta(\log\theta-\log\rho\otimes\sigma) \\ &= S(\rho)+S(\sigma)-S(\theta)=2S(\rho). \end{aligned}$$

That is, for an entangled pure state, the quasi-mutual entropy is twice the von Neumann (reduced) entropy.

From these facts, the following theorem follows.

Theorem 8.27 *For a pure state θ with the marginal states ρ and σ,*

1. *θ is entangled iff $D_{\mathrm{EN}}(\theta;\rho,\sigma)=-\frac{1}{2}\{S(\rho)+S(\sigma)\}<0$, and*
2. *θ is separable iff $D_{\mathrm{EN}}(\theta;\rho,\sigma)=0$.*

Proof (Part 1) In the case of a pure state θ given by $\theta=|\Psi\rangle\langle\Psi|$ where $|\Psi\rangle$ is a normalized vector in $\mathcal{H}\otimes\mathcal{K}$, $D_{\mathrm{EN}}(\theta;\rho,\sigma)$ can be computed as

$$\begin{aligned} D_{\mathrm{EN}}(\theta;\rho,\sigma) &\equiv \frac{1}{2}\bigl\{S(\rho)+S(\sigma)\bigr\}-I_\theta(\rho,\sigma) \\ &= -\frac{1}{2}\bigl\{S(\rho)+S(\sigma)\bigr\} \\ &= -S(\rho)\quad\bigl(\text{or }=-S(\sigma)\bigr). \end{aligned} \tag{8.48}$$

If $D_{\mathrm{EN}}(\theta;\rho,\sigma)<0$, then $S(\rho)=S(\sigma)>0$, which means that ρ and σ are mixture states. When ρ can be written as $\rho=\sum_i\lambda_i|x_i\rangle\langle x_i|$ where $\{|x_i\rangle\}$ is an ONB in $\mathcal{H}$ and $\sum_i\lambda_i=1$, $0\le\lambda_i\le1$, then due to the Schmidt decomposition there exists an ONB $\{|y_i\rangle\}\subset\mathcal{K}$ such that

$$|\Psi\rangle=\sum_i\sqrt{\lambda_i}|x_i\rangle\otimes|y_i\rangle.$$

This implies that ω is a pure entangled state. The converse statement obviously holds.

(Part 2) If $D_{\mathrm{EN}}(\theta;\rho,\sigma)=0$, then $S(\rho)=S(\sigma)=0$, which means that ρ and σ are pure states. When ρ can be written as $\rho=|x\rangle\langle x|$, then there exists a normalized vector $|y\rangle \in \mathcal{K}$ such that

$$|\Psi\rangle = |x\rangle \otimes |y\rangle.$$

This means that ω is a pure separable state. The converse statement obviously holds. □

Theorem 8.28 *$D_{\mathrm{EN}}(\theta; \rho, \sigma)$ is non-negative, i.e., $D_{\mathrm{EN}}(\theta; \rho, \sigma) \geq 0$, if a state θ is separable. Equivalently, θ is entangled if $D_{\mathrm{EN}}(\theta; \rho, \sigma)$ is negative, i.e., $D_{\mathrm{EN}}(\theta; \rho, \sigma) < 0$.*

Proof Let θ be a state on $\mathbf{B}(\mathcal{H}_1 \otimes \mathcal{H}_2)$. If θ is separable, there exist density matrices ρ_n, σ_n in $\mathbf{B}(\mathcal{H}_1)$, $\mathbf{B}(\mathcal{H}_2)$, respectively, such that

$$\theta = \sum_n p_n \rho_n \otimes \sigma_n$$

with

$$p_n \geq 0, \quad \forall n, \sum_n p_n = 1.$$

Let $\{x_n\}$ be an ONB in $\mathcal{H}_1$, and define the completely positive unital $(CP1)$ map $\Lambda_0 : \mathbf{B}(\mathcal{H}_1) \to \mathbf{B}(\mathcal{H}_1)$ by

$$\Lambda_0(A) = \sum_n \mathrm{tr}(A\rho_n) x_n x_n^*, \quad A \in \mathbf{B}(\mathcal{H}_1). \tag{8.49}$$

Then its dual is

$$\Lambda_0^*(\delta) = \sum_n \langle x_n, \delta x_n \rangle \rho_n, \quad \delta \in \mathbf{B}(\mathcal{H}_1)_*, \tag{8.50}$$

so that defining the $CP1$ map

$$\Lambda \equiv \Lambda_0 \otimes \mathrm{id} : \mathbf{B}(\mathcal{H}_1 \otimes \mathcal{H}_2) \to \mathbf{B}(\mathcal{H}_1 \otimes \mathcal{H}_2)$$

and the density matrix

$$\theta_d \equiv \sum_n p_n x_n x_n^* \otimes \sigma_n,$$

one easily verifies that

$$\Lambda^*(\theta_d) = \theta.$$

Moreover, denoting

$$\rho = \sum_n p_n \rho_n \quad \text{and} \quad \sigma = \sum_n p_n \sigma_n,$$

the marginal densities of θ and $\rho_d = \sum_n p_n |x_n\rangle\langle x_n|$, the first marginal density of θ_d, one has:

$$\Lambda^*(\rho_d \otimes \sigma) = \rho \otimes \sigma.$$

Recall now that the monotonicity property of the relative entropy (see [578] for a proof and history) that for any pair of von Neumann algebras $\mathcal{M}$, $\mathcal{M}^0$, for any normal $CP1$ map $\Lambda : \mathcal{M} \to \mathcal{M}^0$, and for any pair of normal states ω_0, φ_0 on $\mathcal{M}^0$, one has

$$R\big(\Lambda^*(\omega_0)|\Lambda^*(\varphi_0)\big) \le R(\omega_0|\varphi_0).$$

Using this property, one finds

$$\begin{aligned} I_\theta(\rho,\sigma) &= R(\theta|\rho\otimes\sigma) \\ &= R\big(\Lambda^*(\theta_d)|\Lambda^*(\rho_d\otimes\sigma)\big) \\ &\le R(\theta_d|\rho_d\otimes\sigma) = I_{\theta_d}(\rho_d,\sigma) \end{aligned}$$

so that

$$S(\sigma) - I_\theta(\rho,\sigma) \ge S(\sigma) - I_{\theta_d}(\rho_d,\sigma) = -\sum_n p_n \operatorname{tr}(\sigma_n \log \sigma_n) \ge 0. \tag{8.51}$$

Introducing the density operator

$$\hat{\theta}_d = \sum_n p_n \rho_n \otimes y_n y_n^*$$

where $\{y_n\}$ is an ONB in $\mathcal{H}_2$, and using a variant of the above argument (in which the density θ_d is replaced by $\hat{\theta}_d$), one proves the analogous inequality

$$S(\rho) - I_\theta(\rho,\sigma) \ge S(\rho) - I_{\hat{\theta}_d}(\rho,\sigma_d) = -\sum_n p_n \operatorname{tr}(\rho_n \log \rho_n) \ge 0. \tag{8.52}$$

Combining (8.51) and (8.52), one obtains

$$\begin{aligned} D_{\mathrm{EN}}(\theta;\rho,\sigma) &= \frac{1}{2}\Big(\big(S(\sigma) - I_\theta(\rho,\sigma)\big) + \big(S(\rho) - I_\theta(\rho,\sigma)\big)\Big) \\ &\ge \frac{1}{2}\bigg(-\sum_n p_n \operatorname{tr}(\rho_n\log\rho_n) - \sum_n p_n \operatorname{tr}(\sigma_n \log\sigma_n)\bigg) \ge 0. \end{aligned}$$ □

Definition 8.29 A state θ (i.e., a density operator) on the tensor product $\mathcal{H}\otimes\mathcal{K}$ of two Hilbert spaces $\mathcal{H}$ and $\mathcal{K}$ is called a product state if it can be represented in the form

$$\theta = \rho \otimes \sigma,$$

where ρ and σ are states in $\mathcal{H}$ and $\mathcal{K}$. A correlated state is a state not belonging to the set of all product states, so a correlated state is simply considered as a non-product state.

Remark 8.30 The above measure $D_{\mathrm{EN}}(\theta;\rho,\sigma)$, in fact, measures the distance between a correlated and a product state.

Now we consider the strength of the correlation.

Definition 8.31

1. θ_1 *has stronger correlation than* θ_2 *iff*

$$D_{\mathrm{EN}}(\theta_1;\rho,\sigma) < D_{\mathrm{EN}}(\theta_2;\rho,\sigma).$$

2. θ is said to be essentially entangled iff

$$D_{\mathrm{EN}}(\theta;\rho,\sigma) < 0.$$

8.6.5 Models of Entanglement in Circulant States

Before we exhibit concrete computations of DEN for some models, we recall the definition of a circulant state suggested by Chruściński and Kossakowski [183].

Consider the finite-dimensional Hilbert space $\mathbb{C}^d$ ($d \in \mathbb{N}$) with the standard basis $\{e_0, e_1, \dots, e_{d-1}\}$. Let Σ_0 be the subspace of $\mathbb{C}^d \otimes \mathbb{C}^d$ generated by $e_i \otimes e_i$ ($i = 0, 1, \dots, d-1$):

$$\Sigma_0 = \overline{\mathrm{span}}\{e_0 \otimes e_0, e_1 \otimes e_1, \dots, e_{d-1} \otimes e_{d-1}\}.$$

For any non-negative integer α, we define the operator S^α on $\mathbb{C}^d$ by extending linearly the map

$$e_k \longmapsto e_{k+\alpha (\mathrm{mod}\ d)} \quad (k = 0, 1, \dots, d-1)$$

and denote by Σ_α the image obtained by applying $\mathrm{I}_d \otimes S^\alpha$ to Σ_0, i.e., $\Sigma_\alpha = (\mathrm{I}_d \otimes S^\alpha)\Sigma_0$.

It can be easily checked that Σ_α and Σ_β ($\alpha \neq \beta$) are orthogonal to each other, and

$$\mathbb{C}^d \otimes \mathbb{C}^d = \Sigma_0 \oplus \Sigma_1 \oplus \cdots \oplus \Sigma_{d-1}.$$

This decomposition is called a circulant decomposition.

Let $\rho_0, \rho_1, \dots, \rho_{d-1}$ be positive $d \times d$ matrices with entries in $\mathbb{C}$ which satisfy

$$\mathrm{tr}(\rho_0 + \cdots + \rho_{d-1}) = 1.$$

For each matrix ρ_α ($\alpha = 0, 1, \ldots, d-1$), we define a new operator $[\rho_\alpha]$ on $\mathbb{C}^d \otimes \mathbb{C}^d$ as

$$[\rho_\alpha] = \sum_{i,j=0}^{d-1} \langle e_i, \rho_\alpha e_j \rangle e_{ij} \otimes S^\alpha e_{ij} (S^\alpha)^*,$$

where e_{ij} denotes $|e_i\rangle\langle e_j|$. Since $S^k e_{ij} (S^k)^* = e_{i+k,j+k}$, $[\rho_\alpha]$ is also written in the form

$$[\rho_\alpha] = \sum_{i,j=0}^{d-1} \langle e_i, \rho_\alpha e_j \rangle e_{ij} \otimes e_{i+\alpha,j+\alpha}.$$

From the simple calculation, we can verify that the operator

$$[\rho] \equiv \sum_{\alpha=0}^{d-1} [\rho_\alpha]$$

is positive, and the trace of $[\rho]$ is equal to 1. That is, $[\rho]$ is a state. We call this state a circulant state. For further details about a circulant state, we refer to [183].

8.6.6 *Computation of DEN*

Circulant State Model 1

The following example of circulant states is given by Horodeckis [351].

Let us consider the finite dimensional Hilbert space $\mathcal{H} = \mathbb{C}^3$ with a standard basis $\{e_0, e_1, e_2\}$. For any $\alpha \in \mathbb{R}$ such that $2 \le \alpha \le 5$, we define the density matrix $\theta(\alpha)$ on $\mathcal{H} \otimes \mathcal{H}$ as

$$\theta(\alpha) = \frac{2}{7} |\psi\rangle\langle\psi| + \frac{\alpha}{7} \theta_+ + \frac{5-\alpha}{7} \theta_-,$$

where

$$\psi = \frac{1}{\sqrt{3}} (e_0 \otimes e_0 + e_1 \otimes e_1 + e_2 \otimes e_2),$$

$$\theta_+ = \frac{1}{3} \bigl\{ |e_0\rangle\langle e_0| \otimes |e_1\rangle\langle e_1| + |e_1\rangle\langle e_1| \otimes |e_2\rangle\langle e_2| + |e_2\rangle\langle e_2| \otimes |e_0\rangle\langle e_0| \bigr\},$$

$$\theta_- = \frac{1}{3} \bigl\{ |e_1\rangle\langle e_1| \otimes |e_0\rangle\langle e_0| + |e_2\rangle\langle e_2| \otimes |e_1\rangle\langle e_1| + |e_0\rangle\langle e_0| \otimes |e_2\rangle\langle e_2| \bigr\}.$$

Then, the marginal states ρ, σ of $\theta(\alpha)$ are given by

$$\rho = \sigma = \frac{1}{3} \bigl\{ |e_0\rangle\langle e_0| + |e_1\rangle\langle e_1| + |e_2\rangle\langle e_2| \bigr\}.$$

Besides, the von Neumann entropy of each $\theta(\alpha)$ can be calculated as follows:

$$S\bigl(\theta(\alpha)\bigr) = -\frac{2}{7}\log\frac{2}{7} - \frac{\alpha}{7}\log\frac{\alpha}{7} - \frac{5-\alpha}{7}\log\frac{5-\alpha}{7} + \frac{5}{7}\log 3,$$
$$S(\rho) = S(\sigma) = \log 3.$$

From these calculations, it follows that

$$D_{\mathrm{EN}}\bigl(\theta(\alpha);\rho,\sigma\bigr) = -\frac{2}{7}\log\frac{2}{7} - \frac{\alpha}{7}\log\frac{\alpha}{7} - \frac{5-\alpha}{7}\log\frac{5-\alpha}{7} - \frac{2}{7}\log 3 > 0.$$

Remark 8.32 The separability of the state $\theta(\alpha)$ is classified by α as follows:

1. $\theta(\alpha)$ is separable iff $2 \le \alpha \le 3$.
2. $\theta(\alpha)$ is both PPT and entangled iff $3 < \alpha \le 4$.
3. $\theta(\alpha)$ is NPT iff $4 < \alpha \le 5$.

Let us compare the above classification with the positivity of DEN of $\theta(\alpha)$. We know that $\theta(\alpha)$ does not have purely quantum correlation in the sense of Definition 8.31 even if $\theta(\alpha)$ is entangled, i.e., $\theta(\alpha)$ is not an essentially entangled state.

Circulant State Model 2

In the same setting as above, we give density matrices on $\mathcal{H}\otimes\mathcal{H}$ by

$$\hat{\theta}_1 = \bigl|\sqrt{3}\psi\bigr\rangle\bigl\langle\sqrt{3}\psi\bigr|, \qquad \hat{\theta}_2 = 3\varepsilon\theta_+, \qquad \hat{\theta}_3 = 3\varepsilon\theta_-.$$

Using these matrices, we define $\vartheta(\varepsilon)$ by

$$\vartheta(\varepsilon) = \frac{1}{1+\varepsilon+1/\varepsilon}\theta_1 + \frac{\varepsilon}{1+\varepsilon+1/\varepsilon}\theta_2 + \frac{1/\varepsilon}{1+\varepsilon+1/\varepsilon}\theta_3,$$

where $\theta_k = \frac{\hat{\theta}_k}{\operatorname{tr}\hat{\theta}_k}$, $k = 1, 2, 3$.

It can be verified that the marginal states of $\vartheta(\varepsilon)$ are the same as in the above example. Noting that θ_1 is pure, we obtain

$$\begin{aligned} D_{\mathrm{EN}}\bigl(\vartheta(\varepsilon);\rho,\sigma\bigr) = &-\frac{1}{1+\varepsilon+1/\varepsilon}\log\frac{1}{1+\varepsilon+1/\varepsilon} \\ &-\frac{\varepsilon}{1+\varepsilon+1/\varepsilon}\log\frac{\varepsilon}{1+\varepsilon+1/\varepsilon} \\ &-\frac{1/\varepsilon}{1+\varepsilon+1/\varepsilon}\log\frac{1/\varepsilon}{1+\varepsilon+1/\varepsilon} > 0. \end{aligned} \tag{8.53}$$

Remark 8.33 The separability of the state $\vartheta(\varepsilon)$ is classified by ε as follows [389]:

1. $\vartheta(\varepsilon)$ is separable if $\varepsilon = 1$.
2. $\vartheta(\varepsilon)$ is both PPT and entangled for $\varepsilon > 1$.

As in Model 1, $\vartheta(\varepsilon)$ is also not an essentially entangled state because of the positivity in (8.53).

We can compare two DEN for the models $\theta(\alpha)$ and $\vartheta(\varepsilon)$ mentioned above because they have the same marginals ρ and σ. If $\varepsilon = 1$ and $\alpha = 3.1$, then we have

$$0.732 \approx D_{\mathrm{EN}}\big(\vartheta\ (\varepsilon = 1)\big) < D_{\mathrm{EN}}\big(\theta\ (\alpha = 3.1)\big) \approx 0.759.$$

This inequality means that the separable state ϑ $(\varepsilon = 1)$ has a stronger correlation than that of the entangled state θ $(\alpha = 3.1)$ in the sense of Definition 8.31.

From the above observations, the concept of "quantum correlation" might be more appropriate than the concepts of "entanglement" and "being separable" to classify quantum states, although we follow the conventional usage in this book.

8.7 Entangled Markov Chains and Their Properties

We discuss here quantum generalization of the classical random walks. The relevance of this problem for quantum information has been emphasized in many papers. However, the given proposals introduce some features which are not satisfactory from the mathematical point of view. Motivated by such a situation, Accardi and Fidaleo introduced the notion of entangled Markov chains, including the quantum random walk [38]. They listed the requirements that should be fulfilled by any candidate definition of a quantum random walk, namely,

1. It should be a quantum Markov chain [3].
2. It should be purely generated in the sense of Fannes, Nachtergaele and Werner [231].
3. Its restriction on at least one maximal abelian subalgebra should be a classical random walk.
4. It should be uniquely determined, up to arbitrary phases, by its classical restriction.

In this section, we discuss entangled Markov chains, and we show that they indeed form entangled states, giving one example in a higher-dimensional Hilbert space of dimension bigger than four.

8.7.1 Entangled Quantum Markov Chains

In order to give an intuitive idea of the connection of their construction with entanglement, let us note that the key characteristic of entanglement is the superposition principle and the corresponding interpretation of the amplitudes as "complex square roots of probabilities". This suggest an approach in which, given a classical Markov chain with finite state space S, a stochastic matrix T, and an initial distribution for it described by a row vector P, one can construct such a quantum Markov chain. At first we review briefly how the entangled Markov chains in [38] can be constructed.

Let us first define a quantum Markov chain for a C^*-algebra $\mathcal{A}$.

Definition 8.34 A state φ on $\otimes\mathcal{A}$ is a quantum Markov chain (QMC for short) with an initial state φ_0 over $\mathcal{A}$ and transition expectation $\mathcal{E}:\mathcal{A}\otimes\mathcal{A}\mapsto\mathcal{A}$ if

$$\begin{aligned}&\varphi(A_0\otimes A_1\otimes\cdots\otimes A_n\otimes 1\otimes 1\otimes\cdots)\\&\quad=\varphi_0\big[\mathcal{E}\big(A_0\otimes\cdots\mathcal{E}\big(A_{n-2}\otimes\mathcal{E}\big(A_{n-1}\otimes\mathcal{E}(A_n\otimes 1)\big)\big)\big)\big].\end{aligned}$$

We will discuss more about the quantum Markov chain in Chap. 20 in terms of the concept of lifting. Let $S=\{e_1,e_2,\dots,e_d\}(\equiv\{1,2,\dots,d\})$ be a state space with cardinality $|S|=d(<\infty)$. We consider a sequence of identical independent distributed (i.i.d. for short) random variables

$$\xi_n:(\Omega,\mathcal{F},P)\to S$$

with distribution $P(\xi_n=j\in S)=p_j$, here $n=1,2,\dots$. Let (S_n) be a classical Markov chain with state space S. For example, in this notation

$$S_n\equiv S_0+\sum_{k=1}^{n}\xi_k=S_{n-1}+\xi_n.$$

Fix an orthonormal basis (ONB for short) $\{|e_i\rangle\}_{i\le d}$ of $\mathbb{C}^{|S|}$ and fix a vector $|e_0\rangle$ in this basis. We consider the infinite tensor product Hilbert space

$$\mathcal{H}_{\mathbb{N}}\equiv\otimes^{\mathbb{N}}\mathbb{C}^{|S|}.$$

Let $T=(t_{ij})$ be any stochastic matrix (i.e., $\sum_j t_{ij}=1$) and let $\sqrt{t_{ij}}$ be any complex square root of t_{ij} (i.e., $t_{ij}\ge 0$, $|\sqrt{t_{ij}}|^2=t_{ij}$). Define the vector

$$|\Psi_n\rangle=\sum_{j_0,\dots,j_n}\sqrt{p_{j_0}}\prod_{\alpha=0}^{n-1}\sqrt{t_{j_\alpha j_{\alpha+1}}}|e_{j_0},\dots,e_{j_n}\rangle$$

where $|e_{j_0},\dots,e_{j_n}\rangle\equiv(\bigotimes_{\alpha\in[0,n]}|e_{j_\alpha}\rangle)$.

Let $M_{|S|}$ denote the $|S|\times|S|$ (i.e., $d\times d$) complex matrix algebra, and let $\mathcal{A}=M_{|S|}\otimes M_{|S|}\otimes\cdots=\otimes^{\mathbb{N}}M_{|S|}$ be the C*-infinite tensor product of $\mathbb{N}$-copies of $M_{|S|}$.

Definition 8.35 An observable A_{Λ^*} is said to be localized in a finite region $\Lambda^*\subseteq\mathbb{N}$ if there exist an operator $\overline{A}_{\Lambda^*}\in\otimes_{\Lambda^*}M_{|S|}$ such that

$$A_{\Lambda^*}=\overline{A}_{\Lambda^*}\otimes I_{\Lambda^{*c}}.$$

We denote by $\mathcal{A}_{\Lambda^*}$ the local algebra at Λ^*, and in the following we will identify $A_{\Lambda^*} = \overline{A}_{\Lambda^*}$.

The basic property of $|\Psi_n\rangle$ is that, although the limit $\lim_{n\to\infty} |\Psi_n\rangle$ does not exist, the following holds:

Lemma 8.36 *For every local observable* $A \in \mathcal{A}_{[0,\,k]}$ $(k \in \mathbb{N})$, *one has*

$$\langle \Psi_{k+1}, A\Psi_{k+1}\rangle = \lim_{n\to\infty} \langle \Psi_n, A\Psi_n\rangle =: \varphi(A). \tag{8.54}$$

Accardi and Fidaleo showed that the state φ defined by (8.54) is a quantum Markov chain in the sense above, and they called a family of quantum Markov chains with the above construction the entangled Markov chains.

For entangled Markov chains the transition expectation $\mathcal{E}$ is expressed in terms of the following linear map:

Definition 8.37 Define the linear map $V_n : \mathcal{H}_n \to \mathcal{H}_n \otimes \mathcal{H}_{n+1}$ by the linear extension of

$$V_n |e_{j_n}\rangle = \sum_{j_{n+1}\in S} \sqrt{t_{j_n j_{n+1}}}\, |e_{j_n}\rangle \otimes |e_{j_{n+1}}\rangle,$$

where $\mathcal{H}_n = \mathcal{H}_{n+1} = \mathbb{C}^{|S|}$ for each $n \in \mathbb{N}$.

It is easy to show that $V_n^* V_n = 1$. Moreover, $\mathcal{E}_n(\cdot) \equiv V_n^* \cdot V_n : \mathcal{A}_n \otimes \mathcal{A}_{n+1} \to \mathcal{A}_n$ becomes a transition expectation, and its dual $\mathcal{E}_n^* : \mathcal{A}_n^* \to (\mathcal{A}_n \otimes \mathcal{A}_{n+1})^*$ becomes a linear lifting in the sense of [19], where $\mathcal{A}_n = \mathcal{A}_{n+1} = M_{|S|}$ for each $n \in \mathbb{N}$.

Now let us extend V_n to an isometry still denoted by the same symbol

$$V_n : \bigotimes_{\alpha\in[0,n]} \mathcal{H}_\alpha \to \bigotimes_{\alpha\in[0,n+1]} \mathcal{H}_\alpha \tag{8.55}$$

by the prescription

$$V_n \bigotimes_{\alpha\in[0,n]} |e_{j_\alpha}\rangle \equiv \Bigl(\bigotimes_{\alpha\in[0,n-1]} |e_{j_\alpha}\rangle\Bigr) \otimes V_n |e_{j_n}\rangle.$$

It is easily shown that for each $j_0 \in S$ one has

$$|\Psi_n\rangle = \sum_{j_0,\dots,j_n} \sqrt{p_{j_0}} \prod_{\alpha=0}^{n-1} \sqrt{t_{j_\alpha j_{\alpha+1}}}\, |e_{j_0},\dots,e_{j_n}\rangle = \sum_{j_0} \sqrt{p_{j_0}}\, V_{n-1}\cdots V_0 |e_{j_0}\rangle. \tag{8.56}$$

We give an initial pure state φ_0 as

$$\varphi_0(\cdot) = \mathrm{tr}_{\mathcal{H}}\bigl(|\Psi_0\rangle\langle\Psi_0|\cdot\bigr) = \langle\Psi_0|\cdot|\Psi_0\rangle. \tag{8.57}$$

Then from (8.56) and (8.57) we define a pure state φ_n over $\bigotimes_{j\in[0,\,n]}\mathcal{A}_j$ by using the extended lifting $\mathcal{E}_n^*$ given by $\mathcal{E}_n^*(|\Psi_n\rangle\langle\Psi_n|)\equiv V_n|\Psi_n\rangle\langle\Psi_n|V_n^*$, namely,

$$\varphi_n(\cdot)\equiv \operatorname{tr}\bigl(\mathcal{E}_{n-1}^*\bigl(\mathcal{E}_{n-2}^*\bigl(\cdots\bigl(\mathcal{E}_1^*\bigl(\mathcal{E}_0^*\bigl(|\Psi_0\rangle\langle\Psi_0|\bigr)\bigr)\bigr)\bigr)\bigr)(\cdot)\bigr).$$

Definition 8.38 An entangled Markov chain (EMC) is a quantum Markov chain $\varphi\equiv\{\varphi_0,\mathcal{E}\}$ over $\mathcal{A}$ where (i) φ_0 is a pure state over $M_{|S|}$, (ii) a transition expectation $\mathcal{E}(\cdot)\equiv V^*\cdot V$ is given by (8.55) for some stochastic matrix $T=(t_{ij})$ and for some fixed ONB $\{|e_i\rangle\}$.

From Lemma (8.54), one has

$$\varphi=\lim_{n\to\infty}\varphi_n.$$

In [38], the entanglements of φ_n and φ are checked. We will discuss the entanglement of EMC following [38]. First, we give three definitions of the entangled compound state:

Definition 8.39 Let $\mathcal{A}_1$ and $\mathcal{A}_2$ be C^*-algebras, then $\omega\in\mathfrak{S}(\mathcal{A}_1\otimes\mathcal{A}_2)$ is entangled if

$$\omega\notin\overline{\operatorname{Conv}}\bigl\{\omega_1\otimes\omega_2;\,\omega_j\in\mathfrak{S}(\mathcal{A}_j),\ j=1,2\bigr\}.$$

Definition 8.40 Let $\mathcal{A}_j$ $(j\in\{1,2,\dots,n\})$ be C^*-algebras, then $\omega\in\mathfrak{S}(\bigotimes_{j=1}^n\mathcal{A}_j)$ is entangled if

$$\omega\notin\overline{\operatorname{Conv}}\left\{\bigotimes_{j=1}^n\omega_j;\ \omega_j\in\mathfrak{S}(\mathcal{A}_j),\ j\in\{1,2,\dots,n\}\right\}.$$

Definition 8.41 Let $\mathcal{A}_j$ $(j\in\{1,2,\dots,\infty\})$ be C^*-algebras, then $\omega\in\mathfrak{S}(\bigotimes_{j=1}^\infty\mathcal{A}_j)$ is entangled if

$$\omega\notin\overline{\operatorname{Conv}}\left\{\bigotimes_{j=1}^\infty\omega_j;\ \omega_j\in\mathfrak{S}(\mathcal{A}_j),\ j\in\{1,2,\dots,\infty\}\right\}.$$

The entanglement of φ_n can be analyzed in the following sense:

Definition 8.42 Let $\mathcal{A}=\bigotimes_{j=1}^n\mathcal{A}_j$ be divided as $\mathcal{A}=\mathcal{A}_{k]}\otimes\mathcal{A}_{(k}\equiv\mathcal{A}_{[1,k]}\otimes\mathcal{A}_{(k,n]}$, then

1. $\omega\in\mathfrak{S}(\mathcal{A})$ is 2-entangled if
$$\omega\notin\overline{\operatorname{Conv}}\bigl\{\omega_{k]}\otimes\omega_{(k};\omega_{k]}\in\mathfrak{S}(\mathcal{A}_{k]}),\ \omega_{(k}\in\mathfrak{S}(\mathcal{A}_{(k)})\bigr\},\quad \forall k\in\{1,2,\dots,n\}.$$
2. $\omega\in\mathfrak{S}(\mathcal{A})$ is 2-separable if
$$\omega\in\overline{\operatorname{Conv}}\bigl\{\omega_{k]}\otimes\omega_{(k};\omega_{k]}\in\mathfrak{S}(\mathcal{A}_{k]}),\ \omega_{(k}\in\mathfrak{S}(\mathcal{A}_{(k)})\bigr\},\quad \forall k\in\{1,2,\dots,n\}.$$

Lemma 8.43 *If $\omega \in \mathfrak{S}(\mathcal{A})$ is 2-entangled, then ω is entangled.*

Proof Clear from the definition. □

The 2-entangled entangled Markov chains can be characterized by computing their degrees of entanglement.

In the next subsection, it will be shown that the 2-separability condition of the EMC generated by a deterministic stochastic matrix is equivalent to the condition of Theorem 8.27 for its separability.

The DEN $D_{\mathrm{EN}}(|\Psi_n\rangle\langle\Psi_n|)$ is that of EMC φ in the interval $[0, n]$, which suggests the following definition of the EMC φ.

Definition 8.44 The DEN of φ is defined by

$$D_{\mathrm{EN}}(\varphi) \equiv \lim_{n\to\infty} D_{\mathrm{EN}}\bigl(|\Psi_n\rangle\langle\Psi_n|\bigr), \tag{8.58}$$

where $D_{\mathrm{EN}}(|\Psi_n\rangle\langle\Psi_n|) = \min_{k\in[1,n]} D_{\mathrm{EN}}(|\Psi_n\rangle\langle\Psi_n|; \rho_{k]}, \sigma_{(k})$ with $\rho_{k]} = \mathrm{tr}_{\mathcal{H}_{(k}} |\Psi_n\rangle\langle\Psi_n|$ and $\sigma_{(k} = \mathrm{tr}_{\mathcal{H}_{k]}} |\Psi_n\rangle\langle\Psi_n|$.

8.7.2 Entangled Markov Chains Generated by Unitary Implementable Matrices

We consider the particular case in which a transition matrix T is given by a unitary implementable matrix, and we will show that the EMC φ of a unitary implementable matrix has the strongest entanglement. We start from an invariant measure $P = (p_i)$ of the stochastic matrix $T = (t_{ij})$ with non-zero elements (i.e., $t_{ij} > 0$ for any i, $j \in S$). Then the following theorem holds.

Theorem 8.45 *To every stochastic matrix T we associate the density matrix σ_T given as*

$$\sigma_T \equiv \sum_i p_i |f_i\rangle\langle f_i|,$$

where $|f_i\rangle = \sum_k \sqrt{t_{ik}}|e_k\rangle$. Then

1. *The state φ_n is a pure 2-separable state for any $n < \infty$ iff $S(\sigma_T) = 0$.*
2. *The state φ_n is a pure 2-entangled state for any $n < \infty$ iff $S(\sigma_T) > 0$.*
3. *There always exists the DEN of φ such that*

$$-S(P) \le D_{\mathrm{EN}}(\varphi) = -S(\sigma_T) \le 0,$$

where $S(P)$ is the Shannon entropy of a probability measure P.

To prove this theorem we need a lemma.

Lemma 8.46 *Put* $\sigma(1) \equiv \sigma_T$ *and define the density matrix* $\sigma(m)$ *as*

$$\sigma(m) \equiv \mathcal{E}_{m-1}^*\big(\mathcal{E}_{m-2}^*\big(\cdots\big(\mathcal{E}_1^*\big(\sigma(1)\big)\big)\big)\big).$$

Then

$$S\big(\sigma(1)\big) = S\big(\sigma(m)\big), \quad \forall m \in [1, \infty).$$

Proof If $\sigma(1)$ is a pure state, then $\sigma(m)$ is a pure state. The claim of the lemma follows.

Assume that $\sigma(1)$ is a mixture state. This assumption means that the rank of $\sigma(1)$ is bigger than 2. Put $k = \text{rank}(\sigma(1))$, then $\sigma(1)$ has a Schatten decomposition

$$\sigma(1) = \sum_{l=1}^{k} \lambda_l \big|g_l(1)\big\rangle\big\langle g_l(1)\big|,$$

where $\lambda_l > 0$ ($l \in [1, k]$), $\sum_{l=1}^{k} \lambda_l = 1$, and $\{|g_l(1)\rangle\}$ is a set of orthonormal vectors. The vector $|g_l(1)\rangle$ can be represented by using the ONB $\{|e_i\rangle\}$

$$\big|g_l(1)\big\rangle = \sum_i \mu(i; l)|e_i\rangle$$

where $\sum_i |\mu(i; l)|^2 = 1$. Due to orthogonality of $\{|g_l(1)\rangle\}$, one has

$$\big\langle g_j(1), g_l(1)\big\rangle = \sum_i \mu(i; j)^* \mu(i; l) = \delta_{j,l}.$$

Using the set $\{|g_l(1)\rangle\}$, the density operator $\sigma(m)$ is given as

$$\sigma(m) = \sum_{l=1}^{k} \lambda_l \mathcal{E}_{m-1}^*\big(\cdots\big(\mathcal{E}_1^*\big(\big|g_l(1)\big\rangle\big\langle g_l(1)\big|\big)\big)\big) = \sum_{l=1}^{k} \lambda_l \big|g_l(m)\big\rangle\big\langle g_l(m)\big|,$$

where

$$\begin{aligned}\big|g_l(m)\big\rangle &= V_{m-1} \cdots V_1 \big|g_l(1)\big\rangle \\ &= \sum_{i_1,\ldots,i_{m+1}} \mu(i_1; l) \prod_{\alpha=1}^{m-1} \sqrt{t_{i_\alpha i_{\alpha+1}}} |e_{i_1}, \ldots, e_{i_m}\rangle.\end{aligned}$$

Then $\{|g_l(m)\rangle\}$ becomes also a set of orthonormal vectors in $\bigotimes_{j=1}^{m} \mathcal{H}_j$. In fact, one can compute

$$\begin{aligned}\big\langle g_j(m), g_l(m)\big\rangle &= \sum_{i_1,\ldots,i_m} \mu(i_1; j)^* \mu(i_1; l) \prod_{\alpha=1}^{m-1} t_{i_\alpha i_{\alpha+1}} \\ &= \sum_{i_1} \mu(i_1; j)^* \mu(i_1; l) = \delta_{j,l}.\end{aligned}$$

Therefore, one has

$$S\big(\sigma(1)\big) = S\big(\sigma(m)\big) = -\sum_{l=1}^{k} \lambda_l \log \lambda_l. \qquad \square$$

Proof of Theorem 8.45 Now φ_n is given by

$$\varphi_n(\cdot) = \operatorname{tr} |\Psi_n\rangle\langle\Psi_n|\cdot,$$

where $|\Psi_n\rangle = \sum_{j_0,\dots,j_n} \sqrt{p_{j_0}} \prod_{\alpha=0}^{n-1} \sqrt{t_{j_\alpha j_{\alpha+1}}} |e_{j_0},\dots,e_{j_n}\rangle$. From the purity of φ_n one has

$$D_{\mathrm{EN}}\big(|\Psi_n\rangle\langle\Psi_n| : \rho_{k]}, \sigma_{(k}\big) = -S(\sigma_{(k}), \quad \forall k \in [1,n].$$

Since P is the invariant measure of T, the marginal density $\sigma_{(k}$ is computed as

$$\begin{aligned}
\sigma_{(k} &= \operatorname{tr}_{\mathcal{H}_{k]}} |\Psi_n\rangle\langle\Psi_n| \\
&= \sum_{j_0,\dots,j_{k-1};l_k,\dots,l_n} p_{j_0} \prod_{\alpha=0}^{k-2} t_{j_\alpha j_{\alpha+1}} \sqrt{t_{j_{k-1}l_k}}^{*} \cdots \sqrt{t_{l_{n-1}l_n}}^{*} \\
&\quad \times \sqrt{t_{j_{k-1}j_k}} \cdots \sqrt{t_{j_{n-1}j_n}} |e_{j_k},\dots,e_{j_n}\rangle\langle e_{l_k},\dots,e_{l_n}| \\
&= \sum_{i;j_k,\dots,j_n;l_k,\dots,l_n} p_i \sqrt{t_{il_k}}^{*} \cdots \sqrt{t_{l_{n-1}l_n}}^{*} \sqrt{t_{ij_k}} \cdots \sqrt{t_{j_{n-1}j_n}} \\
&\quad \times |e_{j_k},\dots,e_{j_n}\rangle\langle e_{l_k},\dots,e_{l_n}| \\
&= \sum_i p_i \big|f_i(n-k)\big\rangle\big\langle f_i(n-k)\big|,
\end{aligned}$$

where $|f_i(n-k)\rangle = \sum_{j_k,\dots,j_n} \sqrt{t_{ij_k}} \cdots \sqrt{t_{j_{n-1}j_n}} |e_{j_k},\dots,e_{j_n}\rangle$.

It is easily checked that the norm of $|f_i(n-k)\rangle$ is equal to 1 but the set $\{|f_i(n-k)\rangle\}$ is not orthogonal, in general. Therefore, one can estimate the entropy of $\sigma_{(k}$ as follows:

$$0 \le S(\sigma_{(k}) \le -\sum p_i \log p_i = H(P),$$

where $S(\sigma_{(k}) = H(P)$ holds if $\{|f_i(n-k)\rangle\}$ is an orthogonal set (i.e., it is an ONB).

In the case of $k = n-1$, one has

$$\sigma_{(n-1} = \sum_i p_i \big|f_i(1)\big\rangle\big\langle f_i(1)\big| = \sum_i p_i |f_i\rangle\langle f_i| = \sigma_T. \tag{8.59}$$

According to the notation of Lemma 8.46, $\sigma_{(n-m}$ can be represented as

$$\sigma_{(n-m} = \sigma(m).$$

Lemma 8.46 means that

$$D_{\mathrm{EN}}\big(|\Psi_n\rangle\langle\Psi_n| : \rho_{k]}, \sigma_{(k}\big) = -S(\sigma_T).$$

From Theorem 8.27 and the definition of $D_{\mathrm{EN}}(\varphi)$, the theorem holds. □

Suppose that there exists a unitary matrix $U = (u_{ij})$ such that $|u_{ij}|^2 = t_{ij}$ for any i and j. Then the stochastic matrix $T = (t_{ij})$ is unitary implementable and we can take $\sqrt{t_{ij}} = u_{ij}$. Under this condition, the set $\{|f_i\rangle\}$ giving the decomposition of σ_T by (8.59) becomes an ONB:

$$\langle f_j, f_i\rangle = \sum_k u_{jk}^* u_{ik} = (UU^*)_{ij} = \delta_{i,j}.$$

Thus we have the following theorem.

Theorem 8.47 *If EMC φ has an invariant measure P of a unitary implementable matrix T, then there exists the D_{EN} of φ such that*

$$D_{\mathrm{EN}}(\varphi) = -S(P).$$

8.8 Notes

The entangled state has been studied recently by several authors [813]. There have been a lot of works discussing the properties of entanglement [108, 324, 350]. In particular, the measures to distinguish the entangled states from separable states have been extensively investigated. One of the computable criterion is the PPT criterion [645] proposed by Peres, which gives a necessary and sufficient condition in the low-dimensional case [349], however, is a necessary condition but not a sufficient one in the high-dimensional case. Belavkin and Ohya [93, 94] have discussed an approach employing a Hilbert–Schmidt operator. Entanglement degree for mixed states has been studied using entropic measures; quantum relative entropy [776] and quantum mutual entropy [94]. This approach can be applied in the infinite-dimensional case, and it is a generalization of the PPT criterion.

Entangled states were introduced by Einstein, Podolsky and Rosen [218] and Schrodinger [692, 693]. Bell's theorem for entangled states of spins without the spatial dependence was proved in [95]. The spatial dependence of entangled states and the modification of the Bell theorem was considered by Volovich [790, 794].

Tomita–Takesaki theory has been applied to the characterized entangled state [504]. The notion of the entanglement witness was introduced by Jamiołkowski [379], and its relation with the entangling operator was discussed in [383]. The various examples of quantum correlation is discussed in [164]. The definition of the entangled Markov chain is given by Accardi and Fidaleo [38] and its development is discussed in [39, 40].

Chapter 9
Quantum Capacity and Coding

We discuss in this chapter the following topics in quantum information: (1) the channel capacity for quantum communication processes by applying the quantum mutual entropy introduced in Chap. 7, (2) formulations of quantum analogues of McMillan's theorem and coding type theorem for entanglement transmission.

9.1 Channel Capacity

As we discussed, it is important to check the ability or efficiency of a channel. It is the channel capacity which mathematically describes this ability. Here we discuss two types of the channel capacity, namely, the capacity of a quantum channel Γ^* and that of a classical (classical–quantum–classical) channel $\tilde{\Xi}^* \circ \Gamma^* \circ \Xi^*$.

9.1.1 Capacity of Quantum Channel

The capacity of a quantum channel is the ability to transmit information by the channel itself, so that it does not depend on how a message being treated as a classical object is coded.

As was discussed in Chap. 1, the main theme of quantum information is to study the information carried by a quantum state and its change associated with a change of the quantum state due to an effect of a quantum channel describing a certain dynamics, in a generalized sense, of a quantum system. So the essential point of quantum communication through a quantum channel is the change of quantum states by the quantum channel, which should be first considered free from any coding of messages. The message is treated as a classical object, so that the information transmission started from messages and their quantum codings is a semi-quantum and is discussed in the next subsection. This subsection treats the pure quantum case, in which the (pure) quantum capacity is discussed as a direct extension of the classical (Shannon's) capacity in Chap. 6.

M. Ohya, I. Volovich, *Mathematical Foundations of Quantum Information and Computation and Its Applications to Nano- and Bio-systems*,
Theoretical and Mathematical Physics,
DOI 10.1007/978-94-007-0171-7_9,

Before starting mathematical discussion, we explain a bit more about what we mean by "pure quantum" for transmission capacity. We have to start from any quantum state and a channel, then compute the supremum of the mutual entropy to define the "pure" quantum capacity. One is often confused by this point, for example, one starts from the coding of a message, computes the supremum of the mutual entropy and then says that the supremum is the capacity of a quantum channel, which is not purely quantum but a classical capacity through a quantum channel.

Even when this coding is a quantum coding and one sends the coded message to a receiver through a quantum channel, if one starts from a classical state, i.e., a probability distribution of messages, then this capacity is not the capacity of the quantum channel itself. In this case, the usual Shannon's theory is applied because one can easily compute the conditional distribution in a usual (classical) way. The supremum is the capacity of a classical–quantum–classical channel, and it is in the second category discussed in the next subsection.

The capacity of a quantum channel $\Gamma^* : \mathfrak{S}(\mathcal{H}) \to \mathfrak{S}(\mathcal{K})$ is defined as follows: Let $\mathcal{S}_0$ ($\subset \mathfrak{S}(\mathcal{H})$) be the set of all states prepared for expression of information. Then the *quantum capacity* of the channel Γ^* with respect to $\mathcal{S}_0$ is defined by

$$C^{\mathcal{S}_0}(\Gamma^*) = \sup\{I(\rho; \Gamma^*); \rho \in \mathcal{S}_0\}.$$

Here $I(\rho; \Gamma^*)$ is the mutual entropy given in Chap. 7 with $\Lambda^* = \Gamma^*$. When $\mathcal{S}_0 = \mathfrak{S}(\mathcal{H})$, $C^{\mathfrak{S}(\mathcal{H})}(\Gamma^*)$ is denoted by $C(\Gamma^*)$ for simplicity. The capacity $C(\Gamma^*)$ is the largest information possibly sent through the channel Γ^*.

We have

Theorem 9.1

$$0 \le C^{\mathcal{S}_0}(\Gamma^*) \le \sup\{S(\rho); \rho \in \mathcal{S}_0\}.$$

Proof From the monotonicity of the relative entropy (i.e., for a channel Λ^* from the state space of any algebra $\mathcal{A}$ into that of $\mathcal{B}$, when $S(\Lambda^*\varphi_1, \Lambda^*\varphi_2) \le S(\varphi_1, \varphi_2)$ for any $\varphi_1 \in \mathfrak{S}(\mathcal{A})$, $\varphi_2 \in \mathfrak{S}(\mathcal{B})$), one has $I(\rho; \Lambda^*) = \sup\{\sum_j \lambda_j S(\Lambda^*\rho_j, \Lambda^*\rho) : \sum_j \lambda_j \rho_j = \rho\} \le \sup\{\sum_j \lambda_j S(\rho_j, \rho) : \sum_j \lambda_j \rho_j = \rho\} = S(\rho)$, which concludes the proof after taking the supremum over $\rho \in \mathcal{S}_0$. □

Remark 9.2 We also considered the pseudo-quantum capacity $C_p(\Gamma^*)$ defined [590] using the pseudo-mutual entropy $I_p(\rho; \Gamma^*)$ where the supremum is taken over all finite decompositions instead of all orthogonal pure decompositions:

$$I_p(\rho; \Gamma^*) = \sup\left\{\sum_k \lambda_k S(\Gamma^*\rho_k, \Gamma^*\rho); \rho = \sum_k \lambda_k \rho_k, \text{ finite decomposition}\right\}.$$

However, the pseudo-mutual entropy is not well-matched to the conditions explained in Sect. 9.2, and it is difficult to compute numerically. It is easy to see that

$$C^{\mathcal{S}_0}(\Gamma^*) \le C_p^{\mathcal{S}_0}(\Gamma^*).$$

It is worth noting that in order to discuss the details of transmission process for a sequence of n messages, we have to consider a channel on the n-tuple space and the average mutual entropy (transmission rate) per message, and such a discussion will be given in Sect. 9.2.

9.1.2 Capacity of Classical–Quantum–Classical Channel

The capacity of C–Q–C channel $\Lambda^* = \tilde{\Xi}^* \circ \Gamma^* \circ \Xi^*$ is the capacity of the information transmission process starting from the coding of messages; therefore, it can be considered as the capacity including a coding (and a decoding). The channel Ξ^* sends a classical state to a quantum one, and the channel $\tilde{\Xi}^*$ transforms a quantum state to a classical one. Note that Ξ^* and $\tilde{\Xi}^*$ can be considered as the dual maps of $\xi : \mathbf{B}(\mathcal{H}) \to \mathbb{C}^n$ $(A \mapsto (\varphi_1(A), \varphi_2(A), \dots, \varphi_n(A)))$ and $\tilde{\xi} : \mathbb{C}^m \to \mathbf{B}(\mathcal{K})$ $((c_1, c_2, \dots, c_m) \mapsto \sum_j c_j A_j)$, respectively.

The capacity of the C–Q–C channel $\Lambda^* = \tilde{\Xi}^* \circ \Gamma^* \circ \Xi^*$ is

$$C^{P_0}(\tilde{\Xi}^* \circ \Gamma^* \circ \Xi^*) = \sup\{I(\rho; \tilde{\Xi}^* \circ \Gamma^* \circ \Xi^*); \rho \in P_0\},$$

where P_0 $(\subset P(\Omega))$ is the set of all probability distributions prepared for input (a-priori) states (distributions or probability measures, that is, classical states). Moreover, the capacity for coding free is found by taking the supremum of the mutual entropy over all probability distributions and all codings Ξ^*:

$$C_c^{P_0}(\tilde{\Xi}^* \circ \Gamma^*) = \sup\{I(\rho; \tilde{\Xi}^* \circ \Gamma^* \circ \Xi^*); \rho \in P_0, \Xi^*\}.$$

The last capacity is both for coding and decoding free and it is given by

$$C_{cd}^{P_0}(\Gamma^*) = \sup\{I(\rho; \tilde{\Xi}^* \circ \Gamma^* \circ \Xi^*); \rho \in P_0, \Xi^*, \tilde{\Xi}^*\}.$$

These capacities $C_c^{P_0}$, $C_{cd}^{P_0}$ do not measure the ability of the quantum channel Γ^* itself, but measure the ability of Γ^* through the coding and decoding.

The above three capacities C^{P_0}, $C_c^{P_0}$, $C_{cd}^{P_0}$ satisfy the following inequalities

$$0 \le C^{P_0}(\tilde{\Xi}^* \circ \Gamma^* \circ \Xi^*) \le C_c^{P_0}(\tilde{\Xi}^* \circ \Gamma^*) \le C_{cd}^{P_0}(\Gamma^*) \le \sup\{S(\rho); \rho \in P_o\}.$$

Here $S(\rho)$ is the Shannon entropy $-\sum p_k \log p_k$ for the initial probability distribution $\{p_k\}$ of the message.

9.1.3 Bound of Mutual Entropy and Capacity

Here we discuss a bound of mutual entropy and capacity. The discussion of this subsection is based on [345, 584, 587, 591, 601].

To each input symbol x_i there corresponds a state σ_i of the quantum communication system; σ_i functions as the codeword of x_i. The coded state is a convex combination

$$\rho = \sum_i p_i \delta_i \to \sigma = \sum_i p_i \Xi^* \delta_i = \sum_i p_i \sigma_i$$

whose coefficients are the corresponding probabilities, p_i is the probability that the letter x_i should be transmitted over the channel. To each output symbol y_j there corresponds a non-negative observable, that is, a self-adjoint operator Q_j on the output Hilbert space $\mathcal{K}$, such that $\sum_j Q_j = I$ ($\{Q_j\}$ is called POVM). In terms of the quantum states, the transition probabilities are $\operatorname{tr} \Gamma^* \sigma_i (Q_j)$, and the probability that x_i was sent and y_j is read is

$$p_{ji} = p_i \operatorname{tr} \Gamma^* \sigma_i (Q_j).$$

On the basis of this joint probability distribution, the classical mutual information is given by

$$I_{cl} = \sum_{i,j} p_{ji} \log \frac{p_{ji}}{p_i q_j}$$

where $q_j = \operatorname{tr} \Gamma^* \sigma (Q_j)$. The next theorem provides a fundamental bound for the mutual information in terms of the quantum von Neumann entropy, suggested by Levitin and Gordon [296, 477] and was proved by Holevo [345] in 1973. Ohya introduced in 1983 the quantum mutual entropy by means of the relative entropy, as discussed in Chap. 7.

Theorem 9.3 *With the above notation,*

$$I_{cl} = \sum_{i,j} p_{ji} \log \frac{p_{ji}}{p_i q_j} \le S(\Gamma^* \sigma) - \sum_i p_i S(\Gamma^* \sigma_i)$$

holds.

Holevo's upper bound can now be expressed as

$$S(\Gamma^* \sigma) - \sum_i p_i S(\Gamma^* \sigma_i) = \sum_i p_i S(\Gamma^* \sigma_i, \Gamma^* \sigma).$$

We shall see that Theorem 9.3 follows from Bogoliubov's inequality and local monotonicity of the relative entropy. For a general quantum case, we have the following inequality according to Theorem 7.15 in Chap. 7.

Theorem 9.4 *When the Schatten decomposition (i.e., one dimensional spectral decomposition) $\rho = \sum_i p_i \rho_i$ is unique,*

$$I_{cl} \le I(\rho; \Gamma^*) = \sum_i p_i S(\Gamma^* \rho_i, \Gamma^* \rho)$$

for any channel Γ^*.

Going back to general discussion, an input state ρ is the probability distribution $\{\lambda_k\}$ of messages, and its Schatten decomposition is unique as $\rho = \sum_k \lambda_k \delta_k$ with delta measures δ_k, so the mutual entropy is written by

$$I(\rho; \tilde{\Xi}^* \circ \Gamma^* \circ \Xi^*) = \sum_k \lambda_k S(\tilde{\Xi}^* \circ \Gamma^* \circ \Xi^* \delta_k, \tilde{\Xi}^* \circ \Gamma^* \circ \Xi^* \rho).$$

If the coding Ξ^* is a quantum coding, then $\Xi^* \delta_k$ is expressed by a quantum state. Denote the coded quantum state by $\sigma_k = \Xi^* \delta_k$ as above and put $\sigma = \Xi^* \rho = \sum_k \lambda_k \Xi^*(\delta_k) = \sum_k \lambda_k \sigma_k$. Then the above mutual entropy in a (classical–quantum–classical) channel $\tilde{\Xi}^* \circ \Gamma^* \circ \Xi^*$ is written as

$$I(\rho; \tilde{\Xi}^* \circ \Gamma^* \circ \Xi^*) = \sum_k \lambda_k S(\tilde{\Xi}^* \circ \Gamma^* \sigma_k, \tilde{\Xi}^* \circ \Gamma^* \sigma). \tag{9.1}$$

This is the expression of the mutual entropy of the whole information transmission process starting from a coding of classical messages.

Remark that if $\sum_k \lambda_k S(\Gamma^* \sigma_k)$ is finite, then (9.1) becomes

$$I(\rho; \tilde{\Xi}^* \circ \Gamma^* \circ \Xi^*) = S(\tilde{\Xi}^* \circ \Gamma^* \sigma) - \sum_k \lambda_k S(\tilde{\Xi}^* \circ \Gamma^* \sigma_k).$$

Further, if ρ is a probability measure having a density function $f(\lambda)$, that is, $\rho(A) = \int_A f(\lambda)\, d\lambda$, where A is an interval in $\mathbb{R}$, and each λ corresponds to a quantum coded state $\sigma(\lambda)$, then

$$\sigma = \int f(\lambda) \sigma(\lambda)\, d\lambda$$

and

$$I(\rho; \tilde{\Xi}^* \circ \Gamma^* \circ \Xi^*) = S(\tilde{\Xi}^* \circ \Gamma^* \sigma) - \int f(\lambda) S\big(\tilde{\Xi}^* \circ \Gamma^* \sigma(\lambda)\big)\, d\lambda.$$

One can prove that this is less than

$$S(\Gamma^* \sigma) - \int f(\lambda) S\big(\Gamma^* \sigma(\lambda)\big)\, d\lambda.$$

This upper bound is a special one of the following inequality

$$I(\rho; \tilde{\Xi}^* \circ \Gamma^* \circ \Xi^*) \le I(\rho; \Gamma^* \circ \Xi^*),$$

which comes from the monotonicity of the relative entropy and gives the proof of Theorem 9.4 above.

Let us discuss a more general bound. If A and B are two positive Hermitian operators (not necessarily states, i.e., not necessarily with unit traces) then we set

$$S(A, B) = \operatorname{tr} A(\log A - \log B).$$

By means of the *Bogoliubov inequality*

$$S(A,B) \geq \operatorname{tr} A(\log \operatorname{tr} A - \log \operatorname{tr} B)$$

which was proved in Sect. 9.5 and the monotonicity of the mutual entropy, we have the following bound giving the bound of the mutual entropy $I(\rho; \Gamma^* \circ \Xi^*)$.

Theorem 9.5 *For a probability distribution $\rho = \{\lambda_k\}$ and a quantum coded state $\sigma = \Xi^* \rho \equiv \sum_k \lambda_k \sigma_k$, $\lambda_k \geq 0$, $\sum_k \lambda_k = 1$, one has the following inequality for any quantum channel decomposed as $\Gamma^* = \Gamma_1^* \circ \Gamma_2^*$ such that $\Gamma_1^* \sigma = \sum_i A_i \sigma A_i^*$, $\sum_i A_i^* A_i = I$:*

$$\begin{aligned}
&\sum_k \lambda_k S(\sigma_k, \sigma) \\
&\quad \geq \sum_{k,i} \lambda_k \operatorname{tr}(A_i \Gamma_2^* \sigma_k A_i^*) \big[\log \operatorname{tr}(A_i \Gamma_2^* \sigma_k A_i^*) - \log \operatorname{tr}(A_i \Gamma_2^* \sigma A_i^*)\big].
\end{aligned}$$

Proof By applying the local monotonicity of the relative entropy,

$$S(\rho,\tau) \geq \sum_i S(A_i \rho A_i^*, A_i \tau A_i^*)$$

where ρ and τ are any states, we have

$$\sum_k \lambda_k S(\sigma_k, \sigma) \geq \sum_k \sum_i \lambda_k S(A_i \Gamma_2^* \sigma_k A_i^*, A_i \Gamma_2^* \sigma A_i^*).$$

Here the second inequality follows from the Bogoliubov inequality. □

In the case that the channel Γ_2^* is identical, $\Gamma_2^* \sigma_k = \sigma_k$, the above inequality reduces to the bound of Theorem 9.3:

$$\sum_k \lambda_k S(\sigma_k, \sigma) \geq \sum_{k,i} \lambda_k \operatorname{tr}(B_i \sigma_k) \big[\log \operatorname{tr}(B_i \sigma_k) - \log \operatorname{tr}(B_i \sigma)\big]$$

where $B_i = A_i^* A_i$.

In fact, it is the classical Shannon's mutual entropy. If we introduce the transition probability $p(i|k) = \operatorname{tr} B_i \sigma_k$ then we can rewrite the last inequality in the form of the bound to the classical mutual entropy

$$\begin{aligned}
\sum_k \lambda_k S(\sigma_k, \sigma) &\geq \sum_{k,i} \lambda_k p(i|k) \left[\log p(i|k) - \log \sum_n p(i|n) \lambda_n\right] \\
&= \sum_{k,i} p_{ik} \log p_{ik} - \sum_k \lambda_k \log \lambda_k - \sum_i \mu_i \log \mu_i,
\end{aligned}$$

where $p_{ik} = p(i|k)\lambda_k$ is the joint probability and

$$\sum_k p_{ik} = \mu_i, \qquad \sum_i p_{ik} = \lambda_k.$$

Note that $\sum_k \lambda_k S(\sigma_k, \sigma)$ and $\sum_{i,k} \lambda_k S(A_i \Gamma_2^* \sigma_k A_i^*, A_i \Gamma_2^* \sigma A_i^*)$ are the quantum mutual entropies $I(\rho; \Gamma^*)$ for special channels Γ^* as above and that the lower bound is equal to the classical mutual entropy which depends on the POVM $\{B_i = A_i^* A_i\}$.

Using the above upper and lower bounds of the mutual entropy, we can compute these bounds of the capacity in many different cases.

9.2 Computation of Capacity

Shannon's communication theory is largely of asymptotic character, the message length N is supposed to be very large. So we consider the N-fold tensor product of the input and output Hilbert spaces $\mathcal{H}$ and $\mathcal{K}$,

$$\mathcal{H}_N = \bigotimes^N \mathcal{H}, \qquad \mathcal{K}_N = \bigotimes^N \mathcal{K}.$$

Note that

$$\mathbf{B}(\mathcal{H}_N) = \bigotimes^N \mathbf{B}(\mathcal{H}), \qquad \mathbf{B}(\mathcal{K}_N) = \bigotimes^N \mathbf{B}(\mathcal{K}).$$

A channel $\Lambda_N^* : \mathfrak{S}(\mathcal{H}_N) \to \mathfrak{S}(\mathcal{K}_N)$ sends density operators acting on $\mathcal{H}_N$ into those acting on $\mathcal{K}_N$. In particular, we take a *memoryless channel* which is the tensor product of the same *single site channels*: $\Lambda_N^* = \Lambda^* \otimes \cdots \otimes \Lambda^*$ (N-fold). In this setting, we compute the quantum capacity and the classical–quantum–classical capacity, denoted by C_{q} and C_{cq} below.

A *pseudo-quantum code* (of order N) is a probability distribution on $\mathfrak{S}(\mathcal{H}_N)$ with finite support in the set of product states. So $\{(p_i), (\varphi_i)\}$ is *a pseudo-quantum code if* (p_i) *is a probability vector and* φ_i *are product states of* $\mathbf{B}(\mathcal{H}_N)$. This code is nothing but a quantum code for a classical input (so a classical–quantum channel) such that $p = \sum_j p_j \delta_j \Rightarrow \varphi = \sum_j p_j \varphi_j$, as discussed in the previous chapter. Each quantum state φ_i is sent over the quantum mechanical media (e.g., optical fiber) and yields the output quantum states $\Lambda_N^* \varphi_i$. The performance of coding and transmission is measured by the quantum mutual entropy

$$I(\varphi; \Lambda_N^*) \bigl(= I\bigl((p_i), (\varphi_i); \Lambda_N^*\bigr)\bigr) = \sum_i p_i S(\Lambda_N^* \varphi_i, \Lambda_N^* \varphi).$$

We regard φ as the quantum state of the n-component quantum system during the information transmission. Taking the supremum over certain classes of pseudo-quantum codes, we obtain various capacities of the channel. The supremum is over product states when we consider memoryless channels, so the capacity is

$$C_{\mathrm{cq}}(\Lambda_N^*) = \sup\bigl\{I\bigl((p_i),(\varphi_i);\Lambda_N^*\bigr);\\ \bigl((p_i),(\varphi_i)\bigr) \text{ is a pseudo-quantum code}\bigr\}.$$

Next we consider a subclass of pseudo-quantum codes. A *quantum code is defined by the additional requirement that* $\{\varphi_i\}$ *is a set of pairwise orthogonal pure states* [559]. This code is pure quantum, namely, we start at a quantum state φ and take *orthogonal extremal decompositions* $\varphi = \sum_i p_i\varphi_i$; this decomposition is not unique. Here the coding is how to take such an orthogonal extremal decomposition. The quantum mutual entropy is

$$I(\varphi;\Lambda_N^*) = \sup\biggl\{\sum_i p_i S(\Lambda_N^*\varphi_i,\Lambda_N^*\varphi); \sum_i p_i\varphi_i = \varphi\biggr\},$$

where the supremum is over all *orthogonal extremal decompositions* $\sum_i p_i\varphi_i = \varphi$ (i.e., $\varphi_i \perp \varphi_j, \varphi_i \in \operatorname{ex}\mathfrak{S}$). Then we arrive at the capacity

$$\begin{aligned} C_{\mathrm{q}}(\Lambda_N^*) &= \sup\bigl\{I(\varphi;\Lambda_N^*) : \varphi\bigr\} \\ &= \sup\bigl\{I\bigl((p_i),(\varphi_i);\Lambda_N^*\bigr) : \bigl((p_i),(\varphi_i)\bigr) \text{ is a quantum code}\bigr\}. \end{aligned}$$

It follows from the definition that

$$C_{\mathrm{q}}(\Lambda_N^*) \le C_{\mathrm{cq}}(\Lambda_N^*) \tag{9.2}$$

holds for every channel.

Proposition 9.6 *For a memoryless channel, the sequences* $C_{\mathrm{cq}}(\Lambda_N^*)$ *and* $C_{\mathrm{q}}(\Lambda_N^*)$ *are subadditive.*

Proof If $((p_i),(\varphi_i))$ and $((q_j),(\psi_j))$ are (pseudo-)quantum codes of order N and M, then $((p_i,q_j),(\varphi_i\otimes\psi_j))$ is a (pseudo-)quantum code of order $N+M$ and

$$I\bigl((p_i,q_j),(\varphi_i\otimes\psi_j);\Lambda_{N+M}^*\bigr) = I\bigl((p_i),(\varphi_i);\Lambda_N^*\bigr) + I\bigl((q_j),(\psi_j);\Lambda_M^*\bigr) \tag{9.3}$$

follows from the additivity of relative entropy when taking tensor product. One can check that if the initial codes are semi-classical (restricted quantum) then the product code is semi-classical (restricted quantum) as well. After taking the supremum, the additivity (9.3) yields the subadditivity of the sequences $C_{\mathrm{cq}}(\Lambda_N^*)$ and $C_{\mathrm{q}}(\Lambda_N^*)$. □

Therefore, the following limits exist and they coincide with the infimum:

$$\widetilde{C_{\mathrm{cq}}} = \lim_{N\to\infty}\frac{1}{N}C_{\mathrm{cq}}(\Lambda_N^*), \qquad \widetilde{C_{\mathrm{q}}^{\infty}} = \lim_{N\to\infty}\frac{1}{N}C_{\mathrm{q}}(\Lambda_N^*). \tag{9.4}$$

(For multiple channels with some memory effect, one may take the limsup in (9.4) to get a good concept of capacity per single use.)

Example 9.7 Let Λ^* be a channel on the 2×2 density matrices such that

$$\Lambda^*\colon \begin{pmatrix} a & b \\ \bar{b} & c \end{pmatrix} \mapsto \begin{pmatrix} a & 0 \\ 0 & c \end{pmatrix}.$$

Consider the input density matrix

$$\rho_\lambda = \frac{1}{2}\begin{pmatrix} 1 & 1-2\lambda \\ 1-2\lambda & 1 \end{pmatrix} \quad (0 < \lambda < 1).$$

For $\lambda \neq 1/2$ the orthogonal extremal decomposition is unique, in fact,

$$\rho_\lambda = \frac{\lambda}{2}\begin{pmatrix} 1 & -1 \\ -1 & 1 \end{pmatrix} + \frac{1-\lambda}{2}\begin{pmatrix} 1 & 1 \\ 1 & 1 \end{pmatrix},$$

and we have

$$I(\rho_\lambda, \Lambda^*) = 0 \quad \text{for } \lambda \neq 1/2.$$

However, $I(\rho_{1/2}, \Lambda^*) = \log 2$. Since $C_q(\Lambda^*) \le C_{\mathrm{cq}}(\Lambda^*) \le \log 2$, we conclude that $C_q(\Lambda^*) = C_{\mathrm{cq}}(\Lambda^*) = \log 2$.

9.2.1 Divergence Center

In order to estimate the quantum mutual entropy, we introduce the concept of a *divergence center*. Let $\{\omega_i : i \in I\}$ be a family of states and consider a constant $r > 0$.

Definition 9.8 We say that the state ω is a divergence center for a family of states $\{\omega_i : i \in I\}$ with radius $\le r$ if

$$S(\omega_i, \omega) \le r \quad \text{for every } i \in I.$$

In the following discussion about the geometry of relative entropy (or divergence as it is called in information theory), the ideas of [184] can be recognized very well.

Lemma 9.9 *Let $((p_i), (\varphi_i))$ be a quantum code for the channel Λ^* and ω a divergence center with radius $\le r$ for $\{\Lambda^*\varphi_i\}$. Then*

$$I\big((p_i), (\varphi_i); \Lambda^*\big) \le r.$$

Proof We assume that the states $\Lambda^*\varphi_i$, $\Lambda^*\varphi = \sum_i p_i \Lambda^*\varphi_i$ and ω have finite entropy, and their densities are denoted by ρ_i, ρ, ρ', respectively. We have

$$-S(\Lambda^*\varphi_i) - \operatorname{tr} \rho_i \log \rho' \le r,$$

hence

$$\sum_i p_i S(\Lambda^*\varphi_i, \Lambda^*\varphi) = -\sum_i p_i S(\Lambda^*\varphi_i) - \operatorname{tr}\rho\log\rho$$
$$\le r - \operatorname{tr}\rho(\log\rho - \log\rho')$$
$$= r - S(\Lambda^*\varphi, \omega). \tag{9.5}$$

The extra assumption we made holds always in the finite-dimensional case. When the entropies are not finite but the relative entropies are such, one has to use more sophisticated methods for the proof. It is quite clear that Inequality (9.5) is close to equality if $S(\Lambda^*\varphi_i, \omega)$ is roughly r and $\sum_i p_i \Lambda^*\varphi_i$ is roughly ω. □

Definition 9.10 Let $\{\omega_i : i \in I\}$ be a family of states. We say that the state ω is an *exact divergence center with radius r* if

$$r = \inf_\varphi \sup_i \{S(\omega_i, \varphi)\}$$

and ω is a minimizer for the right-hand side.

When r is finite, there exists a minimizer because $\varphi \mapsto \sup\{S(\omega_i, \varphi) : i \in I\}$ is lower semicontinuous with compact level sets (cf. Proposition 5.27 in [578]).

Lemma 9.11 *Let ψ_0, ψ_1 and ω be states of $\mathbf{B}(\mathcal{K})$ such that the Hilbert space $\mathcal{K}$ is finite-dimensional, and set $\psi_\lambda = (1-\lambda)\psi_0 + \lambda\psi_1$ $(0 \le \lambda \le 1)$. If $S(\psi_0, \omega)$, $S(\psi_1, \omega)$ are finite and*

$$S(\psi_\lambda, \omega) \ge S(\psi_1, \omega) \quad (0 \le \lambda \le 1)$$

then

$$S(\psi_1, \omega) + S(\psi_0, \psi_1) \le S(\psi_0, \omega).$$

Proof Let the densities of ψ_λ, ω be ρ_λ, ρ. Due to the assumption $S(\psi_\lambda, \omega) < +\infty$, the kernel of ρ is smaller than that of ρ_λ. The function $f(\lambda) = S(\varphi_\lambda, \omega)$ is convex on $[0, 1]$ and $f(\lambda) \ge f(1)$ (cf. Proposition 3.1 in [578]). It follows that $f'(1) \le 0$. Hence we have

$$f'(1) = \operatorname{tr}(\rho_1 - \rho_0)(I + \log\rho_1) - \operatorname{tr}(\rho_1 - \rho_0)\log\rho$$
$$= S(\psi_1, \omega) - S(\psi_0, \omega) + S(\psi_0, \psi_1) \le 0.$$

This is the inequality we had to obtain.

We note that when differentiating the function $f(\lambda)$ the well-known formula

$$\frac{\partial}{\partial t}\operatorname{tr} f(A + tB)|_{t=0} = \operatorname{tr}\bigl(f'(A)B\bigr)$$

can be used. □

Lemma 9.12 *Let $\{\omega_i : i \in I\}$ be a finite set of states of $\mathbf{B}(\mathcal{K})$ such that the Hilbert space $\mathcal{K}$ is finite-dimensional. Then the exact divergence center is unique, and it is in the convex hull of the states ω_i.*

Proof Let $\mathcal{K}$ be the (closed) convex hull of the states $\omega_1, \omega_2, \dots, \omega_n$, and let ω be an arbitrary state such that $S(\omega_i, \omega) < +\infty$. There is a unique state $\omega \in \mathcal{K}$ such that $S(\omega', \omega)$ is minimal (where ω' runs over $\mathcal{K}$) (see Theorem 5.25 in [578]). Then $S(\lambda\omega_i + (1-\lambda)\omega', \omega) \geq S(\omega', \omega)$ for every $0 \leq \lambda \leq 1$ and $1 \leq i \leq n$. It follows from the previous lemma that

$$S(\omega_i, \omega) \geq S(\omega_i, \omega').$$

Hence the divergence center of ω_i's must be in $\mathcal{K}$. The uniqueness of the exact divergence center follows from the fact that the relative entropy functional is strictly convex in the second variable. □

Theorem 9.13 *Let $\Lambda^* : \mathfrak{S}(\mathcal{H}) \to \mathfrak{S}(\mathcal{K})$ be a channel with finite dimensional $\mathcal{K}$. Then the capacity $C_{\mathrm{cq}}(\Lambda^*) = C_{\mathrm{q}}(\Lambda^*)$ is the divergence radius of the range of Λ^*.*

Proof Let $((p_i), (\varphi_i))$ be a quantum code. Then $I((p_i), (\varphi_i); \Lambda^*)$ is at most the divergence radius of $\{\Lambda^*\varphi_i\}$ (according to Lemma 9.9), which is obviously majorized by the divergence radius of the range of Λ^*. Therefore, the capacity does not exceed the divergence radius of the range.

To prove the converse inequality, we assume that the exact divergence radius of $\Lambda^*(\mathfrak{S}(\mathcal{H}))$ is larger than $t \in \mathbb{R}$. Then we can find $\varphi_1, \varphi_2, \dots, \varphi_n \in \mathfrak{S}(\mathcal{H})$ such that the exact divergence radius r of $\Lambda^*(\varphi_1), \dots, \Lambda^*(\varphi_n)$ is larger than t. Lemma 9.12 tells us that the divergence center ω of $\Lambda^*(\varphi_1), \dots, \Lambda^*(\varphi_n)$ lies in their convex hull K. By a possible reordering of the states φ_i, we can achieve

$$S\big(\Lambda^*(\varphi_i), \omega\big) \begin{cases} = r, & \text{if } 1 \leq i \leq k, \\ < r, & \text{if } k < i \leq n. \end{cases}$$

Let K' be the convex hull of $\Lambda^*(\varphi_1), \dots, \Lambda^*(\varphi_n)$. We claim that $\omega \in K'$. Indeed, choose $\omega' \in K'$ such that $S(\omega', \omega)$ is minimal (ω' is running over K'). Then

$$S\big(\Lambda^*\varphi_i, \varepsilon\omega' + (1-\varepsilon)\omega\big) < r$$

for every $1 \leq i \leq k$ and $0 < \varepsilon < 1$, due to Lemma 9.11. However,

$$S\big(\Lambda^*\varphi_i, \varepsilon\omega' + (1-\varepsilon)\omega\big) < r$$

for $k \leq i \leq n$ and for a small ε by a continuity argument. In this way, we conclude that there exists a probability distribution $(p_i, p_2, \dots, p_k)$ such that

$$\sum_{i=1}^{k} p_i \Lambda^*\varphi_i = \omega, \qquad S(\Lambda^*\varphi_i, \omega) = r.$$

Consider now the pseudo-quantum code $((p_i),(\varphi_i))$ in the above and get

$$\sum_{i=1}^{k} p_i S\left(\Lambda^*\varphi_i, \Lambda^*\left(\sum_{j=1}^{k} p_j\varphi_j\right)\right) = \sum_{i=1}^{k} p_i S(\Lambda^*\varphi_i, \omega) = r.$$

So we have found a quantum code which has quantum mutual entropy larger than t. The channel capacity must exceed the entropy radius of the range. □

9.2.2 Comparison of Capacities

Up to now our discussion has concerned the capacities of coding and transmission, which are bounds for the performance of quantum coding and quantum transmission. After a measurement is performed, the quantum channel becomes classical and Shannon's theory is applied. The *total capacity* (or *classical–quantum–classical capacity*) of a quantum channel Λ^* is

$$C_{\mathrm{cqc}}(\Lambda^*) = \sup\bigl\{I\bigl((p_i),(\varphi_i);\ \tilde{\Xi}^* \circ \Lambda^*\bigr)\bigr\},$$

where the supremum is taken over all pseudo-quantum codes $((p_i),(\varphi_i))$ and all measurements $\widetilde{\Xi}^*$. Due to the monotonicity of the mutual entropy,

$$C_{\mathrm{cqc}}(\Lambda^*) \le C_{\mathrm{cq}}(\Lambda^*),$$

and similarly

$$\widetilde{C_{\mathrm{cqc}}}(\Lambda_N^*) \equiv \limsup \frac{1}{N} C_{\mathrm{cqc}}(\Lambda_N^*) \le \widetilde{C_{\mathrm{cq}}}(\Lambda_N^*)$$

holds for the capacities per single use.

Example 9.14 Any 2×2 density operator has the following standard representation

$$\rho_x = \frac{1}{2}(I + x_1\sigma_1 + x_2\sigma_2 + x_3\sigma_3),$$

where $\sigma_1,\sigma_2,\sigma_3$ are the Pauli matrices and $x=(x_1,x_2,x_3)\in\mathbb{R}^3$ with $x_1^2+x_2^2+x_3^2\le 1$. For a positive semi-definite 3×3 matrix A, the application $\Gamma^*:\rho_x \mapsto \rho_{Ax}$ gives a channel when $\|A\|\le 1$. Let us compute the capacities of Γ^*. Since a unitary conjugation does not change capacity, obviously, we may assume that A is diagonal with eigenvalues $1\ge\lambda_1\ge\lambda_2\ge\lambda_3\ge 0$. The range of Γ^* is visualized as an ellipsoid with (Euclidean) diameter $2\lambda_1$. It is not difficult to see that the tracial state τ is the exact divergence center of the segment connected the states $(I\pm\lambda_1\sigma_1)/2$, and hence τ must be the divergence center of the whole range. The divergence radius is

$$S\left(\frac{1}{2}\begin{pmatrix}1 & 0\\ 0 & 0\end{pmatrix} + \frac{\lambda}{2}\begin{pmatrix}1 & 0\\ 0 & -1\end{pmatrix}, \tau\right)$$

$$= \log 2 - S\left(\frac{1}{2}\begin{pmatrix} 1+\lambda & 0 \\ 0 & 1-\lambda \end{pmatrix}\right)$$
$$= \log 2 - \eta\big((1+\lambda)/2\big) - \eta\big((1-\lambda)/2\big).$$

This gives the capacity $C_{\mathrm{cq}}(\Gamma^*)$ according to Theorem 9.13. Inequality (9.2) tells us that the capacity $C_{\mathrm{q}}(\Gamma^*)$ cannot exceed this value. On the other hand, $I(\tau, \Gamma^*) = \log 2 - \eta((1+\lambda)/2) - \eta((1-\lambda)/2)$ and we have $C_{\mathrm{cq}}(\Gamma^*) = C_{\mathrm{q}}(\Gamma^*)$.

The relations among C_{cq}, C_{q} and C_{cqc} form an important problem and are worth studying. For a noiseless channel, $C_{\mathrm{cqc}} = \log n$ was obtained in [264], where n is the dimension of the output Hilbert space (actually identical to the input one). Since the tracial state is the exact divergence center of all density matrices, we have $C_{\mathrm{cq}} = \log n$ and also $C_{\mathrm{q}} = \log n$.

We expect that $C_{\mathrm{cq}} < C_{\mathrm{cqc}}$ for "truely quantum mechanical channels" but $\widetilde{C_{\mathrm{cqc}}} = \widetilde{C_{\mathrm{cq}}} = \widetilde{C_{\mathrm{q}}}$ must hold for a large class of memoryless channels.

One can obtain the following results for the attenuation channel which is discussed in the previous chapter.

Lemma 9.15 *Let Λ^* be the attenuation channel. Then*

$$\sup I\big((p_i), (\varphi_i); \Lambda^*\big) = \log n$$

where the supremum is over all pseudo-quantum codes $((p_i)_{i=1}^n, (\varphi_{f(i)})_{i=1}^n)$ applying n coherent states.

Proof We know that $\Lambda^*\varphi_f = \varphi_{af}$, so the output $\{\Lambda^*\varphi_{f(1)}, \ldots, \Lambda^*\varphi_{f(n)}\}$ consists of n pure states. The corresponding vectors of $\Gamma^*(\mathcal{H})$ span a Hilbert space of dimension $k \le n$. Since the tracial state on that Hilbert space is a divergence center with radius $\le \log k \le \log n$, log$n$ is always a bound for the mutual information according to Lemma 9.9.

In order to show that the bound $\log n$ is really achieved, we choose the vectors $f(k)$ such that

$$f(k) = \lambda k f \quad (1 \le k \le n),$$

where $f \in \mathcal{H}$ is a fixed nonzero vector. Then in the limit $\lambda \to \infty$ the states $\varphi_{f(k)}$ become orthogonal since

$$\big|\langle \Phi_{\lambda k f}, \Phi_{\lambda m f}\rangle\big|^2 = \exp\big(-\lambda^2 (k-m)^2 \|f\|^2/2\big) \to 0$$

whenever $k \ne m$. In the limit $\lambda \to \infty$, the tracial state (of a subspace) becomes the exact divergence center, and we have

$$\lim_{\lambda\to\infty} I\big((1/n), (\varphi_{f(i)}); \Lambda^*\big) = \log n.$$

This proves the lemma. □

The next theorem follows directly from the previous lemma.

Theorem 9.16 *The capacity C_{cq} of the attenuation channel is infinite.*

Since the argument of the proof of the above lemma works for any quasi-free channel, we can conclude $C_{\mathrm{pq}} = \infty$ also in that more general case. Another remark concerns the classical capacity C_{cqc}. Since the states $\varphi_{f(n)}$ used in the proof of Lemma 9.15 commute in the limit $\lambda \to \infty$, the classical capacity C_{cqc} is infinite as well. $C_{\mathrm{cqc}} = \infty$ follows also from the proof of the next theorem.

Theorem 9.17 *The capacity C_{q} of the attenuation channel is infinite.*

Proof We follow the strategy of the proof of the previous theorem, but we use the number states in place of the coherent ones. The attenuation channel sends the number state $|n\rangle\langle n|$ into the binomial mixture of the number states $|0\rangle\langle 0|$ $(\equiv \Phi)$, $|1\rangle\langle 1|, \ldots, |n\rangle\langle n|$. Hence the commuting family of a convex combination of number states is invariant under the attenuation channel, and the channel restricted to those states is classical with obviously infinite capacity. Since C_{q} (as well as C_{cqc}) cannot have a smaller value, the claim follows. □

Let us make some comments on the previous results. The theorems mean that an arbitrarily large amount of information can go through the attenuation channel; however, the theorems do not say anything about the price for it. The expectation value of the number of particles needed in the pseudo-quantum code of Lemma 9.15 tends to infinity. Indeed,

$$\sum_i \frac{1}{n}\varphi_{f(i)}(N) = \frac{1}{n}\sum_{i=1}^{n} \left\| f(i) \right\|^2 = \lambda(n+1)(2n+1)\|f\|^2/6,$$

which increases rapidly with n. (Above N denoted the number operator.) Hence the good question is to ask the capacity of the attenuation channel when some energy constrain is posed:

$$C(E_0) = \sup\left\{ I\big((p_i), (\varphi_i); \Lambda^*\big); \sum_i p_i\varphi_i(N) \le E_0 \right\}.$$

To be more precise, we have posed a bound on the average energy; different constraint is also possible, cf. [153]. Since

$$\Lambda(N) = \eta N$$

for the dual operator Λ of the channel Λ^* and the number operator N, we have

$$C(E_0) = \sup\left\{ \sum_i p_i S\left(\varphi_i, \sum_j p_j\varphi_j\right); S(\varphi) < \sum_i p_i\varphi_i(N) \le \eta E_0 \right\}.$$

The solution of this problem is the same as for

$$S(\varphi) < \sup\big\{ S(\psi) : \psi(N) = \eta E_0 \big\},$$

and the well-known maximizer of this problem is a so-called Gibbs state. Therefore, we have

$$C(E_0) = a^2 E_0 + \log\bigl(a^2 E_0 + 1\bigr).$$

This value can be realized as a classical capacity if the number states can be output states of the attenuation channel.

9.3 Quantum McMillan Type Theorem

McMillan's theorem in classical systems was discussed in Chap. 6. Here we consider how it can be extended to quantum systems. In classical systems, the entropy function and the conditional entropy function with respect to two σ-fields $\tilde{C}$ and $\tilde{D}$, generated by two finite (measurable) partitions of the message space $\mathcal{M}$, are defined as

$$\hat{S}(\tilde{C}) = -\sum_{C\in\tilde{C}} \log\mu(C)1_C,$$

$$\hat{S}(\tilde{C} \mid \tilde{D}) = -\sum_{C\in\tilde{C}, D\in\tilde{D}} \log\mu(C|D)1_{C\cap D}.$$

They are basic quantities of McMillan's theorem. We discuss the quantum analogue of these quantities.

9.3.1 Entropy Operators in Quantum Systems

Let $\mathcal{N}$ be a finite-dimensional von Neumann (matrix) algebra acting on a Hilbert space $\mathcal{H}$ and τ be an α-invariant faithful normal trace on $\mathcal{N}$, where α is an automorphism of $\mathcal{N}$. That is, (i) $\mathcal{N}$ is a subset of $\mathbf{B}(\mathcal{H})$ satisfying $\mathcal{N}'' \equiv (\mathcal{N}')' = \mathcal{N}$ and $\dim\mathcal{H} < +\infty$, (ii) τ is a positive linear functional on $\mathcal{N}$; $\tau(A^*A) \geq 0$, $\tau(\lambda A + B) = \lambda\tau(A) + \tau(B)$ with faithfulness ($\Leftrightarrow \tau(A^*A) = 0$ implies $A = 0$), normality($\Leftrightarrow \sup\tau(A_j) = \tau(\sup A_j)$), trace property ($\Leftrightarrow \tau(AB) = \tau(BA)$) for any $A, B \in \mathcal{N}$.

We take the resolution of unity I in the quantum case to define basic quantities above instead of taking finite (measurable) partitions of $\mathcal{M}$ in the classical case. Let $\mathcal{P}(\mathcal{M})$ be the set of all *minimal finite* resolutions $\widetilde{P}$ of unity I in a subalgebra $\mathcal{M}$ of $\mathcal{N}$, that is, $\widetilde{P} \equiv \{P_j\}$ (i.e., each P_j is a projection in $\mathcal{M}$ such that $P_i \perp P_j$ $(i \neq j)$, $\sum_{j=1}^n P_j = I$ and there is no projection E such as $0 < E < P_j$ for each j).

The entropy operator and the entropy w.r.t. the subalgebra $\mathcal{M}$ and the trace τ are defined by

$$\hat{S}(\mathcal{M}) = -\sum_k P_k \log\tau(P_k) \quad \text{and} \quad S(\mathcal{M}) = \tau\bigl(\hat{S}(\mathcal{M})\bigr).$$

Theorem 9.18 *The entropy operator $\hat{S}(\mathcal{M})$ of a subalgebra $\mathcal{M}$ is uniquely determined (i.e., does not depend on the choice of $\widetilde{P} \equiv \{P_j\}$), so is the entropy $S(\mathcal{M})$.*

Proof (i) When $\mathcal{M}$ is a type I_n factor (i.e., the center $\mathcal{Z} \equiv \{A \in \mathcal{M}; AB = BA, \forall B \in \mathcal{M}\}$ of $\mathcal{M}$ is equal to the set $\mathbb{C}I$, where $\mathbb{C}$ is the set of all complex numbers), $\mathcal{M}$ is isometrically isomorphic to $\mathbf{B}(\mathbb{C}^n)$, the set of all $n \times n$ matrices on the n-dimensional Hilbert space $\mathbb{C}^n$. Then the minimality of the partition implies

$$\hat{S}(\mathcal{M}) = (\log n)I \quad \text{and} \quad S(\mathcal{M}) = \log n.$$

(ii) When $\mathcal{M}$ is not a factor, the center $\mathcal{Z}$ of $\mathcal{M}$ is generated by a minimal finite partition $\{Q_j\} \in \mathcal{Z}$, and $\mathcal{M}$ can be expressed as $\mathcal{M} = \bigoplus_j \mathcal{M}_j$, where $\mathcal{M}_j = Q_j \mathcal{M}$ is a type I_{n_j} factor (i.e., the center $\mathcal{Z}_j$ of $\mathcal{M}_j$ is equal to $\mathbb{C}I$). Then by taking $q_j = \tau(Q_j)^{-1}$ and $\tau_j = q_j \tau \restriction \mathcal{M}_j$ (the restriction of τ to $\mathcal{M}_j$), we have

$$\hat{S}(\mathcal{M}) = \hat{S}(\mathcal{Z}) + \sum_j \hat{S}_j(\mathcal{M}_j),$$

$$S(\mathcal{M}) = S(\mathcal{Z}) + \sum_j q_j S_j(\mathcal{M}_j),$$

where $\hat{S}_j$ and S_j are defined by using τ_j instead of τ, respectively. □

For two von Neumann subalgebras $\mathcal{M}_1$ and $\mathcal{M}_2$, let $\mathcal{M}_1 \vee \mathcal{M}_2$ be the von Neumann subalgebra generated by $\mathcal{M}_1$ and $\mathcal{M}_2$. It is easily seen that a partition $\{Q_k\} \in P(\mathcal{M}_1)$ is not always in $P(\mathcal{M}_1 \vee \mathcal{M}_2)$ but there exists a partition $\{Q_{kj}\}$ in $P(\mathcal{M}_1 \vee \mathcal{M}_2)$ with $Q_k = \sum_j Q_{kj}$. With this fact, we have the conditional entropy operator and the conditional entropy defined as

$$\hat{S}(\mathcal{M}_1|\mathcal{M}_2) = -\sum_{kj} Q_{kj} \{\log \tau(Q_{kj}) - \log \tau(Q_k)\},$$

$$S(\mathcal{M}_1|\mathcal{M}_2) = \tau(\hat{S}(\mathcal{M}_1|\mathcal{M}_2)),$$

where $\{Q_k\} \in P(\mathcal{M}_2)$ and $\{Q_{kj}\} \in P(\mathcal{M}_1 \vee \mathcal{M}_2)$ with $Q_k = \sum_j Q_{kj}$. These entropies are unique as $\hat{S}(\mathcal{M})$ and $S(\mathcal{M})$.

On the basis of the above formulations, we have the following two fundamental propositions.

Proposition 9.19 *For von Neumann subalgebras $\mathcal{M}$, $\mathcal{M}_1$, $\mathcal{M}_2$, the following equalities hold*:

1. $\hat{S}(\mathcal{M}) = \hat{S}(\mathcal{M}_1|\mathbb{C}I)$
2. $\hat{S}(\mathcal{M}_1 \vee \mathcal{M}_2) = \hat{S}(\mathcal{M}_1|\mathcal{M}_2) + \hat{S}(\mathcal{M}_2)$
3. $\hat{S}(\alpha\mathcal{M}_1|\alpha\mathcal{M}_2) = \alpha\hat{S}(\mathcal{M}_1|\mathcal{M}_2)$.

Proof Part 1 is immediate, and Part 2 comes from the equality $Q_k = \sum_j Q_{kj}$. Part 3 is obtained from the α-invariance of τ and the fact that $\tilde{P} = \mathcal{P}(\mathcal{M})$ implies $\alpha\tilde{P} = \mathcal{P}(\alpha\mathcal{M})$. □

Proposition 9.20 *For von Neumann subalgebras* $\mathcal{M}, \mathcal{M}_1, \mathcal{M}_2, \mathcal{M}_3$, *the following equalities and inequalities hold*:

1. $S(\mathcal{M}_1 \vee \mathcal{M}_2) = S(\mathcal{M}_1 \mid \mathcal{M}_2) + S(\mathcal{M}_2)$.
2. $S(\mathcal{M}_1) \le S(\mathcal{M}_1 \vee \mathcal{M}_2)$.
3. *For abelian* $\mathcal{M}_k$ (*i.e.*, $\mathcal{M}_k' = \mathcal{M}_k$), $S(\mathcal{M}_1 \vee \mathcal{M}_2) = S(\mathcal{M}_1) + S(\mathcal{M}_2)$.
4. $S(\mathcal{M}_1 \mid \mathcal{M}_2) \le S(\mathcal{M}_1)$.
5. $\mathcal{M}_2 \subset \mathcal{M}_3$ *implies* $S(\mathcal{M}_2 \mid \mathcal{M}_1) \le S(\mathcal{M}_3 \mid \mathcal{M}_1)$.
6. $S(\alpha\mathcal{M}) = S(\mathcal{M})$.
7. $S(\alpha\mathcal{M}_1 \mid \alpha\mathcal{M}_2) = S(\mathcal{M}_1 \mid \mathcal{M}_2)$.

9.3.2 McMillan's Type Convergence Theorems

The following theorem is obtained from the previous propositions.

Theorem 9.21 *For a von Neumann subalgebra* $\mathcal{M}$,

1. *An* α*-invariant operator* $h \in \mathcal{N}$ (*i.e.*, $\alpha(h) = h$) *exists such that*
$$\lim_{n\to\infty} \hat{S}\left(\bigvee_{k=0}^{n} \alpha^k \mathcal{M}\right) = h.$$
2. *If* α *is ergodic* (*i.e.*, *the set of all* α*-invariant elements is* CI), *then there is an integer* N *satisfying*
$$h = S\left(\bigvee_{k=0}^{N-1} \alpha^k \mathcal{M}\right).$$

Proof (Part 1) Since we have

$$\hat{S}(\mathcal{M} \vee \alpha\mathcal{M}) = \hat{S}(\alpha\mathcal{M} \mid \mathcal{M}) + \hat{S}(\mathcal{M}) \ge \hat{S}(\mathcal{M}),$$
$$\hat{S}\left(\bigvee_{k=0}^{n} \alpha^k \mathcal{M}\right) \le \hat{S}(\mathcal{N})$$

for every n, $\hat{S}(\bigvee_{k=0}^{n} \alpha^k \mathcal{M})$ increasingly converges to a certain operator $h \in \mathcal{N}$ in norm. We thus need to show the α-invariance of this operator h. For each n, we have

$$\hat{S}\left(\bigvee_{k=0}^{n} \alpha^k \mathcal{M}\right) \le \hat{S}\left(\bigvee_{k=1}^{n} \alpha^k \mathcal{M}\right) + \hat{S}\left(\mathcal{M} \mid \bigvee_{k=1}^{n} \alpha^k \mathcal{M}\right),$$

which implies

$$h = \hat{S}\left(\bigvee_{k=1}^{\infty} \alpha^k \mathcal{M}\right) + \hat{S}\left(\mathcal{M} \mid \bigvee_{k=1}^{\infty} \alpha^k \mathcal{M}\right).$$

Therefore, if we can prove the following facts: (i) $\alpha(\mathcal{M} \vee \alpha\mathcal{M}) = \alpha\mathcal{M} \vee \alpha^2\mathcal{M}$, (ii) $\hat{S}(\mathcal{M}_1 \mid \mathcal{M}_2) = 0$ iff $S(\mathcal{M}_1 \mid \mathcal{M}_2) = 0$, and (iii) $S(\mathcal{M} \mid \bigvee_{k=1}^{\infty} \alpha^k \mathcal{M}) = 0$, then the above h is equal to

$$\alpha \hat{S}\left(\bigvee_{k=1}^{\infty} \alpha^k \mathcal{M}\right) + 0 = \alpha(h),$$

that is, h is α-invariant.

Let us show the above facts (i), (ii), and (iii). Fact (i) is simple. To show fact (ii), $\hat{S}(\mathcal{M}_1 \mid \mathcal{M}_2) = 0$ implies $S(\mathcal{M}_1 \mid \mathcal{M}_2) = 0$ from the definition. Conversely, if $S(\mathcal{M}_1 \mid \mathcal{M}_2) = 0$, then the faithfulness of the trace implies $\hat{S}(\mathcal{M}_1 \mid \mathcal{M}_2) = 0$ because of the positivity of $\hat{S}(\mathcal{M}_1 \mid \mathcal{M}_2)$. For Fact (iii), the equality $S(\mathcal{M} \mid \bigvee_{k=1}^{\infty} \alpha^k \mathcal{M}) = 0$ can be proved by the above properties and Part 6 of Proposition 9.20.

(Part 2) Since h is α-invariant and α is ergodic, h is a multiple of identity, $h = \lambda I$. Moreover, it can be shown that, for the set $\mathcal{M}_\alpha \equiv \bigvee_{k=1}^{\infty} \alpha^k \mathcal{M}$, there exists a finite integer N such that $\mathcal{M}_\alpha = \bigvee_{k=1}^{N-1} \alpha^k \mathcal{M}$. These facts imply

$$h = \tau(h) = S\left(\bigvee_{k=0}^{\infty} \alpha^k \mathcal{M}\right) = S\left(\bigvee_{k=0}^{N-1} \alpha^k \mathcal{M}\right).$$

□

In the above theorem, the von Neumann subalgebra $\mathcal{M}_\alpha = \bigvee_{k=0}^{\infty} \alpha^k \mathcal{M}$ is a "space" constructed by moving α over $\mathcal{M}$. This "space" can be considered as a space generated by $\mathcal{M}$ and α. Therefore, the entropy $S(\mathcal{M}_\alpha)/N$ (denoted by $S(\mathcal{M}, \alpha)$) can be read as the entropy rate generated by $\mathcal{M}$ and α. Furthermore, it can readily be seen that (i) $S(\mathcal{M}, \alpha) \le S(\mathcal{M})$ holds and (ii) $\alpha\mathcal{M} = \mathcal{M}$ implies $S(\mathcal{M}, \alpha) = S(\mathcal{M})$.

The above theorem does not contain the averaging w.r.t. time n, so that it is not a complete formulation of McMillan's ergodic theorem in quantum systems. In order to formulate and prove the McMillan ergodic type convergence theorem, we first have to set an infinite tensor product of a Hilbert space $\mathcal{H}$ and that of von Neumann algebra $\mathcal{N}$ with respect to the tracial state τ.

Since τ is faithful normal, τ can be represented by a vector x in $\mathcal{H}$ such that

$$\tau(\cdot) = \langle x, \cdot x \rangle,$$

where $\langle \cdot, \cdot \rangle$ is the inner product of $\mathcal{H}$. Let $\mathcal{K}$ be the infinite tensor product of $\mathcal{H}$ with respect to the above vector x in the sense of von Neumann, which is denoted by

$$\mathcal{K} = \bigotimes_{-\infty}^{\infty} (\mathcal{H}, x).$$

We define the infinite tensor product $\mathcal{A}$ as the von Neumann algebra on $\mathcal{K}$ generated by the following operators $\bar{A}_n$ $(n \in \mathbb{Z})$

$$\bar{A}_n \left(\bigotimes_{-\infty}^{\infty} x_k \right) = \bigotimes_{-\infty}^{\infty} \bar{x}_k \quad \text{with } \bar{x}_k = \delta_{nk} A_n x_n + (1 - \delta_{nk}) x_k$$

with $A_n \in \mathcal{N}$ $(n \in \mathbb{Z})$.

This infinite tensor product $\mathcal{A}$ of von Neumann algebra $\mathcal{N}$ and is denoted by

$$\mathcal{A} = \bigotimes_{-\infty}^{\infty} (\mathcal{N}, \tau).$$

Now, for the n-times tensor product Hilbert space $\mathcal{H}_n = \bigotimes^n \mathcal{H}$, every element Q in $\mathbf{B}(\mathcal{H}_n) = \bigotimes^n \mathbf{B}(\mathcal{H})$ can be canonically embedded into $\mathbf{B}(\mathcal{H}_{n+1})$ in such a way that $Q \subset Q \otimes I$, so that we have the canonical embedding j_n from $\mathcal{N}_n = \bigotimes^n \mathcal{N}$ (the n-times tensor product of von Neumann algebra $\mathcal{N}$) into $\mathcal{A}$.

For a von Neumann subalgebra $\mathcal{M}$ of $\mathcal{N}$, let $\mathcal{M}_n$ and $\mathcal{B}_n$ be

$$\mathcal{M}_n = \bigotimes_1^n \mathcal{M}, \qquad \mathcal{B}_n = j_n(\mathcal{M}_n).$$

Then $\mathcal{B}_n$ becomes a von Neumann subalgebra of $\mathcal{A}$. Using a shift operator α defined as $\alpha(\otimes A_k) = \otimes A_{k+1}$, the above $\mathcal{B}_n$ is expressed by

$$\mathcal{B}_n = \bigvee_{k=0}^{n-1} \alpha^{-k} \mathcal{B}_1.$$

Our "information source" is now described by $(\mathcal{K}, \mathcal{A}, \alpha)$ and an α-invariant faithful state φ on the space $\mathcal{A}$. The state φ controls the transmission of information, so that the McMillan theorem is written in terms of φ, α, and the entropy operator defined in the previous section. We assume that φ_n, the restriction of φ to $\mathcal{M}_n$, is tracial. Then the entropy operator w.r.t. φ and $\mathcal{B}_n$ is given by

$$H_\varphi(\mathcal{B}_n) = -\sum_k Q_k^{(n)} \log \varphi_n\big(Q_k^{(n)}\big),$$

where $\{Q_k^{(n)}\}$ is a minimal finite partition of unity in $\mathcal{M}_n$.

Theorem 9.22 *Under the above settings, we have*

1. *There exists an α-invariant operator h such that $H_\varphi(\mathcal{B}_n)/n$ converges to h φ-almost uniformly as $n \to \infty$.*
2. *If α is ergodic (i.e., $\{A \in \mathcal{A}; \alpha(A) = A\} = \mathbb{C}I$) then $h = \varphi(h)I$.*

Proof (Part 1) For any minimal finite partition $\{P_i;\ i = 1, \dots, N\}$ in $\mathcal{N}$, the family $\{P_{i_1} \otimes \cdots \otimes P_{i_n}\}$ is a minimal finite partition in $\mathcal{N}_n$, where n indices $i_1, \dots, i_n$ run from 1 to N. As $\mathcal{N}_n$ is a finite-dimensional von Neumann algebra, $H_\varphi(\mathcal{B}_n)$ can be expressed by

$$H_\varphi(\mathcal{B}_n) = -\sum_{i_1,\ldots,i_n=1}^{N} P_{i_1} \otimes \cdots \otimes P_{i_n} \log \varphi_n(P_{i_1} \otimes \cdots \otimes P_{i_n}).$$

Let us consider the von Neumann subalgebra $\mathcal{C}_n$ of $\mathcal{M}_n$ generated by the family $\{P_{i_1} \otimes \cdots \otimes P_{i_n}\}$, i.e., $\mathcal{C}_n = \{P_{i_1} \otimes \cdots \otimes P_{i_n}; i_1, \ldots, i_n = 1, \ldots, N\}''$, and let $\mathcal{C}_1 = \{P_i\}'' = \mathcal{C}$. Then $\mathcal{C}_n$ is the n-times tensor product of $\mathcal{C}$, the algebras $\mathcal{C}$, $\mathcal{C}_n$ are commutative von Neumann algebras, and there exist compact Hausdorff spaces Ω_n, Ω and probability measures μ_n, μ such that

$$\mathcal{C}_n = L^\infty(\Omega_n, \mu_n), \qquad \mathcal{C} = L^\infty(\Omega, \mu).$$

Moreover, $\mathcal{C}_n$ is monotonously increasing and generates the infinite tensor product $\tilde{\mathcal{C}}$ of $\mathcal{C}$. Since $\tilde{\mathcal{C}}$ is commutative, there exist a compact Hausdorff space $\tilde{\Omega}$ and a probability measure $\tilde{\mu}$ such that

$$\tilde{\mathcal{C}} = L^\infty(\tilde{\Omega}, \tilde{\mu})$$

and

$$\varphi(P_{i_1} \otimes \cdots \otimes P_{i_n}) = \tilde{\mu}(\Delta_{i_1\cdots i_n}),$$

where $P_{i_1} \otimes \cdots \otimes P_{i_n}$ corresponds to the characteristic function $1_{\Delta_{i_1\cdots i_n}}$ for some measurable set $\Delta_{i_1\cdots i_n}$ in $\prod_i^n \Omega$. Thus the classical McMillan theorem together with the previous theorem (based on the finite-dimensionality of $\mathcal{M}_n$) implies that our entropy operator $H_\varphi(\mathcal{B}_n)/n$ converges to some α-invariant operator h in $L^1(\tilde{\Omega}, \tilde{\mu})$, φ-a.u. because of φ-a.u. $= \tilde{\mu}$-a.e.

The α-invariance of h is clear from the definition of $H_\varphi(\mathcal{B}_n)/n$.

Part 2 is an immediate consequence of Part 1. □

A more general study of the McMillan theorem would be desirable. This will be done by dropping some of the assumptions taken above; for instance, (i) the finite dimensionality of $\mathcal{N}$, (ii) the trace property of $\varphi_n = \varphi \restriction \mathcal{M}_n$.

9.4 Coding Type Theorems

Here we discuss the coding type theorems which are not same as the Shannon coding theorem but a sort of convergence theorems in terms of coding and decoding. These theorems say that if there exist codings and decodings such that some distance between input and output is zero, then the mean information of the input state is equal to the capacity type quantity defined appropriately.

Let $\mathcal{H}$ and $\mathcal{K}$ be the input and the output Hilbert spaces, respectively, so that their N-fold tensor products are denoted by

$$\mathcal{H}_N = \bigotimes^N \mathcal{H}, \qquad \mathcal{K}_N = \bigotimes^N \mathcal{K}.$$

Note that

$$\mathbf{B}(\mathcal{H}_N) = \bigotimes^N \mathbf{B}(\mathcal{H}), \qquad \mathbf{B}(\mathcal{K}_N) = \bigotimes^N \mathbf{B}(\mathcal{K}).$$

A channel $\Lambda_N^* : \mathfrak{S}(\mathcal{H}_N) \to \mathfrak{S}(\mathcal{K}_N)$ sends density operators acting on $\mathcal{H}_N$ into those acting on $\mathcal{K}_N$. We only consider a *memoryless channel* which is the tensor product of the same *single site channels*: $\Lambda_N^* = \Lambda^* \otimes \cdots \otimes \Lambda^*$ (N-fold).

Moreover, let $\mathcal{L}$ be the Hilbert space attached to an information source. Coding and decoding are expressed by the following channel:

$$\begin{aligned} \mathcal{C}_N &: \mathfrak{S}(\mathcal{L}_N) \to \mathfrak{S}(\mathcal{H}_N), \\ \Lambda_N^* &: \mathfrak{S}(\mathcal{H}_N) \to \mathfrak{S}(\mathcal{K}_N), \\ \mathcal{D}_N &: \mathfrak{S}(\mathcal{K}_N) \to \mathfrak{S}(\mathcal{L}_N). \end{aligned}$$

If the Hilbert spaces $\mathcal{L}, \mathcal{H}, \mathcal{K}$ are finite-dimensional, then the channel is called *discrete*. We call $\{\mathcal{L}, \mathcal{H}, \mathcal{K}, \mathcal{C}_N, \Lambda_N^*, \mathcal{D}_N\}$ a *quantum coding scheme*.

Now we give the definition of some rates of transmitted information for a discrete memoryless channel Λ^*. First, we define the rates for transmission of subspaces.

We say that a sequence of subspaces $\mathcal{L}_N^0$ of the source spaces $\mathcal{L}_N$, $N = 1, 2, \ldots$ may be sent reliably over the channel Λ^* if there exists a quantum coding scheme $\{\mathcal{L}, \mathcal{H}, \mathcal{K}, \mathcal{C}_N, \Lambda_N^*, \mathcal{D}_N\}$ such that

$$\lim_{N\to\infty} F_p\big(\mathcal{L}_N^0, \mathcal{D}_N \circ \Lambda_N^* \circ \mathcal{C}_N\big) = 1.$$

Here F_p is the *pure-state fidelity* defined as

$$F_p(\mathcal{V}, \Theta) = \inf_{|\psi\rangle\in\mathcal{V}} \big\langle \psi, \Theta\big(|\psi\rangle\langle\psi|\big)\psi\big\rangle$$

where Θ is a quantum operation on operators in any Hilbert space $\mathcal{H}$ and $\mathcal{V}$ is a subspace in $\mathcal{H}$.

A real number R_s is called the *achievable rate of transmission of subspace dimensions* with the channel Λ^* if there exists a sequence of subspaces $\mathcal{L}_N^0$ of the source spaces $\mathcal{L}_N$ ($N = 1, 2, \ldots$) which can be sent reliably over the channel Λ^*, and it is defined as

$$R_s \equiv \lim_{N\to\infty} \sup \frac{\log \dim(\mathcal{L}_N^0)}{N}.$$

This achievable rate of transmission of subspace dimensions R_s depends on the encoding and decoding maps through the fidelity, and we will write it as

$$R_s = R_s\big(\Lambda^*, \{\mathcal{D}_N, \mathcal{C}_N\}\big).$$

The *maximum subspace rate* $R_s(\Lambda^*)$ of the channel Λ^* is the supremum of achievable rates of subspace dimensions with channel Λ^*,

$$R_s(\Lambda^*) \equiv \sup\big\{R_s\big(\Lambda^*, \{\mathcal{D}_N, \mathcal{C}_N\}\big); \mathcal{D}_N, \mathcal{C}_N\big\}$$

over all encoding and decoding maps.

There is another rate defined for entanglement transmission. Instead of the pure-state fidelity it uses the entanglement fidelity. Let ρ be a state in a Hilbert space $\mathcal{H}$ and ψ_ρ its purification in the Hilbert space $\mathcal{H} \otimes \mathcal{K}$. Then the entanglement fidelity of the state ρ under an operation Θ in $\mathcal{H}$ is defined as

$$F_e(\rho, \Theta) = \langle \psi_\rho | \Theta(\psi_\rho) \otimes I | \psi_\rho \rangle.$$

This is independent on the purification used. One can show that if a quantum operation Θ is represented as

$$\Theta(\rho) = \sum_i A_i \rho A_i^*, \quad \sum_i A_i^* A_i \le 1$$

then the entanglement fidelity of a state ρ under the operation Θ is

$$F_e(\rho, \Theta) = \sum_i |\mathrm{tr}\, A_i \rho|^2.$$

A quantum source $\{\mathcal{L}, \rho_s\}$ is defined as a pair which consists of a Hilbert space $\mathcal{L}$ and a sequence $\rho_s = \{\rho_s^{(1)}, \rho_s^{(2)}, \ldots, \rho_s^{(N)}, \ldots\}$ where $\rho_s^{(N)}$ is a density operator on $\mathcal{L}_N$. One assumes that the density operators in the sequence are consistent with each other in the following sense: for all j and $N > j$, $\mathrm{tr}_{j+1,\ldots,N}(\rho_s^{(N)}) = \rho_s^{(j)}$ where $\mathrm{tr}_{j+1,\ldots,N}$ means the partial trace over the corresponding copies of $\mathcal{L}$. The *entropy rate* for the source Σ is defined as

$$\widetilde{S}(\rho_s) = \lim_{N\to\infty} \sup \frac{S(\rho_s^{(N)})}{N}.$$

We say that a source $\{\mathcal{L}, \rho_s\}$ may be sent reliably over a quantum channel Λ^* if there exists a quantum coding scheme $\{\mathcal{L}, \mathcal{H}, \mathcal{K}, \mathcal{C}_N, \Lambda_N^*, \mathcal{D}_N\}$ such that

$$\lim_{N\to\infty} F_e\big(\rho_s^{(N)}, \mathcal{D}_N \circ \Lambda_N^* \circ \mathcal{C}_N\big) = 1.$$

A real number R_e is called the achievable rate of entanglement transmission with the channel Λ^* if there is a source $\{\mathcal{L}, \rho_s\}$ with entropy rate $\widetilde{S}(\rho_s) = R_e$ which can be sent reliably over the channel Λ^*.

The achievable rate of entanglement transmission R_e depends on the encoding and decoding maps, and we will write it as

$$R_e = R_e\big(\Lambda^*, \{\mathcal{D}_N, \mathcal{C}_N\}\big).$$

The entanglement transmission rate $R_e(\Lambda^*)$ of the channel Λ^* is the supremum of achievable rates of entanglement transmission with channel Λ^*,

$$R_e(\Lambda^*) = \sup\bigl\{R_e\bigl(\Lambda^*, \{\mathcal{D}_N, \mathcal{C}_N\}\bigr); \mathcal{D}_N, \mathcal{C}_N\bigr\}$$

over all encoding and decoding maps.

It was proved in [86] that the two definitions of quantum rates are equivalent, $R_s(\Lambda^*) = R_e(\Lambda^*)$.

Therefore, we will denote

$$R(\Lambda^*) = R_s(\Lambda^*) = R_e(\Lambda^*)$$

and call it simply the entanglement transmission rate of the channel Λ^*.

The coherent information is defined in the previous chapter by

$$I_c(\rho; \Lambda^*) = S\bigl(\Lambda^*(\rho)\bigr) - S_e(\rho, \Lambda^*).$$

There is a *quantum Fano inequality* which relates the entropy exchange and the entanglement fidelity:

$$S_e(\rho, \Lambda^*) \le h\bigl(F_e(\rho, \Lambda^*)\bigr) + \bigl(1 - F_e(\rho, \Lambda^*)\bigr)\log_2\bigl(d^2 - 1\bigr),$$

where $h(x) = -x\log_2 x - (1-x)\log_2(1-x)$ is the dyadic Shannon information and d is the dimension of the Hilbert space with the density matrix ρ.

The following important bound holds.

Theorem 9.23 *Suppose Λ^* is a channel in a Hilbert space H of dimension d, and ρ is a quantum state in H. Then for any channel Γ^* one has*

$$S(\rho) \le I_c(\rho; \Lambda^*) + 2 + 4\bigl(1 - F_e(\rho, \Gamma^* \circ \Lambda^*)\bigr)\log d.$$

This theorem shows that the entropy of the state ρ cannot greatly exceed the coherent information $I_c(\rho, \Lambda^*)$ if the entanglement fidelity is close to one.

Barnum, Nielsen and Schumacher [85] proved the following bound for the entanglement transmission rate:

$$R(\Lambda^*) \le \lim_{n\to\infty} \frac{1}{N} \sup\bigl\{I_c(\rho_N; \Lambda_N^*); \rho_N\bigr\}.$$

Recently, Shor and Devetak [200] proved that, in fact, one has the equality here.

Theorem 9.24

$$R(\Lambda^*) = \lim_{n\to\infty} \frac{1}{N} \sup\bigl\{I_c(\rho_N; \Lambda_N^*); \rho_N\bigr\}.$$

There are several proofs of the formula, but we will not discuss them in this book. This is sometimes called a coding theorem, but it is not so in the sense of Shannon because of two reasons: (i) the one mentioned at the beginning of this section and (ii) I_c does not describe the transmission of information but may describe the transmission of entanglement.

In the discussion of this subsection, it is important that the entanglement fidelity is used. If we were to use another fidelity for information transmission, then we would get a coding type theorem for information transmission. For instance, a candidate of such a fidelity is

$$F(\rho_1, \rho_2) = \sup\{|\langle\psi_1|\psi_2\rangle|^2; \psi_1, \psi_2\},$$

where ψ_1, ψ_2 are certain purifications of states ρ_1, ρ_2. One can conjecture that by using this fidelity one can prove another quantum coding type theorem and estimate a quantum transmission rate by using the quantum mutual entropy $I(\rho; \Lambda^*)$.

Finally, we mention the proper quantum coding theorem which has not been proved so far: Given an input source $\{\mathcal{L}_N, \rho_s \equiv \{\rho_s^{(1)}, \rho_s^{(2)}, \dots, \rho_s^{(N)}, \dots\}\}$ and a channel Λ_N^* (not always memoryless), set the entropy rate $\widetilde{S}(\rho_s)$ and the capacity $C(\Lambda^*) \equiv \sup\{\widetilde{I}(\rho; \Lambda^*); \rho \in \mathfrak{S}(\mathcal{H})\}$. Then if $\widetilde{S}(\rho_s) < C(\Lambda^*)$, there exist a coding $\mathcal{C}$ and decoding $\mathcal{D}$ such that the error probability, properly defined in an information transmission process, is nearly 0.

9.5 Notes

The channel capacity discussed here is mainly due to the authors of [345, 584, 587, 591, 593, 601]. The pseudo-quantum capacity was introduced in [587]. The bound of the capacity has been studied first by Holevo [345] and later by many others [601, 694, 828]. The quantum Macmillan's theorem is discussed in [333, 567]. The coding type theorems discussed here is due to the authors of [85, 86, 200].

Chapter 10
Information Dynamics and Adaptive Dynamics

There exist various approaches to study chaotic behavior of systems by means of several concepts such as entropy, complexity, chaos, fractal, stochasticity. In 1991, the term *Information Dynamics* (ID) was proposed by Ohya with the aim of finding a common framework of treating such chaotic behaviors of different systems altogether. That is, ID is an attempt to synthesize dynamics of state change and complexity of the systems. The basic quantity in ID is called a *chaos degree*, which can be applied to describe chaos phenomena. Since then, ID has been refined and applied to several topics such as communication, chaos, genetics, and finance. Moreover, ID enables us to find mathematics to describe a sort of *subjective* aspects of physical phenomena, for instance, observation, effects of surroundings, which is called *Adaptive Dynamics* (AD).

In this chapter, the time irreversibility problem and a new approach to classical mechanics is also discussed. In classical Newtonian mechanics, the state of the system at some moment of time is represented as a point in the phase space. However, this notion does not have an immediate operational meaning since arbitrary real numbers are not observable. In the new approach to classical non-Newtonian mechanics, suggested by Volovich, the particle is described by a probability distribution in the phase space. The expectation value of the coordinates approximately satisfies the Newton equation, but there are corrections to the Newton trajectories.

Moreover in the last section in this chapter, a new approach to Bell's inequality is discussed in a view of AD which has been called the Chameleon dynamics by Accardi.

10.1 Complex Systems

Let us first review what a complex system is. The discussion leads to the introduction of Information Dynamics in a natural way.

A *complex system* has been considered as follows:

1. A system is composed of several elements called *agents*. The size of the system (the number of the elements) is medium.

M. Ohya, I. Volovich, *Mathematical Foundations of Quantum Information and Computation and Its Applications to Nano- and Bio-systems*, Theoretical and Mathematical Physics,
DOI 10.1007/978-94-007-0171-7_10,

2. Agents have intelligence.
3. Each agent has interaction due to local information. The decision of each agent is determined by not all information but the limited information of the system.

With a small modification, we consider the complex system as follows:

1. A system is composed of several elements. The scale of the system is often large but not always, in some cases, only one element.
2. Some elements of the system have special (self) interactions (relations), which produce dynamics of the system.
3. The system shows a particular character (i.e., structure and dynamics) which is not merely a sum of the characters of all elements.

A system having the above three properties is called a *complex system*. The *complexity* of such a complex system is a quantity measuring the structure of the complex system, and its change (dynamics) describes the appearance of the particular character of the system.

There exist such measures describing the complexity for a system, for instance, variance, correlation, level-statistics, fluctuation, randomness, multiplicity, entropy, fuzzyness, fractal dimension, ergodicity (mixing, flow), bifurcation, localization, computational complexity (Kolmogorov's or Chaitin's), catastrophy, dynamical entropy, Lyapunov exponent, etc. These quantities are used on a case-by-case basis, and they are often difficult to compute. Moreover, the relations among them are lacking (not clear enough). Therefore, it is important to find a common property or expression of these quantities, which is discussed in Information Dynamics and Adaptive Dynamics in the following sections.

We briefly review ID and an axiomatic approach to the complexity in the next section. In Sect. 10.3, various examples of the state changes (channels) and of the complexities are presented, some of which are new expressions of physical and communication processes. In Sect. 10.4, the idea of the adaptive dynamics is explained. The conceptual meaning of AD is discussed in Sect. 10.5. As an illustration of the use of ID and AD, we introduce a certain degree estimating the chaos attached to a dynamical system by means of entropies in Sect. 10.6. In Sect. 10.7, we discuss the algorithms computing the chaos degree. The adaptive dynamics for study of chaos is considered in Sect. 10.8, and in the same section, a new description of chaos observed in various phenomena is discussed.

10.2 Information Dynamics

There are two aspects for the complexity, that is, the complexity of a state describing the system itself and that of a dynamics causing the change of the system (state). The former complexity is simply called the *complexity of the state*, and the later is called the *chaos degree of the dynamics*. Therefore, the examples of the complexity are entropy, fractal dimension as above, and those of the chaos degree are Lyapunov exponent, dynamical entropy, computational complexity. Let us discuss a common quantity measuring the complexity of a system that we can easily handle. The complexity of a general quantum state was introduced in the framework of ID in [359, 573, 590], and the quantum chaos degree was defined in [365, 367], which we will review in Chap. 20.

Information dynamics (ID) is a synthesis of the dynamics of state change and the complexity of states. It is an attempt to provide a new view for the study of chaotic behavior of systems. We briefly review what ID is.

Let $(\mathcal{A}, \mathfrak{S}, \alpha(G))$ be an input (or initial) system and $(\overline{\mathcal{A}}, \overline{\mathfrak{S}}, \overline{\alpha}(\overline{G}))$ be an output (or final) system. Here $\mathcal{A}$ is a set of some objects to be observed and $\mathfrak{S}$ is a set of some means to get the observed value, $\alpha(G)$ describes a certain evolution of system with a set G of parameters. We often have $\mathcal{A} = \overline{\mathcal{A}}$, $\mathfrak{S} = \overline{\mathfrak{S}}$, $\alpha = \overline{\alpha}$. Therefore, we claim

[Giving a mathematical structure to input and output triples

$\equiv$ Having a theory].

The dynamics of state change is described by a channel, which will be explained in the next section, $\Lambda^* : \mathfrak{S} \to \overline{\mathfrak{S}}$ (sometimes $\mathfrak{S} \to \mathfrak{S}$). The information is described by two complexities explained bellow. The fundamental point of ID is that ID contains these two complexities in itself. Let $(\mathcal{A}_{\text{tot}}, \mathfrak{S}_{\text{tot}}, \alpha^{\text{tot}}(G^{\text{tot}}))$ be the total system of $(\mathcal{A}, \mathfrak{S}, \alpha(G))$ and $(\overline{\mathcal{A}}, \overline{\mathfrak{S}}, \overline{\alpha}(\bar{G}))$, and $\mathcal{S}$ be a subset of $\mathfrak{S}$ in which we measure an object (e.g., $\mathcal{S}$ is the set of all KMS or stationary states in a C^*-system). Two complexities are denoted by C and T. C is the complexity of a state φ (the structure of a system) measured from a reference system $\mathcal{S}$, in which we actually observe the objects in $\mathcal{A}$, and T is the transmitted complexity associated with a state change (dynamics) $\varphi \to \Lambda^*\varphi$, both of which should satisfy the following properties:

Axiom 10.1 Complexities

(i) For any $\varphi \in \mathcal{S} \subset \mathfrak{S}$,

$$C^{\mathcal{S}}(\varphi) \geq 0, \qquad T^{\mathcal{S}}(\varphi; \Lambda^*) \geq 0.$$

(ii) $0 \leq T^{\mathcal{S}}(\varphi; \Lambda^*) \leq C^{\mathcal{S}}(\varphi)$.

(iii) $T^{\mathcal{S}}(\varphi; \mathrm{id}) = C^{\mathcal{S}}(\varphi)$, where "id" is the identity map from $\mathfrak{S}$ to $\mathfrak{S}$.

(iv) For any map $j : \mathfrak{S} \to \mathfrak{S}$ such that $j : \operatorname{ex}\mathcal{S} \to \operatorname{ex}\mathcal{S}$ (the set of all extremal points (some elementary points) of $\mathcal{S}$) is a disjoint (in a proper sense) bijection

$$C^{j(\mathcal{S})}\bigl(j(\varphi)\bigr) = C^{\mathcal{S}}(\varphi),$$

$$T^{j(\mathcal{S})}\bigl(j(\varphi); \Lambda^*\bigr) = T^{\mathcal{S}}(\varphi; \Lambda^*).$$

(v) For $\Phi \equiv \varphi \otimes \psi \in \mathcal{S}_{\text{tot}} \subset \mathfrak{S}_{\text{tot}}$, $\psi \in \overline{\mathcal{S}} \subset \overline{\mathfrak{S}}$,

$$C^{\mathcal{S}_{\text{tot}}}(\Phi) = C^{\mathcal{S}}(\varphi) + C^{\overline{\mathcal{S}}}(\psi),$$

where $\otimes$ is a symbolic expression for the combination of two states. When the system is mathematically set, this will be, for instance, a proper tensor product.

Instead of (v), when "(v′) For $\Phi \in \mathcal{S}_{\mathrm{tot}} \subset \mathfrak{S}_{\mathrm{tot}}$, $\varphi \equiv \Phi \upharpoonright \mathcal{A}$, $\psi \equiv \Phi \upharpoonright \overline{\mathcal{A}}$ (i.e., the restriction of Φ to $\mathcal{A}$ and $\overline{\mathcal{A}}$, respectively), $C^{\mathcal{S}_{\mathrm{tot}}}(\Phi) \le C^{\mathcal{S}}(\varphi) + C^{\overline{\mathcal{S}}}(\psi)$." is satisfied, C and T is called a pair of *strong complexity*. Therefore, ID is defined as follows:

Definition 10.2 Information Dynamics is described by

$$\left(\mathcal{A}, \mathfrak{S}, \alpha(G); \overline{\mathcal{A}}, \overline{\mathfrak{S}}, \overline{\alpha}(\overline{G}); \Lambda^*; C^{\mathcal{S}}(\varphi), T^{\mathcal{S}}(\varphi; \Lambda^*)\right)$$

and some relations R among them.

Therefore, in the framework of ID, we have to

(ID-1) Mathematically determine $(\mathcal{A}, \mathfrak{S}, \alpha(G); \overline{\mathcal{A}}, \overline{\mathfrak{S}}, \overline{\alpha}(\overline{G}))$
(ID-2) Choose Λ^* and R, and
(ID-3) Define $C^{\mathcal{S}}(\varphi)$, $T^{\mathcal{S}}(\varphi; \Lambda^*)$.

In ID, several different topics can be treated from a common standing point so that we can find a new clue bridging several fields. For example, we may have the following applications [359]:

1. Study of optical communication processes
2. Formulation of fractal dimensions of states, and the study of complexity for some systems
3. Definition of a genetic matrix for genome sequences and construction of phylogenetic tree for evolution of species
4. Entropic complexities ⇒ Kolmogorov–Sinai type complexities (entropy) ⇒ Classification of dynamical systems
5. Study of optical illusion (psychology)
6. Study of some economic models
7 Study of chaos.

In this chapter, we mainly discuss the applications 4 and 7 above.

10.3 State Change and Complexities

ID contains the dynamics of state change as its part. A state change is mathematically described by a unitary evolution, a semigroup dynamics, generally, a channeling transformation (it is simply called a *channel*) or a bit restricted notion of *lifting*, which have been repeatedly discussed in the previous chapters.

The input and output triple $(\mathcal{A}, \mathfrak{S}, \alpha(G))$ and $(\overline{\mathcal{A}}, \overline{\mathfrak{S}}, \overline{\alpha}(\overline{G}))$ are the above sets, that is, $\mathcal{A}$ is $M(\Omega)$ or $\mathbf{B}(\mathcal{H})$ or $\mathcal{A}$ (C^*-algebra), $\mathfrak{S}$ corresponds to each state space, and $\alpha(G)$ is an inner evolution of $\mathcal{A}$ with a parameter group G (or semigroup), and so is the output system.

A channel is a mapping from $\mathfrak{S}(\mathcal{A})$ to $\overline{\mathfrak{S}}(\overline{\mathcal{A}})$. Almost all physical transformations are described by this mapping, as we have seen.

Although there exist several complexities, one of the most fundamental pairs of C and T in a quantum system is the von Neumann entropy and the mutual entropy, whose C and T are modified to formulate the entropic complexities such as ε-entropy (ε-entropic complexity), Kolmogorov–Sinai type dynamical entropy (entropic complexity).

The concept of entropy was introduced and developed to study the following topics: irreversible behavior, symmetry breaking, amount of information transmission, chaotic properties of states, etc. Here we first show that the quantum entropy and the quantum mutual entropy are examples of our complexities C and T, respectively. Then we give several examples of complexities C and T related to mainly information quantities.

Example 10.3 The first examples of C and T are the entropy S and the mutual entropy I, respectively. Both the classical and quantum S and I satisfy the conditions of the complexities of ID. Here we only discuss the quantum case. For a density operator ρ in a Hilbert space and a channel Λ^*, the entropy $S(\rho)$ and the quantum mutual entropy $I(\rho;\Lambda^*)$ were defined in Chap. 7 as

$$S(\rho) = -\operatorname{tr}\rho\log\rho,$$

$$I(\rho;\Lambda^*) = \sup\Bigl\{\sum_k \lambda_k S(\Lambda^* E_k, \Lambda^*\rho);\ \{E_k\}\Bigr\},$$

where the supremum is taken over all Schatten decompositions $\{E_k\}$ of ρ; $\rho = \sum_k \lambda_k E_k$. According to the fundamental properties of the entropy, $S(\rho)$ satisfies: (i) $S(\rho)\ge 0$, (ii) $S(j(\rho)) = S(\rho)$ for an orthogonal bijection j, that is, it is a map from a set of orthogonal pure states to another set of orthogonal pure states, (iii) $S(\rho_1\otimes\rho_2) = S(\rho_1)+S(\rho_2)$, so that $S(\rho)$ is a complexity C of ID.

The mutual entropy $I(\rho;\Lambda^*)$ satisfies Conditions (i), (ii), (iv) by the fundamental inequality of mutual entropy:

$$0 \le I(\rho;\Lambda^*) \le \min\bigl\{S(\rho), S(\Lambda^*\rho)\bigr\}.$$

Further, for the identity channel, $\Lambda^* = \mathrm{id}$,

$$\begin{aligned} I(\rho;\mathrm{id}) &= \sup\Bigl\{\sum_k \lambda_k S(E_k,\rho);\ \{E_k\}\Bigr\} \\ &= \sup\Bigl\{\sum_k \lambda_k \operatorname{tr} E_k(\log E_k - \log\rho);\ \{E_k\}\Bigr\} \\ &= -\operatorname{tr}\rho\log\rho, \end{aligned}$$

because $S(E_k)=0$, hence it satisfies Condition (v). Thus the quantum entropy and the quantum mutual entropy satisfy all the conditions of the complexity and the transmitted complexity, respectively; $C(\rho) = S(\rho)$, $T(\rho;\Lambda^*) = I(\rho;\Lambda^*)$. Moreover, S satisfies the condition of the strong complexity (subadditivity).

In Shannon's communication theory for classical systems, ρ is a probability distribution $p = (p_k) = \sum_k p_k \delta_k$ and Λ^* is a transition probability $(t_{i,j})$, so that the Schatten decomposition of ρ is unique and the compound state of ρ and its output $\overline{\rho}$ $(\equiv \overline{p} = (\overline{p}_i) = \Lambda^* p)$ is the joint distribution $r = (r_{i,j})$ with $r_{i,j} \equiv t_{i,j} p_j$. Then the above complexities C and T become the Shannon entropy and mutual entropy, respectively:

$$C(p) = S(p) = -\sum_k p_k \log p_k,$$

$$T(p; \Lambda^*) = I(p; \Lambda^*) = \sum_{i,j} r_{i,j} \log \frac{r_{i,j}}{p_j \overline{p}_i}.$$

We can construct several other types of entropic complexities. For instance, one pair of the complexities is

$$T(\rho; \Lambda^*) = \sup\left\{\sum_k p_k S(\Lambda^* \rho_k, \Lambda^* \rho); \rho = \sum_k p_k \rho_k\right\}, \qquad C(\rho) = T(\rho; \mathrm{id})$$

where $\rho = \sum_k p_k \rho_k$ is a finite decomposition of ρ and the supremum is taken over all such finite decompositions.

Example 10.4 Fuzzy entropy has been defined by several authors: Zadeh [829], DeLuca and Termini [192] and Ebanks [217]. Here we take Ebanks' fuzzy entropy and we show that we can use it to construct the complexity C. Let X (this is $\mathcal{A}$ of ID) be a finite set $\{x_1, \ldots, x_n\}$ and f_A be a membership function from X to $[0, 1]$ associated with a subset $A \subset X$. If $f_A = 1_A$, then A is a usual set, which is called a sharp set, and if $f_A \neq 1_A$, then the pair $\{A, f_A\}$ is called a *fuzzy set*. Therefore, the correspondence between a fuzzy set and a membership function is one-to-one. Take a membership function f, and let us denote $f_i = f(x_i)$ for each $x_i \in X$. Then Ebanks' fuzzy entropy $S(f)$ for a membership function f is defined by

$$S(f) \equiv -\sum_{i=1}^n f_i^\nu \log f_i \quad (\nu = \log_2 e).$$

When f is sharp, that is, $f_i = 0$ or 1 for any $x_i \in X$, $S(f) = 0$. When $f_i = \frac{1}{2}$ for any i, $S(f)$ attains the maximum value. Moreover, any two membership functions (or equivalently, fuzzy sets) f and f' have the following order $\prec$:

$$f \prec f' \equiv \begin{cases} f(x) \geq f'(x) & \text{when } f'(x) \geq \frac{1}{2}, \\ f(x) \leq f'(x) & \text{when } f'(x) \leq \frac{1}{2}. \end{cases}$$

Then $f \prec f'$ iff $|f_i - \frac{1}{2}| \geq |f_i' - \frac{1}{2}|$ for any i, which implies

$$S(f) \leq S(f').$$

This fuzzy entropy $S(f)$ defines a complexity $C(f)$ by

$$C(f) = S\left(\frac{f}{n}\right).$$

The positivity of $C(f)$ is proved as follows: From Klein's inequality, $\log \frac{1}{x} \geq 1 - x$ for any $x > 0$, we have

$$\begin{aligned} C(f) &= -\sum_{i=1}^{n}\left(\frac{f_i}{n}\right)^{\nu}\log\frac{f_i}{n} \\ &\geq -\sum_{i=1}^{n}\left(\frac{f_i}{n}\right)^{\nu}\left(1-\frac{n}{f_i}\right) \\ &= \sum_{i=1}^{n}\left(\frac{f_i}{n}\right)^{\nu-1}\left(1-\frac{f_i}{n}\right) \geq 0. \end{aligned}$$

The invariance under a permutation π of indices i of x_i (i.e., $i \to \pi(i)$), directly comes from the invariance of S under π. This $C(f)$ satisfies not only the additivity but also the subadditivity. Let Y be another set $\{y_1, \ldots, y_m\}$ and g be a membership function from Y to $[0, 1]$. Moreover, let h be a membership function on $X \times Y$ to $[0, 1]$ satisfying

$$\sum_{j=1}^{m} h(x_i, y_j) = f(x_i), \qquad \sum_{i=1}^{n} h(x_i, y_j) = g(y_j).$$

What we have to show is the inequality

$$C(h) \leq C(f) + C(g).$$

Without loss of generality, we assume $n \geq m \geq 2$. Put

$$\eta(t) = -t^{\nu}\log t \quad (\nu = \log_2 e).$$

$\eta(t)$ is monotone increasing for $0 \leq t \leq \frac{1}{2}$, so that we have

$$\eta\left(\frac{f_i}{nm}\right) \geq \eta\left(\frac{h_{ij}}{nm}\right) \quad (n \geq m \geq 2)$$

because of $h_{ij} \equiv h(x_i, y_j) \leq f_i \equiv f(x_i)$, and hence $0 \leq h_{ij}/nm \leq f_i/nm \leq \frac{1}{2}$ for any i, j. Thus we have

$$m\eta\left(\frac{f_i}{nm}\right) \geq \sum_{j=1}^{m}\eta\left(\frac{h_{ij}}{nm}\right).$$

Now

$$
\begin{aligned}
&\eta\left(\frac{f_i}{n}\right) - \frac{1}{2} m\eta\left(\frac{f_i}{nm}\right) \\
&\quad = \left(\frac{f_i}{n}\right)^{\nu}\left\{\left(\frac{1}{2m^{\nu-1}} - 1\right)\log f_i + \left(1 - \frac{1}{2m^{\nu-1}}\right)\log n - \frac{1}{2m^{\nu-1}}\log m\right\} \\
&\quad \geq \left(\frac{f_i}{n}\right)^{\nu}\left\{\left(\frac{1}{2m^{\nu-1}} - 1\right)\log f_i + \left(1 - \frac{1}{m^{\nu-1}}\right)\log n\right\},
\end{aligned}
$$

which is positive for any i since $n \geq m \geq 2$. Hence

$$
\eta\left(\frac{f_i}{n}\right) \geq \frac{1}{2}\sum_{j=1}^{m}\eta\left(\frac{h_{ij}}{nm}\right),
$$

which implies

$$
C(f) \geq \frac{1}{2}C(h).
$$

Similarly, we can prove

$$
C(g) \geq \frac{1}{2}C(h).
$$

Therefore, we have the subadditivity

$$
C(h) \leq C(f) + C(g).
$$

Example 10.5 Kolmogorov and Chaitin considered the complexity of sequences [156, 437]. For instance, let us consider the following two sequences a and b composed of 0s and 1s:

$$
\begin{aligned}
&a : 010101010101, \\
&b : 011010000111.
\end{aligned}
$$

In both a and b, the occurrence probabilities $p(0)$ and $p(1)$ are the same. However, the sequence b seems more complicated than a. It is enough for us to know the first two letters to guess the whole a, but one might need the whole sequence of letters to know b. Hence once we consider a computer sending as input a finite sequence of letters 0 and 1 and observing an output sequence, we call the shortest algorithm having the minimum information to produce a proper sequence by a computer the *minimum programming*. When we measure this minimum information by "bit", it is called the *complexity of the sequence*. Let $\mathcal{A}$ be the set of all finite sequences of some letters, say $\{0, 1\}$, and let $\bar{\mathcal{A}}$ be another set. Further, let f be a partial function for $\mathcal{A}$ to $\bar{\mathcal{A}}$ (i.e., f is not always defined on the whole $\mathcal{A}$). The triple $(\mathcal{A}, \bar{\mathcal{A}}, f)$ can be regarded as a language describing certain objects. For an element $a \in \mathcal{A}$, the length of a is denoted by $\ell(a)$. If there exists the minimum length of $a \in \mathcal{A}$ describing

$\bar{a} \in \bar{\mathcal{A}}$ such that $f(a) = \bar{a}$, then the minimum length is called the complexity of description. If there does not exist such an element $a \in \mathcal{H}$, then put $\ell(a) = \infty$. When both $\mathcal{A}$ and $\bar{\mathcal{A}}$ are the set of sequences of 0, 1, we only consider a partial computable function f (i.e., for an input $a \in \mathcal{A}$, there exists a programming that the computation stops with the output $f(a)$ for $a \in \mathrm{dom}(f)$ and it does not stop for $a \notin \mathrm{dom}(f)$). A complexity $H_f(\bar{a})$ determined by $(\mathcal{A}, \bar{\mathcal{A}}, f)$ is defined as

$$H_f(\bar{a}) = \begin{cases} \min\{\ell(a); a \in \mathcal{A}, f(a) = \bar{a}\}, & \exists a \in \mathcal{A} \text{ s.t., } f(a) = \bar{a}, \\ \infty & \text{otherwise.} \end{cases}$$

Add the k symbols 0 and a 1 at the front of $a \in \mathcal{A}$ so that $0\ldots01a$, which is denoted by 0^k1a. Then it is shown by Chaitin [156] that there exist $k \in \mathbb{N}$ and a computable partial function f_U for any f such that $f_U(0^k1a) = f(a)$. This f_U is called the *universal partial function*. Once a partial function f is given, one identifies a computer U with f, and vice-versa. This computer U is called the *universal computer*. Therefore, there exists a universal computer U for f. Important consequences of the universal partial functions f_U are: (i) there exists a constant ε such that $H_{f_U}(\bar{a}) \le H_f(\bar{a}) + \varepsilon$ for any f, and (ii) there exists a constant ε' for two universal partial functions f_U, $f_{U'}$ satisfying $|H_{f_U}(\bar{a}) - H_{f_{U'}}(\bar{a})| \le \varepsilon'$. The above facts imply that H_{f_U} gives the minimum value for H_f if we neglect the constant ε. Kolmogorov and Chaitin introduced the following complexity

$$H(\bar{a}) = H_{f_U}(\bar{a})$$

which does not depend on the choice of f_U because of the property (ii). Moreover, Chaitin introduced the mutual entropy type complexity in the same framework above. His complexity and mutual entropy type complexity can be our complexities C and T, respectively.

Example 10.6 Generalizing the entropy S and the mutual entropy I, we can construct complexities of entropy type: Let $(\mathcal{A}, \mathfrak{S}(\mathcal{A}), \alpha(G))$, $(\overline{\mathcal{A}}, \mathfrak{S}(\overline{\mathcal{A}}), \overline{\alpha}(\overline{G}))$ be C^* systems as before. Let $\mathcal{S}$ be a weak $*$-compact convex subset of $\mathfrak{S}(\mathcal{A})$ and $M_\varphi(\mathcal{S})$ be the set of all maximal measures μ on $\mathcal{S}$ with the fixed barycenter φ

$$\varphi = \int_{\mathcal{S}} \omega \, d\mu.$$

Moreover, let $F_\varphi(\mathcal{S})$ be the set of all measures having finite support with the fixed barycenter φ. The following three pairs C and T satisfy all the conditions of the complexities:

$$T^{\mathcal{S}}(\varphi; \Lambda^*) \equiv \sup\left\{\int_{\mathcal{S}} S(\Lambda^*\omega, \Lambda^*\varphi)\, d\mu;\ \mu \in M_\varphi(\mathcal{S})\right\},$$

$$C_T^{\mathcal{S}}(\varphi) \equiv T^{\mathcal{S}}(\varphi; \mathrm{id}),$$

$$I^{\mathcal{S}}(\varphi; \Lambda^*) \equiv \sup\left\{S\left(\int_{\mathcal{S}} \omega \otimes \Lambda^*\omega\, d\mu, \varphi \otimes \Lambda^*\varphi\right);\ \mu \in M_\varphi(\mathcal{S})\right\},$$

$$C_I^{\mathcal{S}}(\varphi) \equiv I^{\mathcal{S}}(\varphi; \mathrm{id}),$$

$$J^{\mathcal{S}}(\varphi; \Lambda^*) \equiv \sup\left\{ \int_{\mathcal{S}} S(\Lambda^*\omega, \Lambda^*\varphi)\, d\mu;\ \mu \in F_\varphi(\mathcal{S}) \right\},$$

$$C_j^{\mathcal{S}}(\varphi) \equiv J^{\mathcal{S}}(\varphi; \mathrm{id}).$$

Here, the state $\int_{\mathcal{S}} \omega \otimes \Lambda^*\omega\, d\mu$ is the compound state exhibiting the correlation between the initial state and the final state $\Lambda^*\varphi$. As was discussed in Chap. 7, this compound state was introduced as a quantum generalization of the joint probability measure in CDS (classical dynamical system). Note that in the case of $\mathfrak{S} = \mathcal{S}$, $T^{\mathcal{S}}$(resp., $C^{\mathcal{S}}$, $I^{\mathcal{S}}$, $J^{\mathcal{S}}$) is denoted by T (resp., C, I, J) for simplicity.

These complexities and the mixing $\mathcal{S}$-entropy $S^{\mathcal{S}}(\varphi)$, the CNT (Connes–Narnhofer–Thirring) entropy $H_\varphi(\mathcal{A})$ satisfy some relations as shown in the next theorem. Note that the mixing $\mathcal{S}$-entropy and the CNT entropy are discussed in Chap. 7.

Theorem 10.7 *The following relations hold*:

1. $0 \le I^{\mathcal{S}}(\varphi; \Lambda^*) \le T^{\mathcal{S}}(\varphi; \Lambda^*) \le J^{\mathcal{S}}(\varphi; \Lambda^*)$.
2. $C_I(\varphi) = C_T(\varphi) = C_J(\varphi) = S(\varphi) = H_\varphi(\mathcal{A})$.
3. *When* $\mathcal{A} = \overline{\mathcal{A}} = \mathbf{B}(\mathcal{H})$, *for any density operator* ρ

$$0 \le I^{\mathcal{S}}(\rho; \Lambda^*) = T^{\mathcal{S}}(\rho; \Lambda^*) \le J^{\mathcal{S}}(\rho; \Lambda^*).$$

It is possible to construct other complexities not of entropy type in several fields like Genetics, Economics and Computer Sciences, which are beyond the scope of this book.

10.4 Adaptive Dynamics

In the framework of ID, we can propose a special treatment of natural phenomena called Adaptive Dynamics (AD). We start from some general remarks describing a methodological and philosophical approach to adaptive dynamics developed by Ohya. Natural science is not a copy of nature itself, but is a means to understand certain natural phenomena for human beings. Thus it is a sort of a story which we made for recognition of nature, but it is a story beyond each person and personal experience, so that it should have universality in that sense. Following Wilde, "Nature imitates arts". It is the only way for us to come face to face with nature, which is not our conceit but our limit. After discovery of quantum mechanics, we are forced to face with the facts like the above.

In order to understand physical phenomena or other phenomena of human beings, one needs to examine from various viewpoints, not only physical but also observational, the ways how an object exists and how one can recognize the object. It is known that (i) existence itself, (ii) its indicating phenomena, and (iii) their recognition have been extensively studied by philosophers and some physicists. Explaining (i.e., defining and describing) these three is essentially important not only for philosophy but also for physics, information and all other sciences. We should try to explain these three in more rigorous ways beyond usual philosophical and mathematical demonstrations, that is, by finding a method standing on a higher stage, made from dialectic mixing of philosophy, mathematics and something else, although its fulfillment is difficult.

It is appropriate to mention briefly the "phenomenology" of Husserl and the "existentialism" of Sartre.

Before Husserl, in the theory of existence like Kant's or Hegel's, a philosopher could not neglect the existence transcendent, so that he had to distinguish the existence of essence and the existence of phenomenon. An appearance of the essence is a phenomenon and only a description of phenomena is not enough to reach the essence. For instance, Hegel said: "In order to reach the essence, it is necessary for mind to develop itself dialectically". In any case, the dualism of the existence of essence and phenomena has been a basis for several philosophies until materialism of Marx and phenomenology of Husserl appeared.

Husserl was against the idea that the essence of existence is transcendent objects, and he considered that the essence is a chain of phenomena and its integral. The essence is an accurate report of all data (of phenomena) obtained through the stream of consciousness. His consciousness has two characters, "noesis" and "noema". The noesis is the operative part of consciousness to phenomena (objects), in other words, the acting consciousness on objects, and the noema is the object of consciousness experience, i.e., the results obtained by the noesis. His phenomenology is the new dualism of consciousness, but he avoids the existence transcendent, instead he likes to go to the things themselves.

Under a strong influence of Husserl, there appeared several philosophies named "exisitentialism" of Heidegger, of Sartre and of others. Sartre said: "Existence precedes essence". Sartre was affected by "Cogito" of Descartes, and he found two aspects of existence (being) in his famous book "L'Etre et le Neant (Being and Nothingness)", one of which is the "being in itself" and another is the "being for itself". The first one is the being as it is, opaque (nontransparent) being like physical matter itself, being which does not have any connection with another being, being without reason for being, etc. Another one is the being as it is not, being like consciousness, being with cause for being itself, being making any being-in-itself as being, etc. Sartre explained several forms of existence by his new dualism of existence: being-in-itself and being-for-itself. His main concern is being and becoming of human beings, various appearance of emotion, life and ethics, so his expression of philosophy is rather rhetoric and literal. However, we will explain that his idea can be applied for the proper interpretation of quantum entropy and dynamics.

10.4.1 Entropy, Information in Classical and Quantum World

Physics is considered as a "theory of matter", or equivalently, a "theory of existence in itself". Information theory (Entropy theory) is considered as a "theory of events", so that it will be considered as a "theory of changes". Quantum Information can be regarded as a synthesis of these two. The key concept of quantum information bridging between matter and event, so between two modes of being, is "entropy", which was introduced by Shannon in classical systems and by von Neumann in quantum systems. We will discuss how this concept of entropy has a deep connection with the mode of existence considered at the beginning of this section.

According to Shannon, information is related to uncertainty, so it is described by entropy (Information = Uncertainty = Entropy), and dissolution of uncertainty can be regarded as acquisition of information. Historically, the concept of entropy was introduced to describe the flow of heat, then it was recognized that the entropy describes chaos or uncertainty of a system. A system is described by a state such as a probability measure or a density operator, which is a rather abstract concept not belonging to an object (observable) to be measured but a means to get measured values. Thus the entropy is defined through a state of a system, which implies that the entropy is not an object considered in the usual objective classical physics, and it is an existence coming along the action of "observation". It is close to (actually more than) a description of chaos which is a mode taken by consciousness to the being-in-itself. (We will discuss chaos in Sect. 10.3.) Therefore, the entropy can be considered as a representation (formulation) of consciousness involving an observation of a certain object. The concept of entropy is not a direct expression of phenomena associated with a being-in-itself, but is a being having an appearance of consciousness to phenomena of a being-in-itself, so that the mode of existence for the entropy is different from the two modes of being proposed by Sartre, and this third mode is in between being in- and for-itself. The rigorous (mathematical) study with this third mode of being might be important to solving some problems which we face in several fields.

10.4.2 Schematic Expression of Understanding

Metaphysics, idea, feeling, thought are applied to various existence (series of phenomena), which causes understanding (recognition, theory). To understand a physical system, the usual method, often called "Reductionism", is to divide the system into its elements and to study their relations and combinations, which causes the understanding of the whole system.

Our method is the one adding "how to see objects (existence)" to the usual reductionism, so that our method is a mathematical realization of modern philosophy. The fact of "how to see objects" is strongly related to setting the mode for observation, such as selection of phenomena and operation for recognition. Our method is called "*Adaptive dynamics*" or "*Adaptive scheme*" for understanding the existence.

We discuss the conceptual framework of AD and some examples in chaos and quantum algorithms, which are the first steps towards our final aim of making complete mathematics for "adaptivity".

10.5 Adaptive Dynamics—Conceptual Meaning

The adaptive dynamics has two aspects: the "observable-adaptive" and the "state-adaptive".

The idea of observable-adaptive comes from [442, 604, 608] studying chaos. We claimed that any observation will be unrelated or even contradicting the mathematical universalities such as taking limits, sup, inf, etc. An observation is a result due to taking suitable scales of, for example, time, distance or domain, and the observation will not be made in infinite systems such as taking the limits. Such an example will be seen in the sequel; we will consider the appearance of chaos.

The meaning of state-adaptive is that, for instance, the interaction depends on the state at instant time, whose details will be discussed bellow. The idea of the state-adaptive is implicitly started in constructing a compound state for quantum communication [19, 559, 560, 570] and in Accardi's Chameleon dynamics [15]. This adaptivity can be used to solve a pending problem of more than 30 years asking whether there exists an algorithm solving an NP-complete problem in polynomial time. We found such algorithms first by a quantum chaos algorithm [602], and second, by stochastic limit [37] both in state-adaptive dynamics based on the quantum algorithm of the SAT [30, 595].

We will discuss a bit more about the meaning of adaptivity for each of the topics mentioned above.

10.5.1 Description of Chaos

There exist several reports saying that one can observe chaos in nature, which are nothing but reports on how one could observe the phenomena in specified conditions. It has been difficult to find a satisfactory theory (mathematics) to explain such various chaotic phenomena in a unified way.

An idea describing chaos of a phenomenon is to find some divergence of orbits produced by the dynamics explaining the phenomenon. However, to explain such divergence from the differential equation of motion describing the dynamics is often difficult, so that one takes (makes) a difference equation from that differential equation, for which one has to take a certain time interval τ between two steps of dynamics, that is, one needs a processing discretizing time for observing the chaos. In laboratory, any observation is done in finite size for both time and space; however, one believes that natural phenomena do not depend on these sizes or how small they are, so that most of mathematics (theory) has been made free of the sizes taken in laboratory. Therefore, mathematical terminologies such as "lim", "sup", "inf" are very often used to define some quantities measuring chaos, and many phenomena showing chaos have remained unexplained.

In [442, 604, 608], we took the opposite position, that is, any observation is unrelated or even contradicting such limits. Observation of chaos is a result due to taking suitable scales of, for example, time, distance or domain, and it will not be possible in the limiting cases. In other words, as discussed in Sect. 10.1, it is very natural to consider that observation itself plays a similar role of "noesis" of Husserl and the mode of its existence is a "being-for-itself", that is, observation itself cannot exist as is, but it exists only through the results (phenomena) of objects obtained by it. Phenomena cannot be phenomena without observing them, so to explain the phenomena like chaos it is necessary to find a dynamics with observation.

We claimed that most of chaos are scale-dependent phenomena, so the definition of a degree measuring chaos should depend on certain scales taken, and more generally, it is important to find mathematics containing the rules (dynamics) of both object and observation, which is "Adaptive dynamics".

Concerning the definition of a criterion measuring chaos, Information Dynamics [359, 573], a scheme to describe many different types of complex systems, can be applied. We introduced a quantity measuring chaos by means of the complexities of ID, and called it a chaos degree [367, 442, 590]. Using this degree in adaptive dynamics, we can explain or produce many different types of chaos.

10.5.2 Chameleon Dynamics

Accardi considered a problem of whether it is possible to explain quantum effects (e.g., EPR (Einstein–Polodolski–Rosen) correlation) by a sort of classical dynamics [15]. He could find a dynamics positively solving the above problem, and he called it "Chameleon dynamics". He considered two systems having their own particles, initially correlated and later separated. After some time, each particle interacts with a measurement apparatus independently. By the chameleon effect, the dynamical evolution of each particle depends on the setting of the nearby apparatus, but not on the setting of the apparatus interacting with the other particle. Then he reproduced the EPR correlations by this "chameleon dynamics". The explicit construction of the dynamics was done in [32]; see Sect. 10.10 of this chapter. The interaction between a particle and an apparatus depends on the setting of the apparatus, so that the chameleon dynamics is an adaptive dynamics.

10.5.3 Quantum SAT Algorithm

Although the ability of computers has greatly progressed, there are several problems which may not be solved effectively, namely, in polynomial time. Among such problems, NP-problems and NP-complete problems are fundamental. It is known that all

NP-complete problems are equivalent, and an essential question is "*whether there exists an algorithm to solve an NP-complete problem in polynomial time*". Such problems have been studied for decades, and so far all known algorithms have an exponential running time in the length of the input. The P-problem and NP-problem are considered as follows [171, 278].

Let us remind what the P-problem and the NP-problem are (see also Chap. 2). Let n be the size of input.

1. A P-problem is a problem with the time needed for solving it being at worst a polynomial of n. Equivalently, it is a problem which can be recognized in a polynomial in n time by a deterministic Turing machine.
2. An NP-problem is a problem that can be solved in polynomial time by a non-deterministic Turing machine. This can be understood as follows: Consider a problem of finding a solution of $f(x)=0$. We can check in a polynomial in n time whether x_0 is a solution of $f(x)=0$, but we do not know whether we can find a solution of $f(x)=0$ in a time polynomial in n.
3. An NP-complete problem is such an NP-problem to which any other NP problem can be reduced in polynomial time.

To answer an essential question open for more than 30 years, namely, *of the existence of an algorithm to solve an NP-complete problem in polynomial time,* we found two different algorithms [37, 595, 602] (see Chap. 14).

In [595], we discussed the quantum algorithm of the SAT problem and pointed out that the SAT problem, hence all other NP-problems, can be solved in polynomial time by a quantum computer if the superposition of two orthogonal vectors $|0\rangle$ and $|1\rangle$ is physically detected. However, this detection is considered not to be possible in the present technology. The problem to overcome is how to distinguish the pure vector $|0\rangle$ from the superposed one $\alpha|0\rangle+\beta|1\rangle$, obtained by our SAT-quantum algorithm, if β is not zero but very small. If such a distinction is possible, then we can solve the NPC problem in polynomial time.

Chaos SAT Algorithm

It will not be possible to amplify, by a unitary transformation (usual quantum algorithm), the above small positive $q\equiv|\beta|^2$ into a suitably large one to be detected, e.g., $q>1/2$, and if $q=0$ then keeping it as is (see Chap. 14). In [600, 602], we proposed using the output of a quantum computer as an input for another device involving chaotic dynamics, that is, combining a quantum computer with a chaotic dynamics amplifier. We showed that this combination (nonlinear chaos amplifier with the quantum algorithm) provides us with a mathematical algorithm solving NP $=$ P. Using a chaos dynamics to the state computed by quantum unitary operations is one of examples of the state-adaptive approach. This algorithm of Ohya and Volovich is going beyond usual (unitary) quantum Turing algorithm, but there exists a generalized quantum Turing machine in which the OV chaos algorithm can be treated [369, 371, 595].

Adaptive SAT Algorithm

We applied the adaptive dynamics to the OM SAT algorithm. That is, the state adaptive dynamics was applied to the OM SAT algorithm, and if we rescaled the time in the dynamics by the stochastic limit, then we could show that the same amplification (distinction between $q > 0$ and $q = 0$) is possible by unitary adaptive dynamics with the stochastic limit. Its details will be discussed in Sect. 14.5. The AO adaptive algorithm can be treated in the framework of generalized quantum Turing machine as a linear TM.

10.5.4 Summary of Adaptive Dynamics

We summarize our idea on the adaptive dynamics as follows. The mathematical definition of an adaptive system is given in terms of observables and states.

Two adaptivities are characterized (defined) as follows:

The observable-adaptive dynamics is a dynamics characterized by one of the following two statements: (i) *Measurement depends on how one sees an observable to be measured.* (ii) *The interaction between two systems depends on how a fixed observable exists.*

The state-adaptive dynamics is a dynamics characterized by one of the following two statements: (i) *Measurement depends on how the state to be used exists.* (ii) *The correlation between two systems' interaction depends on the state of at least one of the systems at the instant in which the interaction is switched on.*

Examples of the state-adaptive dynamics are seen in compound states [561, 570] (or nonlinear liftings [19]) studying quantum communication and in an algorithm solving NP-complete problem in polynomial time with chaos amplifier or stochastic limit [37, 602].

Examples of the observable-adaptive dynamics are used to understand chaos [442, 590] and examine violation of Bell's inequality [32].

Notice that the definitions of adaptivity make sense both for classical and quantum systems.

The difference between the property (ii) of a state-adaptive system and a nonlinear dynamical system should be remarked here:

(i) In nonlinear dynamical systems (such as those whose evolution is described by the Boltzmann equation, or nonlinear Schrödinger equation, etc.), the interaction depends on the state at any time t: $H_I = H_I(\rho_t)$ $(\forall t)$.
(ii) In state-adaptive dynamical systems, the interaction Hamiltonian depends on the state only at time $t = 0$: $H_I = H_I(\rho_0)$.
(iii) In classical mechanics, the adaptive dynamics is of the type $\ddot{x}(t) = F(x(t), x(0))$, that is, the force depends on the initial state $x(0)$.

The latter class of systems describes the following physical situation: At time $t = -T$ $(T > 0)$, a system S is prepared in a state ψ_{-T} and in the time interval

$[-T,0]$ it evolves according to a fixed (free) dynamics $U_{[-T,0]}$ so that its state at time 0 is $U_{[-T,0]}\psi_{-T} =: \psi_0$. At time $t=0$, an interaction with another system R is switched on and this interaction depends on the state ψ_0: $H_I = H_I(\psi_0)$. If we interpret the system R as environment, we can say that the above interaction describes the response of the environment to the state of the system S. Therefore, the adaptive dynamics can be linear, but it contains the non-linear dynamics in some occasions.

As Darwin said, "It's not the strongest nor the most intelligent that survive but the most adaptable to change". The idea of adaptiveness can be applied to many biological systems which will be discussed mainly in Chap. 21.

10.6 A Use of ID: Chaos Degree

In quantum systems, if we take $C(\rho) = S(\rho) =$ von Neumann entropy, $T(\rho;\Lambda^*) = I(\rho;\Lambda^*) =$ quantum mutual entropy and linear channel Λ^*, then let's consider

$$\begin{aligned} D(\rho;\Lambda^*) &= C(\Lambda^*\rho) - T(\rho;\Lambda^*) \\ &= S(\Lambda^*\rho) - I(\rho;\Lambda^*) \\ &= S(\Lambda^*\rho) - \sup\left\{\operatorname{tr}\left(\sum_n p_n \Lambda^* E_n(\log \Lambda^* E_n - \log \Lambda^*\rho)\right);\ \{E_n\}\right\} \\ &= \inf\left\{\sum_n p_n S(\Lambda^* E_n);\ \{E_n\}\right\} = \inf\left\{\sum_n p_n C(\Lambda^* E_n);\ \{E_n\}\right\} \end{aligned}$$

since $S(\Lambda^*\rho) = -\operatorname{tr}\Lambda^*\rho\log\Lambda^*\rho = -\operatorname{tr}(\sum_n p_n \Lambda^* E_n \log \Lambda^*\rho)$ for any Schatten decomposition $\{E_n\}$ of ρ. *Therefore, the above quantity $D(\rho;\Lambda^*)$ is interpreted as the complexity produced through the channel Λ^**. We apply this quantity $D(\rho;\Lambda^*)$ to study chaos even when the channel describing the dynamics is not linear. $D(\rho;\Lambda^*)$ is called the *entropic chaos degree* in the sequel.

In order to describe more general dynamics such as in continuous systems, we define the entropic chaos degree in C^*-algebraic terminology. This setting will not be much used in the subsequent application, but for mathematical completeness we will discuss the C^*-algebraic setting.

Let $(\mathcal{A},\mathfrak{S})$ be an input C^*-system and $(\overline{\mathcal{A}},\overline{\mathfrak{S}})$ be an output C^*-system; namely, $\mathcal{A}$ is a C^*-algebra with unit I, and $\mathfrak{S}$ is the set of all states on $\mathcal{A}$. We assume $\overline{\mathcal{A}} = \mathcal{A}$ for simplicity. For a weak*-compact convex subset $\mathcal{S}$ (called the reference space) of $\mathfrak{S}$, take a state φ from the set $\mathcal{S}$ and let

$$\varphi = \int_{\mathcal{S}} \omega \, d\mu_\varphi(\omega)$$

be an extremal orthogonal decomposition of φ in $\mathcal{S}$, which describes the degree of mixture of φ in the reference space $\mathcal{S}$. In more details, this formula reads

$$\varphi(A) = \int_{\mathcal{S}} \omega(A)\, d\mu_\varphi(\omega), \quad A \in \mathcal{A}.$$

The measure μ_φ is not uniquely determined unless $\mathcal{S}$ is the Choquet simplex, so that the set of all such measures is denoted by $M_\varphi(\mathcal{S})$.

Definition 10.8 The entropic chaos degree with respect to $\varphi \in \mathcal{S}$ and a channel Λ^* is defined by

$$D^{\mathcal{S}}(\varphi; \Lambda^*) \equiv \inf\left\{ \int_{\mathcal{S}} S^{\mathcal{S}}(\Lambda^*\omega)\, d\mu_\varphi(\omega);\ \mu_\varphi \in M_\varphi(\mathcal{S}) \right\}$$

where $S^{\mathcal{S}}(\Lambda^*\varphi)$ is the mixing entropy of a state φ in the reference space $\mathcal{S}$.

When $\mathcal{S} = \mathfrak{S}$, $D^{\mathcal{S}}(\varphi; \Lambda^*)$ is simply written as $D(\varphi; \Lambda^*)$. This $D^{\mathcal{S}}(\varphi; \Lambda^*)$ contains the classical chaos degree and the quantum one above. The classical entropic chaos degree is the case when $\mathcal{A}$ is abelian and φ is the probability distribution of an orbit generated by a dynamics (channel) Λ^*; $\varphi = \sum_k p_k \delta_k$, where δ_k is the delta measure such as

$$\delta_k(j) \equiv \begin{cases} 1 & (k = j), \\ 0 & (k \neq j). \end{cases}$$

Then the classical entropic chaos degree is

$$D_c(\varphi; \Lambda^*) = \sum_k p_k S(\Lambda^*\delta_k)$$

with the entropy S.

To summarize, the Information Dynamics can be applied to the study of chaos in the following way:

Definition 10.9

1. A state ψ is more chaotic than φ if $C(\psi) > C(\varphi)$.
2. When $\varphi \in \mathcal{S}$ changes to $\Lambda^*\varphi$, the *chaos* degree associated to this state change (dynamics) Λ^* is given by

$$D^{\mathcal{S}}(\varphi; \Lambda^*) = \inf\left\{ \int_{\mathcal{S}} C^{\mathcal{S}}(\Lambda^*\omega)\, d\mu_\varphi(\omega);\ \mu_\varphi \in M_\varphi(\mathcal{S}) \right\}.$$

Definition 10.10 A dynamics Λ^* produces chaos iff $D^{\mathcal{S}}(\varphi; \Lambda^*) > 0$.

Remark 10.11 It is important to note here that the dynamics Λ^* in the definition is not necessarily the same as the original dynamics (channel), but is reduced from the original so that it causes an evolution for a certain observed value like an orbit. However, for simplicity we use the same notation here. In some cases, the above chaos degree $D^{\mathcal{S}}(\varphi; \Lambda^*)$ can be expressed as

$$D^{\mathcal{S}}(\varphi; \Lambda^*) = C^{\mathcal{S}}(\Lambda^*\varphi) - T^{\mathcal{S}}(\varphi; \Lambda^*).$$

10.7 Algorithm Computing Entropic Chaos Degree (A Use of AD)

In order to observe chaos produced by a dynamics, one often looks at the behavior of orbits made by that dynamics, or more generally, one looks at the behavior of a certain observed value. Therefore, in our scheme we directly compute the chaos degree once a dynamics is explicitly given as a state change of system. However, even when the direct calculation does not show chaos, it will appear if one focuses on some aspect of the state change, e.g., a certain observed value which may be called an orbit as usual. In the later case, an algorithm computing the chaos degree for classical or quantum dynamics consists of the following two cases:

1. *Dynamics is given by* $\frac{dx}{dt} = F_t(x)$ *with* $x \in I \equiv [a,b]^N \subset \mathbb{R}^N$. First, find a difference equation $x_{n+1} = F(x_n)$ with a map F on $I \equiv [a,b]^N \subset \mathbb{R}^N$ into itself. Second, let $I \equiv \bigcup_k A_k$ be a finite partition with $A_i \cap A_j = \emptyset$ $(i \neq j)$. Then the state $\varphi^{(n)}$ of the orbit determined by the difference equation is defined by the probability distribution $(p_i^{(n)})$, that is, $\varphi^{(n)} = \sum_i p_i^{(n)} \delta_i$, where for an initial value $x \in I$ and the characteristic function 1_A

$$p_i^{(n)} \equiv \frac{1}{n+1} \sum_{k=m}^{m+n} 1_{A_i}(F^k x).$$

Now when the initial value x is distributed according to a measure ν on I, the above $p_i^{(n)}$ is given as

$$p_i^{(n)} \equiv \frac{1}{n+1} \int_I \sum_{k=m}^{m+n} 1_{A_i}(F^k x)\, d\nu.$$

The joint distribution $(p_{ij}^{(n,n+1)})$ between the time n and $n+1$ is defined by

$$p_{ij}^{(n,n+1)} \equiv \frac{1}{n+1} \sum_{k=m}^{m+n} 1_{A_i}(F^k x) 1_{A_j}(F^{k+1} x),$$

or

$$p_{ij}^{(n,n+1)} \equiv \frac{1}{n+1} \int_I \sum_{k=m}^{m+n} 1_{A_i}(F^k x) 1_{A_j}(F^{k+1} x)\, d\nu.$$

Then the channel Λ_n^* at n is determined by

$$\Lambda_n^* \equiv \left(\frac{p_{ij}^{(n,n+1)}}{p_i^{(n)}} \right) : \text{transition probability} \quad \Longrightarrow \quad \varphi^{(n+1)} = \Lambda_n^* \varphi^{(n)},$$

and the entropic chaos degree is given by

$$D_A(x;F) = D_A\big(p^{(n)};\Lambda_n^*\big) = \sum_i p_i^{(n)} S(\Lambda_n^*\delta_i) = \sum_{i,j} p_{ij}^{(n,n+1)} \log \frac{p_i^{(n)}}{p_{ij}^{(n,n+1)}},$$

which depends on the choice of the partition A (a sort of AD). This partition dependence is particularly important when computing the chaos degree, which will be discussed in more details in Sect. 10.7.

We can judge whether the dynamics causes chaos or not by the value of D as

$$D > 0 \iff \text{chaotic},$$
$$D = 0 \iff \text{stable}.$$

This chaos degree was applied to several dynamical maps such as logistic map, Baker's transformation, and Tinkerbel's map, and it could explain their chaotic character. This chaos degree has several merits compared with usual measures such as Lyapunov exponent as explained below.

Therefore, it is enough to find a partition $\{A_k\}$ such that D is positive for the dynamics to produce chaos.

2. *Dynamics is given by* $\varphi_t = F_t^*\varphi_0$ *on a Hilbert space*. Similarly as writing a difference equation for a (quantum) state, the channel Λ_n^* at n is first deduced from F_t^*, which should satisfy $\varphi^{(n+1)} = \Lambda_n^*\varphi^{(n)}$. By means of this constructed channel (α), we compute the chaos degree D directly according to the definition ($\boldsymbol{\alpha}$) or ($\boldsymbol{\beta}$) we take a proper observable X and put $x_n \equiv \varphi^{(n)}(X)$, then go back to the algorithm 1.

   The entropic chaos degree for quantum systems has been applied to the analysis of quantum spin system and quantum Baker's type transformation, which will be discussed in Chap. 20.

   Note that the chaos degree D above does depend on a partition A taken, which is somehow different from the usual degree of chaos. This is a key point of our understanding of chaos, which will be discussed in Sect. 10.4.

10.7.1 Logistic Map

Let us apply the entropic chaos degree (ECD) to the logistic map. Chaotic behavior in a classical system is often considered as an exponential sensitivity to the initial condition.

The logistic map is defined by

$$x_{n+1} = ax_n(1-x_n), \quad x_n \in [0,1], 0 \le a \le 4.$$

The solution of this equation bifurcates as shown in Fig. 10.1.

In order to compare ECD with other measures describing its chaos, we take Lyapunov exponent for this comparison and remind here its definition.

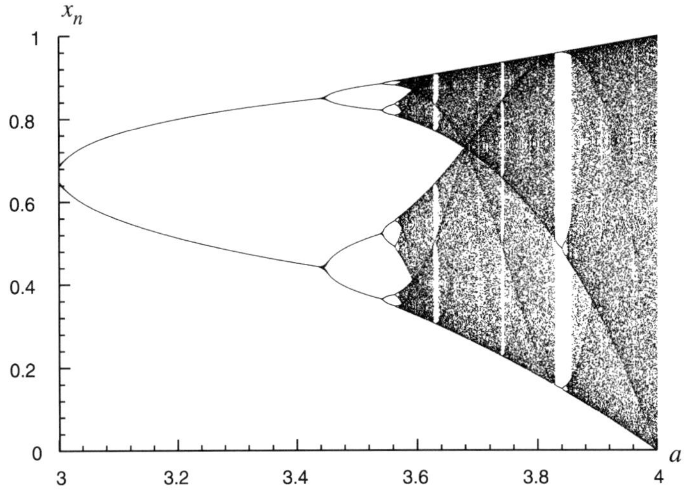

Fig. 10.1 Bifurcation diagram for the logistic map

Lyapunov Exponent $\lambda(f)$

1. Let f be a map on $\mathbb{R}$, and let $x_0 \in \mathbb{R}$. Then the Lyapunov exponent $\lambda_{\mathcal{O}}(f)$ of the orbit $\mathcal{O} \equiv \{f^n(x_0) \equiv f \circ \cdots \circ f(x_0) : n = 0, 1, 2, \ldots\}$ is defined by

$$\lambda_{\mathcal{O}}(f) = \lim_{n\to\infty} \lambda_{\mathcal{O}}^{(n)}(f), \qquad \lambda_{\mathcal{O}}^{(n)}(f) = \frac{1}{n} \log \left| \frac{df^n}{dx}(x_0) \right|.$$

2. Let $f = (f_1, \ldots, f_m)$ be a map on $\mathbb{R}^m$, and let $r_0 \in \mathbb{R}^m$. The Jacobi matrix $J_n = Df^n(r_0)$ at r_0 is defined by

$$J_n = Df^n(r_0) = \begin{pmatrix} \frac{\partial f_1^n}{\partial x_1}(r_0) & \cdots & \frac{\partial f_1^n}{\partial x_m}(r_0) \\ \vdots & & \vdots \\ \frac{\partial f_m^n}{\partial x_1}(r_0) & \cdots & \frac{\partial f_m^n}{\partial x_m}(r_0) \end{pmatrix}.$$

Then, the Lyapunov exponent $\lambda_{\mathcal{O}}(f)$ of f for the orbit $\mathcal{O} \equiv \{f^n(x_0); n = 0, 1, 2, \ldots\}$ is defined by

$$\lambda_{\mathcal{O}}(f) = \log \tilde{\mu}_1, \qquad \tilde{\mu}_k = \lim_{n\to\infty} \left(\mu_k^n\right)^{\frac{1}{n}} \quad (k = 1, \ldots, m).$$

Here, μ_k^n is the kth largest square root of the m eigenvalues of the matrix $J_n J_n^T$.

$$\lambda_{\mathcal{O}}(f) > 0 \quad \Longrightarrow \quad \text{Orbit } \mathcal{O} \text{ is chaotic,}$$
$$\lambda_{\mathcal{O}}(f) \le 0 \quad \Longrightarrow \quad \text{Orbit } \mathcal{O} \text{ is stable.}$$

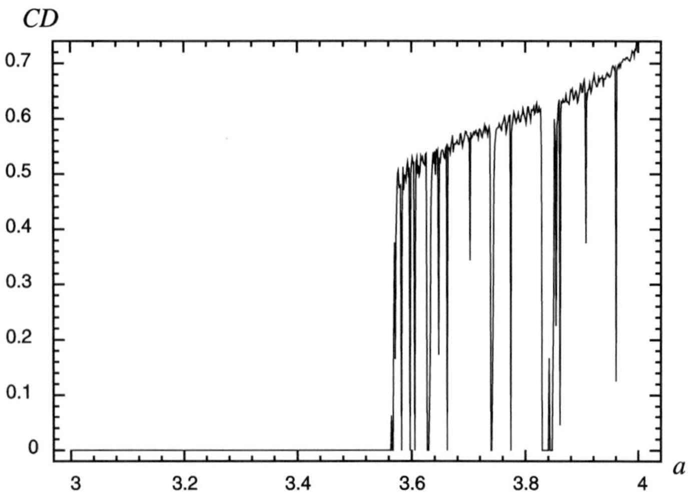

Fig. 10.2 Chaos degree for the logistic map

The properties of the logistic map depend on the parameter a. There exists a critical value $a_0 \simeq 3.57$ such that chaos occurs for $a > a_0$. If we take a particular constant a, for example, $a = 3.71$, then the Lyapunov exponent and the entropic chaos degree are positive (see Figs. 10.2 and 10.3), the trajectory is very sensitive to the initial value and one has chaotic behavior.

We show several applications to some other dynamics.

Bernoulli Shift

Let f be a map from [0, 1] to itself such that

$$f(x_n) = \begin{cases} 2ax^{(n)} & (0 \le x^{(n)} \le 0.5), \\ a(2x^{(n)} - 1) & (0.5 < x^{(n)} \le 1), \end{cases} \tag{10.1}$$

where $x^{(n)} \in [0, 1]$ and $0 \le a \le 1$.

Let us compute the Lyapunov exponent and the entropic chaos degree (ECD for short) for the above Bernoulli shift f.

An orbit of (10.1) is shown in Fig. 10.4.

The Lyapunov exponent $\lambda_n(f)$ is $\log 2a$ for the Bernoulli shift (Fig. 10.5).

On the other hand, the entropic chaos degree of the Bernoulli shift is shown in Fig. 10.6.

Here we took 740 different a's between 0 and 1 with

$$A_i = \left[\frac{i}{2000}, \frac{i+1}{2000}\right] \quad (i = 0, \dots, 1999),$$

$$n = 100000.$$

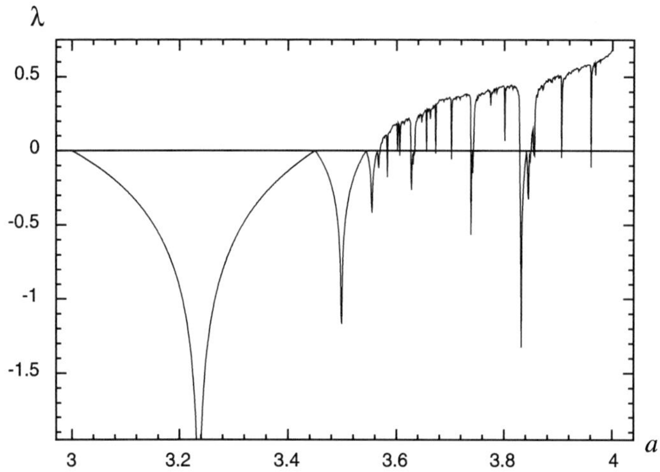

Fig. 10.3 Lyapunov exponent for the logistic map

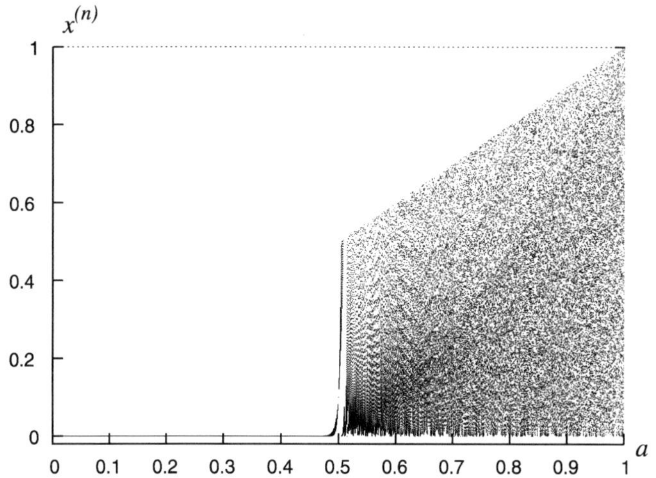

Fig. 10.4 Bifurcation diagram for the Bernoulli shift

Baker's Transformation

We apply the chaos degree to a smooth map on $\mathbb{R}^2$. Let us compute the Lyapunov exponent and the ECD for the following Baker's transformation f_a:

$$f_a\big(x^{(n)}\big) = f_a\big(x_1^{(n)}, x_2^{(n)}\big) = \begin{cases} (2ax_1^{(n)} \frac{1}{2}ax_2^{(n)}) & (0 \le x_1^{(n)} \le 0.5), \\ (a(2x_1^{(n)} - 1), \frac{1}{2}a(x_2^{(n)} + 1)) & (0.5 < x_1^{(n)} \le 1), \end{cases}$$

where $(x_1^{(n)}, x_2^{(n)}) \in [0, 1] \times [0, 1]$ and $0 \le a \le 1$.

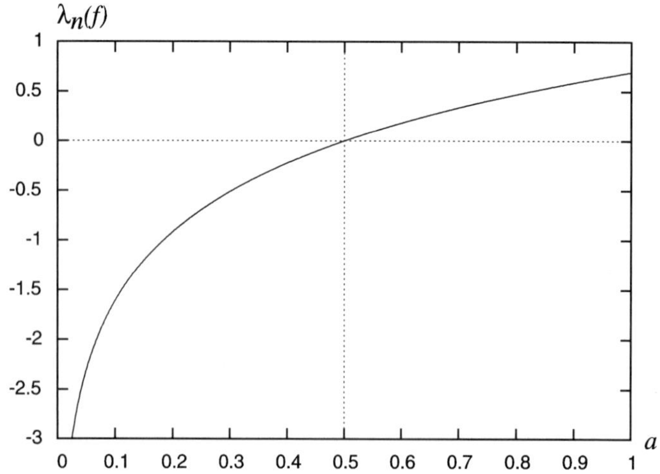

Fig. 10.5 Lyapunov exponent of the Bernoulli shift

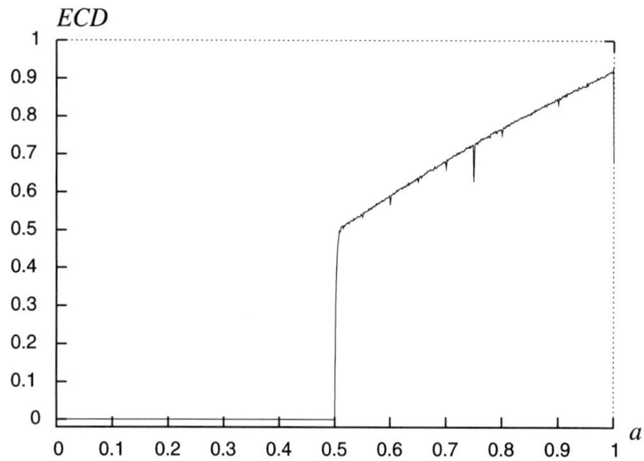

Fig. 10.6 ECD of the Bernoulli shift

Orbits for different a are shown in Figs. 10.7, 10.8, 10.9, 10.10, 10.11, 10.12.

These figures show that the larger a is, the more complicated the orbit is. The maximum Lyapunov exponent $\lambda_n^1(f)$ is $\log 2a$ for the Baker's transformation (Fig. 10.13).

On the other hand, the ECD of the Baker's transformation is shown in Fig. 10.14. Here we took 740 different a's between 0 and 1 with

$$A_{i,j} = \left[\frac{i}{100}, \frac{i+1}{100}\right] \times \left[\frac{j}{100}, \frac{j+1}{100}\right] \quad (i, j = 0, \ldots, 99),$$
$$n = 100000.$$

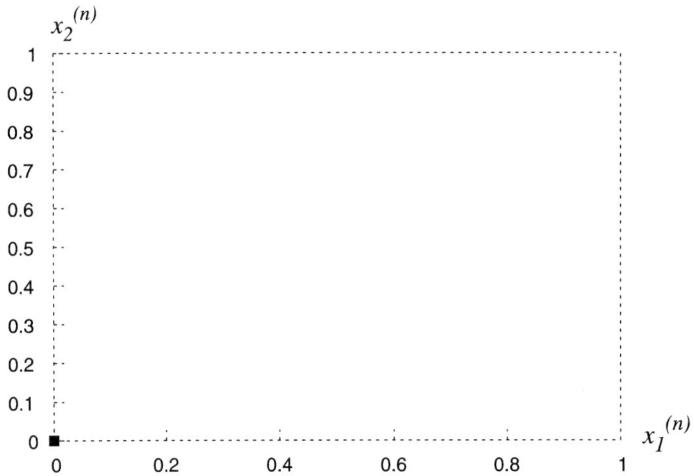

Fig. 10.7 Orbit of f_x for $0 \leq a < 0.5$

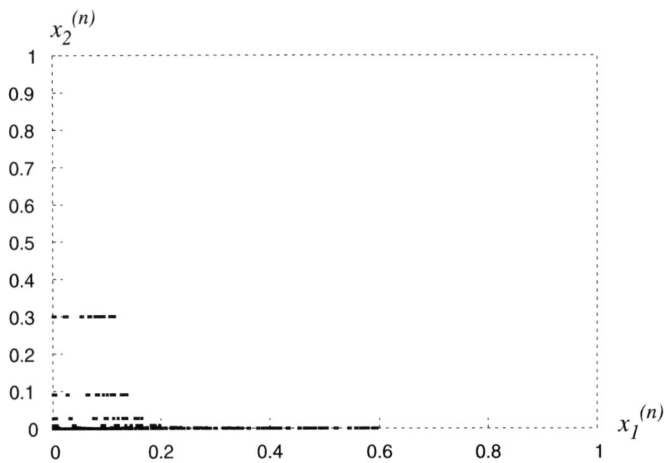

Fig. 10.8 Orbit of f_x for $a = 0.6$

Tinkerbell Map

Let us compute the ECD for the following Tinkerbell maps f_a and f_b on $I = [-1.2, 0.4] \times [-0.7, 0.3]$.

$$\begin{aligned} f_a\big(x^{(n)}\big) &= f_a\big(x_1^{(n)}, x_2^{(n)}\big) \\ &= \Big(\big(x_1^{(n)}\big)^2 - \big(x_2^{(n)}\big)^2 + a x_1^{(n)} + c_2 x_2^{(n)}, 2x_1^{(n)} x_2^{(n)} + c_3 x_1^{(n)} + c_4 x_2^{(n)}\Big), \end{aligned}$$

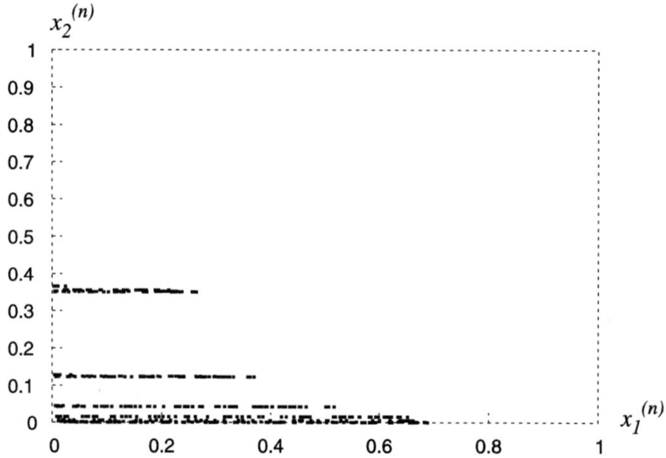

Fig. 10.9 Orbit of f_x for $a = 0.7$

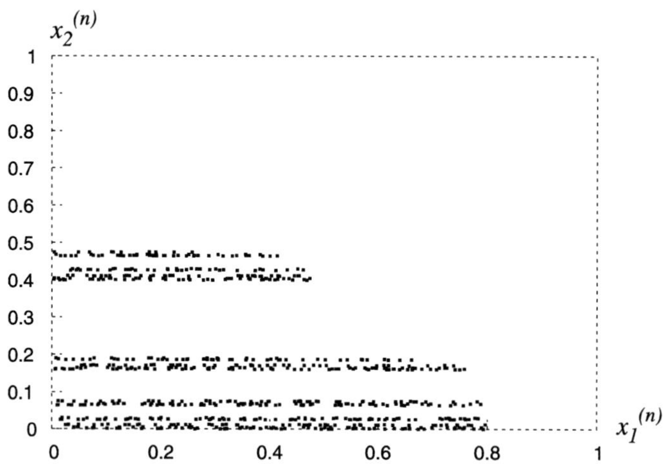

Fig. 10.10 Orbit of f_x for $a = 0.8$

$$
\begin{aligned}
f_b\left(x^{(n)}\right) &= f_b\left(x_1^{(n)}, x_2^{(n)}\right) \\
&= \left(\left(x_1^{(n)}\right)^2 - \left(x_2^{(n)}\right)^2 + c_1 x_1^{(n)} + c_2 x_2^{(n)}, 2x_1^{(n)} x_2^{(n)} + b x_1^{(n)} + c_4 x_2^{(n)}\right),
\end{aligned}
$$

where $(x_1^{(n)}, x_2^{(n)}) \in I, -0.4 \le a \le 0.9, 1.9 \le b \le 2.9, (c_1, c_2, c_3, c_4) = (-0.3, -0.6, 2.0, 0.5)$, and $(x_1^{(0)}, x_2^{(0)}) = (0.1, 0.1)$.

Let us plot the points $(x_1^{(n)}, x_2^{(n)})$ for 3000 different n's between 1001 and 4000.

In a stable domain, the number of the points $(x_1^{(n)}, x_2^{(n)})$ is finite because the point $(x_1^{(n)}, x_2^{(n)})$ periodically appears at time n. Figures 10.15 and 10.16 are examples of the orbits of f_a and f_b in a stable domain.

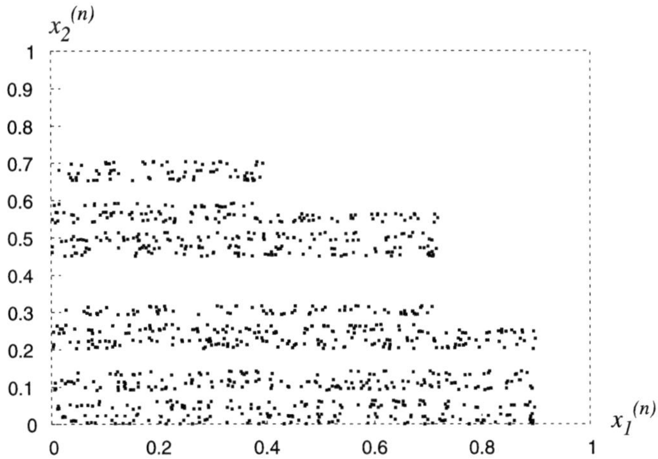

Fig. 10.11 Orbit of f_x for $a = 0.9$

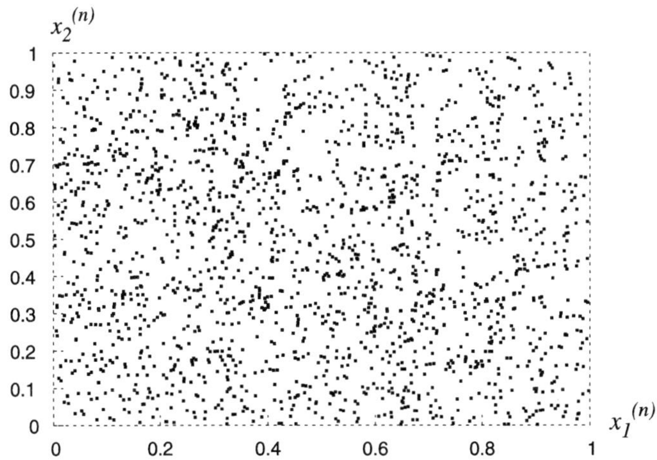

Fig. 10.12 Orbit of f_x for $a = 1.0$

On the other hand, the point $(x_1^{(n)}, x_2^{(n)})$ takes a random value at time n in a chaotic domain. Figures 10.17 and 10.18 are examples of the orbits of f_a and f_b in a chaotic domain.

The ECD of Tinkerbell maps f_a and f_b are shown in Figs. 10.19 and 10.20.

Here we took 740 different a's between -1.2 and 0.9, and 740 different b's between 1.9 and 2.9 with

$$A_{i,j} = \left[\frac{i}{100}, \frac{i+1}{100}\right] \times \left[\frac{j}{100}, \frac{j+1}{100}\right]$$

$$(i = -120, -119, \ldots, -1, 0, 1, \ldots, 38, 39)$$

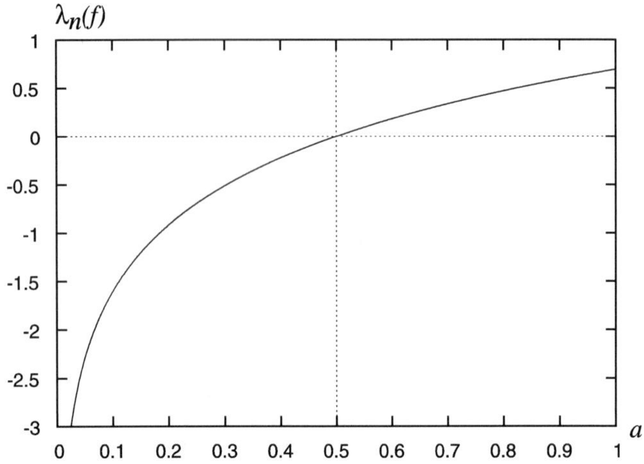

Fig. 10.13 Lyapunov exponent of Baker's transformation

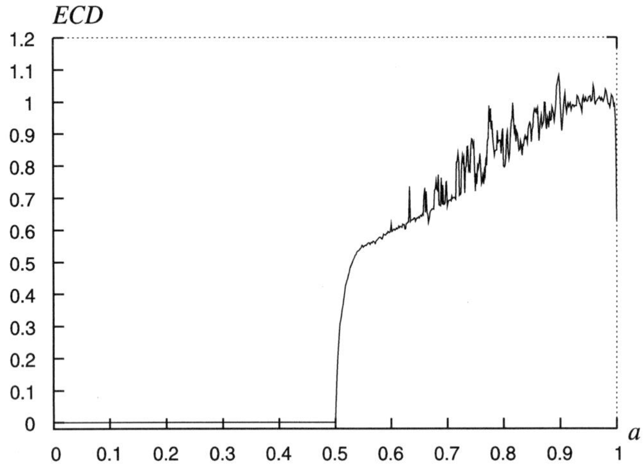

Fig. 10.14 ECD of the Baker's transformation

$$(j = -70, -69, \ldots, -1, 0, 1, \ldots, 28, 29),$$

$$n = 100000.$$

From the above example and some other maps though they are not discussed here (see [360]), the Lyapunov exponent and chaos degree have a clear correspondence, but the ECD can resolve some inconvenient properties of the Lyapunov exponent as follows:

1. Lyapunov exponent takes negative values and is sometimes $-\infty$, but the ECD is always positive for any $a \geq 0$.

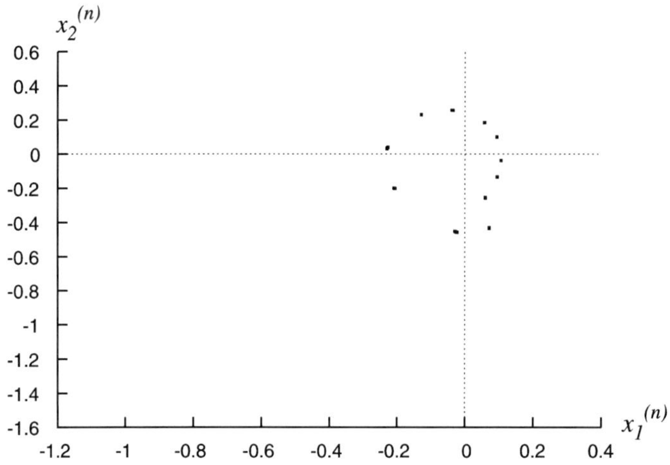

Fig. 10.15 Orbit of f_a for $a = 0.243$

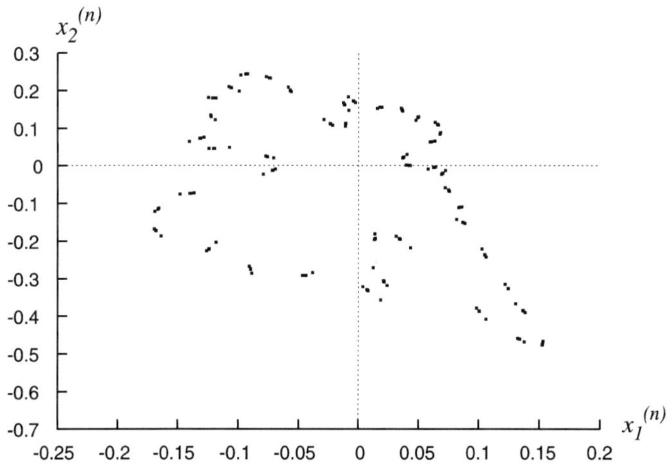

Fig. 10.16 Orbit of f_a for $a = 2.65$

2. It is difficult to compute the Lyapunov exponent for some maps like the Tinkerbell map f because it is difficult to compute f^n for large n. On the other hand, the ECD of f is easily computed.
3. Generally, the algorithm for the ECD is much easier than that for the Lyapunov exponent.

10.7.2 ECD with Memory

Here we generalize the above ECD taking the memory effect into account. Although the original ECD is based upon the choice of the base space $\Sigma \equiv \{1, 2, \ldots, N\}$, we

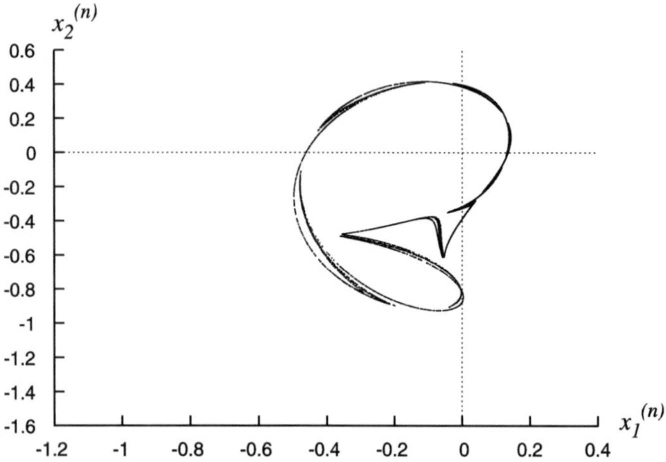

Fig. 10.17 Orbit of f_a for $a = 0.670$

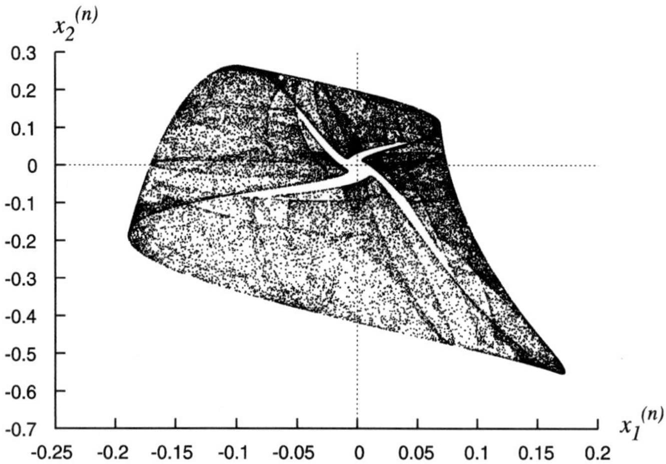

Fig. 10.18 Orbit of f_b for $b = 2.8$

take another choice here: Σ', instead of Σ, will be a new base space. On this base space, a probability distribution is naturally defined as

$$p_{i_0 \cdots i_{n'}}^{(n,\ldots,n+n')} \equiv \frac{1}{n+1} \sum_{k=m}^{m+n} 1_{A_{i_0}}\left(F^k x\right) \cdots 1_{A_{i_{k+n'}}}\left(F^{k+n'} x\right)$$

with its mathematical idealization, $p_{i_0 \cdots i_{n'}} \equiv \lim_{n\to\infty} p_{i_0 \cdots i_{n'}}^{(n,\ldots,n+n')}$. The channel $\Lambda_{n'}^*$ over $\Sigma^{n'}$ is defined by a transition probability,

$$p_{j_0 i_0 \cdots i_{n'+1}} \delta_{i_{n'} j_{n'}} = p(i_1, \ldots, i_{n'+1} | j_0, \ldots, j_{n'}) p_{j_0 \cdots j_{n'}}.$$

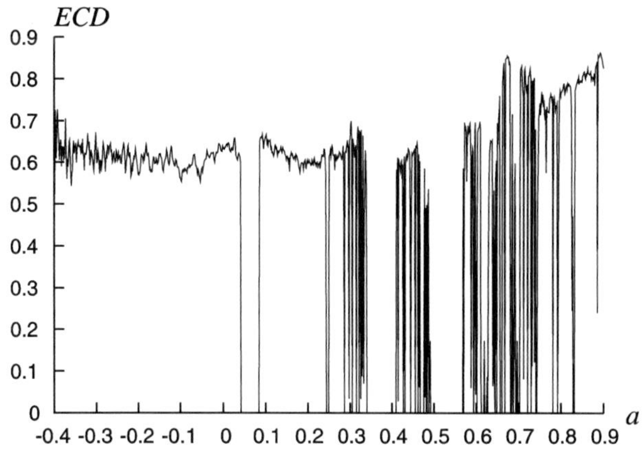

Fig. 10.19 ECD for Tinkerbell map f_a

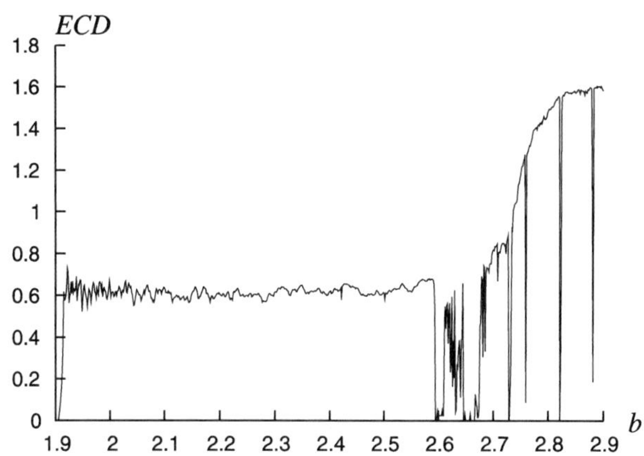

Fig. 10.20 ECD for Tinkerbell map f_b

Thus we derive the ECD with m-step memory effect,

$$D_A^{n'}(x; f) = D_A^{n'}\left(p; \Lambda_{n'}^{*}\right) = \sum_{i_0,\ldots,i_{n'}} p_{i_0\cdots i_{n'}} \log \frac{p_{i_0\cdots i_{n'-1}}}{p_{i_0\cdots i_{n'}}}.$$

It is easy to see that this quantity coincides with the original CD when $n' = 1$.

This memory effect shows an interesting result, namely, the longer the memory, the closer the ECD to the positive part of Lyapunov exponent.

Theorem 10.12 *For given f, x and A, there exist a probability space (Ω, F, ν) and a random variable g depending on f, x, A such that $\lim_{n'\to\infty} D_A^{n'}(x; f) = \int_\Omega g\, d\nu =$ the positive part of Lyapunov exponent.*

Remark 10.13 The use of and a generalization of the ECD to quantum systems have been discussed in [365, 367], some of which are considered in Chap. 19.

10.8 Adaptive Dynamics Describing Chaos

In adaptive dynamics, it is essential to consider in which states and by which ways we see objects. That is, one has to select phenomena and prepare a mode for observation to understand a system as a whole. Typical adaptive dynamics are the dynamics for state-adaptive and that for observable-adaptive as mentioned in the previous section.

We will discuss how such adaptivities appear in dynamics and how they cause chaos.

First of all, we examine carefully what we mean when we say that a certain dynamics produces chaos. Let us take the logistic map as an example. The original differential equation of the logistic map is

$$\frac{dx}{dt} = ax(1-x), \quad 0 \le a \le 4$$

with initial value x_0 in $[0, 1]$. This equation can be easily solved analytically, and the solution (orbit) may not have any chaotic behavior. However, once we make the above equation discrete, i.e., take

$$x_{n+1} = ax_n(1-x_n), \tag{10.2}$$

this difference equation produces chaos.

Taking discrete time is necessary not only for making chaos but also for observing the orbits drawn by the dynamics. Similarly as in quantum mechanics, it is not possible for a human being to understand any object without observing it, for which it will not be possible to trace an orbit continuously in time.

Now let us take a finite partition $A = \{A_k; k = 1, \dots, N\}$ of a proper set $I \equiv [a, b]^N \subset \mathbb{R}^N$ and an equi-partition $B^e = \{B^e_k; k = 1, \dots, N\}$ of I. Here "equi" means that all elements B^e_k are equivalent. We denote the set of all partitions by $\mathcal{P}$ and the set of all equi-partitions by $\mathcal{P}^e$. Such a partition enables observing the orbit of a given dynamics, and moreover, it provides a criterion for observing chaos. There exist several reports saying that one can observe chaos in nature, which are very much related to how one observes the phenomena, for instance, scale, direction and time. It has been difficult to find a satisfactory theory (mathematics) to explain such chaotic phenomena. In the difference equation (10.2), we take some time interval τ between n and $n + 1$. If we let $\tau \to 0$, then we get a completely different dynamics. If we take a coarse graining of the orbit of x_t of τ time units,

$$x_n \equiv \frac{1}{\tau}\int_{(n-1)\tau}^{n\tau} x_t \, dt,$$

we again have a very different dynamics. Moreover, it is important for mathematical consistency to take the limits $n \to \infty$ or N (the number of equi-partitions) $\to \infty$, i.e., making the partition finer and finer, and consider the limits of some quantities as describing chaos, so that mathematical terminologies such as "lim", "sup", "inf" are very often used to define such quantities. *Let us take the opposite position, that is, any observation will be unrelated or even contradicting to such limits. Observation of chaos is a result due to taking suitable scales of, for example, time, distance or domain, and it will not be possible in the limiting cases.*

It is claimed in [442] that most of chaos are scale-dependent phenomena, so the definition of a degree measuring chaos should depend on certain scales taken. Such a scale-dependent dynamics is nothing but adaptive dynamics.

Taking into consideration this view, we modify the definitions of the chaos degree given in the previous section.

Going back to the triple $(\mathcal{A}, \mathfrak{S}, \alpha(G))$ considered in Sect. 10.2, we use this triple both for an input and an output system. Let a dynamics be described by a mapping Γ_t with a parameter $t \in G$ from $\mathfrak{S}$ to $\mathfrak{S}$ and let an observation be described by a mapping $\mathcal{O}$ from $(\mathcal{A}, \mathfrak{S}, \alpha(G))$ to a triple $(\mathcal{B}, \mathfrak{T}, \beta(G))$. The triple $(\mathcal{B}, \mathfrak{T}, \beta(G))$ might be same as the original one or its subsystem, and the observation map $\mathcal{O}$ may contains several different types of observations, that is, it can be decomposed as $\mathcal{O} = \mathcal{O}_m \cdots \mathcal{O}_1$. Let us list some examples of observations.

For a given dynamics $\frac{d\varphi}{dt} = F(\varphi_t)$, or equivalently, $\varphi_t = \Gamma_t \varphi$, one can take several observations.

Example 10.14 (Time scaling (discretizing)) $O_\tau : t \to n$, $\frac{d\varphi}{dt}(t) \to \varphi_{n+1}$, so that $\frac{d\varphi}{dt} = F(\varphi_t) \Rightarrow \varphi_{n+1} = F(\varphi_n)$ and $\varphi_t = \Gamma_t \varphi \Rightarrow \varphi_n = \Gamma_n \varphi$. Here τ is a unit time needed for the observation.

Example 10.15 (Size scaling (conditional expectation, partition)) Let $(\mathcal{B}, \mathfrak{T}, \beta(G))$ be a subsystem of $(\mathcal{A}, \mathfrak{S}, \alpha(G))$, both of which have a certain algebraic structure such as of a C^*-algebra or von Neumann algebra. As an example, the subsystem $(\mathcal{B}, \mathfrak{T}, \beta(G))$ has an abelian structure describing a macroscopic world which is a subsystem of a non-abelian (non-commutative) system $(\mathcal{A}, \mathfrak{S}, \alpha(G))$ describing a micro-world. A mapping $\mathcal{O}_C$ preserving norm (when it is properly defined) from $\mathcal{A}$ to $\mathcal{B}$ is, in some cases, called a conditional expectation. A typical example of this conditional expectation is taken with respect to a projection valued measure

$$\left\{ P_k;\, P_k P_j = P_k \delta_{kj} = P_k^* \delta_{kj} \geq 0, \sum_k P_k = I \right\}$$

associated with a quantum measurement (von Neumann measurement) such that

$$\mathcal{O}_C(\rho) = \sum_k P_k \rho P_k$$

for any quantum state (density operator) ρ. When $\mathcal{B}$ is a von Neumann algebra generated by $\{P_k\}$, it is an abelian algebra isometrically isomorphic to $L^\infty(\Omega)$ for a certain Hausdorff space Ω, so that in this case $\mathcal{O}_C$ sends a general state φ to a probability measure (or distribution) p. A similar example of $\mathcal{O}_C$ comes from a certain representation (selection) of a state such as one Schatten decomposition of ρ,

$$\rho = \mathcal{O}_R \rho = \sum_k \lambda_k E_k,$$

by one-dimensional orthogonal projections $\{E_k\}$ associated to the eigenvalues of ρ with $\sum_k E_k = I$. Another important example of the size scaling is due to a finite partition of an underlining space Ω, e.g., space of orbits defined as

$$\mathcal{O}_P(\Omega) = \left\{ P_k;\ P_k \cap P_j = P_k \delta_{kj}\ (k, j = 1, \dots N), \bigcup_{k=1}^{N} P_k = \Omega \right\}.$$

10.8.1 Chaos Degree with Adaptivity

We go back to the discussion of the entropic chaos degree. Starting from a given dynamics $\varphi_t = \Gamma_t^* \varphi$, it becomes $\varphi_n = \Gamma_n^* \varphi$ after handling the operation $\mathcal{O}_\tau$. Then by taking proper combinations $\mathcal{O}$ of the size scaling operations like $\mathcal{O}_C$, $\mathcal{O}_R$ and $\mathcal{O}_P$, the equation $\varphi_n = \Gamma_n^* \varphi$ changes to $\mathcal{O}(\varphi_n) = \mathcal{O}(\Gamma_n^* \varphi)$, which will be written as $\mathcal{O}\varphi_n = \mathcal{O}\Gamma_n^* \mathcal{O}^{-1} \mathcal{O}\varphi$ or $\varphi_n^{\mathcal{O}} = \Gamma_n^{*\mathcal{O}} \varphi^{\mathcal{O}}$. Then our entropic chaos degree is redefined as follows:

Definition 10.16 The adaptive entropic chaos degree of Γ^* with an initial state φ and observation $\mathcal{O}$ is defined by

$$D^{\mathcal{O}}(\varphi; \Gamma^*) = \int_{\mathcal{O}(\mathfrak{S})} S\left(\Gamma^{*\mathcal{O}} \omega^{\mathcal{O}}\right) d\mu^{\mathcal{O}},$$

where $\mu^{\mathcal{O}}$ is the measure operated by $\mathcal{O}$ to an extremal decomposition measure of φ selected by the observation $\mathcal{O}$ (its part $\mathcal{O}_R$). The adaptive entropic chaos degree of Γ^* with an initial state $\varphi$is defined by

$$D(\varphi; \Gamma^*) = \inf\left\{ D^{\mathcal{O}}(\varphi; \Gamma^*);\ \mathcal{O} \in \mathcal{SO} \right\},$$

where $\mathcal{SO}$ is a proper set of observations naturally determined by a given dynamics.

In this definition, $\mathcal{SO}$ is determined by a given dynamics and some conditions attached to the dynamics, for instance, if we start from a difference equation with a special representation of an initial state, then $\mathcal{SO}$ excludes $\mathcal{O}_\tau$ and $\mathcal{O}_R$.

Then one judges whether a given dynamics causes chaos or not in the following way.

Definition 10.17

1. A dynamics Γ^* is chaotic for an initial state φ in an observation $\mathcal{O}$ iff $D^{\mathcal{O}}(\varphi; \Gamma^*) > 0$.
2. A dynamics Γ^* is totally chaotic for an initial state φ iff $D(\varphi; \Gamma^*) > 0$.

The idea introduced in this section to understand chaos can be applied not only to the entropic chaos degree but also to some other degrees such as dynamical entropy whose applications and the comparison of several degrees will be discussed in [608].

In the case of the logistic map, $x_{n+1} = ax_n(1 - x_n) \equiv F(x_n)$, we obtain this difference equation by taking the observation $\mathcal{O}_\tau$ and take an observation $\mathcal{O}_P$ by equipartition of the orbit space $\Omega = \{x_n\}$ so as to define a state (probability distribution). Thus we can compute the entropic chaos degree in the adaptive sense.

As an example, we consider a circle map

$$\theta_{n+1} = f_v(\theta_n) = \theta_n + \omega \ (\mathrm{mod}\ 2\pi), \tag{10.3}$$

where $\omega = 2\pi v$ $(0 < v < 1)$. If v is a rational number N/M, then the orbit $\{\theta_n\}$ is periodic with the period M. If v is irrational, then the orbit $\{\theta_n\}$ densely fills the unit circle for any initial value θ_0; namely, it is a quasi-periodic motion.

Theorem 10.18 *Let* $\mathbf{I} = [0, 2\pi]$ *be partitioned into L disjoint components of equal length*; $\mathbf{I} = B_1 \cap B_2 \cap \cdots \cap B_L$.

1. *If* v *is rational number* N/M *then the finite equipartition* $P = \{B_k; k = 1, \dots, M\}$ *implies* $D^{\mathcal{O}}(\theta_0; f_v) = 0$.
2. *If* v *is irrational, then* $D^{\mathcal{O}}(\theta_0; f_v) > 0$ *for any finite partition* $P = \{B_k\}$.

Proof (Part 1) Since v is a rational number N/M, for our partition $\{B_k\}$ we choose the M components,

$$B_k \equiv \left\{ x; 2\pi \frac{k-1}{M} \le x < 2\pi \frac{k}{M} \pi \right\}, \quad k = 1, 2, \dots, M.$$

Then, for each component B_k there exists its subset C_k,

$$C_k = \{\theta_j \in B_k;\ j = n+1, \dots, n+m\}$$

where θ_0 is an initial condition, and we have

$$\begin{aligned} f(C_k) &= \left\{ f(\theta_j); 2\pi \frac{k-1}{M} + 2\pi \frac{N}{M} \le f(\theta_j) < 2\pi \frac{k}{M} + 2\pi \frac{N}{M} \ (\mathrm{mod}\ 2\pi), \right. \\ &\qquad \left. j = n+1, \dots, n+m \right\} \\ &= \left\{ \theta_{j+1}; 2\pi \frac{k-1+N}{M} \le \theta_{j+1} < 2\pi \frac{k+N}{M} \ (\mathrm{mod}\ 2\pi), \right. \\ &\qquad \left. j = n+1, \dots, n+m \right\} \end{aligned}$$

$$= \left\{\theta_l;\ 2\pi\frac{k+N}{M} \le \theta_l < 2\pi\frac{k+N+1}{M} \pmod{2\pi},\right.$$
$$\left. l = n+2,\dots,n+m+1\right\}$$
$$\subset B_{k+N \pmod M}.$$

Thus, the map (10.3) maps C_k into $B_{k+N (\bmod M)}$, and we obtain

$$p_{k,k+N \pmod M, B}^{(n,n+1)} = p_{k,B}^{(n)}. \tag{10.4}$$

Equation (10.4) implies that $D(p_B^{(n)}; \Lambda_{n,B}^*) = 0$.

(Part 2) If v is irrational, then the map (10.3) creates a uniform invariant density of the orbit points in the limit as $n \to \infty$. Thus, for any partition $\{B_k\}$, the orbit points are also densely distributed in any component of B_k for sufficiently large m.

Suppose that there exists at least one component B_i of $\{B_k\} = B_1 \cup B_2 \cup \cdots \cup B_L$ such that we can form two nonempty intersections

$$f(B_i) \cap B_j, \qquad f(B_i) \cap B_{j+1}. \tag{10.5}$$

Then we have

$$p_{i,j,B}^{(n,n+1)} > 0, \qquad p_{i,j+1,B}^{(n,n+1)} > 0,$$
$$p_{i,j,B}^{(n,n+1)} + p_{i,j+1,B}^{(n,n+1)} = p_{i,B}^{(n)},$$

so that

$$-\sum_{i,j} p_{i,j,B}^{(n,n+1)} \log p_{i,j,B}^{(n,n+1)} > -\sum_i p_{i,B}^{(n)} \log p_{i,B}^{(n)}.$$

This means that for sufficiently large m, the entropic chaos degree is positive in the case (10.5). Now we randomly choose one component E_1 as

$$E_1 = \left\{x;\ 2\pi a x < 2\pi b \pmod{2\pi}\ 0 < a, b < 1, a \le b\right\}.$$

To overcome situations such as (10.5), we should compose a partition $\{B_k\}$ so that

$$E_2 \equiv f(E_1) \in \{B_k\}$$

and we should take E_1 such that Lebesgue measure of E_1 is less than $2\pi v$. The second condition is needed for the equation

$$f(E_1) \cap E_1 = \emptyset$$

to hold. Continuing in this manner, we should also choose a partition $\{B_k\}$ such that

$$f^n(E_1) \equiv E_{n+1} \in \{B_k\} \quad n = 1, 2, \dots, N-1,$$

where N is the maximum natural number l such that $2\pi l v < 2\pi - 2\pi(b-a)$. Since $2\pi(b-a)$ is the Lebesgue measure of E_1, E_N is the closest left-side component of $\{E_k\}$ to E_1. Now we take the component F_k,

$$\begin{aligned} F_k &= \bigl\{x;\ 2\pi\bigl(b+(k-1)v\bigr) \le x < 2\pi(a+kv) \pmod{2\pi},\\ &\qquad 0 < a, b < 1, a \le b\bigr\}, \quad k = 1, 2, \dots, N-1,\\ F_N &= \bigl\{x;\ 2\pi(a+Nv) \le x \le 2\pi \pmod{2\pi},\\ &\qquad 0 < a, b, < 1, a \le b\bigr\}. \end{aligned}$$

Then for all $k < N-1$, we have

$$\begin{aligned} f(F_k) &= \bigl\{f(x);\ 2\pi\bigl(b+(k-1)v\bigr) + 2\pi v \le f(x) < 2\pi(a+kv)\\ &\quad + 2\pi v \pmod{2\pi}, 0 \le a, b \le 1, a \le b\bigr\}\\ &= \bigl\{x;\ 2\pi(b+kv)+ \le x < 2\pi\bigl(a+(k+v)+2\pi v\bigr) \pmod{2\pi},\\ &\quad 0 \le a, b \le 1, a \le b\bigr\}\\ &= F_{k+1}. \end{aligned}$$

Thus, we have the partition $\{B_k\}$ as

$$\{B_k\} = \left(\bigcup_{k=1}^{N} E_k\right) \cup \left(\bigcup_{k=1}^{N} F_k\right).$$

Since the Lebesgue measure of F_N is less than that of F_1, we form two nonempty intersections,

$$f(E_N) \cap E_1, \quad f(E_N) \cap F_1.$$

From the above discussion, one finds that we cannot choose a partition $\{B_k\}$, in which the entropic chaos degree is equal to 0. □

Note that our entropic chaos degree shows chaos for the quasi-periodic circle dynamics by the observation due to a partition of the orbit, which is different from the usual understanding of chaos. However, the usual belief that quasi-periodic circle dynamics will not cause chaos is not at all obvious, but is realized in a special limiting case as shown in the following theorem.

Theorem 10.19 *For the above circle map, if v is irrational, then $D(\theta_0; f_v) = 0$.*

Proof Let us take an equipartition $P = \{B_k\}$ as

$$B_k \equiv \left\{x;\, 2\pi\frac{k-1}{l} \le x < 2\pi\frac{k}{l}\pi\right\}, \quad k = 1, 2, \dots, l,$$

where l is a certain integer and $B_{k+l} = B_k$. When ν is irrational, put $\nu_0 \equiv [l\nu]$ with Gaussian $[\cdot]$. Then $f_\nu(B_k)$ intersects only two intervals $B_{k+\nu_0}$ and $B_{k+\nu_0+1}$, so that we denote the ratio of the Lebesgue measure of $f_\nu(B_k) \cap B_{k+\nu_0}$ and that of $f_\nu(B_k) \cap B_{k+\nu_0+1}$ by $1-s:s$. This s is equal to $l\nu - [l\nu]$, and the entropic chaos degree becomes

$$D^P = -s\log s - (1-s)\log(1-s).$$

Take the continued fraction expansion of ν and denote its jth approximate by $\frac{b_j}{c_j}$. Then it holds $|\nu - \frac{b_j}{c_j}| \le \frac{1}{c_j^2}$. For the above equipartition $B = \{B_k\}$ with $l = c_j$, we find

$$|l\nu - b_j| \le \frac{1}{k}$$

and

$$[l\nu] = \begin{cases} b_j & \text{when } \nu - \frac{b_j}{c_j} > 0, \\ b_{j-1} & \text{when } \nu - \frac{b_j}{c_j} < 0. \end{cases}$$

It implies

$$D^P \fallingdotseq \frac{\log c_j}{c_j},$$

which goes to 0 as $j \to \infty$. Hence $D = \inf\{D^P; P\} = 0$. □

Such a limiting case will not take place in real observations of natural objects, so that we claim that chaos is a phenomenon depending on observations, environment or periphery, which results in the adaptive definition of chaos as above. The detailed examination of a map of this type is done in [72].

Note here that the chaos degree and the adaptivity can be applied to understand quantum dynamics also [359, 364, 365].

10.9 Time Irreversibility Problem and Functional Mechanics

In this section, we consider the time irreversibility problem and describe a new approach to its solution based on a formulation of classical mechanics which is different form Newton mechanics. This approach, introduced by Volovich [800, 801] fits well with the general idea of adaptive mechanics which says that there are different scales of investigation of determinism and chaos.

The time irreversibility problem is the problem of explaining the irreversible behavior of macroscopic systems from the time-symmetric microscopic laws. The problem has been discussed by Boltzmann, Poincaré, Bogolyubov, Kolmogorov, von Neumann, Landau, Prigogine, Feynman, and many other authors [34, 120, 125,

138, 158, 238, 276, 286, 294, 396, 448, 450, 454, 465, 473, 570, 652, 656, 836], and it deserves a further study.

In particular, in works by Poincaré [652], Landau and Lifshitz [465], Prigogine [656], Ginzburg [286], Feynman [238], it is stressed that the irreversibility problem is still an open problem. Poincaré [652] said that perhaps we will never solve the irreversibility problem. Landau and Lifshitz write about the principle of increasing entropy [465]: "Currently it is not clear whether the law of increasing entropy can be in principle derived from classical mechanics." Landau speculated that to explain the second law of thermodynamics, one has to use quantum mechanical measurement arguments.

From the other side, Lebowitz [473], Goldstein [294] and Bricmont [138] state that the irreversibility problem was basically solved already by Boltzmann by using his notion of macroscopic entropy and the probabilistic approach.

The microscopic-mechanical description of a system assumes that the state of the system at a given moment of time is represented by a point in the phase space with an invariant measure, and the dynamics of the system is described by a trajectory in the phase space, see [58, 71, 213, 439, 465, 718]. It is assumed that the microscopic laws of motion are known (Newton or Schrodinger equations) and there is a problem of derivation from them the macroscopic (Boltzmann, Navier–Stokes, etc.) equations; see, for example, [465, 836].

There are well known critical remarks by Loschmidt and Poincaré and Zermelo on the Boltzmann approach to the irreversibility problem and the H-theorem. Loschmidt remarked that from the symmetry of the Newton equations upon the reverse of time it follows that to every motion of the system on the trajectory towards the equilibrium state one can put into correspondence the motion out of the equilibrium state if we reverse the velocities at some time moment. Such a motion is in contradiction with the tendency of the system to go to the equilibrium state and with the law of increasing of entropy.

Then, there is the Poincaré recurrence theorem which says that a trajectory of a bounded isolated mechanical system will many times come to a very small neighborhood of an initial point. This is also in contradiction with the motion to the equilibrium state. This is the Poincaré–Zermelo paradox.

Boltzmann [126] gave the following answer to the Loschmidt argument: "We do not have to assume a special type of initial condition in order to give a mechanical proof of the second law, if we are willing to accept a statistical viewpoint. While any individual non-uniform state (corresponding to low entropy) has the same probability as any individual uniform state (corresponding to high entropy), there are many more uniform states than non-uniform states. Consequently, if the initial state is chosen at random, the system is almost certain to evolve into a uniform state, and entropy is almost certain to increase."

So, the answer by Boltzmann to the objection of Loschmidt was that, firstly, the probabilistic considerations have been involved, and secondly, he argued that with the overwhelming probability the evolution of the system will occur in the direction of flow of time, corresponding to the increasing entropy, since there are many more uniform states than non-uniform states. The answer by Boltzmann to the Poincaré–Zermelo objection was in pointing out the extremely long Poincaré recurrence time.

These Boltzmann's responses are not very convincing, from our point of view, despite their vigorous support in recent works [138, 294, 473]. Involvement of probability considerations alone does not clarify the issue of irreversibility because if there is symmetry in relation to the direction of time, it remains unclear why the evolution in one direction is more likely than in the other.

Then, the argument that there are many more uniform states than non-uniform states does not clarify the issue of the dynamical evolution since the dynamics does depend on the form of the potential energy between particles, and for many potentials the argument is simply wrong. Therefore, this general Boltzmann's argument does not give a real insight to the irreversibility problem.

Actually, Boltzmann in [126] considered "a large but not infinite number of absolutely elastic spheres, which move in a closed container whose walls are completely rigid and likewise absolutely elastic. No external forces act on our spheres." Even for this simple model it is very difficult to make the Boltzmann argument convincing, i.e., to get a mathematical result, see [71, 718].

Further, an indication to the extremely long Poincaré recurrence time does not remove the contradiction between microscopic reversibility and macroscopic irreversibility, and moreover, no clear mechanism for relaxation to equilibrium is presented.

Lebowitz advanced [473], following Boltzmann, the following arguments to explain irreversibility: (a) the great disparity between microscopic and macroscopic scales, (b) a low entropy state of the early universe, and (c) the fact that what we observe is the behavior of systems coming from such an initial state—not all possible systems.

From our viewpoint these arguments do not lead to explanation of irreversibility even though it is said in [473] that "common alternative explanations, such as those based on the ergodic or mixing properties of probability distribution... are either unnecessary, misguided or misleading."

Boltzmann proposed that we and our observed low-entropy world are a random fluctuation in a higher-entropy universe. These cosmological considerations of the early universe might be entertaining but they should be related with the modern Friedmann [273, 484] gravitational picture of the Big Bang and, what is most important, there is no evidence that the irreversible behavior of gas in a box is somehow related with conditions in the early universe 14 billion years ago.

Notice that in [548] it is shown that the Hawking black hole information paradox is a special case of the irreversibility problem.

Goldstein said in [294]: "The most famous criticisms of Boltzmann's later works on the subject have little merit. Most twentieth century innovations—such as the identification of the state of a physical system with a probability distribution ρ on its phase space, of its thermodynamic entropy with the Gibbs entropy of ρ, and the invocation of the notions of ergodicity and mixing for the justification of the foundations of statistical mechanics—are thoroughly misguided."

And then: "This use of ergodicity is thoroughly misguided. Boltzmann's key insight was that, given the energy of a system, the overwhelming majority of its phase points on the corresponding energy surface are equilibrium points, all of which look macroscopically more or less the same."

The Boltzmann argument about "the overwhelming majority" (i.e., "many more uniform states") was discussed above. Moreover, the main point of the current paper is that we shall use the probability distribution and the Liouville equation not only in statistical mechanics but also in classical mechanics, even for a single particle in empty space.

A powerful method for obtaining kinetic equations from the Newton–Liouville equations was developed by Bogolyubov [120]. He has considered an infinite number of particles in an infinite volume and postulated the condition of weakening of the initial correlations between particles in the distant past, through which the irreversibility entered into the equation for the distribution functions, as well as using a formal expansion in powers of density, which leads to divergences.

Poincaré considered the model of free motion of gas particles in a box with reflecting walls and showed that for solutions of the Liouville equation in this model there is, in some sense, an irreversible diffusion [650]. This result of Poincaré was introduced to modern scientific literature by Kozlov, see [448], where the result of Poincaré was significantly strengthened and consolidated. In the works of Kozlov, a method of the weak limit in the non-equilibrium statistical mechanics has been developed, and in particular, it was proved that for some models the system in the sense of weak convergence tends to one and the same limit in the past and in the future [448, 450]. The method of the weak limit of [448, 450] had a significant influence to the formulation of the approach to the problem of irreversibility through functional formulation of classical mechanics.

Note that the stochastic limit [34] gives a systematic method for investigation of irreversible processes.

Questions about the increase of the fine and coarse entropies are discussed in [158, 396, 449, 450, 465, 651, 836].

In this section, we consider the following approach to the irreversibility problem and to paradoxes of Loschmidt and Poincaré–Zermelo: We describe a *formulation of microscopic dynamics which is irreversible in time*. Thus the contradiction between microscopic reversibility and macroscopic irreversibility of the dynamics disappears, since both microscopic and macroscopic dynamics in the proposed approach are irreversible.

Note that the conventional widely used concept of the microscopic state of the system at some moment in time as the point in phase space, as well as the notion of trajectory and the microscopic equations of motion, has no direct physical meaning, since arbitrary real numbers not observable (observable physical quantities are only presented by rational numbers, cf. the discussion of concepts of space and time in [210, 404, 507, 774, 783, 785, 799, 833]).

In the proposed "functional" approach, the physical meaning is attached not to a single trajectory, but only to a "beam" of trajectories, or the distribution function on the phase space. Individual trajectories are not observable, they could be considered as "hidden variables", if one uses the quantum mechanical notions, see [793, 798].

The fundamental equation of the microscopic dynamics of the proposed functional probabilistic approach is not the Newton's equation, but a Liouville equation for distribution function. It is well known that the Liouville equation is used in statistical mechanics for the description of the motions of gas. Let us stress that we shall use the Liouville equation for the description of a single particle in the empty space.

Although the Liouville equation is symmetric in relation to the reversion of time, its solutions have the property of *delocalization* that, generally speaking, can be interpreted as a manifestation of irreversibility. It is understood that if at some moment in time the distribution function describes a particle, localized to a certain extent, then over time the degree of localization decreases, there is the spreading of distribution function. Delocalization takes place even for a free particle in infinite space where there is no ergodicity and mixing.

In the functional approach to classical mechanics, we do not derive the statistical or chaotic properties of deterministic dynamics, but we suggest that the Laplace's determinism at the fundamental level is absent not only in quantum, but also in classical mechanics.

We show that Newton's equation in the proposed approach appears as an approximate equation describing the dynamics of the average values of coordinates and momenta for not too long time. We calculate corrections to Newton's equation.

In the next subsection, the fundamentals of the functional formulation of classical mechanics are presented. Then we discuss the free movement of particles and Newton's equation for the average coordinates as well as comparison with quantum mechanics. After presenting general comments on the Liouville and Newton equations, we compute corrections to the Newton equation for a nonlinear system. Finally, we discuss the reversibility of motion in classical mechanics and irreversibility in the functional approach to the mechanics. The dynamics of the classical and quantum particle in a box and their interrelationships are summarized in the end of this section.

10.9.1 States and Observables in Functional Classical Mechanics

Usually in classical mechanics, the motion of a point body is described by the trajectory in the phase space, i.e., the values of the coordinates and momenta as functions of time, which are solutions of the equations of Newton or Hamilton.

Note, however, that this mathematical model is an idealization of the physical process, rather far separated from reality. The physical body always has the spatial dimensions, so a mathematical point gives only an approximate description of the physical body. The mathematical notion of a trajectory does not have direct physical meaning, since it uses arbitrary real numbers, i.e., infinite decimal expansions, while the observation is only possible, in the best case, of rational numbers, and even then only with some error. Therefore, in the proposed "functional" approach to classical mechanics, we are not starting from Newton's equation, but form the Liouville equation.

Consider the motion of a classical particle along a straight line in the potential field. The general case of many particles in the three-dimensional space is discussed below. Let (q, p) be the coordinates on the plane $\mathbb{R}^2$ (phase space), $t \in \mathbb{R}$

is time. The state of a classical particle at time t will be described by the function $\rho = \rho(q, p, t)$, it is the density of the probability that the particle at time t has the coordinate q and momentum p.

Note that the description of a mechanical system with the help of probability distribution function $\rho = \rho(q, p, t)$ does not necessarily mean that we are dealing with a set of identically prepared ensembles of particles. Usually in probability theory, one considers an ensemble of events and a sample space [142, 405, 657]. But we can use the description with the function $\rho = \rho(q, p, t)$ also for individual bodies, such as planets in astronomy (the phase space in this case is six-dimensional). In this case, one can think about the "ensemble" of different astronomers which observe the planet, or about the "ensemble" of different models of behavior of a given object. Actually, it is always implicitly dealt with the function $\rho = \rho(q, p, t)$ which takes into account the inherent uncertainty in the coordinates and momentum of the body. An application of these remarks to quantum mechanics will be discussed in a separate work.

The specific type of function ρ depends on the method of preparation of the state of a classical particle at the initial time and the type of potential field. When $\rho = \rho(q, p, t)$ has sharp peaks at $q = q_0$ and $p = p_0$, we say that the particle has the approximate values of coordinate and momentum q_0 and p_0.

We emphasize that the exact derivation of the coordinate and momentum cannot be done, not only in quantum mechanics, where there is the Heisenberg uncertainty relation, but also in classical mechanics. There are always some errors in setting the coordinates and momenta. The concept of arbitrary real numbers, given by the infinite decimal series, is a mathematical idealization, such numbers cannot be measured in the experiment.

Therefore, in the functional approach to classical mechanics, the concept of precise trajectory of a particle is absent, the fundamental concept is a distribution function $\rho = \rho(q, p, t)$ and the δ-function as a distribution function is not allowed.

We assume that the continuously differentiable and integrable function $\rho = \rho(q, p, t)$ satisfies the conditions:

$$\rho \geq 0, \quad \int_{\mathbb{R}^2} \rho(q, p, t)\, dq\, dp = 1, \quad t \in \mathbb{R}. \tag{10.6}$$

The formulation of classical mechanics in the language of states and observables is considered in [229, 431, 494]. The functional approach to classical mechanics differs in the following respects. Because the exact trajectory of a particle in the functional approach does not exist, the function $\rho = \rho(q, p, t)$ cannot be an arbitrary generalized function, it is the usual function of class $L^1(\mathbb{R}^2)$, or even continuously differentiable and integrable function.

In addition, the motion of particles in the functional approach is not described directly by the Newton (Hamilton) equation. Newton's equation in the functional approach is an approximate equation for the average coordinates of the particles, and for nonlinear dynamics there are corrections to the Newton's equations.

As is known, the mathematical description of a moving fluid or gas is given by means of the density distribution functions $\rho(q, t)$, as well as the velocity $v(q, t)$

and pressure $p(q,t)$; see, for example, [463]. Let the function $\rho(q,p,t)$ describe a particle, as proposed in the functional formulation of classical mechanics, and we set $\rho_c(q,t)=\int\rho(q,p,t)\,dp$. We could ask the question if we can determine by the form of functions $\rho(q,t)$ and $\rho_c(q,t)$ whether we are dealing with a continuous medium or with a particle. The general answer is the following: functions $\rho(q,t)$ and $\rho_c(q,t)$ satisfy different equations (the Navier–Stokes or Liouville equation) and different conditions of normalization.

Note, however, that if an error in determining the coordinates and momentum of particles is large enough, it is really not so easy to determine, we have a case of, say, a fast-moving particle in a box with reflecting walls, or a gas of particles.

If $f=f(q,p)$ is a function on the phase space, the average value of f at time t is given by the integral

$$\overline{f}(t)=\int f(q,p)\rho(q,p,t)\,dq\,dp. \tag{10.7}$$

In a sense, we are dealing with a random process $\xi(t)$ with values in the phase space. The motion of a point body along a straight line in the potential field will be described by the equation

$$\frac{\partial\rho}{\partial t}=-\frac{p}{m}\frac{\partial\rho}{\partial q}+\frac{\partial V(q)}{\partial q}\frac{\partial\rho}{\partial p}. \tag{10.8}$$

Here $V(q)$ is the potential field, and mass is $m>0$.

Equation (10.8) looks like the Liouville equation which is used in statistical physics to describe a gas of particles, but here we use it to describe a single particle.

If the distribution $\rho_0(q,p)$ for $t=0$ is known, we can consider the Cauchy problem for (10.8):

$$\rho|_{t=0}=\rho_0(q,p). \tag{10.9}$$

Let us discuss the case when the initial distribution has the Gaussian form:

$$\rho_0(q,p)=\frac{1}{\pi ab}e^{-\frac{(q-q_0)^2}{a^2}}e^{-\frac{(p-p_0)^2}{b^2}}. \tag{10.10}$$

For sufficiently small values of the parameters $a>0$ and $b>0$, the particle has coordinates and momentum close to q_0 and p_0. For this distribution, the average values of the coordinates and momentum are:

$$\overline{q}=\int q\rho_0(q,p)\,dq\,dp=q_0,\qquad \overline{p}=\int p\rho_0(q,p)\,dq\,dp=p_0, \tag{10.11}$$

and dispersion

$$\Delta q^2=\overline{(q-\overline{q})^2}=\frac{1}{2}a^2,\qquad \Delta p^2=\overline{(p-\overline{p})^2}=\frac{1}{2}b^2. \tag{10.12}$$

10.9.2 Free Motion

Consider first the case of the free motion of the particle when $V = 0$. In this case, (10.8) has the form

$$\frac{\partial \rho}{\partial t} = -\frac{p}{m}\frac{\partial \rho}{\partial q}, \tag{10.13}$$

and the solution of the Cauchy problem is

$$\rho(q, p, t) = \rho_0\left(q - \frac{p}{m}t, p\right). \tag{10.14}$$

Using expressions (10.10), (10.14), and

$$\rho(q, p, t) = \frac{1}{\pi ab}\exp\left\{-\frac{(q - q_0 - \frac{p}{m}t)^2}{a^2} - \frac{(p - p_0)^2}{b^2}\right\}, \tag{10.15}$$

we get the time-dependent distribution of coordinates:

$$\rho_c(q, t) = \int \rho(q, p, t)\, dp = \frac{1}{\sqrt{\pi}\sqrt{a^2 + \frac{b^2 t^2}{m^2}}}\exp\left\{-\frac{(q - q_0 - \frac{p_0}{m}t)^2}{(a^2 + \frac{b^2 t^2}{m^2})}\right\}, \tag{10.16}$$

while the distribution of momenta is

$$\rho_m(p, t) = \int \rho(q, p, t)\, dq = \frac{1}{\sqrt{\pi}b}e^{-\frac{(p-p_0)^2}{b^2}}. \tag{10.17}$$

Thus, for the free particle the distribution of the particle momentum with the passage of time does not change, and the distribution of the coordinates changes. There is, as one says in quantum mechanics, the spreading of the wave packet. From (10.16) it follows that the dispersion Δq^2 increases with time:

$$\Delta q^2(t) = \frac{1}{2}\left(a^2 + \frac{b^2 t^2}{m^2}\right). \tag{10.18}$$

Even if the particle was arbitrarily well localized (a^2 is arbitrarily small) at $t = 0$, then at sufficiently large times t the localization of the particle becomes meaningless, there is a *delocalization* of the particle.

10.9.3 Newton's Equation for the Average Coordinate

In the functional approach to classical mechanics, there is no ordinary picture of an individual trajectory of a particle. The starting equation is the dynamic equation (10.8) for the distribution function, rather than the Newton equation.

What role can Newton's equation play in the functional approach? We show that the average coordinate for the free particle in the functional approach satisfies Newton's equation. Indeed, the average coordinate and momentum for the free particles have the form

$$\overline{q}(t) = \int q\rho_c(q,t)\,dq = q_0 + \frac{p_0}{m}t, \qquad \overline{p}(t) = \int p\rho_m(p,t)\,dp = p_0. \tag{10.19}$$

Hence we get

$$\frac{d^2}{dt^2}\overline{q}(t) = 0, \tag{10.20}$$

i.e., we have Newton's equation for the average coordinates.

We also have Hamilton's equations for the average values of the coordinate and momentum:

$$\dot{\overline{q}} = \frac{\partial H}{\partial \overline{p}}, \qquad \dot{\overline{p}} = -\frac{\partial H}{\partial \overline{q}}, \tag{10.21}$$

where the Hamiltonian $H = H(\overline{q}, \overline{p})$ for the free particle has the form

$$H = \frac{\overline{p}^2}{2m}. \tag{10.22}$$

Note that in the functional mechanics the Newton equation for the average coordinates is obtained only for the free particle or for quadratic Hamiltonians with a Gaussian initial distribution function. For a more general case, there are corrections to Newton's equations, as discussed below.

We discussed the spreading of Gaussian distribution functions. Similar results are obtained for the distribution functions of other forms, if they describe, in some sense, localized coordinates and momenta at the initial time.

10.9.4 Comparison with Quantum Mechanics

Compare the evolutions of Gaussian distribution functions in functional classical mechanics and in quantum mechanics for the motion of particles along a straight line. The scene of work for the functional classical mechanics is $L^2(\mathbb{R}^2)$ (or $L^1(\mathbb{R}^2)$), and for quantum mechanics—$L^2(\mathbb{R}^1)$.

The Schrodinger equation for a free quantum particle on a line reads:

$$i\hbar\frac{\partial \psi}{\partial t} = -\frac{\hbar^2}{2m}\frac{\partial^2 \psi}{\partial x^2}. \tag{10.23}$$

Here $\psi = \psi(x,t)$ is the wave function and $\hbar$ is the Planck constant. The density of the distribution function for the Gaussian wave function has the form (see, for example, [261])

$$\rho_q(x,t) = \left|\psi(x,t)\right|^2 = \frac{1}{\sqrt{\pi}\sqrt{a^2 + \frac{\hbar^2 t^2}{a^2 m^2}}} \exp\left\{-\frac{(x - x_0 - \frac{p_0}{m}t)^2}{(a^2 + \frac{\hbar^2 t^2}{a^2 m^2})}\right\}. \tag{10.24}$$

We find that the distribution functions in functional classical and in quantum mechanics (10.16) and (10.24) coincide, if we set

$$a^2 b^2 = \hbar^2. \tag{10.25}$$

We remark that if condition (10.25) is satisfied, then the Wigner function $W(q,p,t)$ [688] for ψ corresponds to the classical distribution function (10.15), $W(q,p,t) = \rho(q,p,t)$.

The problem of spreading of the quantum wave packet in dealing with the potential barrier is considered in [216].

Gaussian wave functions on the line are coherent or compressed states. The compressed states on the interval are considered in [804].

10.9.5 Liouville Equation and the Newton Equation

In the functional classical mechanics, the motion of a particle along the straight line is described by the Liouville equation (10.8). A more general Liouville equation on the manifold Γ with coordinates $x = (x^1, \dots, x^k)$ has the form

$$\frac{\partial \rho}{\partial t} + \sum_{i=1}^{k} \frac{\partial}{\partial x^i}\left(\rho v^i\right) = 0. \tag{10.26}$$

Here $\rho = \rho(x,t)$ is a density function, and $v = v(x) = (v^1, \dots, v^k)$ is a vector field on Γ. The solution of the Cauchy problem for (10.26) with initial data

$$\rho|_{t=0} = \rho_0(x) \tag{10.27}$$

might be written in the form

$$\rho(x,t) = \rho_0\big(\varphi_{-t}(x)\big). \tag{10.28}$$

Here $\varphi_t(x)$ is a phase flow along the solutions of the characteristic equation

$$\dot{x} = v(x). \tag{10.29}$$

In particular, if $k = 2n$, $M = M^n$ is a smooth manifold, the phase space $\Gamma = T^*M$ is a cotangent bundle, and $H = H(q,p)$ is a Hamiltonian function on Γ, then the Liouville equation has the form

$$\frac{\partial \rho}{\partial t} + \sum_{i=1}^{n} \left[\frac{\partial H}{\partial p^i}\frac{\partial \rho}{\partial q^i} - \frac{\partial H}{\partial q^i}\frac{\partial \rho}{\partial p^i}\right] = 0. \tag{10.30}$$

The Liouville measure $d\mu = dq\,dp$ is invariant under the phase flow φ_t.

A classical dynamical system in the functional approach to mechanics is a stochastic process $\xi(t) = \xi(t; q, p) = \varphi_t(q, p)$ which takes values in Γ and with the probabilistic measure $dP(q, p) = \rho_0(q, p)\,dq\,dp$. The correlation functions have the form

$$\langle \xi_{i_1}(t_1) \cdots \xi_{i_s}(t_s) \rangle = \int \xi_{i_1}(t_1; q, p) \cdots \xi_{i_s}(t_s; q, p) \rho_0(q, p)\,dq\,dp. \tag{10.31}$$

Here $i_1, \dots, i_s = 1, \dots, k$.

It is assumed usually that the energy surfaces $\{H = \text{const}\}$ are compact.

A system from N particles in the three-dimensional space has the phase space $\mathbb{R}^{6N}$ with coordinates $q = (\mathbf{q}_1, \dots, \mathbf{q}_N)$, $p = (\mathbf{p}_1, \dots, \mathbf{p}_N)$, $\mathbf{q}_i = (q_i^1, q_i^2, q_i^3)$, $\mathbf{p}_i = (p_i^1, p_i^2, p_i^3)$, $i = 1, \dots, N$, and it is described by the Liouville equation for the function $\rho = \rho(q, p, t)$

$$\frac{\partial \rho}{\partial t} = \sum_{i,\alpha} \left(\frac{\partial V(q)}{\partial q_i^\alpha} \frac{\partial \rho}{\partial p_i^\alpha} - \frac{p_i^\alpha}{m_i} \frac{\partial \rho}{\partial q_i^\alpha} \right). \tag{10.32}$$

Here summation is over $i = 1, \dots, N$, $\alpha = 1, 2, 3$. The characteristic equations for (10.32) are Hamilton's equations

$$\dot{q}_i^\alpha = \frac{\partial H}{\partial p_i^\alpha}, \qquad \dot{p}_i^\alpha = -\frac{\partial H}{\partial q_i^\alpha}, \tag{10.33}$$

where the Hamiltonian is

$$H = \sum_i \frac{\mathbf{p}_i^2}{2m_i} + V(q). \tag{10.34}$$

We emphasize here again that the Hamilton equations (10.33) in the current functional approach to the mechanics do not describe directly the motion of particles, and they are only the characteristic equations for the Liouville equation (10.32) which has a physical meaning. The Liouville equation (10.32) can be written as

$$\frac{\partial \rho}{\partial t} = \{H, \rho\}, \tag{10.35}$$

where the Poisson bracket

$$\{H, \rho\} = \sum_{i,\alpha} \left(\frac{\partial H}{\partial q_i^\alpha} \frac{\partial \rho}{\partial p_i^\alpha} - \frac{\partial H}{\partial p_i^\alpha} \frac{\partial \rho}{\partial q_i^\alpha} \right). \tag{10.36}$$

Criteria for essential self-adjointness of the Liouville operator in the Hilbert space $L^2(\mathbb{R}^{6N})$ are given in [659].

10.9.6 Corrections to Newton's Equations

In the subsection above, it was noted that for the free particle in the functional approach to classical mechanics the averages coordinates and momenta satisfy the Newton equations. However, when there is nonlinear interaction, then corrections to the Newton's equations appear in the functional approach.

Consider the motion of a particle along the line in the functional mechanics. The average value $\overline{f}$ of the function on the phase space $f = f(q,p)$ at time t is given by the integral (10.7)

$$\overline{f}(t) = \langle f(t)\rangle = \int f(q,p)\rho(q,p,t)\,dq\,dp. \tag{10.37}$$

Here $\rho(q,p,t)$ has the form (10.28)

$$\rho(q,p,t) = \rho_0\big(\varphi_{-t}(q,p)\big). \tag{10.38}$$

By making the replacement of variables, subject to the invariance of the Liouville measure, we get

$$\langle f(t)\rangle = \int f(q,p)\rho(q,p,t)\,dq\,dp = \int f\big(\varphi_t(q,p)\big)\rho_0(q,p)\,dq\,dp. \tag{10.39}$$

Let us take

$$\rho_0(q,p) = \delta_\epsilon(q-q_0)\delta_\epsilon(p-p_0), \tag{10.40}$$

where

$$\delta_\epsilon(q) = \frac{1}{\sqrt{\pi}\epsilon}e^{-q^2/\epsilon^2}, \tag{10.41}$$

$q \in \mathbb{R}$, $\epsilon > 0$.

Let us show that in the limit $\epsilon \to 0$ we obtain the Newton (Hamilton) equations:

$$\lim_{\epsilon\to 0}\langle f(t)\rangle = f\big(\varphi_t(q_0,p_0)\big). \tag{10.42}$$

Proposition 10.20 *Let the function* $f(q,p)$ *in the expression* (10.37) *be continuous and integrable, and let* ρ_0 *have the form* (10.40). *Then*

$$\lim_{\epsilon\to 0}\int f(q,p)\rho(q,p,t)\,dq\,dp = f\big(\varphi_t(q_0,p_0)\big). \tag{10.43}$$

Proof Functions $\delta_\epsilon(q)$ form a δ-sequence in $D'(\mathbb{R})$ [781]. Hence we obtain

$$\begin{aligned}\lim_{\epsilon\to 0}\int f\big((q,p)\big)\rho(q,p,t)\,dq\,dp &= \lim_{\epsilon\to 0}\int f\big(\varphi_t(q,p)\big)\delta_\epsilon(q-q_0)\delta_\epsilon(p-p_0)\\ &= f\big(\varphi_t(q_0,p_0)\big),\end{aligned} \tag{10.44}$$

as required. □

We now calculate the corrections to the solution of the Newton equation. In functional mechanics, consider the equation, see (10.8),

$$\frac{\partial \rho}{\partial t} = -p\frac{\partial \rho}{\partial q} + \lambda q^2 \frac{\partial \rho}{\partial p}. \tag{10.45}$$

Here λ is a small parameter, and we set the mass $m = 1$. The characteristic equations have the form of the following Hamilton (Newton) equations:

$$\dot{p}(t) + \lambda q(t)^2 = 0, \qquad \dot{q}(t) = p(t). \tag{10.46}$$

The solution of these equations with the initial data

$$q(0) = q, \qquad \dot{q}(0) = p \tag{10.47}$$

for small t has the form

$$\bigl(q(t), p(t)\bigr) = \varphi_t(q, p) = \left(q + pt - \frac{\lambda}{2}q^2t^2 + \cdots, \ p - \lambda q^2 t + \cdots\right). \tag{10.48}$$

Using the asymptotic expansion $\delta_\epsilon(q)$ in $D'(\mathbb{R})$ as $\epsilon \to 0$, and comparing [34, 639],

$$\delta_\epsilon(q) = \delta(q) + \frac{\epsilon^2}{4}\delta''(q) + \cdots, \tag{10.49}$$

then for $\epsilon \to 0$ we obtain the corrections to the Newton dynamics:

$$\begin{aligned}\bigl\langle q(t)\bigr\rangle &= \int \left(q + pt - \frac{\lambda}{2}q^2t^2 + \cdots\right)\left[\delta(q - q_0) + \frac{\epsilon^2}{4}\delta''(q - q_0) + \cdots\right] \\ &\quad \cdot \left[\delta(p - p_0) + \frac{\epsilon^2}{4}\delta''(p - p_0) + \cdots\right] dq\, dp \\ &= q_0 + p_0 t - \frac{\lambda}{2}q_0^2t^2 - \frac{\lambda}{4}\epsilon^2 t^2.\end{aligned} \tag{10.50}$$

Denoting the Newton solution

$$q_{\text{Newton}}(t) = q_0 + p_0 t - \frac{\lambda}{2}q_0^2 t^2,$$

we obtain for small ϵ, t and λ:

$$\bigl\langle q(t)\bigr\rangle = q_{\text{Newton}}(t) - \frac{\lambda}{4}\epsilon^2 t^2. \tag{10.51}$$

Here $-\frac{\lambda}{4}\epsilon^2 t^2$ is the correction to the Newton solution obtained within the functional approach to classical mechanics with the initial Gaussian distribution function. If we choose a different initial distribution, we get a correction of another form.

We have proved.

Proposition 10.21 *In the functional approach to mechanics, the first correction at* ϵ *to the Newton dynamics for small* t *and* λ *for* (10.46) *has the form* (10.51).

Note that in the functional approach to mechanics, instead of the usual Newton equation

$$m\frac{d^2}{dt^2}q(t) = F(q), \tag{10.52}$$

where $F(q)$ is a force, we obtain

$$m\frac{d^2}{dt^2}\langle q(t)\rangle = \langle F(q)(t)\rangle. \tag{10.53}$$

Indeed, multiplying the equation

$$\frac{\partial\rho}{\partial t} = -\frac{p}{m}\frac{\partial\rho}{\partial q} - F(q)\frac{\partial\rho}{\partial p} \tag{10.54}$$

by q and integrating over p and q, and then integrating by parts, we get

$$\frac{d}{dt}\langle q(t)\rangle = \frac{\langle p(t)\rangle}{m}. \tag{10.55}$$

Similarly, multiplying (10.54) by p and integrating over p and q, and then integrating by parts, we get

$$\frac{d}{dt}\langle p(t)\rangle = \langle F(q)(t)\rangle, \tag{10.56}$$

which gives (10.53).

The task of calculating the corrections at ϵ for Newton's equation for mean values is similar to the problem of calculating semiclassical corrections in quantum mechanics [229, 510, 804].

10.9.7 Time Reversal

Reversibility in Classical Mechanics

Let us present a famous discourse which proves reversibility of the dynamics in classical mechanics. From the symmetry of Newton's equations upon the replacement of the time t with $-t$, it follows that if there exists some motion in the system, then also the reverse motion is possible, i.e. such a motion, in which the system passes same states in the phase space in the reverse order. Indeed, let the function $x(t)$ satisfy the Newton equation

$$\ddot{x}(t) = F\big(x(t)\big) \tag{10.57}$$

with the initial data

$$x(0) = x_0, \qquad \dot{x}(0) = v_0. \tag{10.58}$$

We denote the corresponding solution by

$$x(t) = \Phi(t; x_0, v_0).$$

We fix $T > 0$ and reverse the motion of the particle at some moment in time T by reversing its velocity, i.e., we consider the solution $y(t)$ of the Newton equation

$$\ddot{y}(t) = F\bigl(y(t)\bigr) \tag{10.59}$$

with the following initial data:

$$y(0) = x(T), \qquad \dot{y}(0) = -\dot{x}(T). \tag{10.60}$$

Then it is easy to see that at the time moment T we get

$$y(T) = x_0, \qquad \dot{y}(T) = -v_0, \tag{10.61}$$

i.e., the particle comes back to the initial point with the inverse velocity. To prove the relation (10.61) it is enough to note that the solution of (10.59) with the initial data (10.60) has the form

$$y(t) = \Phi(T - t; x_0, v_0)$$

and use the relations (10.58).

Let us notice that these arguments about reversibility of motion in the classical mechanics used not only symmetry of the Newton equation concerning time reversibility, but also the fact that a state of the particle in the classical mechanics at some instant of time is completely characterized by two parameters: coordinate x and speed v. Reversibility of the motion in classical mechanics means reversibility of the motion along a given trajectory.

As it was discussed above, the notion of an individual trajectory of a particle has no physical sense. In reality, we deal with a bunch of trajectories or a probability distribution. In the functional classical mechanics, the state of the particle is characterized not by the two numerical parameters, but by the distribution function $\rho = \rho(q, p, t)$. In the following subsection, it will be shown how it leads to delocalization and irreversibility.

Irreversibility in the Functional Mechanics

The considered reversibility of motion in classical mechanics deals with an individual trajectory. In the functional mechanics, the concept of the individual trajectory of the particle has no direct physical sense. Instead, the state of the particle is described by the distribution function $\rho = \rho(q, p, t)$ which satisfies the Liouville equation (10.8)

$$\frac{\partial \rho}{\partial t} = -\frac{p}{m}\frac{\partial \rho}{\partial q} + \frac{\partial V(q)}{\partial q}\frac{\partial \rho}{\partial p}. \tag{10.62}$$

The Liouville equation is invariant under the replacement of t with $-t$: If $\rho = \rho(q, p, t)$ is the solution of (10.62), then $\sigma(q, p, t) = \rho(q, -p, -t)$ is its solution also. However, this symmetry does not mean reversibility of the motion of a particle in the functional approach to mechanics, since the state of the particle is described there by the distribution function and the phenomenon of delocalization takes place.

In this way we obtain an answer to the arguments of Loschmidt and Poincaré–Zermelo. Indeed, to reverse the particle motion at the time moment $t = T$ as it is proposed in the Loschmidt argument, it is necessary to make the coordinate and momentum measurement. But it will change the distribution $\rho(q, p, T)$. Further, it is necessary to prepare such a condition of the particle that its evolution back in time would lead to the initial distribution ρ_0 which is difficult since the delocalization takes place. We will need something even better than Maxwell's demon.

For a free particle, the delocalization leads to the increasing of dispersion Δq^2 with time (10.18):

$$\Delta q^2(t) = \frac{1}{2}\left(a^2 + \frac{b^2 t^2}{m^2}\right).$$

Notice that similar phenomenon takes place for the Brownian motion $B(t)$ which has variance t [142, 657].

Concerning the Zermelo argument related with the Poincaré recurrence theorem, we note that this argument cannot be applied to the functional mechanics because this argument is based on the notion of an individual trajectory. In the functional mechanics, the state of the system is characterized by the distribution function, and here the mean values might irreversibly tend to some limits without contradicting the Poincaré theorem as it will be shown in the next section.

The Poincaré theorem is not applicable to the bunch of trajectories or even to two trajectories as it follows from the Lyapunov theory: If two points are situated in some small region of the phase space then they do not necessary come back to this region by moving along their trajectories.

Mixing and Weak Limit

The state $\rho_t = \rho_t(x)$ on the compact phase space Γ is called mixing if its weak limit at $t \to \infty$ is a constant,

$$\lim_{t\to\infty} \rho_t(x) = \text{const}.$$

More precisely, a dynamical system $(\Gamma, \varphi_t, d\mu)$ has the mixing property [401, 718] if

$$\lim_{t\to\infty} \langle f, U_t g\rangle = \int \bar{f}\, d\mu \cdot \int g\, d\mu \tag{10.63}$$

for every $f, g \in L^2(\Gamma)$. Here $U_t g(x) = g(\varphi_t(x))$. For the mixing systems the bunch of trajectories is spreading over the phase space, hence in the functional mechanics we have irreversibility.

The method of the weak limit which generalizes the Poincaré results and which can be applied to a wide class of dynamical systems is developed in [448, 450].

A connection with the irreversibility problem can be explained on the following example. Let us consider the function of two real variables

$$F(t, p) = e^{itp} f(p),$$

where $f(p)$ is an integrable function. It is clear that the function $F(t, p)$ is periodic in t if p is fixed, and it has no limit as $t \to \infty$. However, if we integrate the function $F(t, p)$ over p,

$$F(t) = \int e^{itp} f(p)\, dp,$$

then we get the function $F(t)$ which already has the limit (by the Riemann–Lebesgue lemma):

$$\lim_{t\to\infty} F(t) = 0.$$

10.9.8 Dynamics of a Particle in a Box

Dynamics of collisionless continuous medium in a box with reflecting walls is considered in [448, 450, 650]. This studied asymptotics of solutions of the Liouville equation. In functional approach to mechanics, we interpret the solution of the Liouville equation as describing the dynamics of a single particle. Here we consider this model in the classical and also in the quantum version for the special case of the Gaussian initial data.

Dynamics of a Classical Particle in a Box

Consider the motion of a free particle on the interval with the reflective ends. Using the method of reflections [781], the solution of the Liouville equation (10.13)

$$\frac{\partial \rho}{\partial t} = -\frac{p}{m}\frac{\partial \rho}{\partial q}$$

on the interval $0 \le q \le 1$ with the reflective ends we write as

$$\rho(q, p, t) = \sum_{n=-\infty}^{\infty} \left[\rho_0\left(q - \frac{p}{m}t + 2n, p\right) + \rho_0\left(-q + \frac{p}{m}t + 2n, -p\right)\right], \qquad (10.64)$$

where it is assumed that the function ρ_0 has the Gaussian form (10.10).

One can show that for the distribution for coordinates

$$\rho_c(q,t) = \int \rho(q,p,t)\,dp \tag{10.65}$$

one gets the uniform limiting distribution (pointwise limit):

$$\lim_{t\to\infty} \rho_c(q,t) = 1.$$

For the distribution of the absolute values of momenta $(p > 0)$

$$\rho_a(p,t) = \rho_m(p,t) + \rho_m(-p,t),$$

where

$$\rho_m(p,t) = \int_0^1 \rho(q,p,t)\,dq,$$

as $t \to \infty$ we get the distribution of the Maxwell type (but not the Maxwell distribution):

$$\lim_{t\to\infty} \rho_a(p,t) = \frac{1}{\sqrt{\pi} b}\big[e^{-\frac{(p-p_0)^2}{b^2}} + e^{-\frac{(p+p_0)^2}{b^2}}\big].$$

Dynamics of a Quantum Particle in a Box

The Schrödinger equation for a free quantum particle on the interval $0 \le x \le 1$ with reflecting ends has the form

$$i\hbar\frac{\partial\phi}{\partial t} = -\frac{\hbar^2}{2m}\frac{\partial^2\phi}{\partial x^2} \tag{10.66}$$

with the boundary conditions

$$\phi(0,t) = 0, \qquad \phi(1,t) = 0, \quad t \in \mathbb{R}.$$

The solution of this boundary problem can be written as follows:

$$\phi(x,t) = \sum_{n=-\infty}^{\infty} \big[\psi(x+2n,t) - \psi(-x+2n,t)\big],$$

where $\psi(x,t)$ is some solution of the Schrödinger equation. If we choose the function $\psi(x,t)$ in the form corresponding to the distribution (10.24) then one can show that in the semiclassical limit for the probability density $|\phi(x,t)|^2$ the leading term is the classical distribution $\rho_c(x,t)$ (10.65).

In this section, the functional formulation of classical mechanics is considered which is based not on the notion of an individual trajectory of the particle, but on the distribution function on the phase space.

The fundamental equation of the microscopic dynamics in the proposed functional approach is not the Newton equation but the Liouville equation for the distribution function of a single particle. Solutions of the Liouville equation have the property of delocalization which accounts for irreversibility. It is shown that the Newton equation in this approach appears as an approximate equation describing the dynamics of the average values of the positions and momenta for not too long time intervals. Corrections to the Newton equation are computed.

There are interesting problems related with applications of the functional formulation of mechanics to statistical mechanics, to singularities in cosmology and black holes, and to a new interpretation of quantum mechanics.

10.10 New Interpretation of Bell's Inequality—Chameleon Dynamics

We discuss a recent attempt to violate the Bell's type inequalities in the framework of classical (probability) theory [22, 23, 26]. This attempt is called *Chameleon dynamics* proposed by Accardi and has been studied by Accardi [35], Imafuku, Regoli [32].

The Chameleon dynamics is a classical theory which violates the Bell's inequalities. There is no contradiction with the Bell's theorem because the probability measure in the Chameleon dynamics is not local in the sense of Bell (it depends on the parameters of the apparatus) but it is local in the sense of Accardi, since this dependence has a special form. The key point in Chameleon dynamics is to produce a non-standard statistics corresponding to taking realistic measurement processes into account. A measurement process is nothing but an interaction process between a system and an apparatus followed from von Neumann's proposal. A state of a system to be measured is inevitably effected by a means how to measure it, so that the back effect to the state under the measurement should be carefully taken into account. This situation has occurred even in classical systems. In a real experimental setting, a certain interaction between a system and an apparatus will influence the states of the system afterwards. The notion that the dynamics depends on the observable to be measured is called *Adaptive Dynamics* as already discussed in this.

10.10.1 Dynamical Systems: Passive and Adaptive

Definition 10.22 A classical deterministic (passive) dynamical system is a quadruple:

$$\{\Omega, \mathcal{O}, P, T\}$$

where

- Ω is the state space (more precisely, $\Omega = (\Omega, \mathcal{F})$ is a measurable space);

- $\mathcal{O}$ is a set of observables (measurable maps from Ω to $\mathbb{R}$);
- P is the preparation of the experiment (a probability measure on Ω—initial distribution);
- $T : \Omega \to \Omega$ is a discrete time dynamics (measurable map).

We do not require that the statistics is invariant under the dynamics (i.e., that $P \circ T^{-1} = P$). Thus, if $\langle \cdot \rangle$ denotes the expectation value then for any observable $A \in \mathcal{O}$ one has:

$$\langle A \rangle = \int_\Omega A(Tx)\, dP(x) = \int_\Omega A(y)\, dP \circ T^{-1}(y). \tag{10.67}$$

Definition 10.23 A classical adaptive dynamical system is a quadruple:

$$\left\{\Omega, \mathcal{O}, \{P_A\}_{A\in\mathcal{O}}, \{T_A\}_{A\in\mathcal{O}}\right\}$$

where

- Ω (the state space) and $\mathcal{O}$ (the observables) are as in Definition 10.22.
- for each $A \in \mathcal{O}$:

  (i) P_A is a probability measure (the preparation of an experiment to measure A).
  (ii) $T_A : \Omega \to \Omega$ is an $\mathcal{F}$-measurable map (the adaptive dynamics given that A is measured).

For adaptive dynamical systems, formula (10.67) becomes

$$\langle A \rangle = \int_\Omega A(T_A x)\, dP_A(x) = \int_\Omega A(y)\, dP_A \circ T_A^{-1}(y). \tag{10.68}$$

Consider a classical dynamical system composed of two particles $(1, 2)$ with state spaces S_1, S_2, respectively, and two apparata A_1, A_2 with state spaces M_1, M_2, respectively.

The state space of the composite system will then be

$$\Omega = S_1 \times S_2 \times M_1 \times M_2. \tag{10.69}$$

According to von Neumann's measurement theory, a measurement of the system $(1, 2)$ by means of the apparatus (A_1, A_2) is described, in discrete time, by a reversible dynamical system

$$T : \Omega \to \Omega,$$

and the preparation of the experiment is described by a probability measure P on Ω

$$P \in P(\Omega).$$

Definition 10.24 A dynamics T on the state space Ω, given by (10.69), is called local if it has the form $T = T_1 \otimes T_2$ where

$$T_1 : S_1 \times M_1 \to S_1 \times M_1,$$

$$T_2 : S_2 \times M_2 \to S_2 \times M_2$$

are dynamics. This means

$$T_1 \otimes T_2(\sigma_1, \lambda_1, \sigma_2, \lambda_2) = T_1(\sigma_1, \lambda_1)T_2(\sigma_2, \lambda_2),$$
$$\sigma_1 \in S_1, \lambda_1 \in M_1, \sigma_2 \in S_2, \lambda_2 \in M_2.$$

Definition 10.25 A probability measure P on the space

$$\Omega = S_1 \times S_2 \times M_1 \times M_2$$

is called adaptive local in the sense of Accardi and causal if it has the form

$$P(d\sigma_1, d\sigma_2, d\lambda_1, d\lambda_2) = P_S(d\sigma_1, d\sigma_2)P_1(\sigma_1, d\lambda_1)P_2(\sigma_2, d\lambda_2) \tag{10.70}$$

where

- $P_S(d\sigma_1, d\sigma_2)$ is a probability measure on $S_1 \times S_2$
- $P_1(\sigma_1, d\lambda_1)$ is a positive measure on M_1 for all $\sigma_1 \in S_1$
- $P_2(\sigma_2, d\lambda_2)$ is a positive measure on M_2 for all $\sigma_2 \in S_2$.

We will deal with classical systems composed of two particles $(1, 2)$ and two apparata (M_1, M_2), which measure binary observables, $S_a^{(1)}$, $S_b^{(2)}$, labeled by indices $a, b, \ldots$ and called "spin" to emphasize the analogy with the EPR type experiments.

The apparata can make local independent choices among these labels, and we use the notation (M_a, M_b) to mean that apparatus M_1 has made the choice a and apparatus M_2 the choice b.

A state of the global system is specified by a quadruple $(\sigma_1, \sigma_2, \lambda_1, \lambda_2)$ where (σ_1, σ_2) describe the particle degrees of freedom and (λ_1, λ_2) the apparatus' ones.

Is the emergence of non-Kolmogorovian statistics a specific feature of the quantum world or is it a deeper, more general phenomenon of universal applicability? The analysis of [15] shows that the classical physics of adaptive systems can produce non-Kolmogorovian statistics. This principle was baptized "the chameleon effect". The general analysis of [15] was substantiated in a concrete model in [31], and the simulation of this model on independent classical computers provided an experimental proof of the chameleon local nature of the model [22, 23].

10.10.2 The EPRB–Chameleon Dynamical System

The EPR–Bell–chameleon dynamical system is an adaptive local (in the sense of Accardi), classical dynamical system reproducing the EPRB correlations. It was constructed in [31].

In this construction one considers four classical dynamical systems

$$(1, M_a, 2, M_b).$$

Here 1 and 2 are called particles, M_a and M_b are called measurement apparata.

In the following 1, 2 will be labels for particles, and a, b labels for apparata. We suppose that $a, b \in [0, 2\pi]$,

– The state space of both composite systems $(1, M_a)$ and $(2, M_b)$ is

$$[0, 2\pi] \times \mathbb{R}.$$

– Therefore, the state space of the whole system $(1, M_a, 2, M_b)$ is

$$[0, 2\pi]^2 \times \mathbb{R}^2.$$

Each of the composite systems $(1, M_a)$ and $(2, M_b)$ has a local adaptive dynamics.

For any $a \in [0, 2\pi]$, we define the ± 1-valued maps (observables)

$$S_a^{(1)}, S_a^{(2)} : [0, 2\pi] \times \mathbb{R} \to \{\pm 1\}$$

so that $\forall \sigma \in [0, 2\pi]$ and $\forall \mu \in \mathbb{R}$,

$$S_a^{(1)}(\sigma, \mu) = S_a^{(1)}(\sigma) = \operatorname{sgn}\bigl(\cos(\sigma - a)\bigr),$$
$$S_b^{(2)}(\sigma, \mu) = S_b^{(2)}(\sigma) = \operatorname{sgn}\bigl(\cos(\sigma - b)\bigr) = -S_b^{(1)}(\sigma, \mu).$$

$S_x^{(1)}$ is an observable of particle 1; $S_x^{(2)}$ an observable of particle 2.

Finally, we have to give an initial distribution P of the whole system $(1, M_a, 2, M_b)$.

We define P to be the probability measure on $[0, 2\pi]^2 \times \mathbb{R}^2$:

$$\begin{aligned} &p_S(\sigma_1, \sigma_2) p_{1,a}(\sigma_1, \lambda_1) p_{2,b}(\sigma_2, \lambda_2)\, d\sigma_1 \sigma_2\, d\lambda_1\, d\lambda_2 \\ &\quad := \frac{1}{2\pi} \delta(\sigma_1 - \sigma_2) \delta\left(\frac{4\lambda_1}{\sqrt{2\pi}\,|\cos(\sigma_1 - a)|}\right) \delta\left(\frac{\lambda_2}{\sqrt{2\pi}}\right) d\sigma_1\, d\sigma_2\, d\lambda_1\, d\lambda_2 \end{aligned} \tag{10.71}$$

where $\sigma_1, \sigma_2 \in [0, 2\pi]$, $\lambda_1, \lambda_2 \in \mathbb{R}$. Notice the adaptive local structure of the initial probability measure: $p_S(d\sigma_1, d\sigma_2)$ is the initial preparation, $p_{1,a}(\sigma_1, d\lambda_1)$ and $p_{2,b}(\sigma_2, d\lambda_2)$ are the initial preparations of the local apparata. They are typical "response-type" preparations and must be interpreted in the adaptive sense.

Remark 10.26 The measure (10.71) is adaptive local in the sense of Accardi but it is not local in the sense of Bell since it depends on the parameters a and b.

Theorem 10.27 *The above described dynamical system reproduces the EPR–Bell correlations, i.e.,*

$$\begin{aligned} &\int S_a^{(1)}(\sigma_1) S_b^{(2)}(\sigma_2) \cdot p_S(\sigma_1, \sigma_2)\, d\sigma_1\, d\sigma_2 \\ &\quad = p_{1,a}(\sigma_1, \lambda_1) p_{2,b}(\sigma_2, \lambda_2)\, d\lambda_1\, d\lambda_2 \\ &\quad = -\cos(a - b). \end{aligned} \tag{10.72}$$

Proof Under our assumptions, the left-hand side of (10.72) becomes

$$\begin{aligned} I &= \int\int d\sigma_1\, d\sigma_2 S_a^1(\sigma_1) S_b^2(\sigma_2) \frac{1}{2\pi}\, \delta(\sigma_1 - \sigma_2) \\ &= \int d\lambda_1 \delta\left(\frac{4\lambda_1}{\sqrt{2\pi}\,|\cos(\sigma_1 - a)|}\right) \int d\lambda_2 \delta\left(\frac{\lambda_2}{\sqrt{2\pi}}\right). \end{aligned}$$

Using the identity

$$\int \delta(a\lambda)\, d\lambda = \frac{1}{|a|},$$

one obtains

$$\begin{aligned} I &= \int S_a^1(\sigma_1) S_b^2(\sigma_2) \frac{1}{2\pi}\, \delta(\sigma_1 - \sigma_2)\, d\sigma_1\, d\sigma_2 \cdot \frac{\sqrt{2\pi}}{4} \bigl|\cos(\sigma_1 - a)\bigr| \cdot \sqrt{2\pi} \\ &= \frac{1}{4} \int S_a^1(\sigma_1) S^2(\sigma_1) \bigl|\cos(\sigma_1 - a)\bigr|\, d\sigma_1 \\ &= -\frac{1}{4} \int \operatorname{sgn}\cos(\sigma_1 - u) \bigl|\cos(\sigma_1 - a)\bigr| \cdot \operatorname{sgn}\cos(\sigma_1 - b)\, d\sigma_1 \\ &= -\frac{1}{4} \int \cos(\sigma_1 - a) \operatorname{sgn}\cos(\sigma_1 - b)\, d\sigma_1 = -\cos(b - a). \end{aligned}$$

The theorem is proved. □

There is an experiment realizing this model with three independent computers. It is a classical physics experiment because the personal computers used in it are surely macroscopic classical systems.

10.10.3 Probabilistic Error Model

In [536], the following probabilistic dynamics has been considered. Suppose that there exist two systems, each of which is attached to Alice and Bob. Now we assume that the theory is described by a classical probability theory with a hidden variable $\lambda \in \Delta$, where Δ is assumed to be a discrete set for simplicity. Based on the Chameleon dynamics, it is considered that (i) not every particle clicks the detectors of Alice and Bob, so failed trials occur probabilistically; and (ii) which particle can click the detectors depends on the measurement they perform. Suppose that Alice measures an observable A and Bob measures B, and the hidden variable is denoted by λ. A particle clicks Alice's detector with probability $P_A(\lambda)$, and a particle clicks Bob's with probability $P_B(\lambda)$. In addition, we put $q_A(\mu|\lambda)$ as the probability of obtaining an outcome μ by measuring A when the hidden variable is in λ. $q_B(\nu|\lambda)$ is defined in the same manner. Since Alice and Bob can get rid of the failed trials, the expectation value for the observables should be calculated only by the data in which both experiments, Alice's and Bob's, are successful. Thus, if we denote the probability of the hidden variable as $P(\lambda)$, the expectation value of AB is obtained as

$$\langle AB\rangle = \frac{\sum_{\mu}\sum_{\nu}\sum_{\lambda}\mu\nu q_A(\mu|\lambda)q_B(\nu|\lambda)P_A(\lambda)P_B(\lambda)P(\lambda)}{\sum_{\lambda'}P_A(\lambda')P_B(\lambda')P(\lambda')}.$$

The expression should be compared with the ordinary expectation,

$$\langle AB\rangle = \sum_{\mu}\sum_{\nu}\sum_{\lambda}\mu\nu q_A(\mu|\lambda)q_B(\nu|\lambda)P(\lambda). \tag{10.73}$$

Such a model is called a *probabilistic error model.*

However, this model is a reinterpretation of the Chameleon model. The class of Chameleon dynamics and the class of probabilistic error models are equivalent.

10.10.4 Upper Bounds for CHSH Inequality

A variant of Bell's inequality, the CHSH inequality gives us a bound for the correlation function related with four observables [168]: Alice has two observables A and A', Bob has B and B', all of which take values in $\{-1, 1\}$. It has been shown that for any hidden variable theories (without the Chameleon effect), the expectation value of an observable $C \equiv A(B + B') + A'(B - B')$ is bounded as

$$\left|\langle C\rangle\right| \le 2,$$

but the bound for quantum theory is

$$\left|\langle C\rangle\right| \le 2\sqrt{2}$$

whose maximal value $2\sqrt{2}$ is actually attained by the spin $\frac{1}{2}$ EPR pair. Accardi et al. [32] have proved that the Chameleon dynamics can attain the value $2\sqrt{2}$. The further upper bound can be obtained [524].

Theorem 10.28 *For the Chameleon dynamics, the strict inequality*

$$\left|\langle C\rangle\right| \le 4$$

holds.

Proof Because of the inequality $|\langle C\rangle| \le |\langle AB\rangle| + |\langle AB'\rangle| + |\langle A'B\rangle| + |\langle A'B'\rangle|$, the inequality $|\langle C\rangle| \le 4$ follows. The remaining task is to show that there exists a Chameleon dynamics which attains the maximal value $\langle C\rangle = 4$. Consider the following system. A "hidden variable" takes values in $\{0, 1\}$ with equal probability, i.e., $P(0) = 1/2$ and $P(1) = 1/2$. When Alice measures observable A, the

Chameleon dynamics leads to $P_A(0) = 1$ and $P_A(1) = 0$, that is, the measuring apparatus of A does not accept the hidden variable $\lambda = 1$. In addition, we assume that observable A always takes the value 1 irrespective of the value of λ. Namely, $q_A(1|0) = 1, q_A(-1|0) = 0, q_A(1|1) = 1$, and $q_A(-1|1) = 0$. In terms of the Chameleon dynamics, the apparatus measuring A accepts only the case $\lambda = 0$ and the value of "spin" is always equal to 1. For an observable A' which always takes the value 1, the Chameleon dynamics derives $P_{A'}(0) = 0$ and $P_{A'}(1) = 1$. That is, the apparatus does not accept $\lambda = 0$. For B and B', we have $P_B(0) = P_B(1) = P_{B'}(0) = P_{B'}(1) = 1$. In addition, we assume that the observable B always takes the value 1. The observable B' takes both of -1 and 1 depending upon the value of λ. We put $q_{B'}(1|0) = 1, q_{B'}(-1|0) = 0$, and $q_{B'}(1|1) = 0, q_{B'}(-1|1) = 1$. To explain this situation in terms of the Chameleon dynamics, one must dilate the system. In the above setting, a simple calculation shows $\langle AB\rangle = \langle AB'\rangle = \langle A'B\rangle = -\langle A'B'\rangle = 1$. Thus we obtain the upper bound $\langle C\rangle = 4$. □

10.11 Notes

Information dynamics was introduced by Ohya in [570] as a synthesis of the state change and complex of dynamical systems in order to study chaotic behavior of the systems [359]. There exist several complexities [561]. The complexities in ID provide us a new definition of the quantum dynamical entropy [14, 441, 536, 556]. Connes, Narnhofer and Thirring entropy of a subalgebra was given in [176]. The relative entropy was extensively studied by Araki [63, 64] and Uhlmann [759]. Fuzzy entropy was discussed by several authors like Zadeh [829], DeLuca and Termini [192] and Ebanks [217], and the relative fuzzy entropy was considered by Narituka and Ohya [580]. The complexity of sequences was studied by Kolmogorov [436] and Chaitin [156]. The concept of ID has been applied to several areas such as communication, chaos, genetics, and finance [14, 368, 441, 525, 526, 557, 558, 560, 585]. In particular, a new measure describing chaos was introduced by Ohya [592], which could be applied to several dynamical systems, classical [363] and quantum [364, 365, 367]. In [442], it is discussed how to reach chaos dynamics by starting from general differential dynamics in both classical and quantum systems. That is, it is demonstrated how we can get to chaos dynamics by considering observations introduced in this section. The last section is based on [442] which is an attempt to give a mathematical theory explaining various types of chaos found in several phenomena with finite localized systems, and it gives the origin of adaptivity. Another origin of adaptive dynamics is a work of Accardi, dealing with Chameleon dynamics [15]. Non-Newtonian functional mechanics was introduced by Volovich [800, 801], see also [753, 802].

Chapter 11
Mathematical Models of Quantum Computer

A quantum computer is usually modeled mathematically as a Quantum Turing Machine (QTM) or a uniform family of quantum circuits, which is equivalent to a quantum Turing machine. QTM is a quantum version of the classical Turing machine described in Chap. 2. QTM was introduced by Deutsch and has been extensively studied by Bernstein and Vasirani. The basic properties of the quantum Turing machine and quantum circuits will be described in this chapter. In the last section of the present chapter, we introduce a generalized QTM.

11.1 Quantum Turing Machine

Let us remind that a classical Turing machine M_{cl} is given by a triplet $M_{\mathrm{cl}} = (Q, \Sigma, \delta)$ where Q is a set of states, Σ is a set of finite alphabets with a blank symbol #, Σ^* is the set of all sequences of the elements in Σ and δ is the transition function (program). The classical Turing machine has a processor, a read/write tape head, and an infinite tape indexed by integers. At every moment of time, the state of the Turing machine can be described as a configuration $c = (q, A, i)$ where $q \in Q$ is the state of the processor, an integer i is the location of the tape head, and A is the word on the tape formed by non-blank symbols. We identify A with a tape function $A : \mathbb{Z} \to \Sigma$. The program δ transforms one configuration into another. Computation on the Turing machine is described as a sequence of configurations.

A quantum Turing machine, similarly to the classical Turing machine, has an infinite tape of squares and a read/write tape head that moves along the tape. The work of the quantum Turing machine is the subject to quantum rules.

A quantum Turing machine, as any quantum system, should be described by an appropriate Hilbert space. Computation on the quantum Turing machine is a sequence of unitary transformations. The Hilbert space consists of complex functions defined on the space of classical configurations.

M. Ohya, I. Volovich, *Mathematical Foundations of Quantum Information and Computation and Its Applications to Nano- and Bio-systems*,
Theoretical and Mathematical Physics,
DOI 10.1007/978-94-007-0171-7_11,

Definition 11.1 The usual quantum Turing machine M_{q} is defined by a quadruplet $M_{\mathrm{q}} = (Q, \Sigma, \mathcal{H}, U_\delta)$, where $\mathcal{H}$ is a Hilbert space described below in (11.1) and U_δ is a unitary operator on the space $\mathcal{H}$ of the special form described below in (11.2).

Let $\mathcal{C} = Q \times \Sigma \times \mathbb{Z}$ be the set of all classical configurations of the Turing machine M_{cl}, where $\mathbb{Z}$ is the set of all integers. It is a countable set and one has

$$\mathcal{H} = \left\{ \varphi; \varphi : \mathcal{C} \rightarrow \mathbb{C}, \sum_{c \in \mathcal{C}} |\varphi(c)|^2 < \infty \right\}. \tag{11.1}$$

Since the configuration $c \in \mathcal{C}$ can be written as $c = (q, A, i)$, one can say that the set of functions $\{|q, A, i\rangle\}$ is a basis in the Hilbert space $\mathcal{H}$. Here $q \in Q$, $i \in \mathbb{Z}$, and A is a tape function. We will call this basis the *computational basis*.

Let $\mathcal{H}_Q, \mathcal{H}_\Sigma$ and $\mathcal{H}_Z$ be three Hilbert spaces spanned by canonical bases $\{|q\rangle; q \in Q\}$, $\{|A\rangle; A \in \Sigma^*\}$, and $\{|i\rangle; i \in \mathbb{Z}\}$, respectively, $\mathcal{H}$ is decomposed as

$$\mathcal{H} = \mathcal{H}_Q \otimes \mathcal{H}_\Sigma \otimes \mathcal{H}_Z.$$

By using the computational basis, we now state the conditions for the unitary operator U_δ. We denote the set $\Gamma \equiv \{1, 0, -1\}$. One requires that there is a function $\delta : Q \times \Sigma \times Q \times \Sigma \times \Gamma \rightarrow \tilde{\mathbb{C}}$ which takes values in the field of computable numbers $\tilde{\mathbb{C}}$ and such that the following relation is satisfied:

$$U_\delta |q, A, i\rangle = \sum_{p,b,d \in \Gamma} \delta\big(q, A(i), p, b, d\big) |p, B, i + d\rangle, \tag{11.2}$$

where the sum runs over the states $p \in Q$, the symbols $b \in \Sigma$ and the elements $d \in \Gamma$, and B is a tape function satisfying

$$B(j) = \begin{cases} b, & \text{if } j = i, \\ A(j), & \text{if } j \neq i. \end{cases}$$

Since U_δ is unitary, δ satisfies for any $q \in Q, a \in \Sigma, q'(\neq q) \in Q, a'(\neq a) \in \Sigma$,

$$\begin{aligned} &\sum_{p,b,d} |\delta(q, a, p, b, d)|^2 = 1, \\ &\sum_{p,b,d,d'} \delta(q', a', p, b, d')^* \delta(q, a, p, b, d) = 0. \end{aligned} \tag{11.3}$$

Actually, this is a finite sum. The restriction to the computable number field $\tilde{\mathbb{C}}$ instead of all complex numbers $\mathbb{C}$ is required since otherwise we cannot construct or design a quantum Turing machine.

Let I_Σ and I_Z be identity operators on $\mathcal{H}_\Sigma$ and $\mathcal{H}_Z$, respectively, and let

$$E_Q(q) = |q\rangle\langle q| \otimes I_\Sigma \otimes I_Z$$

be a projection on $\mathcal{H}$.

Definition 11.2 A QTM is called halting at time T if there exists $T < \infty$ such that

$$\begin{aligned} \left\| E_Q(q_F) U_\delta^s |c_0\rangle \right\| &= 0, \quad \forall s < T, \\ \left\| E_Q(q_F) U_\delta^T |c_0\rangle \right\| &= 1, \end{aligned}$$

where c_0 is an initial configuration.

Remark 11.3 Note that this halting is called "stationary" in [114]. All known effective computation schemes halt with probability 1 at some time t and halt with probability 0 before time t. For an arbitrarily constructed quantum Turing machine, however, the different branches of computation might have different numbers of computation steps, in general. In such a case, the halting process or the notion of halting itself may have problems. We will explain this problem in the end of this chapter.

Remark 11.4 For any $q, p \in Q, a, b \in \Sigma, d \in \Gamma$, let $\delta(q, a, p, b, d) \in \{0, 1\}$, then QTM is a reversal TM.

Definition 11.5 A QTM M is said to be in normal form if $\delta(q_F, \sigma, q_0, \sigma, 1) = 1$ for any $\sigma \in \Sigma$.

There are some useful facts from the theory of normal form QTM (NQTM) proved in [114].

Lemma 11.6 (Insertion Lemma) *If M_1 and M_2 are NQTMs with the same alphabets, disjoint state sets, and q is a state of M_1, then there exists a NQTM M which computes as M_1 except that each time when it enters state q, it computes M_2 instead.*

Here, let δ_1 and δ_2 be quantum transition functions of M_1 and M_2, respectively. Using Insertion Lemma, we can construct a NQTM M with a state set $Q = Q_1 \cup Q_2$. M computes M_2 instead of M_1 when the configuration of M denoted by $|\psi\rangle$ becomes

$$|\psi\rangle = \sum_{\sigma \in \Sigma, i \in Z} c_{\sigma,i} |q, \sigma, i\rangle + \sum_{p \in Q, \tau \in \Sigma, j \in Z} c_{p,\tau,j} |p, \tau, j\rangle$$

where $p \neq q$ and

$$\sum_{\sigma \in \Sigma, i \in Z} |c_{\sigma,i}|^2 + \sum_{p \in Q, \tau \in \Sigma, j \in Z} |c_{p,\tau,j}|^2 = 1.$$

Proof Let M_1 and M_2 be NQTMs with the same alphabet denoted by Σ, disjoint state sets denoted by Q_1 and Q_2 with initial and final states $q_{1,0}, q_{1,F} \in Q_1$ and $q_{2,0}, q_{2,F} \in Q_2$, and with q a state of M.

Then we can construct the desired M as follows. First, take $M=(Q_1\cup Q_2,\Sigma,\delta)$ where δ satisfies the following conditions

$$\delta(q_1,\sigma,q_1',\sigma',d)=\delta_1(q_1,\sigma,q_1',\sigma',d),$$
$$\delta(q_2,\sigma,q_2',\sigma',d)=\delta_2(q_2,\sigma,q_2',\sigma',d)$$

for all $\sigma,\sigma'\in\Sigma, d\in\Gamma, q_1,q_1'\neq q\in Q_1$ and $q_2,q_2'\neq q\in Q_2$, and make the initial state $q_{1,0}$ and the final state $q_{1,F}$. Then, swap the incoming transitions of q and $q_{2,0}$ and the outgoing transitions of q and $q_{2,F}$ to construct M. In fact, for all $q_1\in Q_1,\sigma,\sigma',\tau,\tau'\in\Sigma, d\in\Gamma$, we can define the transition function δ by

$$\delta(q_1,\sigma,q,\sigma',d)=\delta_1(q_1,\sigma,q_{2,0},\sigma',d),$$
$$\delta(q_{2,F},\tau,q_{2,0},\tau',d)=\delta_2(q_{2,F},\tau,q,\tau',d).$$

Since M_1 is NQTM, the final state of M leads back to its initial state no matter whether q is the initial state of M_1, the final state of M_1, or neither. □

Lemma 11.7 (Dovetailing Lemma) *If M_1 and M_2 are NQTMs with the same alphabets, then there exists a NQTM M which carries out the computation of M_1 followed by the computation of M_2.*

Let $M_1(x)$ and $M_2(x)$ be the output of M_1 and M_2 for the same input x, respectively. We can construct a NQTM M which outputs $M_1(x)$; $M_2(x)$ for any input x.

Proof Let M_1 and M_2 be NQTMs with the same alphabets and with initial states and final states $q_{1,0},q_{2,0},q_{1,f}$ and $q_{2,f}$, respectively. To construct M, we insert M_2 for the final state of M_1 using *Insertion Lemma*. Then, we show that M carries out the computation of M_1 followed by that of M_2.

Since M_1 and M_2 are two NQTMs, the only transitions into $q_{1,0}$ and $q_{2,0}$ are from $q_{1,F}$ and $q_{2,F}$, respectively. Therefore, since M_1 is NQTM and does not enter $q_{1,F}$, M will compute exactly as M_1 until M_1 halts. At that point M will instead reach a superposition with all configurations in state $q_{2,0}$. Then, since no transitions in M_2 have been changed except for those into or out of $q_{2,F}$, M will proceed exactly as if M_2 had been started in the superposition of outputs computed by M_1. □

11.1.1 Universal Quantum Turing Machine

Deutsch proposed a model of a universal quantum Turing machine which requires exponential time of t to simulate any other QTM with t steps. Bernstein and Vasirani showed the existence of an efficient universal QTM which simulates other QTM with time t in polynomial time by slightly modifying Deutsch's model.

In this section, we construct a novel model of a universal QTM which does not depend on time t in an input data. Our universal QTM M simulates all the steps of a target QTM M for any accuracy ε with a slowdown f (as defined later) which is a polynomial function of t and $1/\varepsilon$.

Here, we define the code of QTM. Considering this code, we can give the information of another QTM to QTM as input data.

Definition 11.8 Let $D = (d_{ij})$ be an $m \times n$ matrix, the code of D is defined to be the following list of finite sequences of numbers

$$\big((x_{11}, y_{11}), (x_{12}, y_{12}), \ldots, (x_{mn}, y_{mn})\big)$$

where $x_{ij} = \mathrm{Re}(a_{ij})$ and $y_{ij} = \mathrm{Im}(a_{ij})$.

Definition 11.9 Suppose a QTM $M = (Q, \Sigma, \mathcal{H}, U_\delta)$, the code of M denoted by $c(M)$ is given as the code of the unitary operator U_δ.

And then, we define the *simulation* in QTM.

Definition 11.10 Suppose that $M = (Q, \Sigma, \delta)$ and $M' = (Q', \Sigma', \delta')$ are quantum Turing machines with the unitary operators U_δ and $U_{\delta'}$, respectively. Let t be a positive integer and $\varepsilon > 0$, we say that a QTM M' with its initial configuration c'_0 simulates M and its initial configuration c_0 for t steps with accuracy ε and slowdown f, which is a polynomial function of t and $1/\varepsilon$, if the following conditions are satisfied: For all $q \in Q$, $T \in \Sigma^*$, $i \in \mathbb{Z}$,

$$\Big| \big|\langle q, T, i|U_\delta^t|c_0\rangle\big|^2 - \big|\langle q, T, i|U_{\delta'}^{t+f(t,\frac{1}{\varepsilon})}|c'_0\rangle\big|^2 \Big| < \varepsilon. \tag{11.4}$$

Bernstein and Vasirani proved [114] that there exists an NQTM $\mathcal{M}_{\mathrm{BV}}$ simulating any NQTM M with any accuracy ε for t steps with slowdown $f(t, \frac{1}{\varepsilon})$ which can be computed in polynomial steps of t and ε. The input data of $\mathcal{M}_{\mathrm{BV}}$ is a quadruplet $(x, \varepsilon, t, c(M))$ where x is an input of M, ε is the accuracy of the simulation, t is a simulation time, and $c(M)$ is a code of M. Note that it is necessary there to give a time t as an input of $\mathcal{M}_{\mathrm{BV}}$.

Theorem 11.11 *There is a normal form QTM $\mathcal{M}_{\mathrm{BV}}$ such that for any NQTM M, any $\varepsilon > 0$, and any t, $\mathcal{M}_{\mathrm{BV}}$ can simulate M with accuracy ε for t steps with slowdown polynomial in $t$and $1/\varepsilon$.*

Proof Suppose $M = (Q, \Sigma, \mathcal{H}, U_\delta)$ is a target NQTM. Using Unitary Transformation Theorem 11.13 showed below, we can construct a NQTM $M'' = (Q'', \Sigma'', \mathcal{H}'', U''_\delta)$ satisfying

$$\|U_\delta - U_{\delta''}\| < \varepsilon$$

with an input data $(A, \varepsilon, c(M))$. Then, we construct a NQTM $M' = (Q', \Sigma', \delta')$ whose algorithm is represented as follows.

1. Transfer the current state and tape alphabet $p;\sigma$ to empty work space near the start cell, leaving a marker in their place.
2. Apply U which is given as an input data to $p;\sigma$ to within ε, transforming $p;\sigma$ into a new state and tape alphabet $q;\tau$.
3. Transfer $q;\tau$ back to the marked cell and empty the work space.
4. Transfer the state to q and move the right or left depending on the data of U.

Using the Synchronization Theorem, we can construct NQTMs for steps 1, 3 and 4 whose time for computation is polynomial in t. Step 2 is executed by NQTM in time polynomial in card(Σ), card(Q) and ε constructed in Unitary Transformation Theorem. Dovetailing these four NQTMs gives us NQTM M''. This M'' simulates one step of the target NQTM M. Therefore, if we insert M'' for the special configuration in NQTM in Looping Lemma, and provide additional input t, the resulting NQTM $M' = (Q', \Sigma', \mathcal{H}', U'_\delta)$ will halt after time polynomial in t and $1/\varepsilon$ after simulating t steps of M with accuracy $t\varepsilon$.

Finally, we can construct the desired universal QTM $\mathcal{M}_{\mathrm{BV}}$ by dovetailing M' after a NQTM which carries out the necessary preparation. □

Now, we consider another model of a universal QTM whose input data is $(A, \varepsilon, c(M))$, that is, we do not need a simulated time t as an input. It suggests that we do not need to know when the given QTM halts. The following theorem is proved in [370].

Theorem 11.12 *There exists an NQTM $\mathcal{M}$ such that for any NQTM M, $\mathcal{M}$ simulates each step of M for an input data $(A, \varepsilon, c(M))$ where A is the input of M, ε is the accuracy of the simulation, and $c(M)$ is the code of M.*

Proof Similarly as $\mathcal{M}_{\mathrm{BV}}$, $\mathcal{M} = (Q, \Sigma, \delta)$ is constructed to have six tracks and moves as follows: The first track of $\mathcal{M}$ is used to represent the result of computation of M. The second track contains a counter of t for $\mathcal{M}_{\mathrm{BV}}$. The third track is used to record the input of M. The fourth and fifth tracks are used to record ε and $c(M)$, respectively. The sixth track is used as a working track. To be precise, for $(x, \varepsilon, c(M))$ as an input data, $\mathcal{M}$ carries out the following algorithm:

(i) $\mathcal{M}$ transfers x, ε and $c(M)$ to the fixed tracks.
(ii) $\mathcal{M}$ sets the counter $t = 1$ and stores the value of t on the second track.
(iii) $\mathcal{M}$ calculates $\frac{6\varepsilon}{\pi^2 t^2}$ and transfers it to the fourth track.
(iv) $\mathcal{M}$ carries out $\mathcal{M}_{\mathrm{BV}}$ with $(x, 6\varepsilon/\pi^2 t^2, t, c(M))$ and writes down the result of $\mathcal{M}_{\mathrm{BV}}$ on the first track. The calculation of $\mathcal{M}$ is carried out on the sixth track, and $\mathcal{M}$ empties the work space, finally.
(v) If the simulated result of M is in a final state, then $\mathcal{M}$ halts, otherwise $\mathcal{M}$ increases the counter by one and repeats (iii) and (iv).

Using the Synchronization theorem, we can construct NQTMs which act steps (i), (ii), and (iii) above, respectively. We can obtain a QTM $\mathcal{M}$ by dovetailing them. We denote the time required to compute the steps from (i) to (v) by $f'(t, \frac{\pi^2 t^2}{6\varepsilon})$, which is polynomial in both variables. Let c_M and $c_{\mathcal{M}}$ be the initial configurations of M and $\mathcal{M}$, respectively, we denote $c_M = |q_0\rangle \otimes |x\rangle \otimes |0\rangle$ and $c_{\mathcal{M}} = |q_0\rangle \otimes |\#, \#, x, \varepsilon, c(M), \#\rangle \otimes |0\rangle$. Since $\mathcal{M}_{\mathrm{BV}}$ simulates M for any ε, x and t, putting $F(t, \frac{1}{\varepsilon}) = \sum_{i=i}^{t} f'(i, \frac{\pi^2 i^2}{6\varepsilon})$, the simulation of t steps for M requires $t + F(t, \frac{1}{\varepsilon})$ steps. For any q, i and T, the following inequality is obtained

$$\left| \left| \langle q, T, i | U_\delta^t | c_M \rangle \right|^2 - \left| \langle q, T, i | U_{\delta'}^{t+F(t,\frac{1}{\varepsilon})} | c_{\mathcal{M}} \rangle \right|^2 \right| < \frac{6\varepsilon}{\pi^2 t^2},$$

where U_δ and $U_{\delta'}$ are the unitary operators corresponding to M and $\mathcal{M}$, respectively. □

Suppose that M halts and gives an outcome with probability p, $\mathcal{M}$ gives the same outcome with probability p' satisfying $|p - p'| \le \varepsilon$. According to a property of the Riemann zeta function, we have

$$|p' - p| \le \sum_{i=1}^{\infty} \frac{6\varepsilon}{\pi^2 i^2} \le \varepsilon.$$

By inserting the process to calculate the proper accuracy for $\mathcal{M}_{\mathrm{BV}}$, we can construct a universal QTM and avoid the halting problem.

11.2 Quantum Gates

A quantum computation is a sequence of unitary transformations. We cannot just assume that any unitary transformation may be efficiently implemented. It must be constructed (using some classical algorithm) from some finite basic set of transformations. We will prove that any unitary matrix can be approximated by means of the product of unitary matrices of a simple form which are called *quantum gates*. Such a representation is the quantum analogue of the representation of recursive functions in terms of primitive functions.

It is known that the set $\{e^{2\pi i n\theta} \mid n \in \mathbb{Z}\}$, where θ is a fixed irrational number, is dense on the unite circle. This can be interpreted as saying that the 1×1-matrix $e^{2\pi i\theta}$ is universal for the set of all unitary 1×1-matrices, i.e., any the complex number on the unit circle can be approximated by the product of $e^{2\pi i\theta}$. The quantity $e^{2\pi i\theta}$ is a universal gate on the unite circle.

A unitary $d \times d$ matrix U is of the simple form, if it has (after possible reordering) a block-diagonal form such that every block is a 2×2-matrix of rotations

$$U_2 = \begin{pmatrix} \cos\theta & -\sin\theta \\ \sin\theta & \cos\theta \end{pmatrix}$$

or it is a number $e^{i\theta}$ ($\equiv U_1$) with some θ. The following theorem demonstrates that any unitary matrix can be effectively approximated by a product of the simple form matrices.

Theorem 11.13 (Unitary Transformation Theorem) *There exists a classical algorithm (i.e., a classical Turing machine) that for a given unitary* 2×2 *matrix* U *and* $\varepsilon > 0$ *it computes in the time polynomial in* $\log \frac{1}{\varepsilon}$ *unitary matrices of the simple form* U_1, U_2 *such that*

$$\left\| U - U_1(\varepsilon) U_2(\varepsilon) \right\| < \varepsilon.$$

Proof If $d = 1$, put $U_1(\varepsilon) \equiv e^{i(\theta + \sqrt{\varepsilon}/2)}$, then $\|U_1 - U_1(\varepsilon)\| < \varepsilon$.

If $d = 2$ and $U_2 = \begin{pmatrix} \cos\theta & -\sin\theta \\ \sin\theta & \cos\theta \end{pmatrix}$, we would take

$$U_2(\varepsilon) = \begin{pmatrix} \cos(\theta + \sqrt{\varepsilon}/2) & -\sin(\theta + \sqrt{\varepsilon}/2) \\ \sin(\theta + \sqrt{\varepsilon}/2) & \cos(\theta + \sqrt{\varepsilon}/2) \end{pmatrix}.$$

In fact,

$$\begin{aligned} \left\| U_2 - U_2(\varepsilon) \right\| &= \left\| \begin{pmatrix} \cos\theta & -\sin\theta \\ \sin\theta & \cos\theta \end{pmatrix} - \begin{pmatrix} \cos(\theta + \sqrt{\varepsilon}/2) & -\sin(\theta + \sqrt{\varepsilon}/2) \\ \sin(\theta + \sqrt{\varepsilon}/2) & \cos(\theta + \sqrt{\varepsilon}/2) \end{pmatrix} \right\| \\ &= \left\| \begin{pmatrix} \sqrt{\varepsilon}/2 & -\sqrt{\varepsilon}/2 \\ \sqrt{\varepsilon}/2 & \sqrt{\varepsilon}/2 \end{pmatrix} \right\| < \varepsilon. \end{aligned}$$

For any unitary U in $\mathbb{C}^2$, the statement holds. □

Therefore, to perform an arbitrary quantum computation one has to build a quantum Turing machine performing the simple unitary transformations [81].

The following theorem shows of how a general d-dimensional unitary transformation can be reduced to the two-dimensional unitary transformations.

Theorem 11.14 *Let* U *be a* $d \times d$ *unitary matrix acting in the space* $\mathbb{C}^d$. *We fix a basis* (e_i) *in* $\mathbb{C}^d$. *Then* U *may be represented as a product of* $2d^2 - d$ *unitary matrices, each of which acts only within a two-dimensional subspace spanned by a pair of the basis vectors.*

Remark 11.15 Such a two dimensional unitary matrix acting in $\mathbb{C}^d$ is often called the two-level unitary matrix.

Proof Consider in $\mathbb{C}^d$ the basis

$$e_1 = \begin{pmatrix} 1 \\ 0 \\ \vdots \\ 0 \end{pmatrix}, \ldots, e_d = \begin{pmatrix} 0 \\ 0 \\ \vdots \\ 1 \end{pmatrix}.$$

Let $z = \sum_{i=1}^{d} z_i e_i$ be an arbitrary vector in $\mathbb{C}^d$. Let us first show that the vector z with components $z = (z_1, \ldots, z_d)$ in our basis may be transformed to the form

$(1, \ldots, 0)$ using a sequence of $(d-1)$ two-dimensional transformations. First, apply the 2×2 unitary transformation

$$V_2 = \begin{pmatrix} \frac{z_1^*}{\sqrt{|z_1|^2+|z_2|^2}} & \frac{z_2^*}{\sqrt{|z_1|^2+|z_2|^2}} & 0 & \cdots & \cdots & 0 \\ -\frac{z_2}{\sqrt{|z_1|^2+|z_2|^2}} & \frac{z_1}{\sqrt{|z_1|^2+|z_2|^2}} & 0 & & & 0 \\ 0 & 0 & 1 & 0 & \cdots & 0 \\ \vdots & & 0 & \ddots & & \vdots \\ \vdots & & \vdots & & \ddots & 0 \\ 0 & \cdots & 0 & \cdots & 0 & 1 \end{pmatrix}$$

in the $\{e_1, e_2\}$ space which reduces z_2 to zero. Indeed, we have

$$V_2 \begin{pmatrix} z_1 \\ z_2 \\ z_3 \\ \vdots \\ z_d \end{pmatrix} = \begin{pmatrix} 1 \\ 0 \\ z_3 \\ \vdots \\ z_d \end{pmatrix}.$$

Similarly use $V_3, \ldots, V_d$ in the span of $\{e_1, e_3\}, \ldots, \{e_1, e_d\}$ to respectively reduce $z_3, \ldots, z_d$ to zero, too. The transformation $V_2 \cdots V_d$ maps $(z_1, \ldots, z_d)$ to $(1, \ldots, 0)$, and $V_2^{-1} \cdots V_d^{-1}$ reverses the action. Now let $\psi_1, \ldots, \psi_d$ be the eigenvectors of U with eigenvalues $e^{i\varphi_1}, \ldots, e^{i\varphi_d}$ so that U can be written as

$$U = \sum_j e^{i\varphi_j} P_{\psi_j}$$

where P_{ψ_j} is the projection operator onto the subspace spanned by the vector ψ_j. The components of these vectors are written by the above z_i $(i = 1, \ldots, d)$. Using the above procedure, transform ψ_1 to e_1, then multiply by $e^{i\varphi_1}$ in the span $\{e_1\}$, then map e_1 back to ψ_1. This requires $(d-1) + 1 + (d-1) = 2d - 1$ two-dimensional transformations. Repeating this for each of the d eigenvectors leads to an expression of U as a product of $d(2d-1)$ two-dimensional transformations.

It is important that this theorem gives a representation for U which is polynomial in d (i.e., $2d^2 - d$) multiplications of the two-dimensional transformations. □

Freudenberg, Ohya and Watanabe generalized the Fredkin–Toffoli–Milburn gate on the Fock space in [271].

11.3 Quantum Circuits

Quantum circuits are quantum analogues of the classical circuits computing Boolean functions. A classical circuit can be represented as a directed acyclic graph. Similarly, a quantum circuit is a sequence of unitary matrices of the special form associated with a (hyper) graph. We will need a special computational basis in the vector space.

Computational basis in n-qubit space. The two-dimensional complex space $\mathbb{C}^2$ is called qubit. We define in a qubit the following computational basis

$$e_0 = \begin{pmatrix} 1 \\ 0 \end{pmatrix}, \qquad e_1 = \begin{pmatrix} 0 \\ 1 \end{pmatrix}.$$

The index $x = 0, 1$ in the basis (e_x) will be interpreted as a Boolean variable. We will use also the Dirac notations

$$e_x = |x\rangle.$$

The n-tuple tensor product of qubits $\mathbb{C}^2 \otimes \cdots \otimes \mathbb{C}^2 = (\mathbb{C}^2)^{\otimes n}$ is called the n-qubit space. It has a computational basis $\{e_{x_1} \otimes \cdots \otimes e_{x_n}\}$ where $x_i = 0, 1$. We will use also the notation

$$e_{x_1} \otimes \cdots \otimes e_{x_n} = |x_1, \ldots, x_n\rangle.$$

Definition 11.16 A quantum circuit QC is defined by the following set of data: $QC = \{\mathcal{H}, U, G, f\}$ where the Hilbert space $\mathcal{H}$ is the n-qubit space $\mathcal{H} = (\mathbb{C}^2)^{\otimes n}$, U is a unitary matrix in $\mathcal{H}$, $G = \{V_1, \ldots, V_r\}$ is a finite set of unitary matrices (quantum gates), and f is a classical Boolean function $f : B^k \to B^m$. Here $B = \{0, 1\}$ and one assumes $k \leq n$ and $m \leq n$. The matrix U should admit a representation as a product of unitary matrices generated by the quantum gates described below (11.5).

The dimension of unitary matrices V_i is normally less then the dimension 2^n of the Hilbert space $\mathcal{H}$, and usually one takes matrices V_i which act in the 2-qubit or in the 3-qubit spaces. We fix the computational basis $\{e_{x_1} \otimes \cdots \otimes e_{x_n}\}$ in $\mathcal{H}$ and define an extension of the matrix V_i to a matrix in the space $\mathcal{H}$. The extension is constructed in the following way. If V_i is an $l \times l$ matrix then we choose l vectors from the computational basis and denote them as $\alpha \equiv \{h_1, \ldots, h_l\}$. Now let us define a unitary transformation $V_i^{(\alpha)}$ in the Hilbert space $\mathcal{H}$ as follows. The action of $V_i^{(\alpha)}$ on the subspace of $\mathcal{H}$ spanned by vectors $\{h_1, \ldots, h_l\}$ is set to be equal to that of V_i, and the action of $V_i^{(\alpha)}$ on the orthogonal subspace to be equal to 0.

The matrix U should be represented in the following product form

$$U = V_{i_1}^{(\alpha_1)} \cdots V_{i_L}^{(\alpha_L)} \tag{11.5}$$

where the matrices V_s are quantum gates, and $V_s^{(\alpha_s)}$ is some extension of V_s to a matrix in the Hilbert space $\mathcal{H}$ described above.

We say that the quantum circuit QC computes the Boolean function $f : B^k \to B^m$ if the following bound is valid

$$\left|\langle \mathbf{0}, f(x_1, \ldots, x_k) \middle| U \middle| x_1, \ldots, x_k, \mathbf{0}\rangle\right|^2 \geq 1 - \varepsilon$$

for all $x_1, \dots, x_{k_1}$ and some fixed $0 \le \varepsilon < 1/2$. Here $|x_1, \dots, x_k, 0\rangle$ is the vector for the computational basis of the form $|x_1, \dots, x_k, 0, \dots, 0\rangle$ ($n - k$ zeros), and $\langle 0, f(x_1, \dots, x_k)|$ is the vector for the computational basis of the form $\langle 0, \dots, 0, f(x_1, \dots, x_k)|$ ($m - k$ zeros).

If there is a quantum circuit QC with the unitary operator U represented as a product of L unitary matrices in the form (11.5) then L is called the *computational time* of the quantum circuit. We are mainly interested in studying the dependence of L on the length k of the input.

11.4 Universal Quantum Gates

We know that one can use some set of gates (e.g., AND, NOT, OR) to compute an arbitrary classical Boolean function. One says that such a set of gates is universal for classical computation. A similar result is valid for quantum computation. We say that a set of gates is *universal for quantum computation* if any unitary operation may be approximated to arbitrary accuracy by a quantum circuit involving only those gates. We will show that any unitary operation can be approximated using C-NOT, Hadamard, and T gates described below.

Let us first recall these gates, although they are briefly discussed in Chap. 1. In terms of the computational basis, the action of C-NOT is given by a unitary matrix $U_{\text{C-NOT}}$:

$$U_{\text{C-NOT}}|x, y\rangle = \big|x, y + x \pmod 2\big\rangle$$

C-NOT is a quantum gate with two input qubits, while Hadamard and T gates are single qubit unitary operations H and T:

$$H = \frac{1}{\sqrt{2}}\begin{pmatrix} 1 & 1 \\ 1 & -1 \end{pmatrix}, \qquad T = \begin{pmatrix} 1 & 0 \\ 0 & e^{i\pi/4} \end{pmatrix}.$$

Let a unitary matrix U which acts on a d-dimensional Hilbert space be given. One can show that U can be decomposed into a product of two-level unitary matrices which act nontrivially only on two-or-fewer vector components. Then, one can show that single qubit and C-NOT gates can be used to implement an arbitrary two-level unitary operation. Finally, one proves that the Hadamard and T gates can be used to approximate any single qubit unitary operation to arbitrary accuracy. We will need the following

Proposition 11.17 *An arbitrary single qubit unitary operation U can be written in the form*

$$U = e^{i\alpha} R_{\mathbf{n}}(\theta)$$

for some real numbers α and θ, and a real three-dimensional unit vector $n = (n_x, n_y, n_z)$. Here

$$R_{\mathbf{n}}(\theta) = \exp(-\mathrm{i}\theta\mathbf{n}\cdot\sigma/2) = \cos\left(\frac{\theta}{2}\right)I - \mathrm{i}\sin\left(\frac{\theta}{2}\right)(\mathbf{n}\cdot\sigma) \tag{11.6}$$

and

$$\mathbf{n}\cdot\sigma = n_x\sigma_x + n_y\sigma_y + n_z\sigma_z,$$

where $\sigma_x, \sigma_y, \sigma_z$ are the Pauli matrices.

Proof To verify (11.6) one uses the relation $(n\cdot\sigma)^2 = I$. A single qubit state $a|0\rangle + b|1\rangle$ can be represented as a point (θ, ϕ) on the unit sphere (i.e., by the unit vector k on the Bloch sphere), where $a = \cos(\theta/2), b = e^{\mathrm{i}\phi}\sin(\theta/2)$. Then the effect of the action of the operator $R_{\mathbf{n}}(\theta)$ on the state $a|0\rangle + b|1\rangle$ represented by the unit vector k is a rotation of the vector k by the angle θ about the n axis of the Bloch sphere. □

Let us define an angle θ_0 by $\cos(\theta_0/2) = \cos^2(\pi/8)$. It can be shown that θ_0 is an irrational multiple of 2π. We denote $R = R_{\mathbf{n}_0}(\theta_0)$ where $n_0 = (\cos\frac{\pi}{8}, \sin\frac{\pi}{8}, \cos\frac{\pi}{8})$. One shows that R is, up to a global phase, equal to $THTH$. One can prove the following

Proposition 11.18 *Let U be any single qubit unitary operator. Then for any $\epsilon > 0$ there are non-negative integers m_1, m_2, m_3 such that $\|U - R^{m_1}HR^{m_2}HR^{m_3}\| < \epsilon$.*

These considerations show that three operators T, H and C-NOT (or R, H and C-NOT) are universal for quantum computations.

11.5 Problem on the Halting Scheme

11.5.1 Destruction of Computation Result

We review a halting scheme for a quantum Turing machine introduced by Deutsch and refined by Bernstein and Vasirani. It seems to us that the Deffinition 11.2 goes well since it is a minimal procedure just to know whether the computation halts or not. However, there still exists a problem if the different branches of quantum computation halt at different steps. Let us see it by a simple example. Suppose that at $t = 0$ the state is in

$$|q_0, A\rangle,$$

where $|A\rangle$ is a state of the tape head and tape cells. By unitary transformation, at $t = 1$ it evolves into a superposition of the following two branches:

$$\frac{1}{\sqrt{2}}\big(|q_F, B\rangle + |q_1, C\rangle\big).$$

Next we suppose that each branch evolves as follows:

$$\begin{aligned} |q_F, B\rangle &\mapsto \frac{1}{\sqrt{2}}\big(|q_F, D\rangle + |q_F, E\rangle\big), \\ |q_1, C\rangle &\mapsto \frac{1}{\sqrt{2}}\big(-|q_F, D\rangle + |q_F, E\rangle\big). \end{aligned} \tag{11.7}$$

Then, as a whole, (11.7) evolves into

$$|q_F, E\rangle$$

at $t = 2$ if no observation was performed at $t = 1$. However, if we observe the internal state at $t = 1$ whether it halts or not then at $t = 2$ the state becomes $|q_F, B\rangle$ with probability $1/2$ (computation halts) and $|q_1, C\rangle$ with probability $1/2$ and at $t = 2$ it evolves into $\frac{1}{\sqrt{2}}(-|q_F, D\rangle + |q_F, E\rangle)$. As a whole, the outcome is B with probability $1/2$, D with probability $1/4$ and E with probability $1/4$. Thus in the simple example we have seen that even the minimal observation to know the halting destroys the computation result.

11.5.2 QND (Quantum Non-demolition Monitoring)-Type Quantum Turing Machine

Once the internal state drops into the halting state $|q_F\rangle$, the quantum Turing machine will never change the internal data or the tape [623]. We here call such a quantum Turing Machine a QND-type quantum Turing machine. It is easily seen that the previous simple example violates this condition. It is proved that for the QND-type quantum Turing machines the minimal halting protocol does not affect the result of the computation. More precise explanation is the following. Let $P\{\text{output} = A_j|\text{ monitored}\}$ be the probability of finding the output A_j up to T steps by the halting protocol. Let $P\{\text{output} = A_j|\text{ not-monitored}\}$ be the probability of finding the output A_j by the single measurement after T steps. One can show

$$P\{\text{output} = A_j \mid \text{monitored}\} = P\{\text{output} = A_j \mid \text{not-monitored}\}.$$

That is, the probability distribution of the output is not affected by monitoring of the sign of halting.

11.5.3 Problem for Halting

The above QND-type quantum Turing machine scheme gives a possible solution to the problem but it still remains a problem. Even for the above restricted class of quantum Turing machines, the notion of halting is ambiguous due to its probabilistic character. We will show, in the following, the restriction is not realistic, unfortunately.

Definition 11.19 We call a pair of a quantum Turing machine Q and its input x *conventional halting* iff one of the following conditions is satisfied:

(i) There exists a $t_0 \in \{1, 2, \dots\}$ such that at step t_0, Q under x halts with probability one and for $s < t_0$ it halts with probability zero.
(ii) For all the steps, Q under x halts with probability zero.

Definition 11.20 A pair of Q and x is called *probabilistic halting* when it is not conventional halting.

The following theorem is proved in [526].

Theorem 11.21 *There does not exist a QTM to judge whether a given QTM is conventional halting or not.*

Proof Assume the existence of a classical TM M_0 that determines conventional halting or not for any input $\langle M, x\rangle$ where M is a TM (or QTM) and x is an input for M. Let M_a and M_b be two reversible TMs, construct a special QTM $M_1(M_a, M_b)$ which runs M_a and M_b at same time without interference by Branching Lemma.

Let S be a set of QTMs given as

$$S \equiv \big\{M(M_a, M_b) \mid M_a \text{ and } M_b \text{ are reversal TMs}\big\}.$$

Since S is a subset of whole set of QTMs, M_0 could determine whether or not M_1 with the input x is a probabilistic halting. Then we can determine that for any reversible TMs M_a and M_b their computational steps for an input x are same or not. And we obtain a TM M_1' which reads input $\langle M_a, M_b, x\rangle$ to compare their computational steps, whose output is *accept* if their steps are same and otherwise *reject*. Here we can construct the following TM M_2 with its input $\langle M_a, x\rangle$.

1. Read $\langle M_a, x\rangle$.
2. Construct a TM M_c which never halts under any inputs.
3. Run M_1' with $\langle M_a, M_c, x\rangle$.
4. Output the result of M_1'.

One can see the result of M_2 as Table 11.1.

According to the classical halting problem (see Chap. 2), the existence of M_0 is a contradiction. □

Here it is proved that for arbitrarily constructed quantum Turing machine one cannot say whether it is conventional halting or not. The result will suggest that considering quantum Turing machines with different computation steps for each branch is necessary and the notion of halting in a quantum Turing machine should be re-examined again.

Table 11.1 Outputs of M_2

Output of M_2	M_a with x
Accept	Halts
Reject	Does not halt

11.6 Generalized Quantum Turing Machine

In this section, we define the generalized quantum Turing machine (GQTM) by using a quantum channel (see below) instead of a unitary operator.

Definition 11.22 A generalized quantum Turing machine M_{gq} (GQTM) is defined by a quadruplet $M_{\mathrm{gq}} = (Q, \Sigma, \mathcal{H}, \Lambda_{\delta_1}^*)$, where $\mathcal{H} \equiv \mathcal{H}_Q \otimes \mathcal{H}_\Sigma \otimes \mathcal{H}_Z$ is a Hilbert space and $\Lambda_{\delta_1}^*$ is a quantum transition channel on the space of states on $\mathcal{H}$ of the special form described below.

Here, we define the transition function

$$\delta_1 : \mathcal{R} \times Q \times \Sigma \times Q \times \Sigma \times Q \times \Sigma \times \Gamma \times Q \times \Sigma \times \Gamma \to \mathbb{C}.$$

Let $\mathfrak{S}(\mathcal{H})$ be the set of all density operators in the Hilbert space $\mathcal{H}$, a quantum transition function is given by a quantum channel

$$\Lambda_{\delta_1}^* : \mathfrak{S}(\mathcal{H}) \to \mathfrak{S}(\mathcal{H}),$$

satisfying the following condition.

Definition 11.23 $\Lambda_{\delta_1}^*$ is called a quantum transition channel if there exists a transition function δ_1 such that for all quantum configurations $\rho = \sum_k \lambda_k |\psi_k\rangle\langle\psi_k|$, $|\psi_k\rangle = \sum_l \alpha_{k,l} |q_{k,l}, A_{k,l}, i_{k,l}\rangle$, $\sum_k \lambda_k = 1, \forall \lambda_k \geq 0$, $\sum_l |\alpha_{k,l}|^2 = 1, \forall \alpha_{k,l} \in \mathbb{C}$, it holds

$$\Lambda_{\delta_1}^*(\rho) \equiv \sum_{k,l,m,n,p,b,d,p',b',d'} \delta_1\bigl(\lambda_k, q_{k,l}, A_{k,l}(i_{k,l}), q_{m,n}, A_{m,n}(i_{m,n}), p, b, d, p', b', d'\bigr) |p, B, i_{k,l} + d\rangle\langle p', B', i_{m,n} + d'|,$$

$$B(j) = \begin{cases} b, & j = i_{k,l}, \\ A_{k,l}(j), & \text{otherwise}, \end{cases}$$

$$B'(j) = \begin{cases} b', & j = i_{m,n}, \\ A_{m,n}(j), & \text{otherwise}, \end{cases}$$

so that RHS of the first equation is a state.

A configuration ρ of GQTM M_{gq} is described by a density operator on $\mathcal{H}$. For instance, given a configuration $\rho \equiv \sum_k \lambda_k |\psi_k\rangle\langle\psi_k|$, where $\sum \lambda_k = 1, \lambda_k \geq 0$ and $\psi_k = |q_k\rangle \otimes |A_k\rangle \otimes |i_k\rangle$ $(q_k \in Q, A_k \in \Sigma^*, i_k \in Z)$ is a vector in a basis of $\mathcal{H}$. This configuration changes to a new configuration ρ' by a one step transition as $\rho' = \Lambda^*_{\delta_1}(\rho) = \sum_k \mu_k |\psi_k\rangle\langle\psi_k|$ with $\sum \mu_k = 1, \mu_k \geq 0$.

One requirement on the GQTM $M_{\mathrm{gq}} = (Q, \Sigma, \mathcal{H}, \Lambda^*)$ is the correspondence with a QTM. If the channel Λ^* in the GQTM is a unitary operator U then the GQTM $M_{\mathrm{gq}} = (Q, \Sigma, \mathcal{H}, \Lambda^* = U \cdot U^*)$ reduces to the QTM $M_{\mathrm{q}} = (Q, \Sigma, \mathcal{H}, U)$.

Definition 11.24 $M_{\mathrm{gq}} = (Q, \Sigma, \mathcal{H}, \Lambda^*_{\delta_1})$ is called a LQTM (Linear Quantum Turing Machine) if there exists a transition function

$$\delta_2 : Q \times \Sigma \times Q \times \Sigma \times Q \times \Sigma \times \Gamma \times Q \times \Sigma \times \Gamma \to \mathbb{C}$$

such that for all quantum configuration ρ_k, $\Lambda^*_{\delta_1}$ is written as

$$\Lambda^*_{\delta_1}(\rho_k) \equiv \sum_{k,l,m,n,p,b,d,p',b',d'} \delta_2\big(q_{k,l}, A_{k,l}(i_{k,l}), q_{m,n}, A_{m,n}(i_{m,n}), p, b, d, p', b', d'\big) \times |p, B, i_{k,l} + d\rangle\langle p', B', i_{m,n} + d'|$$

so that the RHS is a state.

For all quantum configuration $\sum_k \lambda_k \rho_k$, $\Lambda^*_{\delta_1}$ is affine, that is,

$$\Lambda^*_{\delta_1}\left(\sum_k \lambda_k \rho_k\right) = \sum_k \lambda_k \Lambda^*_{\delta_1}(\rho_k).$$

Definition 11.25 A GQTM $M_{\mathrm{gq}} = (Q, \Sigma, \mathcal{H}, \Lambda^*_{\delta_1})$ is called a unitary QTM (UQTM), if the quantum transition channel $\Lambda^*_{\delta_1}$ is a unitary channel implemented as $\Lambda^*_{\delta_1} = Ad_{U_{\delta_3}}$ where U_{δ_3}, for $|\psi\rangle = |q, A, i\rangle$, is given by

$$\begin{aligned} U_{\delta_3}|\psi\rangle &= U_{\delta_3}|q, A, i\rangle \\ &= \sum_{p,b,d} \delta_3\big(q, A(i), p, b, d\big)|p, B, i + d\rangle \end{aligned}$$

with the condition (11.3).

We show the differences of delta functions in GQTM, LQTM and UQTM in Table 11.2. Then one can easily see that a quantum Turing machine is classified by its transition function.

Remark 11.26 A classical Turing machine is represented as a LQTM with a transition channel that has diagonal part only. Moreover, for any $q, p \in Q, a, b \in \Sigma, d \in \{0, \pm 1\}$, define $\delta_3(q, a, p, b, d) = 0$ or 1. Then a UQTM is a reversal TM.

Table 11.2 Transition functions for classes

Classes	Transition function
GQTM	$\mathbb{R} \times Q \times \Sigma \times Q \times \Sigma \times Q \times \Sigma \times \Gamma \times Q \times \Sigma \times \Gamma \to \mathbb{C}$
LQTM	$Q \times \Sigma \times Q \times \Sigma \times Q \times \Sigma \times \Gamma \times Q \times \Sigma \times \Gamma \to \mathbb{C}$
UQTM	$Q \times \Sigma \times Q \times \Sigma \times \Gamma \to \mathbb{C}$

We will discuss this in Chap. 14 by constructing a GQTM which solves the SAT problem in polynomial time for the size of input data.

11.7 Notes

Deutsch formulated a precise model of a quantum-physical computer as a quantum Turing machine [197] and quantum circuits [198]. Bernstein and Vasirani [114] refined the halting scheme which was introduced by Deutsch [197], and showed the existence of an efficient universal quantum Turing machine in Deutsch's model. In [709], several issues regarding a quantum Turing machine, universal quantum Turing machine, and programmable quantum gate array are discussed. In [538], Myers drew attention to the halting problem between the different branches of quantum computation, Ozawa gave a possible solution of this problem [623]. Miyadera and Ohya proved that the restriction of a quantum Turing machine whose branches halt at a same time or none of them halt, introduced by Bernstein and Vasirani [114], is not realistic [526]. Shor proposed efficient quantum algorithms for the factoring problem and the discrete logarithm problem [713]. Grover showed that the search problem could be speeded up on a quantum computer [305]. Ohya and Volovich, using a previous work by Ohya and Masuda, proved that the NP-complete SAT problem can be solved in polynomial time by using quantum computation with a chaos amplifier [595, 597, 600, 602]. The generalized QTM and its application to the chaos SAT algorithm was discussed in [369, 371].

Chapter 12
Quantum Algorithms I

In this chapter, the discrete quantum Fourier transform, the Deutsch–Jozsa algorithm for balanced functions, and the Grover algorithm for database search are exposed.

12.1 Discrete Quantum Fourier Transform and Principle of Quantum Algorithm

12.1.1 Hadamard Gate

We define here an important quantum gate which is called the Hadamard gate. It is a unitary transformation on the qubit space $\mathbb{C}^2$. It is defined as follows.

Let $|0\rangle = \binom{1}{0}$ and $|1\rangle = \binom{0}{1}$ be a basis of $\mathbb{C}^2$. Then

$$H = \frac{1}{\sqrt{2}} \begin{pmatrix} 1 & 1 \\ 1 & -1 \end{pmatrix}$$

is called the Hadamard matrix which gives a transformation:

$$H|0\rangle = \frac{1}{\sqrt{2}}\big(|0\rangle + |1\rangle\big), \qquad H|1\rangle = \frac{1}{\sqrt{2}}\big(|0\rangle - |1\rangle\big).$$

One can easily check that this is a unitary operator, so that we call this H the *Hadamard gate*.

12.1.2 Discrete Quantum Fourier Transformation

Consider the Hilbert space $\mathbb{C}^2 \otimes \cdots \otimes \mathbb{C}^2 = (\mathbb{C}^2)^{\otimes n}$ of dimension 2^n. Let the basis of the Hilbert space $\mathcal{H} = (\mathbb{C}^2)^{\otimes n}$ be

$$e_0\big(=|\mathbf{0}\rangle\big) = |0\rangle \otimes \cdots \otimes |0\rangle \otimes |0\rangle,$$

M. Ohya, I. Volovich, *Mathematical Foundations of Quantum Information and Computation and Its Applications to Nano- and Bio-systems*,
Theoretical and Mathematical Physics,
DOI 10.1007/978-94-007-0171-7_12,

$$
\begin{aligned}
e_1 \big(= |\mathbf{1}\rangle\big) &= |0\rangle \otimes \cdots \otimes |0\rangle \otimes |1\rangle, \\
e_2 \big(= |2\rangle\big) &= |0\rangle \otimes \cdots \otimes |1\rangle \otimes |0\rangle, \\
&\vdots \\
e_{2^n-2} \big(= \big|2^n - 2\big\rangle\big) &= |1\rangle \otimes \cdots \otimes |1\rangle \otimes |0\rangle, \\
e_{2^n-1} \big(= \big|2^n - 1\big\rangle\big) &= |1\rangle \otimes \cdots \otimes |1\rangle \otimes |1\rangle.
\end{aligned}
$$

We extend the action of the Hadamard gate to the n-qubit space as

$$H_j = I \otimes \cdots \otimes H \otimes \cdots \otimes I, \quad j = 1, \ldots, n.$$

Any number $a \in [0, 2^n - 1]$ can be expressed as

$$a = \sum_{k=0}^{n-1} a_k 2^k, \quad a_k = 0 \text{ or } 1,$$

so that the associated vector is written as

$$|a\rangle (= e_a) = \bigotimes_{k=0}^{n-1} |a_k\rangle \equiv |a_{n-1}, \ldots, a_0\rangle.$$

And applying n-tuples of Hadamard matrix to the vector $|\mathbf{0}\rangle$, we get

$$\bigotimes^n H |\mathbf{0}\rangle = \bigotimes^n \frac{1}{\sqrt{2}} \big(|0\rangle + |1\rangle\big).$$

Put

$$W(a) \equiv \bigotimes_{j=0}^{n-1} \begin{pmatrix} 1 & 0 \\ 0 & \exp(\frac{2\pi \mathrm{i} a}{2^n} 2^j) \end{pmatrix}.$$

Applying $W(a) \otimes^n H$ to the vector $|\mathbf{0}\rangle$ we have

$$|a_F\rangle = W(a) \bigotimes^n H |\mathbf{0}\rangle = \frac{1}{\sqrt{2^n}} \sum_{b=0}^{2^n-1} \exp\left(\frac{2\pi \mathrm{i} ab}{2^n}\right) |b\rangle.$$

The above operations altogether yield a unitary operator $U_F(a) \equiv W(a) \otimes^n H$ and the vector

$$|a_F\rangle = U_F(a) |\mathbf{0}\rangle.$$

This vector transformation is called the *discrete quantum Fourier transformation* (DFT). For simplicity, this $|a_F\rangle$ can be expressed as

$$U_F |a\rangle = \frac{1}{\sqrt{2^n}} \sum_{b=0}^{2^n-1} \exp\left(\frac{2\pi \mathrm{i} ab}{2^n}\right) |b\rangle.$$

The discrete quantum Fourier transformation is multiplication by an $N \times N$ unitary matrix so that the (x, y)-matrix element is $e^{2\pi \mathrm{i} xy/N}$, where $N \equiv 2^n$. Naively, this multiplication requires $O(n^2)$ elementary operations. However, we will show that due to special properties of the discrete quantum Fourier transformation, it can be implemented asymptotically by means of only $O((\log N)^2)$ elementary operations.

It is important to notice that the action of the discrete quantum Fourier transform can be written in the factorized (unentangled) form:

$$U_F|a_{n-1}, \ldots, a_0\rangle = \frac{1}{\sqrt{2^n}}\big(|0\rangle + e^{\mathrm{i}\phi_a 2^{n-1}}|1\rangle\big) \otimes \big(|0\rangle + e^{\mathrm{i}\phi_a 2^{n-2}}|1\rangle\big) \otimes \cdots$$
$$\otimes\big(|0\rangle + e^{\mathrm{i}\phi_a}|1\rangle\big)$$

where $\phi_a = 2\pi a/2^n$.

We will prove that the discrete quantum Fourier transform can be written as a product of matrices generated by Hadamard gates and by the following 4×4 matrix B,

$$B|a_1, a_0\rangle = \begin{cases} e^{\mathrm{i}\pi/2}|a_1, a_0\rangle, & \text{if } a_1 = a_0 = 1, \\ |a_1, a_0\rangle, & \text{otherwise.} \end{cases}$$

We denote $B_{j,k}$, $j < k$ the following extension of the matrix B:

$$B_{j,k}|a_{n-1}, \ldots, a_k, \ldots, a_j, \ldots, a_0\rangle = e^{\mathrm{i}\theta_{k-j}}|a_{n-1}, \ldots, a_k, \ldots, a_j, \ldots, a_0\rangle$$

where

$$e^{\mathrm{i}\theta_{k-j}} = \begin{cases} (e^{\mathrm{i}\pi/2})^{(k-j)}, & \text{if } a_1 = a_0 = 1, \\ 1, & \text{otherwise.} \end{cases}$$

The computational complexity of the discrete Fourier transform is also described by the following theorem.

Theorem 12.1 *The discrete quantum Fourier transform in the space* $\mathbb{C}^{2^n}$ *can be represented as a product of* $O(n^2)$ *operators* H_j *and* $B_{j,k}$.

Proof To explain the proof of the theorem, we define the *reversal* Fourier transform

$$U_F^{\mathrm{Rev}}|a_{n-1}, \ldots, a_0\rangle = \frac{1}{\sqrt{2^n}} \sum_{b=0}^{2^n-1} e^{2\pi \mathrm{i} ab/2^n}|b_0, b_1, \ldots, b_{n-1}\rangle.$$

In particular, one has

$$U_{F(n=2)}^{\mathrm{Rev}} = H_0 B_{0,1} H_1.$$

One can easily show an important formula

$$U_F^{\mathrm{Rev}} = H_0 B_{0,1} \cdots B_{0,n-1} H_1 \cdots B_{n-4,n-3} B_{n-4,n-2} \cdot B_{n-4,n-1} H_{n-3} B_{n-3,n-2} B_{n-3,n-1} H_{n-2} B_{n-2,n-1} H_{n-1}.$$

In this formula, one has n matrices H_j and $n(n-1)/2$ matrices $B_{j,k}$. Now since $U_F = U_F^{\mathrm{Rev}} T$, where T is the transposition operator, the theorem follows. □

Therefore, there is a quantum algorithm for implementation of the discrete quantum Fourier transform which is polynomial as a function of the input size.

12.1.3 Principle of Quantum Algorithm

The quantum algorithm is essentially made of the following three steps.

1. Take an input state $|\phi\rangle$ in the total Hilbert space $\mathcal{H}$ attached to a quantum Turing machine, $|\phi\rangle = \sum_k c_k |\phi_k\rangle \otimes |\psi\rangle$, where $\{|\phi_k\rangle\}$ is a basis of the Hilbert space for the input and $|\psi\rangle$ is a vector state for the register.
2. Apply a unitary operator U made of some gates designed by a program and get a resulting vector, $U|\phi\rangle (\equiv |\phi'\rangle) = \sum_k c'_k |\phi'_k\rangle \otimes |\psi'\rangle$.
3. After a long time (i.e., when the Turing machine halts), observe the state vector $|\phi'\rangle$ (or $|\psi'\rangle$).

This computation process can be written in the terminology of channels as discussed in Chap. 7. Several examples for quantum computations will be discussed in the subsequent sections and chapters.

12.2 Deutsch–Jozsa Algorithm

Let N be a positive integer and Z_{2N} be the set of all nonnegative integers less than $2N$. Then, for any function defined on Z_{2N} with values in $Z_2 \equiv \{0, 1\}$,

- f is said to satisfy *Condition A* if f is not a constant function, and
- f is said to satisfy *Condition B* if the cardinality of the set $\{l; 0 \le l \le 2N-1, f(l) = 0\}$ is not equal to the cardinality of the set $\{l; 0 \le l \le 2N-1, f(l) = 1\}$.

Problem 12.2 (Deutsch–Jozsa) Determine whether one of the above Conditions A or B is satisfied for any given function f.

For example, let f, g and h be the functions defined as

$$f(l) = 0, \qquad g(l) = \frac{1}{2} + \frac{(-1)^l}{2}, \qquad h(l) = 1_{\{0,\dots,2N-2\}}(l),$$

where $l \in Z_{2N}$. Then, f does not satisfy Condition A but satisfies Condition B, g does not satisfy Condition B but satisfies Condition A, and h satisfies both Condition A and Condition B. It should be remarked that any function defined on Z_{2N} with values in Z_2 satisfies at least one of these conditions.

Here let f be a function prepared as an *Oracle*, which means that we do not know an algorithm to compute f.

Remark 12.3 Note that there is also another formulation of the problem when one has to determine whether a function f is constant or balanced.

If a classical computer is used to obtain the solution to the Deutsch–Jozsa problem, then the required running time is (at worst) $N+1$, so we say *the complexity in a classical algorithm of the Deutsch–Jozsa problem is* $N+1$.

In 1992, Deutsch and Jozsa gave a quantum algorithm solving the problem more effectively.

Let U_f and W be the two unitary operators on $\mathcal{H}\otimes\mathbb{C}^2$, where $\mathcal{H}$ is a Hilbert space of dimension $2N$, which are respectively defined as

$$U_f|i\rangle\otimes|\xi\rangle\equiv|i\rangle\otimes\big|\xi\oplus f(i)\big\rangle \quad\text{and}\quad W|i\rangle\otimes|\xi\rangle\equiv(-1)^{\xi}|i\rangle\otimes|\xi\rangle,$$

where $\xi\oplus f(i)\equiv\xi+f(i)\ (\mathrm{mod}\,2)$ and $|i\rangle\in\mathcal{H}$ $(i=0,\dots,2N-1)$, $|\xi\rangle\in\mathbb{C}^2$ $(\xi=0,1)$. The above U_f and W can be represented as

$$U_f=\frac{1}{\sqrt{2N}}\sum_{i=0}^{2N-1}|i\rangle\langle i|\otimes\begin{pmatrix}0&1\\1&0\end{pmatrix}^{f(i)}$$

and

$$W=\frac{1}{\sqrt{2N}}\sum_{i=0}^{2N-1}|i\rangle\langle i|\otimes\begin{pmatrix}1&0\\0&-1\end{pmatrix}^{f(i)}.$$

The Deutsch–Jozsa quantum algorithm is characterized by a unitary operator $U_{\mathrm{DJ}}(f)\equiv U_fWU_f$ as follows: Let $|x_{\mathrm{in}}\rangle$ be the input vector defined as

$$|x_{\mathrm{in}}\rangle=\sum_{i=0}^{2N-1}\frac{1}{\sqrt{2N}}|i\rangle\otimes|0\rangle,$$

so the output vector $|x_{\mathrm{out}}\rangle$ is obtained as

$$\begin{aligned}|x_{\mathrm{out}}\rangle=U_{\mathrm{DJ}}(f)|x_{\mathrm{in}}\rangle&=U_fWU_f\sum_{i=0}^{2N-1}\frac{1}{\sqrt{2N}}|i\rangle\otimes|0\rangle\\&=U_fW\sum_{i=0}^{2N-1}\frac{1}{\sqrt{2N}}|i\rangle\otimes\big|f(i)\big\rangle\\&=U_f\sum_{i=0}^{2N-1}\frac{1}{\sqrt{2N}}|i\rangle\otimes(-1)^{f(i)}\big|f(i)\big\rangle\end{aligned}$$

$$= \sum_{i=0}^{2N-1} \frac{1}{\sqrt{2N}} |i\rangle \otimes (-1)^{f(i)} |0\rangle.$$

Let us compute the inner product $\langle x_{\text{in}} | x_{\text{out}} \rangle$. Then we obtain

$$\begin{aligned} \langle x_{\text{in}} | x_{\text{out}} \rangle &= \langle x_{\text{in}} | U_{\text{DJ}}(f) | x_{\text{in}} \rangle \\ &= \left[\sum_{i=0}^{2N-1} \frac{1}{\sqrt{2N}} \langle i | \otimes \langle 0 | \right] \left[\sum_{i=0}^{2N-1} \frac{1}{\sqrt{2N}} |i\rangle \otimes (-1)^{f(i)} |0\rangle \right] \\ &= \frac{1}{2N} \sum_{i=0}^{2N-1} (-1)^{f(i)}. \end{aligned}$$

Let $(Z_2)^{Z_{2N}}$ be the set of all functions defined on Z_{2N} with values in Z_2. Then the above equalities enable us to classify $(Z_2)^{Z_{2N}}$ into three categories:

1. The subset consisting of all functions that do not satisfy Condition A but satisfy Condition B: $\{f;\, |\langle x_{\text{in}} | x_{\text{out}} \rangle|^2 = 1\}$.
2. The subset consisting of all functions that do not satisfy Condition B but satisfy Condition A: $\{f;\, |\langle x_{\text{in}} | x_{\text{out}} \rangle|^2 = 0\}$.
3. The subset consisting of all functions that satisfy both Condition A and Condition B: $\{f; 0 < |\langle x_{\text{in}} | x_{\text{out}} \rangle|^2 < 1\}$.

Here we define the computational complexity of D–J quantum algorithm by the total number of unitary gates. So then, the computational complexity of Deutsch–Jozsa quantum algorithm is 3 because $U_{\text{DJ}} = U_f W U_f$ contains three unitary gates. On the other hand, the computational complexity of the classical algorithm is $N+1$. Therefore, the computational complexity of the quantum algorithm is the logarithm of the classical one.

12.3 Grover's Search Algorithm

In this section, we discuss the quantum algorithm for database search developed by Grover. This searching problem is the problem of finding one data-file among N data-files. Generally, let A be a certain subset of $\{0, 1, \ldots, N-1\}$, and one considers a problem of finding any element of A.

If we follow the classical computational procedure, then it is clear that it takes $O(N)$ steps to find an element of A. In a quantum algorithm, Grover proved that it is sufficient to take $O(\sqrt{N})$ steps. In particular, when A consists of only one element, one takes about $\frac{\pi}{4}\sqrt{N}$ steps.

Throughout this section, we assume that

$$N = 2^n$$

for a certain positive integer n, without loss of generality. That is, we suppose there exist N data-files, and we seek for one file in A. Let f_A be the function defined on $\{0, 1, \dots, N-1\}$ with values in $\{0, 1\}$ satisfying the following condition:

$$f_A(x) = \begin{cases} 1, & x \in A, \\ 0, & x \notin A. \end{cases}$$

This function is called the *Oracle* corresponding to A. The "*Oracle*" means that we do not know where such an A exists and take it as granted.

Let us take two vectors $|x\rangle \equiv |x_0, \dots, x_{n-1}\rangle$ and $|y\rangle \equiv |y_0, \dots, y_{n-1}\rangle$ in $\otimes^n \mathbb{C}^2 \equiv \mathbb{C}^N$, that is, x is any number in $\{0, 1, \dots, N-1\}$ and its binary representation is $x = x_0 + x_1 2 + \cdots + x_{n-1} 2^{n-1}$, $x_i = 0, 1$. Then the Walsh-Hadamard transformation W on $\mathbb{C}^N$ is defined as

$$W \equiv \begin{pmatrix} W(0,0) & \cdots & W(0, N-1) \\ \vdots & & \vdots \\ W(N-1, 0) & \cdots & W(N-1, N-1) \end{pmatrix},$$

where $W(x, y)$ is given as

$$W(x, y) = \frac{1}{\sqrt{N}} (-1)^{\sum_{k=0}^{n-1} x_k y_k},$$

and it can be easily proved that W is a unitary operator. We define a selective unitary operator with respect to f_A, denoted by R_{f_A}, as

$$R_{f_A} = I - 2 \sum_{z \in A} |z\rangle\langle z|,$$

where I is the identity operator on $\mathbb{C}^{2^n}$. This operator flips the direction of the vectors in A. One more operator we need is R_0 denoting another selective unitary operator defined by

$$R_0 = -I + 2|0\rangle\langle 0|. \tag{12.1}$$

The component in the xth row and the yth column of R_0 is represented as

$$R_0(x, y) = (-1)^{1-\delta_{x0}} \delta_{xy}.$$

We denote

$$U_{f_A} = W R_0 W R_{f_A},$$

and call U_{f_A} the Grover's unitary operator.

Let us discuss the structure and the work of the operator U_{f_A}. Let $|\varphi_0\rangle$ be the vector of the initial database (N-files) defined as

$$|\varphi_0\rangle = \frac{1}{\sqrt{N}} \sum_{x=0}^{N-1} |x\rangle.$$

Proposition 12.4 *One has*

$$WR_0W = -I + 2|\varphi_0\rangle\langle\varphi_0|.$$

Proof It is sufficient to show that the component in the xth row and the yth column of WR_0W is equal to the corresponding component of $-I + 2|\varphi_0\rangle\langle\varphi_0|$. We have

$$\begin{aligned} -I + 2|\varphi_0\rangle\langle\varphi_0| &= -I + 2\left(\frac{1}{\sqrt{N}}\sum_{x=0}^{N-1}|x\rangle\right)\left(\frac{1}{\sqrt{N}}\sum_{y=0}^{N-1}\langle y|\right) \\ &= -I + \frac{2}{N}\sum_{x=0}^{N-1}\sum_{y=0}^{N-1}|x\rangle\langle y|, \end{aligned}$$

which implies

$$\big(-I + 2|\varphi_0\rangle\langle\varphi_0|\big)(x, y) = -\delta_{xy} + \frac{2}{N}.$$

Moreover, we have

$$\begin{aligned} (WR_0W)(x, y) &= \sum_{u=0}^{N-1}\sum_{v=0}^{N-1} W(x, u)R_0(u, v)W(v, y) \\ &= \sum_{u=0}^{N-1}\sum_{v=0}^{N-1}\left[\frac{1}{\sqrt{N}}(-1)^{\sum_{k=0}^{n-1} x_k u_k}\right]\left[(-1)^{1-\delta_{u0}}\delta_{uv}\right] \\ &\quad\times\left[\frac{1}{\sqrt{N}}(-1)^{\sum_{k=0}^{n-1} v_k y_k}\right] \\ &= \frac{1}{N}\sum_{v=0}^{N-1}\left[\sum_{u=0}^{N-1}(-1)^{\sum_{k=0}^{n-1} x_k u_k}(-1)^{1-\delta_{u0}}\delta_{uv}\right] \\ &\quad\times\left[(-1)^{\sum_{k=0}^{n-1} v_k y_k}\right] \end{aligned}$$

and

$$\begin{aligned} \sum_{u=0}^{N-1}(-1)^{\sum_{k=0}^{n-1} x_k u_k}(-1)^{1-\delta_{u0}}\delta_{uv} &= \delta_{0v} + \sum_{u=1}^{N-1}(-1)^{\sum_{k=0}^{n-1} x_k u_k}(-1)\delta_{uv} \\ &= 2\delta_{0v} - \sum_{u=0}^{N-1}(-1)^{\sum_{k=0}^{n-1} x_k u_k}\delta_{uv} \\ &= 2\delta_{0v} - \prod_{k=0}^{n-1}\left[\sum_{u_k=0}^{1}(-1)^{x_k u_k}\delta_{u_k v_k}\right], \end{aligned}$$

which imply

$$\begin{aligned}
&(WR_0W)(x,y)\\
&= \frac{1}{N}\sum_{v=0}^{N-1}\left[2\delta_{0v} - \prod_{k=0}^{n-1}\left(\sum_{u_k=0}^{1}(-1)^{x_k u_k}\delta_{u_k v_k}\right)\right]\left[(-1)^{\sum_{k=0}^{n-1} v_k y_k}\right]\\
&= \frac{2}{N} - \frac{1}{N}\sum_{v=0}^{N-1}\left[\prod_{k=0}^{n-1}\left(\sum_{u_k=0}^{1}(-1)^{x_k u_k}\delta_{u_k v_k}\right)\right]\left[\prod_{k=0}^{n-1}(-1)^{v_k y_k}\right]\\
&= \frac{2}{N} - \frac{1}{N}\sum_{v_0=0}^{1}\cdots\sum_{v_{n-1}=0}^{1}\left[\prod_{k=0}^{n-1}\left(\sum_{u_k=0}^{1}(-1)^{x_k u_k + v_k y_k}\delta_{u_k v_k}\right)\right]\\
&= \frac{2}{N} - \frac{1}{N}\left[\prod_{k=0}^{n-1}\left(\sum_{u_k=0}^{1}(-1)^{(x_k+y_k)u_k}\right)\right]\\
&= \frac{2}{N} - \frac{1}{N}\prod_{k=0}^{n-1}\left(1+(-1)^{x_k+y_k}\right) = \frac{2}{N} - \frac{1}{N}\prod_{k=0}^{n-1}(2\delta_{x_k y_k})\\
&= \frac{2}{N} - \frac{1}{N}(2^n\delta_{xy}) = \frac{2}{N} - \delta_{xy}.
\end{aligned}$$

Thus we conclude the proof. □

Any vector $|\varphi\rangle$ in $\mathbb{C}^N$ can be written as

$$|\varphi\rangle = \sum_{x=0}^{N-1} w_x|x\rangle,$$

where $\{w_x; 0 \le x \le N-1\}$ are real numbers. Let $\overline{w}$ be the mean value which is defined as

$$\overline{w} = \frac{1}{N}\sum_{x=0}^{N-1} w_x.$$

Then WR_0W is represented in terms of $\{w_x; 0 \le x \le N-1\}$ and $\overline{w}$ as follows.

Proposition 12.5 *One has*

$$WR_0W|\varphi\rangle = \sum_{x=0}^{N-1}(2\overline{w} - w_x)|x\rangle.$$

Proof Since

$$|\varphi_0\rangle = \frac{1}{\sqrt{N}} \sum_{x=0}^{N-1} |x\rangle,$$

and due to Proposition 12.4, we have

$$\begin{aligned}
WR_0W|\varphi\rangle &= \left(-I + 2|\varphi_0\rangle\langle\varphi_0|\right)\left(\sum_{x=0}^{N-1} w_x|x\rangle\right) \\
&= -\sum_{x=0}^{N-1} w_x|x\rangle + 2\sum_{x=0}^{N-1} w_x|\varphi_0\rangle\langle\varphi_0|x\rangle \\
&= -\sum_{x=0}^{N-1} w_x|x\rangle + 2\sum_{x=0}^{N-1} \frac{w_x}{\sqrt{N}}|\varphi_0\rangle \\
&= -\sum_{x=0}^{N-1} w_x|x\rangle + 2\left[\sum_{x=0}^{N-1} \frac{w_x}{N}\right]\left[\sqrt{N}\sum_{y=0}^{N-1} \frac{1}{\sqrt{N}}|y\rangle\right] \\
&= -\sum_{x=0}^{N-1} w_x|x\rangle + 2\overline{w}\sum_{y=0}^{N-1} |y\rangle,
\end{aligned}$$

which concludes the proof. □

In the same way as above, we can prove the following corollary.

Corollary 12.6 *One has*

$$R_{f_A}|\varphi\rangle = \sum_{x\notin A} w_x|x\rangle - \sum_{x\in A} w_x|x\rangle.$$

The Grover's unitary operator $U_{f_A} \equiv WR_0WR_{f_A}$ changes the state $|\varphi\rangle$ as follows:

Proposition 12.7 *One has*

$$U_{f_A}|\varphi\rangle = \sum_{x\notin A}(2\widehat{w} - w_x)|x\rangle + \sum_{z\in A}(2\widehat{w} + w_z)|z\rangle,$$

where $\widehat{w}$ is defined as

$$\widehat{w} = \frac{1}{N}\left(\sum_{x\notin A} w_x - \sum_{z\in A} w_z\right).$$

Proof The previous corollary assures that the following equalities hold:

$$U_{f_A}|\varphi\rangle = (WR_0W)R_{f_A}|\varphi\rangle = (WR_0W)\left[\sum_{x\notin A} w_x|x\rangle - \sum_{z\in A} w_z|z\rangle\right]$$
$$= \sum_{x\notin A}(2\widehat{w} - w_x)|x\rangle + \sum_{z\in A}\bigl(2\widehat{w} - (-w_z)\bigr)|z\rangle.$$

Therefore, we have the conclusion. □

Let us study U_{f_A} more. Assume that a and b are any two real numbers satisfying $|A|a^2 + (N - |A|)b^2 = 1$, where $|A|$ is a cardinal number of A.

Lemma 12.8 *For a and b above, one has*

$$U_{f_A}\left(a\sum_{z\in A}|z\rangle + b\sum_{x\notin A}|x\rangle\right)$$
$$= \left(\frac{N-2|A|}{N}a + \frac{2(N-|A|)}{N}b\right)\sum_{z\in A}|z\rangle + \left(-\frac{2|A|}{N}a + \frac{N-2|A|}{N}b\right)\sum_{x\notin A}|x\rangle.$$

Proof Let $\overline{\overline{w}}$ be the number defined as

$$\overline{\overline{w}} = \frac{1}{N}\bigl[(N - |A|)b - |A|a\bigr].$$

Proposition 12.7 assures two equalities

$$2\overline{\overline{w}} - b = -\frac{2|A|}{N}a + \frac{N-2|A|}{N}b$$

and

$$2\overline{\overline{w}} + a = \frac{N-2|A|}{N}a + \frac{2(N-|A|)}{N}b,$$

which imply the conclusion. □

When we apply the Grover's operator U_{f_A} k times to the initial vector (state) $|\varphi_0\rangle$ equally distributed on all files, the resulting vector can be written as

$$U_{f_A}^k|\varphi_0\rangle = a_k\sum_{z\,\in A}|z\rangle + b_k\sum_{z\,\notin A}|x\rangle,$$

where a_k and b_k are some real numbers. For $k = 0$,

$$U_{f_A}^{k=0}|\varphi_0\rangle = |\varphi_0\rangle = \frac{1}{\sqrt{N}}\sum_{x=0}^{N-1}|x\rangle$$

so that

$$a_0 = b_0 = \frac{1}{\sqrt{N}}.$$

Next task is to find these a_k and b_k for all k.

Proposition 12.9 *The above a_k and b_k satisfy the relations*

$$a_k = \frac{N - 2|A|}{N} a_{k-1} + \frac{2(N - |A|)}{N} b_{k-1}, \qquad b_k = -\frac{2|A|}{N} a_{k-1} + \frac{N - 2|A|}{N} b_{k-2}.$$

These equations can be solved as

$$a_k = \frac{1}{\sqrt{|A|}} \sin(2k + 1)\theta_A, \qquad b_k = \frac{1}{\sqrt{N - |A|}} \cos(2k + 1)\theta_A,$$

where θ_A is the smallest positive number satisfying

$$\sin \theta_A = \sqrt{\frac{|A|}{N}}.$$

Proof The relations satisfied by a_k and b_k are an immediate consequence of the above lemma. So we have

$$\begin{pmatrix} a_k \\ b_k \end{pmatrix} = \begin{pmatrix} \frac{N-2|A|}{N} & \frac{2(N-|A|)}{N} \\ -\frac{2|A|}{N} & \frac{N-2|A|}{N} \end{pmatrix} \begin{pmatrix} a_{k-1} \\ b_{k-1} \end{pmatrix}.$$

Put c_k as

$$c_k = \frac{\sqrt{N - |A|}}{\sqrt{|A|}} b_k.$$

Then the relation between (a_k, c_k) and (a_{k-1}, c_{k-1}) is

$$\begin{aligned} \begin{pmatrix} a_k \\ c_k \end{pmatrix} &= \begin{pmatrix} \frac{N-2|A|}{N} & \frac{2\sqrt{|A|}\sqrt{N-|A|}}{N} \\ -\frac{2\sqrt{|A|}\sqrt{N-|A|}}{N} & \frac{N-2|A|}{N} \end{pmatrix} \begin{pmatrix} a_{k-1} \\ c_{k-1} \end{pmatrix} \\ &= \begin{pmatrix} \cos 2\theta_A & \sin 2\theta_A \\ -\sin 2\theta_A & \cos 2\theta_A \end{pmatrix} \begin{pmatrix} a_{k-1} \\ c_{k-1} \end{pmatrix}, \end{aligned}$$

where we take θ_A as

$$\sin 2\theta_A = \frac{2\sqrt{|A|}\sqrt{N - |A|}}{N}, \qquad \cos 2\theta_A = \frac{N - 2|A|}{N}.$$

Thus

$$\begin{pmatrix} a_k \\ c_k \end{pmatrix} = \begin{pmatrix} \cos 2\theta_A & \sin 2\theta_A \\ -\sin 2\theta_A & \cos 2\theta_A \end{pmatrix}^k \begin{pmatrix} a_0 \\ c_0 \end{pmatrix}$$

$$= \begin{pmatrix} \cos 2k\theta_A & \sin 2k\theta_A \\ -\sin 2k\theta_A & \cos 2k\theta_A \end{pmatrix} \begin{pmatrix} \frac{1}{\sqrt{N}} \\ \frac{1}{\sqrt{N}} \frac{\sqrt{N-|A|}}{\sqrt{|A|}} \end{pmatrix}.$$

From the above equality the result follows easily. □

According to this proposition, under the operator $U_{f_A}^k$, the initial vector $|\varphi_0\rangle$ is transformed into

$$U_{f_A}^k|\varphi_0\rangle = \frac{1}{\sqrt{|A|}} \sin(2k+1)\theta_A \sum_{z\in A} |z\rangle + \frac{1}{\sqrt{N-|A|}} \cos(2k+1)\theta_A \sum_{x\notin A} |x\rangle.$$

Let $P_{f_A}(k)$ be the probability of succeeding in the database search after applying $U_{f_A}^k$ to $|\varphi_0\rangle$, that is,

$$P_{f_A}(k) = \left| \sum_{z\in A} \frac{1}{\sqrt{|A|}} \langle z|U_{f_A}^k|\varphi_0\rangle \right|^2.$$

It holds that

$$P_{f_A}(k) = \sum_{z\in A} \left(\frac{1}{\sqrt{|A|}} \sin(2k+1)\theta_A \right)^2 = \sin^2(2k+1)\theta_A.$$

Since it is assumed that the cardinal number of A is much smaller than N, we can prove the following theorem established in [304].

12.3.1 Complexity of Grover's Algorithm

Theorem 12.10 *Let m_{θ_A} be*

$$m_{\theta_A} = \left[\frac{\pi}{4\theta_A} \right],$$

where $[\cdot]$ is the Gauss symbol. Then

$$P_{f_A}(m_{\theta_A}) \geq 1 - \frac{|A|}{N} \quad \textit{and} \quad m_{\theta_A} \simeq \frac{\pi}{4}\sqrt{\frac{N}{|A|}}.$$

Proof Since $m_{\theta_A} = [\frac{\pi}{4\theta_A}]$, we have

$$\left| (2m_{\theta_A}+1)\theta_A - \frac{\pi}{2} \right| \leq \theta_A.$$

The equality $\sin\theta_A = \sqrt{\frac{|A|}{N}}$ and $|A| \ll N$ imply that θ_A is sufficiently small, so that

$$-\sin\theta_A = \cos\left(\frac{\pi}{2} + \theta_A\right) \le \cos\bigl((2m_{\theta_A} + 1)\theta_A\bigr) \le \cos\left(\frac{\pi}{2} - \theta_A\right) = \sin\theta_A.$$

From these inequalities, it follows that

$$\cos^2\bigl((2m_{\theta_A} + 1)\theta_A\bigr) \le \sin^2\theta_A = \frac{|A|}{N}.$$

Therefore, we have

$$P_{f_A}(m_{\theta_A}) \ge 1 - \frac{|A|}{N}.$$

Moreover, m_{θ_A} can be estimated from above as

$$m_{\theta_A} \le \frac{\pi}{4\theta_A} \le \frac{\pi}{4\sin\theta_A} = \frac{\pi}{4}\sqrt{\frac{N}{|A|}}.$$

□

We summarize the characteristic points of Grover's algorithm for database search in the following two sentences:

1. If the cardinal number of A is not sufficiently smaller than N, the probability of succeeding in database search might decrease.
2. m_{θ_A} can be regarded as the number of the optimal steps to search the database, and it is $\frac{\pi}{4}\sqrt{\frac{N}{|A|}}$. So we proved that the upper bound of complexity of Grover's algorithm is $O(\sqrt{N})$.

In the rest of this section, we will show that the lower bound of the complexity for the Grover algorithm is $O(\sqrt{N})$. Let $|\varphi_0\rangle$ be the initial state vector as before, i.e.,

$$|\varphi_0\rangle = \frac{1}{\sqrt{N}} \sum_{j=0}^{N-1} |x\rangle.$$

Let f be an *Oracle* and let U_f be the unitary operator constructed from f as in the previous section. In particular, U_0 denotes the unitary operator corresponding to the *Oracle* at $|\varphi_0\rangle$, and the image of $|\varphi_0\rangle$ under the operator U_f^k (resp., U_0^k) denotes $|\varphi_k^f\rangle$ (resp., $|\varphi_k^0\rangle$). Note that $|\varphi_0^f\rangle = |\varphi_0^0\rangle = |\varphi_0\rangle$. Let $\{c_x^f(k) \in \mathbb{C}; k = 0, \dots, N-1\}$ and $\{c_x^0(k) \in \mathbb{C}; k = 0, \dots, N-1\}$ be the coefficient sequences obtained by

$$|\varphi_k^f\rangle = \sum_{x=0}^{N-1} c_x^f(k)|x\rangle, \qquad |\varphi_k^0\rangle = \sum_{x=0}^{N-1} c_x^0(k)|x\rangle.$$

We proved in the previous section that the computational complexity T is bounded by $\frac{\pi\sqrt{N}}{4}$. We assume that $|A|=1$ in the sequel. Take $k=T$ and put

$$P^f \equiv \sum_{x\in J^f} \left|c_x^f(T)\right|^2, \qquad P\equiv \frac{1}{N}\sum_{f\in F} P^f,$$

where $J^f = \{x; 0\le x\le N-1,\ f(x)=1\}$ and F is the set of all functions from $\{0,1,\dots,N-1\}$ to $\{0,1\}$ such that it takes the value 1 at only one point in $\{0,1,\dots,N-1\}$ and the value 0 at other $N-1$ points in $\{0,1,\dots,N-1\}$.

Remark 12.11 The cardinality of the set F is N.

The above P is the average probability of succeeding in database search after T steps, that is,

$$\frac{1}{N}\sum_{f\in F} P^f = P.$$

In the rest of this section, we will prove the lower bound $T\ge O(\sqrt{N})$.

Now we first prove the following proposition.

Theorem 12.12 *The following inequality holds*

$$2N-2\sqrt{N}\sqrt{P}-2\sqrt{N}\sqrt{N-1}\sqrt{1-P}\le \sum_{f\in F}\left\| |\varphi_T^f\rangle - |\varphi_T^0\rangle\right\|^2.$$

Proof Since $\{|x\rangle; x=0,1,\dots,N-1\}$ is an orthonormal base, one has

$$\begin{aligned}\left\| |\varphi_T^f\rangle - |\varphi_T^0\rangle\right\|^2 &= \sum_{x=0}^{N-1} \left|c_x^f(T)-c_x^0(T)\right|^2\\ &\ge 2-2\sum_{x=0}^{N-1}\left|c_x^f(T)\right|\cdot\left|c_x^0(T)\right|.\end{aligned}$$

The sequence $\{c_x^f(T); x=0,1,\dots,N-1\}$ has the following constraints:

$$\sum_{x\in J^f}\left|c_x^f(T)\right|^2 = P^f \quad\text{and}\quad \sum_{x=0}^{N-1}\left|c_x^f(T)\right|^2=1.$$

Thus we use Lagrange method to determine the coefficients as follows: Put

$$\begin{aligned}&L\left(\left|c_0^f(T)\right|,\dots,\left|c_{N-1}^f(T)\right|\right)\\ &\equiv 2\sum_{x=0}^{N-1}\left|c_x^f(T)\right|\cdot\left|c_x^0(T)\right|\end{aligned}$$

$$-\lambda\left[\sum_{x=0}^{N-1}\left|c_x^f(T)\right|^2-1\right]-\mu\left[\sum_{x\in J^f}\left|c_x^f(T)\right|^2-P^f\right].$$

Then we compute

$$\begin{cases}\dfrac{\partial L}{\partial|c_x^f(T)|}=2|c_x^0(T)|-2\lambda|c_x^f(T)|-2\mu|c_x^f(T)|=0, & x\in J^f,\\ \dfrac{\partial L}{\partial|c_x^f(T)|}=2|c_x^0(T)|-2\lambda|c_x^f(T)|=0, & x\notin J^f.\end{cases}$$

From here we have

$$\begin{cases}|c_x^f(T)|=\frac{|c_x^0(T)|}{\lambda+\mu}, & x\in J^f,\\ |c_x^f(T)|=\frac{|c_x^0(T)|}{\lambda}, & x\notin J^f.\end{cases}$$

By substituting these relations into the above constraints, we obtain

$$\sum_{x\in J^f}\left|c_x^f(T)\right|^2=\left(\frac{1}{\lambda+\mu}\right)^2\sum_{x\in J^f}\left|c_x^0(T)\right|^2,$$
$$\sum_{x=0}^{N-1}\left|c_x^f(T)\right|^2=\left(\frac{1}{\lambda+\mu}\right)^2\sum_{x\in J^f}\left|c_x^0(T)\right|^2+\left(\frac{1}{\lambda}\right)^2\sum_{x\notin J^f}\left|c_x^0(T)\right|^2.$$

Here, let c^f be the positive number defined as

$$c^f=\sum_{x\in J^f}\left|c_x^0(T)\right|^2.$$

Then $\lambda+\mu$ and λ are given by

$$\frac{1}{\lambda+\mu}=\sqrt{\frac{P^f}{c^f}}\quad\text{and}\quad\frac{1}{\lambda}=\sqrt{\frac{1-P^f}{1-c^f}},$$

and the solutions $|c_0^f(T)|,\dots,|c_{N-1}^f(T)|$ are expressed in terms of $|c_0^0(T)|,\dots,$ $|c_{N-1}^0(T)|$ as

$$\left|c_x^f(T)\right|=\sqrt{\frac{P^f}{c^f}}\left|c_x^0(T)\right|,\quad x\in J^f,$$
$$\left|c_x^f(T)\right|=\sqrt{\frac{1-P^f}{1-c^f}}\left|c_x^0(T)\right|,\quad x\notin J^f.$$

Thus we can estimate $\||\varphi_T^f\rangle-|\varphi_T^0\rangle\|^2$ from below as

$$\left\||\varphi_T^f\rangle-|\varphi_T^0\rangle\right\|^2\geq 2-\sum_{x\in J^f}\sqrt{\frac{P^f}{c^f}}\left|c_x^0(T)\right|^2-2\sum_{x\notin J^f}\sqrt{\frac{1-P^f}{1-c^f}}\left|c_x^0(T)\right|^2.$$

From this inequality, we have

$$\sum_{f\in F} \left\| |\varphi_T^f\rangle - |\varphi_T^0\rangle \right\|^2 \ge 2N - 2\sum_{f\in F} \sqrt{P^f}\sqrt{c^f} - 2\sum_{f\in F} \sqrt{1-P^f}\sqrt{1-c^f}.$$

Now we have two more constraints:

$$\frac{1}{N}\sum_{f\in F} P^f = P \quad \text{and} \quad \sum_{f\in F} c^f = 1.$$

Let $J^f = \{x\}$, and let f_x be the corresponding oracle. Let use Lagrange method again to the function

$$\begin{aligned} &L\big(P^{f_0},\dots,P^{f_{N-1}},c^{f_0},\dots,c^{f_{N-1}}\big)\\ &\quad = 2\sum_{f\in F}\sqrt{P^f}\sqrt{c^f} + 2\sum_{f\in F}\sqrt{1-P^f}\sqrt{1-c^f}\\ &\qquad - \lambda\left(\frac{1}{N}\sum_{f\in F} P^f - P\right) - \mu\left(\sum_{f\in F} c^f - 1\right). \end{aligned}$$

Then we obtain the $2N$ equations with $2N$ variables $P^{f_0},\dots,P^{f_{N-1}}$ and $c^{f_0},\dots,c^{f_{N-1}}$ such that

$$\begin{cases} \dfrac{\partial L}{\partial P^{f_0}} = \dfrac{\sqrt{a^{f_0}}}{\sqrt{P^{f_0}}} - \dfrac{\sqrt{1-a^{f_0}}}{\sqrt{1-P^{f_0}}} - \frac{\lambda}{N} = 0,\\ \vdots\\ \dfrac{\partial L}{\partial P^{f_{N-1}}} = \dfrac{\sqrt{a^{f_{N-1}}}}{\sqrt{P^{f_{N-1}}}} - \dfrac{\sqrt{1-a^{f_{N-1}}}}{\sqrt{1-P^{f_{N-1}}}} - \frac{\lambda}{N} = 0,\\ \dfrac{\partial L}{\partial a^{f_0}} = \dfrac{\sqrt{P^{f_0}}}{\sqrt{a^{f_0}}} - \dfrac{\sqrt{1-P^{f_0}}}{\sqrt{1-a^{f_0}}} - \mu = 0,\\ \vdots\\ \dfrac{\partial L}{\partial a^{f_{N-1}}} = \dfrac{\sqrt{P^{f_{N-1}}}}{\sqrt{a^{f_{N-1}}}} - \dfrac{\sqrt{1-P^{f_{N-1}}}}{\sqrt{1-a^{f_{N-1}}}} - \mu = 0. \end{cases}$$

The solution of the above equations is

$$\begin{aligned} P^{f_0} &= \cdots = P^{f_{N-1}} = P,\\ a^{f_0} &= \cdots = a^{f_{N-1}} = \frac{1}{N}, \end{aligned}$$

from which we complete the proof of the theorem. □

Theorem 12.13 *For any positive integer T, the following inequality holds*:

$$\sum_{f\in F} \left\| |\varphi_T^f\rangle - |\varphi_T^0\rangle \right\|^2 \leqq 4T^2.$$

Proof For any positive integer T, we will prove the following equality

$$|\varphi_T^0\rangle = |\varphi_T^f\rangle + \sum_{k=0}^{T-1} U_f^k \Delta U_f |\varphi_{T-1-k}^0\rangle$$

by mathematical induction, where ΔU_f denotes $U_0 - U_f$.

In the case of $T = 1$, $\varphi_0^0 = \varphi_0^f$ implies

$$\begin{aligned}|\varphi_1^f\rangle + \sum_{k=0}^{0} U_f^0 \Delta U_f |\varphi_{0-k}^0\rangle &= |\varphi_1^f\rangle + (U_0 - U_f)|\varphi_0^0\rangle \\ &= |\varphi_1^f\rangle + |\varphi_1^0\rangle - |\varphi_1^f\rangle = |\varphi_1^0\rangle.\end{aligned}$$

Assume that the equality holds in the case of T. Then the vector $|\varphi_{T+1}^0\rangle$ becomes

$$\begin{aligned}|\varphi_{T+1}^0\rangle &= U_0|\varphi_T^0\rangle = (U_f + \Delta U_f)|\varphi_T^0\rangle \\ &= U_f\left[|\varphi_T^f\rangle + \sum_{k=0}^{T-1} U_f^k \Delta U_f |\varphi_{T-1-k}^0\rangle\right] + \Delta U_f|\varphi_T^0\rangle \\ &= U_f|\varphi_T^f\rangle + \sum_{k=0}^{T-1} U_f^{k+1} \Delta U_f |\varphi_{T-1-k}^0\rangle + \Delta U_f|\varphi_T^0\rangle \\ &= |\varphi_{T+1}^f\rangle + \sum_{k=1}^{(T+1)-1} U_f^k \Delta U_f |\varphi_{(T+1)-1-k}^0\rangle + U_f^0 \Delta U_f|\varphi_T^0\rangle \\ &= |\varphi_{T+1}^f\rangle + \sum_{k=0}^{(T+1)-1} U_f^k \Delta U_f |\varphi_{(T+1)-1-k}^0\rangle,\end{aligned}$$

which is the equality to be proved.

Next we estimate $\||\varphi_T^f\rangle - |\varphi_T^0\rangle\|$ from above. The above equality and the unitarity of U_f imply

$$\begin{aligned}\||\varphi_T^f\rangle - |\varphi_T^0\rangle\| &= \left\|\sum_{k=0}^{T-1} U_f^k \Delta U_f |\varphi_{T-1-k}^0\rangle\right\| \le \sum_{k=0}^{T-1} \|U_f^k \Delta U_f |\varphi_{T-1-k}^0\rangle\| \\ &= \sum_{k=0}^{T-1} \|\Delta U_f |\varphi_{T-1-k}^0\rangle\| = \sum_{k=0}^{T-1} \|\Delta U_f |\varphi_k^0\rangle\|.\end{aligned}$$

Since we have

$$|\varphi_k^0\rangle = \sum_{x=0}^{N-1} c_x^0(k)|x\rangle = \sum_{x\in J^f} c_x^0(k)|x\rangle + \sum_{x\notin J^f} c_x^0(k)|x\rangle,$$

we obtain

$$\begin{aligned}
\Delta U_f\left|\varphi_k^0\right\rangle &= (U_0 - U_f)\left|\varphi_k^0\right\rangle \\
&= \left[\sum_{x\in J^f} c_x^0(k)U_0|x\rangle + \sum_{x\notin J^f} c_x^0(k)U_0|x\rangle\right] \\
&\quad - \left[\sum_{x\in J^f} c_x^0(k)U_f|x\rangle + \sum_{x\notin J^f} c_x^0(k)U_f|x\rangle\right] \\
&= \left[\sum_{x\in J^f} c_x^0(k)U_0|x\rangle - \sum_{x\in J^f} c_x^0(k)U_f|x\rangle\right] \\
&\quad + \left[\sum_{x\notin J^f} c_x^0(k)U_0|x\rangle - \sum_{x\notin J^f} c_x^0(k)U_f|x\rangle\right].
\end{aligned}$$

As $U_0|x\rangle = U_f|x\rangle$ for any $x \notin J^f$, we have

$$\begin{aligned}
\Delta U_f\left|\varphi_k^0\right\rangle &= (U_0 - U_f)\sum_{x\in J^f} c_x^0(k)|x\rangle \\
&= (U_0 - U_f)E_f\left|\varphi_k^0\right\rangle,
\end{aligned}$$

where E_f is the projection to the subspace spanned by $\{|x\rangle; x \in J^f\}$. Since U_0 and U_f are unitary, we have

$$\left\|\Delta U_f\left|\varphi_k^0\right\rangle\right\| = \left\|U_0E_f\left|\varphi_k^0\right\rangle - U_fE_f\left|\varphi_k^0\right\rangle\right\| \le 2\left\|E_f\left|\varphi_k^0\right\rangle\right\|.$$

Thus it holds that

$$\sum_{f\in F}\left\|\left|\varphi_T^f\right\rangle - \left|\varphi_T^0\right\rangle\right\|^2 \le 4\sum_{f\in F}\left[\sum_{k=0}^{T-1}\left\|E_f\left|\varphi_k^0\right\rangle\right\|\right]^2.$$

Applying the Schwartz inequality to the right-hand side of the above inequality (i.e., $|\sum a_k b_k|^2 \leqq \sum |a_k|^2 \sum |b_k|^2$), we obtain

$$\begin{aligned}
&= 4\sum_{f\in F}\left[\sum_{k=0}^{T-1}\left(1\cdot\left\|E_f\left|\varphi_k^0\right\rangle\right\|\right)\right]^2 \\
&\le 4\sum_{f\in F}\left[\sum_{k=0}^{T-1}1^2\right]\left[\sum_{k=0}^{T-1}\left\|E_f\left|\varphi_k^0\right\rangle\right\|^2\right] \\
&= 4T\sum_{f\in F}\left[\sum_{k=0}^{T-1}\left\|E_f\left|\varphi_k^0\right\rangle\right\|^2\right].
\end{aligned}$$

It is clear that E_{f_1} is orthogonal to E_{f_2} if $f_1 \neq f_2$. Therefore, Pythagorean theorem assures

$$\sum_{f\in F} \|E_f|\varphi_k^0\rangle\|^2 = \left\|\sum_{f\in F} E_f|\varphi_k^0\rangle\right\|^2,$$

and we obtain

$$\begin{aligned}\sum_{f\in F} \||\varphi_T^f\rangle - |\varphi_T^0\rangle\|^2 &\le 4T\sum_{k=0}^{T-1}\left\|\sum_{f\in F} E_f|\varphi_k^0\rangle\right\|^2 \\ &\le 4T\sum_{k=0}^{T-1}\||\varphi_k^0\rangle\|^2 = 4T^2.\end{aligned}$$

□

Now, the rapidity of the Grover algorithm can be estimated below by the following theorem.

Theorem 12.14 *A lower bound of T can be estimated by*

$$2N - 2\sqrt{N}\sqrt{P} - 2\sqrt{N}\sqrt{N-1}\sqrt{1-P} \le 4T^2.$$

Proof Clear from the previous two theorems. □

Thus this theorem together with Theorem 12.10 tells that the complexity of the database search quantum algorithm is $O(\sqrt{N})$ because of $2N - 2\sqrt{N} \le 4T^2$. Note that even when $P \neq 1$, this P is close to 1 for large N ($\gg |A|$) because of the above theorem, so that T is bounded below by $\sqrt{N}$.

12.3.2 Accardi–Sabadini's Unitary Operator

There exist several applications and extensions of the Grover algorithm. We mention here the work done by Accardi and Sabbadini [33]. They presented a general method to choose the unitary operator which can amplify any components of a given vector, and it was shown that the Grover algorithm is a particular case of the general method.

The Grover algorithm is the construction of a unitary operator U which increases the probability of one of the components of a given unit vector at the expense of the remaining ones. A preliminary step to solve this problem is to determine the most general unitary operator which performs the same task of Grover's unitarity operator. Accardi and Sabbadini found that up to the five parameters taking the value 1 or -1, there exists exactly one class of such unitary operators, labeled by an arbitrary parameter in the interval [0, 1] (Theorem 12.15).

In this section, we assume that the set A has only one element, and it is "0" for notational simplicity.

Theorem 12.15 *Given the linear functions with γ_i and η_i real,*

$$\eta(a) = \sum_{i=0}^{N-1} \eta_i a_i, \qquad c(a) = \sum_{i=0}^{N-1} \gamma_i a_i,$$

the operator U defined by

$$U \sum_i a_i |i\rangle = \varepsilon_1 \big(a_0 + \eta(a)\big)|0\rangle + \varepsilon_2 \sum_{i \neq 0} \big(a_i + c(a)\big)|i\rangle$$

is unitary iff there exists a real number β_0 satisfying $|\beta_0| \leq 1$ such that

$$\begin{aligned} \gamma_0 &= \varepsilon_3 \frac{\sqrt{1-\beta_0^2}}{\sqrt{N-1}}, \\ \eta_i &= \varepsilon_4 \varepsilon_5 \gamma_0, \\ \gamma_i &= -\frac{1+\varepsilon_4 \beta_0}{N-1}, \\ \eta_0 &= -1 + \varepsilon_4 \varepsilon_5 \beta_0, \end{aligned} \tag{12.2}$$

where $\varepsilon_1, \varepsilon_2, \varepsilon_3, \varepsilon_4, \varepsilon_5 = \pm 1$.

Proof Put $\eta(a) =: \eta$ and $c(a) =: c$. In finite dimension, the unitarity is equivalent to isometry. Therefore, the following isometry condition is a necessary condition:

$$\begin{aligned} \sum_i a_i^2 &= (a_0 + \eta)^2 + \sum_{i \neq 0} (a_i + c)^2 \\ &= a_0^2 + \eta^2 + 2a_0\eta + \sum_{i \neq 0} a_i^2 + (N-1)c^2 + 2c \sum_{i \neq 0} a_i, \end{aligned}$$

which equals to

$$\eta^2 + 2a_0\eta + (N-1)c^2 + 2c \sum_{i \neq 0} a_i = 0. \tag{12.3}$$

We rewrite this equation as

$$\eta^2 + 2a_0\eta + \gamma = 0$$

with

$$\gamma = (N-1)c^2 + 2c \sum_{i \neq 0} a_i.$$

Thus a possible solution is

$$\eta\big(= \eta(a)\big) = -a_0 + \varepsilon_5 \sqrt{a_0^2 - \gamma}.$$

The given $\eta(a)$ will be linear in a iff for any $a_0, \dots, a_N$

$$a_0^2 - \gamma = \Big(\sum_j \beta_j a_j\Big)^2$$

with the β_j independent from a. The further linearity condition of $c(a)$ leads to:

$$c(a) = \sum_j \gamma_j a_j$$

with the γ_j independent from a. From (12.3), one has

$$\begin{aligned}
&-a_0^2 + (N-1)\Big(\sum_j \gamma_j a_j\Big)^2 + \Big(\sum_j \beta_j a_j\Big)^2 + 2\sum_j \gamma_j a_j \sum_{i\neq 0} a_i = 0 \\
\iff\quad & -a_0^2 + 2\sum_j \gamma_j a_j \sum_{i\neq 0} a_i + \sum_{i,j}\big[(N-1)\gamma_i\gamma_j + \beta_i\beta_j\big]a_i a_j = 0 \\
\iff\quad & a_0^2\big[(N-1)\gamma_0^2 + \beta_0^2 - 1\big] + \sum_{i,j\neq 0}\big[2\gamma_j + (N-1)\gamma_i\gamma_j + \beta_i\beta_j\big]a_i a_j \\
& + 2\sum_{i\neq 0}\big[\gamma_0 + (N-1)\gamma_0\gamma_i + \beta_0\beta_i\big]a_0 a_i = 0.
\end{aligned}$$

This identity holds for any $a_0, \dots, a_N$, iff

$$\begin{cases}
\text{(a) } (N-1)\gamma_0^2 + \beta_0^2 - 1 = 0, \\
\text{(b) } 2\gamma_j + (N-1)\gamma_i\gamma_j + \beta_i\beta_j = 0 \quad \forall i, j \neq 0, \\
\text{(c) } 2\gamma_i + (N-1)\gamma_i^2 + \beta_i^2 = 0 \quad \forall i \neq 0, \\
\text{(d) } \gamma_0 + (N-1)\gamma_0\gamma_i + \beta_0\beta_i = 0 \quad \forall i \neq 0.
\end{cases}$$

Hence one obtains

$$\gamma_0 = \varepsilon_3 \frac{\sqrt{1-\beta_0^2}}{\sqrt{N-1}}, \qquad \gamma_i = -\frac{\gamma_0 + \beta_0\beta_i}{\gamma_0(N-1)} \quad (i \neq 0).$$

Substituting these into (c), one has

$$-\frac{2(\gamma_0 + \beta_0\beta_i)}{\gamma_0(N-1)} + \frac{(\gamma_0 + \beta_0\beta_i)^2}{\gamma_0^2(N-1)} + \beta_i^2 = 0$$

or equivalently,

$$\big[(N-1)\gamma_0^2 + \beta_0^2\big]\beta_i^2 = \gamma_0^2.$$

Then from (a) this is equivalent to

$$\beta_i = \varepsilon_4 \gamma_0 = \varepsilon_4 \varepsilon_3 \frac{\sqrt{1-\beta_0^2}}{\sqrt{N-1}}$$

with $\varepsilon_3 = \pm 1$, so that every β_i $(i \neq 0)$ is written by β_0.

Using $c(a)$ instead of $\eta(a)$ one arrives at

$$\gamma_i = -\frac{1+\varepsilon_4 \beta_0}{N-1}.$$

Put $\beta_0 = \cos\theta$, then $\sqrt{N-1}\gamma_0 = \sin\theta$. From all discussions above, one finally obtains

$$\begin{aligned}\eta(a) &= (-1+\varepsilon_5\beta_0)a_0 + \varepsilon_5\varepsilon_4\gamma_0 \sum_{k\neq 0} a_k \\ &= (-1+\varepsilon_5\beta_0)a_0 + \varepsilon_4\varepsilon_3\varepsilon_5 \frac{\sqrt{1-\beta_0^2}}{\sqrt{N-1}} \sum_{k\neq 0} a_k \\ &= (-1+\varepsilon_5\cos\theta)a_0 + \varepsilon_3\varepsilon_4\varepsilon_5 \frac{\sin\theta}{\sqrt{N-1}} \sum_{k\neq 0} a_k\end{aligned}$$

and

$$\begin{aligned}c(a) &= \gamma_0 a_0 - \frac{1+\varepsilon_4\beta_0}{N-1} \sum_{k\neq 0} a_k \\ &= \varepsilon_3 \frac{\sqrt{1-\beta_0^2}}{\sqrt{N-1}} a_0 - \frac{1+\varepsilon_4\beta_0}{N-1} \sum_{k\neq 0} a_k \\ &= \varepsilon_3 \frac{\sin\theta}{\sqrt{N-1}} a_0 - \frac{1+\varepsilon_4\cos\theta}{N-1} \sum_{k\neq 0} a_k.\end{aligned}$$

It is a similar exercise to verify that conditions (12.2) are sufficient for the operator U to be unitary. □

Corollary 12.16 *Grover's method corresponds to the case* $\varepsilon_1\varepsilon_3 = 1$, $\varepsilon_2 = -1$, $\beta_0 = \frac{N-2}{N}$, $\gamma_0 = \frac{2}{N}$, *i.e., for such a choice of parameters the corresponding unitary operator* U *is Grover's unitary operator.*

Proof Grover's unitary operator is characterized by

$$a_0 \quad \mapsto \quad \frac{N-2}{N} a_0 + \frac{2}{N} \sum_{k\neq 0} a_k = \varepsilon_1 \big(a_0 + \eta(a)\big),$$

$$a_i \quad \mapsto \quad -a_i + \frac{2}{N}\Big(-a_0 + \sum_{k\neq 0} a_k\Big) = \varepsilon_2\big(a_i + c(a)\big).$$

From $\eta(a)$ and $c(a)$ obtained in the above theorem, one has

$$\begin{aligned}\varepsilon_1\big(a_0 + \eta(a)\big) &= \varepsilon_1\varepsilon_4\Big(\beta_0 a_0 + \varepsilon_3\gamma_0 \sum_{k\neq 0} a_k\Big),\\ \varepsilon_2\big(a_i + c(a)\big) &= \varepsilon_2\Big(a_i + \gamma_0 a_0 - \frac{1+\varepsilon_3\beta_0}{N-1}\sum_{k\neq 0} a_k\Big)\end{aligned} \tag{12.4}$$

with

$$\beta_i = \varepsilon_3 \frac{\sqrt{1-\beta_0^2}}{\sqrt{N-1}}.$$

Thus

$$\varepsilon_1\varepsilon_5\beta_0 = \frac{N-2}{N}.$$

Now choose $\varepsilon_1\varepsilon_4 = 1$ and $\beta_0 = \frac{N-2}{N}$, then

$$\gamma_0 = \sqrt{\frac{\frac{1-(N-2)^2}{N^2}}{N-1}}\left(\varepsilon_3\frac{2}{N}\right)$$

as in (12.4) with $\varepsilon_2 = -1$. And finally,

$$-\varepsilon_2\frac{1+\varepsilon_5\beta_0}{N-1} = \frac{1+\varepsilon_5\frac{N-2}{N}}{N-1} = \frac{N+\varepsilon_5 N - 2\varepsilon_5}{N(N-1)}$$

gives the parameter $\gamma_0 = \frac{2}{N}$ iff $\varepsilon_5 = 1$. □

Thus the unitary operator U of Accardi–Sabbadini generalizes that of Grover's.

12.4 Notes

Deutsch and Jozsa showed that quantum computer can efficiently solve the problem of whether a function is constant or balanced [199]. Grover showed the search problem could be speeded up on a quantum computer [305]. Accardi and Sabbadini proved that there exists a one step solution of Grover's algorithm under some conditions [33].

Chapter 13
Quantum Algorithms II

In this chapter, the Shor's quantum algorithm for factoring integers is described. Factoring integers plays an important role in modern cryptography. We start by reviewing some fundamental facts of number theory.

13.1 Elements of Number Theory

In this section, we collect some relevant material from number theory. It is used in the Shor's factoring algorithm and in quantum cryptography.

Given two integers a and b, not both zero, the *greatest common divisor* of a and b, denoted $\gcd(a, b)$ is the biggest integer d dividing both a and b. For example, $\gcd(9, 12) = 3$.

13.1.1 Euclid's Algorithm

There is the well known *Euclid's algorithm* of finding the greatest common divisor. Although it was discussed in Chap. 2, we will discuss it more for the subsequent uses in this section. Euclid's algorithm proceeds as follows.

Problem: Find $\gcd(a, b)$ where $a > b > 0$.

1. Divide b into a and write down the quotient q_1 and the remainder r_1:
$$a = q_1 b + r_1, \quad 0 < r_1 < b.$$
2. Next, perform a second division with b playing the role of a and r_1 playing the role of b:
$$b = q_2 r_1 + r_2, \quad 0 < r_2 < r_1.$$
3. Next write
$$r_1 = q_3 r_2 + r_3, \quad 0 < r_3 < r_2.$$

M. Ohya, I. Volovich, *Mathematical Foundations of Quantum Information and Computation and Its Applications to Nano- and Bio-systems*,
Theoretical and Mathematical Physics,
DOI 10.1007/978-94-007-0171-7_13,

Continue in this way until we finally obtain a remainder that divides the previous remainder, then we are done, that is, that final nonzero remainder is the gcd of a and b:

$$r_t = q_{t+2}r_{t+1} + r_{t+2},$$
$$r_{t+1} = q_{t+3}r_{t+2}.$$

We obtain: $r_{t+2} = d = \gcd(a, b)$.

Example 13.1 Find $\gcd(128, 24)$: $128 = 5 \cdot 24 + 8$, $24 = 3 \cdot 8$, so we obtain $\gcd(128, 24) = 8$.

Let us prove that Euclid's algorithm indeed gives the greatest common divisor. Note first that $b > r_1 > r_2 > \cdots$ is a sequence of decreasing positive integers which cannot be continued indefinitely. Consequently, Euclid's algorithm must end.

Let us go up throughout Euclid's algorithm. $r_{t+2} = d$ divides $r_{t+1}, r_t, \ldots, r_1, b, a$. Thus d is a common divisor of a and b.

Now let c be any common divisor of a and b. Go downward throughout Euclid's algorithm. c divides $r_1, r_2, \ldots, r_{t+2} = d$. Thus d, being a common divisor of a and b, is divisible by any common divisor of these numbers. Consequently, d is the greatest common divisor of a and b.

Another (but similar) proof is based on the formula

$$\gcd(qb + r, b) = \gcd(b, r).$$

Note that from Euclid's algorithm it follows (going up) that if $d = \gcd(a, b)$ then there are integers u and v such that

$$d = ua + vb.$$

In particular, one has

$$ua \equiv d \pmod{b}. \tag{13.1}$$

One can estimate the efficiency of Euclid's algorithm. By *Lame's theorem* the number of divisions required to find the greatest common divisor of two integers is never greater than five-times the number of digits in the smaller integer.

Congruences: An integer a is *congruent to b modulo m*,

$$a \equiv b \pmod{m},$$

iff m divides $(a - b)$. In this case, $a = b + km$ where $k = 0, \pm 1, \pm 2, \ldots$.

Proposition 13.2 *Let us be given two integers a and m. The following are equivalent*:

(i) *There exists a u such that* $au \equiv 1 \pmod{m}$.
(ii) $\gcd(a, m) = 1$.

Proof From (i), follows $ab - mk = 1$. Therefore, the $\gcd(a, m) = 1$, i.e., we get (ii).

Now if (ii) is true then one has the relation (13.1) for $d = 1, b = m$, i.e., $au \equiv 1 \pmod{m}$, which gives (i). □

Let us find integer solutions of the equation

$$ax \equiv c \pmod{m}. \tag{13.2}$$

We suppose that $\gcd(a, m) = 1$. Then by the previous proposition there exists a b such that

$$ab \equiv 1 \pmod{m}.$$

Multiplying (13.2) by b, we obtain the solution

$$x \equiv bc \pmod{m},$$

or more explicitly,

$$x = bc + km, \quad k = 0, \pm 1, \pm 2, \ldots.$$

Exercise 13.3 Find all the solutions of the congruence $3x \equiv 4 \pmod 7$.

Continued fractions: Euclid's algorithm is closely related to continued fractions. If a and b are two integers then by using Euclid's algorithm we write

$$\begin{aligned}
a &= q_1 b + r_1; & \frac{a}{b} &= q_1 + \frac{1}{b/r_1}, \\
b &= q_2 r_1 + r_2; & \frac{b}{r_1} &= q_2 + \frac{1}{r_1/r_2}, \\
r_1 &= q_3 r_2 + r_3; & \frac{r_1}{r_2} &= q_3 + \frac{1}{r_1/r_2}, \\
&\vdots \\
r_t &= q_{t+2} r_{t+1} + r_{t+2}; & \frac{r_t}{r_{t+1}} &= q_{t+2} + \frac{1}{r_{t+1}/r_{t+2}}, \\
r_{t+1} &= q_{t+3} r_{t+2}; & \frac{r_{t+1}}{r_{t+2}} &= q_{t+3}.
\end{aligned}$$

Therefore, we obtain a representation of a/b as a continued fraction

$$\frac{a}{b} = q_1 + \frac{1}{q_2 + \frac{1}{q_3 + \cdots \frac{1}{q_{t+3}}}}.$$

Hence any positive rational number can be represented by a continued fraction. Fractions

$$\delta_1 = q_1, \qquad \delta_2 = q_1 + \frac{1}{q_2}, \qquad \delta_3 = q_1 + \frac{1}{q_2 + \frac{1}{q_3}}, \dots$$

are called convergents. We will use the following theorems.

Theorem 13.4 *If x is a real number and a and b are positive integers satisfying*

$$\left| \frac{a}{b} - x \right| < \frac{1}{2b^2},$$

then a/b is a convergent of the continued fraction of x.

Theorem 13.5 (Chinese remainder theorem) *Suppose there is a system of congruences with respect to different moduli*:

$$\begin{aligned} x &\equiv a_1 \pmod{m_1}, \\ x &\equiv a_2 \pmod{m_2}, \\ &\vdots \\ x &\equiv a_t \pmod{m_t}. \end{aligned}$$

Suppose $\gcd(m_i, m_j) = 1$ *for* $i \neq j$. *Then there exists a solution x to all of the congruences, and any two solutions are congruent to one another modulo*

$$M = m_1 m_2 \cdots m_t.$$

Proof Let us denote $M_i = M/m_i$. There exists N_i such that $M_i N_i \equiv 1 \pmod{m_i}$.
Let us set $x = \sum_i a_i M_i N_i$. This is the solution. Indeed, we have

$$\sum_i a_i M_i N_i = a_1 M_1 M_1 + \cdots \equiv a_1 + a_2 + \cdots \equiv a_1 \pmod{m_1},$$

and similarly for other congruences. □

We will also need

Theorem 13.6 (Fermat's little theorem) *Let p be a prime number. Any integer a satisfies*

$$a^p \equiv a \pmod{p},$$

and any integer a not divisible by p satisfies

$$a^{p-1} \equiv 1 \pmod{p}.$$

Proof Suppose a is not divisible by p. Then $\{0a, 1a, 2a, \ldots, (p-1)a\}$ forms a complete set of residues modulo p, i.e., $\{a, 2a, \ldots, (p-1)a\}$ is a rearrangement of $\{1, 2, \ldots, p-1\}$ when considered modulo p. Hence the product of the numbers in the first sequence is congruent modulo p to the product of the members in the second sequence, i.e.,

$$a^{p-1}(p-1)! \equiv (p-1)! \pmod{p}.$$

Thus p divides $a^{p-1} - 1$. □

13.1.2 The Euler Function

The Euler function $\varphi(n)$ is the number of nonnegative integers a less than n which are prime to n:

$$\varphi(n) = \#\{0 \le a < n; \gcd(a, n) = 1\}.$$

In particular, $\varphi(1) = 1,\ \varphi(2) = 1, \ldots, \varphi(6) = 2, \ldots$. One has $\varphi(p) = p - 1$ for any prime p.

Exercise 13.7 Prove that $\varphi(p^n) = p^n - p^{n-1}$ for any n and prime p.

The Euler function is multiplicative in the sense that

$$\varphi(mn) = \varphi(m)\varphi(n)$$

whenever $\gcd(m, n) = 1$.

If

$$n = p_1^{\alpha_1} p_2^{\alpha_2} \cdots p_k^{\alpha_k}$$

then

$$\varphi(n) = n\left(1 - \frac{1}{p_1}\right)\cdots\left(1 - \frac{1}{p_k}\right).$$

In particular, if n is the product of two primes, $n = pq$, then

$$\varphi(n) = \varphi(p)\varphi(q) = (p-1)(q-1).$$

There is the following generalization of Fermat's Little Theorem.

Theorem 13.8 (Euler's theorem) *If* $\gcd(a, m) = 1$ *then*

$$a^{\varphi(m)} \equiv 1 \pmod{m}.$$

Proof Let $r_1, r_2, \ldots, r_{\varphi(m)}$ be classes of integers relatively prime to m. Such a system is called a reduced system of residues mod m. Then $ar_1, ar_2, \ldots, ar_{\varphi(m)}$ is another reduced system since $\gcd(a, m) = 1$. Therefore,

$$ar_1 \equiv r_{\pi(1)}, \qquad ar_2 \equiv r_{\pi(2)}, \ldots, ar_{\varphi(m)} \equiv r_{\pi(m)} \pmod{m}.$$

On multiplying these congruences, we get

$$a^{\varphi(m)} r_1 r_2 \cdots r_{\varphi(m)} \equiv r_1 r_2 \cdots r_{\varphi(m)} \pmod{m}.$$

Now since $r_1 r_2 \cdots r_{\varphi(m)}$ is relatively prime to m the theorem is proved. □

We will use the following result on the asymptotic behavior of the Euler function.

Theorem 13.9 *There is a constant $C > 0$ such that for sufficiently large n one has*

$$\frac{\varphi(n)}{n} \geq \frac{C}{\log\log n}.$$

13.1.3 Modular Exponentiation

Sometimes it is necessary to do classical computations on a quantum computer. Since quantum computation is reversible, a deterministic classical computation is performable on a quantum computer only if it is reversible. It was shown that any deterministic computation can be made reversible for only a constant factor cost in time and by using as much space as time.

In this section, we discuss the modular exponential problem. The problem is: Given N, a and m, $m \leq N$, $a \leq N$ find $m^a \pmod{N}$.

Theorem 13.10 *There exists a classical algorithm for computation $m^a \pmod{N}$ which requires asymptotically $O(n^2 \log n \log\log n)$ arithmetical operations with bits in the binary representation of the numbers where $n = \log N$.*

Proof The algorithm proceeds as follows:

1. Write the binary representation

$$a = a_0 + 2a_1 + 2^2 a_2 + \cdots + 2^s a_s$$

where $a_i = 0, 1$ and $a_0 = 1$.

2. Set $m_0 = m$ and then for $i = 1, \ldots, s$ compute

$$m_i \equiv m_{i-1}^2 m^{a_{s-i}} \pmod{N}.$$

3. The final result is m_s given by

$$m_s = m^a \pmod{N}.$$

The validity of the algorithm follows from the relation

$$m_i \equiv m^{a_0+2a_1+\cdots+2^i a_i} \pmod{N}.$$

Computation at the third step requires no more than three multiplications, and it is repeated no more than $s \le n = \log N$ times. There is the Schönhage–Strassen algorithm for integer multiplication that uses asymptotically $O(n \log n \log\log n)$ operations on bits. This proves the theorem. □

13.2 Shor's Quantum Factoring

Let us discuss the problem of factoring. It is known that every integer N is uniquely decomposable into a product of prime numbers. However, we do not know *efficient* (i.e., polynomial in the number of operations) classical algorithms for factoring. Given a large integer N, one has to find efficiently such integers p and q that $N = pq$, or to prove that such a factoring does not exist. It is assumed that p and q are not equal to 1.

An algorithm of factoring the number N is called efficient if the number of elementary arithmetical operations which are used for large N is bounded by a polynomial in n where $n = \log N$ is the number of digits in N.

The most naive factoring method would be just to divide N by each number from 1 to $\sqrt{N}$. This requires at least $\sqrt{N}$ operations. Since $\sqrt{N} = 2^{\frac{1}{2}\log N}$ is exponential in the number of digits $n = \log N$ in N, this method is not an efficient algorithm. There is no known efficient classical algorithm for factoring, but the quantum polynomial algorithm does exist.

The best classical factoring algorithm which is currently known is the number field sieve. It requires asymptotically

$$\exp\bigl(cn^{1/3}(\log n)^{2/3}\bigr)$$

operations for some constant c, i.e., it is exponential in $n^{1/3}$. Shor has found a quantum algorithm which takes asymptotically

$$O\bigl(n^2 \log n \log\log n\bigr),$$

i.e., only a polynomial number, of operations on a quantum computer along with a polynomial amount of time on a classical computer.

In the description of Shor's algorithm, we essentially follow his original presentation. It is known that using randomization the factorization of N can be reduced to finding the *order* of an arbitrary element m in the multiplicative group of residues modulo N, that is, the least integer r such that

$$m^r \equiv 1 \pmod{N}.$$

The reduction will be discussed below. Therefore, to factorize N it is enough to find the order r of m.

Shor's algorithm for finding the order consists of five steps:

1. Preparation of quantum state
2. Modular exponentiation
3. Quantum Fourier transform
4. Measurement
5. Computation of the order at the classical computer.

These steps will be discussed in details in the next section.

Moreover, in the subsequent sections, we have the following discussions: Shor's algorithm for finding the order is exposed. Then the computational complexity of Shor's algorithm is considered. Finally, the reduction of the problem of factorization to finding the order is discussed.

The main results of the quantum algorithm for finding the order are given in Theorem 13.11 on the lower bound for the probability of measurement and in Theorem 13.12 on the derivation of the order. Theorem 13.13 describes the computational complexity of the algorithm. The main result of the quantum algorithm for factoring is presented in Theorem 13.16.

13.2.1 Finding the Order

For a given N, choose a random (with the uniform distribution) m $(1 < m \leq N)$. We assume $\gcd(m, N) = 1$, otherwise we would already know a divisor of N. We want to find the order of m, i.e., the least integer r such that

$$m^r \equiv 1 \pmod{N}.$$

Fix some q of the form $q = 2^s$ with $N^2 \leq q < 2N^2$. The algorithm will use the Hilbert space

$$\mathcal{H} = \mathbb{C}^q \otimes \mathbb{C}^{N_1} \otimes \mathbb{C}^k$$

where $\mathbb{C}^q$ and $\mathbb{C}^{N_1}$ are two quantum registers which hold integers represented in binary form. Here N_1 is an integer of the form $N_1 = 2^l$ for some l such that $N \leq N_1$. There is also the work space $\mathbb{C}^k$ to make arithmetical operations. We will not indicate it explicitly. If

$$a = a_0 + 2a_1 + 2^2 a_2 + \cdots + 2^s a_s$$

is the binary representation $(a_i = 0, 1)$ of an integer a then we write

$$|a\rangle = |a_0\rangle \otimes \cdots \otimes |a_s\rangle$$

where

$$|0\rangle = \begin{pmatrix} 1 \\ 0 \end{pmatrix}, \qquad |1\rangle = \begin{pmatrix} 0 \\ 1 \end{pmatrix}$$

form the basis in the two dimensional complex space $\mathbb{C}^2$.

We have the data (N, m, q). The algorithm for finding the order r of m consists of five steps stated as above: 1. Preparation of quantum state; 2. Modular exponentiation; 3. Quantum Fourier transform; 4. Measurement; 5. Computation of the order on a classical computer.

13.2.2 Description of the Algorithm

1. (Preparation of quantum state) Put the first register in the uniform superposition of states representing numbers $a \pmod q$. The quantum computer will be in the state

$$|\psi_1\rangle = \frac{1}{\sqrt{q}} \sum_{a=0}^{q-1} |a\rangle \otimes |0\rangle.$$

2. (Modular exponentiation) Compute $m^a \pmod N$ in the second register. This leaves the quantum computer in the state

$$|\psi_2\rangle = \frac{1}{\sqrt{q}} \sum_{a=0}^{q-1} |a\rangle \otimes \big|m^a \pmod N\big\rangle.$$

3. (DFT) Perform the discrete quantum Fourier transform on the first register, mapping $|a\rangle$ to

$$\frac{1}{\sqrt{q}} \sum_{c=0}^{q-1} e^{2\pi iac/q} |c\rangle.$$

The quantum computer will be in the state

$$|\psi_3\rangle = \frac{1}{q} \sum_{a=0}^{q-1} \sum_{c=0}^{q-1} e^{2\pi iac/q} |c\rangle \otimes \big|m^a \pmod N\big\rangle.$$

4. (Measurement) Make the measurement on both registers $|c\rangle$ and $|m^a \pmod N\rangle$. To find the period r, we will need only the value of $|c\rangle$ in the first register, but for clarity of computations we make the measurement on the both registers. The probability $P(c, m^k \pmod N)$ that the quantum computer ends in a particular state

$$\big|c; m^k \pmod N\big\rangle = |c\rangle \otimes \big|m^k \pmod N\big\rangle$$

is

$$P\big(c, m^k \pmod N\big) = \big|\big\langle m^k \pmod N; c|\psi_3\big\rangle\big|^2$$

where we can assume $0 \le k < r$.

We will use the following theorem which shows that the probability $P(c, m^k \pmod N)$ is large if the residue of $rc \pmod q$ is small. Here r is the order of m in the group $(Z/NZ)^\times$ of residues of modulo N.

Theorem 13.11 *If there is an integer d such that*

$$-\frac{r}{2} \le rc - dq \le \frac{r}{2} \tag{13.3}$$

and N is sufficiently large then

$$P\big(c, m^k \pmod N\big) \ge \frac{1}{3r^2}. \tag{13.4}$$

The theorem is proved below.

5. (Computation of the order on a classical computer) We know N, c and q and we want to find the order r. Because $q > N^2$, there is at most one fraction d/r with $r < N$ that satisfies inequality (13.3). We can obtain the fraction d/r in lowest terms by rounding c/q to the nearest fraction having a denominator smaller than N. To this end, we can use the continued fraction expansion of c/q and Theorem 13.4.

We will prove the following theorem which summarizes the main results for the quantum algorithm for finding the order.

Theorem 13.12 *If the integer N is sufficiently large then by repeating the first four steps of the algorithm for finding the order $O(\log\log N)$ times one can obtain the value of the order r with the probability $\gamma > 0$ where the constant γ does not depend on N.*

Proof of Theorem 13.11 First, let us notice the relation

$$\big\langle m^k \pmod N \big| m^a \pmod N\big\rangle = \begin{cases} 1, & \text{if } a \equiv k \pmod r, \\ 0, & \text{otherwise.} \end{cases}$$

Hence the amplitude is

$$\big\langle m^k \pmod N; c\big|\psi_3\big\rangle = \frac{1}{q}\sum_a e^{2\pi i ac/q}$$

where the summation on a runs on the subset $a \equiv k \pmod r$ of the set $\{0, 1, \dots, q-1\}$. One sets

$$a = br + k$$

to get

$$\sum_a e^{2\pi i ac/q} = \sum_{b=0}^{f} e^{2\pi i c(br+k)/q} = \frac{1 - e^{2\pi i cr(f+1)/q}}{1 - e^{2\pi i cr/q}} e^{2\pi i ck/q}$$

where f is the integer part

$$f = \left[\frac{q-1-k}{r}\right].$$

Therefore, the probability is

$$P\big(c, m^k \ (\mathrm{mod}\ N)\big) = \left|\frac{1}{q}\frac{1-e^{2\pi icr(f+1)/q}}{1-e^{2\pi icr/q}}\right|^2,$$

which is equal to

$$P\big(c, m^k \ (\mathrm{mod}\ N)\big) = \frac{1}{q^2}\frac{\sin^2\frac{\pi cr(f+1)}{q}}{\sin^2\frac{\pi cr}{q}}.$$

Now to prove the theorem we use condition (13.3) and the relation

$$\sin x > \frac{2}{\pi}x, \quad 0 < x < \frac{\pi}{2}.$$

□

Proof of Theorem 13.12 If we know the fraction d/r in lowest terms and if d is relatively prime to r then we can derive r. There are $r\varphi(r)$ states $|c; m^k \ (\mathrm{mod}\ N)\rangle$ which enable us to compute r because there are $\varphi(r)$ values of d relatively prime to r and also there are r possible values for $m^k \ (\mathrm{mod}\ N)$. By Theorem 13.9, each of these states occurs with the probability at least $1/3r^2$. Therefore, we can get r with probability at least $\varphi(r)/3r$. Now the theorem follows from Theorem 13.11. □

13.2.3 Computational Complexity of Shor's Algorithm

Let us estimate the number of operations (or gates) needed to implement the first three steps of the Shor's algorithm for finding the order.

Theorem 13.13 *Shor's algorithm for finding the order of an element in the group of residues modulo N requires*

$$O\big((\log N)^2(\log\log N)(\log\log\log N)\big) \tag{13.5}$$

operations (gates) on a quantum computer.

Proof Let us estimate the number of operations (gates) needed to implement the first three steps of the algorithm on a quantum computer. To prepare the state $|\psi_1\rangle$, one needs $s = \log q = O(\log N)$ Hadamard's gates. Then let us consider the modular exponentiation. It is the most time consuming part of the algorithm. As it is discussed in Sect. 13.1.3, asymptotically, modular exponentiation requires

$$O\big(n^2 \log n \log\log n\big) \tag{13.6}$$

operations, where $n = O(\log N)$. The computation can be made reversible for only a constant factor cost in time and the same amount in space. Finally, it is shown in Sect. 13.2.2 that to make the third step of the algorithm, quantum Fourier transform, one takes

$$O\big((\log N)^2\big) \tag{13.7}$$

quantum gates. Actually, this is the key ingredient in the factoring algorithm. Just because of the polynomial bound (13.7) we obtain the polynomial efficiency of the factoring algorithm. Now the theorem follows from the estimates (13.6) and (13.7). □

13.3 Factoring Integers

In this section, the factoring algorithm will be described. The factoring algorithm solves the following problem: Given an integer N, find such integers p and q that $N = pq$ or prove that such a factoring does not exist. It is assumed that p and q are not equal to 1. We shall use the algorithm for finding the order described in Sect. 13.2.1.

13.3.1 Factoring Algorithm

1. Choose a random m, $1 \le m \le N$ (with uniform distribution) and find its order r by using the factoring algorithm from Sect. 13.2.1.
2. If r is even, compute

   $$\gcd\big(m^{r/2} - 1, N\big)$$

   by using Euclid's algorithm.
3. If $\gcd(m^{r/2} - 1, N) > 1$ then it gives a factor of N. In the case if $\gcd(m^{r/2} - 1, N) = 1$ or when the order r of m is odd, one has to repeat steps 1 and 2 for another integer m.

Let us explain why the algorithm works. Consider the equation

$$y^2 \equiv 1 \pmod N.$$

There are the trivial solutions

$$y \equiv \pm 1 \pmod N.$$

Suppose there is also a nontrivial solution $y = b$,

$$b^2 \equiv 1 \pmod N; \quad b \not\equiv \pm 1 \pmod N.$$

Then

$$(b+1)(b-1) \equiv 0 \pmod N,$$

i.e.,

$$(b+1)(b-1) = kN$$

and neither of the factors $b+1$ or $b-1$ is 0 (mod N). Thus, $(b+1)$ must contain one factor of N, and $(b-1)$ another.

Now, if r is the order of m (mod N) and r is even, then $b = m^{r/2}$ is the solution of the equation $b^2 \equiv 1 \pmod{N}$. If $m^{r/2} \not\equiv \pm 1 \pmod{N}$ then $\gcd(m^{r/2}-1, N) > 1$. We have proved the following:

Lemma 13.14 *If the order r of m* (mod N) *is even and*

$$m^{r/2} \not\equiv \pm 1 \pmod{N}$$

then

$$\gcd\bigl(m^{r/2} - 1, N\bigr) > 1.$$

The above process may fail if r is odd or if r is even but $m^{r/2}$ is a trivial solution. However, due to the following theorem, these situations can arise only with small probability.

Theorem 13.15 *Let N be an odd natural number with prime factorization*

$$N = p_1^{\alpha_1} p_2^{\alpha_2} \cdots p_k^{\alpha_k}.$$

Suppose m is chosen at random, $1 \le m \le N$ *(with uniform distribution), satisfying* $\gcd(m, N) = 1$. *Let r be the order of m (mod N). Then*

$$P\bigl\{r : r \text{ is even and } m^{r/2} \not\equiv \pm 1 \pmod{N}\bigr\} \ge 1 - \frac{1}{2^{k-1}}.$$

The probability is positive if $k \ge 2$.

Proof Since r is the order, we never have $m^{r/2} \equiv -1 \pmod{N}$. One can prove that

$$P\bigl\{r; r \text{ is odd or } m^{r/2} \equiv -1 \pmod{N}\bigr\} \le \frac{1}{2^{k-1}},$$

by using the Chinese remainder theorem. □

Theorem 13.16 *If an integer N is sufficiently large and if it is a product of at least two prime numbers then the factoring algorithm finds the factors with the probability greater than* $\gamma/2$ *where* γ *is the constant defined in Theorem* 13.12. *One needs asymptotically*

$$O\bigl((\log N)^2 (\log\log N)(\log\log\log N)\bigr)$$

quantum gates to implement the quantum circuit for the factoring algorithm.

Proof The conclusion of the theorem follows from the description of the factoring algorithm and from Theorem 13.8 and Theorem 13.11. □

13.4 Notes

Shor proposed efficient quantum algorithms for the factoring problem and the discrete logarithm problem [713]. The presentation in this chapter is based on [710] and [220, 795]. For the basic mathematical results used in this chapter related to the number theory, we refer to [779]. The problem of computational complexity of quantum modular exponentiation (Step 2 in the description of the algorithm) is discussed in [371].

Chapter 14
Quantum Algorithm III

In this chapter we will consider a new approach to quantum computations. We shall discuss an algorithm introduced by Ohya and Volovich which can solve the NP-complete satisfiability (SAT) problem in polynomial time. The algorithm goes beyond the quantum Turing machine paradigm.

As we discussed in Chaps. 2 and 10, it is known that all NP-complete (NPC, for brevity) problems are equivalent and an essential question is whether there exists an algorithm to solve an NP-complete problem in polynomial time. There are examples of the NP-complete problems such as the knapsack problem, the traveling salesman problem, the integer programming problem, the subgraph isomorphism problem, the satisfiability problem, which have been studied for decades and for which all known algorithms have a running time that is exponential in the size of input.

It is widely believed that quantum computers are more efficient than classical computers. In particular, Shor gave a quantum polynomial-time algorithm for the factoring problem, as was discussed in Chap. 13. However, it is known that this problem is not NP-complete, but is NP-intermediate.

In Sect. 14.1, the quantum algorithm of SAT problem, due to Ohya and Masuda, is discussed (OM algorithm). Accardi and Sabbadini showed that OM algorithm is a combinatorial one, and they rewrote the OM algorithm in the language of combinatorics. In Sect. 14.2, it is shown that the SAT problem can be solved in polynomial time by means of chaos dynamics with the quantum algorithm given in Sect. 14.1, which goes beyond the usual quantum Turing machine algorithm. The use of chaos dynamics depends on the state computed by quantum unitary operators, so that our method is an example of the adaptive dynamics explained in Chap. 10. This algorithm is called OV-SAT algorithm here. In Sect. 14.3, the channel expression of algorithm is discussed as a possible extension of usual quantum algorithm to a more general quantum algorithm pointed out in Sect. 14.2. The general QTM is used to discuss the OV-SAT algorithm in Sect. 14.4. In Sect. 14.5, applying the stochastic limit and the state adaptive dynamics, the SAT algorithm is reconsidered.

M. Ohya, I. Volovich, *Mathematical Foundations of Quantum Information and Computation and Its Applications to Nano- and Bio-systems*, Theoretical and Mathematical Physics,
DOI 10.1007/978-94-007-0171-7_14, © Springer Science+Business Media B.V. 2011

14.1 Quantum Algorithm of SAT Problem

In this section, we discuss the quantum algorithm of the SAT problem and point out that this problem, hence all other NPC problems, can be solved in polynomial time by a quantum computer if the superposition of two orthogonal vectors $|0\rangle$ and $|1\rangle$ is physically detected. However, *this detection is considered not to be possible in the present technology*, so that in the following sections we discuss what detection is possible to answer the question of the NPC problem.

Let $X \equiv \{x_1, \dots, x_n\}$ be a set. x_k and its negation $\overline{x}_k$ $(k = 1, \dots, n)$ are called literals. Let $\bar{X} \equiv \{\bar{x}_1, \dots, \bar{x}_n\}$ be a set, then the set of all literals is denoted by $X' \equiv X \cup \bar{X} = \{x_1, \dots, x_n, \bar{x}_1, \dots, \bar{x}_n\}$. We write the set of all subsets of X and $\bar{X}$ as $F(X)$ and $F(\bar{X})$, respectively. An element $C \in F(X')$ is called a clause. A map $t : X' \to \{0, 1\}$ is called a truth assignment if for all literals $x_k \in X'$ it satisfies the following equation

$$t(x_k) + t(\bar{x}_k) = 1.$$

Definition 14.1 A clause C is satisfiable if there exists a truth assignment t such that

$$\begin{aligned} t(C) &\equiv \bigvee_{x \in C} t(x) \\ &= 1. \end{aligned}$$

For $X = \{x_1, \dots, x_n\}$ and a truth assignment t, we put

$$t(x_1) = \varepsilon_1, \dots, t(x_n) = \varepsilon_n,$$

where $\varepsilon_1, \dots, \varepsilon_n \in \{0, 1\}$, and we write t as a sequence of binary symbols:

$$t = \varepsilon_1 \cdots \varepsilon_n.$$

Definition 14.2 Let $\mathcal{C} = \{C_1, \dots, C_m\}$ be a set of clauses, $\mathcal{C}$ is satisfiable iff the meet of all truth values of C_j is 1, that is,

$$t(\mathcal{C}) \equiv \bigwedge_{j=1}^{m} t(C_j) = 1.$$

Thus the SAT problem is written as follows:

Problem 14.3 (SAT problem) *Given a Boolean set $X \equiv \{x_1, \dots, x_n\}$ and a set $\mathcal{C} = \{C_1, \dots, C_m\}$ of clauses, determine whether $\mathcal{C}$ is satisfiable or not.*

In other words, this problem is asking whether there exists a truth assignment to make $\mathcal{C}$ satisfiable. It is known in usual algorithms that it takes polynomial time to check the satisfiability only when a specific truth assignment is given, but we cannot determine the satisfiability in polynomial time when an assignment is not specified.

Since it is necessary to compute $t(C)$ for all assignments in the worst case to determine whether C is SAT or not, it is known that the SAT problem belongs to NP when m is a polynomial of n [278].

Ohya and Masuda pointed out [595] that the SAT problem, hence all other NP-complete problems, can be solved in polynomial time by a quantum computer if the superposition of two orthogonal vectors $|0\rangle$ and $|1\rangle$ is physically detected. However, this detection is considered not to be possible in the present technology. The problem to be overcome is how to distinguish the pure vector $|0\rangle$ from the superposed one $\alpha|0\rangle + \beta|1\rangle$, obtained by the OM SAT-quantum algorithm, if β is not zero, but very small. If such a distinction is possible, then we can solve the NPC problem in polynomial time. In [600, 602], it is shown that it can be possible by combining nonlinear chaos amplifier with the quantum algorithm, which implies the existence of a mathematical algorithm solving NP = P.

14.1.1 Quantum Computing

In this section, we review the fundamentals of quantum computation discussed in Chap. 12. We describe a qubit as a unit vector on the Hilbert space $\mathbb{C}^2$ such that $|\psi\rangle = \alpha|0\rangle + \beta|1\rangle$, where $|0\rangle = \binom{1}{0}$ and $|1\rangle = \binom{0}{1}$, α and β are complex numbers with the property $|\alpha|^2 + |\beta|^2 = 1$. We represent two or more sequences of qubits as $|x,y\rangle \equiv |x\rangle \otimes |y\rangle$, $|x^n\rangle = \underbrace{|x\rangle \otimes \cdots \otimes |x\rangle}_{n}$. The n qubit sequence is denoted in Sect. 12.1 of Chap. 12.

$$\begin{aligned}
|e_0\rangle &= |0, 0, \ldots, 0\rangle, \\
|e_1\rangle &= |1, 0, \ldots, 0\rangle, \\
|e_2\rangle &= |0, 1, \ldots, 0\rangle, \\
&\vdots \\
|e_{2^n-1}\rangle &= |1, 1, \ldots, 1\rangle.
\end{aligned}$$

In quantum computing, we apply unitary operators to qubits; the fundamental gates such as the NOT gate, C-NOT gate, and the C-C-N gate are defined as

$$\begin{aligned}
U_{\mathrm{NOT}} &\equiv |1\rangle\langle 0| + |0\rangle\langle 1|, \\
U_{\mathrm{CN}} &= |0\rangle\langle 0| \otimes I + |1\rangle\langle 1| \otimes U_{\mathrm{NOT}}, \\
U_{\mathrm{CCN}} &= |0\rangle\langle 0| \otimes I \otimes I + |1\rangle\langle 1| \otimes |0\rangle\langle 0| \otimes I + |1\rangle\langle 1| \otimes |1\rangle\langle 1| \otimes U_{\mathrm{NOT}}.
\end{aligned}$$

Note that these gates act on the Hilbert space $\mathcal{H} = \mathbb{C}^2$, $(\mathbb{C}^2)^{\otimes 2}$ and $(\mathbb{C}^2)^{\otimes 3}$, respectively.

The quantum computation is performed by the unitary gates constructed from several fundamental gates such as the NOT gate, Controlled–NOT gate, Controlled–Controlled–NOT gate. It is convenient for computation to construct the AND gate U_{AND}, OR gate U_{OR}, and the COPY gate U_{COPY}. The AND gate and the OR gate are two unitary operators on $\mathcal{H} = (\mathbb{C}^2)^{\otimes 3}$. For any state vectors of the form $|\varepsilon_1, \varepsilon_2, 0\rangle$ in $(\mathbb{C}^2)^{\otimes 3}$ for $\varepsilon_1, \varepsilon_2 \in \{0, 1\}$, we obtain

$$U_{\text{AND}}|\varepsilon_1, \varepsilon_2, 0\rangle = |\varepsilon_1, \varepsilon_2, \varepsilon_1 \wedge \varepsilon_2\rangle,$$
$$U_{\text{OR}}|\varepsilon_1, \varepsilon_2, 0\rangle = |\varepsilon_1, \varepsilon_2, \varepsilon_1 \vee \varepsilon_2\rangle.$$

The COPY gate U_{COPY} is a unitary operator acting on $(\mathbb{C}^2)^{\otimes 2}$, which copies the information of the first qubit to the second qubit. For instance, applying U_{COPY} to any state vectors of the form $|\varepsilon_1, 0\rangle$, we have

$$U_{\text{COPY}}|\varepsilon_1, 0\rangle = |\varepsilon_1, \varepsilon_1\rangle.$$

Let us write these unitary gates more precisely, then we can easily check the unitarity of each gate:

- AND gate

$$\begin{aligned} U_{\text{AND}} \equiv &\sum_{\varepsilon_1, \varepsilon_2 \in \{0,1\}} |\varepsilon_1, \varepsilon_2, \varepsilon_1 \wedge \varepsilon_2\rangle\langle\varepsilon_1, \varepsilon_2, 0| + |\varepsilon_1, \varepsilon_2, 1 - \varepsilon_1 \wedge \varepsilon_2\rangle\langle\varepsilon_1, \varepsilon_2, 1| \\ &= |0,0,0\rangle\langle 0,0,0| + |0,0,1\rangle\langle 0,0,1| + |1,0,0\rangle\langle 1,0,0| + |1,0,1\rangle\langle 1,0,1| \\ &\quad + |0,1,0\rangle\langle 0,1,0| + |0,1,1\rangle\langle 0,1,1| + |1,1,1\rangle\langle 1,1,0| \\ &\quad + |1,1,0\rangle\langle 1,1,1|. \end{aligned}$$

- OR gate

$$\begin{aligned} U_{\text{OR}} \equiv &\sum_{\varepsilon_1, \varepsilon_2 \in \{0,1\}} |\varepsilon_1, \varepsilon_2, \varepsilon_1 \vee \varepsilon_2\rangle\langle\varepsilon_1, \varepsilon_2, 0| + |\varepsilon_1, \varepsilon_2, 1 - \varepsilon_1 \vee \varepsilon_2\rangle\langle\varepsilon_1, \varepsilon_2, 1| \\ &= |0,0,0\rangle\langle 0,0,0| + |0,0,1\rangle\langle 0,0,1| + |1,0,1\rangle\langle 1,0,0| + |1,0,0\rangle\langle 1,0,1| \\ &\quad + |0,1,1\rangle\langle 0,1,0| + |0,1,0\rangle\langle 0,1,1| + |1,1,1\rangle\langle 1,1,0| \\ &\quad + |1,1,0\rangle\langle 1,1,1|. \end{aligned}$$

- COPY gate

$$\begin{aligned} U_{\text{COPY}} \equiv &\sum_{\varepsilon_1 \in \{0,1\}} |\varepsilon_1, \varepsilon_1\rangle\langle\varepsilon_1, 0| + |\varepsilon_1, 1 - \varepsilon_1\rangle\langle\varepsilon_1, 1| \\ &= |0,0\rangle\langle 0,0| + |0,1\rangle\langle 0,1| + |1,1\rangle\langle 1,0| + |1,0\rangle\langle 1,1|. \end{aligned}$$

Here we extend the above expressions to the case of N qubits. Let N be a positive integer greater than 3, and s, t, u be three numbers in $\{1, 2, \ldots, N\}$ such that

$s < t < u$. We rewrite the above four unitary gates into those acting on $(\mathbb{C}^2)^{\otimes N}$.

$$U_{\mathrm{NOT}}(s) \equiv \underbrace{I \otimes \cdots \otimes I}_{s-1} \otimes U_{\mathrm{NOT}} \otimes \underbrace{I \otimes \cdots \otimes I}_{N-s},$$

$$U_{\mathrm{NOT}}(1) \equiv U_{\mathrm{NOT}} \otimes I^{\otimes N-1}, \qquad U_{\mathrm{NOT}}(N) \equiv I^{\otimes N-1} \otimes U_{\mathrm{NOT}}.$$

The unitary gate $U_{\mathrm{NOT}}(s)$ applies U_{NOT} to the sth qubit. In the subsequent discussion, put

$$\underbrace{I \otimes \cdots \otimes I}_{m} = I^{\otimes m}, \quad m \in \mathbb{N}$$

so that

$$U_{\mathrm{NOT}}(s) = I^{\otimes s-1} \otimes U_{\mathrm{NOT}} \otimes I^{\otimes N-s}.$$

Let us define the following unitary operators:

$$\begin{aligned} U_{\mathrm{OR}}(s,t,u) \equiv \sum_{\varepsilon_1,\varepsilon_2 \in \{0,1\}} & \big(I^{\otimes s-1} \otimes |\varepsilon_1\rangle\langle\varepsilon_1| \otimes I^{\otimes t-s-1} \otimes |\varepsilon_2\rangle\langle\varepsilon_2| \\ & \otimes I^{\otimes u-t-1} \otimes |\varepsilon_1 \vee \varepsilon_2\rangle\langle 0| \otimes I^{\otimes N-u}. \\ & + I^{\otimes s-1} \otimes |\varepsilon_1\rangle\langle\varepsilon_1| \otimes I^{\otimes t-s-1} \otimes |\varepsilon_2\rangle\langle\varepsilon_2|. \\ & \otimes I^{\otimes u-t-1} \otimes |1-\varepsilon_1 \vee \varepsilon_2\rangle\langle 1| \otimes I^{\otimes N-u}\big), \end{aligned}$$

$$\begin{aligned} U_{\mathrm{AND}}(s,t,u) \equiv \sum_{\varepsilon_1,\varepsilon_2 \in \{0,1\}} & \big(I^{\otimes s-1} \otimes |\varepsilon_1\rangle\langle\varepsilon_1| \otimes I^{\otimes t-s-1} \otimes |\varepsilon_2\rangle\langle\varepsilon_2| \\ & \otimes I^{\otimes u-t-1} \otimes |\varepsilon_1 \wedge \varepsilon_2\rangle\langle 0| \otimes I^{\otimes N-u} \\ & + I^{\otimes s-1} \otimes |\varepsilon_1\rangle\langle\varepsilon_1| \otimes I^{\otimes t-s-1} \otimes |\varepsilon_2\rangle\langle\varepsilon_2| \\ & \otimes I^{\otimes u-t-1} \otimes |1-\varepsilon_1 \wedge \varepsilon_2\rangle\langle 1| \otimes I^{\otimes N-u}\big), \end{aligned}$$

$$\begin{aligned} U_{\mathrm{COPY}}(s,t) \equiv \sum_{\varepsilon_1,\varepsilon_2 \in \{0,1\}} & \big(I^{\otimes s-1} \otimes |\varepsilon_1\rangle\langle\varepsilon_1| \otimes I^{\otimes t-s-1} |\varepsilon_1\rangle\langle 0| \otimes I^{\otimes N-t} \\ & + I^{\otimes s-1} \otimes |\varepsilon_1\rangle\langle\varepsilon_1| \otimes I^{\otimes t-s-1} \otimes |1-\varepsilon_1\rangle\langle 1| \otimes I^{\otimes N-t}\big). \end{aligned}$$

For the state vector on the Hilbert space $\mathcal{H} = (\mathbb{C}^2)^{\otimes N}$,

$$|\psi\rangle = \big|\varepsilon_1, \ldots, \varepsilon_s, \ldots, \varepsilon_t, \ldots, \underbrace{0}_{u\text{th qubit}}, \ldots, \varepsilon_N\big\rangle,$$

we have

$$U_{\mathrm{OR}}(s,t,u)|\psi\rangle = |\varepsilon_1, \ldots, \varepsilon_s, \ldots, \varepsilon_t, \ldots, \varepsilon_s \vee \varepsilon_t, \ldots, \varepsilon_N\rangle,$$

$$U_{\mathrm{AND}}(s,t,u)|\psi\rangle = |\varepsilon_1,\dots,\varepsilon_s,\dots,\varepsilon_t,\dots,\varepsilon_s\wedge\varepsilon_t,\dots,\varepsilon_N\rangle.$$

One can consider $U_{\mathrm{OR}}(s,t,u)$ and $U_{\mathrm{AND}}(s,t,u)$ as the OR gate and AND gates with these conditions, respectively. $U_{\mathrm{COPY}}(s,t)$ copies the sth qubit to the tth qubit.

Unitarity of these gates is shown easily because these gates are made of U_{NOT}, U_{OR}, U_{AND}, U_{COPY}, and the identity operators.

14.1.2 Quantum Algorithm of SAT Problem

In this section, we explain the quantum algorithm of SAT problem based on [595]. Here, we estimate the total number of qubits more rigorously and show how to construct the unitary gates more precisely for every case.

We first calculate the total number of qubits and show that this number depends on the input data only, and is calculated in polynomial time. Then we decide on the Hilbert space for the quantum algorithm and the initial state vector on it.

Let $\mathcal{C}=\{C_1,\dots,C_m\}$ be a set of clauses on $X'\equiv\{x_1,\dots,x_n,\bar{x}_1,\dots,\bar{x}_n\}$. The computational basis of this algorithm is on the Hilbert space $\mathcal{H}=(\mathbb{C}^2)^{\otimes n+\mu+1}$ where μ is a number of dust qubits (i.e., the qubits added for constructing unitary gates) written linearly in mn.

Let

$$|v_{\mathrm{in}}\rangle \equiv \left|0^n,0^\mu,0\right\rangle$$

be an initial state vector. A unitary operator $U_{\mathcal{C}}:\mathcal{H}\to\mathcal{H}$ computes $t(\mathcal{C})$ for the truth assignment e_i $(i=1,\dots,2^{n-1})$ as follows

$$\begin{aligned} U_{\mathcal{C}}|v_{\mathrm{in}}\rangle &= U_{\mathcal{C}}\left|0^n,0^\mu,0\right\rangle \\ &= \frac{1}{\sqrt{2^n}}\sum_{i=0}^{2^n-1}\left|e_i,d^\mu,t(\mathcal{C})\right\rangle \equiv |v_{\mathrm{out}}\rangle, \end{aligned}$$

where $|d^\mu\rangle$ stands for the dust qubits denoted by the μ strings of binary symbols.

Theorem 14.4 *For a set of clauses $\mathcal{C}=\{C_1,\dots,C_m\}$ on $X'\equiv\{x_1,\dots,x_n,\bar{x}_1,\dots,\bar{x}_n\}$, the number μ of dust qubits for an algorithm of the SAT problem is*

$$\begin{aligned} \mu &= s_f-1-n \\ &= \sum_{k=1}^{m}\left(i_k+i_k'+\delta_{1,i_k+i_k'}\right)-2 \\ &\le 2nm, \end{aligned}$$

where i_k is the number of literals in C_k, i_k' is the number of negations, s_f is the index of the last qubit given by

$$s_f = s_m - 1 + i_m + i_m' + \delta_{1,i_m+i_m'},$$

and s_k is the index of workspace for the calculation of $t(C_k)$ such that

$$s_k = s_{k-1} - 1 + i_{k-1} + i'_{k-1} + \delta_{1,i_{k-1}+i'_{k-1}} \tag{14.1}$$

and where

$$s_1 = n + 1,$$
$$s_2 = s_1 - 1 + i_1 + i'_1 + \delta_{1,i_1+i'_1} - 1.$$

Proof In the case of $m = 1$, the numbers of dust qubits to calculate OR and that of OR with negation are $i_1 - 1$ and i'_1, respectively. If $i_1 + i'_1 = 1$, then we add one more dust qubit to preserve $t(C_k)$. Therefore, the number of dust qubits μ is

$$\begin{aligned}\mu &= s_f - 1 - n \\ &= s_1 - 1 + i_1 + i'_1 + \delta_{1,i_1+i'_1} - 1 - 1 - n \\ &= i_1 + i'_1 + \delta_{1,i_1+i'_1} - 2\end{aligned}$$

where $s_1 = n + 1$.

In the case of $m = 2$, the number of dust qubits to calculate $t(C_2)$ is $i_2 + i'_2 + \delta_{1,i'_2+i'_2} - 1$, and one needs to add 1 qubit to calculate $t(C_1) \wedge t(C_2)$. Thus, μ becomes

$$\begin{aligned}\mu &= s_f - 1 - n \\ &= s_2 - 1 + i_2 + i'_2 + \delta_{1,i_2+i'_2} - 1 - n \\ &= s_1 - 1 + i_1 + i'_1 + \delta_{1,i_1+i'_1} - 1 + i_2 + i'_2 + \delta_{1,i_2+i'_2} - 1 - n \\ &= \sum_{k=1}^{2} (i_k + i'_k + \delta_{1,i_k+i'_k}) - 2.\end{aligned}$$

In the case of $m = k \geq 3$, we prove by mathematical induction. Suppose that (14.1) holds. Since the number of dust qubits to calculate $t(C_{k+1})$ and $\bigwedge_{i=1}^{k} t(C_i)$ is $i_{k+1} + i'_{k+1} + \delta_{1,i'_{k+1}+i'_{k+1}}$, we obtain

$$\begin{aligned}\mu &= s_f - 1 - n \\ &= s_{k+1} - 1 + i_{k+1} + i'_{k+1} + \delta_{1,i'_{k+1}+i'_{k+1}} - 1 - n \\ &= s_k - 1 + i_k + i'_k + \delta_{1,i'_k+i'_k} - 1 - 1 - n + i_{k+1} + i'_{k+1} + \delta_{1,i'_{k+1}+i'_{k+1}} \\ &= \sum_{k=1}^{m} (i_k + i'_k + \delta_{1,i_1+i'_1}) - 2.\end{aligned}$$

□

Theorem 14.5 *For a set of clauses $\mathcal{C}=\{C_1,\dots,C_m\}$, we can construct the unitary operators $U_{\mathrm{AND}}(i)$, $U_{\mathrm{OR}}(i)$ and $U_{\mathcal{C}}$ to calculate the truth value of $\mathcal{C}$ as*

$$U_{\mathcal{C}} \equiv \prod_{i=1}^{m-1} U_{\mathrm{AND}}(i) \prod_{j=1}^{m} U_{\mathrm{OR}}(j) H(n)$$

where $H(k)$ is a unitary operator applying Hadamard transformation to the first k qubits, that is,

$$H(k) = H^{\otimes k} I^{\otimes N-k}.$$

Proof Following the algorithm below, we construct the unitary operator $U_{\mathcal{C}}$ to compute $t(\mathcal{C})$.

1. Compute the indices s_i $(i=1,\dots,m)$ and s_f.
2. Construct the unitary gates computing $t(C_k)$ from the elements of C_k.
3. Construct the unitary gate computing $t(\mathcal{C})$.

First, we compute the indices s_k as

$$\begin{aligned} s_1 &= n+1, \\ s_2 &= n+i_1+i_1'-2+\delta_{1,i_1+i_1'}, \\ s_k &= s_{k-1}+i_{k-1}+i_{k-1}'-1+\delta_{1,i_k+i_k'} \quad (k\ge 3). \end{aligned}$$

Moreover, $s_f = s_m + i_m + i_m' - 1 + \delta_{1,i_m+i_m'}$. Then we construct the unitary gates to compute $t(C_k)$. Here, we give an expression relating $U_{\mathrm{OR}}(s,t,u)$ with $x_s, x_t, x_u \in X$ and $U_{\mathrm{OR}}(\bar{s},t,u)$ with $x_t, x_u \in X$, $x_s \in \bar{X}$:

$$U_{\mathrm{OR}}(\bar{s},t,u) \equiv U_{\mathrm{NOT}}(s) U_{\mathrm{OR}}(s,t,u) U_{\mathrm{NOT}}(s).$$

Similarly,

$$\begin{aligned} U_{\mathrm{OR}}(s,\bar{t},u) &\equiv U_{\mathrm{NOT}}(t) U_{\mathrm{OR}}(s,t,u) U_{\mathrm{NOT}}(t), \\ U_{\mathrm{OR}}(\bar{s},\bar{t},u) &\equiv U_{\mathrm{NOT}}(s) U_{\mathrm{NOT}}(t) U_{\mathrm{OR}}(s,t,u) U_{\mathrm{NOT}}(s) U_{\mathrm{NOT}}(t). \end{aligned}$$

We define an arrangement map $J(i)$ such that for x_j, the ith literal in an ordered sequence S made from C,

$$J(i) \equiv \begin{cases} j, & x_j \in S \cap X, \\ \bar{j}, & x_j \in S \cap \overline{X}, \end{cases} \qquad 1 \le i \le |S|.$$

For instance, if $S=\{x_3, x_5, \bar{x}_8\}$ then

$$J(1)=3, \qquad J(2)=5, \qquad J(3)=\bar{8}.$$

The unitary operator $U_{\mathrm{OR}}(k)$ is constructed using the previous U_{OR} and U_{COPY} with J as follows:

$$U_{\mathrm{OR}}(k)=\begin{cases} U_{\mathrm{COPY}}(J(1),s_k), & |C_k|=1,\\ U_{\mathrm{OR}}(J(1),J(2),s_k), & |C_k|=2,\\ \prod_{i=1}^{|C_k|-2}U_{\mathrm{OR}}(J(i+2),s_k+i-1,s_k+i) & \\ \quad\times U_{\mathrm{OR}}(J(1),J(2),s_k), & |C_k|\geq 3, \end{cases}$$

where $|C|$ is the cardinal number of C.

Moreover, the unitary gate $U_{\mathrm{AND}}(k)$ is represented as follows: In the case of $m=1$, U_{AND} copies $t(C_1)$ to the last qubit

$$U_{\mathrm{AND}}(k)=U_{\mathrm{COPY}}(s_f-1,s_f).$$

In the case of $m\geq 2$,

$$U_{\mathrm{AND}}(k)=\begin{cases} U_{\mathrm{AND}}(s_{k+1}-1,s_{k+2}-2,s_{k+2}-1), & 1\leq k\leq m-2,\\ U_{\mathrm{AND}}(s_m-1,s_f-1,s_f), & k=m-1. \end{cases}$$

Therefore, for $\mathcal{C}=\{C_1,\dots,C_m\}$ on $X=\{x_1,\dots,x_n\}$, $U_{\mathcal{C}}$ is given by the following

$$U_{\mathcal{C}}\equiv\prod_{i=1}^{m-1}U_{\mathrm{AND}}(i)\prod_{j=1}^{m}U_{\mathrm{OR}}(j)H(n).$$

Obviously, for all k and i we can compute s_k and $J(i)$ in a polynomial time in n, so we can construct $U_{\mathcal{C}}$ in a polynomial time in n. □

The last qubit is the state representing the result of the computation obtained by applying a unitary operator to the initial state. Let r be the cardinality of the set $\mathcal{C}$ such that

$$t(\mathcal{C})=1$$

and put $q=\sqrt{\frac{r}{2^n}}$. After the quantum computation, the final state vector is

$$|v_{\mathrm{out}}\rangle=\sqrt{1-q^2}|\varphi_0\rangle\otimes|0\rangle+q|\varphi_1\rangle\otimes|1\rangle$$

where $|\varphi_0\rangle$ and $|\varphi_1\rangle$ are normalized n qubit states.

Let $\rho=|v_{\mathrm{in}}\rangle\langle v_{\mathrm{in}}|$ be the initial state, the computation is described by a channel $\Lambda_U^*=\mathrm{Ad}_{U_{\mathcal{C}}}$ (unitary channel). Then the final state $\rho'=|v_{\mathrm{out}}\rangle\langle v_{\mathrm{out}}|$ is written as

$$\Lambda_U^*(\rho)=U_{\mathcal{C}}\rho U_{\mathcal{C}}^*\equiv\rho'.$$

Therefore, if we have an experiment to detect the above vector $|v_{\mathrm{out}}\rangle$ or the state ρ' directly, then we can conclude that the SAT is solved. However, this will not be in the case that we are now, so that we need further elaborated treatment, which is discussed in the next section. The difficulty comes from the following observation.

Proposition 14.6 *$\mathcal{C}$ is SAT if and only if*

$$P_{n+\mu,1}U_{\mathcal{C}}|v_{\text{in}}\rangle \neq 0$$

where $P_{n+\mu,1}$ denotes the projector

$$P_{n+\mu,1} \equiv I^{\otimes n+\mu} \otimes |1\rangle\langle 1|$$

onto the subspace of $\mathcal{H}$ spanned by the vectors $|\varepsilon^n, \varepsilon^\mu, 1\rangle$.

After a measurement of the event $P_{n+\mu,1}$, the state is changed according to the standard theory of quantum measurement. This measurement is expressed by a CP channel Λ_M^* as

$$\begin{aligned}\Lambda_M^*(\rho') &= \frac{P_{n+\mu,1}\rho' P_{n+\mu,1}}{\text{tr}\rho' P_{n+\mu,1}} = q^2|1\rangle\langle 1| + \left(1-q^2\right)|0\rangle\langle 0| \\ &\equiv \bar{\rho}\end{aligned}$$

where $0 \leq q^2 \leq 1$.

The final step to check the satisfiability of $\mathcal{C}$ is to apply the projection $P_{n+\mu,1} \equiv \bigotimes_1^{n+\mu} I \otimes |1\rangle\langle 1|$ to the state $|v_{\text{out}}\rangle = U_{\mathcal{C}}|v_{\text{in}}\rangle$, which is mathematically equivalent to computing the value $\langle v_{\text{out}}|P_{n+\mu,1}|v_{\text{out}}\rangle$. If the vector $P_{n+\mu,1}|v_{\text{out}}\rangle$ exists or the value $\langle v_{\text{out}}|P_{n+\mu,1}|v_{\text{out}}\rangle$ is not 0, then we conclude that $\mathcal{C}$ is satisfiable. The value of $\langle v_{\text{out}}|P_{n+\mu,1}|v_{\text{out}}\rangle$ is often very small and is difficult to detect even when it is not zero, so that we desire to have some way of amplifying the value, which is considered in the next section.

14.1.3 Example

In this subsection, we give an example of the SAT computation in the case $n = 4$. Let the literals and the clauses be

$$\begin{aligned}X &= \{x_1, x_2, x_3, x_4\}, \quad n = 4, \\ \mathcal{C} &= \{C_1, C_2, C_3, C_4\}, \\ C_1 &= \{x_1, x_4, \bar{x}_2\}, \qquad C_2 = \{x_2, x_3, x_4\}, \\ C_3 &= \{x_1, \bar{x}_3\}, \qquad C_4 = \{x_3, \bar{x}_1, \bar{x}_2\}.\end{aligned}$$

We first calculate the indices of the work space:

$$\begin{aligned}s_1 &= n + 1 = 5, \\ s_2 &= s_1 + i_1 + i_1' + \delta_{1,i_1+i_1'} - 1 \\ &= 5 + 2 + 1 + 0 - 1 = 7,\end{aligned}$$

$$\begin{aligned} s_3 &= s_2 + i_2 + i_2' + \delta_{1,i_2+i_2'} \\ &= 7 + 3 = 10, \\ s_4 &= s_3 + i_3 + i_3' + \delta_{1,i_3+i_3'} \\ &= 10 + 1 + 1 = 12, \\ s_f &= s_4 + i_4 + i_4' + \delta_{1,i_4+i_4'} - 1 \\ &= 12 + 1 + 2 - 1 = 7. \end{aligned}$$

Second, we construct the OR and AND gates:

$$\begin{aligned} U_{\mathrm{OR}}(1) &= U_{\mathrm{OR}}(\bar{2}, 5, 6) U_{\mathrm{OR}}(1, 4, 5), \\ U_{\mathrm{OR}}(2) &= U_{\mathrm{OR}}(4, 7, 8) U_{\mathrm{OR}}(2, 3, 7), \\ U_{\mathrm{OR}}(3) &= U_{\mathrm{OR}}(1, \bar{3}, 10), \\ U_{\mathrm{OR}}(4) &= U_{\mathrm{OR}}(\bar{2}, 12, 13) U_{\mathrm{OR}}(3, \bar{1}, 12), \\ U_{\mathrm{AND}}(1) &= U_{\mathrm{AND}}(6, 8, 9), \\ U_{\mathrm{AND}}(2) &= U_{\mathrm{AND}}(9, 10, 11), \\ U_{\mathrm{AND}}(3) &= U_{\mathrm{AND}}(11, 13, 14). \end{aligned}$$

Thus, we obtain a unitary gate U_C

$$\begin{aligned} U_C =& U_{\mathrm{AND}}(11, 13, 14) U_{\mathrm{AND}}(9, 10, 11) U_{\mathrm{AND}}(6, 8, 9) \\ & \cdot U_{\mathrm{OR}}(\bar{2}, 12, 13) U_{\mathrm{OR}}(3, \bar{1}, 12) U_{\mathrm{OR}}(1, \bar{3}, 10) \\ & \cdot U_{\mathrm{OR}}(4, 7, 8) U_{\mathrm{OR}}(2, 3, 7) U_{\mathrm{OR}}(\bar{2}, 5, 6) U_{\mathrm{OR}}(1, 4, 5). \end{aligned}$$

Let $|v_{\mathrm{in}}\rangle = |0^4, 0^{10}, 0\rangle$ be an initial state, applying $H(4)$, we have

$$\begin{aligned} |v\rangle &\equiv H(4)\big|0^4, 0^9, 0\big\rangle \\ &= \frac{1}{(\sqrt{2})^4} \sum_{i=0}^{2^4-1} \big|e_i, 0^9, 0\big\rangle. \end{aligned}$$

Applying $\prod U_{\mathrm{OR}}(k)$, we have

$$\begin{aligned} & U_{\mathrm{OR}}(4) U_{\mathrm{OR}}(3) U_{\mathrm{OR}}(2) U_{\mathrm{OR}}(1) |v\rangle \\ &= \frac{1}{(\sqrt{2})^4} U_{\mathrm{OR}}(4) U_{\mathrm{OR}}(3) U_{\mathrm{OR}}(2) U_{\mathrm{OR}}(1) \sum_{\varepsilon_i \in \{0,1\}} \big|\varepsilon_1, \varepsilon_2, \varepsilon_3, \varepsilon_4, 0^{10}\big\rangle \\ &= \frac{1}{(\sqrt{2})^4} U_{\mathrm{OR}}(4) U_{\mathrm{OR}}(3) U_{\mathrm{OR}}(2) \sum_{\varepsilon_i \in \{0,1\}} \big|\varepsilon_1, \varepsilon_2, \varepsilon_3, \varepsilon_4, \varepsilon_1 \vee \varepsilon_4, \varepsilon_1 \vee \varepsilon_4 \vee \bar{\varepsilon}_2, 0^9\big\rangle \end{aligned}$$

$$= \frac{1}{(\sqrt{2})^4} \sum_{\varepsilon_i \in \{0,1\}} |\varepsilon_1, \varepsilon_2, \varepsilon_3, \varepsilon_4, \varepsilon_1 \vee \varepsilon_4, \varepsilon_1 \vee \varepsilon_4 \vee \bar{\varepsilon}_2, \varepsilon_2 \vee \varepsilon_3, \varepsilon_2 \vee \varepsilon_3 \vee \varepsilon_4, 0,$$

$$\varepsilon_1 \vee \bar{\varepsilon}_3, 0, \varepsilon_3 \vee \bar{\varepsilon}_1, \varepsilon_3 \vee \bar{\varepsilon}_1 \vee \bar{\varepsilon}_2, 0\rangle \equiv |v'\rangle.$$

Applying AND gates to the state $|v'\rangle$, we obtain

$$\begin{aligned} U_{\mathrm{AND}}(m)|v'\rangle &= U_{\mathrm{AND}}(11, 14, 15) U_{\mathrm{AND}}(9, 10, 11) U_{\mathrm{AND}}(6, 8, 9)|v'\rangle \\ &= \frac{1}{(\sqrt{2})^4} \sum_{\varepsilon_i \in \{0,1\}} \big|\varepsilon_1, \varepsilon_2, \varepsilon_3, \varepsilon_4, \varepsilon_1 \vee \varepsilon_4, \varepsilon_1 \vee \varepsilon_4 \vee \bar{\varepsilon}_2, \\ &\qquad \varepsilon_2 \vee \varepsilon_3, \varepsilon_2 \vee \varepsilon_3 \vee \varepsilon_4, \\ &\qquad t(C_1) \wedge t(C_2), \varepsilon_1 \vee \bar{\varepsilon}_3, t(C_1) \wedge t(C_2) \wedge t(C_3), \\ &\qquad \varepsilon_3 \vee \bar{\varepsilon}_1, \varepsilon_3 \vee \bar{\varepsilon}_1 \vee \bar{\varepsilon}_2, t(C)\big\rangle \equiv |v_{\mathrm{out}}\rangle. \end{aligned}$$

At last, we obtain the final state $\bar{\rho} = |v_{\mathrm{out}}\rangle\langle v_{\mathrm{out}}|$.

14.2 Quantum Chaos Algorithm

As we discussed in the previous section, the solvability of the SAT problem is reduced to checking that $\bar{\rho} \neq 0$. The difficulty is that the probability of $P_{n+\mu,1}$ is

$$\operatorname{tr} \bar{\rho} P_{n+\mu,1} = \big\| P_{n+\mu,1} |v_{\mathrm{out}}\rangle \big\|^2 = \frac{r}{2^n}$$

where r is the cardinality of the set C such that

$$t(C) = 1.$$

If r is suitably large to detect it, then the SAT problem is solved in polynomial time. However, for small r, the probability is very small, and this means that we, in fact, do not get information about the existence of the solution of the equation $t(C) = 1$, so that in such a case we need further consideration.

Let us simplify our notations. As shown above, after quantum computation the final state vector becomes

$$|v_{\mathrm{out}}\rangle = \sqrt{1 - q^2} |\varphi_0\rangle \otimes |0\rangle + q|\varphi_1\rangle \otimes |1\rangle$$

where $|\varphi_1\rangle$ and $|\varphi_0\rangle$ are normalized n qubit states and $q = \sqrt{r/2^n}$. Effectively, our problem is reduced to the following 1 qubit problem. We have the state

$$|\phi\rangle = \sqrt{1 - q^2} |0\rangle + q|1\rangle$$

and we want to distinguish between the cases $q = 0$ and $q > 0$ (small positive number).

It is argued that a quantum computer can speed up NP-problems quadratically, but not exponentially. The *no-go* theorem states that if the inner product of two quantum states is close to 1, then the probability that a measurement distinguishes between the two states is exponentially small. And one could claim that amplification of this distinguishability is not possible.

At this point, we emphasize that we do not propose to make a measurement which will overwhelmingly likely fail. Instead, we propose using the output of the quantum computer as an input for another device which uses chaotic dynamics.

The amplification would be impossible if we use the standard model of quantum computations with a unitary evolution. Ohya and Volovich proposed combining a quantum computer with a chaotic dynamics amplifier. Such a quantum chaos computer is a new model of computations, and we demonstrate that the amplification is possible in polynomial time.

One could object that we do not suggest a practical realization of the new model of computations. But at the moment, nobody knows how to make a practically useful implementation of the standard model of quantum computing at all. A quantum circuit or a quantum Turing machine is a mathematical model, though a convincing one. It seems to us that the quantum chaos computer deserves an investigation and has potential to be realizable.

14.2.1 Amplification Process in SAT Algorithm

If q^2 is very small such that q^2 cannot be distinguished from the value 0, then it is necessary to amplify q^2 so as to detect efficiently, for instance, $q^2 \geq \frac{1}{2}$. For this purpose, we have a few attempts of amplification. Here we take the chaos amplification proposed in [600, 602].

We will argue that chaos can play a constructive role in computations.

Chaotic behavior in a classical system is usually considered as an exponential sensitivity to initial conditions. It is this sensitivity we would like to use to distinguish between the cases $q = 0$ and $q > 0$ from the previous section.

Consider the so-called logistic map which is given by the equation

$$x_{n+1} = ax_n(1 - x_n) \equiv g_a(x), \quad x_n \in [0, 1].$$

The properties of the map depend on the parameter a. If we take, for example, $a = 3.71$ (see Sect. 10.7.1), then the Lyapunov exponent is positive, the trajectory is very sensitive to the initial value, and one has chaotic behavior. It is important to notice that if the initial value $x_0 = 0$, then $x_n = 0$ for all n.

Our quantum chaos computer will be consisting from two blocks. One block is the ordinary quantum computer performing computations with the output $|\phi\rangle = \sqrt{1-q^2}|0\rangle + q|1\rangle$. The second block is a computer performing computations of the *classical* logistic map. These two blocks should be connected in such a way that the state $|\phi\rangle$ first be transformed into the density matrix of the form

$$\bar{\rho} = q^2 P_1 + (1 - q^2) P_0$$

where P_1 and P_0 are the projectors to the state vectors $|1\rangle$ and $|0\rangle$. This connection is, in fact, nontrivial and actually it should be considered as the third block. One has to notice that P_1 and P_0 generate an abelian algebra which can be considered as a classical system. In the second block, the density matrix ρ above is interpreted as the initial data ρ_0, and we apply the logistic map as

$$\Lambda_{\mathrm{CA}}^*(\bar{\rho}) = \frac{(I + g_a(\bar{\rho})\sigma_3)}{2},$$

where I is the identity matrix and σ_3 is the z-component of Pauli matrices:

$$\bar{\rho}_k = (\Lambda_{\mathrm{CA}}^*)^k(\bar{\rho}).$$

To find a proper value k we finally measure the value of σ_3 in the state ρ_k such that

$$M_k \equiv \mathrm{tr}\, \bar{\rho}_k \sigma_3.$$

We obtain [600, 602]

Theorem 14.7

$$\bar{\rho}_k = \frac{(I + g_a^k(q^2)\sigma_3)}{2} \quad \textit{and} \quad M_k = g_a^k(q^2).$$

Thus the question is whether we can find such a number k in polynomially many steps of n satisfying the inequality $M_k \geq \frac{1}{2}$ for very small but non-zero q^2. Here we have to remark that if one has $q = 0$ then $\bar{\rho} = P_0$, and we obtain $M_k = 0$ for all k. If $q \neq 0$, the chaotic dynamics leads to the amplification of the small magnitude q in such a way that it can be detected. The transition from $\bar{\rho}$ to $\bar{\rho}_k$ is nonlinear and can be considered as a classical evolution because our algebra generated by P_0 and P_1 is abelian. The amplification can be done within at most $2n$ steps due to the following propositions. Since $g_a^k(q^2)$ is the x_k of the logistic map $x_{k+1} = g_a(x_k)$ with $x_0 = q^2$, we use the notation x_k in the logistic map for simplicity.

Theorem 14.8 *For the logistic map $x_{n+1} = ax_n(1 - x_n)$ with $a \in [0, 4]$ and $x_0 \in [0, 1]$, let x_0 be $\frac{1}{2^n}$ and a set J be $\{0, 1, 2, \ldots, n, \ldots, 2n\}$. If a is 3.71, then there exists an integer k in J satisfying $x_k > \frac{1}{2}$.*

Proof Suppose that there does not exist such an m in J. Then $x_m \leq \frac{1}{2}$ for any $m \in J$. The inequality $x_m \leq \frac{1}{2}$ implies

$$x_m = 3.71(1 - x_{m-1})x_{m-1} \geq \frac{3.71}{2} x_{m-1}.$$

Thus we have

$$\frac{1}{2} \geq x_m \geq \frac{3.71}{2} x_{m-1} \geq \cdots \geq \left(\frac{3.71}{2}\right)^m x_0 = \left(\frac{3.71}{2}\right)^m \frac{1}{2^n},$$

from which we get

$$2^{n+m-1} \geq (3.71)^m.$$

According to the above inequality, we obtain

$$m \leq \frac{n-1}{\log_2 3.71 - 1}.$$

Since $\log_2 3.71 \doteqdot 1.8912$, we have

$$m \leq \frac{n-1}{\log_2 3.71 - 1} < \frac{5}{4}(n-1),$$

which is definitely less than $2n - 1$, and it is contradictory to the statement "$x_m \leq \frac{1}{2}$ for any $m \in J$". Thus there exists an m in J satisfying $x_m > \frac{1}{2}$. □

Theorem 14.9 *Let a and n be the same as in the above theorem. If there exists k in J such that* $x_k > \frac{1}{2}$, *then* $k > \frac{n-1}{\log_2 3.71 - 1}$.

Proof Since $0 \leq x_m \leq 1$, we have

$$x_m = 3.71(1 - x_{m-1})x_{m-1} \leq 3.71 x_{m-1},$$

which reduces to

$$x_m \leq (3.71)^m x_0.$$

For m_0 in J satisfying $x_{m_0} > \frac{1}{2}$, it holds

$$x_0 \geq \frac{1}{(3.71)^{m_0}} x_{m_0} > \frac{1}{2 \times (3.71)^{m_0}}.$$

Then for $x_0 = \frac{1}{2^n}$ we obtain

$$\log_2 2 \times (3.71)^{m_0} > n,$$

which implies

$$m_0 > \frac{n-1}{\log_2 3.71}.$$

□

Corollary 14.10 *If* $x_0 \equiv \frac{r}{2^n}$ *with* $r \equiv |T(\mathcal{C})|$ *and there exists k in J such that* $x_k > \frac{1}{2}$, *then there exists k satisfying the following inequality if* $\mathcal{C}$ *is SAT*:

$$\left[\frac{n - 1 - \log_2 r}{\log_2 3.71 - 1}\right] \leq k \leq \left[\frac{5}{4}(n-1)\right].$$

From these theorems, for all k, it holds

$$M_k \begin{cases} = 0 & \text{iff } \mathcal{C} \text{ is not SAT}, \\ > 0 & \text{iff } \mathcal{C} \text{ is SAT}. \end{cases}$$

14.2.2 Computational Complexity of SAT Algorithm

The computational complexity of quantum computation depends on the number of unitary operators in the quantum circuit. Let U be a unitary operator which is written as

$$U = U_n U_{n-1} \cdots U_1$$

where $U_n, \ldots, U_1$ are fundamental gates. Thus the computational complexity $T(U)$ is considered as n.

We need to combine some fundamental gates such as U_{NOT}, U_{CN}, and U_{CCN} to construct,in fact, the quantum circuit. U_{AND} and U_{OR} can be written as a combination of fundamental gates. Here we obtain the computational complexity of SAT algorithm by counting the numbers of U_{NOT}, U_{AND} and U_{OR}.

Theorem 14.11 *For a set of clauses $\mathcal{C} = \{C_1, \ldots, C_m\}$ and the set of literals $X' = \{x_1, \ldots, x_n, \bar{x}_1, \ldots, \bar{x}_n\}$, the complexity $T(U_{\mathcal{C}})$ is*

$$\begin{aligned} T(U_{\mathcal{C}}) &= m - 1 + \sum_{k-1}^{m} \left(|C_k| + 2i'_k - 1\right) \\ &\leq 4mn - 1. \end{aligned}$$

Proof For a set of clauses $\mathcal{C} = \{C_1, \ldots, C_k\}$ and the set of literals $X' = \{x_1, \ldots, x_n, \bar{x}_1, \ldots, \bar{x}_n\}$, we can construct a unitary operator $U_{\mathcal{C}}$ in polynomial time (see Theorem 14.5). From the form of $U_{\mathcal{C}}$, we calculate the number of fundamental gates in it as follows: Let i'_k be the number of negations in C_k. For all $k = 1, \ldots, m$, U_{OR} is applied $|C_k| - 1$ times and U_{NOT} is applied $2i'_k$ times to compute $t(C_k)$. U_{AND} is applied $m - 1$ times to compute $t(\mathcal{C})$:

$$U_{\text{OR}}\colon \sum_{k=1}^{m} \left(|C_k| - 1\right) \leq m(2n - 1),$$

$$U_{\text{NOT}}\colon \sum_{k-1}^{m} 2i'_k \leq 2mn,$$

$$U_{\text{AND}}\colon m - 1.$$

Hence, the total number of fundamental gates in $U_{\mathcal{C}}$ is

$$m-1+\sum_{k-1}^{m}\left(|C_k|+2i'_k-1\right)\le 4mn-1.$$

□

Consider the amplifier process and let s be the number of the steps of amplification. According to Corollary 14.10, there exists an $s\le\frac{5}{4}(n-1)$ such that $q^2>\frac{1}{2}$.

The computational complexity $T(SAT)$ is

$$\begin{aligned}T(SAT)&=T(U_{\mathcal{C}})\times\frac{5}{4}(n-1)\\&=(4mn-1)\times\frac{5}{4}(n-1)=\mathcal{O}\big(\mathrm{poly}(n)\big)\end{aligned}$$

where $\mathrm{poly}(n)$ is a polynomial in n. Note that m is a polynomial of n defined in Sect. 14.2.

14.3 Channel Expression of Quantum Algorithm

The quantum algorithm discussed so far is a sort of idealistic one written by unitary operations (computation). Unitary operation is rather difficult to realize in physical processes, a more realistic operation is the one allowing some dissipation like semi-group dynamics. However, such dissipative dynamics reduces the ability of quantum computation very much because the ability is based on preserving the entanglement of states and dissipativity destroys the entanglement. To keep high ability of quantum computation and good entanglement, some kinds of amplification will be necessary in the course of real physical processes in physical devices, which will be similar as amplification processes in quantum communication processes. In this section, in the search for more realistic operations in a quantum computer, the channel expression will be useful, at least in the sense of a mathematical scheme of quantum computation, because the channel is not always unitary and represents many different types of dynamics.

We explain the channel expression of quantum algorithm.

14.3.1 Channel Expression of Unitary Algorithm

We explained several quantum algorithms in Chaps. 2, 11, 13 and Sects. 14.1, 14.2 of this chapter; all algorithms can be written as the following three steps:

1. (Preparation of state) Take a state ρ (e.g., $\rho=|0\rangle\langle 0|$) and apply $\Lambda_F^*=\mathrm{Ad}_{U_F}$

$$\rho\implies\Lambda_F^*\rho=U_F\rho U_F^*,$$

where U_F is the discrete quantum Fourier transformation given in Sect. 12.1.

2. (Computation) Let U be a unitary operator on $\mathcal{H}$ representing the computation followed by a suitable programming of a certain problem, then the computation is described by a channel

$$\Lambda_U^* = \mathrm{Ad}_U \quad \text{(unitary channel)}.$$

After the computation, the final state ρ_{out} will be

$$\rho_{\text{out}} = \Lambda_U^* \Lambda_F^* \rho.$$

For the SAT problem, we use $U = U_{\mathcal{C}}$ defined in Sect. 14.2.

3. (Register and Measurement) This stage will again remain; however, for registration of the computed result and its measurement we might need an additional system $\mathcal{K}$ (e.g., register), so that the lifting $\mathcal{E}_m^*$ from $\mathfrak{S}(\mathcal{H})$ to $\mathfrak{S}(\mathcal{H} \otimes \mathcal{K})$ discussed in Chap. 7 is useful to describe this stage. Thus the whole process is written as

$$\rho_{\text{out}} = \mathcal{E}_m^*(\Lambda_U^* \Lambda_F^* \rho).$$

Finally, we measure the state in $\mathcal{K}$: for instance, let $\{P_k; k \in J\}$ be a projection valued measure (PVM) on $\mathcal{K}$

$$\Lambda_m^* \rho_{\text{out}} = \sum_{k \in J} I \otimes P_k \rho_{\text{out}} I \otimes P_k,$$

after which we can get a desired result by observations in finite times if the size of the set J is small.

14.3.2 Channel Expression of Arbitrary Algorithm

The above three steps are generalized so that dissipative nature is involved. Such a generalization can be expressed by means of a suitable channel, not necessarily unitary.

1. (Preparation of state) We may use the same channel $\Lambda_F^* = \mathrm{Ad}_{U_F}$ in this first step, but if the number of qubits N is large so that it will not be built physically, then Λ_F^* should be modified. Let us denote it by Λ_P^*.
2. (Computation) This stage is certainly modified to a channel Λ_C^* reflecting the physical device for a computer.
3. (Register and Measurement) This stage will again remain; however, for registration of the computed result and its measurement we might need an additional system $\mathcal{K}$ (e.g., register), so that the lifting $\mathcal{E}_m^*$ from $\mathfrak{S}(\mathcal{H})$ to $\mathfrak{S}(\mathcal{H} \otimes \mathcal{K})$ discussed in Chap. 7 is useful to describe this stage.

   Thus the whole process is written as

$$\rho_{\text{out}} = \mathcal{E}_m^*(\Lambda_C^* \Lambda_P^* \rho).$$

14.4 SAT Algorithm in GQTM

The generalized quantum Turing machine (GQTM) was discussed in Chap. 11. In this section, we will write the SAT chaos algorithm in the terminology of GQTM. We construct the 3 multi-track GQTM $M_{\mathrm{SAT}} = (Q, \Sigma^3, \mathcal{H}, \Lambda_\delta^*)$ that realizes the OVM SAT algorithm. This GQTM does not belong to LQTM and UQTM because the chaos amplification process is described by a nonlinear CP channel, not a unitary and linear one. The OVM algorithm runs from an initial state $\rho_0 \equiv |v_{\mathrm{in}}\rangle\langle v_{\mathrm{in}}|$ to $\bar{\rho}_k$ through $\rho \equiv |v_f\rangle\langle v_f|$ explained above. The computation from $\rho_0 \equiv |v_{\mathrm{in}}\rangle\langle v_{\mathrm{in}}|$ to $\rho \equiv |v_f\rangle\langle v_f|$ is due to a unitary channel $\Lambda_C^* \equiv U_C \bullet U_C$, and that from $\rho \equiv |v_f\rangle\langle v_f|$ to $\bar{\rho}_k$ is due to a non-unitary channel $(\Lambda_{\mathrm{CA}}^*)^k \circ \Lambda_I^*$, so that all computation can be done by $(\Lambda_{\mathrm{CA}}^*) \circ \Lambda_I^* \circ \Lambda_C^*$, which is completely positive, so the whole computation process is deterministic.

Let us explain our computation by a multi-track GQTM. The first track stores the input data and literals. The second track is used for the computation of $f(C_i)$ $(i = 1, \ldots, m)$, and the third track is used for the computation of $f(\mathcal{C})$ denoting the result. This algorithm is represented by the following five steps:

Step 1. Apply the Hadamard matrix to the state $|0\rangle$ in Track 1.
Step 2. Calculate $f(C_1), \ldots, f(C_m)$ and store them in Track 2.
Step 3. Calculate $f(\mathcal{C})$ and store it in Track 3.
Step 4. Empty the working space.
Step 5. Apply the chaos amplifier to the result state and repeat this step.

Let us explain how to construct the initial state and transition functions for this algorithm. Let $X' \equiv X \cup \bar{X} = \{x_1, \ldots, x_n, \bar{x}_1, \ldots, \bar{x}_n\}$ be a set of literals and $\mathcal{C} = \{C_1, \ldots, C_m\}$ be a set of clauses on X', we write $\mathcal{C}$ in the form

$$L_{\mathcal{C}} = V C_S B(C_1) C_E C_S B(C_2) C_E \cdots C_S B(C_m) C_E$$

where V indicates the top of the sequence $L_{\mathcal{C}}$, C_S and C_E mean that the information of each clause begins and ends, respectively, and $B(C_i)$ is a binary form of clause C_i such that

$$B(C_i) = \varepsilon_1 \cdots \varepsilon_n W \bar{\varepsilon}_1 \cdots \bar{\varepsilon}_n,$$

$$\varepsilon_k = \begin{cases} 0, & x_k \notin X, \\ 1, & x_k \in X, \end{cases}$$

$$\bar{\varepsilon}_k = \begin{cases} 0, & \bar{x}_k \notin \bar{X}, \\ 1, & \bar{x}_k \in \bar{X}, \end{cases}$$

where W is a partition to split up Boolean variables and their negations.

Then we construct the initial tape state as

$$\left(0^n L_C, \#, \#\right).$$

Table 14.1 DFT

	#	0	1	V
q_0		$q_a, 0, +1$	$q_a, 1, +1$	
q_a		$q_a, 0, +1$	$q_a, 1, +1$	$q_b, V, -1$
q_b	$q_{0,2}, \#, +1$	$\frac{1}{\sqrt{2}}q_b, 0, L + \frac{1}{\sqrt{2}}q_b, 1, -1$	$\frac{1}{\sqrt{2}}q_b, 0, -1 - \frac{1}{\sqrt{2}}q_b, 1, -1$	

For example, given $X = \{1, 2, 3\}$, $\mathcal{C} = \{C_1, C_2, C_3\}$, $C_1 = \{x_1, x_2, \bar{x}_3\}$, $C_2 = \{x_3, \bar{x}_2\}$, $C_3 = \{x_1, \bar{x}_2, \bar{x}_3\}$, we represent L_c as

$$0^n L_C = 000VC_S110W001C_EC_S001W010C_EC_S100W011C_E.$$

Therefore, we prepare the initial state of M_{SAT} as

$$\rho_0 = |v_{\text{in}}\rangle\langle v_{\text{in}}|,$$
$$|v_{\text{in}}\rangle = \big|q_0, (L_C, \#, \#), 0\big\rangle.$$

In Step 1, GQTM applies the DFT (discrete Fourier transform) to a part of literals in Track 1. The transition function for DFT is written by Table 14.1. Write the vector in $\mathcal{H}_Q$ by $q.$ instead of $|q.\rangle$ and denote the direction of moving the tape head to the right by $+1$, to the left by -1, and staying put by 0. Note that O is the starting position.

The tape head moves to the right until it reads a symbol C_S. When the tape head reads C_S, GQTM increases a program counter by one, which is indicated by the processor state, while moving to the right until it reads 1. Then GQTM stops increasing the counter, and the tape head moves to the top of the tape. According to the program counter, the tape head moves to the right and reduces the counter by one. When the counter becomes zero, GQTM reads the 0 or 1 and calculates OR, then GQTM writes the result on Track 2. GQTM goes back to the top cell of Track 1 and repeats the above processes until it reads W.

When GQTM reads W, it calculates OR with the negation and repeats the processes as above. When it reads C_E, it writes down $f(C_k)$ on Track 2 and cleans the workspace for the next calculation. Then GQTM reads the blank symbol #, and it begins to calculate AND. The calculation of AND is done on Track 3. GQTM calculates them while moving to the left because the position of the tape head is at the end of Track 2 when the OR calculation is finished. Then the result of the calculation $f(\mathcal{C})$ will be showed in the top cell of Track 3.

The transition function of OR calculation is described, similarly as for the classical TM, by Tables 14.2, 14.3, 14.4.

The transition function of AND calculation is described by Table 14.5.

After Step 3, we obtain the following state

$$\rho_5 = |\psi_5\rangle\langle\psi_5|,$$
$$|\psi_5\rangle = \frac{1}{\sqrt{2^n}}\sum_{i=0}^{2^n-1}\big|q_5, (A_{1,i}, A_{2,i}, A_{3,i}), 0\big\rangle$$

Table 14.2 OR calculation

	0	1	C_S	V	W	#
$q_{0,2}$	$q_a, 0, +1$		$q_b, C_S, +1$		$q_{\bar{\text{OR}}}, W, +1$	
q_a	$q_a, 0, +1$	$q_a, 1, +1$	$q_b, C_S, +1$	$q_a, V, +1$	$q_{\bar{\text{OR}}}, W, +1$	$q_{\text{AND}}, \#, 0$
q_b	$q_{b,1}, 0, +1$	$q_{c,1}, 0, -1$				
$q_{b,1}$	$q_{b,2}, 0, +1$	$q_{c,1}, 0, -1$				
$\vdots$						
$q_{b,k}$	$q_{b,k+1}, 0, +1$	$q_{c,k+1}, 0, -1$				
$\vdots$						
$q_{b,n-1}$	$q_{b,n}, 0, +1$	$q_{c,n}, 0, -1$				
$q_{b,n}$					$q_{\bar{\text{OR}}}, W, +1$	
$q_{c,1}$	$q_{c,1}, 0, -1$	$q_{c,1}, 1, -1$	$q_{c,1}, C_S, -1$	$q_{c,1}, V, -1$		$q_{d,1}, \#, +1$
$\vdots$						
$q_{c,n}$	$q_{c,n}, 0, -1$	$q_{c,n}, 1, -1$	$q_{c,n}, C_S, -1$	$q_{c,n}, V, -1$		$q_{d,n}, \#, +1$
$q_{d,1}$	$q_{t2,0}, 0, 0$	$q_{t2,1}, 1, 0$				
$q_{d,2}$	$q_{d,1}, 0, +1$	$q_{d,1}, 1, +1$				
$\vdots$						
$q_{d,k}$	$q_{d,k-1}, 0, +1$	$q_{d,k-1}, 1, +1$				
$\vdots$						
$q_{d,n}$	$q_{d,n-1}, 0, +1$	$q_{d,n-1}, 0, +1$				

where $A_{1,i} = e_i L_{\mathcal{C}}$, $A_{2,i}$ contains $f(C_1), \dots, f(C_m)$ and some binary symbols used for the computation, and $A_{3,i}$ has $f(\mathcal{C})$ on the top cell and the other cells are used for the computation. In Step 4, we empty each tape except for $A_{3,i}(0)$ in order to fix the state for the next computation.

This step is represented as a linear channel. First, we empty Track 1 according to Table 14.6.

Then, we empty Track 2 like Track 1 as in Table 14.7.

Moreover, we empty Track 3 except for the cell storing $f(\mathcal{C})$ as in Table 14.8.

Let A_i, B_i, $i = 1, 2, 3$ be the tape states of the ith track. After Step 4, we obtain the state ρ_6

$$\begin{aligned}\rho_6 &= q^2 \big|q_f, (A_1, A_2, A_3), 0\big\rangle\big\langle q_f, (A_1, A_2, A_3), 0\big| \\ &\quad + \left(1 - q^2\right) \big|q_6, (B_1, B_2, B_3), 0\big\rangle\big\langle q_6, (B_1, B_2, B_3), 0\big|,\end{aligned}$$

where for all $i \in \mathbb{Z}$

$$\begin{aligned}A_1(i) &= A_2(i) = \#, \\ B_1(i) &= B_2(i) = \#\end{aligned}$$

Table 14.3 OR calculation

	0	1	W	C_S	C_E	V	#
$q_{\bar{OR}}$	$q_{g,1,}, 0, +1$	$q_{h,1}, 0, -1$					
q_e	$q_e, 0, +1$		$q_g, W, +1$	$q_e, C_S, +1$			
q_g	$q_{g,1}, 0, +1$	$q_{h,1}, 0, -1$					
$q_{g,1}$	$q_{g,2}, 0, +1$	$q_{h,2}, 0, -1$					
$\vdots$							
$q_{g,k}$	$q_{g,k+1}, 0, +1$	$q_{h,k+1}, 0, -1$					
$\vdots$							
$q_{g,n-1}$	$q_{g,n}, 0, +1$	$q_{h,n}, 0, -1$					
$q_{g,n}$					$q_j, 0, -1$		
$q_{h,1}$	$q_{h,1}, 0, -1$	$q_{h,1}, 1, -1$	$q_{h,1}, W, -1$	$q_{h,1}, C_S, -1$		$q_{h,1}, V, -1$	$q_{i,1}, \#, +1$
$\vdots$							
$q_{h,n}$	$q_{h,n}, 0, -1$	$q_{h,n}, 1, -1$	$q_{h,n}, W, -1$	$q_{h,n}, C_S, -1$		$q_{h,n}, V, -1$	$q_{i,n}, \#, +1$
$q_{i,1}$	$q_{t2,1}, 0, 0$	$q_{t2,0}, 1, 0$					
$q_{i,2}$	$q_{i,1}, 0, +1$	$q_{i,1}, 1, +1$					
$\vdots$							
$q_{i,k}$	$q_{i,k-1}, 0, +1$	$q_{i,k-1}, 1, +1$					
$\vdots$							
$q_{i,n}$	$q_{i,n-1}, 0, +1$	$q_{i,n-1}, 0, +1$					
q_j	$q_j, 0, -1$		$q_j, 0, -1$	$q_{t2,a}, 0, 0$			

Table 14.4 OR calculation

	0	1	#
$q_{t3,0}$	$q_{t3,0}, 0, +1$	$q_{t3,1}, 1, +1$	$q_a, 0, 0$
$q_{t3,1}$	$q_{t3,1}, 0, +1$	$q_{t3,1}, 1, +1$	$q_a, 1, 0$
$q_{t3,a}$	$q_{t4,0}, 0, 0$	$q_{t4,1}, 1, 0$	
$q_{t3,b}$	$q_{t3,b}, \#, -1$	$q_{t3,b}, \#, -1$	$q_{t3,c}, \#, +1$
$q_{t3,c}$			$q_a, 0, 0$

and

$$A_3(i) = \begin{cases} 1, & i = 0, \\ \#, & \text{otherwise}, \end{cases}$$

$$B_3(i) = \begin{cases} 0, & i = 0, \\ \#, & \text{otherwise}. \end{cases}$$

Table 14.5 AND

	0	1	#
$q_{t4,0}$			$q_{t3,b}, 0, +1$
$q_{t4,1}$			$q_{t3,b}, 1, +1$
q_{AND}			$q_{t4,a}, \#, -1$
$q_{t4,a}$	$q_{t4,a}, \#, -1$	$q_{t4,b}, \#, -1$	$q_{t4,c}, \#, +1$
$q_{t4,b}$	$q_{t4,a}, \#, -1$	$q_{t4,b}, \#, -1$	$q_{t4,d}, \#, +1$
$q_{t4,c}$			$q_5, 0, -1$
$q_{t4,d}$			$q_5, 1, +1$

Table 14.6 Erasing Track 1

	0	1	C_S	V	W	#
q_5	$q_{5,a}, \#, +1$	$q_{5,a}, \#, +1$				
$q_{5,a}$	$q_{5,a}, \#, +1$	$q_{5,a}, \#, +1$	$q_{5,a}, \#, +1$	$q_{5,a}, \#, +1$	$q_{5,a}, \#, +1$	$q_{5,b}, \#, -1$

Table 14.7 Erasing Track 2

	0	1	#
$q_{5,b}$	$q_{5,c}, \#, -1$	$q_{5,c}, \#, -1$	
$q_{5,c}$			$q_{5,d}, \#, +1$

Table 14.8 Output

	0	1	#
$q_{5,d}$	$q_{5,e}, 0, +1$	$q_{5,e}, 1, +1$	
$q_{5,e}$	$q_{5,e}, \#, +1$	$q_{5,e}, \#, +1$	$q_{5,f}, \#, -1$
$q_{5,f}$	$q_{5,g,0}, 0, -1$	$q_{5,g,1}, 1, -1$	$q_{5,f}, \#, -1$
$q_{5,g,0}$			$q_6, \#, +1$
$q_{5,g,1}$			$q_f, \#, +1$

We put

$$P_1 = \big|q_6, (A_1, A_2, A_3), 0\big\rangle\big\langle q_6, (A_1, A_2, A_3), 0\big|,$$
$$P_0 = \big|q_f, (B_1, B_2, B_3), 0\big\rangle\big\langle q_f (B_1, B_2, B_3), 0\big|$$

then P_0 and P_1 are abelian. For any state in the form $\rho = (1-q^2)P_0 + q^2 P_1$ where $q^2 \in \{0, \frac{1}{2^n}, \frac{2}{2^n}, \ldots, \frac{2^n-1}{2^n}, 1\}$, define the delta function δ_1 as

$$\delta_1\big(1-q^2, q_6, 0, q_6, 0, q_6, 0, 0, q_6, 0, 0\big) = 1 - g_a\big(q^2\big),$$

$$\delta_1\big(q^2, q_f, 1, q_f, 1, q_f, 1, 0, q_f, 1, 0\big) = g_a\big(q^2\big)$$

where g_a is the logistic map.

Using this δ, the transition function of Step 5 denoted by the chaos amplifier is formally written as

$$\begin{aligned}\Lambda_{\mathrm{CA}}^{*}(\rho_6) &= \delta_1\big(q^2, q_f, 1, q_f, 1, q_f, 1, 0, q_f, 1, 0\big) P_1\\ &\quad + \delta_1\big(1 - q^2, q_6, 0, q_6, 0, q_6, 0, 0, q_6, 0, 0\big) P_0\\ &= g_a\big(q^2\big) P_1 + \big(1 - g_a\big(q^2\big)\big) P_0,\\ \big(\Lambda_{\mathrm{CA}}^{*}\big)^k(\rho_6) &= g_a^k\big(q^2\big) P_1 + \big(1 - g_a^k\big(q^2\big)\big) P_0.\end{aligned}$$

According to Corollary 14.10, GQTM halts in at most $[\frac{5}{4}(n-1)]$ steps with the probability $p \geq \frac{1}{2}$, by which we can claim that $\mathcal{C}$ is SAT.

14.5 SAT Algorithm with Stochastic Limits

We illustrate the general scheme described in the previous section in the simplest case when the state space of the system is $\mathcal{H}_S = \mathbb{C}^2$. Let us use some simple notations in this section. We fix an orthonormal basis of $\mathcal{H}_S$ as $\{e_0, e_1\}$. The unknown state (vector) of the system at time $t = 0$ is denoted

$$|\psi\rangle = \sum_{\varepsilon\in\{0,1\}} \alpha_\varepsilon e_\varepsilon = \alpha_0 e_0 + \alpha_1 e_1, \qquad \|\psi\| = 1.$$

In the sections above, α_1 corresponds to q, and e_j to $|j\rangle$ $(j = 0, 1)$. This vector after quantum computation of the SAT problem is taken as input and defines the interaction Hamiltonian (adaptive dynamics) in an external field

$$\begin{aligned}H_I &= \lambda|\psi\rangle\langle\psi| \otimes \big(A_g^* + A_g\big)\\ &= \sum_{i,j=0}^{1} \lambda\alpha_i\overline{\alpha}_j |e_i\rangle\langle e_j| \otimes \big(A_g^* + A_g\big)\end{aligned}$$

where λ is a small coupling constant. Here and in the following, summation over repeated indices is understood.

The free system Hamiltonian is taken to be diagonal in the e_ε-basis

$$H_S \equiv \sum_{i=0}^{1} E_i |e_i\rangle\langle e_i| = E_0|e_0\rangle\langle e_0| + E_1|e_1\rangle\langle e_1|$$

and the energy levels are ordered so that $E_0 < E_1$. Thus there is a single Bohr frequency $\omega_0 \equiv E_1 - E_0 > 0$. The 1-particle field Hamiltonian is

$$S_t g(k) = e^{it\omega(k)} g(k)$$

where $\omega(k)$ is a function satisfying the basic analytical assumption of the stochastic limit. Its second quantization is the free field evolution

$$e^{\mathrm{i}tH_0} A_g e^{-\mathrm{i}tH_0} = A_{S_t g}.$$

We can distinguish two cases described below, corresponding to the two cases of Sect. 14.3, i.e., $q > 0$ and $q = 0$.

Case 1. If $\alpha_0, \alpha_1 \neq 0$ then, according to the general theory of stochastic limit (i.e., $t \to t/\lambda^2$) [34], the interaction Hamiltonian H_I is in the same universality class as

$$\tilde{H}_I = D \otimes A_g^* + D^* \otimes A_g$$

where $D \equiv |e_0\rangle\langle e_1|$. The interaction Hamiltonian at time t is then

$$\tilde{H}_I(t) = e^{-\mathrm{i}t\omega_0} D \otimes A_{S_t g}^* + h.c. = D \otimes A^*\big(e^{\mathrm{i}t(\omega(p)-\omega_0)} g\big) + h.c.$$

and the white noise ($\{b_t\}$) Hamiltonian equation associated, via the stochastic golden rule, to this interaction Hamiltonian is

$$\partial_t U_t = \mathrm{i}\big(D b_t^* + D^* b_t\big) U_t.$$

Its causally normal ordered form is equivalent to the stochastic differential equation

$$dU_t = \big(\mathrm{i}D\, dB_t^* + \mathrm{i}D^*\, dB_t - \gamma D^+ D\, dt\big) U_t,$$

where $dB_t \equiv b_t\, dt$ and γ is a constant.

The causally ordered inner Langevin equation for $j_t(\rho) \equiv U_t^* \rho U_t$ is

$$\begin{aligned}
dj_t(\rho) &= dU_t^* \rho U_t + U_t^* \rho\, dU_t + dU_t^* \rho\, dU_t \\
&= U_t^*\big(-\mathrm{i}D^* \rho\, dB_t - \mathrm{i}D\rho\, dB_t^* - \bar{\gamma} D^* D \rho\, dt + \mathrm{i}\rho D\, dB_t^* \\
&\quad + \mathrm{i}\rho D^*\, dB_t - \gamma \rho D^* D\, dt + \gamma D^* \rho D\, dt\big) U_t \\
&= \mathrm{i} j_t\big(\big[\rho, D^*\big]\big)\, dB_t + \mathrm{i} j_t\big([\rho, D]\big)\, dB_t^* \\
&\quad - (\mathrm{Re}\,\gamma) j_t\big(\big\{D^* D, \rho\big\}\big)\, dt + \mathrm{i}(\mathrm{Im}\,\gamma) j_t\big(\big[D^* D, \rho\big]\big)\, dt \\
&\quad + j_t\big(D^* \rho D\big)(\mathrm{Re}\,\gamma)\, dt.
\end{aligned}$$

Therefore, the master equation is

$$\begin{aligned}
\frac{d}{dt} P^t(\rho) &= (\mathrm{Im}\,\gamma)\mathrm{i}\big[D^* D, P^t(\rho)\big] - (\mathrm{Re}\,\gamma)\big\{D^* D, P^t(\rho)\big\} \\
&\quad + (\mathrm{Re}\,\gamma) D^* P^t(\rho) D
\end{aligned}$$

where $D^* D = |e_1\rangle\langle e_1|$ and $D^* \rho D = \langle e_0, x e_0\rangle |e_1\rangle\langle e_1|$.

The dual Markovian evolution P_*^t acts on density matrices, and its generator is

$$L_* \rho = (\operatorname{Im} \gamma) \mathrm{i}\big[\rho, D^* D\big] - (\operatorname{Re} \gamma)\big\{\rho, D^* D\big\} + (\operatorname{Re} \gamma) D \rho D^*.$$

For $\rho = |e_0\rangle\langle e_0|$ one has

$$L_* \rho = 0,$$

so ρ is an invariant measure. From the Fagnola–Rebolledo criterion [36, 230], it is the unique invariant measure and *the semigroup* $\exp(tL_*)$ *converges exponentially to it.*

Case 2. If $\alpha_1 = 0$, then the interaction Hamiltonian H_I is

$$H_I = \lambda |e_0\rangle\langle e_0| \otimes \big(A_g^* + A_g\big)$$

and, according to the general theory of stochastic limit, the reduced evolution has no damping and corresponds to the pure Hamiltonian

$$H_S + |e_0\rangle\langle e_0| = (E_0 + 1)|e_0\rangle\langle e_0| + E_1 |e_1\rangle\langle e_1|;$$

therefore, if we choose the eigenvalues E_1, E_0 to be integers (in appropriate units) then *the evolution will be periodic*.

Since the eigenvalues E_1, E_0 can be chosen a-priori, by fixing the system Hamiltonian H_S, it follows that the period of the evolution can be known a-priori. This gives a simple criterion for the solvability of the SAT problem because, by waiting a sufficiently long time, one can experimentally detect the difference between a damping and an oscillating behavior.

A precise estimate of this time can be achieved either by theoretical methods or by computer simulation. Both methods will be analyzed in [37].

Czachor [187] gave an example of a nonlinear Schrödinger equation to distinguish two cases, similar to $\alpha_1 \neq 0$ and $\alpha_1 = 0$ given above, in a certain oracle computation. We used the resulting (flag) state after quantum computation of the truth function of SAT to couple the external field and took the stochastic limit, then our final evolution became "linear" for the state ρ described as above. The stochastic limit is historically important to realize macroscopic (time) evolution, and it is now rigorously established as explained in [34], and we gave a general protocol to study the distinction of the two cases $\alpha_1 \neq 0$ and $\alpha_1 = 0$ by this rigorous mathematics. The macro-time enables us to measure the behavior of the outcomes practically. Thus our approach is conceptually different from [187]. Moreover, in [187] it is discussed that some expectation value is constant for the case $\alpha_1 = 0$ and oscillating for $\alpha_1 \neq 0$, and ours gives the detailed behavior of the state w.r.t. the macro-time: damping ($\alpha_1 \neq 0$ case) and oscillating ($\alpha_1 = 0$ case).

In the previous sections, we discussed an algorithm for solving the SAT problems in polynomial steps by combining a quantum algorithm with a chaos dynamics (one adaptive method). In this section, we pointed out that it is possible to distinguish the two different states, $\sqrt{1-q^2}|0\rangle + q|1\rangle$ $(q \neq 0)$ and $|0\rangle$, by means of the adaptive interaction and the stochastic limit (another adaptive method). Finally, we remark that these algorithms can be described by a generalized quantum Turing machine, discussed in Chap. 11.

14.6 Notes

Ordinary approach to quantum algorithm is based on a quantum Turing machine or quantum circuits [114, 171, 197]. It is known that this approach is not powerful enough to solve NP-complete problems [109, 278]. Ohya and Volovich [597, 600, 602] considered an approach to these problems based on quantum computers and chaotic dynamics. It is shown that the satisfiability problem as an example of NP-complete problems [301], in principle, can be solved in polynomial time by using their new quantum algorithm. This approach goes beyond the quantum Turing machine paradigm. Moreover, an alternative solution of the SAT problem is given in the stochastic limit in [37].

Accardi and Sabbadini showed that this algorithm is a combinatoric one, and they discussed its combinatoric representation [30]. It was shown in [372, 595] that the SAT problem can be solved in polynomial time by using a quantum computer with a chaos amplifier. A channel expression of an arbitrary algorithm is a generalization of a quantum algorithm by means of a CP channel and lifting [19].

Ohya and Iriyama showed that there exists a classical algorithm to create the unitary operator for the SAT problem in polynomial time in [372], and discussed the computational complexity of OMV SAT algorithm rigorously by using a generalized quantum Turing machine [373].

Chapter 15
Quantum Error Correction

In this chapter, we discuss the basic theory of quantum error-correcting codes, fault-tolerant quantum computation, and the threshold theorem.

Noise and decoherence are the major obstacles for the experimental implementation of quantum processing. One uses error-correcting codes to protect against the effects of noise. The basic principle is that to protect a message against the effects of noise we should *encode* the message by adding *redundant information* to the message. If there is enough redundancy, then one can *decode* or recover the original message even if it is corrupted by noise.

Noises affect each of the elements of a quantum circuit, the state preparation procedure, quantum logic gates, and measurement of the output. Actually, the state of quantum circuit is not a pure state but a mixed state, and it should be described by a density matrix. Moreover, quantum gates should not be just unitary operators but more general quantum operations (channels).

15.1 Three Qubit Code and Fidelity

15.1.1 Three Qubit Code

As a simple example, consider sending a classical bit through a classical noisy communications channel. We assume that the effect of noise in the channel is to flip the bit with probability p, while the bit is transmitted without error with probability $1 - p$. To protect the bit against the effects of noise, we replace the bit with three copies of itself:

$$0 \to 000,$$

$$1 \to 111.$$

We encode the message to be sent by repeating it three times. Such a code is called a *repetition code*. Bob knows that he should get 000 or 111. After sending all three

M. Ohya, I. Volovich, *Mathematical Foundations of Quantum Information and Computation and Its Applications to Nano- and Bio-systems*, Theoretical and Mathematical Physics,
DOI 10.1007/978-94-007-0171-7_15,

bits through the noisy channel which flips each bit with probability p the receiver has to decide what the value of the original bit was. Suppose 011 was the output. If the probability p of a bit flip is not too high, then it is likely that the first bit was flipped by the channel, and that 1 was the bit that was sent. Notice here that if one were to use a simpler coding like

$$0 \to 00,$$
$$1 \to 11,$$

then it would be impossible to correct errors.

Let us show that the code makes the transmission more reliable if $p < 1/2$. Indeed, the code fails if two or all three of the bits sent through the channel were flipped, and succeeds otherwise. The probability of error, i.e., that two or all three of the bits were flipped, is $p_0 = 3p^2(1-p) + p^3 = 3p^2 - 2p^3$. Without encoding, the probability of an error was p, so the encoding makes the transmission more reliable if $p_0 < p$, i.e., if $p < 1/2$.

There are important differences between classical and quantum communication channels. In particular, a continuum of different errors may occur on a single qubit for the quantum case. That is, the type of errors is not restricted to the bit flip. We have to take into account also the features of quantum measurement. Moreover, the no-cloning theorem forbids the copying of the quantum state. Fortunately, these problems can be overcome, and a theory of quantum error-correcting codes has been developed.

One can try to encode the single qubit state $a|0\rangle + b|1\rangle$ in three qubits, by analogy with the classical encoding, as

$$a|000\rangle + b|111\rangle.$$

Let each of the three qubits be transmitted through an independent copy of the bit flip channel. One can use the following procedure to recover the original quantum state in this case. It involves:

1. Error syndrome measurement
2. Recovery.

The *error syndrome* measurement is such a measurement which tells us what error, if any, occurred in a quantum state. There are four projection operators describing the error syndrome measurements on the bit flip quantum channel. They are

$$P_0 = |000\rangle\langle 000| + |111\rangle\langle 111|,$$
$$P_1 = |100\rangle\langle 100| + |011\rangle\langle 011|,$$
$$P_2 = |010\rangle\langle 010| + |101\rangle\langle 101|,$$
$$P_3 = |001\rangle\langle 001| + |110\rangle\langle 110|.$$

The operator P_0 describes the measurement when no error occurs. The operators P_i describe the measurements with a bit flip on qubit i, $i = 1, 2, 3$. One has

$$\sum_{i=0}^{3} P_i = 1.$$

For example, suppose that a bit flip occurs on the second qubit, so the transmitted state is $|\Psi\rangle = a|010\rangle + b|101\rangle$. If we make the measurement of the projection operator P_2 then the outcome of the measurement result will be 1, since $P_2|\Psi\rangle = |\Psi\rangle$. So, the syndrome measurement in this case is P_2. Moreover, the state is not perturbed since the syndrome measurement does not change the state. The syndrome contains only information about what error has occurred, and does not allow us to derive the value of a or b.

Now if we know the error syndrome we can recover the initial state. In the case when the syndrome measurement is P_2, we just have to flip the second qubit again.

15.1.2 Fidelity and Error Correction

One can characterize quantum error correction by using fidelity. Recall that fidelity measures the distance between quantum states. The fidelity of states ρ and σ is

$$F(\rho, \sigma) = \mathrm{tr}\sqrt{\sigma^{1/2}\rho\sigma^{1/2}}.$$

The fidelity is a symmetric function, $0 \le F(\rho, \sigma) \le 1$, with equality in the first inequality if ρ and σ have orthogonal supports, and equality in the second inequality if and only if $\rho = \sigma$. The fidelity between a pure state $|\psi\rangle$ and a mixed state σ is given by

$$F\big(|\psi\rangle\langle\psi|, \sigma\big) = \sqrt{\langle\psi|\sigma|\psi\rangle}.$$

The aim of quantum error correction is to increase the fidelity with which quantum information is transmitted up near the maximum possible fidelity of 1.

Let us compare the minimum fidelity which can be achieved by the three qubit bit flip code with the fidelity when no error correction is performed. The minimum fidelity without using the error correction code is $F = \sqrt{1-p}$. Indeed, if the one qubit state which we want to send through the qubit flip channel is $|\psi\rangle = a|0\rangle + b|1\rangle$ then the state of the qubit after being sent through the channel is

$$\sigma = (1-p)|\psi\rangle\langle\psi| + p\sigma_x|\psi\rangle\langle\psi|\sigma_x$$

where σ_x is a Pauli matrix. The fidelity in this case is

$$F = \sqrt{\langle\psi|\sigma|\psi\rangle} = \sqrt{(1-p) + p\langle\psi|\sigma_x|\psi\rangle^2}.$$

The minimum fidelity is $F = \sqrt{1-p}$ since the second term under the square root is non-negative, and it vanishes when $|\psi\rangle = |0\rangle$.

Now suppose that we use the three qubit error correcting code

$$|\psi\rangle = a|0\rangle + b|1\rangle \to |\Psi\rangle = a|000\rangle + b|111\rangle.$$

The probability that the state $|\psi\rangle$ will be not flipped through the channel is $1 - p_0 = 1 - 3p^2 + 2p^3$. Therefore, the state of the qubit after being sent through the channel is

$$\sigma = \left(1 - 3p^2 + 2p^3\right)|\Psi\rangle\langle\Psi| + T$$

where T is a positive operator representing contributions from bit flips on two and three qubits. We obtain a lower bound on the fidelity as

$$F = \sqrt{\langle\Psi|\sigma|\Psi\rangle} \geq \sqrt{1 - 3p^2 + 2p^3}.$$

Hence encoding improves the fidelity of the transmission, i.e., $\sqrt{1 - 3p^2 + 2p^3} \geq \sqrt{1-p}$, if $p < 1/2$. This conclusion is consistent with that obtained earlier by a more elementary analysis.

Here we discuss general properties of fidelity.

Theorem 15.1 *Fidelity has the following properties*:

$$\begin{aligned} 0 &\leq F(\rho,\sigma) \leq 1, \\ F(\rho,\sigma) = 1 &\iff \rho = \sigma, \\ F(\rho,\sigma) &= F(\sigma,\rho). \end{aligned} \tag{15.1}$$

To prove this theorem, we need the following lemma and theorem.

Lemma 15.2 *For any compact operator A and any unitary operator U, we have*

$$\left|\mathrm{tr}(AU)\right| \leq \mathrm{tr}\left(|A|\right).$$

Proof For any two compact operators B, C, define an inner product by $\langle B, C\rangle \equiv \mathrm{tr}(B^*C)$. By the Cauchy–Schwarz inequality, one has

$$\left|\mathrm{tr}(BC)\right|^{1/2} \leq \mathrm{tr}(B^*B)\mathrm{tr}(C^*C).$$

Let $A = |A|V$ be the polar decomposition of A. We have

$$\left|\mathrm{tr}(AU)\right| = \left|\mathrm{tr}\left(|A|^{1/2}|A|^{1/2}VU\right)\right|.$$

The Cauchy–Schwarz inequality implies

$$\left|\mathrm{tr}(AU)\right| \leq \sqrt{\mathrm{tr}\left(|A|\right)\mathrm{tr}\left(U^*V^*|A|VU\right)} = \mathrm{tr}\left(|A|\right). \qquad \square$$

Theorem 15.3 *Let ρ, σ be two states (density operators) on a Hilbert space H and $|\phi\rangle\langle\phi|$, $|\psi\rangle\langle\psi|$ be their purification with additional Hilbert space $\mathcal{K}$ (i.e., $\mathrm{tr}_{\mathcal{K}}(|\phi\rangle\langle\phi|) = \rho$, $\mathrm{tr}_{\mathcal{K}}(|\psi\rangle\langle\psi|) = \sigma$). Then it holds that*

$$F(\rho,\sigma) = \sup_{|\phi\rangle,|\psi\rangle} \left|\langle\phi,\psi\rangle\right|.$$

Proof Let $\{|e_n\rangle\}$ and $\{|f_n\rangle\}$ be CONS of $\mathcal{H}$ and $\mathcal{K}$, respectively. Then the purification vector $|\phi\rangle$ of ρ can be expressed by unitary operators $U \in \mathbf{B}(\mathcal{H})$ and $U_R \in \mathbf{B}(\mathcal{K})$ as

$$|\phi\rangle = \sum_n \left(\rho^{1/2} U \otimes U_R\right)|e_n\rangle \otimes |f_n\rangle.$$

The purification vector $|\psi\rangle$ of σ is also written as

$$|\psi\rangle = \sum_n \left(\sigma^{1/2} V \otimes V_R\right)|e_n\rangle \otimes |f_n\rangle$$

with unitary operators $V \in \mathbf{B}(\mathcal{H})$ and $V_R \in \mathbf{B}(\mathcal{K})$. Thus

$$\left|\langle\phi|\psi\rangle\right| = \left|\mathrm{tr}\left(V_R^* U_R U^* \rho^{1/2}\sigma^{1/2} V\right)\right|.$$

Then we use the lemma above for the unitary operator $V V_R^* U_R U^*$ and we get

$$\left|\langle\phi|\psi\rangle\right| \le \mathrm{tr}\left|\rho^{1/2}\sigma^{1/2}\right| = F(\rho,\sigma).$$

Note that taking the polar decomposition $\rho^{1/2}\sigma^{1/2} = |\rho^{1/2}\sigma^{1/2}|W$ and $U_R = I$, $U = I$, $V_R = I$, $V = W^*$ reduces the equality. □

This theorem is essentially proved by Uhlmann [759], from which the inequality (15.1) follows.

Another important property of fidelity is the following:

Theorem 15.4 *For any probability distributions $\{p_i\}$, $\{q_i\}$ and any states ρ_i, σ_i, we have*

$$F\left(\sum_i p_i\rho_i, \sum_i q_i\sigma_i\right) \ge \sum_i \sqrt{p_i q_i}\, F(\rho_i,\sigma_i).$$

Proof Let $|\phi_i\rangle$ and $|\psi_i\rangle$ be the purification vectors of ρ_i and σ_i, respectively, such that $F(\rho_i,\sigma_i) = |\langle\phi_i|\psi_i\rangle|$. Let $\mathcal{L}$ be another Hilbert space with CONS $\{|i\rangle\}$. We define two vectors on the Hilbert space $\mathcal{H}\otimes\mathcal{K}\otimes\mathcal{L}$ by

$$|\phi\rangle \equiv \sum_i \sqrt{p_i}|\phi_i\rangle \otimes |i\rangle,$$

$$|\psi\rangle \equiv \sum_i \sqrt{q_i}|\psi_i\rangle \otimes |i\rangle.$$

It is easily seen that these vectors are the purification vectors of $\sum_i p_i \rho_i$ and $\sum_i q_i \sigma_i$, so that the Uhlmann theorem concludes this proof. □

Remark 15.5 When $p_j = 1$ $(j = 1, 2, \ldots)$, one has

$$F\left(\rho, \sum_i q_i \sigma\right) \geq \sqrt{q_j} F(\rho, \sigma_j).$$

It is not easy to compute the fidelity for general states, but for the case when one of the states ρ or σ is pure, i.e., $\sigma = |\phi\rangle\langle\phi|$, the fidelity becomes

$$F(\rho, \sigma) = \sqrt{\langle\phi|\rho|\phi\rangle}.$$

15.2 The Shor Code

Unfortunately, this simple encoding does not protect against errors other than bit flips. Let us consider the *phase flip error* which occurs when the state $a|0\rangle + b|1\rangle$ is taken to the state $a|0\rangle - b|1\rangle$. For protection against the phase flip error, one can turn the phase flip error channel into a bit flip error channel. Note that the phase flip error channel takes the state $|+\rangle = (|0\rangle + |1\rangle)/\sqrt{2}$ into the state $|-\rangle = (|0\rangle - |1\rangle)/\sqrt{2}$, and vice versa. Therefore, if we work not with the original qubit basis $\{|0\rangle, |1\rangle\}$ but with the basis $\{|+\rangle, |-\rangle\}$, then the phase flip error channel acts just like a bit flip error channel. So, one can use the states $|+++\rangle$, $|---\rangle$ to encode the states $|0\rangle$, $|1\rangle$ in the case of the phase flip error channel.

It is remarkable that there is a code which can protect against the effects of an arbitrary error on a single qubit. The code is a combination of the three bit flip and $|\pm\rangle$ flip codes. The code was suggested by Shor. First, we encode the qubit using the phase flip code: $|0\rangle$ is encoded as $|+++\rangle$ and $|1\rangle$ as $|---\rangle$. Then, we encode each of these qubits using the three bit flip code: $|+\rangle \to (|000\rangle + |111\rangle)/\sqrt{2}$ and $|-\rangle \to (|000\rangle - |111\rangle)/\sqrt{2}$. The result is a nine qubit code:

$$|0\rangle \to \big(|000\rangle + |111\rangle\big)\big(|000\rangle + |111\rangle\big)\big(|000\rangle + |111\rangle\big)/2\sqrt{2},$$
$$|1\rangle \to \big(|000\rangle - |111\rangle\big)\big(|000\rangle - |111\rangle\big)\big(|000\rangle - |111\rangle\big)/2\sqrt{2}.$$

The Shor code protects not only against the flip and phase flip errors on a single qubit. One can show that the code protects against *arbitrary* errors on a single qubit. It follows from the superposition principle in quantum mechanics.

An operator A_k acts only on one qubit, say the first one. Hence it can be represented as a linear combination of the identity I, the bit flip σ_x, the phase flip σ_z, and the combined bit and phase flip $\sigma_x\sigma_z$:

$$A_k = \alpha_k I + \beta_k \sigma_x + \gamma_k \sigma_z + \delta_k \sigma_x \sigma_z$$

where α_k, β_k, γ_k and δ_k are properly chosen complex numbers. Let us define $A_k^{(i)}$ by

$$A_k^{(i)} \equiv I \otimes I \otimes \cdots \otimes \underbrace{A_k}_{i\text{th qubit}} \otimes \cdots \otimes I.$$

Suppose that we encode a qubit as $|\Psi\rangle = a|+++\rangle + b|---\rangle$ and the noise is described by a channel (quantum operation) Λ^* expressed by operation elements $\{A_k\}$. Then, after the noise has acted, we obtain the state

$$\Lambda^{(i)*}\big(|\Psi\rangle\langle\Psi|\big) = \sum_k A_k^{(i)}|\Psi\rangle\langle\Psi|A_k^{(i)*}.$$

Therefore, we can write

$$\Lambda^{(i)*}\big(|\Psi\rangle\langle\Psi|\big) = \sum_{k,l,m} c_{klm}^{(i)} B_l|\Psi\rangle\langle\Psi|B_m^*$$

where $c_{klm}^{(i)}$ are complex numbers determined by $\alpha_k, \beta_k, \gamma_k$ and δ_k, and B_l has the form of the tensor product of one of the qubit operations I, σ_x, σ_z, $\sigma_x\sigma_z$. Here l, m take four values. Measuring the error syndrome reduces this density operator to the one of the form $B_l|\Psi\rangle\langle\Psi|B_m^*$. Then one can recover the initial state by applying the appropriate inversion operations.

15.3 Calderbank–Shor–Steane Codes

We describe now a large class of quantum error-correction codes which were invented by Calderbank, Shor and Steane, called the CSS codes.

First, let us briefly review linear codes. A linear code C encoding k bits of information into an m bit code space is a k-dimensional subspace in the vector space $\mathbb{Z}_2^m$. Such a code is called an $[m,k]$-code. It is defined by an $m \times k$ generator matrix A whose entries are zeros and ones. The $m-k$ bits correspond to redundancy of a code, which will be used to correct an error in information sent through a channel. By using the $[m,k]$-code with the generator matrix A, we encode the k bit message x into Ax, where the message x is treated as a column vector and the arithmetic operations are done modulo 2.

One of results on the existence of linear codes is given by the *Gilbert–Varshamov bound* which states that for large m there exists an $[m,k]$-code protecting against errors on l bits for some k such that

$$\frac{k}{m} \geq 1 - S\left(\frac{l}{m}\right)$$

where $S(x) = -x\log x - (1-x)\log(1-x)$.

Another definition of linear codes can be given in terms of a *parity check* matrix. In this formulation, an $[m,k]$-code is defined to consist of all m element vectors such that $Hx = 0$ where H is an $(m-k) \times m$ matrix and all entries of H and x are zeros and ones. The matrix H is called the parity check matrix. The parity check matrix H and the generator matrix A for the same linear code satisfy $HA = 0$. If a generator matrix A is given then we can find the corresponding parity check matrix H, and vice versa.

Let us describe the so-called dual error correction code. If C is an $[m,k]$-code with the generator matrix A and the parity check matrix H, then the *dual* of C, denoted $C^{\perp}$, is the code with generator matrix H' (transpose of H) and parity check matrix G'.

Now let us describe the quantum CSS codes. Assume that C_1 and C_2 are $[m,k_1]$ and $[m,k_2]$ linear codes (i.e., subspaces of $\mathbb{Z}_2^m$) such that $\{0\} \subset C_2 \subset C_1 \subset \mathbb{Z}_2^m$. *The quantum code* $\mathrm{CSS}(C_1, C_2)$, the CSS code of C_1 over C_2, is defined to be the vector space spanned by the states $|\theta_x\rangle$ for all $x \in C_1$. Here the quantum state $|\theta_x\rangle$ is

$$|\theta_x\rangle = \frac{1}{\sqrt{|C_2|}} \sum_{y \in C_2} |x+y\rangle$$

where the addition is done modulo 2. The state $|\theta_x\rangle$ depends only on the coset space C_1/C_2. The number of cosets of C_2 in C_1 is $|C_1|/|C_2|$, therefore $\mathrm{CSS}(C_1, C_2)$ is an $[m, k_1 - k_2]$ linear code.

One can prove that for CSS codes there is the following quantum Gilbert–Varshamov bound, i.e., for large m there exists an $[m,k]$-code protecting against errors on l bits for some k such that

$$\frac{k}{m} \geq 1 - 2S\left(\frac{2l}{m}\right).$$

In the next section, the punctured self dual doubly even CSS codes will be used. This means that one deletes one coordinate form the code $C_1^{\perp}$, $C_1' = C_1'^{\perp}$, the weight of each word in the code is divisible by 4 and also $C_2 = C_1^{\perp}$. One can prove that in this case these are only two cosets of $C_1^{\perp}$ in C_1. One can write them as $|\theta_0\rangle$ and $|\theta_1\rangle$, and they can be thought of as encoding $|0\rangle$ and $|1\rangle$, respectively.

15.4 General Theory of Quantum Error-Correcting Codes

The basic idea of the general theory of quantum error-correcting codes is the following. The Hilbert space of quantum states is encoded by a unitary transformation into a quantum error-correcting code, which is a subspace C of some larger Hilbert space. Then, the noise is described by a quantum operation Γ (not necessarily trace preserving), and the error-correction procedure is performed by a trace-preserving quantum operation $\mathcal{R}$, which is called the error-correction operation. It is required that for any density operator ρ with support in the code C one has the relation

$$(\mathcal{R} \circ \Gamma)(\rho) = c\rho \tag{15.2}$$

where c is a certain constant. We can write the relation (15.2) also in the form

$$(\mathcal{R}\circ\Gamma)(P\rho P)=cP\rho P \tag{15.3}$$

where P is the projector onto C. In this last form, the relation should hold for any density operator ρ.

The following theorem gives a set of equations which can be checked to determine whether a quantum error-correcting code protects against the noise Λ.

Theorem 15.6 *Let C be a quantum code, and let P be the projector onto C. Suppose that Γ is a quantum operation with operation elements $\{N_k\}$. Then an error-correcting operation $\mathcal{R}$ correcting Γ on C exists if and only if there exists an Hermitian matrix (α_{kn}) such that the following conditions hold*

$$PN_k^*N_nP=\alpha_{kn}P.$$

The conditions are called quantum error-correction conditions.

Proof If $\mathcal{R}$ is an operation with operation elements $\{R_i\}$ then one can write the relation (15.3) in the form

$$\sum_{i,k}R_iN_kP\rho PN_k^*R_i^*=cP\rho P.$$

Therefore, the quantum operation with operation elements $\{R_iN_k\}$ is the same as the quantum operation with a single operation element $\sqrt{c}P$. One can prove that there exist complex numbers c_{ik} such that

$$R_iN_kP=c_{ik}P.$$

Then we derive the quantum error-correction conditions

$$PN_k^*N_nP=\alpha_{kn}P$$

where $\alpha_{kn}=\sum_i c_{ik}^*c_{in}$ is the knth element of a Hermitian matrix. □

15.5 Depolarizing Channel

As an example of application of the previous theorem, let us consider the so-called *depolarizing channel* which is an important type of quantum noise. Suppose we have a state ρ on a single qubit, and with probability p the state is replaced by the completely mixed state, $I/2$. With probability $1-p$, the state is left untouched. One gets the channel sending ρ to the state defined above such that

$$\Lambda^*(\rho)=p\frac{I}{2}+(1-p)\rho.$$

One can write it in the canonical Kraus form. It follows from the canonical representation of a state in the Hilbert space $\mathbb{C}^2$:

$$\rho = \frac{1}{2}\left[I + \sum_{i=1}^{3} \sigma_i a_i\right], \qquad |a| \leq 1,$$

that for arbitrary ρ one has

$$\frac{I}{2} = \frac{1}{4}\sum_{i=0}^{3} \sigma_i \rho \sigma_i$$

where $\sigma_0 = I$, σ_1, σ_2, and σ_3 are the Pauli matrices. Then we get the depolarizing channel in the canonical form

$$\Lambda^*(\rho) = \left(1 - \frac{3p}{4}\right)\rho + \frac{p}{4}\sum_{i=1}^{3} \sigma_i \rho \sigma_i$$

with the operation elements $\{\sqrt{1-3p/4}I, \sqrt{p}\sigma_1/2, \sqrt{p}\sigma_2/2, \sqrt{p}\sigma_3/2\}$.

Let us prove the following

Proposition 15.7 *If a quantum code corrects the depolarizing channel then it corrects an arbitrary single qubit operation.*

Proof Note that the quantum error-correction conditions for the depolarizing channel read

$$P\sigma_k\sigma_n P = \alpha_{kn} P, \quad k, n = 0, 1, 2, 3. \tag{15.4}$$

Suppose now that Γ is an *arbitrary* quantum operation acting on a single qubit, say, the first one. Since its operation elements $\{N_k\}$ can be written as a linear combination of the Pauli matrices $\sigma_0 = I$, σ_1, σ_2, and σ_3, we obtain that the quantum error-correction conditions for the quantum operation Γ are the same as the conditions (15.4). The proposition is proved. □

15.6 Fault-Tolerant Quantum Computation and the Threshold Theorem

One of important applications of quantum error correction is to quantum computations. Noisy quantum systems are not isolated from the environment. Noises affect each of the elements of a quantum circuit: the state preparation procedure, quantum logic gates, and measurement of the output. Actually, the state of a quantum circuit is not a pure state but a mixed state, and it should be described by a density matrix, and quantum gates should be not just unitary operators but more general quantum operations (channels).

The main idea of the fault-tolerant quantum computation is to compute on encoded data. If there is a quantum circuit with unitary gates then we replace each gate in the original circuit with a *procedure* for performing an encoded gate acting on the encoded state. The procedure is defined by a quantum operation (channel). We have to make error corrections on the encoded states. We follow [49] in this section.

Since the procedures (encoded gates) can cause errors to propagate, we have to design them carefully. The procedures should be *fault-tolerant*, i.e., if only one component in the procedure fails then the failure causes at most one error in each encoded block of qubits output from the procedure. It is shown that it is possible to perform a universal set of quantum gates—controlled-NOT, Hadamard, phase, Toffoli—using fault-tolerant procedures. Similarly, we define a fault-tolerant measurement operation and fault-tolerant state preparation.

A quantum circuit with mixed states is defined as follows (see Chap. 11). Let $\mathcal{G}$ be a set of quantum operations (channels). A quantum circuit with gates from $\mathcal{G}$ is a directed acyclic (i.e., without cycles) graph with each vertex in the graph being labeled with a gate from $\mathcal{G}$.

CSS codes will be used to perform computations fault-tolerantly. Each unitary gate in the original quantum circuit is replaced by a procedure (i.e., a quantum operation) which imitates the operation of the gates on the encoded states. If $|\alpha\rangle$ is a state in the original circuit and g is a unitary gate then the *procedure* $O(g)$ *encodes the gate* g if it acts on the state $|\theta_{|\alpha\rangle}\rangle$ which encodes $|\alpha\rangle$ as follows:

$$O(g)|\theta_{|\alpha\rangle}\rangle = |\theta_{g|\alpha\rangle}\rangle.$$

Let us denote by $\mathcal{G}$ the following set of gates (here $a, b, c \in \mathbb{Z}_2$ and all operation are made mod 2):

1. (NOT) $|a\rangle \to |a+1\rangle$
2. (Controlled NOT) $|a, b\rangle \to |a, a+b\rangle$
3. (Phase rotation) $|a\rangle \to i^a|a\rangle$
4. (Controlled phase rotation) $|a, b\rangle \to (-1)^{ab}|a, b\rangle$
5. (Hadamard) $|a\rangle \to \frac{1}{\sqrt{2}}\sum_b(-1)^{ab}|b\rangle$
6. (Toffoli) $|a, b, c\rangle \to |a, b, c+ab\rangle$
7. (Swap) $|a, b\rangle \to |b, a\rangle$
8. Adding a qubit in the state $|0\rangle$
9. Discarding a qubit.

The set $\mathcal{G}$ is a universal set of gates (though not minimal). The following theorem shows that one can perform the gates form $\mathcal{G}$ on the encoded states fault-tolerantly.

Theorem 15.8 *There exist fault-tolerant procedures which simulate all the operations of gates from $\mathcal{G}$ on states encoded by punctured self-dual doubly-even CSS codes such that one error in a qubit or gate effects at most four qubits in each block at the end of the procedure. There exist also such procedures for encoding, decoding, and error correction. Moreover, all these procedures use only gates from $\mathcal{G}$.*

The proof will not be given here (see [49]) we only comment on the bitwise procedures. A bitwise procedure of a gate on k qubits is defined by labeling the qubits in each one of k blocks from 1 to m, and then applying the gate m times. In particular, we have for the NOT gate ($a, b \in C/C^{\perp}$ and we omit normalization factors):

$$|\theta_a\rangle = \sum_{x \in C^{\perp}} |a_1 + x_1\rangle \otimes \cdots \otimes |a_m + x_m\rangle$$

$$\longrightarrow \sum_{x \in C^{\perp}} |a_1 + 1 + x_1\rangle \otimes \cdots \otimes |a_m + 1 + x_m\rangle = |\theta_{a+1}\rangle.$$

For C-NOT one has:

$$\begin{aligned}
&|\theta_a\rangle \otimes |\theta_b\rangle \\
&= \sum_{x \in C^{\perp}} |a_1 + x_1\rangle \otimes \cdots \otimes |a_m + x_m\rangle \otimes \sum_{y \in C^{\perp}} |b_1 + y_1\rangle \otimes \cdots \otimes |b_m + y_m\rangle \\
&\longrightarrow \sum_{x \in C^{\perp}} |a_1 + x_1\rangle \otimes \cdots \otimes |a_m + x_m\rangle \\
&\quad \otimes \sum_{y \in C^{\perp}} |a_1 + b_1 + x_1 + y_1\rangle \otimes \cdots \otimes |a_m + b_m + x_m + y_m\rangle \\
&= |\theta_a\rangle \otimes |\theta_{a+b}\rangle.
\end{aligned}$$

The last equality follows form the relation

$$\begin{aligned}
&\sum_{y \in C^{\perp}} |a_1 + b_1 + x_1 + y_1\rangle \otimes \cdots \otimes |a_m + b_m + x_m + y_m\rangle \\
&\quad = \sum_{y \in C^{\perp}} |a_1 + b_1 + y_1\rangle \otimes \cdots \otimes |a_m + b_m + y_m\rangle
\end{aligned}$$

since $C^{\perp}$ is a linear space.

Now let us discuss noise in computation. One uses the following model for noise. Each time step, each qubit or gate undergoes a fault (i.e., an arbitrary quantum operation) which is at most η-far from the identity, in some metric on operators. The η is called the *error rate*. Let a quantum circuit be represented as

$$\Xi(\Lambda^*) = \Lambda_{i_T}^{(\alpha_T)*} \cdots \Lambda_{i_2}^{(\alpha_2)*} \Lambda_{i_1}^{(\alpha_1)*}$$

where $\{\Lambda_k^*\}$ are channels from some fixed set $\mathcal{G}$ of channels and $\Lambda_{i_k}^{(\alpha_k)*}$ is an α_k-extension of the channel Λ_k^*. It transforms a density matrix ρ into $\Lambda_{i_T}^{(\alpha_T)*} \circ \cdots \circ \Lambda_{i_2}^{(\alpha_2)*} \circ \Lambda_{i_1}^{(\alpha_1)*} \circ \rho$.

Gates are applied at integer time steps. Faults occur in between time steps, in the *locations* of the circuit. A set $(q_1,\dots,q_r,t)$ is a location of the quantum circuit if the qubits $q_1,\dots,q_r$ are all gates participating in the same gate, in time step t. If a qubit q did not participate in any gate at time t, then (q,t) is also considered as a location.

Each location in the circuit exhibits a fault with independent probability η. A *fault* at a certain location at time t means that a quantum operation (channel, noise operator) $\mathcal{E}_t$ is applied on the faulty qubits after time t. The list of locations where faults have occurred, in a specific run of the computation, is called a fault path. Each fault path, F, is assigned a probability $P(F)$. Denote by $\mathcal{E}(F)=(\mathcal{E}_1,\mathcal{E}_2,\dots,\mathcal{E}_T)$ the choice of faults for the fault path and

$$\Xi\bigl(\Lambda^*,\mathcal{E}(F)\bigr)=\mathcal{E}_T\Lambda_{i_T}^{(\alpha_T)*}\cdots\mathcal{E}_2\Lambda_{i_2}^{(\alpha_2)*}\mathcal{E}_1\Lambda_{i_1}^{(\alpha_1)*}.$$

The output density operator of the circuit is defined as

$$\Xi\circ\rho=\sum_{\Gamma}P(F)\Xi\bigl(\Lambda^*,\mathcal{E}(F)\bigr)\circ\rho.$$

This circuit is called a quantum circuit with probabilistic errors. A more general noise model is given by the formula

$$\Xi(\Lambda^*,\mathcal{E})=\mathcal{E}_T\Lambda_{i_T}^{(\alpha_T)*}\cdots\mathcal{E}_2\Lambda_{i_2}^{(\alpha_2)*}\mathcal{E}_1\Lambda_{i_1}^{(\alpha_1)*}$$

where $\mathcal{E}_t$ are now channels (noise operators) of the form

$$\mathcal{E}_t=\mathcal{E}_{B_{1,t}}(t)\otimes\mathcal{E}_{B_{2,t}}(t)\otimes\cdots\otimes\mathcal{E}_{B_{s,t}}(t).$$

Here $B_{i,t}$ runs over all possible locations at time t and for each of them, one has

$$\bigl\|\mathcal{E}_{B_{i,t}}(t)-I\bigr\|\le\eta$$

in some norm on operators.

This model includes, in particular, the probabilistic errors, decoherence, phase and amplitude damping. Note that this definition assumes independence between different fault in space, and independence in time (Markovianity).

Now we can formulate the threshold theorem [49] for quantum computation.

Theorem 15.9 *Let C be a computational code with gates G. Let $\varepsilon>0$. There exists a threshold $\eta_0>0$ and a constant $\gamma>0$ such that the following holds. Let M be a quantum circuit with n input qubits, which operates T time steps, uses s gates from G, and has v locations, and we denote $L=\log^{\gamma}(\frac{v}{\varepsilon})$. Then there exists a quantum circuit M' which operates on nL qubits, for time tL and uses vL gates from G such that in the presence of general noise with error rate $\eta<\eta_0$, the quantum circuit M' computes a function which is ε-close to that computed by M.*

A similar theorem is valid for quantum circuits with probabilistic errors. We present some comments on the proof of the theorem for the last case. We will use the computational code recursively. It encodes one qubit on m qubits. Consider a fault-tolerant procedure for a gate in $\mathcal{G}$ preceded by fault-tolerant error corrections on each block participating in the procedure. The *spread* of the code is l if each fault which occurs during this sequence of gates effects at most l qubits in each block at the end of the procedure. It is required that the number of errors that the code can correct, d, is larger than the spread l, $l \leq d$.

We denote the original quantum circuit $M = M_0$ and simulate M_0 by a more reliable circuit M_1, as follows. Each qubit is replaced by a block of qubits, and each time step in M_0 transforms in M_1 to a working period which includes an error correction procedure, and then the operation of each gate in M_0 is replaced in M_1 by its procedure. We repeat this r levels to get M_r, an r-simulating circuit of M_0. The output of M_r is defined by taking recursive majority on the outputs. The number of levels will be the polynomial order of $\log(V(M_0))$ where $V(M_0)$ is the number of locations in M_0.

Every qubit transforms to a block of m qubits in the next level and so on. One qubit in M_{r-s} transforms to m^s qubits in M_r. This set of qubits is called an s-block. The recursive simulation induces a partition of the set of locations in M_r into generalized rectangles. An *r-rectangle* in M_r is the set of locations which originated from one location on M_0. Let B be the set of qubits in some r-blocks. An (r, k)-*sparse* set of qubits A in B is a set of qubits in which for every r-block in B, there are at most k $(r-1)$-blocks such that the set A in these blocks is not $(r-1, k)$-sparse. An $(0, k)$-sparse set of qubits A is an empty set of qubits. Two density operators ρ_1, ρ_2 of the set of qubits B are (r, k)-deviated if there exists an (r, k)-sparse set of qubits $A \subseteq B$, with $\rho_1|_{B-A} = \rho_2|_{B-A}$. A computation is successful if the error at the end of each r-working period is sparse enough. A set of locations in an r-rectangle is called (r, k)-sparse if there are no more than k $(r-1)$-rectangles in which the set is not $(r-1, k)$-sparse. A fault path in M_r is (r, k)-sparse if in each r-rectangle, the set is (r, k)-sparse. For the proof of the theorem, the following three lemmas are used.

Lemma 15.10 *Consider a computation subjected to an (r, k)-sparse fault path with $kl \leq d$. Then at the end of each r-working period the error is (r, d)-sparse.*

This and the next lemmas show that if the fault path is sparse enough, then the error correction keeps the deviation small.

Lemma 15.11 *Let $2d + 1 \leq m$. If the final density matrix of M_r is (r, d)-deviated for the correct one, then the distribution on the strings which are obtained when taking recursive majority on each r-block of its output is correct.*

Then it is shown that non-sparse fault paths are rare below the threshold. Let N be the maximal number of locations in a rectangle. It is required that the probability for more than $[d/l] = k$ faults in N locations is smaller than η. This requirement is satisfied if one has the following threshold condition:

$$\binom{N}{k+1}\eta^{k+1} < \eta.$$

The *threshold* for probabilistic noise for the code C is defined as

$$\eta_c = \binom{N}{k+1}^{-k}.$$

Any $\eta < \eta_c$ satisfies the threshold condition. The threshold was estimated as 10^{-6} in some cases. The next lemma shows that the threshold bad fault paths are rare.

Lemma 15.12 *Let $P(r)$ be the probability for the faults in an r-rectangle to be (r,k)-sparse. If $\eta < \eta_c$ then there exists a $\delta > 0$ such that*

$$P(r) > 1 - \eta^{(1+\delta)^r}.$$

To prove the threshold theorem for probabilistic noise, one chooses r such that

$$\upsilon\eta^{(1+\delta)^r} < \varepsilon.$$

Then from the previous three lemmas the theorem follows.

To prove the threshold theorem for general noise, one writes

$$\mathcal{E}_{B_{i,t}}(t) = (1-\eta)I + \mathcal{E}'_{B_{i,t}}(t)$$

(this is the definition of $\mathcal{E}'_{B_{i,t}}(t)$). The operators $\mathcal{E}'_{B_{i,t}}(t)$ satisfy the bound $\|\mathcal{E}'_{B_{i,t}}(t)\| \le 2\eta$. By replacing the error operators in the formula for $\mathcal{E}_t$ by the product of operators of the last form, one can get the sum of terms corresponding to fault paths. The threshold theorem for general noise is basically reduced to the theorem for the probabilistic noise.

The threshold theorem was derived also for the error model with exponentially decaying correlations. Moreover, a threshold result was obtained for some more realistic non-Markovian error models admitting a Hamiltonian description. A non-Markovian noise model is familiarly formulated in terms of a Hamiltonian H that governs the joint evolution of the system (quantum circuit) and the "bath" (noise). We write

$$H = H_S + H_B + H_{SB}$$

where H_S is the (time-dependent) Hamiltonian of the ideal quantum circuit, H_B is an Hamiltonian of the bath, and H_{SB} is the interacting Hamiltonian coupling the system to the bath. A specific form of the Hamiltonian is assumed. In particular, it is assumed that every qubit of the system has its own bath. The results depend on two parameters t_0 and λ_0. t_0 is the time to do a one- or two-qubit gate by a unitary evolution $U(t+t_0, t)$ during the time interval from t to $t+t_0$. λ_0 is a constant which describes the strength of the coupling,

$$\|H_{SB}\| \leq \lambda_0.$$

Instead of η, in this model one has $\lambda_0 t_0$, and the critical error threshold is

$$(\lambda_0 t_0)_c = \frac{1}{eN(N-1)}.$$

Here N is the number of locations in a 1-rectangle that can correct two errors and has spread $s=1$, and e is the base of the natural logarithm. The following theorem was proved.

Theorem 15.13 *Let an error free quantum circuit M output samples from a probability distribution P. Let $\varepsilon > 0$. There exists a quantum circuit M' subjected to noise according to the Hamiltonian H_{SB} and the bath Hamiltonian H_B, and if $\lambda_0 t_0 \leq (\lambda_0 t_0)_c$ then it outputs the probability distribution P' such that*

$$\|P' - P\| \leq \varepsilon.$$

Theorem 15.14 *The number of locations in M' is N_0 polylog$(N_0/\sqrt{\varepsilon})$ where N_0 is the number of locations of M.*

15.7 Notes

Examples of quantum error correcting code were first found by Shor [711] and Steane [733]. The 9-qubit encoding (Shor code) explained in this chapter was introduced in [711]. The theory of quantum codes is presented in [547]. The notion of fidelity was introduced by Uhlmann [759] and has often been employed to measure the difference of states in the various places [393]. The Calderbank–Shor–Steane code was first introduced in [148, 149, 734]. The general theory of quantum error-correcting code was discussed in [434]. The fault-tolerant method was proposed by [205, 712] and was extensively investigated in [298]. The quantum threshold theorems were proved in [49, 50, 56, 297, 429, 435, 654, 746]. We follow the Aharonov and Ben-Or [49] and the Therhal and Burkard papers [746]. Hamiltonian models of noise were considered by Alicky et al. [54, 55].

Chapter 16
Quantum Field Theory, Locality and Entanglement

In this chapter, some basic notions of quantum field theory will be exposed and properties of entanglement and locality will be considered again in this new context. The relativistic corrections to the EPR–Bell type correlation functions for entangled states of the Dirac particles will be computed and it will be shown that the spatially depending correlations are consistent with locality.

The central notion in modern fundamental physics is the notion of a *quantum field* (and its extension to string theory). Therefore, quantum information theory should be based on quantum field theory, and first of all, on quantum electrodynamics. One of the future goals is an attempt to develop quantum information theory starting from quantum field theory.

A quantum field is an operator-valued function on the Minkowski space–time (or, more generally, on the curved space–time). In the quantum field theory, there is a fundamental *property of locality* (local commutativity) which for the bosonic quantum field $\Phi(x)$ reads

$$\big[\Phi(x),\Phi(y)\big]=0$$

if the points x and y are space-like separated. We will discuss how this locality is consistent with the properties of entangled states and with the collapse of the wave function in the measurement procedure.

The modern standard model of the elementary particles theory is the theory of electro-weak interactions and quantum chromodynamics. The list of quantum fields includes scalar fields with spin 0, Fermi fields with spin $1/2$ and gauge fields with spin 1. The prototype theory for the standard model is quantum electrodynamics which is the most important for practical applications.

16.1 Quantum Electrodynamics (QED)

In this section, some basic formulae from quantum electrodynamics (QED) are collected. They are used in particular in quantum optics, the cavity QED, and in the theory of quantum dots which will be discussed in Chap. 19.

M. Ohya, I. Volovich, *Mathematical Foundations of Quantum Information and Computation and Its Applications to Nano- and Bio-systems*,
Theoretical and Mathematical Physics,
DOI 10.1007/978-94-007-0171-7_16,

16.1.1 Maxwell Equations

An electric field $\mathbf{E}=(E_1, E_2, E_3)$ and a magnetic field $\mathbf{B}=(B_1, B_2, B_3)$ are mappings from $\mathbf{G}\times\mathbb{R}$ to $\mathbb{R}^3$ where $\mathbf{G}$ is a region in $\mathbb{R}^3$. The Maxwell equations for the free electromagnetic field have the form

$$\begin{aligned} \operatorname{rot}\mathbf{E}(\mathbf{r},t) &= -\frac{\partial}{\partial t}\mathbf{B}(\mathbf{r},t), \quad \operatorname{div}\mathbf{E}(\mathbf{r},t)=0, \\ \operatorname{rot}\mathbf{B}(\mathbf{r},t) &= \frac{1}{c^2}\frac{\partial}{\partial t}\mathbf{E}(\mathbf{r},t), \quad \operatorname{div}\mathbf{B}(\mathbf{r},t)=0, \end{aligned} \tag{16.1}$$

where c is the speed of light, $\operatorname{rot}\mathbf{E}$ is a vector defined by

$$\operatorname{rot}\mathbf{E}=\nabla\times\mathbf{E}, \quad \nabla=\left(\frac{\partial}{\partial x_1}, \frac{\partial}{\partial x_2}, \frac{\partial}{\partial x_3}\right),$$

$$(\operatorname{rot}\mathbf{E})_i \equiv \frac{1}{2}\sum_{j,k=1}^{3}\varepsilon_{ijk}\partial_j\mathbf{E}_k, \quad i=1,2,3,$$

where ε_{ijk} is an antisymmetric tensor such that

$$\varepsilon_{ijk}=-\varepsilon_{jik}=\varepsilon_{kij}, \quad \varepsilon_{123}=1,$$

and $\operatorname{div}\mathbf{E}$ is the divergence of $\mathbf{E}$, that is,

$$\operatorname{div}\mathbf{E}=\nabla\cdot\mathbf{E}=\sum_{i=1}^{3}\partial_i\mathbf{E}_i,$$

$$\partial_i=\frac{\partial}{\partial x_i}.$$

Here $\mathbf{E}(\mathbf{r},t)$ and $\mathbf{B}(\mathbf{r},t)$ are the electric and magnetic fields at the space–time point $(\mathbf{r},t)$, $\mathbf{r}\in\mathbf{G}$, $t\in\mathbb{R}$.

The energy of the electromagnetic field is given by

$$H=\frac{1}{2}\int\left[\mathbf{E}^2(\mathbf{r},t)+\mathbf{B}^2(\mathbf{r},t)\right]d^3\mathbf{r}.$$

One can represent the fields by using the vector-potential $\mathbf{A}(\mathbf{r},t)$ in the Coulomb gauge as

$$\mathbf{E}(\mathbf{r},t)=-\frac{\partial}{\partial t}\mathbf{A}(\mathbf{r},t), \qquad \mathbf{B}(\mathbf{r},t)=\operatorname{rot}\mathbf{A}(\mathbf{r},t)$$

where the vector-potential satisfies the equations

$$\Box\mathbf{A}(\mathbf{r},t)\equiv\frac{1}{c^2}\frac{\partial^2}{\partial t^2}\mathbf{A}(\mathbf{r},t)-\Delta\mathbf{A}(\mathbf{r},t)=0, \quad \operatorname{div}\mathbf{A}(\mathbf{r},t)=0. \tag{16.2}$$

Here Δ is the Laplace operator,

$$\Delta = \sum_{i=1}^{3} \frac{\partial^2}{\partial x_i^2}, \qquad \mathbf{r} = (x_1, x_2, x_3).$$

It follows from (16.2) that the vector potential $\mathbf{A}(\mathbf{r}, t)$ can be represented as

$$\mathbf{A}(\mathbf{r}, t) = \sum_k c_k(t) u_k(\mathbf{r})$$

where $u_k(\mathbf{r})$ is an eigenfunction of the Laplace operator with the eigenvalue ω_k

$$-\Delta u_k = \omega_k u_k, \tag{16.3}$$

and $c_k(t)$ is a vector satisfying

$$\left(\frac{\partial^2}{\partial t^2} + \omega_k^2\right) c_k(t) = 0,$$

from which we can get

$$c_k(t) = c_k \exp(-\mathrm{i}\omega_k t) + c_k^* \exp(\mathrm{i}\omega_k t).$$

16.1.2 Quantization of Electromagnetic Field

We consider the system in a box $\mathbf{G}$ of volume $V = L^3$ with the periodic boundary conditions. Let us solve (16.3) as

$$u_k(\mathbf{r}) = \exp(\mathrm{i}\mathbf{kr})$$

where $\mathbf{k} = (k_1, k_2, k_3)$, $k_i = 2\pi n_i / L$ $(n_i = 0, \pm 1, \pm 2, \ldots)$, $\mathbf{kr}$ means the inner product of $\mathbf{k}$ and $\mathbf{r}$, and the solution of (16.2) is written as

$$\mathbf{A}(\mathbf{r}, t) = \sum_k \big(c_k \exp(-\mathrm{i}\omega_k t) + c_k^* \exp(\mathrm{i}\omega_k t)\big) \exp(\mathrm{i}\mathbf{kr}).$$

Using (16.2), the above formula can be written in the form

$$\mathbf{A}(\mathbf{r}, t) = \sum_{\mathbf{k}} \sum_{s} \left(\frac{\hbar}{2V\omega_k}\right)^{1/2} \left[a_{\mathbf{k}s} \varepsilon_{\mathbf{k}s} e^{-\mathrm{i}\omega_k t + \mathrm{i}\mathbf{kr}} + a_{\mathbf{k}s}^* \varepsilon_{\mathbf{k}s} e^{\mathrm{i}\omega_k t - \mathrm{i}\mathbf{kr}}\right].$$

Here $a_{\mathbf{k}s}$ is a complex-valued function and $a_{\mathbf{k}s}^*$ is its complex conjugate, $\omega_k = c|\mathbf{k}|$, and $\hbar$ is the Planck constant. Remark that c_k corresponds to $\sum_{s=1}^{2} a_{\mathbf{k}s} \varepsilon_{\mathbf{k}s}$. The factor

$(\hbar/2V\omega_k)^{1/2}$ is inserted for further convenience. The polarization vectors $\varepsilon_{\mathbf{k}s}$ are three-dimensional vector functions which satisfy

$$\mathbf{k}\varepsilon_{\mathbf{k}s}=0, \qquad \varepsilon_{\mathbf{k}s}\varepsilon_{\mathbf{k}s'}=\delta_{ss'}, \quad s,s'=1,2; \qquad \varepsilon_{\mathbf{k}_1}\times\varepsilon_{\mathbf{k}_2}=\mathbf{k}/|\mathbf{k}|.$$

We quantize the electromagnetic field by declaring $a_{\mathbf{k}s}$ and $a^*_{\mathbf{k}s}$ to be the annihilation and creation operators which obey the canonical commutation relations

$$[a_{\mathbf{k}s}, a^*_{\mathbf{k}'s'}]=\delta_{\mathbf{k}\mathbf{k}'}\delta_{ss'}.$$

The Hamiltonian of a quantum electromagnetic field is

$$H=\frac{1}{2}\int_{\mathbf{G}}\left[\mathbf{E}^2(\mathbf{r},t)+\mathbf{B}^2(\mathbf{r},t)\right]d^3\mathbf{r}=\sum_{\mathbf{k}}\sum_{s}\hbar\omega_k\left[a^*_{\mathbf{k}s}a_{\mathbf{k}s}+\frac{1}{2}\right].$$

The sum $\sum_{\mathbf{k}}\frac{1}{2}\hbar\omega_k$ is divergent. This is a particular case of the so-called ultraviolet divergences in the quantum field theory. To remove this divergence, one has to apply the renormalization theory. In the given case, we just replace the divergent vacuum energy $\sum_{\mathbf{k}}\frac{1}{2}\hbar\omega_k$ with zero. So the renormalized Hamiltonian will have the form

$$H=\sum_{\mathbf{k}}\sum_{s}\hbar\omega_k a^*_{\mathbf{k}s}a_{\mathbf{k}s}.$$

The n photon basis Fock state can be written as

$$|\mathbf{k}_1 s_1,\ldots,\mathbf{k}_n s_n\rangle=a^*_{\mathbf{k}_1 s_1}\cdots a^*_{\mathbf{k}_n s_n}|0\rangle,$$

where $|0\rangle$ is the vacuum vector, $a_{\mathbf{k}s}|0\rangle=0$. Here $\mathbf{k}_i$ is the momentum of the ith photon and s_i gives its polarization.

16.1.3 Casimir Effect

As an illustration of the renormalization procedure we consider the Casimir effect.

The Casimir effect is the attractive force between two uncharged metal plates that are placed very near to each other in a vacuum. A typical example is of two uncharged metallic plates in a vacuum, placed a few micrometers apart, without any external electromagnetic field. In a classical description, no force would be measured between them. When this field is instead studied using quantum electrodynamics, it is seen that the plates do affect the virtual photons which constitute the field, and generate a net force—either an attraction or a repulsion depending on the specific arrangement of the two plates. The attraction arises due to a reduction in the energy of the ground state of the electromagnetic field between the two plates. Because fluctuations in the field between the plates can only have wavelengths equal to or smaller than the distance between the plates, the vacuum electromagnetic field has less energy between the plates than outside of them.

It has been suggested that the Casimir forces have application in nanotechnology, in particular, in silicon integrated circuit technology based micro- and nano-electromechanical systems, and the so-called Casimir oscillators.

The Casimir effect can be expressed in terms of virtual particles interacting with the objects. It is calculated in terms of the zero-point energy of a quantized field in the intervening space between the objects.

Consider a quantum electromagnetic field in the space between a pair of conducting metal plates at a distance a apart. In this case, the transverse component of the electric field and the normal component of the magnetic field must vanish on the surface of a conductor. Assuming the parallel plates lie in the x–y plane, the standing waves are

$$F(x,y,z,t) = e^{-\omega_n t + ik_x x + ik_y y} \sin(k_n z)$$

where F denotes a component of the electromagnetic field. Here, k_x and k_y are the wave vectors in directions parallel to the plates, and $k_n = \pi n/a$ is the wave-vector perpendicular to the plates, $n = 1, 2, \ldots$. The energy is

$$\omega_n = c\sqrt{k_x^2 + k_y^2 + \frac{\pi^2 n^2}{a^2}}$$

where c is the speed of light. The vacuum energy is then the sum over all possible excitation modes

$$E = \frac{\hbar}{2} 2\sigma \int \frac{dk_x\,dk_y}{(2\pi)^2} \sum_{n=1}^{\infty} \omega_n .$$

Here σ is the area of the metallic plates, and a factor of 2 is introduced for the two possible polarizations of the wave. This expression is infinite, and it is required to introduce a regularization. The regulator will serve to make the expression finite, and in the end will be removed. One shall use the zeta regularization by introducing a complex parameter s:

$$\frac{E(s)}{\sigma} = \hbar \int \frac{dk_x\,dk_y}{(2\pi)^2} \sum_{n=1}^{\infty} \omega_n^{1-s} .$$

We first compute the integral and the sum for $\mathrm{Re}\, s > 3$, then make the analytical continuation, and finally take the limit $s \to 0$. We obtain

$$\begin{aligned}\frac{E(s)}{\sigma} &= \frac{\hbar c^{1-s}}{4\pi^2} \sum_{n=1}^{\infty} \int_0^{\infty} 2\pi k\,dk \left(k^2 + \frac{\pi^2 n^2}{a^2}\right)^{\frac{1-s}{2}} \\ &= -\frac{\hbar c^{1-s} \pi^{2-s}}{2a^{3-s}} \zeta(s-3).\end{aligned}$$

Here $\zeta(s)$ is the Riemann zeta function,

$$\zeta(s) = \sum_{n=1}^{\infty} n^{-s}$$

which admits the analytic continuation to the complex plane. Now we take $s = 0$, and since $\zeta(-3) = 1/120$ we get

$$\frac{E}{\sigma} = -\frac{\hbar c \pi^2}{3 \cdot 240 a^3}.$$

Then the Casimir force per unit area F_C/σ for conducting plates is

$$\frac{F_C}{\sigma} = -\frac{d}{da}\frac{E}{\sigma} = -\frac{\hbar c \pi^2}{240 a^4}.$$

The force is negative, indicating that the force is attractive.

16.1.4 Correlation Functions and Photo-Detection

Coherence and other properties of light are described by using the correlation functions

$$\begin{aligned} &\Gamma^{(N,M)}_{i_1,\dots,i_N;j_1,\dots,j_M}(\mathbf{r}_1, t_1, \dots, \mathbf{r}_N, t_N; \mathbf{r}'_M, t'_M, \dots, \mathbf{r}'_1, t'_1) \\ &\quad = \langle F^{(-)}_{i_1}(\mathbf{r}_1, t_1) \cdots F^{(-)}_{i_N}(\mathbf{r}_N, t_N) F^{(+)}_{j_M}(\mathbf{r}'_M, t'_M) \cdots F^{(+)}_{j_1}(\mathbf{r}'_1, t'_1) \rangle. \end{aligned}$$

Here N, M are natural numbers, and the brackets mean the expectation value in the state with a density operator ρ, $\langle A \rangle = \mathrm{tr}[A\rho]$, and $F^{(\pm)}_j(\mathbf{r}, t)$, $j = 1, 2, 3$ are negative (positive) frequency components of the electromagnetic field, $F^{(\pm)}_j(\mathbf{r}, t) = E^{(\pm)}_j(\mathbf{r}, t)$ or $B^{(\pm)}_j(\mathbf{r}, t)$. For example, the components of the electric field are

$$E^{(+)}_j(\mathbf{r}, t) = \sum_{\mathbf{k}} \sum_s \mathrm{i} \left(\frac{\hbar\omega}{2V} \right)^{1/2} a_{\mathbf{k}s} \varepsilon_{j\mathbf{k}s} e^{-\mathrm{i}\omega t + \mathrm{i}\mathbf{k}\mathbf{r}},$$

$$E^{(-)}_j(\mathbf{r}, t) = -\sum_{\mathbf{k}} \sum_s \mathrm{i} \left(\frac{\hbar\omega}{2V} \right)^{1/2} a^*_{\mathbf{k}s} \varepsilon_{j\mathbf{k}s} e^{\mathrm{i}\omega t - \mathrm{i}\mathbf{k}\mathbf{r}}.$$

Here $\varepsilon_{j\mathbf{k}s}$ is the jth component of the polarization vector $\varepsilon_{\mathbf{k}s}$, $j = 1, 2, 3$.

The theory of photon detection was discussed by Glauber. As a model of the detector, he has considered an atom which can change its state due to interaction with photons. In the dipole approximation, the interaction Hamiltonian in the interaction representation reads

$$-e\sum_{k}\mathbf{q}_k(t)\mathbf{E}(\mathbf{r},t).$$

Here $\mathbf{q}_k$ is the position of the kth electron.

In the first order of the perturbation theory, the probability density of the single-photon counting during the time (t_0,t) at the space point $\mathbf{r}$ is

$$p^{(1)}(\mathbf{r},t)=\int_{t_0}^{t}dt_1\int_{t_0}^{t}dt_1' S_{ij}(t_1-t_1')\Gamma_{i;j}^{(1,1)}(\mathbf{r},t_1;\mathbf{r},t_1')$$

where the summation over the repeating indices $i,j=1,2,3$ is assumed. Here $S_{ij}(t_1-t_1')$ is the window function of the detector.

If the period of time $\Delta t=t-t_0$ is small, we can set approximately

$$p^{(1)}(\mathbf{r},t)=C\Gamma_{i;i}^{(1,1)}(\mathbf{r},t;\mathbf{r},t)=C\big\langle F_i^{(-)}(\mathbf{r},t)F_i^{(+)}(\mathbf{r},t)\big\rangle$$

where C is a constant characterizing the detector.

16.1.5 Interference: Two-Slits Experiment

The last formula can be used to describe the interference in the two-slits experiment when the light emitted from two space points $\mathbf{r}_1$ and $\mathbf{r}_2$ is detected at the point $\mathbf{r}$. We can obtain the relation between corresponding solutions of the Maxwell equations which are valid in the quantum case

$$\mathbf{F}^{(+)}(\mathbf{r},t)=K_1\mathbf{F}^{(+)}(\mathbf{r}_1,t-\tau_1)+K_2\mathbf{F}^{(+)}(\mathbf{r}_2,t-\tau_2).$$

Here $c\tau_1=|\mathbf{r}-\mathbf{r}_1|$, $c\tau_2=|\mathbf{r}-\mathbf{r}_2|$, and K_1, K_2 are some constants in $\mathbb{C}$.

Now if we denote $\langle I(\mathbf{r},t)\rangle=\langle F_i^{(-)}(\mathbf{r},t)F_i^{(+)}(\mathbf{r},t)\rangle$, we get

$$\begin{aligned}\big\langle I(\mathbf{r},t)\big\rangle&=|K_1|^2\big\langle I(\mathbf{r}_1,t-\tau_1)\big\rangle+|K_2|^2\big\langle I(\mathbf{r}_2,t-\tau_2)\big\rangle\\&\quad+2\,\mathrm{Re}\big[K_1^*K_2\big\langle\mathbf{F}^{(-)}(\mathbf{r}_1,t-\tau_1)\mathbf{F}^{(+)}(\mathbf{r}_2,t-\tau_2)\big\rangle\big].\end{aligned}$$

For appropriate coherent states, one can take the expectation values of quantum operators equal to the classical plane wave values

$$\mathbf{F}^{(+)}(\mathbf{r}_1,t-\tau_1)=\mathbf{A}_1e^{\mathrm{i}(\mathbf{k}\mathbf{r}_1-\omega(t-\tau_1))},\qquad \mathbf{F}^{(+)}(\mathbf{r}_2,t-\tau_2)=\mathbf{A}_2e^{\mathrm{i}(\mathbf{k}\mathbf{r}_2-\omega(t-\tau_2))}$$

to get the familiar interference formula. Here $\mathbf{A}_1$ and $\mathbf{A}_2$ are constant amplitudes.

The probability density of the N-photon counting during the time (t_0,t) at the space points $\mathbf{r}_1,\ldots,\mathbf{r}_N$ is

$$p^{(N)}(\mathbf{r}_1, \ldots, \mathbf{r}_n; t)$$
$$= \int_{t_0}^{t} \cdots \int_{t_0}^{t} S^{(1)}_{i_1 j_1}(t_1 - t'_1) \cdots S^{(N)}_{i_N j_N}(t_N - t'_N)$$
$$\times \Gamma^{(N,N)}_{i_1,\ldots,i_N; j_1,\ldots,j_N}(\mathbf{r}_1, t_1, \ldots, \mathbf{r}_N, t_N; \mathbf{r}_N, t'_N, \ldots, \mathbf{r}_1, t'_1)\, dt_1\, dt'_1 \cdots dt_N\, dt'_N.$$

16.1.6 Lorentz-Invariant Form of Maxwell Equations

The Lorentz-invariant form of the Maxwell equations is

$$\partial^\mu F_{\mu\nu} = 0.$$

Remark that $\partial^\mu F_{\mu\nu}$ means $\sum_{\mu=0}^{3} \partial^\mu F_{\mu\nu}$ (Einstein's summation rule). Here we use relativistic notations:

$$\partial^\mu = \eta^{\mu\nu} \partial_\nu, \qquad \partial_\nu = \frac{\partial}{\partial x^\nu}, \qquad \mu, \nu = 0, 1, 2, 3$$

where $x = (x^\mu)$ are the coordinates in the Minkowski space–time ($x^0 = t$ is time and $x^k, k = 1, 2, 3$, are spatial coordinates) with the metric ($\eta^{\mu\nu} = \mathrm{diag}(-1, 1, 1, 1)$) and the summation over the repeating indices is assumed. The tensor $F_{\mu\nu}$ is expressed in terms of the vector-potential $A_\mu = A_\mu(x)$ as

$$F_{\mu\nu} = \partial_\mu A_\nu - \partial_\nu A_\mu.$$

The tensor $F_{\mu\nu}$ describes the components of the electric E_j and magnetic B_j fields:

$$F_{0j} = E_j, \qquad F_{jk} = \varepsilon_{jkl} B_l, \quad j, k, l = 1, 2, 3.$$

In the Lorentz gauge $\partial^\mu A_\mu = 0$, the Maxwell equations take the form

$$\Box A_\mu = 0, \quad \Box = -\eta^{\mu\nu} \partial_\mu \partial_\nu.$$

We have set $c = \hbar = 1$. The solution of the Maxwell equations can be written in the form

$$A_\mu(x) = \sum_{k\alpha} \frac{1}{\sqrt{2V\omega}} e^\alpha_\mu \left[a_{k\alpha} e^{ikx} + a^*_{k\alpha} e^{-ikx}\right].$$

Here $k = (k_\mu) = (|\mathbf{k}|, \mathbf{k})$, $\mu = 0, 1, 2, 3$; $kx = -|\mathbf{k}|t + \mathbf{kx}$ and $e^\alpha_\mu = e^\alpha_\mu(k)$ are polarization vectors, $\alpha = 1, 2, 3, 4$. We set $\hbar = c = 1$.

In the quantum theory of electromagnetic fields, the annihilation and creation operators $a_{k\alpha}, a^*_{k\alpha}$ obey the canonical commutation relations

$$[a_{k\alpha}, a^*_{k'\alpha'}] = \delta_{kk'} \delta_{\alpha\alpha'}.$$

The Hamiltonian has the form

$$H_{\mathrm{EM}} = \sum_{k\alpha} |\mathbf{k}| a^*_{k\alpha} a_{k\alpha}.$$

In the covariant Gupta–Bleuler formulation of quantum electromagnetic field, an indefinite metric in the Fock space is used.

16.1.7 Dirac Equation

Particles with spin 1/2 in the Minkowski space–time are described by the Dirac equation:

$$\left[\gamma^\mu(\partial_\mu - \mathrm{i}eA_\mu) + m\right]\psi = 0.$$

Here γ^μ, $\mu = 0, 1, 2, 3$, are Dirac 4×4 matrices, $\{\gamma^\mu, \gamma^\nu\} \equiv \gamma^\mu\gamma^\nu + \gamma^\nu\gamma^\mu = 2\eta^{\mu\nu} I$ where I is the unit 4×4 matrix, A_μ is the vector-potential of the electromagnetic field, e is electric charge, m is mass, and $\psi = (\psi^\alpha(x))$, $\alpha = 1, 2, 3, 4$, is the 4-component wave function.

To quantize the Dirac field, let us consider a set of stationary solutions of the Dirac equations in a constant weak electromagnetic field $\psi_r(x) = \psi_r(\mathbf{x})\exp(-\mathrm{i}\omega_r t)$ where the index r is a label for different solutions. We assume that the functions $\psi_r(\mathbf{x})$ form a complete orthonormal system,

$$\int \psi^*_r(\mathbf{x})\psi_{r'}(\mathbf{x})\, d^3x = \delta_{rr'},$$

and any solution of the Dirac equation can be represented by using $\psi_r(\mathbf{x})$:

$$\psi(\mathbf{x}) = \sum_{\omega_r > 0} b_r \psi_r(\mathbf{x}) + \sum_{\omega_r < 0} c^*_r \psi_r(\mathbf{x})$$

where b_r, c_r are the coefficients of the representation. The Dirac field is quantized by postulating the anticommutation relations for creation and annihilation operators (see Chap. 4)

$$\{b_r, b^*_{r'}\} = \delta_{rr'}, \qquad \{c_r, c^*_{r'}\} = \delta_{rr'}.$$

The Hamiltonian operator of the quantum Dirac field is

$$H = \sum_r |\omega_r|(b^*_r b_r + c^*_r c_r) + E_0$$

and the electric charge

$$Q = e\sum_r (b^*_r b_r - c^*_r c_r) + Q_0,$$

where E_0 and Q_0 are infinite constants as before for the electromagnetic field.

In particular, for a free Dirac particle in a cube of volume $V = L^3$ with the periodic boundary conditions, one can write

$$\psi(x) = \sum_{p\lambda} \frac{1}{\sqrt{2V\varepsilon(\mathbf{p})}} \big[b_{p\lambda} u_\lambda(p) e^{\mathrm{i}px} + c^*_{p\lambda} u_\lambda(-p) e^{-\mathrm{i}px}\big].$$

Here $p = (p_\mu) = (\varepsilon(\mathbf{p}), \mathbf{p})$, $\mu = 0, 1, 2, 3$,

$$\varepsilon(\mathbf{p}) = \sqrt{|\mathbf{p}|^2 + m^2},$$
$$\mathbf{p} = (p_1, p_2, p_3),\ p_i = 2\pi k_i/L,\ i = 1, 2, 3;\ k_i = 0, \pm 1, \pm 2, \ldots,$$

and $px = -\varepsilon(\mathbf{p})t + \mathbf{px}$. The 4-spinors $\mathbf{u}(p) = (u_\lambda(\pm p))$ are solutions of the system of equations

$$-\mathrm{i}\big(p_\mu \gamma^\mu\big)\mathbf{u}(\alpha p) = m\alpha \mathbf{u}(\alpha p),$$
$$\big(p_\mu n^\mu\big)\gamma_4 \mathbf{u}(\alpha p) = \alpha \mathbf{u}(\alpha p).$$

Here the polarization index α takes values $\alpha = +1, -1$, $n = (n_0, \mathbf{n})$ is a unit 4-vector orthogonal to the 4-momentum,

$$n^\mu n_\mu = -n_0^2 + \mathbf{nn} = 1, \qquad n^\mu p_\mu = 0,$$

and $\gamma_4 = \gamma_0\gamma_1\gamma_2\gamma_3$.

The operator $b_{p\lambda}$ is called the annihilation operator of the particle (electron) with momentum $\mathbf{p}$ and polarization λ, and the operator $c^*_{p\lambda}$ the creation operator of the antiparticle (positron) with momentum $\mathbf{p}$ and polarization λ.

The Hamiltonian operator of the free quantum Dirac field has the form

$$H_\mathrm{D} = \sum_{p\lambda} \varepsilon(\mathbf{p})[b^*_{p\lambda} b_{p\lambda} + c^*_{p\lambda} c_{p\lambda}].$$

16.1.8 Pauli Equation

In the non-relativistic limit, one can get the Pauli equation from the Dirac equation as follows: We insert back the speed of light c. Then if we write the 4-component wave function ψ as a pair of two 2-component wave functions

$$\psi = e^{imc^2 t}(\phi, \xi),$$

we obtain in the first order over $1/c$ the Pauli equation for the 2-component spinor ϕ:

$$\mathrm{i}\frac{\partial \phi}{\partial t} = \left[\frac{1}{2m}\left(\mathbf{p} - \frac{e}{c}\mathbf{A}\right)^2 + eA_0 - \frac{e}{2mc}\boldsymbol{\sigma} \boldsymbol{B}\right]\phi.$$

Here $\mathbf{p} = -\mathrm{i}\,\mathrm{grad}$ and $\boldsymbol{\sigma}$ are the Pauli matrices.

16.1.9 Equations of Quantum Electrodynamics

The fundamental equations of quantum electrodynamics describing interaction of the Maxwell and the Dirac fields are

$$[\gamma^{\mu}(\partial_{\mu} - \mathrm{i}eA_{\mu} + m]\psi = 0,$$
$$\partial^{\mu}(\partial_{\mu}A_{\nu} - \partial_{\nu}A_{\mu}) = \mathrm{i}e\bar{\psi}\gamma_{\nu}\psi,$$

where $\bar{\psi} = \psi^{*}\gamma^{0}$.

The Hamiltonian operator H_{QED} of quantum electrodynamics is the sum of the Hamiltonian for the free electromagnetic field H_{EM}, the Hamiltonian for free Dirac field H_{D}, and the interaction Hamiltonian H_{int},

$$H_{\mathrm{QED}} = H_{\mathrm{EM}} + H_{\mathrm{D}} + H_{\mathrm{int}}$$

where the interaction Hamiltonian in the Lorentz gauge is

$$H_{\mathrm{int}} = \mathrm{i}e \int \bar{\psi}(\mathbf{x})\gamma^{\mu}\psi(\mathbf{x})A_{\mu}(\mathbf{x})\, d^{3}x.$$

Here $\bar{\psi}(\mathbf{x})$, $\psi(\mathbf{x})$, and $A_{\mu}(\mathbf{x})$ are represented in terms of the creation and annihilation operators given above and taken at $t = 0$.

16.2 Quantum Fields and Locality

16.2.1 Wightman Axioms

A quantum electromagnetic field and a Dirac field are operator-valued generalized functions on the Minkowski space–time. The basic principles of quantum theory were exposed in Chap. 5. In the quantum field theory, in addition to the seven principles of quantum mechanics, one has to add relativistic invariance, positivity of energy and locality. A rigorous formulation of the general properties of a quantum field are given by *Wightman axioms*. One works in the infinite volume. The following axioms are assumed.

1. It is assumed that there is a Hilbert space $\mathcal{H}$ and for each space–time function f belonging to a certain class of test functions one defines a finite number of operators $\Phi_j(f),\ j = 1, 2, \ldots$ (quantum fields), which are formally written as

$$\Phi_j(f) = \int \Phi_j(x) f(x)\, d^{4}x.$$

2. One assumes that in the Hilbert space there is a unitary representation $U(L)$ of the Poincaré group. A Poincaré transformation $L = (a, \Lambda)$ of the Minkowski

space–time is given by

$$x^\mu \to x'^\mu = (Lx)^\mu = \Lambda^\mu_\nu x^\nu + a^\mu.$$

Here $\mu, \nu = 0, 1, 2, 3$, and Λ^μ_ν is a real 4×4 matrix satisfying $\Lambda^T \eta \Lambda = \eta$ where η is diagonal with entries $(-1, 1, 1, 1)$, and the translations a^μ are real. One has the property of spectrality which means that the generator of translations along the time directions (Hamiltonian) is positively defined. There exists a unique vector, vacuum $|0\rangle$, which is invariant under $U(L)|0\rangle = |0\rangle$.

3. One assumes the transformation law (relativistic covariance)

$$U(L)\Phi_j(x)U^*(L) = \sum_k D_{jk}\Phi_k(Lx).$$

4. The property of relativistic causality or locality for bosonic fields is formulated as follows:

$$\big[\Phi_j(x), \Phi_k(y)\big] = 0$$

if the points x and y are space-like separated: $-(x^0 - y^0)^2 + (\mathbf{x} - \mathbf{y})^2 > 0$. This requirement rests on the principle that no physical effect can propagate in space-like directions. If the field is Hermitian, it expresses also the interpretation of $\Phi_j(f)$ as a measurement within the space–time region where f does not vanish (see the discussion of locality in Chap. 8).

A quantum field can be characterized by the set of its vacuum expectation values, the Wightman functions,

$$W_{j_1,\dots,j_n}(x_1,\dots,x_n) = \langle 0|\Phi_{j_1}(x_1)\cdots\Phi_{j_n}(x_n)|0\rangle,$$

properties of which can be derived from axioms 1–4.

Let us note that for bosonic fields in the standard model of elementary particles the property of locality is not postulated but follows from equations of motion.

Fermi fields are anticommuting for space-like arguments.

Note, however, that the quantum electrodynamics which is the most important quantum field theory is not included into the framework of Wightman's axioms since it requires indefinite metric, i.e., its norm is not always positive-definite.

16.2.2 Algebraic Quantum Theory and Local Channels

More invariant though less practically useful formulation of the quantum theory is given in the algebraic approach developed by Haag, Araki, and many others [130, 131, 578, 740]. We suppose that to any region $\mathcal{O}$ in Minkowski space–time we are given a von Neumann algebra $\mathcal{N}(\mathcal{O})$ (there is a similar approach with C^*-algebras). The discussions of von Neumann and C^*-algebras are given in Chap. 4. Often one considers not arbitrary regions but some more restrictive set, for example, open double cones. The correspondence shall satisfy the following requirements:

1. $\mathcal{N}(\mathcal{O}_1) \subset \mathcal{N}(\mathcal{O}_2)$ when $\mathcal{O}_1 \subset \mathcal{O}_2$.
2. (Locality) $\mathcal{N}(\mathcal{O}_1)$ commutes with $\mathcal{N}(\mathcal{O}_2)$ if $\mathcal{O}_1$ and $\mathcal{O}_2$ are space-like separated.
3. There is a representation of the Poincaré group by automorphisms of algebras, $L \to \alpha_L$ such that

$$\alpha_L\big(\mathcal{N}(\mathcal{O})\big) = \mathcal{N}(\mathcal{O}_L)$$

where $\mathcal{O}_L$ is the transformed region. Note that

$$\mathcal{N} = \overline{\bigcup_{\mathcal{O}} \mathcal{N}(\mathcal{O})}^{\|\cdot\|_{uw}}$$

is a von Neumann algebra, where $\|\cdot\|_{uw}$ is the ultra-weak norm.

Using this framework we formulate the notion of a *local channel.* A channel is local to a space–time region $\mathcal{O}$, denoted by $\Lambda^*_{\mathcal{O}}$ if in the Kraus–Sudarschan representation it is given by

$$\Lambda^*_{\mathcal{O}}(\rho) = \sum_i A_i \rho A_i^*, \quad \text{with} \sum_i A_i^* \rho A_i \leq I,$$

where the operators both A_i and ρ belong to $\mathcal{N}(\mathcal{O})$. One could compare this definition of a local channel with the discussion of the Bogolyubov local causality condition in Chap. 8.

16.3 Quantum Field Theory in Quantum Probability Scheme

Ultimately, quantum information theory should become a part of quantum field theory (perhaps, in the future, a part of the superstring theory) since the quantum field theory is our most fundamental physical theory.

The quantum field theory is not just an abstract mathematical theory of operators in a Hilbert space. Basic equations of the quantum field theory such as the Maxwell, Dirac, Yang–Mills equations are differential equations for operator functions defined on the space–time. The nonrelativistic Schrödinger equation is also a differential equation in space–time. Therefore, a future relativistic quantum information theory should be based on the study of the solutions of these equations propagating in space–time.

One could suggest defining a context described in a boundary condition for a differential equation. Then we would derive the contextual dependence of probabilities from the study of the dependence of solutions of the equation on the boundary conditions.

In the modern quantum information theory, the basic notion is of the two-dimensional Hilbert space, i.e., a qubit. We suggest that in a relativistic quantum information theory, when the existence of space–time is taken into account, the basic notion should be a notion of an elementary quantum system, i.e., according to Wigner, an infinite dimensional Hilbert space H invariant under an irreducible representation of the Poincaré group labeled by $[m, s]$ where $m \geq 0$ is mass and $s = 0, 1/2, 1, \ldots$ is spin (helicity).

In quantum probability, we are given a $*$-algebra $\mathcal{A}$ and a state (i.e., a linear positive normalized functional) ω on $\mathcal{A}$. Elements from $\mathcal{A}$ are called random variables. Two random variables A and B are called (statistically) independent if $\omega(AB) = \omega(A)\omega(B)$.

Consider the free scalar quantum field $\varphi(x)$ which satisfies the Klein–Gordon–Fock equation

$$\left(\Box + m^2\right)\varphi(x) = 0$$

where $\Box = -\eta^{\mu\nu}\partial_\mu\partial_\nu$ is as above.

We prove the following:

Proposition 16.1 *There is a statistical dependence between two spacelike separated regions in the theory of free scalar quantum field.*

Proof One represents free scalar quantum field $\varphi(x)$ in the infinite volume as:

$$\varphi(x) = \frac{1}{(2\pi)^{3/2}} \int_{R^3} \frac{d\mathbf{k}}{\sqrt{2k^0}} \left(e^{ikx} a^*(\mathbf{k}) + e^{-ikx} a(\mathbf{k})\right).$$

Here $kx = k^0x^0 - \mathbf{kx}$, $k^0 = \sqrt{|\mathbf{k}|^2 + m^2}$, $m \geq 0$, and $a(\mathbf{k})$ and $a^*(\mathbf{k})$ are annihilation and creation operators,

$$\left[a(\mathbf{k}), a^*(\mathbf{k}')\right] = \delta(\mathbf{k} - \mathbf{k}').$$

It is an operator-valued distribution acting in the Fock space $\mathcal{F} = \bigoplus_n L^2(\mathbb{R}^3)^{\otimes_s n}$ with the vacuum $|0\rangle$,

$$a(\mathbf{k})|0\rangle = 0.$$

The vacuum expectation value of two fields is

$$\omega_0\big(\varphi(x)\varphi(y)\big) = \langle 0|\varphi(x)\varphi(y)|0\rangle = W_0\big(x - y, m^2\big)$$

where

$$W_0\big(x - y, m^2\big) = \frac{1}{(2\pi)^3} \int_{R^3} \frac{d\mathbf{k}}{2k^0} e^{-ik(x-y)}.$$

The statistical independence of two spacelike separated regions in particular would lead to the relation

$$\omega_0\big(\varphi(x)\varphi(y)\big) - \omega_0\big(\varphi(x)\big)\omega_0\big(\varphi(y)\big) = 0$$

if

$$(x - y)^2 = -(x_0 - y_0)^2 + (x_1 - y_1)^2 + (x_2 - y_2)^2 + (x_3 - y_3)^2 > 0.$$

But since $\omega_0(\varphi(x)) = 0$, in fact, we have

$$\omega_0\big(\varphi(x)\varphi(y)\big) - \omega_0\big(\varphi(x)\big)\omega_0\big(\varphi(y)\big) = W_0\big(x - y, m^2\big) \neq 0.$$

So there is violation of statistical independence of spacelike separated regions. However, the violation of the statistical independence vanishes exponentially with the spacial separation of x and y since for large $\lambda = m\sqrt{x^2}$ the function $W_0(x, m^2)$ behaves like

$$\frac{m^2}{4\pi\lambda}\left(\frac{\pi}{2\lambda}\right)^{1/2} e^{-\lambda}.$$

□

Let us prove that any polynomial state is asymptotically *disentangled* (factorized) for large spacelike distances. Let $\mathcal{A}$ be the algebra of polynomials in the Fock space $\mathcal{F}$ at the field $\varphi(f)$ with the test functions f. Let $C \in \mathcal{A}$ and $|\psi\rangle = C|0\rangle$. Denote the state $\omega(A) = \langle\psi|A|\psi\rangle/\|\psi\|^2$ for $A \in \mathcal{A}$.

Theorem 16.2 *One has the following asymptotic disentanglement property*

$$\lim_{|l|\to\infty}\big[\omega\big(\alpha_l(A)B\big) - \omega\big(\alpha_l(A)\big)\omega(B)\big] = 0.$$

Here A and B belong to $\mathcal{A}$ and $\alpha_l(A)$ is the translation of A along the three-dimensional vector l. One has also

$$\lim_{|l|\to\infty}\big[\omega\big(\alpha_l(A)\big) - \langle 0|\alpha_l(A)|0\rangle\big] = 0.$$

The proof of the theorem is based on the Wick theorem and the Riemann–Lebesgue lemma.

Similar theorems take place also for the Dirac and the Maxwell fields. In particular, for the Dirac field $\psi(x)$ one can prove the asymptotic factorization for the local spin operator

$$\mathbf{S}(\mathcal{O}) = \int_{\mathcal{O}} \psi^* \boldsymbol{\Sigma} \psi \, dx,$$

which will be discussed in Sect. 16.5.

Finally, let us show that some correlation functions in the relativistic quantum field theory can be represented as mathematical expectations of the classical (generalized) random fields.

Theorem 16.3 *If $\Phi(x)$ is a scalar complex quantum field (in this case $[\Phi(x), \Phi(y)] = 0$ for all x, y) then one has a representation*

$$\langle 0|\Phi(x_1)\cdots\Phi(x_n)\Phi^*(y_1)\cdots\Phi^*(y_n)|0\rangle = E\xi(x_1)\cdots\xi(x_n)\xi^*(y_1)\cdots\xi^*(y_n).$$

Here $\xi(x)$ is a complex random field. More explicitly, there exists a probability space (Ω, F, P) and random field $\xi(x) = \xi(x, \lambda)$ such that

$$\langle 0|\Phi(x_1)\cdots\Phi(x_n)\Phi^*(y_1)\cdots\Phi^*(y_n)|0\rangle = \int_\Omega \xi(x_1, \lambda)\cdots\xi^*(y_n, \lambda)\, dP(\lambda).$$

The proof of the theorem follows from the positivity of the quantum correlation functions. It is interesting that we have obtained a functional integral representation for the quantum correlation functions in real time. Similar representation is valid also for the 2-point correlation function of a Hermitian scalar field. It follows from the Kallen–Lehmann representation.

This theorem shows that there exists a hidden variable representation for the correlation functions of scalar quantum field. Similar representation also exists for the Glauber correlation function for the free electromagnetic field:

$$\langle\psi|F_{i_1}^{(-)}(\mathbf{r}_1, t_1)\cdots F_{i_N}^{(-)}(\mathbf{r}_N, t_N)F_{j_M}^{(+)}(\mathbf{r}'_M, t'_M)\cdots F_{j_1}^{(+)}(\mathbf{r}'_1, t'_1)|\psi\rangle$$
$$= \int_\Omega \xi_{i_1}(\mathbf{r}_1, t_1, \lambda)\cdots\xi^*_{j_1}(\mathbf{r}'_1, t'_1, \lambda)\, dP(\lambda).$$

Here $\xi_i(\mathbf{r}, t, \lambda)$ is a random field, $i = 1, 2, 3$.

16.4 Expansion of Wave Packet

Let us remind that there is a well known effect of expansion of wave packets due to the free time evolution. If ϵ is the characteristic length of the Gaussian wave packet describing a particle of mass M at time $t = 0$ then at time t the characteristic length ϵ_t will be

$$\epsilon_t = \epsilon\sqrt{1 + \frac{\hbar^2 t^2}{M^2\epsilon^4}}.$$

It tends to $(\hbar/M\epsilon)t$ as $t \to \infty$. Therefore, the locality criterion is always satisfied for non-relativistic particles if regions $\mathcal{O}_A$ and $\mathcal{O}_B$ are far enough from each other.

16.4.1 Relativistic Particles

We cannot immediately apply the previous considerations to the case of relativistic particles such as photons and the Dirac particles because in these cases the wave function cannot be represented as a product of the spin part and the space–time part. Let us show that the wave function of a photon cannot be represented in the product form. Let $A_i(k)$ be the wave function of a photon, where $i = 1, 2, 3$ and $k \in \mathbb{R}^3$. One has the gauge condition $k^i A_i(k) = 0$ [46]. If one supposes that the wave function has a product form $A_i(k) = \phi_i f(k)$ then from the gauge condition one gets $A_i(k) = 0$. Therefore, the case of relativistic particles requires a separate investigation. Here we consider the Dirac relativistic particles.

We study the spin correlation function for two relativistic Dirac particles. We compute the spin correlation function and show its dependence on the distance between the particles and the anisotropic dependence on the angular position of detectors. The principal term in the obtained expansion is reduced to the non-relativistic Bell's correlation, but in our case it has an extra factor, depending on the distance between two particles.

In the next section, we introduce the notations used in the Dirac equation. Then we describe the form of the spin operator and the singlet state. It allows us to define the space-dependent correlation function. After that we discuss the possibility of experimental measurement of the space-dependent correlation function. In the following subsections, the scheme of calculations is described, and the main result, i.e., the correlation function for the spin 1/2 particles with relativistic corrections is presented.

16.5 Space Dependence of the Dirac Correlation Function

We consider two spin 1/2 particles (electrons, protons) with total angular momentum 0 (the so-called singlet pairs). Let us remind some notations for a single fermion. A fermion in the space–time is described by the Dirac equation. Further on, we will use the coordinate realizations of the spinor as the set of functions $\psi = (\psi^\alpha(x))$, $\alpha = 1, 2, 3, 4$. Here $x = (x^\mu)$ are the coordinates in the Minkowski space–time, $\mu = 0, 1, 2, 3$, $x^0 = ct$. The spinor satisfies the Dirac equation [111]:

$$\left(i\gamma^\mu \frac{\partial}{\partial x^\mu} - M\right)\psi(x) = 0. \tag{16.4}$$

Here M is the mass, γ^μ are the Dirac matrices satisfying the condition

$$\gamma^\mu\gamma^\nu + \gamma^\nu\gamma^\mu = 2\eta^{\mu\nu}$$

where $\mu, \nu = 0, 1, 2, 3$, and $(\eta^{\mu\nu}) = \mathrm{diag}(1, -1, -1, -1)$ is the Minkowski metric.

In the sequel, we use the following set of the Dirac γ matrices [111]:

$$\begin{aligned} \gamma^0 &= \begin{pmatrix} I & 0 \\ 0 & -I \end{pmatrix}, \qquad \gamma^1 = \begin{pmatrix} 0 & \sigma_1 \\ -\sigma_1 & 0 \end{pmatrix}, \\ \gamma^2 &= \begin{pmatrix} 0 & \sigma_2 \\ -\sigma_2 & 0 \end{pmatrix}, \qquad \gamma^3 = \begin{pmatrix} 0 & \sigma_3 \\ -\sigma_3 & 0 \end{pmatrix}, \end{aligned} \tag{16.5}$$

where σ are the 2×2 Pauli matrices, I is the identity 2×2 matrix, and γ_μ is defined by $\gamma_\mu = \eta_{\mu\nu}\gamma^\nu$.

Let Ψ be a state vector of two particles and suppose there are two detectors in spatial regions $\mathcal{O}_1$ and $\mathcal{O}_2$ where one makes measurements of the projection of spins to the direction of the unit vectors **a** and **b**, respectively. We consider the correlation function of the form

$$f(\mathcal{O}_1, \mathcal{O}_2, \mathbf{a}, \mathbf{b}) = \langle \Psi | \mathrm{Spin}(\mathcal{O}_1, \mathbf{a}) \otimes \mathrm{Spin}(\mathcal{O}_2, \mathbf{b}) | \Psi \rangle. \tag{16.6}$$

Here $\mathrm{Spin}(\mathcal{O}_1, \mathbf{a})$ is an operator describing the projection of the spin along the **a**-axis in the detector $\mathcal{O}_1$, and similarly $\mathrm{Spin}(\mathcal{O}_2, \mathbf{b})$ for detector $\mathcal{O}_2$.

In EPR–Bohm model the spatial dependence is neglected, and nonrelativistic limit is considered. In such a model, the spin projection operator is

$$\mathrm{Spin}(\mathcal{O}_1, \mathbf{a}) = \boldsymbol{\sigma}\mathbf{a}; \tag{16.7}$$

here $\boldsymbol{\sigma} = (\sigma_1, \sigma_2, \sigma_3)$ are the Pauli matrices.

The Hilbert space for the model is just $\mathbb{C}^2 \otimes \mathbb{C}^2$, with two basis vectors in $\mathbb{C}^2$, denoted by $|0\rangle$ and $|1\rangle$. The singlet state is

$$|\Phi_0\rangle = \frac{1}{\sqrt{2}}\big(|0\rangle \otimes |1\rangle - |1\rangle \otimes |0\rangle\big).$$

If one takes $|\Phi_0\rangle$ as our state vector $|\Psi\rangle$ and the spin operator (16.7), the correlation function (16.6) will be

$$f(\mathcal{O}_1, \mathcal{O}_2, \mathbf{a}, \mathbf{b}) = -\mathbf{ab}. \tag{16.8}$$

One can study the spatial dependence in the nonrelativistic case if one considers the operators of the form

$$\mathrm{Spin}(\mathcal{O}_1, \mathbf{a}) = P_{\mathcal{O}_1}\boldsymbol{\sigma}\mathbf{a},$$

where $P_{\mathcal{O}_1}$ is the projection operator to the region $\mathcal{O}_1$.

We want to study the correlation function (16.6) in the relativistic case, using the Dirac equation. Therefore, we have to define the spin operator and the entangled singlet state in the relativistic case. In further computations, we take

$$\mathrm{Spin}(\mathcal{O}_1, \mathbf{a}) = P_{\mathcal{O}_1}\boldsymbol{\Sigma} \cdot \mathbf{a}, \tag{16.9}$$

where $P_{\mathcal{O}_1}$ is the projection operator to the area $\mathcal{O}_1$, and $\boldsymbol{\Sigma}$ is

$$\boldsymbol{\Sigma} = \begin{pmatrix} \boldsymbol{\sigma} & 0 \\ 0 & \boldsymbol{\sigma} \end{pmatrix}.$$

As discussed in [747], it is more rigorous instead of $\boldsymbol{\Sigma} \cdot \mathbf{a}$ to use the so-called relativistic spin operator $\mathbf{O} \cdot \mathbf{a}$,

$$\mathbf{O} = -\gamma^0\Sigma - c\gamma^5\frac{\mathbf{p}}{\varepsilon} - \gamma^0\frac{c^2\mathbf{p}(\boldsymbol{\Sigma}, \mathbf{p})}{\varepsilon(\varepsilon + Mc^2)}, \tag{16.10}$$

where ε is the energy and γ_5 is defined by

$$\gamma_5 = -i\gamma_0\gamma_1\gamma_2\gamma_3 = \begin{pmatrix} 0 & I \\ I & 0 \end{pmatrix}.$$

The operator $\mathbf{O}$ is approximately equal to $\boldsymbol{\Sigma}$, when

$$\frac{|p|}{Mc} \ll 1, \qquad \left(\mathbf{a}, \frac{\mathbf{p}}{|p|}\right) \ll 1.$$

The singlet state of two Dirac particles is the state with the total angular momentum equal to zero. To describe the singlet state, we shall introduce the *basis of spherical waves* in the set of solutions of the Dirac equation (16.4).

An element of the basis $\psi_{\varepsilon jlm}$ is labeled by

- energy $\varepsilon = \sqrt{M^2c^4 + P^2c^2}$ $(P^2 = \sum_{i=1,2,3} p_i^2)$
- total angular momentum $\mathbf{j} = \mathbf{p} \times \mathbf{r} + \frac{1}{2}\boldsymbol{\Sigma}$, $\mathbf{j}^2\psi = j(j+1)\psi$
- projection m of total angular momentum to the Oz axis
- parity $(-1)^l$, $l \in \{j \pm \frac{1}{2}\}$.

In other words, ψ is the eigenvector of the operators of energy, total angular momentum, projection of the total angular momentum, and parity with the eigenvalues ε, j, m, and $(-1)^l$.

The corresponding solution of (16.4) has the form [111]:

$$\psi_{\varepsilon jlm}(r,\theta,\varphi,t) = \frac{1}{\sqrt{2\varepsilon}} \begin{pmatrix} \sqrt{\varepsilon + Mc^2}\, R_{Pl}(r)\Omega_{jlm}(\theta,\varphi) \\ -\sqrt{\varepsilon - Mc^2}\, R_{Pl'}(r)\Omega_{jl'm}(\theta,\varphi) \end{pmatrix} e^{-\mathrm{i}\varepsilon t}. \tag{16.11}$$

Here (r,θ,φ) are the spherical coordinates of $\mathbf{x}$,

$$R_{Pl}(r) = \sqrt{\frac{2\pi P}{r}}\, J_{l+1/2}(Pr),$$

$J_{l+1/2}(x)$ is the Bessel function, $l' = l \pm 1$, and

$$\Omega_{l+1/2,l,m}(\theta,\varphi) = \begin{pmatrix} \sqrt{\frac{j+m}{2j}}\, Y_{l,m-1/2}(\theta,\varphi) \\ \sqrt{\frac{j-m}{2j}}\, Y_{l,m+1/2}(\theta,\varphi) \end{pmatrix}, \tag{16.12}$$

$$\Omega_{l-1/2,l,m}(\theta,\varphi) = \begin{pmatrix} -\sqrt{\frac{j-m+1}{2j+2}}\, Y_{l,m-1/2}(\theta,\varphi) \\ \sqrt{\frac{j+m+1}{2j+2}}\, Y_{l,m+1/2}(\theta,\varphi) \end{pmatrix} \tag{16.13}$$

where $Y_{lm}(\theta,\varphi)$ is the usual spherical function. Such an Ω_{jlm} is called the spherical spinor.

Here are the exact forms of the important spherical spinors:

$$\Omega_{\frac{1}{2},0,\frac{1}{2}} = \begin{pmatrix} \frac{1}{2\sqrt{\pi}} \\ 0 \end{pmatrix}, \qquad \Omega_{\frac{1}{2},0,-\frac{1}{2}} = \begin{pmatrix} 0 \\ \frac{1}{2\sqrt{\pi}} \end{pmatrix},$$

$$\Omega_{\frac{1}{2},1,\frac{1}{2}} = \begin{pmatrix} -\frac{\cos\theta}{2\sqrt{\pi}} \\ -\frac{e^{\mathrm{i}\varphi}\sin\theta}{2\sqrt{\pi}} \end{pmatrix}, \qquad \Omega_{\frac{1}{2},1,-\frac{1}{2}} = \begin{pmatrix} -\frac{e^{-\mathrm{i}\varphi}\sin\theta}{2\sqrt{\pi}} \\ \frac{\cos\theta}{2\sqrt{\pi}} \end{pmatrix},$$

and spherical functions

$$Y_{0,0}=\frac{1}{\sqrt{4\pi}},\qquad Y_{1,0}=\mathrm{i}\sqrt{\frac{3}{4\pi}}\cos\theta,\qquad Y_{1,\pm1}=\mp\mathrm{i}\sqrt{\frac{3}{8\pi}}e^{\pm\mathrm{i}\varphi}\sin\theta.$$

One can prove, that the set of (16.11) spinors, with $\varepsilon\geq Mc^2$; $j=0,\frac{1}{2},1,\frac{3}{2},\dots$; $m=-j,-j+1,\dots,j$; $l=j\pm\frac{1}{2}$, forms an orthonormal basis.

The "singlet" state with the total angular momentum $J=0$ has the form:

$$\begin{aligned}\Phi^{\alpha\beta}(\mathbf{x}_1,\mathbf{x}_2)=&\frac{1}{\sqrt{2}}\big(\psi^{\alpha}_{\varepsilon,\frac{1}{2},0,\frac{1}{2}}(\mathbf{x}_1)\psi^{\beta}_{\varepsilon,\frac{1}{2},0,-\frac{1}{2}}(\mathbf{x}_2)\\&-\psi^{\alpha}_{\varepsilon,\frac{1}{2},0,-\frac{1}{2}}(\mathbf{x}_1)\psi^{\beta}_{\varepsilon,\frac{1}{2},0,\frac{1}{2}}(\mathbf{x}_2)\big).\end{aligned}\tag{16.14}$$

In the given form of the spin operator (16.9), the correlation function can be represented in a more elegant way. Let us define a partial matrix element $\langle\langle\psi,\xi\rangle\rangle(\mathbf{x})$ by

$$\big\langle\langle\psi,\xi\rangle\big\rangle(\mathbf{x})\equiv\sum_{\alpha}\psi^{\alpha}(\mathbf{x})\xi^{\alpha}(\mathbf{x}).\tag{16.15}$$

The partial matrix element

$$\big\langle\langle\Psi|A\otimes B|\Psi\rangle\big\rangle(\mathbf{x}_1,\mathbf{x}_2)\tag{16.16}$$

is called a *space-dependent correlation function* of the operators A and B in the state Ψ. One can give a physical meaning to the space-dependent correlation function. Let us denote balls with the center at the point x and radius r by $V_r(x)$. Then,

$$\begin{aligned}&\big\langle\langle\Psi(\mathbf{x}_1,\mathbf{x}_2)\big|A\otimes B\big|\Psi(\mathbf{x}_1,\mathbf{x}_2)\rangle\big\rangle(\mathbf{x}_1,\mathbf{x}_2)\\&\quad=\lim_{\varepsilon\to0}\frac{\langle\Psi|P_{V_\varepsilon(\mathbf{x}_1)}A\otimes P_{V_\varepsilon(\mathbf{x}_2)}B|\Psi\rangle}{(4/3\pi\varepsilon^3)^2}.\end{aligned}\tag{16.17}$$

Let us define the *spin correlation matrix* g_{ij} by

$$g_{ij}(\mathbf{x}_1,\mathbf{x}_2)\equiv\big\langle\langle\Phi|\Sigma_i\otimes\Sigma_j|\Phi\rangle\big\rangle(\mathbf{x}_1,\mathbf{x}_2)\tag{16.18}$$

where Φ is defined by (16.14), and $i,j=1,2,3$.

Using the spin correlation matrix, one can obtain space-dependent correlation function (16.16).

Thus, the spin correlation matrix is the basic object in studying the correlation function. It is a natural generalization of the Bell's correlation function (16.8).

16.6 Wave Packets

What can we really measure? A common answer is the following: We can obtain a sequence of the pairs (r_{ai},t_{ai}), $a\in\{1,2\}$, $i\in\{1,2,\dots,N_a\}$. Here $r_{ai}=\pm1$ is the result of the measurement, and t_{ai} is the moment of "clicking" the detector.

Using these data we can find all simultaneous events. Let us denote by $Y[j]$ and $Z[j]$ the number of the jth simultaneous event in the first and second detector sequence, so $t_{1,Y[j]}$ is in some sense near $t_{2,Z[j]}$. Then, we can compute an experimental value of the correlation function defined by

$$K_{\text{exp}} = \frac{1}{\#\text{Particles}} \sum_{1<i<\#\text{simult.detected}} r_{1,Y[i]} r_{2,Z[i]}. \tag{16.19}$$

Possibly, we can omit the #Particles factor if one can prove that the measurement is not selective, in other words, the probability of loss of the particle is independent of the spin value. This is the subject of further discussion.

We have to measure individual events, corresponding to the single particles. Before the measurement, a particle wave function is somehow distributed, and that distribution depends on time. Classically, we identify the center of the area, where the wave function is distributed, with the location of the particle, and the speed of that center with the speed of the particle. Let us denote the effective size of the wave function by l and the number of the particles emitted in 1 second by τ^{-1}. The term "simultaneously" means that we have to demand the following inequality: $\tau \gg l/v$. Otherwise, the wave functions of the different particles will intersect. The singlet wave function (16.14) is not a wave function of the single particle because the effective size of (16.14) is infinite. A particle, localized in space, can be described by the wave packet [119]

$$\int \rho(P)\Phi_P(\mathbf{x}_1, \mathbf{x}_2)\, dP,$$

where $\rho(P)$ describes the prepared entangled state, and $\Phi_P(x_1, x_2)$ is the singlet state with the momentum P.

How can we estimate the answer without exactly knowing $\rho(P)$ distribution? We propose the following. Let the wave function be localized in the spherical layer $S(t) = \{x = (r, \theta, \phi) | V_{\min} t < r < V_{\max} t\}$. Let us replace the original wave function with

$$\int \rho(P)\Phi_P(\mathbf{x}_1, \mathbf{x}_2)\, dP \to N\Phi_{P_0}(\mathbf{x}_1, \mathbf{x}_2)\chi_{S(t)},$$

where $\chi_{S(t)}(x_1, x_2)$ is the characteristic function of the area $S(t)$, P_0 is the effective center of the wave packet in the momentum space, and we choose the factor N such that the norm of wave function is invariant under replacement. Let us denote the spin correlation matrix for the singlet with the momentum P by $g_{ij\,P}(x_1, x_2)$.

Suppose detectors are localized in space areas $\mathcal{O}_1$ and $\mathcal{O}_2$, P_0 is the center of the wave packet in the momentum space, and $g_{ij\,P}$ is very slow on P. Then,

$$K_{\text{exp}} \approx \int_{(\mathcal{O}_1 \cap S(t_0)) \times (\mathcal{O}_2 \cap S(t_0))} g_{ij\,P_0}(\mathbf{x}_1, \mathbf{x}_2)\, d\mathbf{x}_1\, d\mathbf{x}_2. \tag{16.20}$$

Therefore, the exact answer for the K_{exp} can be obtained only with $\rho(P)$ distribution. That distribution strongly depends on how the entangled pair was prepared.

However, a very nice estimate is given with the (16.20) formula. Such an estimate is based on the spin correlation matrix (16.18). We see that the relation between experimentally measured quantities and correlation function is not trivial, and it should be a subject of further discussion.

16.6.1 Calculations and Results

In the previous section, we showed that the spin correlation matrix (16.18) is the basic object in studying the spin correlations. Here we briefly describe the main steps in its computation. The calculation of the g_{ij} matrix proceeds in five steps:

Step 1. Decomposing the basis wave functions $\psi_{\varepsilon jlm}$ (16.11) into the sum over $\frac{P}{mc}$ powers. We denote

$$\psi_{\varepsilon(P)jlm} = \psi_{Pjlm}. \tag{16.21}$$

Then

$$\psi_{Pjlm} = \begin{pmatrix} \varphi_{Pjlm} \\ \chi_{Pjlm} \end{pmatrix} = \begin{pmatrix} \sqrt{\frac{\varepsilon+Mc^2}{2\varepsilon}} R_{Pl} \Omega_{jlm} \\ -\sqrt{\frac{\varepsilon-Mc^2}{2\varepsilon}} R_{Pl'} \Omega_{jl'm} \end{pmatrix}. \tag{16.22}$$

Let us set $\psi_\pm = \psi_{P,\frac{1}{2},0,\pm\frac{1}{2}}$, $\varphi_\pm = \varphi_{P,\frac{1}{2},0,\pm\frac{1}{2}}$, and $\chi_\pm = \chi_{P,\frac{1}{2},0,\pm\frac{1}{2}}$. Then,

$$\Phi = \frac{1}{\sqrt{2}}(\psi_+ \otimes \psi_- - \psi_- \otimes \psi_+)$$

and

$$\psi_\pm = \begin{pmatrix} \varphi_\pm \\ \chi_\pm \end{pmatrix}.$$

Step 2. Decomposing (16.18) into the sum

$$g_{ij} = \langle \Phi | \Sigma_i \otimes \Sigma_j | \Phi \rangle = \frac{1}{2} \sum_{A,B=\pm} \sigma(A,B) G_{A,B,i,j}$$

where

$$\sigma(A,B) = \begin{cases} +1, & \text{if } A = B, \\ -1, & \text{if } A \neq B, \end{cases}$$

and

$$\begin{aligned} G_{A,B,i,j} &= \langle \psi_A \otimes \psi_{-A} | \Sigma_i \otimes \Sigma_j | \psi_B \otimes \psi_{-B} \rangle \\ &= \langle \psi_A | \Sigma_i | \psi_{-B} \rangle \langle \psi_{-A} | \Sigma_j | \psi_B \rangle \end{aligned}$$

$$= (\varphi_A^* \sigma_i \varphi_B + \chi_A^* \sigma_i \chi_B) \\ \times (\varphi_{-A}^* \sigma_j \varphi_{-B} + \chi_{-A}^* \sigma_j \chi_{-B}). \tag{16.23}$$

Step 3. Denoting

$$V_{l_1,l_2,A,i,B} \equiv R_{Pl_1} R_{Pl_2} \Omega^*_{\frac{1}{2},l_1,A\frac{1}{2}} \sigma_i \Omega_{\frac{1}{2},l_2,B\frac{1}{2}}.$$

Then

$$\varphi_A^* \sigma_i \varphi_B = \frac{\varepsilon + Mc^2}{2\varepsilon} V_{0,0,A,i,B},$$

$$\chi_A^* \sigma_i \chi_B = \frac{\varepsilon - Mc^2}{2\varepsilon} V_{1,1,A,i,B}.$$

Step 4. Now representing g_{ij} as the sum

$$g_{ij} = \frac{1}{2} \sum_{A,B=\pm} \sigma(A,B) \Delta_{A,B,i,j} \tag{16.24}$$

where

$$\Delta_{A,B,i,j} = \frac{1}{4}\Bigg(\left(\frac{\varepsilon + Mc^2}{\varepsilon}\right)^2 V_{0,0,A,i,B} V_{0,0,-A,j,-B} + \frac{\varepsilon^2 - (Mc^2)^2}{\varepsilon^2} \\ \times (V_{1,1,A,i,B} V_{0,0,-A,j,-B} + V_{0,0,A,i,B} V_{1,1,-A,j,-B}) \\ + \left(\frac{\varepsilon - Mc^2}{\varepsilon}\right)^2 V_{1,1,A,i,B} V_{1,1,-A,j,-B}\Bigg). \tag{16.25}$$

Step 5. Computer calculations.

Now one can see that the calculation of g_{ij} is the direct application of formulae (16.8), (16.12), (16.13), (16.24), and (16.25). These calculation were made using a computer program for symbolic mathematics.

Results

Let us expand g_{ij} into the series at P/Mc:

$$g_{ij} = g_{ij}^{(0)} + \left(\frac{P}{Mc}\right)^2 g_{ij}^{(1)} + \cdots.$$

We are interested in the leading term and in the first non-vanishing correction to the leading term. To obtain correct results for higher orders one should use the operator **O** (16.10). Denote

$$R_0(r) = R_{P0}(r), \qquad R_1(r) = R_{P1}(r).$$

Computation of $g_{ij}^{(0)}$

The following computation explains the correspondence with the original nonrelativistic Bell's result. Denote $|+\rangle = \binom{1}{0}$, $|-\rangle = \binom{0}{1}$. We have

$$V_{0,0,A,i,B} = \frac{1}{4\pi} R_0^2(r)\langle A|\sigma_i|B\rangle,$$

$$\begin{aligned}
g_{ij}^{(0)} &= \frac{1}{2}\sum_{A,B=\pm} \sigma(A,B) V_{0,0,A,i,B} V_{0,0,-A,j,-B} \\
&= \left(\frac{1}{16\pi^2} R_0(r_1)^2 R_0(r_2)^2\right)\frac{1}{2}\sum_{A,B=\pm} \sigma(A,B)\langle A|\sigma_i|B\rangle\langle -A|\sigma_j|-B\rangle \\
&= \frac{1}{2}\left(\frac{1}{16\pi^2} R_0(r_1)^2 R_0(r_2)^2\right)\big(\langle +|\sigma_i|+\rangle\langle -|\sigma_j|-\rangle - \langle +|\sigma_i|-\rangle\langle -|\sigma_j|+\rangle \\
&\quad - \langle -|\sigma_i|+\rangle\langle +|\sigma_j|-\rangle + \langle -|\sigma_i|-\rangle\langle +|\sigma_j|+\rangle\big) \\
&= \left(\frac{1}{16\pi^2} R_0(r_1)^2 R_0(r_2)^2\right)\frac{1}{\sqrt{2}}\big(\langle +-| - \langle -+|\big)\sigma_i \otimes \sigma_j \frac{1}{\sqrt{2}}\big(|+-\rangle - |-+\rangle\big).
\end{aligned}$$

We define $\Psi_0 = \frac{1}{\sqrt{2}}(|+-\rangle - |-+\rangle)$. Then

$$\begin{aligned}
g_{ij}^{(0)} &= \left(\frac{1}{16\pi^2} R_0(r_1)^2 R_0(r_2)^2\right)\langle \Psi_0|\sigma_i \otimes \sigma_j|\Psi_0\rangle \\
&= -\left(\frac{1}{16\pi^2} R_0(r_1)^2 R_0(r_2)^2\right)\delta_{ij}.
\end{aligned}$$

We obtain the non-relativistic result:

$$g_{ij}^{(0)} = -\frac{1}{16\pi^2} R_0(r_1)^2 R_0(r_2)^2 \delta_{ij}. \tag{16.26}$$

Computation of $g_{ij}^{(1)}$

We define $Z_{ij}(\theta,\varphi)$ by

$$\frac{1}{2}\sum_{A,B=\pm} \sigma(A,B) V_{1,1,A,i,B} V_{0,0,-A,j,-B} = \frac{1}{16\pi^2} R_1^2(r_1) R_0^2(r_2) Z_{ij}(\theta_1,\varphi_1).$$

Then, using computer computations we obtain

$$\big(Z_{ij}(\theta,\varphi)\big) \equiv \begin{pmatrix} \cos^2\theta - \cos 2\varphi \sin^2\theta & -\sin 2\varphi \sin^2\theta & -\cos\varphi \sin 2\theta \\ -\sin 2\varphi \sin^2\theta & \cos^2\theta + \cos 2\varphi \sin^2\theta & -\sin\varphi \sin 2\theta \\ -\cos\varphi \sin 2\theta & -\sin\varphi \sin 2\theta & -\cos 2\theta \end{pmatrix}. \tag{16.27}$$

The first order relativistic correction is

$$g_{ij}^{(1)} = \frac{1}{64\pi^2}\big(R_1^2(r_1)R_0^2(r_2)Z_{ij}(\theta_1,\varphi_1) + R_0^2(r_1)R_1^2(r_2)Z_{ij}(\theta_2,\varphi_2) + R_0^2(r_1)R_0^2(r_2)\delta_{ij}\big). \quad (16.28)$$

The main result of this section is the formulae (16.27) and (16.28) for the spin space-dependent correlation function of two relativistic entangled fermions.

One has a nontrivial dependence on angular coordinates and on the distance between particles. It would be interesting to study such dependence experimentally.

16.7 Noncommutative Spectral Theory and Local Realism

As a generalization of the previous discussion, we would like to suggest here a general relation between the quantum theory and the theory of classical stochastic processes [343] which expresses the condition of local realism. Let $\mathcal{H}$ be a Hilbert space, ρ the density operator, and let $\{A_\alpha\}$ be a family of self-adjoint operators in $\mathcal{H}$. One says that the family of observables $\{A_\alpha\}$ and the state ρ satisfy *the condition of local realism* if there exists a probability space $(\Omega, \mathcal{F}, d\rho(\lambda))$ and a family of random variables $\{\xi_\alpha\}$ such that the range of ξ_α belongs to the spectrum of A_α and for any subset $\{A_i\}$ of mutually commutative operators one has a representation

$$\operatorname{tr}(\rho A_{i_1}\cdots A_{i_n}) = E\xi_{i_1}\cdots\xi_{i_n}.$$

The physical meaning of the representation is that it describes the quantum–classical correspondence. If the family $\{A_\alpha\}$ were a maximal commutative family of self-adjoint operators then for pure states the previous representation could be reduced to the von Neumann spectral theorem [540]. In our case, the family $\{A_\alpha\}$ consists of not necessary commuting operators. Hence we will call such a representation a *noncommutative spectral representation*. Of course, one can ask for which families of operators and states a *noncommutative spectral theorem* is valid, i.e., When can we write the noncommutative spectral representation? We need a noncommutative generalization of von Neumann's spectral theorem.

It would be helpful to study the following problem: Describe the class of functions $f(t_1,\dots,t_n)$ which admits the representation of the form

$$f(t_1,\dots,t_n) = Ex_{t_1}\cdots z_{t_n}$$

where $x_t,\dots,z_t$ are random processes which obey the bounds $|x_t|\le 1,\dots,|z_t|\le 1$.

From the previous discussion, we know that there are families of operators and states which do not admit the noncommutative spectral representation, and therefore they do not satisfy the condition of local realism. Indeed, let us take the Hilbert space

$\mathcal{H} = \mathbb{C}^2 \otimes \mathbb{C}^2$ and four operators A_1, A_2, A_3, A_4 of the form (we denote $A_3 = B_1$, $A_4 = B_2$)

$$A_1 = \begin{pmatrix} \sin\alpha_1 & \cos\alpha_1 \\ \cos\alpha_1 & -\sin\alpha_1 \end{pmatrix} \otimes I, \qquad A_2 = \begin{pmatrix} \sin\alpha_2 & \cos\alpha_2 \\ \cos\alpha_2 & -\sin\alpha_2 \end{pmatrix} \otimes I,$$

and

$$B_1 = I \otimes \begin{pmatrix} -\sin\beta_1 & -\cos\beta_1 \\ -\cos\beta_1 & \sin\beta_1 \end{pmatrix}, \qquad B_2 = I \otimes \begin{pmatrix} -\sin\beta_2 & -\cos\beta_2 \\ -\cos\beta_2 & \sin\beta_2 \end{pmatrix}.$$

Here operators A_i correspond to operators $\sigma \cdot a$, and operators B_i correspond to operators $\sigma \cdot b$ where $a = (\cos\alpha, 0, \sin\alpha)$, $b = (-\cos\beta, 0, -\sin\beta)$. Operators A_i commute with operators B_j, $[A_i, B_j] = 0$, $i, j = 1, 2$, and one has

$$\langle\psi_{\text{spin}}|A_i B_j|\psi_{\text{spin}}\rangle = \cos(\alpha_i - \beta_j), \quad i, j = 1, 2.$$

We know from the discussion of Bell's theorem in Chap. 8 that this function cannot be represented as the expected value $E\xi_i\eta_j$ of random variables with the bounds $|\xi_i| \leq 1$, $|\eta_j| \leq 1$.

However, as it was discussed above, the space part of the wave function was neglected in the previous consideration. We suggest that *in physics one could prepare only such states and observables which satisfy the condition of local realism.* Perhaps we should restrict ourself in this proposal to the consideration of only such families of observables which satisfy the condition of relativistic local causality. If there are physical phenomena which do not satisfy this proposal then it would be important to *describe quantum processes which satisfy the above formulated condition of local realism and also processes which do not satisfy this condition.*

We have discussed in this section some problems in quantum information theory which require the inclusion of space–time variables. In particular, entangled states in space and time were considered. A modification of Bell's equation which includes the space–time variables will be studied more.

There are many interesting open problems in the approach to quantum information in space and time. Some of them related to the noncommutative spectral theory and the theory of classical stochastic processes have been discussed above.

In quantum cryptography, there are important open problems which require further investigations. In quantum cryptographic protocols with two entangled photons to detect the eavesdropper's presence by using Bell's inequality, we have to estimate the function $g(\mathcal{O}_A, \mathcal{O}_B)$. To increase the detectability of the eavesdropper one has to do a thorough investigation of the process of preparation of the entangled state and then its evolution in space and time towards Alice and Bob. One has to develop a proof of the security of such a protocol.

In the next chapter, Eve is interpreted as an abstract hidden variable. However, one can assume that more information about Eve is available. In particular, one can assume that she is located somewhere in space, in a region $\mathcal{O}_E$. It seems one has to study a generalization of the function $g(\mathcal{O}_A, \mathcal{O}_B)$ which depends not only on the Alice's and Bob's locations $\mathcal{O}_A$ and $\mathcal{O}_B$, but also on the Eve's location $\mathcal{O}_E$, and try to find a strategy which leads to an optimal value of this function.

16.8 Contextual Approach

A general contextual approach to the probabilistic scheme of the quantum theory was proposed by Khrennikov in [409]. It is based on the transformation rules induced by context transitions. A context is a complex of physical conditions used for the preparation of quantum or classical states. The idea of the contextual dependence of probabilistic results of observations is a very general one. It can be used and developed in various directions. In particular, it was suggested in [797] to treat boundary conditions for quantum mechanical differential equations as an appropriate context.

A context describes a measure of idealization which we use to construct a mathematical model for a physical process. For example, in some approximation one can deal with models of quantum phenomena when the spatial characteristics are neglected as it was done by Bell in his consideration of the EPR paradox. However, if we want to speak about fundamental properties of the quantum theory then the principal role of the space–time picture should not be overlooked. In the axiomatic approach to the quantum theory after von Neumann [806], one often postulates only the formalism of a Hilbert space, its statistical interpretation, and the abstract Schrödinger evolution equation, but without indication of the spatial properties of a quantum system. The necessity of including into the list of basic axioms of quantum mechanics the property of covariance of the physical system under the spatial translation and rotation and moreover under the Galilei or Poincaré group was considered in [793].

16.8.1 Contextual Classical and Quantum Probability

The contextual probabilistic approach is nothing than a probabilistic formalization of Bohr's idea that the whole experimental arrangement must be taken into account. The basic postulate of the contextual probabilistic approach to general statistical measurements is that probability distributions for physical variables depend on complexes of experimental physical conditions. Such complexes are called (experimental) contexts. Mathematically, contextualism means providing a context to an abstract (e.g., Kolmogorov) probability. Mathematical formalization of the notion of a context in the general case is a problem of large complexity. Here we use the following definition of a *quantum context*.

Definition 16.4 Every family $\mathcal{A} = \{A_1, A_2, \dots\}$ (finite or infinite) of self-adjoint commutative operators is said to be a quantum context.

Example 16.5 (Space–time context) Let $\mathcal{A} = \{A_1, A_2, A_3, A_4\}$ be the system of generators of the unitary group of translations. Then $\mathcal{A}$ is said to be the space–time context.

Example 16.6 (Internal symmetry) Let G be a compact Lie group of internal symmetries (for example, the gauge group $U(1)$ which describes the electric charge). Then the generators of the unitary representation of the group G define the internal symmetry context.

16.9 Notes

Quantum electrodynamics and more general quantum field theory is exposed in [122, 698, 811]. Axiomatic approach to the quantum field theory is considered in [123, 736]. Quantum optics and the Glauber correlation functions are discussed in [505, 699, 808]. The space dependence of the correlation function for the spin $1/2$ particles is studied by Roschin and Volovich [663].

Chapter 17
Quantum Cryptography

Cryptography is the art of code-making, code-breaking, and secure communication. It has a long history of military, diplomatic, and commercial applications dating back to ancient societies. In this chapter, an introduction to basic notions of classical and quantum cryptography is given.

In 1976, Diffie and Hellman [202] discovered a new type of cryptosystem and invented *public key cryptography*. In this method, the problem of key distribution was solved. A public key cryptosystem has the property that someone who knows only how to encipher cannot use the *enciphering key* to find the deciphering key without a prohibitively lengthy computation. The best-known public key cryptosystem, RSA [661], is widely used, and it relies on the difficulty of factoring large integers.

In the 1980s, Wiesner [818] and Bennett and Brassard [102] (their method is called the BB84 protocol) have proposed the idea of quantum cryptography. They used the sending of single quantum particles. The method of quantum cryptography also can solve the key distribution problem. Moreover, it can detect the presence of an eavesdropper. In 1991, Ekert [219] proposed using in quantum cryptography the phenomena of entanglement and Bell's inequalities.

Experimental quantum key distribution was demonstrated for the first time in 1989, and since then tremendous progress has been made. Several groups have shown that quantum key distribution is possible, even outside the laboratory. In particular, the creation of a key over the distance of several dozens kilometers was reported [288].

First, we will discuss Caesar's cryptosystem, and then, in Sect. 17.3, the elements of the number theory needed for cryptography are discussed. In Sect. 17.4, the public key distribution and the RSA cryptosystem are considered. The BB84 quantum cryptographic protocol is discussed in Sect. 17.5. Some useful notions of the mutual information and Shannon's entropy are included and proofs of security of the protocol are discussed. In Sect. 17.6, the Einstein–Podolsky–Rosen–Bell–Ekert (EPRBE) quantum cryptographic protocol is considered. The security of the protocol is based on Bell's theorem describing nonlocal properties of entangled states. The importance of considering entangled states in space and time is stressed. A modification of Bell's equation which includes the space–time variables is given, and the problem of security of the EPRBE protocol in real space–time is discussed.

M. Ohya, I. Volovich, *Mathematical Foundations of Quantum Information and Computation and Its Applications to Nano- and Bio-systems*,
Theoretical and Mathematical Physics,
DOI 10.1007/978-94-007-0171-7_17,

17.1 Private Key Cryptosystems

Cryptography is the art of sending messages in disguised form. We shall use the following notions.

- Alphabet—a set of letters
- Plaintext—the message we want to send
- Ciphertext—the disguised message.

The plaintext and ciphertext are broken up into *message units*. A message unit might be a single letter, a pair of letters, or a block of k letters.

An *enciphering* transformation is a function f from the set $\mathcal{P}$ of all possible plaintext message units to the set $\mathcal{C}$ of all possible ciphertext units. We assume that f is a one-to-one correspondence, $f:\mathcal{P}\to\mathcal{C}$. The *deciphering* transformation is the map f^{-1} which goes back and recovers the plaintext from the ciphertext. Schematically one has the diagram

$$\mathcal{P}\overset{f}{\to}\mathcal{C}\overset{f^{-1}}{\to}\mathcal{P}.$$

Any such set-up is called a *cryptosystem*.

17.1.1 Julius Caesar's Cryptosystem

Let us discuss Caesar's cryptosystem in more detail. Suppose we use the 26-letter Latin alphabet $A, B, \dots, Z$ with numerical equivalents $0, 1, \dots, 25$. Let the letter $x \in \{0, 1, \dots, 25\}$ stands for a plaintext message unit. Define a function

$$f:\{0,\dots,25\}\to\{0,\dots,25\}$$

by the rule

$$f(x)=\begin{cases} x+3, & \text{if } x<23,\\ x+3-26=x-23, & \text{if } x\geq 23.\end{cases}$$

In other words, $f(x)\equiv x+3 \pmod{26}$.

To decipher a message one subtracts 3 modulo 26.

Exercise 17.1 According to the Caesar's cryptosystem, the word "COLD" reads "FROG".

More generally, consider the congruence (see Sect. 17.3 about the properties of congruences)

$$f(x)=x+b \pmod{N},$$

i.e.,

$$\begin{cases} x+b, & \text{if } x < N-b, \\ x-(N-b) = x+b-N, & \text{if } x \geq N-b. \end{cases}$$

Here N is bigger than the cardinality of an alphabet set. In the case of Caesar's cryptosystem $N = 26$, $b = 3$. To decipher a message one subtracts b modulo N.

We could use a more general *affine map*, i.e., $f(x) = ax + b \pmod{N}$. To decipher a message $y = ax + b \pmod{N}$ one solves for x in terms of y obtaining

$$x = a'y + b' \pmod{N},$$

where a' is the inverse of a modulo N and $b' = -a^{-1}b \pmod{N}$. Assume a is relatively prime to N, then there exists a^{-1} (see Sect. 17.3).

In this example, the enciphering function f depends upon the choice of parameters a and b. The values of parameters are called the *enciphering key* $K_E = (a, b)$. In order to compute f^{-1} (decipher), we need a *deciphering key* K_D. In our example $K_D = (a', b')$ where $a' = a^{-1} \pmod{N}$ and $b' = -a^{-1}b \pmod{N}$.

17.1.2 Symmetric Cryptosystems—DES and GOST

Suppose that the algorithm of the cryptosystem is publicly known but the keys are kept secret. It is a *private key cryptography*. Examples of such cryptosystems are Data Encryption Standard (DES), with 56-bit private key (USA, 1980) and a more secure GOST-28147-89 which uses 256-bit key (Russia, 1989). In such cryptosystems, anyone who knows an enciphering key can determine the deciphering key. Such cryptosystems are called *symmetric cryptosystems*.

17.2 Public Key Cryptography and RSA Cryptosystem

First, let us define some extra notions that we will use along with ones defined in the previous sections.

- Information channel—a way to transmit information from one endpoint to another.
- Trusted channel—an information channel where it is believed that it is impossible to eavesdrop the transmitted information. For example, military optical communication channels.
- Public channel—an information channel where the transmitted information could be quite easily overheard. An example is the Internet.

Let us introduce our main characters: Alice, Bob, and Eve. Alice wants to send ciphertext to Bob. Eve, the eavesdropper, wants to catch the ciphertext and break it, i.e., decipher without knowing the deciphering key. In our scheme, in order to produce a ciphertext from the plaintext, Alice has to have an *enciphering key*. In turn, Bob to read (decipher) the Alice's ciphertext needs a *deciphering key*. If Alice and Bob use a *private key cryptosystem*, i.e., a cryptosystem where enciphering and deciphering keys could be easily produced one from another, they come to the *key distribution problem*. Indeed, Alice and Bob should use a trusted channel to share the keys.

At first glance, it seems to be impossible to get rid of the need for a secret channel. However, in 1976 Diffie and Hellman [202] discovered a new type of cryptosystem called *public key cryptosystem* where there is no key distribution problem at all. A public key cryptosystem has the property that having the enciphering key one cannot find the deciphering key without a prohibitively lengthy computation. In other words, the enciphering function $f:\mathcal{P}\to\mathcal{C}$ is easy to compute if the enciphering key K_E is known, but it is very hard to compute the inverse function $f^{-1}:\mathcal{C}\to\mathcal{P}$ without knowing the *deciphering* key K_D even having the *enciphering* key K_E.

One of the most widely used public key cryptosystem is RSA—a cryptosystem named after the three inventors, Ron Rivest, Adi Shamir, and Leonard Adleman [661]. The RSA cryptosystem is based on the fact that in order to factorize a big natural number with N digits any classical computer needs at least a number of steps that grows faster than any polynomial in N. Honestly speaking, there is no rigorous proof of this fact but all known factoring algorithms obey this fact.

Let us describe RSA cryptosystem in more detail. First, we describe the *protocol*, i.e., the steps Alice and Bob should perform in order to allow Alice send enciphered messages to Bob. The mathematical basis of the RSA cryptosystem will be described in the next section.

17.2.1 The RSA Protocol

The RSA protocol solves the following problem. Bob wants to announce publicly a public key such that Alice using this key could send to him an enciphered message and nobody but Bob would be able to decipher it.

Bob generates public and private keys—each of them is a pair of two natural numbers, (e,n) and (d,n). Here $K_e=(e,n)$ is the enciphering key (public) and $K_d=(d,n)$ is the deciphering key (private).

In order to generate public and private keys, Bob does the following:

(a) He takes any two big prime numbers p and q and computes $n=pq$ with the value of the Euler function $\varphi(n)=(p-1)(q-1)$. In modern cryptosystems, one uses $\log_2 p\approx\log_2 q\approx 1000$.
(b) He then takes any $e<n$ such that $\gcd(e,\varphi(n))=1$.
(c) And he computes $d=e^{-1}\pmod{\varphi(n)}$, i.e., finds a natural number d such that

$$ed\equiv 1\ \big(\mathrm{mod}\ \varphi(n)\big),\quad 1\le d<\varphi(n). \tag{17.1}$$

(d) Then Bob sends a public key (n,e) to Alice via a public channel.

(e) Alice, having Bob's public key (n, e) and a plaintext m (assume m is a natural number and $m < n$) that she wants to send to Bob, computes

$$c = m^e \pmod{n},$$

and sends c (ciphertext) to Bob.

(f) When Bob receives c from Alice, he computes

$$c^d \pmod{n},$$

and gets the Alice's plaintext m because $m = c^d \pmod{n}$.

Nobody but Bob will be able to decipher Alice's message as explained below.

17.2.2 Mathematical Basis of the RSA Protocol

In this section, we will show why the RSA cryptosystem works. Then we will discuss the *security* of the protocol, i.e., how hard it is for Eve, the eavesdropper, to decipher the Alice's message without knowing the private key.

In order to prove that RSA cryptosystem works, we have to prove that the computations that Bob does in the step (f) of the protocol is the inverse to the computations that Alice does in the step (e). That is,

$$c^d \equiv m \pmod{n}.$$

From (17.1), we have

$$ed = 1 + k\varphi(n), \quad k \in \mathbb{Z}.$$

We have

$$c^d = m^{ed} = m \cdot m^{k\varphi(n)}. \tag{17.2}$$

Finally, using the Euler's theorem for the r.h.s. of (17.2), we obtain

$$c^d \equiv m \pmod{n}.$$

Now let us investigate the security of the RSA cryptosystem. At first glance, it seems to be rather straightforward for Eve to obtain the Bob's *private* key having his *public* key. The only thing she has to do is, having n and e, to solve the congruence and find d

$$ed \equiv 1 \big(\text{mod } \varphi(n)\big), \quad 1 \le d < \varphi(n).$$

The problem that Eve faces here is computing $\varphi(n)$. To this end, she has to know p and q, i.e., she has to solve the factoring problem. The practical solution of this problem is not possible with modern technology since factoring large numbers take too much time. For a discussion of this problem, see, for example [793].

17.3 Entropic Uncertainty Relations

The fundamental Heisenberg uncertainty relation is a particular case of the Robertson inequality

$$\Delta(A,\psi)\Delta(B,\psi) \geq \frac{1}{2}\big|\langle\psi,[A,B]\psi\rangle\big|,$$

where A and B are two observables and

$$\Delta(A,\psi) = \sqrt{\big\langle\psi,\big(A-\langle\psi,A\psi\rangle\big)^2\psi\big\rangle}.$$

Here we discuss a generalization of the uncertainty relation which uses the notions of entropy and mutual entropy. The quantum entropy of an observable A in the state ρ is given by

$$S(A,\rho) = -\sum_i p(i,\rho)\log_2 p(i,\rho), \tag{17.3}$$

where $p(\cdot,\rho)$ is the probability distribution of an observable A in the state ρ. That is, for the spectral decomposition of A, $A=\sum_i a_i P_i$, $p(i,\rho)$ is $\operatorname{tr}\rho P_i$. If the state ρ is pure, i.e., $\rho=|\varphi\rangle\langle\varphi|$, and P_i is one-dimensional, $P_i=|\xi_i\rangle\langle\xi_i|$, where φ and ξ_i are unit vectors in a Hilbert space, one can rewrite (17.3) as

$$S(A,\varphi) = -\sum_i \big|\langle\xi_i,\varphi\rangle\big|^2\log_2\big|\langle\xi_i,\varphi\rangle\big|^2. \tag{17.4}$$

Theorem 17.2 *For any nondegenerate observables A and B in a finite dimensional Hilbert space the entropic uncertainty relation holds* [493, 578]:

$$S(A,\rho)+S(B,\rho) \geq -2\log_2 c, \tag{17.5}$$

where c is defined as the maximum possible overlap of the eigenstates of A and B,

$$c \equiv \max_{a,b}\big|\langle a,b\rangle\big|. \tag{17.6}$$

Here $\{|a\rangle\}$ *and* $\{|b\rangle\}$ *are orthonormal bases consisting of the eigenvectors of A and B, respectively.*

One can check that for any nondegenerate observable A in an N-dimensional Hilbert space there exists an upper bound on the entropy

$$S(A,\rho) \leq \log_2 N. \tag{17.7}$$

Let us illustrate the entropic uncertainty relation on a simple spin $\frac{1}{2}$ particle. Taking the Pauli matrices

$$\sigma_x = \begin{pmatrix}0&1\\1&0\end{pmatrix}, \qquad \sigma_z = \begin{pmatrix}1&0\\0&-1\end{pmatrix} \tag{17.8}$$

as observables with eigenstates

$$h_1 = \frac{1}{\sqrt{2}}\begin{pmatrix}1\\1\end{pmatrix}, \qquad h_2 = \frac{1}{\sqrt{2}}\begin{pmatrix}1\\-1\end{pmatrix}, \qquad e_1 = \begin{pmatrix}0\\1\end{pmatrix}, \qquad e_2 = \begin{pmatrix}1\\0\end{pmatrix}, \tag{17.9}$$

we compute $c = 1/\sqrt{2}$. Now taking 2 as the base of the logarithm, the relation (17.5) states that for any unit vector $\varphi \in \mathbb{C}^2$,

$$\sum_{i=1,2} \left(\left|\langle e_i, \varphi\rangle\right|^2 \log_2 \left|\langle e_i, \varphi\rangle\right|^2 + \left|\langle h_i, \varphi\rangle\right|^2 \log_2 \left|\langle h_i, \varphi\rangle\right|^2 \right) \le -1. \tag{17.10}$$

Now we will formulate the uncertainty relation using the mutual entropy. Consider a quantum system which is described by a density operator ρ_i with probability p_i. Then the density operator of the whole ensemble $\mathcal{E} = \{\rho_i\}$ of all possible states of the system is given by

$$\rho = \sum_i p_i \rho_i.$$

The mutual information corresponding to a measurement of an observable A is given by

$$I(A, \mathcal{E}) = S(A, \rho) - \sum_i p_i S(A, \rho_i).$$

From (17.5), using (17.7) one can obtain the following theorem (information exclusion relation [313]).

Theorem 17.3 *Let A and B be arbitrary observables in an N-dimensional Hilbert space, then*

$$I(A, \mathcal{E}) + I(B, \mathcal{E}) \le 2 \log_2 Nc,$$

where c is defined by (17.6).

Remark 17.4 The proofs of the above two theorems are just common computation at the level of the classical theory of entropy, so that we leave them as exercises for the readers.

17.4 No-cloning Theorem

The eavesdropper, Eve, wants to have a perfect copy of Alice's message. However, Wootters and Zurek [821] proved that perfect copying is impossible in the quantum world.

It is instructive to start with the following:

Proposition 17.5 *If $\mathcal{H}$ is a Hilbert space and ϕ_0 is a vector from $\mathcal{H}$ then there is no a linear map $M:\mathcal{H}\otimes\mathcal{H}\to\mathcal{H}\otimes\mathcal{H}$ with the property $M(\psi\otimes\phi_0)=\psi\otimes\psi$ for any ψ.*

Proof Indeed, we would have

$$M(2\psi\otimes\phi_0)=2\psi\otimes 2\psi=4\psi\otimes\psi.$$

But because of linearity, we should have

$$M(2\psi\otimes\phi_0)=2M(\psi\otimes\phi_0)=2\psi\otimes\psi.$$

This contradiction proves the claim. □

Now let us prove the no-cloning theorem.

Theorem 17.6 *Let $\mathcal{H}$ and $\mathcal{K}$ be two Hilbert spaces,* $\dim\mathcal{H}\geq 2$. *Let M be a linear map (copy machine)*

$$M:\mathcal{H}\otimes\mathcal{H}\otimes\mathcal{K}\to\mathcal{H}\otimes\mathcal{H}\otimes\mathcal{K}$$

with the property

$$M(\psi\otimes\phi_0\otimes\xi_0)=\psi\otimes\psi\otimes\eta_\psi$$

for any $\psi\in\mathcal{H}$ and some nonzero vectors $\phi_0\in\mathcal{H}$ and $\xi_0\in\mathcal{K}$ where $\eta_\psi\in\mathcal{K}$ can depend on ψ. Then M is a trivial map, $M=0$ (i.e., $\eta_\psi=0$ for any ψ).

Proof Let $\{e_i\}$ be an orthonormal basis in $\mathcal{H}$. We have

$$M(e_i\otimes\phi_0\otimes\xi_0)=e_i\otimes e_i\otimes\eta_i$$

where η_i are some vectors in $\mathcal{K}$. To prove the theorem, we prove that $\eta_i=0$. If $i\neq j$ then $(e_i+e_j)/\sqrt{2}$ is a unit vector (here we use that $\dim\mathcal{H}\geq 2$). We have the equality

$$\frac{1}{\sqrt{2}}(e_i+e_j)\otimes\phi_0\otimes\xi_0=\frac{1}{\sqrt{2}}e_i\otimes\phi_0\otimes\xi_0+\frac{1}{\sqrt{2}}e_j\otimes\phi_0\otimes\xi_0.$$

Let us apply the map M to both sides of this equality. Then we get

$$\frac{1}{\sqrt{2}}(e_i+e_j)\otimes\frac{1}{\sqrt{2}}(e_i+e_j)\otimes\eta_{ij}=\frac{1}{\sqrt{2}}e_i\otimes\frac{1}{\sqrt{2}}e_i\otimes\eta_i+\frac{1}{\sqrt{2}}e_j\otimes\frac{1}{\sqrt{2}}e_j\otimes\eta_j \tag{17.11}$$

where η_{ij} is a vector in $\mathcal{K}$. We can rewrite (17.11) as

$$e_i\otimes e_i\otimes(\eta_{ij}-\eta_i)+e_i\otimes e_j\otimes\eta_{ij}+e_j\otimes e_i\otimes\eta_{ij}+e_j\otimes e_j\otimes(\eta_{ij}-\eta_j)=0.$$

Now taking into account that e_i and e_j belong to a basis in $\mathcal{H}$, we get

$$\eta_{ij}-\eta_i=0,\qquad \eta_{ij}=0,\qquad \eta_{ij}-\eta_j=0.$$

Hence $\eta_i=0$ for any i, and the theorem is proved. □

Remark 17.7 If $\dim \mathcal{H} = 1$, i.e., $\mathcal{H} = \mathbb{C}$, then Theorem 17.6 is not valid. For $\phi_0 = 1$ and $\psi \in \mathbb{C}$ one can set $M(\psi \xi_0) = \psi \xi_0 = \psi^2 \eta_\psi$ where $\eta_\psi = \xi_0/\psi$ for $\psi \neq 0$.

We proved that Eve cannot get a perfect quantum copy because perfect quantum copy machines cannot exist. The possibility to copy classical information is one of the most crucial features of information needed for eavesdropping. The quantum no-cloning theorem prevents Eve from perfect eavesdropping, and hence makes quantum cryptography potentially secure.

Note, however, that though there is no perfect quantum cloning machine, there are cloning machines that achieve the optimal approximate cloning transformation compatible with the no-cloning theorem, see [155, 287].

17.5 The BB84 Quantum Cryptographic Protocol

Quantum cryptographic protocols differ from the classical ones in that their security is based on the laws of quantum mechanics, rather than the conjectured computational difficulty of certain functions. In this section, we will describe the Bennett and Brassard (BB84) quantum cryptographic protocol [102].

17.5.1 The BB84 Protocol

First, let us describe the physical devices used by Alice and Bob.

Alice has a *photon emitter*—a device which is capable to emit single photons that are linearly polarized in one of four directions. The polarizations are described by the four unit vectors in $\mathbb{C}^2$ here they are e_1, e_2, h_1, h_2 given in (17.9). We will call the polarizations vertical, horizontal, diagonal, and anti-diagonal, and denote them respectively (| , — , \ , /). We have two bases in $\mathbb{C}^2$. One basis, $G_z = \{e_1, e_2\}$, describes the vertical and horizontal polarizations. Another basis, $G_x = \{h_1, h_2\}$, describes the diagonal and anti-diagonal polarizations. Note that one has

$$\left|(e_i, h_j)\right| = 1/\sqrt{2}, \quad i, j = 1, 2. \tag{17.12}$$

Bases with such a property are called conjugate. Note also that the vectors e_1, e_2 from the basis G_z and h_1, h_2 from the basis G_x are the eigenvectors of the Pauli matrices σ_z and σ_x, respectively; see (17.8).

Bob has a *photon detector*—a device that detects single photons in one of the two bases.

Alice can send photons emitted by the photon emitter to Bob, and Bob detects the photons with the photon detector.

The Protocol

1. Alice chooses a random polarization basis and prepares photons with a random polarization that belongs to the chosen basis. She sends the photons to Bob.
2. For each photon, Bob chooses at random which polarization basis he will use, and measures the polarization of the photon. (If Bob chooses the same basis as Alice, he can identify the polarization of the photon for sure.)
3. Alice and Bob use the public channel to compare the polarization bases they used. They keep only the polarization data for which the polarization bases are the same. In the absence of errors and eavesdropping, these data should be the same on both sides, it is called a *raw key*.
4. At the last step, Alice and Bob use methods of classical information theory to check whether their raw keys are the same. For example, they choose a random subset of the raw key and compare it using a public channel. They compute *the error rate D* (that is, the fraction of data for which their values disagree). If the error rate is unreasonably high—above, say, 10%—they abort the protocol and maybe try it again later. If the error rate is not that high they could use error correction codes.

As a result of the protocol Alice and Bob share the same random data. This data could now be used as a private key in the symmetric cryptosystems.

Instead of polarized photons one can use any two-level quantum system. One can consider also a generalized quantum key distribution protocol using a d-dimensional Hilbert space with k bases, each basis having d states [88, 128, 141, 154].

17.5.2 The Security of BB84

In transmitting information, there are always some errors, and Alice and Bob must apply some classical information processing protocols to improve their data. They can use *error correction* to obtain identical keys and *privacy amplification* to obtain a secret key. To solve the problem of eavesdropping, one has to find a protocol, which assuming that Alice and Bob can only measure the error rate of the received data, either provides Alice and Bob with a secure key, or aborts the protocol and tells the parties that the key distribution has failed. There are various eavesdropping problems, depending in particular on the technological power which Eve could have and on the assumed fidelity of Alice and Bob's devices, [155, 274, 288].

There is a simple eavesdropping strategy, called intercept–resend. Eve measures each qubit in one of the two basis and resends to Bob a qubit in the state corresponding to the result of her measurement. This attack belongs to the class of the so-called individual attacks. In this way, Eve will get 50% information. However, Alice and Bob can detect the actions of Eve because they will have 25% of errors in their sifted key. But it would be not so easy to detect eavesdropping if Eve applies the intercept–resend strategy to only a fraction of the Alice's sending.

In this case, one can use methods of classical cryptography. We suppose that once Alice, Bob, and Eve have made their measurements, they will get classical random variables α, β and ϵ respectively, with a joint probability distribution $p(x, y, z)$. Let $I(\alpha, \beta)$ be the mutual information of Alice and Bob, and $I(\alpha, \epsilon)$ and $I(\beta, \epsilon)$ the mutual information of Alice and Eve and Bob and Eve, respectively. Intuitively, it is clear that only if Bob has more information on Alice's bits than Eve then it could be possible to establish a secret key between Alice and Bob. In fact, one can prove (see [185, 288]) the following:

Theorem 17.8 *Alice and Bob can establish a secret key* (*using error correction and privacy amplification*) *if and only if*

$$I(\alpha, \beta) \geq I(\alpha, \epsilon) \quad or \quad I(\alpha, \beta) \geq I(\beta, \epsilon).$$

Let D be the error rate. Then one can prove that the BB84 protocol is secure against individual attacks if one has the following bound

$$D < D_0 \equiv \frac{1 - 1/\sqrt{2}}{2} \approx 15\%.$$

More general coherent or joint attacks when Eve measures several qubits simultaneously have also been discussed. An important problem of the eavesdropping analysis is to find quantum cryptosystems for which one can prove its *ultimate security*. Ultimate security means that the security is guaranteed against the whole class of eavesdropping attacks, even if Eve uses any conceivable technology of the future.

We assume that Eve has perfect technology which is only limited by the laws of quantum mechanics. This means she can use any unitary transformation between any number of qubits and an arbitrary auxiliary system. But Eve is not allowed to come to Alice's lab and read all her data.

17.5.3 Ultimate Security Proofs

The main ideas of how to prove the security of the BB84 protocol were presented by Mayers [515] in 1996. The security issues were considered in recent papers [117, 141, 488, 514, 515, 714, 835]. We describe here a simple and general method proposed in [128, 154, 288]. The method is based on Theorem 17.8 from Sect. 17.5.2 on classical cryptography and on Theorem 17.3 from Sect. 17.3 on information uncertainty relations.

The argument runs as follows. Suppose Alice sends out a large number of qubits and Bob receives n of them in the correct basis. The relevant Hilbert space dimension is then 2^n. Let us relabel the bases used for each of the n qubits in such a way that Alice used n times the x-basis. Hence, Bob's observable is the n-time tensor product $\sigma_x \otimes \cdots \otimes \sigma_x$. Since Eve had no way to know the correct bases, her optimal information on the correct ones is precisely the same as her optimal information on the incorrect ones. Hence one can bound her information assuming she measures $\sigma_z \otimes \cdots \otimes \sigma_z$. Therefore, $c = 2^{-n/2}$ and Theorem 17.3 from Sect. 17.3 implies:

$$I(\alpha, \epsilon) + I(\alpha, \beta) \le 2\log_2\bigl(2^n 2^{-n/2}\bigr) = n. \tag{17.13}$$

Next, combining the bound (17.13) with Theorem 17.8 from Sect. 17.5.2, one deduces that a secret key is achievable if $I(\alpha, \beta) \ge n/2$. Using

$$I(\alpha, \beta) = n\bigl(1 - D\log_2 D - (1 - D)\log_2(1 - D)\bigr),$$

one obtains the sufficient condition on the error rate D:

$$-D\log_2 D - (1 - D)\log_2(1 - D) \le \frac{1}{2},$$

which implies $D \le 0.11$. This bound was obtained in Mayers proof (after an improvement by Shor and Preskill [714]). It is compatible with the 15% bound found for individual attacks.

One can argue, however, that previous arguments lead, in fact, to another result: $c = 2^{-n/4}$. Indeed, Bob's observable is the n-time tensor product $\sigma_x \otimes \cdots \otimes \sigma_x$. Now, since Eve had no way to know the correct basis, it was assumed that she measures $\sigma_z \otimes \cdots \otimes \sigma_z$. However, it seems that if Eve does not know the correct basis then her observables σ_i will be complementary observables to σ_x only in a half of the cases. In the other half of cases, her observables σ_i will be the same as Bob's, i.e., σ_x. Therefore, one gets: $c = (1/\sqrt{2})^{n/2} = 2^{-n/4}$. This leads to a lower error rate, instead of 11% one gets 4%.

17.6 The EPRBE Quantum Cryptographic Protocol

17.6.1 Quantum Nonlocality and Cryptography

Bell's theorem [95] states that there are quantum correlation functions that cannot be represented as classical correlation functions of separated random variables. It has been interpreted as incompatibility of the requirement of locality with the statistical predictions of quantum mechanics [95]. For a recent discussion of Bell's theorem see, for example, [48, 791] and references therein. It is now widely accepted, as a result of Bell's theorem and related experiments, that "local realism" must be rejected.

Bell's theorem constitutes an important part in quantum cryptography [219]. It is now generally accepted that techniques of quantum cryptography can allow secure communications between distant parties. The promise of some secure cryptographic quantum key distribution schemes is based on the use of quantum entanglement in the spin space and on quantum no-cloning theorem. An important contribution of quantum cryptography is a mechanism for detecting eavesdropping.

Let us stress that the very formulation of the problem of locality in quantum mechanics is based on ascribing a special role to the position in the ordinary three-dimensional space. However, the space dependence of the wave function is neglected in many discussions of the problem of locality in relation to Bell's inequalities. Actually, it is the space part of the wave function which is relevant to the consideration of the problem of locality.

It was pointed out in [790] that the space part of the wave function leads to an extra factor in quantum correlation which changes the Bell's equation. A criterion of locality (or nonlocality) of quantum theory in a realist model of hidden variables was suggested. In particular, predictions of quantum mechanics can be consistent with Bell's inequalities for some Gaussian wave functions.

If one neglects the space part of the wave function in a cryptographic scheme then such a scheme could be insecure in the real three-dimensional space.

We will discuss how one can try to improve the security of quantum cryptography schemes in space by using a special preparation of the space part of the wave function, see [791].

17.6.2 Bell's Inequality and Localized Detectors

It was shown in the previous chapter that if one takes into account the space part of the wave function then the quantum correlation in the simplest case will take the form $g\cos(\alpha-\beta)$ instead of just $\cos(\alpha-\beta)$ where the parameter g describes the location of the system in space and time. In this case, one can get the representation [790]

$$g\cos(\alpha-\beta)=E(\xi_\alpha\eta_\beta), \tag{17.14}$$

if g is small enough (see below). The factor g gives a contribution to visibility or efficiency of detectors that are used in the phenomenological description of detectors.

In the previous section, the space part of the wave function of the particles was neglected. However, exactly the space part is relevant to the discussion of locality. The complete wave function is $\psi=(\psi_{\alpha\beta}(\mathbf{r}_1,\mathbf{r}_2))$ where α and β are spinor indices and $\mathbf{r}_1$ and $\mathbf{r}_2$ are vectors in the three-dimensional space.

We suppose that Alice and Bob have detectors which are located within the two localized regions $\mathcal{O}_A$ and $\mathcal{O}_B$, respectively, well separated from one another.

Quantum correlation describing the measurements of spins by Alice and Bob at their localized detectors is

$$G(a,\mathcal{O}_A,b,\mathcal{O}_B)=\langle\psi|\sigma\cdot aP_{\mathcal{O}_A}\otimes\sigma\cdot bP_{\mathcal{O}_B}|\psi\rangle, \tag{17.15}$$

where $P_{\mathcal{O}}$ is the projection operator onto the region $\mathcal{O}$.

As we discussed in the previous chapter, consider the case when the wave function has the form of the product of the spin function and the space function $\psi=\psi_{\mathrm{spin}}\phi(\mathbf{r}_1,\mathbf{r}_2)$. Then one has

$$G(a,\mathcal{O}_A,b,\mathcal{O}_B)=g(\mathcal{O}_A,\mathcal{O}_B)D_{\mathrm{spin}}(a,b), \tag{17.16}$$

where the function

$$g(\mathcal{O}_A, \mathcal{O}_B) = \int_{\mathcal{O}_A \times \mathcal{O}_B} \left|\phi(\mathbf{r}_1, \mathbf{r}_2)\right|^2 d\mathbf{r}_1\, d\mathbf{r}_2 \tag{17.17}$$

describes the correlation of particles in space. It is the probability to find one particle in the region $\mathcal{O}_A$ and another particle in the region $\mathcal{O}_B$. One has

$$0 \leq g(\mathcal{O}_A, \mathcal{O}_B) \leq 1. \tag{17.18}$$

Remark 17.9 In the relativistic quantum field theory, there is no nonzero strictly localized projection operator that annihilates the vacuum. It is a consequence of the Reeh–Schlieder theorem. Therefore, apparently, the function $g(\mathcal{O}_A, \mathcal{O}_B)$ should be always strictly smaller than 1.

Now one inquires whether one can write the representation

$$g(\mathcal{O}_A, \mathcal{O}_B) D_{\text{spin}}(a, b) = \int \xi(a, \mathcal{O}_A, \lambda) \eta(b, \mathcal{O}_B, \lambda)\, d\rho(\lambda). \tag{17.19}$$

Note that if we are interested in the conditional probability of finding the projection of spin along vector a for the particle 1 in the region $\mathcal{O}_A$ and the projection of spin along the vector b for the particle 2 in the region $\mathcal{O}_B$ then we have to divide both sides of (17.19) by $g(\mathcal{O}_A, \mathcal{O}_B)$. The factor g was written as (see Chap. 15) for $0 \leq g \leq 1/2$:

$$g \cos(\alpha - \beta) = \int_0^{2\pi} \sqrt{2g} \cos(\alpha - \lambda) \sqrt{2g} \cos(\beta - \lambda) \frac{d\lambda}{2\pi}. \tag{17.20}$$

Let us now apply these considerations to quantum cryptography.

17.6.3 The EPRBE Quantum Key Distribution

Ekert [219] showed that one can use the Einstein–Podolsky–Rosen correlations to establish a secret random key between two parties ("Alice" and "Bob"). Bell's inequalities are used to check the presence of an intermediate eavesdropper ("Eve"). We will call this method the Einstein–Podolsky–Rosen–Bell–Ekert (EPRBE) quantum cryptographic protocol. There are two stages to the EPRBE protocol: the first stage over a quantum channel, the second over a public channel.

The quantum channel consists of a source that emits pairs of spin 1/2 particles in a singlet state. The particles fly apart towards Alice and Bob, who, after the particles have separated, performs measurements on spin components along one of three directions, given by unit vectors a and b. In the second stage, Alice and Bob communicate over a public channel. They announce in public the orientation of the detectors they have chosen for particular measurements. Then they divide the measurement results into two separate groups: the first group for which they used different orientation of the detectors, and the second group for which they used the same orientation of the detectors. Now Alice and Bob can reveal publicly the results they obtained but within the first group of measurements only. This allows them, by using Bell's inequality, to establish the presence of an eavesdropper (Eve). The results of the second group of measurements can be converted into a secret key. One supposes that Eve has a detector which is located within the region $\mathcal{O}_E$ and she is described by hidden variables λ.

We will interpret Eve as a hidden variable in a realist theory and will study whether the quantum correlation (17.16) can be represented in the form (17.15). From (17.15), (17.18), and (17.19), one can see that if the following inequality

$$g(\mathcal{O}_A, \mathcal{O}_B) \leq 1/\sqrt{2}, \tag{17.21}$$

is valid for regions $\mathcal{O}_A$ and $\mathcal{O}_B$ which are well separated from one another then there is no violation of the CHSH inequalities (17.10), and therefore Alice and Bob cannot detect the presence of an eavesdropper. On the other side, if for a pair of well separated regions $\mathcal{O}_A$ and $\mathcal{O}_B$ one has

$$g(\mathcal{O}_A, \mathcal{O}_B) > 1/\sqrt{2}, \tag{17.22}$$

then it could be a violation of the realist locality in these regions for a given state. Then, in principle, one can hope to detect an eavesdropper in these circumstances.

Note that if we set $g(\mathcal{O}_A, \mathcal{O}_B) = 1$ in (17.19) as it was done in the original proof of Bell's theorem, then it means we did a special preparation of the states of particles to be completely localized inside of detectors. There exist such well localized states (see, however, the previous remark) but there exist also another states, with the wave functions which are not very well localized inside the detectors, and still particles in such states are also observed in detectors. The fact that a particle is observed inside the detector does not mean, of course, that its wave function is strictly localized inside the detector before the measurement. Actually one has to perform a thorough investigation of the preparation and the evolution of our entangled states in space and time if one needs to estimate the function $g(\mathcal{O}_A, \mathcal{O}_B)$.

17.6.4 Gaussian Wave Functions

Now let us consider the criterion of locality for Gaussian wave functions. We will show that with a reasonable accuracy there is no violation of locality in this case. Let us take the wave function ϕ of the form $\phi = \psi_1(\mathbf{r}_1)\psi_2(\mathbf{r}_2)$ where the individual wave functions have the moduli

$$|\psi_1(\mathbf{r})|^2 = \left(\frac{m^2}{2\pi}\right)^{3/2} e^{-m^2\mathbf{r}^2/2}, \qquad |\psi_2(\mathbf{r})|^2 = \left(\frac{m^2}{2\pi}\right)^{3/2} e^{-m^2(\mathbf{r}-\mathbf{l})^2/2}. \tag{17.23}$$

We suppose that the length of the vector $\mathbf{l}$ is much larger than $1/m$. We can make measurements of $P_{\mathcal{O}_A}$ and $P_{\mathcal{O}_B}$ for any well separated regions $\mathcal{O}_A$ and $\mathcal{O}_B$. Let us suppose a rather unfavorable case for the criterion of locality when the wave functions of the particles are almost localized inside the regions $\mathcal{O}_A$ and $\mathcal{O}_B$, respectively. In such a case, the function $g(\mathcal{O}_A, \mathcal{O}_B)$ can take values near its maximum. We suppose that the region $\mathcal{O}_A$ is given by $|r_i| < 1/m, \mathbf{r} = (r_1, r_2, r_3)$, and the region $\mathcal{O}_B$ is obtained from $\mathcal{O}_A$ by translation on $\mathbf{l}$. Hence $\psi_1(\mathbf{r}_1)$ is a Gaussian function with modules appreciably different from zero only in $\mathcal{O}_A$, and similarly $\psi_2(\mathbf{r}_2)$ is localized in the region $\mathcal{O}_B$. Then we have

$$g(\mathcal{O}_A, \mathcal{O}_B) = \left(\frac{1}{\sqrt{2\pi}} \int_{-1}^{1} e^{-x^2/2}\, dx \right)^6. \tag{17.24}$$

One can estimate (17.24) as

$$g(\mathcal{O}_A, \mathcal{O}_B) < \left(\frac{2}{\pi} \right)^3, \tag{17.25}$$

which is smaller than $1/2$. Therefore, the locality criterion (17.21) is satisfied in this case.

Let us remind that there is a well known effect of expansion of wave packets due to the free time evolution. If ϵ is the characteristic length of the Gaussian wave packet describing a particle of mass M at time $t = 0$ then at time t the characteristic length ϵ_t will be

$$\epsilon_t = \epsilon \sqrt{1 + \frac{\hbar^2 t^2}{M^2 \epsilon^4}}. \tag{17.26}$$

It tends to $(\hbar/M\epsilon)t$ as $t \to \infty$. Therefore, the locality criterion is always satisfied for nonrelativistic particles if regions $\mathcal{O}_A$ and $\mathcal{O}_B$ are far enough from each other.

In quantum cryptography there are many interesting open problems which require further investigations. In quantum cryptographic protocols with two entangled photons (such as the EPRBE protocol) to detect the eavesdropper's presence by using Bell's inequality, we have to estimate the function $g(\mathcal{O}_A, \mathcal{O}_B)$. In order to increase the detectability of the eavesdropper, one has to do a thorough investigation of the process of preparation of the entangled state and then its evolution in space and time towards Alice and Bob. One has to develop a proof of the security of such a protocol.

In the previous section, Eve was interpreted as an abstract hidden variable. However, one can assume that more information about Eve is available. In particular, one can assume that she is located somewhere in space in a region $\mathcal{O}_E$. It seems that one has to study a generalization of the function $g(\mathcal{O}_A, \mathcal{O}_B)$, which depends not only on the Alice and Bob locations $\mathcal{O}_A$ and $\mathcal{O}_B$ but also on Eve's location $\mathcal{O}_E$. Then one can try to find a strategy which leads to an optimal value of this function.

In quantum cryptographic protocols with single photons (such as the BB84 protocol), further investigation of the security under various types of attacks, including attacks from real space, would be desirable.

17.7 Notes

Public key distribution was proposed by Diffie and Hellman in 1976 [202]. The widely used public key distribution RSA was proposed in [661]. The no-cloning theorem was discussed by Wootters and Zurek [821]. The idea of quantum cryptography was proposed by Wiesner [818]. The BB84 protocol was proposed by Bennett and Brassard [102]. Nambu et al. experimentally demonstrated the BB84 protocol [542]. Ekert proposed the QKD protocol with EPR nonlocality [219]. Other protocols of quantum key distribution were discussed in B92 [103], DPS-QKD (Differential Phase Shift-Quantum Key Distribution) [361, 362] and BBM92 [104]. Yuen proposed a protocol of classical key distribution with a quantum state [827]. The presentation of this chapter uses [792]. Quantum cryptography in real space was considered by Volovich [791, 792]. A general mathematical approach to quantum cryptography is presented in [803]. Specifications and standards of security in quantum cryptographic ptotocols are discussed by Trushechkin and Volovich [754].

Chapter 18
Quantum Teleportation

In quantum communication theory, one looks for the most efficient way to code information and construct a physical device (channel) in order to send information as completely as possible. There "quantum" means that we code information by quantum states and send it through a properly designed quantum device. If one can send any quantum state from an input system to an output system as it is, then it will be an ultimate way of information transmission. Such an ultimate method is not only ultimate for information transmission but also considered to enable sending matter existing in real world to other place without destroying itself which was a dream as in science fiction. In this chapter some protocols for quantum teleportation of states are discussed.

Although it needs an assumption that quantum mechanics can describe all aspects of existence in our world, it is in quantum teleportation that we can discuss such an ultimate communication system. More precisely, quantum teleportation is to send a quantum state itself containing all information of a certain system from one place to another. The problem of quantum teleportation is whether there exist a physical device and a key (or a set of keys) by which a quantum state attached to a sender (Alice) is completely transmitted, and a receiver (Bob) can reconstruct the state sent by Alice. It has been considered that such a teleportation would not be realistic because usual quantum state contains information which cannot be observed simultaneously. Bennett, Brassard, Crepeau, Jozsa, Peres, and Wootters (BBCJPW) showed that such a teleportation is possible by means of a device (channel) made from proper entangled states (i.e., Einstein, Podolsky and Rosen (EPR), see Chap. 8) of Bell's basis. The basic idea behind their discussion is to divide the information encoded in the state into two parts, classical and quantum, and send them through different channels, a classical channel and an EPR channel. The classical channel is nothing but a simple correspondence between the sender and the receiver, and the EPR channel is constructed by using a certain entangled state. However, the EPR channel is not so stable due to decoherence. Fichtner and Ohya studied the quantum teleportation by means of general beam splitting processes so that it contains the EPR channel, and they constructed a more stable teleportation process with coherent entangled states.

M. Ohya, I. Volovich, *Mathematical Foundations of Quantum Information and Computation and Its Applications to Nano- and Bio-systems*,
Theoretical and Mathematical Physics,
DOI 10.1007/978-94-007-0171-7_18, © Springer Science+Business Media B.V. 2011

In Sect. 18.1 of this chapter, we discuss the channel expression of quantum teleportation. In Sect. 18.2, we explain original treatment of quantum teleportation done by Bennett et al. within the channel expression. In Sect. 18.3, the weak type of quantum teleportation and the uniqueness of the set of keys given to Bob are discussed. In Sect. 18.4 and Sect. 18.5, we discuss a general treatment of quantum teleportation process in bosonic Fock space with basic techniques of beam splitting processes, often used in usual quantum communication. In Sect. 18.6, we present a calculation of fidelity with respect to the teleportation model using the beam splitter. In Sect. 18.7, we discuss a continuous teleportation model based on a paper by Braunstein and Kimble and some related mathematics. Finally, we close this chapter with a new scheme of the quantum teleportation due to Kossakowski and Ohya [444], where the teleportation channel is always linear, and perfect (complete) teleportation is possible for non-maximally entangled states.

18.1 Channel Expression of Quantum Teleportation

As was discussed in Chap. 6, we have to prepare at least two dynamical systems, input and output, for information transmission. Since every system can be described by a state, it is important to know the relation between the input state and the output state for the study of information transmission. Such a relation is described by a channel bridging between two systems, namely, providing the state change in the course of information transmission.

In the classical communication theory, a state of input or output system is described by a probability distribution (or measure), so that a channel causes a change of this probability distribution. On the other hand, in quantum communication theory, a state of input or output system should be described by a certain noncommutative state (quantum state) such as a density operator or a positive normalized linear functional, more generally. Here we restrict our discussion to the former case, the case of density operators.

We need three separable Hilbert spaces $\mathcal{H}_1$, $\mathcal{H}_2$, and $\mathcal{H}_3$. $\mathcal{H}_1$ and $\mathcal{H}_2$ are attached to Alice, and $\mathcal{H}_3$ is attached to Bob. Let $\mathbf{B}(\mathcal{H}_i)$ $(i = 1, 2, 3)$ be the set of all bounded linear operators on $\mathcal{H}_i$, and let $\mathfrak{S}(\mathcal{H}_i)$ be the set of all states (density operators) on the Hilbert spaces $\mathcal{H}_i$.

The goal of quantum teleportation is to faithfully send a state from Alice to Bob, the state being unknown to Alice. The quantum teleportation was studied by Bennett et al. and its a bit more general scheme is expressed in the following steps: Alice is provided with a device expressed by the set of projections $\{F_k^{(12)}\}$ in her Hilbert space $\mathcal{H}_1 \otimes \mathcal{H}_2$ and Bob is provided with a set of unitary keys $\{U_k\}$.

Step 0. Alice has an unknown to her quantum state $\rho^{(1)}$ (on the Hilbert space $\mathcal{H}_1$) and she was asked to teleport it to Bob.

Step 1. Prearrange a state $\sigma^{(23)} \in \mathfrak{S}(\mathcal{H}_2 \otimes \mathcal{H}_3)$ having a certain correlation between Alice and Bob, which is an entangled state between two systems $\mathcal{H}_2$ and $\mathcal{H}_3$.

Step 2. Prepare the set of projections $\{F_k^{(12)}\}$ and an observable $F^{(12)} \equiv \sum_k \lambda_k F_k^{(12)}$ $(k \neq j \Rightarrow \lambda_k \neq \lambda_j)$ on a tensor product Hilbert space (system) $\mathcal{H}_1 \otimes \mathcal{H}_2$. Alice performs the joint measurement of the observable $F^{(12)}$, and she obtains the outcome λ_k.

Step 3. After the measurement by Alice, Bob obtained a state $\rho^{(3)}$ due to the reduction of wave packet and he is informed which outcome λ_k was obtained by Alice. This information is completely transmitted from Alice to Bob without disturbance (for instance, by telephone).

Step 4. Bob reconstructs $\rho^{(1)}$ from $\rho^{(3)}$ by using the key U_k which corresponds to the outcome λ_k Bob got from Alice in the above *Step 3*.

The above steps can be exhibited by a channel $\Lambda_k^* : \mathfrak{S}(\mathcal{H}_1) \to \mathfrak{S}(\mathcal{H}_3)$ constructed by the following three maps (channels):

1. $\gamma^* : \mathfrak{S}(\mathcal{H}_1) \to \mathfrak{S}(\mathcal{H}_1 \otimes \mathcal{H}_2 \otimes \mathcal{H}_3)$,

$$\gamma^*(\rho^{(1)}) = \rho^{(1)} \otimes \sigma^{(23)} \quad \forall \rho^{(1)} \in \mathfrak{S}(\mathcal{H}_1).$$

This channel γ expresses a coupling of an initial state $\rho^{(1)}$ with the entangled state $\sigma^{(23)}$.

2. $\pi_k^* : \mathfrak{S}(\mathcal{H}_1 \otimes \mathcal{H}_2 \otimes \mathcal{H}_3) \to \mathfrak{S}(\mathcal{H}_1 \otimes \mathcal{H}_2 \otimes \mathcal{H}_3)$ is a state change describing the nonclassical effect of the teleportation determined by such as Alice's measurement.

   Since $F_k^{(12)}$ is a projection (Alice's measurement) on $\mathcal{H}_1 \otimes \mathcal{H}_2$, that is, $F_k^{(12)} = (F_k^{(12)})^* = (F_k^{(12)})^2$, the map $\pi_k^* : \mathfrak{S}(\mathcal{H}_1 \otimes \mathcal{H}_2 \otimes \mathcal{H}_3) \to \mathfrak{S}(\mathcal{H}_1 \otimes \mathcal{H}_2 \otimes \mathcal{H}_3)$ is given by

$$\pi_k^*(\rho^{(123)}) = \frac{1}{L}(F_k^{(12)} \otimes I^{(3)})\rho^{(123)}(F_k^{(12)} \otimes I^{(3)}),$$
$$\forall \rho^{(123)} \in \mathfrak{S}(\mathcal{H}_1 \otimes \mathcal{H}_2 \otimes \mathcal{H}_3).$$

Here $L = \mathrm{tr}_{123}(F_k^{(12)} \otimes I^{(3)})\rho^{(123)}(F_k^{(12)} \otimes I^{(3)})$.

3. $a^* : \mathfrak{S}(\mathcal{H}_1 \otimes \mathcal{H}_2 \otimes \mathcal{H}_3) \to \mathfrak{S}(\mathcal{H}_3)$,

$$\rho^{(3)} = a^*(\rho^{(123)}) = \mathrm{tr}_{12}\, \rho^{(123)}, \quad \forall \rho^{(123)} \in \mathfrak{S}(\mathcal{H}_1 \otimes \mathcal{H}_2 \otimes \mathcal{H}_3).$$

This channel a^* represents a reduction from the state $\rho^{(123)}$ to Bob's state $\rho^{(3)}$ after the Alice's measurement, where tr_{12} is the partial trace with respect to $\mathcal{H}_1 \otimes \mathcal{H}_2$.

4. In the teleportation scheme, there exists a unitary channel $u_k^* : \mathfrak{S}(\mathcal{H}_3) \to \mathfrak{S}(\mathcal{H}_1)$,

$$u_k^*(\rho^{(3)}) = U_k \rho^{(3)} U_k^* \quad \forall \rho^{(3)} \in \mathfrak{S}(\mathcal{H}_3).$$

u_k^* is Bob's unitary operator (key) corresponding to Alice's measurement $F_k^{(12)}$.

Therefore, we obtain the channel $\Lambda_k^* : \mathfrak{S}(\mathcal{H}_1) \to \mathfrak{S}(\mathcal{H}_3)$

$$\Lambda_k^* = a^* \circ \pi_k^* \circ \gamma^*,$$

or more concretely,

$$\Lambda_k^* \rho^{(1)} = \mathrm{tr}_{12}\, \pi_k^* \big(\rho^{(1)} \otimes \sigma^{(23)}\big), \quad \forall \rho^{(1)} \in \mathfrak{S}(\mathcal{H}_1),$$

where the subscript "k" means that the channel Λ_k^* depends on the choice of Alice's measurement $F_k^{(12)}$.

Thus, the whole teleportation channel $\Lambda_k^* : \mathfrak{S}(\mathcal{H}_1) \to \mathfrak{S}(\mathcal{H}_3)$ is written as

$$\Lambda_k^* \rho^{(1)} \equiv \mathrm{tr}_{12}\left[\frac{(F_k^{(12)} \otimes I^{(3)})(\rho^{(1)} \otimes \sigma^{(23)})(F_k^{(12)} \otimes I^{(3)})}{\mathrm{tr}_{123}(F_k^{(12)} \otimes I^{(3)})(\rho^{(1)} \otimes \sigma^{(23)})(F_k^{(12)} \otimes I^{(3)})}\right], \quad \forall \rho^{(1)} \in \mathfrak{S}(\mathcal{H}_1).$$

Note that this channel is sometimes called the von Neumann–Lüders projection postulate, and Λ_k^* *is nonlinear unless the entangled state* σ *and the projection* P_α are maximally entangled (see Sect. 18.3).

The quantum teleportation is stated as follows:

Definition 18.1 Quantum teleportation is realized if there exist the set of operators $\{F_k^{(12)}\}$ on $\mathcal{H}_1 \otimes \mathcal{H}_2$, an entangled state $\sigma^{(23)}$ on $\mathcal{H}_2 \otimes \mathcal{H}_3$, and the set of keys $\{U_k\}$ such that $U_k \Lambda_k^* \rho^{(1)} U_k^* = \rho^{(1)}$ for any state $\rho^{(1)}$ in $\mathcal{H}_1$ and Λ_k^* above.

When such teleportation is realized, the state $\rho^{(3)}$ transferred to Bob from Alice is unitarily equivalent to the original state $\rho^{(1)}$ sent by Alice, so that all information stored in $\rho^{(1)}$ is completely transmitted to Bob; $I(\rho^{(1)}; \Lambda_k^*) = S(\rho^{(1)}) = S(\rho^{(3)})$ for any k.

18.2 BBCJPW Model of Teleportation

In this section, we will notice that the original BBCJPW (Bennett, Brassard, Crepeau, Jozsa, Peres and Wootters) scheme provides nice examples to the described framework. Let the initial state given to Alice be $\rho^{(1)} = |\zeta\rangle\langle\zeta|$ where $|\zeta\rangle = \alpha|0^{(1)}\rangle + \beta|1^{(1)}\rangle$ with $|\alpha|^2 + |\beta|^2 = 1$. In their scheme, $\sigma^{(23)}$ is given by an EPR spin pair in a singlet state such as

$$\sigma^{(23)} = |\xi\rangle\langle\xi|,$$

where

$$|\xi\rangle = \sqrt{\frac{1}{2}}\big|0^{(2)}\big\rangle \otimes \big|1^{(3)}\big\rangle - \sqrt{\frac{1}{2}}\big|1^{(2)}\big\rangle \otimes \big|0^{(3)}\big\rangle$$

with the spin-up vector $|0\rangle \equiv \binom{1}{0}$ and the spin-down vector $|1\rangle \equiv \binom{0}{1}$. There, Alice's measurement $F_k^{(12)}$ is one of the projections $\{F_k^{(12)}; k = 1, 2, 3, 4\}$,

$$F_1^{(12)} = \big|\psi^{(-)}\big\rangle\big\langle\psi^{(-)}\big|, \qquad F_2^{(12)} = \big|\psi^{(+)}\big\rangle\big\langle\psi^{(+)}\big|,$$

$$F_3^{(12)} = |\varphi^{(-)}\rangle\langle\varphi^{(-)}|, \qquad F_4^{(12)} = |\varphi^{(+)}\rangle\langle\varphi^{(+)}|$$

with the Bell's CONS

$$|\psi^{(-)}\rangle = \sqrt{\frac{1}{2}}(|0^{(1)}\rangle \otimes |1^{(2)}\rangle - |1^{(1)}\rangle \otimes |0^{(2)}\rangle),$$

$$|\psi^{(+)}\rangle = \sqrt{\frac{1}{2}}(|0^{(1)}\rangle \otimes |1^{(2)}\rangle + |1^{(1)}\rangle \otimes |0^{(2)}\rangle),$$

$$|\varphi^{(-)}\rangle = \sqrt{\frac{1}{2}}(|0^{(1)}\rangle \otimes |0^{(2)}\rangle - |1^{(1)}\rangle \otimes |1^{(2)}\rangle),$$

$$|\varphi^{(+)}\rangle = \sqrt{\frac{1}{2}}(|0^{(1)}\rangle \otimes |0^{(2)}\rangle + |1^{(1)}\rangle \otimes |1^{(2)}\rangle).$$

The unitary (key) operators U_k ($k = 1, 2, 3, 4$) are given as follows

$$U_1 \equiv |0^{(1)}\rangle\langle 0^{(3)}| + |1^{(1)}\rangle\langle 0^{(3)}|,$$

$$U_2 \equiv |0^{(1)}\rangle\langle 0^{(3)}| - |1^{(1)}\rangle\langle 0^{(3)}|,$$

$$U_3 \equiv |0^{(1)}\rangle\langle 1^{(3)}| + |1^{(1)}\rangle\langle 1^{(3)}|,$$

$$U_4 \equiv |0^{(1)}\rangle\langle 1^{(3)}| - |1^{(1)}\rangle\langle 1^{(3)}|.$$

If Alice measures $F_k^{(12)}$, then Bob uses U_k. The channels constructed by the above states became linear, and the teleportation is done completely.

18.3 Weak Teleportation and Uniqueness of Key

As we discussed in Sect. 18.1, the teleportation problem is to find the following objects 1–3:

1. An entangled state $\sigma^{(23)}$ on $\mathcal{H}_2 \otimes \mathcal{H}_3$.
2. A family of mutually orthogonal projections $\{F_k^{(12)}\}$ acting on $\mathcal{H}_1 \otimes \mathcal{H}_2$.
3. For each k, a unitary operator U_k such that the associated unitary channel satisfies the identity $\Lambda_k^* \rho^{(1)} = U_k^* \rho^{(1)} U_k$ for any k and for any state $\rho^{(1)} \in \mathfrak{S}(\mathcal{H}_1)$, or at least for $\rho^{(1)}$ in a preassigned subset of $\mathfrak{S}(\mathcal{H}_1)$.

If the above conditions are replaced by the weaker ones:

1'. An entangled state $\sigma^{(23)}$ acting on $\mathcal{H}_1 \otimes \mathcal{H}_2$
2'. A single projection $F^{(12)}$ acting on $\mathcal{H}_1 \otimes \mathcal{H}_2$
3'. A single unitary operator U such that the identity $\Lambda^* \rho^{(1)} = U^* \rho^{(1)} U$ holds for any state $\rho^{(1)} \in \mathfrak{S}(\mathcal{H}_1)$

then we speak of the *weak teleportation problem*.

The connection between the weak and the general teleportation problem is the following.

Given a family $\{\sigma^{(23)}, F_k^{(12)}, U_k\}$ *of solutions of the weak teleportation problem for each k such that the projections* $F_k^{(12)}$ *are mutually orthogonal, this family provides a solution of the general teleportation problem.*

We shall solve the weak teleportation problem, and then we show the uniqueness of the key.

In the notations above, we shall assume that

$$N = \dim \mathcal{H}_1 < +\infty.$$

Under this assumption we shall look for a solution of the weak teleportation problem in which

$$\begin{aligned} N &= \dim \mathcal{H}_1 = \dim \mathcal{H}_2 = \dim \mathcal{H}_3, \\ \sigma &= |\xi\rangle\langle\xi|, \\ F &\equiv |\psi\rangle\langle\psi|, \end{aligned} \tag{18.1}$$

where $\xi \in \mathcal{H}_2 \otimes \mathcal{H}_3$ and $\psi \in \mathcal{H}_1 \otimes \mathcal{H}_2$ are unit vectors. We look for a unitary transformation $U : \mathcal{H}_3 \to \mathcal{H}_1$ such that for any density matrix $\rho \in \mathfrak{S}(\mathcal{H}_1)$ one has

$$U\left\{\frac{1}{L}\mathrm{tr}_{12}\big(F \otimes I\big(\rho \otimes |\xi\rangle\langle\xi|\big)F \otimes I\big)\right\}U^* = \rho.$$

Here $L = \mathrm{tr}(F \otimes I^{(3)}(\rho \otimes |\xi\rangle\langle\xi|)F \otimes I^{(3)})$.

Let us fix three arbitrary orthonormal bases: $\{|x_l\rangle\}$ of $\mathcal{H}_1$, $\{|x'_h\rangle\}$ of $\mathcal{H}_2$, and $\{|x''_k\rangle\}$ of $\mathcal{H}_3$. Fix an arbitrary $N \times N$ complex unitary matrix (λ_{hk}) and let

$$|\xi\rangle \equiv \frac{1}{\sqrt{N}} \sum \lambda_{hk} |x'_h\rangle \otimes |x''_k\rangle \in \mathcal{H}_2 \otimes \mathcal{H}_3,$$

$$|\psi\rangle = \frac{1}{\sqrt{N}} \sum |x_l\rangle \otimes |x'_l\rangle \in \mathcal{H}_1 \otimes \mathcal{H}_2.$$

Then we have [20]:

Theorem 18.2 *Consider a unitary operator* $U : \mathcal{H}_3 \to \mathcal{H}_1$ *such that* $\sum_h \bar{\lambda}_{hk}|x_h\rangle = U|x''_k\rangle$, *then the triple* $(|\xi\rangle, |\psi\rangle, U)$ *satisfies*

$$\frac{1}{L}\mathrm{tr}_{12}\big(F \otimes I(\rho \otimes \sigma)F \otimes I\big) = U^*\rho U$$

for any density operator $\rho \in \mathfrak{S}(\mathcal{H}_1)$.

Theorem 18.3 *Let* $\rho = \sum p_\gamma P_\gamma$ *be the spectral decomposition of* $\rho \in \mathcal{H}_1$. *If* U_1 *and* U_2 *satisfy condition* 3 *of a key with the above* ρ, *then there exists a unitary operator* V *from* $\mathcal{H}_1$ *to* $\mathcal{H}_1$ *such that* $U_2U_1^* = \sum V_\gamma$ *with* $V_\gamma \equiv P_\gamma V P_\gamma$. *Moreover, the equality* $V_{\gamma'} V_\gamma^* = \delta_{\gamma\gamma'} P_\gamma$ *is satisfied.*

Proof Suppose U and V are two solutions of the equation in condition 3, then $U_1^* \rho U_1 = U_2^* \rho U_2$, or equivalently, $U_2 U_1^* \rho = \rho U_2 U_1^*$. This means that $U_2 U_1^* \equiv W : \mathcal{H}_1 \to \mathcal{H}_1$ is in the commutant of ρ. Since V is a unitary operator commuting with ρ, $V P_\gamma = P_\gamma V$ is satisfied. Therefore, $V = \sum P_\gamma V P_\gamma = \sum V_\gamma : V_\gamma P_\gamma = P_\gamma V_\gamma = V_\gamma$. The equalities $1 = \sum_{\gamma\gamma'} V_\gamma^* V_{\gamma'} = \sum_{\gamma\gamma'} V_\gamma^* P_\gamma P_{\gamma'} V_\gamma = \sum_\gamma V_\gamma^* V_\gamma$ imply $V_\gamma V_{\gamma'}^* = \delta_{\gamma\gamma'} P_\gamma$. □

In the notations and assumptions above, let us suppose that the normalized state vectors $|\xi\rangle$ and $|\widetilde{\psi}\rangle$ have the form below for some constants $\{s_k\}$ and $\{t_k\}$

$$|\xi\rangle = \sum_l t_l |x_l\rangle \otimes |x_l'\rangle,$$

$$|\widetilde{\psi}\rangle = \sum_k s_k \sum_h \lambda_{hk} |x_h'\rangle \otimes |x_k''\rangle,$$

and let us look for the conditions under which the teleportation map Λ^* becomes linear.

Theorem 18.4 *Given* $|\xi\rangle$, $|\widetilde{\psi}\rangle$ *and* U *as above, the teleportation channel* Λ^*, $\mathrm{tr}_{123} F \otimes I^{(3)}(\rho \otimes |\widetilde{\psi}\rangle\langle\widetilde{\psi}|) F \otimes I^{(3)}$ *is independent of* ρ *if and only if the coefficients* t_k *have the following form* $s_l = e^{i\theta_l}/\sqrt{N}$, $t_k = e^{i\theta_k}/\sqrt{N}$, *which means that the entangled state* σ *is maximal.*

Proof One has

$$\begin{aligned}
&F \otimes I\big(\rho \otimes |\widetilde{\psi}\rangle\langle\widetilde{\psi}|\big) F \otimes I \\
&= \sum \lambda_{h\alpha} \overline{\lambda}_{k\beta} F \otimes I\big(\rho \otimes |x_h'\rangle\langle x_k'|\big) F \otimes I \otimes |x_\alpha''\rangle\langle x_\beta''| \\
&= \sum \lambda_{h\alpha} \overline{\lambda}_{k\beta} t_\mu \overline{t}_{\mu'} t_\nu \overline{t}_{\nu'} |x_\mu\rangle\langle x_{\mu'}, \rho, x_\nu\rangle\langle x_{\nu'}| \\
&\quad \otimes |x_\mu'\rangle\langle x_{\mu'}', x_h'\rangle\langle x_k', x_\nu'\rangle\langle x_{\nu'}'| \otimes |x_\alpha''\rangle\langle x_\beta''| \\
&= \sum \lambda_{h\alpha} \overline{\lambda}_{k\beta} t_\mu \overline{t}_h t_k \overline{t}_\nu \langle x_h, \rho x_k\rangle |x_\mu\rangle\langle x_{\nu'}| \otimes |x_\mu'\rangle\langle x_{\nu'}'| \otimes |x_\alpha''\rangle\langle x_\beta''|.
\end{aligned}$$

Tracing over $\mathcal{H}_1 \otimes \mathcal{H}_2$, it becomes

$$\sum \lambda_{h\alpha} \overline{\lambda}_{k\beta} t_\mu \overline{t}_h t_k \overline{t}_\mu \langle x_h, \rho x_k\rangle |x_\alpha''\rangle\langle x_\beta''|.$$

Since $\sum_\mu |t_\mu|^2 = 1$, this is equal to

$$\sum \lambda_{h\alpha} \overline{\lambda}_{k\beta} \overline{t}_h t_k \langle x_h, \rho x_k\rangle |x_\alpha''\rangle\langle x_\beta''|.$$

Taking the $\mathcal{H}_3$-trace, we find

$$\sum \lambda_{h\alpha} \overline{\lambda}_{k\alpha} \overline{t}_h t_k \langle x_h, \rho x_k\rangle = |t_h|^2 \langle x_h, \rho x_h\rangle$$

because of the unitarity of $(\lambda_{\alpha,\beta})$. It follows that the problem linearizes if and only if $\sum_k |t_k|^2 |x_k\rangle\langle x_k| = cI$ (c is a constant), and this is equivalent to: $|t_k|^2 = c$, $\forall k = 1, \ldots, N$. Consequently, $c = N$, and (18.1) follows. □

18.4 Perfect Teleportation in Bose Fock Space

Bennett and others used EPR spin pair to construct a teleportation model as we discussed in Sect. 18.2. In order to have a more handy model, in the subsequent sections we use coherent states to construct a model based on the works by Fichtner and Ohya. One of the main points for such a construction is how to prepare the entangled state. The EPR entangled state used by Bennett et al. can be identified with the splitting of a one-particle state, so that their teleportation model can be described in terms of Fock spaces and the beam splittings, which makes it possible for the whole teleportation process to work in a general beam splitting scheme. Moreover, to work with beams having a fixed number of particles seems to be not realistic, especially in the case of large distance between Alice and Bob because we have to take into account that the beams will lose particles (or energy). For that reason, one should use a class of beams being insensitive to this loss of particles. That and other arguments lead to superpositions of coherent beams.

Before starting the discussion, we slightly generalize the teleportation scheme, namely, the steps starting with Step 2 of the teleportation, described in the first section. Remark that we drop the indices (1), (12), (23) for $\rho^{(1)}$, $F_k^{(12)}$, $\sigma^{(23)}$ and others for notational simplicity.

Step 2. One then fixes a family of mutually orthogonal projections $(F_{nm})_{n,m=1}^N$ on the Hilbert space $\mathcal{H}_1 \otimes \mathcal{H}_2$ corresponding to an observable $F \equiv \sum_{n,m} z_{nm} F_{nm}$, and for a fixed pair of indices n, m, Alice performs a measurement of the observable F, involving only the $\mathcal{H}_1 \otimes \mathcal{H}_2$ part of the system in the state $\rho \otimes \sigma$. When Alice obtains an outcome z_{nm}, the state becomes

$$\rho_{nm}^{(123)} \equiv \frac{1}{L}(F_{nm} \otimes I)\rho \otimes \sigma(F_{nm} \otimes I),$$

where $L \equiv \mathrm{tr}_{123}(F_{nm} \otimes I)\rho \otimes \sigma(F_{nm} \otimes I)$. Note that in the first section we used the notation F_k instead of F_{nm} above, but in the sequel we use F_{nm} with the double indices n, m because one can easily discuss the beam-splitters, and so on.

Step 3. Bob is informed about which outcome was obtained by Alice. This is equivalent to transmitting the information that the eigenvalue z_{nm} was detected. This information is transmitted from Alice to Bob without disturbance and by means of classical tools.

Step 4. Making only partial measurements on the third part on the system in the state $\rho_{nm}^{(123)}$ means that Bob will control a state $\Lambda_{nm}^*(\rho)$ on $\mathcal{H}_3$ given by the partial trace on $\mathcal{H}_1 \otimes \mathcal{H}_2$ of the state $\rho_{nm}^{(123)}$ (after Alice's measurement)

$$\begin{aligned}\Lambda_{nm}^{*}(\rho) &= \operatorname{tr}_{12} \rho_{nm}^{(123)} \\ &= \operatorname{tr}_{12} \frac{1}{L}(F_{nm} \otimes I)\rho \otimes \sigma (F_{nm} \otimes I).\end{aligned}$$

Thus the whole teleportation scheme given by the family (F_{nm}) and the entangled state σ can be characterized by the family (Λ_{nm}^{*}) of channels from the set of states on $\mathcal{H}_1$ into the set of states on $\mathcal{H}_3$, and the family (p_{nm}) given by

$$p_{nm}(\rho) \equiv \operatorname{tr}_{123}(F_{nm} \otimes I)\rho \otimes \sigma (F_{nm} \otimes I)$$

of the probabilities that Alice's measurement according to the observable F will show the value z_{nm}.

The teleportation scheme works perfectly with respect to a certain class $\mathfrak{S}$ of states ρ on $\mathcal{H}_1$ if the following conditions are fulfilled.

(E1) For each n, m there exists a unitary operator $U_{nm} : \mathcal{H}_3 \to \mathcal{H}_1$ such that

$$U_{nm} \Lambda_{nm}^{*}(\rho) U_{nm}^{*} = \rho, \quad \forall \rho \in \mathfrak{S}.$$

(E2)

$$\sum_{n,m} p_{nm}(\rho) = 1, \quad \forall \rho \in \mathfrak{S}.$$

Condition (E1) means that Bob can reconstruct the original state ρ by unitary keys $\{U_{nm}\}$ provided to him. Condition (E2) means that Bob will succeed in finding a proper key with certainty.

In this section, we construct a teleportation model being perfect in the sense of conditions 1 and 2 above, where we take the Bose Fock space $\Gamma(L^2(G)) \equiv \mathcal{H}_1 = \mathcal{H}_2 = \mathcal{H}_3$ over a configuration space G with a certain class ρ of states on this Fock space. Before stating the main results on perfect teleportation, we review basic notations and facts on Bose Fock space.

18.4.1 Basic Notations and Facts

We collect some basic facts concerning the (symmetric) Fock space.

Symmetric (Bose) Fock Space

Let us recall the definition of the symmetric Fock space (see Chap. 4) Let $\mathcal{L}_1$ be a Hilbert space, which is called a one-particle Hilbert space. Then the n-particle Hilbert space is defined by

$$\mathcal{L}_n \equiv S_+ \mathcal{L}_1^{\otimes^n},$$

where S_+ is the symmetrizing operator on n-tuple tensor product Hilbert space $\mathcal{L}_1^{\otimes n}$;

$$S_+ \equiv \frac{1}{n!}\sum_P S_P,$$

$$S_P(f_1 \otimes \cdots \otimes f_n) = f_{P(1)} \otimes \cdots \otimes f_{P(n)}.$$

The above P indicates a permutation of indices, and note that $\mathcal{L}_0 \equiv \mathbb{C}$ describes the zero-particle Hilbert space. Indistinguishable Bose particles are described in the symmetric (Boson) Fock space

$$\Gamma(\mathcal{L}_1) \equiv \bigoplus_{n=0}^{\infty} \mathcal{L}_n.$$

Let Ω be the vacuum vector in $\Gamma(\mathcal{L}_1)$; $\Omega = (1, 0, 0, \ldots)$. Note that we have an important equality $\Gamma(\mathcal{L}_1 \oplus \mathcal{L}_1') = \Gamma(\mathcal{L}_1) \otimes \Gamma(\mathcal{L}_1')$ for two Hilbert spaces $\mathcal{L}_1$ and $\mathcal{L}_1'$.

Fichtner–Freudenberg Expression of Fock Space

We discuss Fichtner–Freudenberg expression of Fock space in a way adapted to the language of counting measures. Their expression looks difficult, but it is very useful to prove some statements.

Let G be an arbitrary complete separable metric space. Further, let μ be a locally finite diffuse measure on G, i.e., $\mu(B) < +\infty$ for bounded measurable subsets of G and $\mu(\{x\}) = 0$ for all singletons $x \in G$. In order to describe the teleportation of states on a finite-dimensional Hilbert space through the k-dimensional space $\mathbb{R}^k$, especially we are concerned with the case

$$G = \mathbb{R}^k \times \{1, \ldots, N\},$$
$$\mu = l \times \#,$$

where l is the k-dimensional Lebesgue measure and # denotes the counting measure on $\{1, \ldots, N\}$, we denote the set of all finite counting measures on G by $M = M(G)$. Since $\varphi \in M$ can be written in the form $\varphi = \sum_{j=1}^n \delta_{x_j}$ for some $n = 0, 1, 2, \ldots$ and $x_j \in G$ with the Dirac measure δ_x corresponding to $x \in G$, the elements of M can be interpreted as finite (symmetric) point configurations in G. We equip M with its canonical σ-algebra $\mathfrak{M}$ and we consider the σ-finite measure F by setting

$$F(Y) \equiv 1_Y(O) + \sum_{n\geq 1} \frac{1}{n!}\int_G 1_Y\left(\sum_{j=1}^n \delta_{x_j}\right)\mu^n\left(d[x_1, \ldots, x_n]\right), \quad Y \in \mathfrak{M},$$

where 1_Y denotes the indicator function of a set Y, and O represents the empty configuration, i.e., $O(G) = 0$.

Since μ was assumed to be diffuse, one easily checks that F is concentrated on the set of a simple configurations (i.e., without multiple points)

$$\hat{M} \equiv \{\varphi \in M; \varphi(\{x\}) \leq 1 \text{ for all } x \in G\}.$$

Definition 18.5 $\mathcal{M} = \mathcal{M}(G) \equiv L^2(M, \mathfrak{M}, F)$ is called the (symmetric) Fock space over G.

It was proved by Fichtner and Freudenberg [242, 243] that $\mathcal{M}$ and the Boson Fock space $\Gamma_+(L^2(G))$ in the usual definition are isomorphic.

Basic Facts in Symmetric Fock Space

For each vector Φ in the symmetric Fock space $\mathcal{M}$ with $\Phi \neq 0$, we denote by $|\Phi\rangle$ the corresponding normalized vector

$$|\Phi\rangle \equiv \frac{\Phi}{\|\Phi\|}.$$

Further, $|\Phi\rangle\langle\Phi|$ denotes the corresponding one-dimensional projection describing a pure state given by the normalized vector $|\Phi\rangle$. Now, for each $n \geq 1$ let $\mathcal{M}^{\otimes n}$ be the n-fold tensor product of the Hilbert space $\mathcal{M}$, which can be identified with $L^2(M^n, F^n)$.

Definition 18.6 For a given function $g : G \to \mathbb{C}$ the function $\exp(g) : M \to \mathbb{C}$ defined by

$$\exp(g)(\varphi) \equiv \begin{cases} 1, & \text{if } \varphi = 0, \\ \prod_{x \in G, \varphi(\{x\})>0} g(x), & \text{otherwise} \end{cases}$$

is called the exponential vector generated by g.

Those not familiar with the above expression of a coherent vector have only to understand the following definition.

Definition 18.7 For any $g \in \mathcal{L}_1$ (one-particle state vector), the coherent vector with g is defined by

$$\exp(g) \equiv \bigoplus_{n=0}^{\infty} \frac{1}{\sqrt{n!}} g^{\otimes^n} \in \Gamma_+(\mathcal{L}_1).$$

Observe that $\exp(g) \in \mathcal{M}$ if and only if $g \in L^2(G)$, and one has in this case $\|\exp(g)\|^2 = e^{\|g\|^2}$, where $\|\cdot\|$ is the norm deduced from the inner product $(\cdot, \cdot)$ of $L^2(G)$, so that the normalized vector is $|\exp(g)\rangle = \frac{\exp(g)}{\|\exp(g)\|} = e^{-\frac{1}{2}\|g\|^2} \exp(g)$. The projection $|\exp(g)\rangle\langle\exp(g)|$ is called the coherent state corresponding to $g \in L^2(G)$. In the special case $g \equiv 0$, we get the vacuum state

$$\big|\exp(0)\big\rangle = 1_{\{0\}} = \Phi_0.$$

The linear span of the exponential vectors of $\mathcal{M}$ is dense in $\mathcal{M}$, so that bounded operators and certain unbounded operators can be characterized by their actions on exponential vectors.

Definition 18.8 The operator $D : \mathrm{dom}(D) \to \mathcal{M}^{\otimes 2}$ on a dense domain $\mathrm{dom}(D) \subset \mathcal{M}$ containing the exponential vectors from $\mathcal{M}$ given by

$$(D\psi)(\varphi_1, \varphi_2) \equiv \psi(\varphi_1 + \varphi_2) \quad \big(\psi \in \mathrm{dom}(D),\ \varphi_1, \varphi_2 \in M\big)$$

is called the compound Hida–Malliavin derivative.

On exponential vectors $\exp(g)$ with $g \in L^2(G)$, one gets immediately

$$D\exp(g) = \exp(g) \otimes \exp(g).$$

Definition 18.9 The operator $S : \mathrm{dom}(S) \to \mathcal{M}$ on a dense domain $\mathrm{dom}(S) \subset \mathcal{M}^{\otimes 2}$ containing tensor products of exponential vectors given by

$$S\Phi(\varphi) \equiv \sum_{\tilde{\varphi} \le \varphi} \Phi(\tilde{\varphi}, \varphi - \tilde{\varphi}) \quad \big(\Phi \in \mathrm{dom}(S), \varphi \in M\big)$$

is called the compound Skorohod integral.

After some calculations, one gets

$$\langle D\psi, \Phi\rangle_{\mathcal{M}^{\otimes 2}} = \langle \psi, S\Phi\rangle_{\mathcal{M}} \quad \big(\psi \in \mathrm{dom}(D), \Phi \in \mathrm{dom}(S)\big),$$
$$S\big(\exp(g) \otimes \exp(h)\big) = \exp(g + h) \quad \big(g, h \in L^2(G)\big).$$

Definition 18.10 Let T be a linear operator on $L^2(G)$ with $\|T\| \le 1$. Then the operator $\Gamma(T)$ called the second quantization of T is the (uniquely determined) bounded operator on $\mathcal{M}$ fulfilling

$$\Gamma(T)\exp(g) = \exp(Tg) \quad \big(g \in L^2(G)\big).$$

Clearly,

$$\Gamma(T_1)\Gamma(T_2) = \Gamma(T_1T_2),$$
$$\Gamma(T^*) = \Gamma(T)^*.$$

It follows that $\Gamma(T)$ is a unitary operator on $\mathcal{M}$ if T is a unitary operator on $L^2(G)$. This second quantization is expressed as a unitary operator

$$\Gamma(T) \equiv \bigoplus_n T^{\otimes^n}.$$

The following lemma is useful.

Lemma 18.11 *Let K_1, K_2 be linear operators on $L^2(G)$ with the property*

$$K_1^* K_1 + K_2^* K_2 = 1. \tag{18.2}$$

Then there exists exactly one isometry ν_{K_1,K_2} from $\mathcal{M}$ to $\mathcal{M}^{\otimes 2} \equiv \mathcal{M} \otimes \mathcal{M}$ with $\nu_{K_1,K_2} \exp(g) = \exp(K_1 g) \otimes \exp(K_2 g)$ $(g \in L^2(G))$. Thus defined isometry ν_{K_1,K_2} is called the generalized beamsplitting. Furthermore, $\nu_{K_1,K_2} = (\Gamma(K_1) \otimes \Gamma(K_2))D$.

Proof First, we define an operator B on $\mathcal{M}$ as

$$B \equiv S\big(\Gamma(K_1^*) \otimes \Gamma(K_2^*)\big)\big(\Gamma(K_1) \otimes \Gamma(K_2)\big)D.$$

This operator is well defined and its domain is dense. For any $g \in L^2(G)$ and the vector $\exp(g)$, we calculate

$$\begin{aligned}
&B \exp(g) \\
&\quad = S\big(\Gamma(K_1^*) \otimes \Gamma(K_2^*)\big)\big(\Gamma(K_1) \otimes \Gamma(K_2)\big)D \exp(g) \\
&\quad = S\big(\Gamma(K_1^*) \otimes \Gamma(K_2^*)\big)\big(\exp(K_1 g) \otimes \exp(K_2 g)\big) \\
&\quad = S\big(\exp(K_1^* K_1 g) \otimes \exp(K_2^* K_2 g)\big) \\
&\quad = \exp\big((K_1^* K_1 + K_2^* K_2)g\big) = \exp(g).
\end{aligned}$$

Therefore, we can see

$$B = I.$$

Here, one has an operator ν_{K_1,K_2} from $\mathcal{M}$ to $\mathcal{M}^{\otimes 2}$ defined as

$$\nu_{K_1,K_2} \equiv \Gamma(K_1) \otimes \Gamma(K_2)D.$$

Then, we obtain

$$\begin{aligned}
\|\nu_{K_1,K_2}\psi\|^2 &= \langle \nu_{K_1,K_2}\psi, \nu_{K_1,K_2}\psi \rangle \\
&= \big\langle \big(\Gamma(K_1) \otimes \Gamma(K_2)\big)D\psi, \big(\Gamma(K_1) \otimes \Gamma(K_2)\big)D\psi \big\rangle.
\end{aligned}$$

By Definitions 18.8 and 18.9, it becomes

$$\begin{aligned}&\big\langle\big(\Gamma(K_1)\otimes\Gamma(K_2)\big)D\psi,\big(\Gamma(K_1)\otimes\Gamma(K_2)\big)D\psi\big\rangle\\&\quad=\big\langle\psi,S\big(\Gamma(K_1^*)\otimes\Gamma(K_2^*)\big)(\Gamma(K_1)\otimes\Gamma(K_2))D\psi\big\rangle\\&\quad=\langle\psi,B\psi\rangle=\langle\psi,\psi\rangle=\|\psi\|^2,\end{aligned}$$

so that ν_{K_1,K_2} is an isometry. Note that ν_{K_1,K_2} can be expanded into a bounded operator on $\mathcal{M}$ such that

$$\|\nu_{K_1,K_2}\psi\|=\|\psi\|.$$

Moreover, it holds

$$\begin{aligned}\nu_{K_1,K_2}\exp(g)&=\big(\Gamma(K_1)\otimes\Gamma(K_2)\big)D\exp(g)\\&=\big(\Gamma(K_1)\otimes\Gamma(K_2)\big)\big(\exp(g)\otimes\exp(g)\big)\\&=\Gamma(K_1)\exp(g)\otimes\Gamma(K_2)\exp(g)\\&=\exp(K_1g)\otimes\exp(K_2g).\end{aligned}$$

□

Remark 18.12 υ_{K_1,K_2} is one of the liftings (see Chap. 7).

Above we defined the *generalized* beam splitting. Here we explain that the ordinary scheme of beam splitting is an example of the generalized splitting. Let $K_1\equiv\alpha I$ and $K_2\equiv\beta I$ with $|\alpha|^2+|\beta|^2=1$. Then we obtain

$$\nu_{K_1,K_2}\exp(g)=\exp(\alpha g)\otimes\exp(\beta g),$$

which is the well-known beam splitting often used in optical communication and quantum measurements.

Example 18.13 ($\alpha=\beta=1/\sqrt{2}$ above) Let $K_1=K_2$ be the following operator of multiplication on $L^2(G)$

$$K_1g=\frac{1}{\sqrt{2}}g=K_2g\quad\big(g\in L^2(G)\big).$$

We put

$$\nu\equiv\nu_{K_1,K_2},$$

and obtain

$$\nu\exp(g)=\exp\bigg(\frac{1}{\sqrt{2}}g\bigg)\otimes\exp\bigg(\frac{1}{\sqrt{2}}g\bigg)\quad\big(g\in L^2(G)\big).$$

Example 18.14 Let $L^2(G)=\mathcal{H}_1\oplus\mathcal{H}_2$ be the orthogonal sum of the subspaces $\mathcal{H}_1,\mathcal{H}_2$. K_1 and K_2 denote the corresponding projections.

We use the above example in order to describe a teleportation model where Bob performs his experiments on the same ensemble of the systems as Alice.

Remark 18.15 The property (18.2) implies

$$\|K_1 g\|^2 + \|K_2 g\|^2 = \|g\|^2 \quad \big(g \in L^2(G)\big).$$

Remark 18.16 Let U, V be unitary operators on $L^2(G)$. If operators K_1, K_2 satisfy the equality of (18.2), then the pair $\hat{K}_1 = UK_1,\ \hat{K}_2 = VK_2$ fulfills the same equalities.

18.4.2 A Perfect Teleportation

The state of Alice asked to teleport is of the type

$$\rho = \sum_{s=1}^{N} \lambda_s |\Phi_s\rangle\langle\Phi_s|, \tag{18.3}$$

where $|\Phi_s\rangle$ is constructed as

$$|\Phi_s\rangle = \sum_{j=1}^{N} c_{sj} \big|\exp(aK_1 g_j) - \exp(0)\big\rangle, \quad \sum_j |c_{sj}|^2 = 1; s = 1, \ldots, N, \tag{18.4}$$

with an ONS $\{g_j\}_{j=1,2,\ldots,N}$ (i.e., $\langle g_i, g_j\rangle = \delta_{ij}$ holds) of the one-particle space $\mathcal{L}_1$ and $a = \sqrt{d}$. One easily checks that $(|\exp(aK_1 g_j) - \exp(0)\rangle)_{j=1}^N$ and $(|\exp a \times K_2 g_j) - \exp(0)\rangle)_{j=1}^N$ are ONS in $\mathcal{M}$. The set $\{\Phi_s; s = 1, \ldots, N\}$ makes the N-dimensional Hilbert space $\mathcal{H}_1$ defining an input state teleported by Alice. Although we may include the vacuum state $|\exp(0)\rangle$ to define $\mathcal{H}_1$, here we take the N-dimensional Hilbert space $\mathcal{H}_1$ as above because of computational simplicity.

In order to achieve that $(|\Phi_s\rangle)_{s=1}^N$ is still an ONS in $\mathcal{M}$, we assume

$$\sum_{j=1}^{N} \bar{c}_{sj} c_{kj} = 0 \quad (j \neq k;\ j, k = 1, \ldots, N). \tag{18.5}$$

Denote $c_s = [c_{s1}, \ldots, c_{sN}] \in \mathbb{C}^N$, then $(c_s)_{s=1}^N$ is a CONS in $\mathbb{C}^N$.

Let $(b_n)_{n=1}^N$ be a sequence in $\mathbb{C}^N$,

$$b_n = [b_{n1}, \ldots, b_{nN}]$$

with properties

$$|b_{nk}| = 1 \quad (n, k = 1, \ldots, N), \tag{18.6}$$

$$\langle b_n, b_j \rangle = 0 \quad (n \neq j; n, j = 1, \dots, N). \tag{18.7}$$

Now, for each m, n $(= 1, \dots, N)$, we have unitary operators U_m, B_n on $\mathcal{M}$ given by

$$\begin{aligned} B_n \big|\exp(aK_1 g_j) - \exp(0)\big\rangle &= b_{nj} \big|\exp(aK_1 g_j) - \exp(0)\big\rangle \quad (j = 1, \dots, N), \\ U_m \big|\exp(aK_1 g_j) - \exp(0)\big\rangle &= \big|\exp(aK_1 g_{j\oplus m}) - \exp(0)\big\rangle \quad (j = 1, \dots, N), \end{aligned} \tag{18.8}$$

where $j \oplus m \equiv j + m \pmod N$.

Then Alice's measurements are performed using the projection

$$F_{nm} = |\xi_{nm}\rangle\langle\xi_{nm}| \quad (n, m = 1, \dots, N)$$

given by

$$|\xi_{nm}\rangle = \frac{1}{\sqrt{N}} \sum_{j=1}^{N} b_{nj} \big|\exp(aK_1 g_j) - \exp(0)\big\rangle \otimes \big|\exp(aK_1 g_{j\oplus m}) - \exp(0)\big\rangle.$$

One easily checks that $(|\xi_{nm}\rangle)_{n,m=1}^{N}$ is an ONS in $\mathcal{M}^{\otimes 2}$. Further, the state vector $|\xi\rangle$ of the entangled state $\sigma = |\xi\rangle\langle\xi|$ is given by

$$|\xi\rangle = \frac{1}{\sqrt{N}} \sum_{k} \big|\exp(aK_1 g_k) - \exp(0)\big\rangle \otimes \big|\exp(aK_2 g_k) - \exp(0)\big\rangle.$$

By using the above facts, it can be easily seen that the model is unitary equivalent with the original perfect teleportation model proposed by Bennett et al. Thus the following theorem is proved.

Theorem 18.17 *For each $n, m = 1, \dots, N$, define a channel Λ_{nm}^* by*

$$\Lambda_{nm}^*(\rho) \equiv \operatorname{tr}_{12} \frac{(F_{nm} \otimes 1)(\rho \otimes \sigma)(F_{nm} \otimes \mathbf{1})}{\operatorname{tr}_{123}(F_{nm} \otimes 1)(\rho \otimes \sigma)(F_{nm} \otimes \mathbf{1})} \quad (\rho \text{ normal state on } \mathcal{M}). \tag{18.9}$$

Then we have for all states ρ on M

$$\Lambda_{nm}^*(\rho) = \big(\Gamma(T) U_m B_n^*\big) \rho \big(\Gamma(T) U_m B_n^*\big)^*.$$

If Alice performs a measurement according to the following self-adjoint operator

$$F = \sum_{n,m=1}^{N} z_{nm} F_{nm}$$

with $\{z_{nm} | n, m = 1, \dots, N\} \subseteq \mathbb{R} - \{0\}$, then she will obtain the value z_{nm} with probability $1/N^2$. The sum over all this probabilities is 1, so that the teleportation model works perfectly.

Before stating some fundamental results for the non-perfect case, we note that our perfect teleportation is obviously treated in general finite-dimensional Hilbert spaces $\mathcal{H}_k$ $(k = 1, 2, 3)$ in the same way as the usual one. Moreover, our teleportation scheme can be a bit generalized by introducing the entangled state σ_{12} on $\mathcal{H}_1 \otimes \mathcal{H}_2$ defining the projections $\{F_{nm}\}$ by the unitary operators B_n, U_m. We here discuss the perfect teleportation on general Hilbert spaces $\mathcal{H}_k$ $(k = 1, 2, 3)$. Let $\{\xi_j^k; j = 1, \dots, N\}$ be a CONS of the Hilbert space $\mathcal{H}_k$ $(k = 1, 2, 3)$. Define the entangled states σ_{12} and σ_{23} on $\mathcal{H}_1 \otimes \mathcal{H}_2$ and $\mathcal{H}_2 \otimes \mathcal{H}_3$, respectively, by

$$\sigma_{12} = |\xi_{12}\rangle\langle\xi_{12}|, \qquad \sigma_{23} = |\xi_{23}\rangle\langle\xi_{23}|$$

with $\xi_{12} \equiv \frac{1}{\sqrt{N}} \sum_{j=1}^{N} \xi_j^1 \otimes \xi_j^2$ and $\xi_{23} \equiv \frac{1}{\sqrt{N}} \sum_{j=1}^{N} \xi_j^2 \otimes \xi_j^3$. Using a sequence $\{b_n = [b_{n1}, \dots, b_{nN}]; n = 1, \dots, N\}$ in $\mathbb{C}^N$ with the properties (18.6) and (18.7), we define the unitary operators B_n and U_m such as

$$B_n \xi_j^1 \equiv b_{nj} \xi_j^1 \quad (n, j = 1, \dots, N) \quad \text{and} \quad U_m \xi_j^2 \equiv \xi_{j \oplus m}^2 \quad (n, j = 1, \dots, N)$$

with $j \oplus m \equiv j + m \pmod N$. Then the set $\{F_{nm}; n, m = 1, \dots, N\}$ of the projections of Alice is given by

$$F_{nm} = (B_n \otimes U_m)\sigma_{12}(B_n \otimes U_m)^*,$$

and the teleportation channels $\{\Lambda^*; n, m = 1, \dots, N\}$ are defined as

$$\Lambda_{nm}^*(\rho) \equiv \operatorname{tr}_{12} \frac{(F_{nm} \otimes 1)(\rho \otimes \sigma_{23})(F_{nm} \otimes \mathbf{1})}{\operatorname{tr}_{123}(F_{nm} \otimes 1)(\rho \otimes \sigma_{23})(F_{nm} \otimes \mathbf{1})}.$$

Finally, the unitary keys $\{W_{nm}; n, m = 1, \dots, N\}$ of Bob are given as

$$W_{nm} \xi_j^1 = \overline{b}_{nj} \xi_{j \oplus m}^3 \quad (n, m = 1, \dots, N),$$

by which we obtain the perfect teleportation

$$\Lambda_{nm}^*(\rho) = W_{nm} \rho W_{nm}^*.$$

The above perfect teleportation is unique in the sense of unitary equivalence.

18.5 Non-perfect Teleportation in Bose Fock Space

In this section, we consider a teleportation model where the entangled state σ is given by the splitting of a superposition of certain coherent states instead of subtracting the vacuum. Unfortunately, this model does not work perfectly, that is, neither condition (E2) nor condition (E1) holds. However, this model is more realistic than that in the previous section, and we show that this model provides a nice approximation of the perfect case. To estimate the difference between the perfect teleportation and non-perfect teleportation, we add a further step in the teleportation scheme:

Step 5. Bob will perform a measurement on his part of the system according to the projection

$$F_+ \equiv I - \big|\exp(0)\big\rangle\big\langle\exp(0)\big|,$$

where $|\exp(0)\rangle\langle\exp(0)|$ denotes the vacuum state (the coherent state with density 0).

Then our new teleportation channels (we denote them again by Λ^*) have the form

$$\Lambda^*(\rho) \equiv \mathrm{tr}_{12} \frac{(F_{nm} \otimes F_+)\rho \otimes \sigma(F_{nm} \otimes F_+)}{\mathrm{tr}_{123}(F_{nm} \otimes F_+)\rho \otimes \sigma(F_{nm} \otimes F_+)},$$

and the corresponding probabilities are

$$p_{nm}(\rho) \equiv \mathrm{tr}_{123}(F_{nm} \otimes F_+)\,\rho \otimes \sigma(F_{nm} \otimes F_+).$$

For this teleportation scheme, condition (E1) is fulfilled. Furthermore, we get

$$\sum_{n,m} p_{nm}(\rho) = \frac{(1 - e^{-\frac{d}{2}})^2}{1 + (N-1)e^{-d}} \quad \big(\to 1\ (d \to +\infty)\big),$$

in which the probability of teleporting the state from Alice to Bob is less than 1, and it depends on the density parameter d (may be the energy of the beams) of the coherent vector. Here N denotes the dimension of the Hilbert space and d is the expectation value of the total number of particles (or energy) of the beam, so that in the case of high density (or energy) "$d \to +\infty$" of the beam the model works perfectly. Thus the model can be considered as asymptotically perfect.

We discuss the above facts and explain that such a teleportation scheme can be understood as "quantum teleportation with a test".

Take the normalized vector which is a superposition of coherent states,

$$|\eta\rangle \equiv \frac{\gamma}{\sqrt{N}} \sum_{k=1}^{N} \big|\exp(ag_k)\big\rangle$$

$$\text{with } \gamma \equiv \left(\frac{1}{1 + (N-1)e^{-d}}\right)^{\frac{1}{2}} = \left(\frac{1}{1 + (N-1)e^{-a^2}}\right)^{\frac{1}{2}},$$

and we employ it as the input of the beam splitter to obtain the entangled state

$$\tilde{\xi} \equiv \nu_{K_1,K_2}(\eta) = \frac{\gamma}{\sqrt{N}} \sum_{k=1}^{N} \big|\exp(aK_1 g_k)\big\rangle \otimes \big|\exp(aK_2 g_k)\big\rangle.$$

We hence replace in the entangled state σ of the perfect teleportation by

$$\tilde{\sigma} \equiv |\tilde{\xi}\rangle\langle\tilde{\xi}|.$$

Then for each $n, m = 1, \ldots, N$, we get the channels on any normal state ρ on $\mathcal{M}$ by setting

$$\Lambda_{nm}^*(\rho) \equiv \mathrm{tr}_{12} \frac{(F_{nm} \otimes \mathbf{1})(\rho \otimes \tilde{\sigma})(F_{nm} \otimes \mathbf{1})}{\mathrm{tr}_{123}(F_{nm} \otimes \mathbf{1})(\rho \otimes \tilde{\sigma})(F_{nm} \otimes \mathbf{1})},$$
$$\Theta_{nm}^*(\rho) \equiv \mathrm{tr}_{12} \frac{(F_{nm} \otimes F_+)(\rho \otimes \tilde{\sigma})(F_{nm} \otimes F_+)}{\mathrm{tr}_{123}(F_{nm} \otimes F_+)(\rho \otimes \tilde{\sigma})(F_{nm} \otimes F_+)}, \tag{18.10}$$

where $F_+ = I - |\exp(0)\rangle\langle\exp(0)|$, i.e., F_+ is the projection onto the space $\mathcal{M}_+$ of configurations having no vacuum part,

$$\mathcal{M}_+ \equiv \big\{\psi \in \mathcal{M} \big| \big\| |\exp(0)\rangle\langle\exp(0)|\psi \big\| = 0\big\}.$$

One easily checks that

$$\Theta_{nm}^*(\rho) = \frac{F_+ \Lambda_{nm}^*(\rho) F_+}{\mathrm{tr}(F_+ \Lambda_{nm}^*(\rho) F_+)},$$

that is, after receiving the state $\Lambda_{nm}^*(\rho)$ from Alice, Bob has to omit the vacuum.

From Theorem 18.17, it follows that for all ρ with (18.3) and (18.4)

$$\Lambda_{nm}^*(\rho) = \frac{F_+ \Lambda_{nm}^*(\rho) F_+}{\mathrm{tr}\,(F_+ \Lambda_{nm}^*(\rho) F_+)}.$$

This is not true if we replace Λ_{nm}^* by Λ_{nm}^*, namely, in general it does not hold that

$$\Theta_{nm}^*(\rho) = \Lambda_{nm}^*(\rho).$$

The following theorem is true.

Theorem 18.18 *For all states ρ on $\mathcal{M}$ with* (18.3) *and* (18.4) *and each pair n, m $(= 1, \ldots, N)$, we have*

$$\Theta_{nm}^*(\rho) = \big(\Gamma(T) U_m B_n^*\big) \rho \big(\Gamma(T) U_m B_n^*\big)^* \quad \text{or} \quad \Theta_{nm}^*(\rho) = \Lambda_{nm}^*(\rho) \tag{18.11}$$

and

$$\sum_{n,m} p_{nm}(\rho) = \sum_{n,m} \mathrm{tr}_{123}(F_{nm} \otimes F_+)(\rho \otimes \tilde{\sigma})(F_{nm} \otimes F_+) = \frac{(1 - e^{-\frac{d}{2}})^2}{1 + (N-1)e^{-d}}. \tag{18.12}$$

That is, the model works only asymptotically perfectly in the sense of condition (E2). In other words, the model works perfectly for the case of high density (or energy) of the considered beams. In order to prove Theorem 18.18, we fix ρ with (18.3) and (18.4) and start with a lemma.

Lemma 18.19 *For each* $n, m, s\ (=1,\dots,N)$,

$$(F_{nm}\otimes I)\big(|\Phi_s\rangle\otimes|\tilde{\xi}\rangle\big)=\frac{\gamma}{N}\big(1-e^{-\frac{d}{2}}\big)|\xi_{nm}\rangle\otimes\big(\Gamma(T)U_m B_n^*|\Phi_s\rangle\big)$$
$$+\frac{\gamma}{N}\left(\frac{e^{\frac{d}{2}}-1}{e^d}\right)^{\frac{1}{2}}\langle b_n, c_s\rangle_{\mathbb{C}^N}\xi_{nm}\otimes\big|\exp(0)\big\rangle.$$

Proof For all $k, j, r=1,\dots,N$, we get

$$\begin{aligned}\alpha_{k,j,r}&\equiv\big\langle\big|\exp(aK_1g_r)-\exp(0)\big\rangle\otimes\big|\exp\left(aK_1g_{r\oplus m}\right)-\exp(0)\big\rangle,\\&\qquad\big|\exp(aK_1g_j)-\exp(0)\big\rangle\otimes\big|\exp(aK_1g_k)\big\rangle\big\rangle\\&=\begin{cases}(\frac{e^{\frac{a^2}{2}}-1}{e^{\frac{a^2}{2}}}), & \text{if } r=j \text{ and } k=r\oplus m,\\ 0, & \text{otherwise}\end{cases}\end{aligned}$$

and

$$\big|\exp(aK_2g_{j\oplus m})\big\rangle=e^{-\frac{a^2}{2}}\big(e^{\frac{a^2}{2}}-1\big)^{\frac{1}{2}}\big|\exp(aK_2g_{j\oplus m})-\exp(0)\big\rangle+e^{-\frac{a^2}{2}}\big|\exp(0)\big\rangle.$$

On the other hand, we have

$$(F_{nm}\otimes I)\big(|\Phi_s\rangle\otimes|\tilde{\xi}\rangle\big)=\frac{\gamma}{N}\sum_k\sum_j\sum_r c_{sj}\bar{b}_{nr}\alpha_{k,j,r}\xi_{nm}\otimes\big|\exp(aK_2g_k)\big\rangle.$$

It follows with $a^2=d$

$$\begin{aligned}&(F_{nm}\otimes I)(\Phi_s\otimes\tilde{\xi})\\&\quad=\frac{\gamma}{N}\big(e^{\frac{d}{2}}-1\big)e^{-\frac{d}{2}}\xi_{nm}\otimes\Big(\sum_j c_{sj}\bar{b}_{nj}\big|\exp(aK_2g_{j\oplus m})-\exp(0)\big\rangle\Big)\\&\qquad+\frac{\gamma}{N}\big(e^{\frac{d}{2}}-1\big)^{\frac{1}{2}}e^{-\frac{d}{2}}\sum_j c_{sj}\bar{b}_{nj}\xi_{nm}\otimes\big|\exp(0)\big\rangle\\&\quad=\frac{\gamma}{N}\big(1-e^{-\frac{d}{2}}\big)\xi_{nm}\otimes\big(\Gamma(T)U_mB_n^*\Phi_s\big)\\&\qquad+\frac{\gamma}{N}\left(\frac{e^{\frac{d}{2}}-1}{e^d}\right)^{\frac{1}{2}}\langle b_n,c_s\rangle_{\mathbb{C}^N}\xi_{nm}\otimes\big|\exp(0)\big\rangle.\end{aligned}$$

□

If ρ is a pure state,

$$\rho=|\Phi_s\rangle\langle\Phi_s|,$$

then we obtain from Lemma 18.19

$$\begin{aligned}
&\mathrm{tr}_{123}(F_{nm}\otimes I)(\rho\otimes\tilde{\sigma})(F_{nm}\otimes I)\\
&=\frac{\gamma^2}{N^2}\left(\left(1-e^{-\frac{d}{2}}\right)^2+\frac{e^{\frac{d}{2}}-1}{e^d}\left|\langle b_n,c_s\rangle\right|^2\right)\\
&=\frac{1}{N^2(1+(N-1)e^{-d})}\left(\left(1-e^{-\frac{d}{2}}\right)^2+\frac{e^{\frac{d}{2}}-1}{e^d}\left|\langle b_n,c_s\rangle\right|^2\right)
\end{aligned}$$

and

$$\Lambda^*_{nm}(\rho)\neq\left(\Gamma(T)U_mB^*_n\right)\rho\left(\Gamma(T)U_mB^*_n\right)^*.$$

Now we have

$$\Gamma(T)U_mB^*_n\Phi_s\in\mathcal{M}_+,\qquad \left|\exp(0)\right\rangle\in\mathcal{M}_+^{\perp}.$$

Hence, Lemma 18.19 implies

$$(I\otimes I\otimes F_+)(F_{nm}\otimes I)(\Phi_s\otimes\tilde{\xi})=\frac{\gamma}{N}\left(1-e^{-\frac{d}{2}}\right)\xi_{nm}\otimes\left(\Gamma(T)U_mB^*_n\Phi_s\right),$$

that is, we have the following lemma

Lemma 18.20 *For each $n,m,s=1,\dots,N$,*

$$(F_{nm}\otimes F_+)(\Phi_s\otimes\tilde{\xi})=\frac{\gamma}{N}\left(1-e^{-\frac{d}{2}}\right)\xi_{nm}\otimes\left(\Gamma(T)U_mB^*_n\Phi_s\right).\tag{18.13}$$

Remark 18.21 Let K_2 be a projection of the type

$$K_2h=h1_X;\quad h\in L^2(G),$$

where $X\subseteq G$ is measurable. Then (18.13) also holds if we replace F_+ by the projection $F_{+,X}$ onto the subspace $\mathcal{M}_{+,X}$ of $\mathcal{M}$ given by

$$\mathcal{M}_{+,X}\equiv\left\{\psi\in\mathcal{M}\mid\psi(\varphi)=0\text{ if }\varphi(X)=0\right\}.$$

Observe that $\mathcal{M}_{+,G}=\mathcal{M}_+$.

Proof of Theorem 18.18 We have assumed that $(|\Phi_s\rangle)_{s=1}^N$ is an ONS in $\mathcal{M}$, which implies that $(|\xi_{nm}\rangle\otimes(\Gamma(T)U_mB^*_n|\Phi_s\rangle))_{s=1}^N$ is an ONS in $\mathcal{M}^{\otimes3}$. Hence we obtain (18.11) and (18.12) by Lemma 18.20. This proves Theorem 18.18. □

We can further generalize the above teleportation schemes, namely, replacing the projectors F_{nm} by projectors $\tilde{F}_{nm}$ defined as follows:

$$\tilde{F}_{nm}\equiv\left(B_n\otimes U_m\Gamma(T)^*\right)\tilde{\sigma}\left(B_n\otimes U_m\Gamma(T)^*\right)^*.$$

In order to make this definition precise, we assume, in addition to (18.8), that

$$U_m\big|\exp(0)\big\rangle = \big|\exp(0)\big\rangle \quad (m = 1, \ldots, N).$$

Together with (18.11) this implies

$$U_m\big|\exp(aK_1 g_j)\big\rangle = \big|\exp(aK_1 g_{j\oplus m})\big\rangle \quad (m, j = 1, \ldots, N).$$

Formally, we have the same relation between $\tilde{\sigma}$ and $\tilde{F}_{nm}$ like the relation between σ and F_{nm} (cf. Remark 18.21). Further for each pair $n, m = 1, \ldots, N$ we define channels on normal states on $\mathcal{M}$ by

$$\tilde{\Theta}^*_{nm}(\rho) \equiv \mathrm{tr}_{12} \frac{(\tilde{F}_{nm} \otimes F_+)(\rho \otimes \tilde{\sigma})(\tilde{F}_{nm} \otimes F_+)}{\tilde{p}_{nm}(\rho)},$$

where

$$\tilde{p}_{nm}(\rho) \equiv \mathrm{tr}_{123}(\tilde{F}_{nm} \otimes F_+)(\rho \otimes \tilde{\sigma})(\tilde{F}_{nm} \otimes F_+).$$

The following theorem can be proved.

Theorem 18.22 *For each state ρ on $\mathcal{M}$ satisfying* (18.3), *and* (18.4), *each pair n, m $(= 1, \ldots, N)$, and each bounded operator A on $\mathcal{M}$,*

$$\big|\mathrm{tr}\,\big(\tilde{\Theta}^*_{nm}(\rho)A\big) - \mathrm{tr}\,\big(\Lambda^*_{nm}(\rho)A\big)\big| \le \frac{2e^{-\frac{d}{2}}}{(1 - e^{-\frac{d}{2}})}\big(N^2 + N\sqrt{N} + N\big)\|A\|,$$

$$\left|\tilde{p}_{nm}(\rho) - \frac{1}{N^2}\right| \le e^{-\frac{d}{2}}\left(\frac{14}{N^2} + 2 + \frac{2}{\sqrt{N}}\right).$$

The theorem is proved in [248]. (We omit the proof.)

From the earlier theorem and the fact that $e^{-\frac{d}{2}} \to 0$ $(d \to +\infty)$, the above theorem means that our modified teleportation model works asymptotically perfectly (the case of high density or energy) in the sense of conditions (E1) and (E2).

The non-perfect teleportation scheme can be understood as the perfect teleportation with a "test" of Alice and Bob in the following sense: When Alice performs a measurement of the observable F, there is a possibility to obtain an outcome 0, that is, none of $\{z_{nm}\}$ is obtained. In such a case, Alice quits the experiment and tries again from the first procedure. And at the final step, Bob performs a measurement with $F_+ = I - |\exp(0)\rangle\langle\exp(0)|$. If his result $=0$, then he asks Alice to try again, and if his result $=1$, then he continues. The state he obtained is

$$\Theta^*_{nm}(\rho) \equiv \frac{\mathrm{tr}_{12}(F_{nm} \otimes F_+)(\rho \otimes \tilde{\sigma})(F_{nm} \otimes F_+)}{\mathrm{tr}_{123}[(F_{nm} \otimes F_+)(\rho \otimes \tilde{\sigma})(F_{nm} \otimes F_+)]},$$

on which he applies the proper key provided to get the state sent by Alice.

18.6 Fidelity in Teleportation

In this section, we investigate the imperfect teleportation model without test but with natural entangled state constructed by the beam splitting. To discuss the (non-) perfectness of channels, we need some proper quantity to measure how close two states are. We explain the *fidelity* of the non-perfect teleportation model. The notion of fidelity is frequently used in the context of quantum information and quantum optics. The fidelity of a state ρ with respect to another state σ is defined in Chap. 15 as

$$F(\rho,\sigma) \equiv \operatorname{tr}\sqrt{\sigma^{1/2}\rho\sigma^{1/2}},$$

which possesses some nice properties (Theorem 13.11, in Chap. 13).

$$\begin{aligned} &0 \le F(\rho,\sigma) \le 1, \\ &F(\rho,\sigma) = 1 \iff \rho = \sigma, \\ &F(\rho,\sigma) = F(\sigma,\rho). \end{aligned}$$

Thus we can say that two states ρ and σ are close when the fidelity between them is close to unity. Moreover, it satisfies a kind of concavity relation:

$$F\biggl(\sum_i p_i\rho_i, \sum_i q_i\sigma_i\biggr) \ge \sum_i \sqrt{p_i q_i}\, F(\rho_i,\sigma_i),$$

where ρ_i and σ_i are states, and p_i and q_i are nonnegative numbers satisfying $\sum_i p_i = \sum_i q_i = 1$. In particular, by putting $p_j = 1$, one obtains

$$F\biggl(\rho, \sum_i q_i\sigma\biggr) \ge \sqrt{q_j}\, F(\rho,\sigma_j)$$

for $j = 1, 2, \ldots$.

Now let us begin with the teleportation model having the entangled state $\tilde{\sigma}$. Since in this section Bob is not allowed to subtract the vacuum, put

$$\varXi_{nm}^{*}(\rho) \equiv \frac{\operatorname{tr}_{12}[(F_{nm}\otimes W_{nm})(\rho\otimes\tilde{\sigma})(F_{nm}\otimes W_{nm}^{*})]}{\operatorname{tr}_{123}[(F_{nm}\otimes 1)(\rho\otimes\tilde{\sigma})(F_{nm}\otimes 1)]},$$

which takes place with probability $p_{nm} \equiv \operatorname{tr}_{123}[(F_{nm}\otimes 1)(\rho\otimes\tilde{\sigma})(F_{nm}\otimes 1)]$.

In this section, we do not perform any tests, and therefore even if the outcome of Alice's measurement is 0, the procedure is not stopped. We put the key of Bob corresponding to the outcome 0 as W_0. Then the nonlinear channel with the result 0 is

$$\varXi_{0}^{*}(\rho) = \frac{\operatorname{tr}_{12}[(F_0\otimes W_0)(\rho\otimes\tilde{\sigma})(F_0\otimes W_0^{*})]}{\operatorname{tr}_{123}[(F_0\otimes 1)(\rho\otimes\tilde{\sigma})(F_0\otimes 1)]},$$

which takes place with probability

$$p_0 \equiv \mathrm{tr}_{123}\big[(F_0 \otimes 1)(\rho \otimes \tilde{\sigma})(F_0 \otimes 1)\big].$$

Without knowing the result, the expected state is due to a linear channel (=unital completely positive map)

$$\begin{aligned}
\Lambda^*(\rho) &\equiv \sum_{n,m} p_{nm} \Xi^*_{nm}(\rho) + p_0 \Xi^*_0(\rho) \\
&= \sum_{n,m} \mathrm{tr}_{12}\big[(F_{nm} \otimes W_{nm})(\rho \otimes \tilde{\sigma})(F_{nm} \otimes W^*_{nm})\big] \\
&\quad + \mathrm{tr}_{12}\big[(F_0 \otimes W_0)(\rho \otimes \tilde{\sigma})(F_0 \otimes W^*_0)\big].
\end{aligned}$$

Note that $\{F_{nm} \otimes W_{nm}\}_{nm}$ and $F_0 \otimes W_0$ form a partition of unity.

To estimate $F(\rho, \Xi^*(\rho))$, one needs to begin with a calculation of $\Xi^*(\rho) = \sum_{nm} \Xi^*_{nm}(\rho) + \Xi^*_0(\rho)$, for which we need a lemma:

Lemma 18.23 *For a general state* $\rho = \sum_s \lambda_s |\Phi_s\rangle\langle\Phi_s|$,

$$\begin{aligned}
\Xi^*_{nm}(\rho) &= \frac{\gamma^2}{N^2}\big(1 - e^{-d/2}\big)^2 \rho + \frac{\gamma^2}{N^2}\left(\frac{e^{d/2} - 1}{e^d}\right) \sum_s \lambda_s \big|\langle b_n, c_s\rangle\big|^2 \big|\exp(0)\big\rangle\big\langle\exp(0)\big| \\
&\quad + \frac{\gamma^2}{N^2}\left(\frac{e^{d/2} - 1}{e^d}\right)^{1/2} \big(1 - e^{-d/2}\big) \\
&\quad \times \left(\sum_s \langle b_n, c_s\rangle^* \lambda_s |\Phi_s\rangle\big\langle\exp(0)\big| + \sum_s \langle b_n, c_s\rangle \lambda_s \big|\exp(0)\big\rangle\langle\Phi_s|\right).
\end{aligned}$$

Proof It is a direct consequence of Lemma 18.19. □

Next we will obtain an expression of $\Xi^*_0(\rho)$.

Lemma 18.24 *For a general state* $\rho = \sum_s \lambda_s |\Phi_s\rangle\langle\Phi_s|$,

$$\Xi^*_0(\rho) = e^{-|a|^2/2} \frac{\gamma^2}{N} \sum_{k=1}^{N} \sum_{l=1}^{N} W_0 \big|\exp(aK_2 g_k)\big\rangle\big\langle\exp(aK_2 g_l)\big| W^*_0.$$

Proof Let $\mathcal{L}$ be a subspace spanned by an ONS $\{|\exp(aK_1 g_k) - \exp(0)\rangle\}$ ($k = 1, \dots, N$), and suppose $\sum_{nm} F_{nm}$ is a projection onto $\mathcal{L} \otimes \mathcal{L}$. Then, we obtain

$$(F_0 \otimes I)(\Phi_s \otimes \tilde{\xi}) = \big(I \otimes \big|\exp(0)\big\rangle\big\langle\exp(0)\big| \otimes I\big)(\Phi_s \otimes \tilde{\xi}).$$

Here we used the fact that $|\exp(0)\rangle$ is orthogonal to the set $\{|\exp(aK_1 g_k) - \exp(0)\rangle\}$. Hence we easily see that the following holds

$$(F_0 \otimes I)\big(|\Phi_s\rangle \otimes |\tilde{\xi}\rangle\big) = |\Phi_s\rangle \otimes \big|\exp(0)\big\rangle \otimes e^{-|a|^2/4} \frac{\gamma}{\sqrt{N}} \sum_{k=1}^{N} \big|\exp(aK_2 g_k)\big\rangle.$$

Taking convex combinations of them yields the lemma. □

Hence we obtained the formula for $\Xi^*(\rho)$. Now let us estimate the fidelity between $\Xi^*(\rho)$ and ρ.

Theorem 18.25 *For any input state* ρ *of type* (18.3) *and* (18.4),

$$F\big(\rho, \Xi^*(\rho)\big) \geq \sqrt{\frac{(1-e^{-d/2})^2}{1+(N-1)e^{-d}}}.$$

Proof We must first compute

$$\begin{aligned}
\rho^{1/2}\Xi^*(\rho)\rho^{1/2} &= \sum_{nm} \rho^{1/2}\Xi^*_{nm}(\rho)\rho^{1/2} + \rho^{1/2}\Xi_0\rho^{1/2} \\
&= \gamma^2\big(1-e^{-d/2}\big)^2 \rho^{1/2}\rho\rho^{1/2} + \rho^{1/2}e^{-|a|^2/2} \\
&\quad \times \frac{\gamma^2}{N} \sum_{k=1}^{N}\sum_{l=1}^{N} W_0 \big|\exp(aK_2 g_k)\big\rangle\big\langle \exp(aK_2 g_l)\big| W_0^* \rho^{1/2},
\end{aligned}$$

where we used the relation $\langle \exp(0)|\Phi_s\rangle = 0$. Because $\Xi_0^*(\rho)$ is positive operator, $\Xi_0^*(\rho)/\mathrm{tr}[\Xi_0^*(\rho)]$ becomes a state and we can rearrange the expression of fidelity as

$$F\big(\rho, \Xi^*(\rho)\big) = F\big(\rho, \gamma^2\big(1-e^{-d/2}\big)^2\rho + \mathrm{tr}\big[\Xi_0^*(\rho)\big]\Xi_0^*(\rho)/\mathrm{tr}\big[\Xi_0^*(\rho)\big]\big).$$

Thanks to the concavity property of fidelity, we obtain

$$F\big(\rho, \Xi^*(\rho)\big) \geq \gamma\big(1-e^{-d/2}\big)F(\rho,\rho) = \gamma\big(1-e^{-d/2}\big).$$ □

Therefore, the teleportation protocol approaches the perfect one as the parameter $|a|$ goes to infinity.

With some additional condition, one can strengthen the above estimate to an equality.

Proposition 18.26 *Let* $L^2(G) = \mathcal{H}_1 \oplus \mathcal{H}_2$ *be the orthogonal sum of the subspaces* $\mathcal{H}_1, \mathcal{H}_2$. *Let* K_1 *and* K_2 *denote the corresponding projections, and* $W_0 = I$. *Then*

$$F\big(\rho, \Xi^*(\rho)\big) = \sqrt{\frac{(1-e^{-d/2})^2}{1+(N-1)e^{-d}}}$$

holds for any input state ρ.

Proof Because $\langle \exp(aK_1 g_k) - \exp(0) | \exp(aK_2 g_l) \rangle = 0$ holds, $\rho^{1/2} \Xi^*(\rho) \rho^{1/2} = \gamma^2 (1 - e^{-d/2})^2 \rho^2$ follows. □

We have considered the fidelity of a teleportation scheme with beam splittings. We showed that as the parameter $|a|$ goes to infinity, the fidelity approaches unity, and the teleportation scheme also approaches the perfect scheme as does the teleportation scheme with tests. In fact, the fidelity can be bounded from below by the square root of the probability to complete a successful teleportation with tests.

The above result can be extended to the case of general inputs for the beam splitter which produces an entangled state. To get an insight for the generalization, it is instructive to consider the following simple splitting model. Here we put $\mathcal{H}_j \simeq \mathbb{C}^N$ for all $j = 1, 2, 3$ and take $\{e_n^{(j)}\}_{n=1}^N$ as its basis. In addition, we put the input system of the splitter as $\mathcal{H}_0$ which has $\{e_n\}_{n=1}^N$ as its basis. Now we define a simple splitting isometry $J : \mathcal{H}_0 \to \mathcal{H}_2 \otimes \mathcal{H}_3$ by $e_k \mapsto e_k^{(2)} \otimes e_k^{(3)}$ for all k. Then if the input of the simple splitting is $|\xi_0\rangle \equiv \frac{1}{\sqrt{N}} \sum_k e_k$, the resulted state is perfectly entangled, otherwise not. We put a general input state for the splitter as τ hereafter, and the ideal one as $\tau_0 \equiv |\xi_0\rangle\langle\xi_0|$. Let us consider the following teleportation scheme:

Step 0. Alice has an unknown quantum state $\rho^{(1)}$ (on Hilbert space $\mathcal{H}_1$) and she was asked to teleport it to Bob.

Step 1. For this purpose, we need three additional Hilbert spaces $\mathcal{H}_0$, $\mathcal{H}_2$, and $\mathcal{H}_3$. $\mathcal{H}_2$ is attached to Alice, and $\mathcal{H}_3$ is given to Bob. Prearrange a state $J\tau J^* \in \mathfrak{S}(\mathcal{H}_2 \otimes \mathcal{H}_3)$ having a certain correlation between Alice and Bob by use of the input $\tau \in \mathfrak{S}(\mathcal{H}_0)$ for the simple splitter.

Step 2. Prepare the set of projections $\{F_{nm}^{(12)}\}$ and an observable $F^{(12)} \equiv \sum_{nm} z_{nm} F_{nm}^{(12)}$ on a tensor product Hilbert space (system) $\mathcal{H}_1 \otimes \mathcal{H}_2$ defined by $F_{nm}^{(12)} \equiv |\xi_{nm}\rangle\langle\xi_{nm}|$ with

$$\left|\xi_{nm}{}^{(12)}\right\rangle \equiv \frac{1}{\sqrt{N}} \sum_{j=1}^N b_{nj} e_j^{(1)} \otimes e_{j\oplus m}^{(2)} \quad \left(\sum_{j=1}^N b_{nj}^* b_{lj} = \delta_{nl}, |b_{nj}| = 1\right).$$

Alice performs the joint measurement of the observable $F^{(12)}$.

Step 3. Bob obtained a state $\rho^{(3)}$ due to the reduction of wave packet, and he is informed which outcome was obtained by Alice. This information is completely transmitted from Alice to Bob without disturbance (for instance, by telephone).

Step 4. $\rho^{(1)}$ is reconstructed from $\rho^{(3)}$ by using the key which corresponds to the outcome Bob got from Alice in the above Step 3. That is, for z_{nm}, the operation W_{nm}, $W_{nm} e_{j\oplus m}^{(3)} \equiv b_{nj}^* e_j^{(1)}$, is employed.

In the above scheme, Alice and Bob believe that the teleportation setting (entangled state between them) is perfect and do their own jobs; however, if the prepared entangled state is not an ideal one ($\tau \neq \tau_0$), the experiment yields error. Now the expected state obtained by Bob is written as

$$\widetilde{\Lambda}^*{}_\tau(\rho) = \sum_{nm} \mathrm{tr}_{12}\big(\big(F_{nm}^{(12)} \otimes W_{nm}\big)(\rho \otimes J\tau J^*)\big(F_{nm}^{(12)} \otimes W_{nm}^*\big)\big).$$

Let us estimate the fidelity. We obtain the following theorem.

Theorem 18.27 *The following inequality for the fidelity holds*:

$$(1 \geq) F\big(\rho, \widetilde{\Lambda}^*{}_\tau(\rho)\big) \geq F(\tau_0, \tau),$$

where $\tau = \tau_0 = |\xi_0\rangle\langle\xi_0|$ *with* $|\xi_0\rangle \equiv \frac{1}{\sqrt{N}} \sum_{k=1}^N e_k \in \mathcal{H}_0$ *represents the ideal input state for the splitter to produce a perfect entangled state. In addition, the inequality is strict, that is, there exists an unknown state to be sent which makes the inequality into an equality.*

Proof For simplicity, for the unknown state of Alice, we assume pure state $\rho = |\Phi\rangle\langle\Phi|$. The fidelity is written as

$$F\big(\rho, \widetilde{\Lambda}^*{}_\tau(\rho)\big)^2 = \langle\Phi|\widetilde{\Lambda}^*{}_\tau\big(|\Phi\rangle\langle\Phi|\big)|\Phi\rangle.$$

For $\tau = \sum_k \lambda_k |\varphi_k\rangle\langle\varphi_k|$, it is expressed in a simple form by using f^k, $g \in L^2(\{1, 2, \dots N\})$ defined by

$$f^k(j) \equiv \sum_t \langle e_{t \oplus j}, \varphi_k\rangle \big|\langle e_t, \Phi\rangle\big|^2,$$

$$g(j) \equiv \frac{1}{\sqrt{N}} \quad (j = 1, 2, \dots, N).$$

We obtain

$$F\big(\rho, \widetilde{\Lambda}^*{}_\tau(\rho)\big)^2 = \sum_k \lambda_k \|f_k\|^2 \|g\|^2.$$

Thanks to the Cauchy–Schwarz inequality, it is estimated as

$$\begin{aligned} F\big(\rho, \widetilde{\Lambda}^*{}_\tau(\rho)\big)^2 &\geq \sum_k \lambda_k \big|(f^k, g)\big|^2 \\ &= F(\tau_0, \tau)^2. \end{aligned}$$

For the general mixed input case, the proof goes similarly. Since we have just only used Cauchy-Schwarz inequality, the equality is attained with the choice $|\Phi\rangle = \sum_k \frac{e^{i\theta_k}}{\sqrt{N}} |e_k^{(1)}\rangle$. Thus the proof is completed. □

The theorem shows that the closer the input state for the splitter to the perfect one, the closer the fidelity to one.

Finally, we present the generalized result to the generalized beam splitting case:

Theorem 18.28 *For an arbitrary input τ of the generalized beam splitter, the fidelity is bounded by*

$$F\big(\rho, \Lambda^*(\rho)\big) \geq \frac{1-e^{-a^2/2}}{\sqrt{1+e^{-a^2}}} F(\tau_0, \tau),$$

where $\tau_0 \equiv |\xi_0\rangle\langle\xi_0|$ with

$$|\xi_0\rangle \equiv \frac{1}{\sqrt{N}} \sum_{k=1}^{N} \big|\exp(ag_k) - \exp(0)\big\rangle.$$

The theorem is proved by combining the above two theorems. (We omit the proof.) It is seen that when τ is close to τ_0 and $|a|$ is large, the fidelity becomes close to one.

18.7 Spatially Separated Teleportation

Specializing this model, we consider the teleportation of all states on a finite-dimensional Hilbert space (through the space $\mathbb{R}^k$). Further specialization leads to a teleportation model where Alice and Bob are spatially separated, that is, we have to teleport the information given by the state of our finite-dimensional Hilbert space from one region $X_1 \subseteq \mathbb{R}^k$ into another region $X_2 \subseteq \mathbb{R}^k$ with $X_1 \cap X_2 = \varnothing$, and Alice can only perform local measurements (inside of region X_1) as well as Bob (inside of X_2).

To discuss these, we start by discussing the teleportation inside the Euclidean space $\mathbb{R}^k$.

18.7.1 Teleportation of States Inside $\mathbb{R}^k$

Let $\mathcal{H}$ be a finite-dimensional Hilbert space. We consider the case $\mathcal{H} = \mathbb{C}^N = L^2(\{1, \dots, N\}, \#)$, without loss of generality, where # denotes the counting measure on the set $\{1, \dots, N\}$. We want to teleport states on $\mathcal{H}$ with the aid of the constructed channels $(\Lambda^*_{nm})^N_{n,m=1}$ or $(\Theta^*_{nm})^N_{n,m=1}$. We fix

- A CONS $(|j\rangle)^N_{j=1}$ of $\mathcal{H}$,
- $f \in L^2(\mathbb{R}^k)$, $\|f\| = 1$,
- $d = a^2 > 0$,
- $\hat{K}_1, \hat{K}_2$ linear operators on $L^2(\mathbb{R}^k)$,
- $\hat{T}$ unitary operator on $L^2(\mathbb{R}^k)$

with two properties

$$\hat{K}_1^* \hat{K}_1 f + \hat{K}_2^* \hat{K}_2 f = f,$$
$$\hat{T} \hat{K}_1 f = \hat{K}_2 f.$$

We put

$$G = \mathbb{R}^k \times \{1, \ldots, N\}, \quad \mu = l \times \#,$$

where l is the Lebesgue measure on $\mathbb{R}^k$. Then $L^2(G) = L^2(G, \mu) = L^2(\mathbb{R}^k) \otimes \mathcal{H}$. Further, put

$$g_j \equiv f \otimes |j\rangle \quad (j = 1, \ldots, N).$$

Then $(g_j)_{j=1}^N$ is an ONS in $L^2(G)$. We consider linear operators K_1, K_2 on $L^2(G)$ with $K_1^* K_1 + K_2^* K_2 = 1$ and

$$K_r g_j = (\hat{K}_r f) \otimes |j\rangle \quad (j = 1, \ldots, N; r = 1, 2),$$

which determine operators K_1, K_2 on the subspace of $\mathcal{M}$ spanned by the ONS $(g_j)_{j=1}^N$. On the orthogonal complement, one can put, for instance,

$$K_r \psi = \frac{1}{\sqrt{2}} \psi.$$

Then K_1, K_2 are well defined and fulfill $K_1^* K_1 + K_2^* K_2 = 1$. Further, one checks that $\langle K_i g_k, K_i g_j \rangle = 0$ $(i = 1, 2; k \neq j = 1, \ldots, N)$ hold.

Now let T be a unitary operator on $L^2(G)$ with

$$T(K_1 g_j) = (\hat{T} \hat{K}_2 f) \otimes |j\rangle.$$

From $\langle K_i g_k, K_i g_j \rangle = 0$ one can prove the existence of T. Further, we get $T K_1 g_k = K_2 g_k$ from $\hat{T} \hat{K}_1 f = \hat{K}_2 f$.

Summarizing, we obtain that $\{g_1, \ldots, g_N\}$, K_1, K_2, T fulfill all the assumptions required in Sect. 18.5. Thus we have the corresponding channels $\Lambda_{n,m}^*$ and Θ_{nm}^*. It follows that we are able to teleport a state ρ on $\mathcal{M} = \mathcal{M}(G)$ the same as before.

In order to teleport states on $\mathcal{H}$ through the space $\mathbb{R}^k$ using the above channels, we have to consider:

First a "lifting" $\mathcal{E}^*$ of the states on $\mathcal{H}$ into the set of states on the bigger state space on $\mathcal{M}$ such that $\rho = \mathcal{E}^*(\hat{\rho})$ can be described by (18.3), (18.4), and (18.5).

Second a "reduction" $\mathcal{R}$ of (normal) states on $\mathcal{M}$ to states on $\mathcal{H}$ such that for all states $\hat{\rho}$ on $\mathcal{H}$

$$\begin{aligned} &\mathcal{R}\big(\big(\Gamma(T) U_m B_n^*\big) \mathcal{E}^*(\hat{\rho}) \big(\Gamma(T) U_m B_n^*\big)^*\big) \\ &\quad = V_{nm} \hat{\rho} V_{nm}^* \quad (n, m = 1, \ldots, N), \end{aligned} \tag{18.14}$$

where $(V_{nm})_{n,m=1}^N$ are unitary operators on $\mathcal{H}$.

This we can obtain as follows: We have already stated in Sect. 18.2 that

$$\left(\left|\exp\left(aK_1(g_j)\right) - \exp(0)\right\rangle\right)_{j=1}^{N} \quad (r = 1, 2)$$

are ONS in $\mathcal{M}$. We denote by $\mathcal{M}_r$ $(r = 1, 2)$ the corresponding N-dimensional subspaces of $\mathcal{M}$. Then for each $r = 1, 2$, there exists exactly one unitary operator W_r from $\mathcal{H}$ onto $\mathcal{M}_r \subseteq \mathcal{M}$ with

$$W_r|j\rangle = \left|\exp(aK_r g_j) - \exp(0)\right\rangle \quad (j = 1, \ldots, N). \tag{18.15}$$

We put

$$\mathcal{E}^*(\hat{\rho}) \equiv W_1 \hat{\rho} W_1^* \Pi_{\mathcal{M}_1} \quad (\hat{\rho} \text{ state on } \mathcal{H}), \tag{18.16}$$

where $\Pi_{\mathcal{M}_r}$ denotes the projection onto $\mathcal{M}_r$ $(r = 1, 2)$. Describing the state $\hat{\rho}$ on $\mathcal{H}$ by

$$\hat{\rho} = \sum_{s=1}^{N} \lambda_s |\hat{\Phi}_s\rangle\langle\hat{\Phi}_s| \tag{18.17}$$

with

$$|\hat{\Phi}_s\rangle = \sum_{j=1}^{N} c_{sj}|j\rangle,$$

where $(c_{sj})_{s,j=1}^{N}$ fulfills (18.5), we obtain that $\rho = \mathcal{E}^*(\hat{\rho})$ is given by (18.3) and (18.4).

Now, for each state ρ on $\mathcal{M}$ we put

$$\mathcal{R}^*(\rho) \equiv \frac{W_2^* \Pi_{\mathcal{M}_2} \rho W_2}{\operatorname{tr}_{\mathcal{M}} W_2^* \Pi_{\mathcal{M}_2} \rho W_2}. \tag{18.18}$$

Since

$$\Pi_{\mathcal{M}_2} \Gamma(T) U_m B_n^* \mathcal{E}^*(\hat{\rho}) \left(\Gamma(T) U_m B_n^*\right)^* = \Gamma(t) U_m B_n^* \mathcal{E}^*(\hat{\rho}) \left(\Gamma(T) U_m B_n^*\right)^*$$

we get

$$\operatorname{tr}_{\mathcal{M}} W_2^* \Pi_{\mathcal{M}_2} \Gamma(t) U_m B_n^* \mathcal{E}^*(\hat{\rho}) \left(\Gamma(T) U_m B_n^*\right)^* = 1,$$

and

$$\begin{aligned}
&\mathcal{R}^*\left(\Gamma(T) U_m B_n^* \mathcal{E}^*(\hat{\rho}) \left(\Gamma(T) U_m B_n^*\right)^*\right) \\
&\quad = W_2^* \Gamma(T) U_m B_n^* W_1 \hat{\rho} W_1^* \Pi_{\mathcal{M}_1} \left(\Gamma(T) U_m B_n^*\right)^* W_2.
\end{aligned}$$

So we have the equality

$$\Pi_{\mathcal{M}_1} \left(\Gamma(T) U_m B_n^*\right)^* W_2 = \left(\Gamma(T) U_m B_n^*\right)^* W_2$$

which implies

$$\mathcal{R}^*\big(\Gamma(T)U_m B_n^* \mathcal{E}^*(\hat{\rho})\big(\Gamma(T)U_m B_n^*\big)^*\big)$$
$$= W_2^* \Gamma(T)U_m B_n^* W_1 \hat{\rho} W_1^* \big(\Gamma(T)U_m B_n^*\big)^* W_2.$$

Put

$$V_{nm} \equiv W_2^* \Gamma(T) V_m B_n^* W_1 \quad (n, m = 1, \dots, N), \tag{18.19}$$

then V_{nm} $(n, m = 1, \dots, N)$ is a unitary operator on $\mathcal{H}$ and (18.14) holds. One easily checks

$$V_{nm}|j\rangle = \bar{b}_{nj}|j \otimes m\rangle \quad (j, m, n = 1, \dots, N).$$

Summarizing the above, we have the following theorem:

Theorem 18.29 *Consider the channels on the set of states on* $\mathcal{H}$

$$\hat{\Lambda}_{nm}^* \equiv \mathcal{R}^* \circ \Lambda_{nm}^* \circ \mathcal{E}^* \quad (n, m = 1, \dots, N), \tag{18.20}$$

$$\hat{\Theta}_{nm}^* \equiv \mathcal{R}^* \circ \Theta_{nm}^* \circ \mathcal{E}^* \quad (n, m = 1, \dots, N), \tag{18.21}$$

where $\mathcal{R}^*$, $\mathcal{E}^*$, Λ_{nm}^*, Θ_{nm}^* *are given by* (18.18), (18.16), (18.9), *and* (18.10), *respectively. Then for all states* $\hat{\rho}$ *on* $\mathcal{H}$,

$$\hat{\Lambda}_{nm}^*(\hat{\rho}) = V_{nm} \hat{\rho} V_{nm}^* = \hat{\Theta}_{nm}^*(\hat{\rho}) \quad (n, m = 1, \dots, N), \tag{18.22}$$

where V_{nm} $(n, m = 1, \dots, N)$ *are the unitary operators on* $\mathcal{H}$ *given by* (18.19).

Remark 18.30 Remember that the teleportation model according to $(\Lambda_{nm}^*)_{n,m=1}^N$ works perfectly in the sense described before, and the model dealing with $(\Theta_{nm}^*)_{n,m=1}^N$ was only asymptotically perfect for large d (i.e., high density or high energy of the beams). They can transfer to $(\hat{\Lambda}_{n,m}^*)$, $(\hat{\Theta}_{nm}^*)$.

Example 18.31 We specialize

$$\hat{K}_1 h = \hat{K}_2 h = \frac{1}{\sqrt{2}} h \quad \big(h \in L^2\big(\mathbb{R}^k\big)\big), \qquad \hat{T} = I.$$

Realizing the teleportation in this case means that Alice has to perform measurements (F_{nm}) in the whole space $\mathbb{R}^k$ and so does Bob (concerning F_+).

18.7.2 Alice and Bob Are Spatially Separated

We specialize the situation in the previous section as follows:

- We fix a $t \in \mathbb{R}^k$.

– Take $X_1, X_2, X_3 \subseteq \mathbb{R}^k$ to be a measurable decomposition of $\mathbb{R}^k$ such that $l(X_1) \neq 0$ and

$$X_2 = X_1 + t \equiv \{x + t \mid x \in X_1\}.$$

Put

$$\hat{T}h(x) \equiv h(x - t) \quad \left(x \in \mathbb{R}^k, h \in L^2(\mathbb{R}^k)\right),$$
$$\hat{K}_r h \equiv h\mathbf{1}_{X_r} \quad \left(r = 1, 2, h \in L^2(\mathbb{R}^k)\right),$$

and assume that the function $f \in L^2(\mathbb{R}^d)$ has the properties

$$f\mathbf{1}_{X_2} = \hat{T}(f\mathbf{1}_{X_1}), \quad f\mathbf{1}_{X_3} \equiv 0.$$

Then $\hat{T}$ is a unitary operator on $L^2(\mathbb{R}^k)$. Using the assumption that X_1, X_2, X_3 form a measurable decomposition of $\mathbb{R}^k$, we get immediately that

$$G_s \equiv X_s \times \{1, \ldots, N\} \quad (s = 1, 2, 3)$$

is a measurable decomposition of G. It follows that $\mathcal{M} = \mathcal{M}(G)$ is decomposed into the tensor product

$$\mathcal{M}(G) = \mathcal{M}(G_1) \otimes \mathcal{M}(G_2) \otimes \mathcal{M}(G_3).$$

According to this representation, the local algebras $\mathcal{A}(X_s)$ corresponding to regions $X_s \subseteq \mathbb{R}^d$ $(s = 1, 2, 3)$ are given by

$$\mathcal{A}(X_1) \equiv \left\{A \otimes \mathbf{1} \otimes \mathbf{1};\ A \text{ bounded operator on } \mathcal{M}(G_1)\right\},$$
$$\mathcal{A}(X_2) \equiv \left\{\mathbf{1} \otimes A \otimes \mathbf{1};\ A \text{ bounded operator on } \mathcal{M}(G_2)\right\}.$$

One easily checks in our special case that

$$F_{nm} \in \mathcal{A}(X_1) \otimes \mathcal{A}(X_1) \quad (n, m = 1, \ldots, N),$$

and $\mathcal{E}^*(\hat{\rho})$ gives a state on $\mathcal{A}(X_1)$ (the number of particles outside of G_1 is 0 with probability 1). That is, Alice has to perform only local measurements inside of the region X_1 in order to realize the teleportation processes described in Sect. 18.4 or measure the state $\mathcal{E}^*(\hat{\rho})$. On the other hand, $\Lambda^*_{nm}(\mathcal{E}^*(\hat{\rho}))$ and $\Theta_{nm}(\mathcal{E}^*(\hat{\rho}))$ give local states on $\mathcal{A}(X_2)$ such that by measuring these states Bob has to perform only local measurements inside of the region X_2. The only problem could be that, according to the definition of the channels Θ_{nm}, Bob has to perform the measurement by F_+ which is not local. However, as we have already stated, this problem can be avoided if we replace F_+ by $F_{+,X_2} \in \mathcal{A}(X_2)$.

Therefore, we can describe the special teleportation process as follows: We have a beam in the pure state $|\eta\rangle\langle\eta|$ $(|\eta\rangle \equiv \frac{\gamma}{\sqrt{N}} \sum_{k=1}^{N} |\exp(ag_k)\rangle)$. After splitting, one part of the beam is located in the region X_1 or will go to X_1, and the other part is located in the region X_2 or will go to X_2. Further, there is a state $\mathcal{E}^*(\hat{\rho})$ localized in the region X_1. Now Alice will perform the local measurement inside of X_1 according to $F = \sum_{n,m} z_{nm} F_{nm}$ involving the first part of the beam and the state $\mathcal{E}^*(\rho)$. This leads to a preparation of the second part of the beam located in the region X_2 which can be controlled by Bob, and the second part of the beam will show the behavior of the state $\Lambda^*_{nm}(\mathcal{E}^*(\hat{\rho})) = \Theta_{nm}(\mathcal{E}^*(\hat{\rho}))$ if Alice's measurement shows the value z_{nm}. Thus we have teleported the state $\hat{\rho}$ on $\mathcal{H}$ from the region X_1 into the region X_2.

18.8 Model of Continuous Teleportation

In this section, we discuss another model of teleportation, namely continuous variable teleportation.

18.8.1 Scheme of Continuous Variable Teleportation

Braunstein and Kimble proposed a model of teleportation which uses continuous variable measurement with respect to a photon. As discussed above, what is important to realize in a teleportation scheme is to produce a well-entangled state and perform a well-controlled measurement which distinguishes the entangled basis used. Braunstein and Kimble proposed a slightly modified model from the original one. They employed the measurement of the continuous instead of discrete spectrum.

To simplify we explain the one-mode version of their model. Thus each system attached with Alice and Bob is a one-mode electromagnetic field which can be identified with a one-particle quantum harmonic oscillator. We write the "position" operator of the system as $\hat{x}$ and the "momentum" as $\hat{y}$. The ordinary commutation relation $[\hat{x}, \hat{y}] = \mathrm{i}$ holds. (We put the Planck constant $\hbar = 1$.) We put for simplicity the angular velocity of all oscillators as $\omega = 1$, that is, we assume all oscillators have a same angular velocity. For each system, the energy eigenvector can be written as $|n\rangle$, which has $n + \frac{1}{2}$ as its energy eigenvalue. (Since what we indeed treat is a photon and not harmonic oscillator, the renormalization constant $\frac{1}{2}$ can be subtracted, but it does not matter at all.) It may be instructive to mention the same thing in Fock-space language. The "one-particle" Hilbert space $\mathcal{L}_1$ is one-dimensional with a normalized vector $|1\rangle$. Then the "n-particle" Hilbert space has one-dimensional structure whose basis is written as $|n\rangle$. We have a natural definition of the creation and anihilation operators given by

$$a|n\rangle = \sqrt{n}|n-1\rangle,$$
$$a^*|n\rangle = \sqrt{n+1}|n+1\rangle.$$

One can easily see that they satisfy the commutation relation

$$[a, a^*] = 1,$$

and the position and momentum operators are represented by

$$\hat{x} = \sqrt{\frac{1}{2}}(a + a^*),$$

$$\hat{y} = \sqrt{\frac{-i}{2}}(a - a^*).$$

Three systems of Alice and Bob are specified with the indices (1), (2) and (3): (1) and (2) are attached to Alice, and (3) corresponds to Bob. For instance, we write the nth eigenstate of the system (1) as $|n\rangle^{(1)}$. The entangled state over (2) and (3) is given by the so-called two-mode squeezed state which can be realized by nondegenerate optical parametric amplifier. The state is specified by a parameter $r > 0$ and constructed by applying the two-mode squeezed operator, $S^{(23)}(r) \equiv \exp(r(a_{(2)}a_{(3)} - a^*_{(2)}a^*_{(3)}))$, to the vacuum. That is, the state is

$$|r\rangle^{(23)} \equiv S^{(23)}(r)|0\rangle^{(2)} \otimes |0\rangle^{(3)}.$$

It is difficult to see in the above expression how strongly the state is entangled, so the following Fock basis representation is useful:

$$\begin{aligned} |r\rangle^{(23)} &= \frac{1}{\cosh(r)} \sum_{n=0}^{\infty} (\tanh(r))^n |n\rangle^{(2)} \otimes |n\rangle^{(3)} \\ &= \sqrt{1 - q^2} \sum_{n=0}^{\infty} q^n |n\rangle^{(2)} \otimes |n\rangle^{(3)}. \end{aligned} \tag{18.23}$$

Here we put $q \equiv \tanh(r)$ to simplify the notation. Thus the states are parameterized by a positive number $q \in (0, 1)$. When the parameter q is closer to 1, it represents more perfectly entangled state. According to the experiment, the value $q = 0.33$ has already been realized. To understand how strongly the state (18.23) is entangled, it should be noted how q^n behaves as n becomes large. q^n can be estimated to have sufficiently large nonzero value for $0 \le n \le \frac{1}{1-q^2}$. Therefore, the effectively entangled subspace has $\frac{1}{(1-q^2)^2}$ dimension in the tensor product Hilbert space of (2) and (3).

Another way to characterize the state is to redefine the creation and anihilation operators using a squeezing operator:

$$b_{(2)} \equiv S_{(23)}(r)^* a_{(2)} S_{(23)} = \cosh(r) a_{(2)} - \sinh(r) a^*_{(3)},$$

$$b_{(3)} \equiv S_{(23)}(r)^* a_{(3)} S_{(23)} = \cosh(r) a_{(3)} - \sinh(r) a^*_{(2)}.$$

The state $|r\rangle^{(23)}$ is the vacuum state for these new operators $b_{(2)}$ and $b_{(3)}$. In the position representation, the wave function is written by means of the Gaussian one as

$$\Psi_q(x_2, x_3) = \frac{1}{\pi} \exp\left(-\frac{1-q}{4(1+q)}(x_2 - x_3)^2 - \frac{1+q}{4(1-q)}(x_2 + x_3)^2 \right).$$

In the next subsection, we will discuss the state obtained by letting the parameter q to 1, which reveals a mathematically interesting structure.

Now let us see what kind of measurements Alice performs. Alice mixes her input state with the reference EPR beam by a 50% beam-splitter and she performs an entanglement measurement of the complex field value $\hat{\beta} = \hat{x}_- + \mathrm{i}\hat{y}_+$, where

$$\hat{x}_- = \hat{x}_1 - \hat{x}_2,$$
$$\hat{y}_+ = \hat{y}_1 + \hat{y}_2.$$

An important point is that the commutativity of $\hat{x}_-$ and $\hat{x}_+$ yields the measurement of two different observables at the same time. It is easily seen that both of them have continuous spectrum. The measurement of an observable with a continuous spectrum cannot be done precisely, but can be done approximately depending on one's resolution ability. The unnormalized "eigenvector" of the operator $\hat{\beta}$ is written as

$$|\beta\rangle^{(12)} = \frac{1}{\sqrt{\pi}} \sum_{n=0}^{\infty} \hat{D}_{(1)}(\beta)|n\rangle^{(1)} \otimes |n\rangle^{(2)},$$

where $\hat{D}_{(1)}(\beta)$ is a displacement operator acting on the mode (1) with a displacement amplitude of $\beta \in \mathbb{C}$, that is, it is defined by

$$D_{(1)}(\beta) \equiv \exp(\beta a^*_{(1)} - \beta^* a_{(1)}).$$

When Alice obtains an outcome β, the resulted state is not $|\beta\rangle_{A,R}$ but the smeared normalized state

$$|\beta, \Delta\rangle^{(12)} \equiv \int d^2\beta' f_\Delta(\beta' - \beta)|\beta'\rangle^{(12)},$$

where f_Δ is a function whose support is only around the origin with the width Δ (resolution ability) and it satisfies $\int d^2\beta' |f_\Delta(\beta')|^2 = 1$. Thus the state of (3) is changed into

$$\rho(\beta) \equiv \frac{|\psi(\beta)\rangle^{(3)(3)}\langle\psi(\beta)|}{{}^{(3)}\langle\psi(\beta)|\psi(\beta)\rangle^{(3)}}$$

with

$$\begin{aligned}
|\psi(\beta)\rangle^{(3)} &= \left({}^{(1,2)}\langle\beta, \Delta| \otimes I^{(3)}\right)\left(|\psi\rangle^{(1)} \otimes I^{(23)}\right)\left(I^{(1)} \otimes |q\rangle^{(23)}\right) \\
&= \left({}^{(1,2)}\langle\beta, \Delta| \otimes I^{(3)}\right)\left(|\psi\rangle^{(1)} \otimes |q\rangle^{(23)}\right) \\
&= \int d'\beta \sqrt{\frac{1-q^2}{\pi}} \sum_{n=0}^{\infty} q^n |n\rangle^{(3)} f^*_\Delta(\beta' - \beta)\langle n|\hat{D}_{(1)}(-\beta')|\psi\rangle^{(1)}.
\end{aligned}$$

The probability obtaining the field measurement value β is given by $P_q(\beta) = {}^{(3)}\langle\psi(\beta)|\psi(\beta)\rangle^{(3)}$.

After Bob gets information about the field measurement value β from Alice, Bob applies a corresponding displacement unitary operator to the output state by mixing the coherent field of a local oscillator with the output EPR beam B. Thus the output state is $|\psi_{\text{out}}(\beta)\rangle^{(3)} = \hat{D}_{(3)}(\beta)|\psi(\beta)\rangle^{(3)}$. The above mentioned scheme is called the continuous teleportation, which is proposed by Kimble et al. After some calculations, one can show that by letting $q \to 1$ and $\Delta \to 0$ the scheme approaches the perfect one, that is, it teleports exactly all the states of one-mode photons.

18.8.2 Entangled State Employed by Continuous Teleportation

Let us go back to the entangled state employed by the above scheme and close this section by a comment on its mathematical property. As we mentioned, each system attached to Alice and Bob is identified with a one-particle system. The observable algebra of a one-particle system is a so-called CCR algebra defined as follows. Although ordinary quantum mechanics textbooks begin with commutation relation (we put $\hbar = 1$)

$$[x, y] = \mathrm{i},$$

here note that the momentum is written as y, this equation is somewhat singular since the operators x and y cannot be realized by bounded operators but can be realized by unbounded ones. To treat it more soundly, it is nice to begin with unitary operators whose generators are x and y. Thus we consider the unitary objects, $W(k, k')$, $((k, k') \in \mathbb{R}^2)$ with CCR relation,

$$W(k, k')W(p, p') = W(k + p, k' + p')e^{-\frac{\mathrm{i}}{2}k\wedge p},$$

where $k \wedge p \equiv kp' - k'p$ represents a wedge product. Intuitively, $W(k, k')$ represents $\exp[\mathrm{i}(kx + k'y)]$. A C^*-algebra $\mathcal{A}$ generated by $\{W(k, k')\}$'s is called CCR algebra. The well-known von Neumann uniqueness theorem says that the representation π of CCR yielding *continuous* unitary group of $\pi(W(k, k'))$ with respect to (k, k') is essentially unique. Almost all standard textbooks treat things in the so-defined *regular* representation, Shrödinger representation. The above discussion of continuous teleportation scheme also can be written in this representation. The Hilbert space is spanned by a complete orthonormal system $\{|n\rangle\}$. The observable algebra of composite system attached to Alice and Bob, (2) and (3), is written by a tensor product of two such CCR systems. There exists subtle problems when one treats the tensor product of C^*-algebras since the norm on it is not uniquely defined; however, we can here define the norm in a natural way. We obtain the observable algebra of the composite system $\mathcal{A}_2 \otimes \mathcal{A}_3$, in which the situation is not changed from the one-particle case above and the von Neumann uniqueness theorem still holds here. Thus we can play the same game in the regular representation Hilbert space $\mathcal{H}_2 \otimes \mathcal{H}_3$.

Let us now consider the entangled state $\langle q| \cdot |q\rangle$. As long as $q < 1$ holds, the vector $|q\rangle$ is well-defined in $\mathcal{H}_2 \otimes \mathcal{H}_3$, but we have a question "how it behaves as

$q \to 1$". The vector becomes closer and closer to the basis $|n\rangle^{(2)} \otimes |n\rangle^{(3)}$, and it does *not* converge to any vector in $\mathcal{H}_2 \otimes \mathcal{H}_3$. However, this fact does not mean that the limiting state is a mathematically illegal one. Let us remind the algebraic formulation of the quantum theory. We have a C^*-algebra and states over it. A state ω is nothing but a normalized positive linear functional over the algebra. In our CCR algebra case, a state is exactly defined by specifying a so-called generating functional $\omega(W(k_2, k_2') \otimes W(k_3, k_3'))$. Before considering the entangled state, let us begin with an exercise with respect to one mode CCR. For instance, let us see the generating functional $\omega_0(\cdot) = \langle 0| \cdot |0\rangle$, the ground state. In Shrödinger representation, the representation and the creation–anihilation a, a^* operators are related by

$$\pi\left(W(k, k')\right) = e^{-\frac{1}{4}(k^2+k'^2)} e^{\mathrm{i}\frac{1}{\sqrt{2}}(k+\mathrm{i}k')a^*} e^{\mathrm{i}\frac{1}{\sqrt{2}}(k-\mathrm{i}k')a},$$

and by using the relation $a|0\rangle = 0$, we obtain

$$\omega_0\left(W(k, k')\right) = e^{-\frac{1}{4}(k^2+k'^2)}.$$

Now let us consider the entangled state employed in continuous teleportation model, $\omega_q(\cdot) \equiv \langle q| \cdot |q\rangle$ over the tensored observable algebra. By a long calculation, one obtains generating functional:

$$\begin{aligned}
&\omega_q\left(W(k_2, k_2') \otimes W(k_3, k_3')\right) \\
&\quad = \exp\left(-\frac{1+q}{8(1-q)}\left\{(k_2-k_3)^2 + (k_2'+k_3')^2\right\}\right) \\
&\qquad \times \exp\left(-\frac{1-q}{8(1+q)}\left\{(k_2+k_3)^2 + (k_2'-k_3')^2\right\}\right).
\end{aligned}$$

By letting q to 1 in the above equation, as the weak limit we obtain a state over CCR tensored algebra such that

$$\omega_{q=1}\left(W(k_2, k_2') \otimes W(k_3, k_3')\right) = \begin{cases} 1, & \text{if } k_2 = k_3,\ k_2' = -k_3', \\ 0, & \text{otherwise} \end{cases}$$

for which one can easily verify that it indeed satisfies the requirement for a state (positivity, normality, etc.). It is nothing but the state that Einstein, Podolsky and Rosen originally considered. The state does not stay in the regular representation space anymore, but it is a state over the CCR C^*-algebra. One can show that this state maximally violates Bell's inequality and has a strong entanglement property.

18.9 Quantum Teleportation with Non-maximally Entangled States

In some models [20, 105, 246] *perfect teleportation is possible if the entangled state σ used for the teleportation and the projection P_α are maximally entangled.*

18.9.1 Basic Setting

Let $\mathcal{H} = \mathbb{C}^n$ be a finite-dimensional complex Hilbert space in which the scalar product $\langle \cdot, \cdot \rangle$ is defined as usual. Let e_n $(n = 1, \ldots, n)$ be a fixed orthonormal basis (ONB) in $\mathcal{H}$, and let $\mathbf{B}(\mathcal{H})$ be the set of all bounded linear operators on $\mathcal{H}$, which is simply denoted by M_n. In M_n, the scalar product $(\cdot, \cdot)$ is defined by

$$(A, B) \equiv \operatorname{tr} A^* B = \sum_{i=1}^{n} \langle Ae_i, Be_i \rangle.$$

Note that $e_{ij} \equiv |e_i\rangle\langle e_j|$ $(i, j = 1, \ldots, n)$ is an ONB in M_n with respect to the above scalar product. The mappings

$$A \in M_n \to A^L \equiv \sum_{i=1}^{n} Ae_i \otimes e_i \in \mathcal{H} \otimes \mathcal{H},$$

$$A \in M_n \to A^R \equiv \sum_{i=1}^{n} e_i \otimes Ae_i \in \mathcal{H} \otimes \mathcal{H}$$

define the inner product isomorphisms from M_n into $\mathcal{H} \otimes \mathcal{H}$ such that

$$(A, B) = \langle\!\langle A^L, B^L \rangle\!\rangle = \langle\!\langle A^R, B^R \rangle\!\rangle,$$

where the inner products in $\mathcal{H} \otimes \mathcal{H}$ is denoted by $\langle\!\langle \cdot, \cdot \rangle\!\rangle$.

Let $L(M_n, M_n)$ be the vector space of all linear maps $\Phi : M_n \to M_n$. $M_n \otimes M_n$ is the set of all linear maps from $\mathcal{H} \otimes \mathcal{H}$ to $\mathcal{H} \otimes \mathcal{H}$. By analogy between M_n and $\mathcal{H} \otimes \mathcal{H}$, one can construct the inner product isomorphisms between $L(M_n, M_n)$ and $M_n \otimes M_n$ such as

$$\Phi \in L(M_n, M_n) \to \Phi^L \equiv \sum_{i,j=1}^{n} \Phi e_{ij} \otimes e_{ij} \in M_n \otimes M_n,$$

$$\Phi \in L(M_n, M_n) \to \Phi^R \equiv \sum_{i,j=1}^{n} e_{ij} \otimes \Phi e_{ij} \in M_n \otimes M_n.$$

The inner products in $L(M_n, M_n)$ is defined as follows:

$$\big((\Phi, \Psi)\big) \equiv \operatorname{tr} \Phi^* \Psi = \sum_{i,j=1}^{n} (\Phi e_{ij}, \Psi e_{ij}).$$

One can easily verify that it is equal to

$$\operatorname{tr}_{12} \Phi^{L*} \Psi^L = \operatorname{tr}_{12} \Phi^{R*} \Psi^R,$$

where tr_{12} is the trace over the space $M_n \otimes M_n$, whose ONB is $\{e_{ij} \otimes e_{kl}\}$.

Let $\{f_\alpha; \alpha = 1, \ldots, n^2\}$ be another ONB in M_n so that one has $\operatorname{tr} f_\alpha^* f_\beta = \delta_{\alpha\beta}$. It is easy to check that the maps $\Phi_{\alpha\beta} \in L(M_n, M_n)$ defined by $\Phi_{\alpha\beta}(A) \equiv f_\alpha A f_\beta^*$ for any $A \in M_n$ can be written as $\Phi_{\alpha\beta} = |f_\alpha\rangle\langle f_\beta|$ and the set $\{\Phi_{\alpha\beta}\}$ is a ONB of $M_n \otimes M_n$. Moreover, the corresponding elements $\Phi_{\alpha\beta}^L, \Phi_{\alpha\beta}^R \in M_n \otimes M_n$ form ONBs of $M_n \otimes M_n$. The explicit expressions of $\Phi_{\alpha\beta}^L$ and $\Phi_{\alpha\beta}^R$ are

$$\Phi_{\alpha\beta}^L \equiv \sum_{i,j=1}^n f_\alpha e_{ij} f_\beta^* \otimes e_{ij} \quad \text{and} \quad \Phi_{\alpha\beta}^R \equiv \sum_{i,j=1}^n e_{ij} \otimes f_\alpha e_{ij} f_\beta^*.$$

There exist some important consequences for the above isomorphisms:

1. Any map $\Phi \in L(M_n, M_n)$ is uniquely written as

$$\Phi(A) = \sum c_{\alpha\beta} \Phi_{\alpha\beta}(A) = \sum c_{\alpha\beta} f_\alpha A f_\beta^* \quad \text{with some } c_{\alpha\beta} \in \mathbb{C}.$$

2. If $\Phi(A^*) = \Phi(A)^*$, then $c_{\alpha\beta} = \overline{c_{\alpha\beta}} \in \mathbb{R}$ and Φ^L, Φ^R are self-adjoint in $\mathcal{H} \otimes \mathcal{H}$.
3. If $\Phi(A) = \Phi(A)^*$, that is, the matrix $C \equiv (c_{\alpha\beta})$ is Hermitian, then Φ and Φ^L, Φ^R can be written in the following canonical forms:

$$\Phi(A) = \sum_\alpha c_\alpha g_\alpha A g_\alpha^*,$$

$$\Phi^L = \sum_{\alpha,i,j} c_\alpha g_\alpha e_{ij} g_\alpha^* \otimes e_{ij},$$

$$\Phi^R = \sum_{\alpha,i,j} c_\alpha e_{ij} \otimes g_\alpha e_{ij} g_\alpha^*,$$

where $\{g_\alpha; \alpha = 1, \ldots, n^2\}$ is some ONB in M_n and $c_\alpha \in \mathbb{R}$.
4. From part 3, it follows that for any ONB $\{f_\alpha\}$

$$P_\alpha \equiv \Phi_{\alpha\alpha}^L = \sum_{i,j=1}^n f_\alpha e_{ij} f_\alpha^* \otimes e_{ij}, \qquad Q_\alpha \equiv \Phi_{\alpha\alpha}^R = \sum_{i,j=1}^n e_{ij} \otimes f_\alpha e_{ij} f_\alpha^*$$

are mutually orthogonal projections in $\mathcal{H} \otimes \mathcal{H}$ satisfying

$$\sum_{\alpha=1}^{n^2} P_\alpha = \sum_{\alpha=1}^{n^2} Q_\alpha = I \otimes I \quad (I \text{ is the unity of } M_n).$$

5. A any state (density operator) σ_{12} on $\mathcal{H} \otimes \mathcal{H}$ can be written in the form

$$\sigma_{12} = \sum_{\alpha=1}^{n^2} \lambda_\alpha Q_\alpha = \sum_{\alpha=1}^{n^2} \lambda_\alpha \sum_{i,j=1}^n e_{ij} \otimes f_\alpha e_{ij} f_\alpha^*$$

with $\sum_{\alpha=1}^{n^2} \lambda_\alpha = 1$ and $\lambda_\alpha \geq 0$. Put

$$\Theta^*(A) \equiv \sum_{\alpha=1}^{n^2} \lambda_\alpha f_\alpha A f_\alpha^*$$

for any $A \in M_n$. Then Θ^* is a completely positive (CP) map on M_n, and σ_{12} is written as

$$\sigma_{12} = \sum_{i,j=1}^{n} e_{ij} \otimes \Theta^*(e_{ij}).$$

Let us take $A \in M_n$ with $\operatorname{tr} A^* A = 1$, then A^L (A^R) is a normalized vector in $\mathcal{H} \otimes \mathcal{H}$, and it defines a state σ in $\mathcal{H} \otimes \mathcal{H}$ as $\sigma \equiv |A^L\rangle\langle A^L|$.

Definition 18.32 The above state σ is maximally entangled if $A^* A = A A^* = \frac{I}{n}$, equivalently, $A = \frac{1}{\sqrt{n}} U$ for some unitary operator U in $\mathcal{H}$.

Remark 18.33 One can construct an ONB $\{f_\alpha = U_\alpha / \sqrt{n}; \alpha = 1, \ldots, n^2\}$ with unitary U_α. Then the corresponding projections P and Q given above in part 4 are maximally entangled states.

Definition 18.34 The map $\Phi \in L(M_n, M_n)$ is said to be normalized if $\Phi(I) = I$, base-preserving if $\operatorname{tr} \Phi(A) = \operatorname{tr} A$ for all $A \in M_n$, self-adjoint if $\Phi(A)^* = \Phi(A^*)$ for all $A \in M_n$, positive if $\Phi(A^* A) \geq 0$ for all $A \in M_n$, and completely positive if $\sum_{i,j=1}^{n} \langle x_i, \Phi(A_i^* A_j) x_j \rangle \geq 0$ for any x_i $(i = 1, \ldots, n) \in \mathcal{H}$ and any A_i $(i = 1, \ldots, n) \in M_n$. Note that the canonical form of completely positive map is given by Θ^* above.

18.9.2 New Scheme of Teleportation

We propose a new protocol for quantum teleportation. Let us take the conditions that all three Hilbert spaces $\mathcal{H}_1$, $\mathcal{H}_2$, and $\mathcal{H}_3$ are $\mathbb{C}^n$. Let the state σ in $\mathcal{H}_2 \otimes \mathcal{H}_3 = \mathbb{C}^n \otimes \mathbb{C}^n$ be

$$\sigma = \sum_{i,j=1}^{n} e_{ij} \otimes \Theta^*(e_{ij}).$$

Here e_{ij}, Θ^* are those given in the previous subsection with an ONB $\{f_\alpha; \alpha = 1, \ldots, n^2\}$, but are defined on $\mathcal{H}_2$ and $\mathcal{H}_3$. We set an observable F in $\mathcal{H}_1 \otimes \mathcal{H}_2$ to be measured by Alice as follows:

$$F = \sum_\alpha z_\alpha P_\alpha \equiv \sum_\alpha z_\alpha \sum_{i,j=1}^{n} g_\alpha^* e_{ij} g_\alpha \otimes e_{ij},$$

where $\{g_\alpha; \alpha = 1, \dots, n^2\}$ is another ONB of M_n. Then we define the unnormalized teleportation map for an input state ρ in $\mathcal{H}_1$ and the measured value z_α of Alice by

$$T_\alpha^*(\rho) \equiv \mathrm{tr}_{12}(P_\alpha \otimes I)\rho \otimes \sigma (P_\alpha \otimes I).$$

Lemma 18.35 *The teleportation map T_α^* has the form $T_\alpha^*(\rho) = \Theta^*(g_\alpha \rho g_\alpha^*)$ for any ρ in $\mathcal{H}_1$.*

Proof One can write $T_\alpha^*(\rho)$ as

$$\begin{aligned}
T_\alpha^*(\rho) &= \sum_{i,j=1}^n \sum_{k,l=1}^n \sum_{t,s=1}^n \mathrm{tr}\,(g_\alpha^* e_{ij} g_\alpha \rho g_\alpha^* e_{ts} g_\alpha)\mathrm{tr}\,(e_{ij} e_{kl} e_{ts})\Theta^*(e_{kl}) \\
&= \sum_{i,j,s=1}^n \mathrm{tr}\,(g_\alpha^* e_{ij} g_\alpha \rho g_\alpha^* e_{ti} g_\alpha)\Theta^*(e_{jt}) \\
&= \sum_{i=1}^n \langle g_\alpha^* e_i, g_\alpha e_i\rangle \sum_{j,t=1}^n \langle e_j, g_\alpha \rho g_\alpha^* e_t\rangle \Theta^*(e_{jt}) \\
&= \sum_{j,t=1}^n \mathrm{tr}\,(g_\alpha \rho g_\alpha^* e_{jt})\Theta^*(e_{jt}) \\
&= \Theta^*(g_\alpha \rho g_\alpha^*).
\end{aligned}$$

□

It is easily seen that T_α^* is completely positive but not trace-preserving. In order to consider the trace-preserving map from T_α^*, let us consider the dual map T_α of T_α^*, i.e., $\mathrm{tr}\, A T_\alpha^*(\rho) =: \mathrm{tr}\, T_\alpha(A)\rho$. Indeed, it is

$$T_\alpha(A) = g_\alpha^* \Theta(A) g_\alpha, \qquad A \in M_n,$$

where Θ is the dual to Θ^*:

$$\Theta(A) = \sum_{\alpha=1}^{n^2} \lambda_\alpha f_\alpha^* A f_\alpha.$$

The map T_α is normalizable iff $\mathrm{rank}\, T_\alpha(I) = n$, that is, the operator $T_\alpha(I)$ is invertible. Put

$$\kappa_\alpha \equiv T_\alpha(I).$$

In this case, the dual teleportation map T_α is normalized as

$$\Upsilon_\alpha \equiv \kappa_\alpha^{-\frac{1}{2}} T_\alpha \kappa_\alpha^{-\frac{1}{2}}.$$

The dual map Υ_α^* of Υ_α is trace-preserving and it has the form

$$\Upsilon_\alpha^*(\rho) = \Theta^*\left(g_\alpha \kappa_\alpha^{-\frac{1}{2}} \rho \kappa_\alpha^{-\frac{1}{2}} g_\alpha^*\right) = \sum_{\beta=1}^{n^2} \lambda_\beta f_\beta g_\alpha \kappa_\alpha^{-\frac{1}{2}} \rho \kappa_\alpha^{-\frac{1}{2}} (f_\beta g_\alpha)^*.$$

Definition 18.36 We call the map Υ_α^* a quantum teleportation channel.

It is important to note that this teleportation channel Υ_α^* is linear with respect to all initial states ρ.

Let us consider a special case of σ such that

$$\sigma = \sum_{i,j=1}^{n} e_{ij} \otimes \Theta^*(e_{ij}) \quad \text{with } \Theta^*(\cdot) \equiv f \cdot f^* \text{ and } \operatorname{tr} f^* f = 1.$$

That is, σ is a pure state. In this case, one has

$$T_\alpha^*(\rho) = (g_\alpha f)\rho(g_\alpha f)^*$$

and

$$\kappa_\alpha = (g_\alpha f)^*(g_\alpha f).$$

Remark 18.37 If $g_\alpha = U_\alpha/\sqrt{n}$ and $f = V/\sqrt{n}$, where U_α and V are unitary operators, then $\kappa_\alpha = 1/n^2$, which corresponds to the usual discussion.

Further, it follows that the teleportation Υ_α^* is trace-preserving iff $\operatorname{rank}(g_\alpha) = \operatorname{rank}(f) = n$, and in such a case one has

$$\Upsilon_\alpha^*(\rho) = (f g_\alpha)\kappa_\alpha^{-\frac{1}{2}} \rho \kappa_\alpha^{-\frac{1}{2}} (f g_\alpha)^*.$$

Put

$$W_\alpha \equiv f g_\alpha \kappa_\alpha^{-\frac{1}{2}},$$

which is easily seen to be unitary. Thus we proved the following theorem.

Theorem 18.38 *Given an ONB $\{g_\alpha; \alpha = 1, \ldots, n^2\}$ and a vector f in M_n on the n-dimensional Hilbert space, if* $\operatorname{rank}(g_\alpha) = \operatorname{rank}(f) = n$ *is satisfied, then one can construct an entangled state σ and the set of keys $\{W_\alpha\}$ such that the teleportation channel is linear, and perfect teleportation occurs.*

That is, our teleportation protocol does not require that the entangled state be maximal for the linear perfect teleportation. In other protocols, teleportation channel becomes nonlinear if the entangled state is not maximal; moreover, perfect teleportation can occur if the entangled state is maximal.

We will give examples of our teleportation scheme with non-maximally entangled states in the next subsection.

18.9.3 Examples

Let us construct an example. That is, we construct an entangled state given in the form: $\sigma = \sum_{i,j=1}^{n} e_{ij} \otimes \Theta^*(e_{ij})$ with $\Theta^*(\cdot) \equiv f \cdot f^*$ and $\operatorname{tr} f^* f = 1$. Then it is possible in our protocol to teleport completely by means of a non-maximally entangled state σ. The above state σ is pure, so that σ is maximally entangled iff $f = U/\sqrt{n}$ with some unitary operator u. Therefore, if $\operatorname{rank}(f) = n$ and $f \neq U/\sqrt{n}$, then σ is not maximally entangled.

We will consider a bit more general question: For an ONB $\{f_\alpha\}$ $(\alpha = 1, \dots, n^2)$ in M_n, can we construct n^2 projections $Q_\alpha = \sum_{i,j=1}^{n} e_{ij} \otimes f_\alpha e_{ij} f_\alpha^*$ such that all Q_α are mutually orthogonal and not maximally entangled? This question is reduced to finding the basis $\{f_\alpha\}$ such that $\operatorname{rank}(f_\alpha) = n$ for any α and $f_\alpha \neq \frac{U}{\sqrt{n}}$ with unitary U.

1. A positive answer for the above question is given in the case $n = 2$, that is, M_2. Let S_α $(\alpha = 0, 1, 2, 3)$ be spin matrices,

$$S_0 = I, \qquad S_1 = \begin{pmatrix} 0 & 1 \\ 1 & 0 \end{pmatrix}, \qquad S_2 = \begin{pmatrix} 0 & -\mathrm{i} \\ \mathrm{i} & 0 \end{pmatrix}, \qquad S_3 = \begin{pmatrix} 1 & 0 \\ 0 & -1 \end{pmatrix},$$

and put

$$\omega_\alpha \equiv \frac{S_\alpha}{\sqrt{2}} \quad (\alpha = 0, 1, 2, 3).$$

Now we consider an orthogonal transformation $C : \mathbb{R}^4 \to \mathbb{R}^4$. In terms of $C \equiv (C_{\alpha\beta})$ one defines a new basis $\{f_\alpha\}$ in M_2:

$$f_\alpha = \sum_{\beta=0}^{3} C_{\alpha\beta} \omega_\beta. \tag{18.24}$$

Since $C_{\alpha\beta}$ is real and $\omega_\alpha = \omega_\alpha^*$, it implies that $f_\alpha = f_\alpha^*$ and the equality

$$\det f_\alpha = \frac{1}{2} \left(C_{\alpha 0}^2 - \sum_{\beta=1}^{3} C_{\alpha\beta}^2 \right),$$

so that all f_α have rank 2 iff $\det f_\alpha \neq 0$. Such f_α $(\alpha = 0, 1, 2, 3)$ generate the corresponding projection $Q_\alpha = \sum_{i,j=1}^{n} e_{ij} \otimes f_\alpha e_{ij} f_\alpha^*$ on mutually orthogonal subspaces of $\mathbb{C}^2 \otimes \mathbb{C}^2$ such that Q_α $(\alpha = 0, 1, 2, 3)$ are non-maximally entangled states iff the transformation $\{\omega_\alpha\}$ to $\{f_\alpha\}$ cannot be generated by unitary U such as $U \omega_\alpha U^* = f_\alpha$.

From the orthogonality relation to C, it follows that

$$C_{\alpha 0}^2 + \sum_{\beta=1}^{3} C_{\alpha\beta}^2 = 1 \quad \text{and} \quad \sum_{\alpha=0}^{3} C_{\alpha 0}^2 = 1.$$

These relations tell us that $\det f_\alpha \neq 0$ iff $C_{\alpha 0}^2 \neq \frac{1}{2}$. Thus the relation $\sum_{\alpha=0}^{3} C_{\alpha 0}^2 = 1$ implies that $\det f_\alpha \neq 0$ iff $C_{\alpha 0}^2 > \frac{1}{2}$.

As an example, let us take the matrix C of the form

$$C \equiv R_{01}(\theta_1) R_{02}(\theta_2) R_{03}(\theta_3),$$

where $R_{ab}(\theta)$ is the rotation in (a, b)-plane with the angle θ. Then one finds

$$C = \begin{pmatrix} c_1 c_2 c_3 & -s_1 & -c_1 s_2 & -c_1 c_2 s_3 \\ s_1 c_2 c_3 & c_1 & -s_1 s_2 & -s_1 c_2 s_3 \\ s_2 c_3 & 0 & c_2 & -s_2 s_3 \\ s_3 & 0 & 0 & c_3 \end{pmatrix},$$

where $c_i \equiv \cos\theta_i$ and $s_i \equiv \sin\theta_i$. It is easy to check that f_α generate the projections Q_α, whose corresponding states are non-maximally entangled if $|s_3| > \frac{1}{2}$. This inequality can be realized by taking θ_3 properly, e.g., $\frac{\pi}{6} < \theta_3 < \frac{5\pi}{6}$.

2. We can construct even simpler ONB $\{f_\alpha; \alpha = 0, 1, 2, 3\}$ generating non-maximally entangled states such as

$$f_0 = \begin{pmatrix} \cos\theta_1 & 0 \\ 0 & \sin\theta_1 \end{pmatrix}, \qquad f_1 = \begin{pmatrix} -\sin\theta_1 & 0 \\ 0 & \cos\theta_1 \end{pmatrix},$$

$$f_2 = \begin{pmatrix} 0 & \cos\theta_2 \\ \sin\theta_2 & 0 \end{pmatrix}, \qquad f_3 = \begin{pmatrix} 0 & -\sin\theta_2 \\ \cos\theta_2 & 0 \end{pmatrix}.$$

These are the matrices with rank $= 2$ for $0 < \theta_1, \theta_2 < \pi/2$, and they generate a non-maximally entangled state when $\theta_1, \theta_2 \neq \pi/4$.

18.9.4 Perfect Teleportation for Non-maximally Entangled State

In this subsection, we briefly explain the physical model of the Kossakowski–Ohya teleportation scheme. The details are given in [743].

This model is dependent on an entangled state generated by a photon number state with N photons through the half beamsplitter. The entangled state is written on $\mathcal{H}_2 \otimes \mathcal{H}_3$ as

$$|\xi\rangle = \sum_{n=0}^{N} d_n |n\rangle \otimes |N - n\rangle,$$

where $|n\rangle$ is the n-photon number state and

$$d_n = (-1)^{N-n} e^{-\mathrm{i}\phi(N-n)} \sqrt{2^{-N} \binom{N}{n}}.$$

The above ϕ is the phase difference between the reflected beam and the transmitted beam [427]. Remark that the $|\xi\rangle$ is a non-maximally entangled state, in general.

In the model, we take a superposition $|\psi\rangle$ of Schrödinger's cat state vectors $|\alpha_{\mathrm{even}}\rangle$ and $|\alpha_{\mathrm{odd}}\rangle$ as an input state vector being sent from Alice to Bob which is written as

$$|\psi\rangle = c_1|\alpha_{\mathrm{even}}\rangle + c_2|\alpha_{\mathrm{odd}}\rangle.$$

Here, $|\alpha_{\mathrm{even}}\rangle$ and $|\alpha_{\mathrm{odd}}\rangle$ are defined by

$$|\alpha_{\mathrm{even}}\rangle = \frac{1}{\sqrt{C_+}}\sum_{n=0}^{\infty}\frac{\alpha^n + (-\alpha)^n}{\sqrt{n!}}|n\rangle,$$

$$|\alpha_{\mathrm{odd}}\rangle = \frac{1}{\sqrt{C_-}}\sum_{n=0}^{\infty}\frac{\alpha^n - (-\alpha)^n}{\sqrt{n!}}|n\rangle,$$

where $C_\pm \equiv 2\exp(|\alpha|^2) \pm 2\exp(-|\alpha|^2)$ [384].

Alice measures *the phase difference of beams* and *the sum of photon number* on $\mathcal{H}_1 \otimes \mathcal{H}_2$ [491]. The operator expressing the sum of photon number is

$$\begin{aligned}\hat{N}_+ &= \hat{N}_1 + \hat{N}_2\\ &\equiv \sum_{q=0}^{\infty} q\left(\sum_{t=0}^{q}|q-t\rangle\langle q-t| \otimes |t\rangle\langle t|\right),\end{aligned}$$

where $\hat{N}_1$ (resp., $\hat{N}_2$) is the number operator on $\mathcal{H}_1$ (resp., $\mathcal{H}_2$). The operator expressing the phase difference of beam is

$$\hat{\Phi}_- = \sum_{m=0}^{q}\phi_m^-\left(\sum_{t=0}^{q}\left|\phi_{t+m}^{(1)}\right\rangle\left\langle\phi_{t+m}^{(1)}\right| \otimes \left|\phi_t^{(2)}\right\rangle\left\langle\phi_t^{(2)}\right|\right)$$

where

$$\left|\phi_m^{(k)}\right\rangle = \sum_{n=0}^{q}\frac{\exp(in\phi_m^{(k)})}{\sqrt{q+1}}|n\rangle,$$

$$\phi_m^{(k)} = \phi_0^{(k)} + \frac{2\pi m}{q+1}$$

are Pegg–Barnett phase state vectors on $\mathcal{H}_k$, and $\phi_m^- = \phi_0^{(1)} - \phi_0^{(2)} + \frac{2\pi m}{q+1}$ [549, 640]. There exist common eigenvectors of both $\hat{N}_+$ and $\hat{\Phi}_-$, which are given by

$$|q, \phi_m^-\rangle = \sum_{n=0}^{q}\frac{\exp(-in\phi_m^-)}{\sqrt{q+1}}|q-n\rangle \otimes |n\rangle$$

for the eigenvalue q of $\hat{N}_+$ and that ϕ_m^- of $\hat{\Phi}_-$.

Let us rewrite $|\xi\rangle$ and $|q,\phi_m^-\rangle$ with the Kossakowski–Ohya formalism as

$$|\xi\rangle = \sum_{n=0}^{N} |n\rangle \otimes f|n\rangle,$$

$$|q,\phi_m^-\rangle = \sum_{n=0}^{q} g_{q,m}^* |n\rangle \otimes |n\rangle,$$

where

$$f = \sum_{n=0}^{N} d_n |N-n\rangle\langle n|,$$

$$g_{q,m}^* = \sum_{n=0}^{q} \frac{\exp(-in\phi_m^-)}{\sqrt{q+1}} |q-n\rangle\langle n|.$$

Obviously, the $g_{q,m}$ satisfy $\mathrm{tr}(g_{q,m}^* g_{q',m'}) = \delta_{q,q'}\delta_{m,m'}$.

In the same way discussed above, we can construct Bob's key by f and $g_{q,m}$ through which complete teleportation is achieved when $|\alpha|^2 = \frac{N}{2}$.

18.10 Notes

The first model of quantum teleportation was given by Bennett et al. [105, 107], whose model is discussed in Sect. 18.2. The channel expression of the teleportation process is studied in [363]. Weak teleportation was considered in [20], in which the uniqueness of keys and the maximality of entanglement are proved. Similar discussion was given in [348, 813]. A teleportation model in Fock space was given by Fichtner and Ohya [246, 248], in which incomplete teleportation is rigorously discussed and its importance is pointed out. A model using squeezed states was considered in [134]. The fidelity [392, 759] of teleportation process has been computed in [247]. The Fichtner–Freutenberg expression of Fock space was given in [242–244, 486]. Mathematically rigorous study of beam splitting was given in [20, 244]. Spatially separated teleportation discussed here is taken from [246]. The mathematical structure of Einstein–Podolsky–Rosen state using C^*-algebra is presented in [315, 813]. The complete teleportation is realized mostly for maximally entangled states, Kossakowski and Ohya found [444] a new teleportation scheme so that the complete teleportation is possible for non-maximally entangled states.

Almost all discussions of quantum teleportation have been based on finite dimensionality of the Hilbert spaces attached to Alice and Bob. As is well-known, success of quantum mechanics is due to the discovery that nature is described by infinite-dimensional Hilbert spaces, so that it is desirable to demonstrate the quantum teleportation process in a certain infinite-dimensional Hilbert space. A recent paper [251] was an attempt to describe the teleportation process in an infinite-dimensional Hilbert space by giving simple examples.

Chapter 19
Physical Nanosystems

In this chapter, certain experimental realizations of quantum information schemes are briefly discussed. Some nanosystems used for experimental implementation of quantum computation such as quantum dots, ion traps, and nuclear magnetic resonance are considered. Parametric down-conversion for producing entangled photons is described. Full exposition of recent experimental works is beyond the scope of this book, so we will discuss only some fundamental topics.

19.1 Quantum Dots

19.1.1 Quantum Wells, Quantum Wires, Quantum Dots

A quantum dot is a semiconductor whose excitons are confined in all three spatial dimensions ranging from 2–10 nanometers (10–50 atoms) in diameter. They have properties that are between those of bulk semiconductors and those of discrete molecules. They were discovered at the beginning of the 1980s by Ekimov in a glass matrix and by Brus in colloidal solutions.

Quantum dots are quasi-zero-dimensional systems that contain a small and controllable number of electrons, see [385]. One has the possibility of controlling their properties, for example, their shape, dimensions, and the number of confined electrons.

Quantum dots are similar to atoms and often referred to as the artificial atoms. However, quantum dots do not have nuclei, and the potential of a quantum dot differs from the Coulomb potential binding electrons in an atom. In a good approximation, an electron has the Fock–Darwin energy levels.

From the 1970s, the electronic structures of quasi-two-dimensional structures, quantum wells, are explored. A *quantum well* is a very thin, of a few nanometer thick, flat layer of semiconductor sandwiched between two layers of another semiconductor with a higher conduction-band energy. The material used commonly for creating quantum wells is gallium arsenide, GaAs. Remarkable physical properties

M. Ohya, I. Volovich, *Mathematical Foundations of Quantum Information and Computation and Its Applications to Nano- and Bio-systems*,
Theoretical and Mathematical Physics,
DOI 10.1007/978-94-007-0171-7_19, © Springer Science+Business Media B.V. 2011

of quantum wells such as the integer and fractional quantum Hall effects were observed. Quantum wells are implemented in numerous devices, for example, laser diodes used in CD players.

Quasi-one-dimensional structures, *quantum wires*, were produced at the beginning of the 1980s in the form of miniature strips, etched in a sample containing a quantum well.

Quantum dots were first created at the end of 1980s by etching them in quantum wells by means of lithography.

Quantum dots are also created by using modulated electric field from miniature electrodes over the surface of a quantum well by means of lithographic techniques, through the self-crystallization, and by using other methods. Both single quantum dots and large arrays (matrices) of dots were produced.

19.1.2 Fock–Darwin Hamiltonian and Interacting Electrons in Quantum Dot

The energy of an electron in quantum dot is quantized as a result of the confinement in a small area. Quantum dots are created usually through producing the potential $V(x, y)$ restricting the motion of the electrons, which are confined in a very narrow quantum well. Therefore, they have the shape of flat discs. In a good approximation, one can use the parabolic well $V(\mathbf{r}) \sim \mathbf{r}^2$, $\mathbf{r} = (x, y)$. The Hamiltonian operator of an electron in the parabolic well in a perpendicular magnetic field has the form:

$$\begin{aligned} H &= \frac{1}{2m}\left(\mathbf{p} - \frac{e}{c}\mathbf{A}\right)^2 + \frac{m}{2}\omega_0^2\mathbf{r}^2 \\ &= \frac{\mathbf{p}^2}{2m} + \frac{m}{2}\left(\omega_0^2 + \frac{\omega_c^2}{4}\right)\mathbf{r}^2 - \frac{1}{2}\omega_c l_z \end{aligned}$$

where c is the speed of light in vacuum. Here m is the effective mass, $\mathbf{r} = (x, y)$ is the position, $\mathbf{p} = (p_x, p_y) = (-\mathrm{i}\partial_x, -\mathrm{i}\partial_y)$ is the momentum, $l_z = xp_y - yp_x$ is the projection of the angular momentum onto the field direction. Here note that

$$\mathbf{p}^2 \equiv \mathbf{p} \cdot \mathbf{p} = p_x^2 + p_y^2 = -\partial_x^2 - \partial_y^2.$$

A is the vector potential of the magnetic field, $\mathbf{A} = \frac{1}{2}B(y, -x)$ (see Sect. 16.1.1), and $\omega_c = eB/mc$ is the cyclotron frequency. The eigenstates and the energy levels of the Hamiltonian were determined by Fock and Darwin by using the transformation to a pair of harmonic oscillators.

Using the complex variables $z = x + \mathrm{i}y$, $\overline{z} = x - \mathrm{i}y$ and differential operators $\partial = (\partial_x - \mathrm{i}\partial_y)/2$, $\overline{\partial} = (\partial_x + \mathrm{i}\partial_y)/2$, we define two pairs of annihilation–creation operators:

$$a = \frac{1}{\sqrt{2}}\left(\frac{1}{2l_0}\overline{z} + 2l_0\partial\right), \qquad a^* = \frac{1}{\sqrt{2}}\left(\frac{1}{2l_0}z + 2l_0\overline{\partial}\right)$$

and

$$b = \frac{1}{\sqrt{2}}\left(\frac{1}{2l_0} z + 2l_0 \bar{\partial}\right), \qquad b^* = \frac{1}{\sqrt{2}}\left(\frac{1}{2l_0} \bar{z} - 2l_0 \partial\right)$$

where

$$l_0 = \frac{l_B}{\sqrt[4]{1 + 4\omega_0^2/\omega_c^2}}, \quad l_B = \sqrt{\frac{\hbar c}{eB}}.$$

The operators satisfy the following commutation relations

$$[a, a^*] = [b, b^*] = 1, \qquad [a, b] = [a, b^*] = 0.$$

The Hamiltonian above takes the form

$$H = \hbar\omega_+ \left(a^* a + \frac{1}{2}\right) + \hbar\omega_- \left(b^* b + \frac{1}{2}\right)$$

i.e., represents a pair of independent harmonic oscillators, where

$$\omega_\pm = \sqrt{\omega_0^2 + \frac{1}{4}\omega_c^2} \pm \frac{1}{2}\omega_c.$$

Therefore, the eigenstates of the Hamiltonian have the form

$$|n_+, n_-\rangle = \frac{1}{\sqrt{n_+! n_-!}} a^{*n_+} b^{*n_-} |0, 0\rangle, \quad n_+, n_- = 0, 1, 2, \ldots,$$

where $|0, 0\rangle$ is the vacuum state, $a|0, 0\rangle = b|0, 0\rangle = 0$. The eigenenergies are

$$E(n_+, n_-) = \hbar\omega_+ \left(n_+ + \frac{1}{2}\right) + \hbar\omega_- \left(n_- + \frac{1}{2}\right).$$

They are called the Fock–Darwin energy levels. In the strong magnetic field when $\omega_c \gg \omega_0$, one can take $\omega_0 \simeq 0$ and one gets the Landau energy levels:

$$E_n = \hbar\omega_c \left(n + \frac{1}{2}\right).$$

The Hamiltonian describing an N-electron quantum dot is

$$H = \sum_{i=1}^{N} \left[\frac{1}{2m}\left(\mathbf{p}_i - \frac{e}{c}\mathbf{A}(\mathbf{r}_i)\right)^2 + V(\mathbf{r}_i) + \alpha \mathbf{L}_i \cdot \boldsymbol{\sigma}_i - g\mu_B \boldsymbol{\sigma}_i \cdot \mathbf{B}\right] + \frac{1}{2}\sum_{i \neq j} \frac{e^2}{|\mathbf{r}_i - \mathbf{r}_j|}.$$

Here $\mathbf{L}_i = \mathbf{r}_i \times \mathbf{p}_i$ is the orbital angular momentum and $\boldsymbol{\sigma}_i$ is the spin (one of Pauli matrices) of the ith electron. The term $\alpha \mathbf{L}_i \cdot \boldsymbol{\sigma}_i$ describes the spin–orbit interaction with the coupling constant α, and the Zeeman coupling $g\mu_B \boldsymbol{\sigma}_i \cdot \mathbf{B}$ describes the coupling with the magnetic field, where g is the effective g-factor and μ_B the Bohr magneton.

The potential $V(\mathbf{r})$ in a good approximation can be taken to be a Gaussian or a parabolic well. For applications to quantum computing, the potential was taken to be a quartic polynomial which was used to model the coupling of two quantum dots. The Hamiltonian was studied by using known quantum mechanical methods and numerical simulation.

19.1.3 Properties of Quantum Dots

Many interesting properties of quantum dots have been investigated theoretically and in experiments. For the single particle states, the use of strong magnetic fields enables a transition for the regime of spatial quantization of the Fock–Darwin energy levels to the regime of magnetic field Landau quantization.

In a quantum dot, not only the conduction-band electrons but also the another type of carriers, the valence-band holes, can be bound. The basic tool for the investigation of discrete energy levels of quantum dots are the photoluminescence measurements. Electron–hole pairs (excitons) can be created by means of a laser beam.

In large quantum dots, the Coulomb interaction between electrons is important, in particular, it determines the ground state of the system. It leads to the formation of the so-called incompressive magic states of a few electron systems. The magic states of *strongly* interacting electrons were described as some states of *weakly* interacting *composite fermions*, i.e., free fermions with attached fluxes of a magnetic field.

There are considerations of possible applications of quantum dots in the construction of new quantum-dot-based lasers and in quantum computers.

19.1.4 Quantum Computation with Quantum Dots

There are several proposals of how to use quantum dots for quantum computation. Loss and DiVincenzo [490] proposed to consider the spin of the electron on a single-electron quantum dot as the qubit and coupled quantum dots as quantum gates. In this model, the two-qubit quantum gate operates by an electrical gating of the tunneling barrier between neighboring quantum dots.

The exchange coupling between two spins $\boldsymbol{\sigma}_1$ and $\boldsymbol{\sigma}_2$ is described by the Heisenberg Hamiltonian

$$H(t) = J(t)\boldsymbol{\sigma}_1 \cdot \boldsymbol{\sigma}_2.$$

Here $J(t)$ is the time-dependent exchange coupling that is produced by turning on and off of the tunneling matrix elements by using, for example, a split-gate technique.

If the coupling is pulsed such that $\int J(t)\,dt = J_0\tau_0 = \pi \pmod{2\pi}$, then for the specific duration τ_0 the unitary evolution operator $U(t)$ corresponds to the swap operator U_{sw} which exchanges the quantum states of qubit 1 and 2, $U(\tau_0) = U_{\mathrm{sw}}$. The quantum XOR operation U_{XOR} can be obtained by applying a sequence of square-roots of swap and single-qubit operations:

$$U_{\mathrm{XOR}} = \exp\left[\frac{\mathrm{i}\pi}{4}\sigma_1^z\right]\exp\left[-\frac{\mathrm{i}\pi}{4}\sigma_2^z\right]U_{\mathrm{sw}}^{1/2}\exp\left[\frac{\mathrm{i}\pi}{2}\sigma_1^z\right]U_{\mathrm{sw}}^{1/2}.$$

It is known that U_{XOR} is a universal quantum gate if it is combined with single qubit rotations. Therefore, the described quantum dot model can, in principle, be used to implement any quantum algorithm.

Also the decoherence in this model was considered by using the spin-boson Hamiltonian.

19.2 Quantum Communication Experiments

19.2.1 Quantum Cryptography and Teleportation

There are several quantum communication experiments. In a demonstration of entanglement-based key distribution, the sources uses type II parametric down-conversion in β-barium borate ($\beta - \mathrm{B}_a\mathrm{B}_2\mathrm{O}_4$; we call it BBO), pumped with an argon laser. The photons, with a wavelength of 702 nm, are each coupled into 500 m long optical fibers and transmitted to Alice and Bob, who are separated by 400 m. Quantum key distribution is started by a single light pulse from the source to Alice and Bob. After a run of about 5 s duration has been completed, and Alice and Bob compare their lists of detections to extract the coincidences. After a measurement run, the quantum keys are established by Alice and Bob through classical communication over a standard computer network. The system has a measured rate of total coincidence of about 1700 per second, and the collection efficiency of each photon path of 5%. For more recent experiments on quantum cryptography, see [751, 822].

In experiments on quantum teleportation, polarization-entangled photons were produced by type II down-conversion in a nonlinear BBO crystal. Here the UV beam was pulsed, the pulses had a duration of about 200 fs (1 fs $= 10^{-15}$ s) and $\lambda = 294$ nm.

The entangled pair of photons 2 and 3 is produced in a Bell state in the first passage of the UV pulses through the crystal which is distributed to Alice and Bob. The pulse is reflected at a mirror back through the crystal and produces another pair of photons, one of which is prepared in the initial state to be teleported (photon 1), and the other one (photon 4) serves as a trigger indicating that a photon to be teleported is on its way. Alice then looks for coincidences behind her beam splitter, where the initial photon and one of the ancillaries are superposed. Bob, after receiving the classical information that Alice has obtained a coincidence count identifying the Bell state, knows that his photon 3 is in the initial state of photon 1, which then can be verified using polarization analysis e.g., [129]. Therefore, the initial state of photon 1 is teleported onto the state of photon 3.

For more recent experiments on quantum teleportation including teleportation of continuous variables, see [135, 613, 824, 834].

Here we mention that there exists a serious question of where the photon number states exist so that we can control them for a special purpose. Fichtner and Ohya proposed a mathematical model by means of coherent states as was discussed in Chap. 18 with a new concept "non-perfect teleportation".

19.3 Experimental Realizations of Quantum Computations

Several experiments demonstrating the basic properties of quantum computations were performed. These experiments showed that small-scale quantum logic operations are conceivable. However, whether or not it is possible to scale them up to practical quantum computation remains to be seen. Experimental methods based on trapped ions, cavity quantum electrodynamics (cavity QED), nuclear magnetic resonances (NMR), single atoms, and solid state devices were proposed among others to implement quantum computations.

Here we briefly discuss some of these experiments and proposals. To implement quantum computations, we need to encode the information then to process it by means of quantum gates, and finally to read out the result. We will discuss how these steps could be implemented in various proposed methods.

19.3.1 Ion Traps

In the realization of a quantum computer with trapped ions, each qubit can be implemented as a superposition of the ground electronic state and the excited state of an ion. It is shown that a set of ions interacting with laser light and moving in a linear trap provides a physical system to realize a quantum computer.

Consider N ions which are confined in a linear Paul trap by means of a time dependent inhomogeneous electric field. The ions basically move only in one dimension and interact with laser fields. Let us assume that the ions have been laser cooled in all three dimensions so that they undergo very small oscillations around the equilibrium positions. To implement a quantum computer, one has to implement single and two-qubit gates. Single qubit gates can be implemented by using the Rabi flopping between the internal states of the qubit. To implement two-qubit gates, one uses the external degree of freedom associated with the string of ions and, in particular, the center-of-mass motion which is the lowest quantized mode.

The Hamiltonian describing the interaction of a given ion i with a standing laser wave in the Lamb–Dicke limit and when the laser beam acts on one of the ions is given by $H = H_{\text{ex}} + H_{\text{int}} + H_{\text{las}}$, where

$$H_{\text{ex}} = \sum_{k=1}^{N} \nu_k a_k^* a_k, \qquad H_{\text{int}} = -\frac{\delta_i}{2}\sigma_z^i,$$

$$H_{\text{las}} = \frac{\Omega_i^a}{2}\left(\sigma_i^+ + \sigma_i^-\right) + \frac{\Omega_i^b}{2}\frac{\eta_{cm}}{\sqrt{N}}\left(a\sigma_i^+ + a^*\sigma_i^-\right).$$

Here ν_k is the frequency of the laser normal modes, δ_i is the laser detuning, Ω_i^a and Ω_i^b are Rabi frequencies not modifying, or modifying the motion of the ions, η_{cm} is the Lamb–Dicke parameter associated with the center of mass (CM) mode and a, a^* are annihilation and creation operators of the CM mode and $\sigma^{\pm} \equiv \sigma_x \pm \mathrm{i}\sigma_y$.

A suitable light field can couple two internal electronic levels of the ions, the ground state $|g\rangle$ giving the minimum energy and the excited state $|e\rangle$ to the external vibrational motion.

Quantum gates between one or two qubits can be realized by using this interaction as follows. Single-qubit quantum gates imply only individual rotations of a single ion. They can be realized using a laser at resonance with the internal transition frequency $\delta_i = 0$ with the ion localized at the antinode of the standing wave laser beam. The evolution in this case is given by the Hamiltonian

$$H_a^i = \frac{\Omega_i^a}{2}\left(\sigma_i^+ + \sigma_i^-\right)$$

inducing the rotations

$$|g\rangle_i \to \cos(k_L\pi/2)|g\rangle_i - \mathrm{i}e^{\mathrm{i}\phi}\sin(k_L\pi/2)|e\rangle_i,$$
$$|e\rangle_i \to \cos(k_L\pi/2)|e\rangle_i - \mathrm{i}e^{-\mathrm{i}\phi}\sin(k_L\pi/2)|g\rangle_i.$$

To implement two-qubit gates first chose the laser frequency in such a way that $\delta_i = -\nu_z$, i.e., it excites only the CM mode and the ion localized at the node of the standing wave laser beam. Then the interaction with the laser is given by the Hamiltonian

$$H_b^i = \frac{\Omega_i^b}{2}\frac{\eta_{cm}}{\sqrt{N}}\left(a\sigma_i^+ + a^*\sigma_i^-\right).$$

If one applies the laser for a fixed time $t = k\pi/(\Omega_i^b\eta_z/\sqrt{N})$ (a $k\pi$ pulse) the states will evolve as

$$|g\rangle_i|1\rangle \to \cos(k_L\pi/2)|g\rangle_i|1\rangle - \mathrm{i}e^{\mathrm{i}\phi}\sin(k_L\pi/2)|e'\rangle_i|0\rangle,$$
$$|e'\rangle_i|0\rangle \to \cos(k_L\pi/2)|e'\rangle_i|0\rangle - \mathrm{i}e^{-\mathrm{i}\phi}\sin(k_L\pi/2)|g\rangle_i|1\rangle,$$
$$|g\rangle|0\rangle \to |g\rangle|0\rangle.$$

We note that the tensor product $|x\rangle \otimes |y\rangle$ is denoted by $|x\rangle|y\rangle$ in this chapter.

Here $|0\rangle$ ($|1\rangle$) denotes the CM mode with zero (one) phonon, ϕ is the phase of the laser and $|e'\rangle$ can be either the state $|1\rangle$ of the qubit considered (denoted $|e\rangle$) or an auxiliary state selectively excited. Now a two-qubit quantum logic gate can be implemented as follows:

(i) Swap the internal state of the first ion to the motional state of the CM mode by using a π pulse focused on the first ion.
(ii) Introduce a conditional sign flip by means of a 2π pulse on the second ion using the auxiliary level $|e'\rangle_i$.
(iii) Swap back the quantum state of CM mode to the internal state of the first ion by using a π pulse.

We assume that before and after the gate the CM mode is in the vacuum state $|0\rangle$. The complete evolution will be

$$|g\rangle_1|g\rangle_2|0\rangle \to |g\rangle_1|g\rangle_2|0\rangle,$$
$$|g\rangle_1|e\rangle_2|0\rangle \to |g\rangle_1|e\rangle_2|0\rangle,$$
$$|e\rangle_1|g\rangle_2|0\rangle \to |e\rangle_1|g\rangle_2|0\rangle,$$
$$|e\rangle_1|e\rangle_2|0\rangle \to -|e\rangle_1|e\rangle_2|0\rangle.$$

The net effect of the interaction is a sign flip only when both ions are in the internal excited state.

Dynamical trapping of charged particles was first experimentally verified by W. Paul in 1958 [634]. A radiofrequency (rf) electric field, generated by an electrode structure, creates a pseudo-potential confining a charged particle. For the trapping of single atomic ions, the electrodes have typical dimensions of a few millimeters down to about 100 μm. The rf fields are in the 10–300 MHz range. The motion of a particle confined in such a field involves a fast component synchronous to the applied driving frequency and the slow secular motion in the dynamically created pseudo-potential. For a quadrupole field geometry, the pseudo-potential is harmonic, and the quantized secular motion of the trapped ion is very accurately described by quantum harmonic oscillator.

To prepare the initial state, one cools ions into their motional ground state and hyperfine ground state. To readout the results of computations, one measures populations of hyperfine states.

Some quantum logical gates have been experimentally demonstrated in ion traps in [167, 312, 533, 619].

19.3.2 Nuclear Magnetic Resonance

Nuclear magnetic resonance (NMR) studies transitions between the Zeeman levels of an atomic nucleus in a magnetic field. It is one of important spectroscopic techniques available in the molecular sciences, and moreover commercial spectrometers are widely available. There are well developed techniques of manipulation and detection of nuclear spin states using radio-frequency electromagnetic waves. A pulsed NMR system which is used for implementing quantum computation consists of a liquid sample and an NMR spectrometer. A molecule which might be used contains N protons which have spin $1/2$. The molecules are dissolved in a solvent giving an ensemble of N qubit quantum computers.

Let us describe the theory of NMR for a model of one and two spins. First, we consider the single spin dynamics. The Hamiltonian describing the interaction of a classical magnetic field with a two-state spin has the form

$$H = H(t) = \frac{\omega_0}{2}\sigma_z + g(\sigma_x \cos\omega t + \sigma_y \sin\omega t).$$

Here ω_0 is related to the strength of the static magnetic field, and g is related to the strength of the alternating magnetic field. The solution to the Schrödinger equation

$$\mathrm{i}\partial_t \big|\varphi(t)\big\rangle = H\big|\varphi(t)\big\rangle$$

is

$$\big|\varphi(t)\big\rangle = e^{-\mathrm{i}\frac{\omega}{2}\sigma_z t} e^{\mathrm{i}[\frac{\omega_0-\omega}{2}\sigma_z + g\sigma_x]t}\big|\varphi(0)\big\rangle.$$

The solution can be understood as a single qubit rotation about the axis

$$\mathbf{m} = \frac{\mathbf{n}_z + \frac{2g}{\omega_0-\omega}\mathbf{n}_x}{\sqrt{1+(\frac{2g}{\omega_0-\omega})^2}}$$

by an angle

$$t\sqrt{\left(\frac{\omega_0-\omega}{2}\right)^2 + g^2}.$$

Here $\mathbf{n}_x = (1,0,0)$, $\mathbf{n}_y = (0,1,0)$, and $\mathbf{n}_z = (0,0,1)$. When ω is far from ω_0, the axis of the rotation of the spin is nearly parallel with $\mathbf{z}$, and its time evolution is nearly exactly that of the static Hamiltonian. On the other hand, when $\omega_0 \approx \omega$, i.e., if there is a resonance, then even a small alternating field can cause large changes in the state.

The spin–spin coupling in a good approximation can be described by the Hamiltonian

$$H = J\sigma_z^{(1)} \otimes \sigma_z^{(2)}$$

where J is the coupling constant.

Now let us show how one can build the controlled-NOT operation U_{CNOT} by using the one and two spin coupling Hamiltonians. One has the representation

$$U_{\mathrm{CNOT}} = (I \otimes H)K(I \otimes H)$$

where

$$U_{\mathrm{CNOT}} = \begin{pmatrix} 1 & 0 & 0 & 0 \\ 0 & 1 & 0 & 0 \\ 0 & 0 & 0 & 1 \\ 0 & 0 & 1 & 0 \end{pmatrix}, \qquad K = \begin{pmatrix} 1 & 0 & 0 & 0 \\ 0 & 1 & 0 & 0 \\ 0 & 0 & 1 & 0 \\ 0 & 0 & 0 & -1 \end{pmatrix},$$

and H is the Hadamard gate,

$$H = \frac{1}{\sqrt{2}} \begin{pmatrix} 1 & 1 \\ 1 & -1 \end{pmatrix}.$$

One can check that

$$K = \sqrt{i} e^{i\sigma_z^{(1)} \otimes \sigma_z^{(2)} \pi/4} \left(e^{-i\sigma_z^{(1)} \pi/4} \otimes e^{-i\sigma_z^{(2)} \pi/4} \right)$$

giving a controlled-NOT operation from one evolution period of time $\pi/4$ together with several single qubit operations.

Grover's algorithm and various other quantum algorithms have been realized on small molecules by using NMR approach [772].

19.3.3 Atomic Quantum Computer

Quantum computers [98, 197, 220, 240, 487] have an information processing capability much greater than that of the classical computers. Considerable progress in quantum computing has been made in recent years.

The proposed technologies for realization of a quantum computer have serious intrinsic limitations [655]. In particular, NMR devices suffer from an exponential attenuation of signal to noise as the number of qubits increases, and an ion trap computer is limited by the frequencies of the vibrational modes in the trap. Here we discuss a possible realization of a quantum computer which perhaps can help to avoid these limitations.

Basic elements of a quantum computer are qubits and logic elements (quantum gates). A qubit is a two-state quantum system with a prescribed computational basis. The current proposals for the experimental realization of a quantum computer are based on the implementation of the qubit as a two-state atom or an ion. A quantum computer in these schemes is a *molecular machine* because it is built up from a number of coupled atoms or quantum dots. In [788], it was proposed to do quantum computations using a single atom. In that scheme, the atomic quantum computer is a single atom. It is interesting to study such an *atomic machine* theoretically, but it could have also some advantages for the practical realization with respect to the molecular machines.

It is well known that in atomic physics the concept of the individual state of an electron in an atom is accepted and one proceeds from the self-consistent field approximation; see, for example, [727]. The state of an atom is determined by the set of the states of the electrons. Each state of an electron is characterized by a definite value of its orbital angular momentum l, by the principal quantum number n, and by the values of the projections of the orbital angular momentum m_l and of the spin m_s on the z-axis. In the Hartree–Fock central field approximation, the energy of an atom is completely determined by the assignment of the electron configuration, i.e., by the assignment of the values of n and l for all the electrons.

One can implement a single qubit in an atom as a one-particle electron state in the self-consistent field approximation, and multi-qubit states as the corresponding multi-particle states represented by the Slater determinant.

Almost all real spectra can be systematized with respect to *LS* or *JJ* coupling schemes. Every stationary state of the atom in the *LS* coupling approximation is characterized by a definite value of the orbital angular momentum L and the total spin S of the electrons. Under the action of the relativistic effects, a degenerate level with given L and S is split into a number of distinct levels (the fine structure of the level), which differ in the value of the total angular momentum J. The relativistic terms in the Hamiltonian of an atom include the spin–orbit and spin–spin interaction. There is also the further splitting of atomic energy levels as a result of the interaction of electrons with the spin of the nucleus. This is the hyperfine structure of the levels.

One can use these interactions to build quantum logic gates.

As a simple example let us discuss how the hyperfine splitting can be used to do quantum computations on a single atom. Let us consider the Hamiltonian which includes both nucleus and electron for a case of quenched orbital angular momentum. If one assumes that the electron spin Zeeman energy is much bigger than the hyperfine coupling energy then one gets an approximate Hamiltonian [725]

$$H = g\beta B S_z - \gamma_n \hbar B I_z + A S_z I_z.$$

Here S_z and I_z are the electron and nuclear spin operators, B is the magnetic field which is parallel to the z-axis, β is the Bohr magneton, γ_n is the nuclear gyromagnetic ratio, A is the hyperfine coupling energy, and g is the g-factor.

Let us consider the simplest case of nuclear and electron spins of $1/2$. Then a single qubit is a nuclear spin $|m_I\rangle$ and electron spin $|m_S\rangle$ function, where m_I and m_S stand for the eigenvalues of I_z and S_z. The two-qubit states are the eigenfunctions of the Hamiltonian H, and they are given by the product of the nuclear spin and electron spin functions

$$|m_I, m_S\rangle = |m_I\rangle |m_S\rangle.$$

The coupling used to produce magnetic resonances is an alternating magnetic field applied perpendicularly to the static field. The possible transitions produced by an alternating field are found by considering a perturbing term in the Hamiltonian

$$H_m(t) = (\gamma_e \hbar S_x - \gamma_n \hbar I_x) B_x \cos \omega t.$$

Many of the basic principles of nuclear magnetic resonance apply to electromagnetic spin resonance (ESR). However, there are some special features of spin echoes that arise for electron spin resonance which are not encountered in nuclear magnetic resonance. This is because in many cases the nuclear quantization direction depends on the electron spin orientation.

As was explained in Chaps. 1 and 11, any quantum algorithm can be implemented with one-qubit rotations and two-qubit controlled-NOT gate. The implementation of the controlled-NOT gate by using pulse sequences is well known in NMR [179, 281, 391]. For example, it can be represented as a network which includes one-qubit Hadamard gates and a 4×4 matrix which can be implemented as the following pulse sequence

$$\left(R\left(\frac{\pi}{2}\right)_{I_z}\right)\left(R\left(\frac{\pi}{2}\right)_{S_z}\right)\left(R\left(-\frac{\pi}{2}\right)_{2I_zS_z}\right).$$

Two-qubit realizations of the Deutsch–Jozsa algorithm and the Grover algorithm have been accomplished using NMR spectroscopy of spin $1/2$ nuclei of appropriate molecules in solution [179, 281, 391]. One can use the similar technique in the case of ESR.

If computers are to become much smaller in the future, the miniaturization might lead to the atomic quantum computer. One of advantages of the atomic quantum computer is that quantum state of a single atom can be stable against decoherence; for a discussion of the decoherence problem in quantum computing, see [487, 655, 787] and references therein. Recent experimental and theoretical advances on quantum state engineering with a natural and artificial (quantum dots) atoms [617, 665, 690, 771, 825] and the development of methods for completely determining the quantum state of an atom [775] show that quantum computations with a single atom should be possible.

Thus Volovich proposed using a single atom to do quantum computations. Such an atom can be also used, of course, as a part of a computational network. He discussed the simple realization of the two-qubit atomic quantum computer based on ESR and hyperfine splitting. However, the idea of an atomic quantum computer is more general. To build a multi-qubit atomic quantum computer one has to use the fine and hyperfine splitting of energy levels to process the information encoded in the multi-electron states. In principle, one can build an atomic quantum computer based on a natural or artificial (quantum dot) atom.

19.4 Parametric Down Conversion

Entanglement between two photons was first observed in measurements of the polarization correlation between photon emitted in positron annihilation and then in experiments with a two-photon cascade emission from calcium.

Current experiments are performed by using the spontaneous parametric down-conversion in nonlinear crystals [75, 201, 532]. In an optically nonlinear medium, a light quantum from the incident pump E_p can convert into a pair of photons in the "signal" E_s and "idler" E_i modes. The three-wave mixing requires an appropriately anisotropic medium. In an anisotropic medium, light propagates as ordinary or as extraordinary waves with opposite polarizations. There are two types of down-converters. In a type I down-converter, the modes E_s and E_i are both ordinary or extraordinary waves. In a type II down-converter, E_s and E_i have opposite polarizations. Also the pump has the extraordinary polarization and the down-converted photons are emitted into two cones, one with the ordinary polarization and the other with the extraordinary polarization. One has the conservation of energy and conservation of the wave vectors

$$\omega_p = \omega_s + \omega_i, \qquad \mathbf{k}_p = \mathbf{k}_s + \mathbf{k}_i,$$

where $\mathbf{k}_p$, $\mathbf{k}_s$, and $\mathbf{k}_i$ are three-dimensional wave vectors explained below. The two cones intersect along two lines, and along the lines the light is in an entangled state with a relative phase arising from the crystal birefringence.

Let us present some details of the formalism of entangled-photon state generated via type II spontaneous down conversion (for more details, see [75]). Due to the weak interaction in the nonlinear crystal, we can restrict ourselves to the two-photon state generated within the first-order time dependent perturbation theory:

$$\left|\Psi^{(2)}\right\rangle \sim \frac{\mathrm{i}}{\hbar}\int_{t_0}^{t_1} dt\, H_{\mathrm{int}}(t)|0\rangle.$$

Here $H_{\mathrm{int}}(t)$ is the interaction Hamiltonian, $[t_0, t_1]$ is the duration of the interaction, and $|0\rangle$ is the initial vacuum state. The interaction Hamiltonian is

$$H_{\mathrm{int}}(t) \sim \chi^{(2)} \int_V d\mathbf{r} E_p^{(+)}(\mathbf{r},t) E_o^{(-)}(\mathbf{r},t) E_e^{(-)}(\mathbf{r},t) + h.c.$$

where $\chi^{(2)}$ is the second-order susceptibility, and V is the volume of the nonlinear medium in which the interaction takes place. The operator $E_j^{(\pm)}(\mathbf{r},t)$ represents the positive-(negative) frequency portion of the jth electric field operator at position $\mathbf{r}$ and time t, with the subscript j representing the pump (p), ordinary (o), and extraordinary (e) waves, and $h.c.$ stands for Hermitian conjugate.

We decompose the three-dimensional wave-vector $\mathbf{k}$ into a two-dimensional transverse wave-vector $\mathbf{q}$ and frequency ω, and write the pump field in the form

$$E_p^{(+)}(\mathbf{r},t) = \int d\mathbf{q}_p\, d\omega_p \tilde{E}^{(+)}(\mathbf{q}_p;\omega_p) e^{\mathrm{i}\kappa_p z} e^{\mathrm{i}\mathbf{q}_p \mathbf{x}} e^{-\mathrm{i}\omega_p t}$$

where $\mathbf{x}$ spans the transverse plane perpendicular to the propagation direction z. Similarly, the ordinary and extraordinary fields can be expressed in terms of the quantum-mechanical creation operators as

$$E_j^{(+)}(\mathbf{r},t) = \int d\mathbf{q}_j\, d\omega_j a_j^*(\mathbf{q}_j,\omega_j) e^{-\mathrm{i}\kappa_j z} e^{-\mathrm{i}\mathbf{q}_j \mathbf{x}} e^{\mathrm{i}\omega_j t}$$

where $j = o, e$. The longitudinal component of $\mathbf{k}$, denoted κ, is

$$\kappa = \sqrt{\left(n(\omega,\theta)\omega/c\right)^2 - |\mathbf{q}|^2}$$

where $n(\omega,\theta)$ is the index of refraction in the nonlinear medium, θ is the angle between $\mathbf{k}$ and the optical axis of the nonlinear crystal, and c is the speed of light in vacuum.

The quantum state at the output of the crystal will have the form

$$\left|\Psi^{(2)}\right\rangle \sim \int d\mathbf{q}_o\, d\mathbf{q}_e\, d\omega_o\, d\omega_e \Phi(\mathbf{q}_o,\mathbf{q}_e;\omega_o,\omega_e) a_o^*(\mathbf{q}_o,\omega_o) a_e^*(\mathbf{q}_e,\omega_e)|0\rangle$$

with the function

$$\Phi(\mathbf{q}_o,\mathbf{q}_e;\omega_o,\omega_e) = \tilde{E}^{(+)}(\mathbf{q}_o+\mathbf{q}_e;\omega_o+\omega_e) L \operatorname{sinc}\left(\frac{L\Delta}{2}\right) e^{-\mathrm{i}L\Delta/2}.$$

Here L is the thickness of the crystal, $\Delta = \kappa_p - \kappa_o - \kappa_e$, and $\operatorname{sinc}(x) = \sin(x)/x$. The quantum state $|\Psi^{(2)}\rangle$ is entangled for nonseparable function $\Phi(\mathbf{q}_o,\mathbf{q}_e;\omega_o,\omega_e)$.

The joint probability amplitude of detecting the photon pair at the space–time coordinates $(\mathbf{x}_A, t_A)$ and $(\mathbf{x}_B, t_B)$ is given by

$$A(\mathbf{x}_A,t_A;\mathbf{x}_B,t_B) = \langle 0|E_A^{(+)}(\mathbf{x}_A,t_A) E_B^{(+)}(\mathbf{x}_B,t_B)\left|\Psi^{(2)}\right\rangle$$

where $E_A^{(+)}$ and $E_B^{(+)}$ are the positive-frequency components of the electric fields at points A and B:

$$E_A^{(+)}(\mathbf{x}_A,t_A) = \sum_{j=o,e} \int d\mathbf{q}\, d\omega e^{-\mathrm{i}\omega t_A} \mathcal{H}_{Aj}(\mathbf{x}_A,\mathbf{q},\omega) a_j^*(\mathbf{q},\omega),$$

$$E_B^{(+)}(\mathbf{x}_B,t_B) = \sum_{j=o,e} \int d\mathbf{q}\, d\omega e^{-\mathrm{i}\omega t_B} \mathcal{H}_{Bj}(\mathbf{x}_B,\mathbf{q},\omega) a_j^*(\mathbf{q},\omega).$$

Here the transfer functions $\mathcal{H}_{ij}(\mathbf{x}_i,\mathbf{q},\omega)$ ($i = A, B$ and $j = o, e$) describe the propagation of a mode $(\mathbf{q},\omega)$ through the optical system from the output plane of the nonlinear medium to the detection plane. An example of the transfer functions in the paraxial approximation is given by [75],

$$\mathcal{H}_{ij}(\mathbf{x}_i,\mathbf{q},\omega) = (\mathbf{e}_i\cdot\mathbf{e}_j) e^{-\mathrm{i}\omega\tau\delta_{ej}} H(\mathbf{x}_i,\mathbf{q},\omega).$$

The unit vector $\mathbf{e}_i$ describes the orientation of the polarization analyzers, $\mathbf{e}_j$ is the unit vector that describes the polarization of the down-converted photons, and δ_{ej} is the Kronecker delta ($\delta_{ee} = 1, \delta_{eo} = 0$). The function H has the form

$$H(\mathbf{x}_i,\mathbf{q},\omega) = P\left(\frac{\omega}{cf}\mathbf{x} - \mathbf{q}\right) F(\omega) \\ \times \exp\left\{\mathrm{i}\left[\frac{\omega}{c}(d_1+d_2+f) - \frac{\omega|\mathbf{x}|^2}{2cf}\left(\frac{d_2}{f}-1\right) - \frac{d_1 c}{2\omega}|\mathbf{q}|^2\right]\right\}$$

where $F(\omega)$ is a spectral filter profile, P is the Fourier transform of the aperture function, and parameters d_1, d_2, and f characterize the optical system.

The formulation of the detection process depends on the scheme of detection apparatus. Slow detectors, for example, impart temporal integration while detectors of finite area impart spatial integration. If the temporal response of a point detector is spread negligibly with respect to the characteristic time scale of spontaneous parametric down conversion then the coincidence rate R reduces to

$$R = \left|A(\mathbf{x}_A, t_A; \mathbf{x}_B, t_B)\right|^2.$$

The more general coincidence rate is given by

$$R = \lim_{T\to\infty} \frac{\zeta_A \zeta_B}{T} \int_{-T}^{T} dt_A \int_{-T}^{T} dt_B W(t_A - t_B)\left|A(\mathbf{x}_A, t_A; \mathbf{x}_B, t_B)\right|^2$$

where $W(t_A - t_B)$ is a coincidence window function and ζ_A and ζ_B describe the efficiency of detectors.

Working with the β-barium borate (BBO) crystal and by using two extra birefringent elements, it was possible to produce each of the four orthogonal Bell's states and observe violations of Bell's inequalities. Note, however, that the violation of Bell's inequalities was obtained only under some auxiliary assumptions. Therefore, up to now the local realism was not refuted without auxiliary assumption.

As another remark note that the conversion process in the parametric down conversion is random and there is no control or prior knowledge of when the event will occur. Moreover, there is a possibility of producing more than one pair at a time. Therefore, as currently implemented, single-photon sources cannot produce single photons on demand.

Note also that not only one has the limited efficiency of detectors, but for geometrical reasons most of created pairs do not reach the detectors. This means that the detected photons in all Bell's inequalities tests were actually in a highly mixed states.

19.4.1 Notes

Quantum dots were discovered by Ekimov and Onushchenko [221] and by Brus. Quantum dots in life sciences are discussed in [522]. For a further discussion, see [191, 211]. Experimental realization of quantum algorithms are discussed in [165, 166]. Atomic quantum computer using a single atom was discussed by Volovich [788].

Chapter 20
Miscellaneous in Quantum Theory and Information

This chapter is devoted to miscellaneous topics related to quantum information and quantum probability. They haven't been either completed or used much for quantum information, yet, but they will be important for the future development in these fields. Topics discussed in this chapter are more or less conceptual, so that we will not always provide complete proofs to some of the statements, but we will indicate where each proof can be found. In particular such topics as lifting, possible decreasing of entropy, stochastic limit, Janes–Cummings model, Kolmogorov–Sinai complexities, decoherence, chaos degree, and quantum baker's map are discussed.

20.1 Lifting and Conditional Expectation

As we pointed out, the conditional probability is not suitable for the purposes of quantum probability. In order to obtain a more useful generalization, a more general point of view should be adopted and this naturally leads to the notion of "lifting", which was introduced in Sect. 7.8. We are going to discuss more about the lifting here.

20.1.1 Lifting

The probabilistic and physical meaning of "conditioning" is that one acquires new information on a system and one would like to know how the probabilities of other events related to the system should be changed.

In the classical probability, the set of all the events is represented by a σ-algebra $\mathcal{F}$ and the acquired information by a sub σ-algebra $\mathcal{F}_0 \subseteq \mathcal{F}$. The classical conditional probability from $\mathcal{F}$ to $\mathcal{F}_0$ is a way of "lifting" probability measures on $\mathcal{F}_0$ to probability measures on $\mathcal{F}$.

Since this is an important notion, we again discuss it in great detail in the simplest case, when $\mathcal{F}_0$ is generated by a finite partition $E_1, \ldots, E_n$ of Ω such that

M. Ohya, I. Volovich, *Mathematical Foundations of Quantum Information and Computation and Its Applications to Nano- and Bio-systems*,
Theoretical and Mathematical Physics,
DOI 10.1007/978-94-007-0171-7_20,

$E_i \cap E_j = \emptyset$; $\bigcup_{j=1}^n E_j = \Omega$. Then the conditional probability from $\mathcal{F}$ to $\mathcal{F}_0$, associated to a probability measure μ on $\mathcal{F}$, uniquely determines n probability measures on $\mathcal{F}$, indexed by the sets E_j such that

$$P(\cdot|E_j), \quad j=1,\dots,n \tag{20.1}$$

by the Bayes formula

$$P(A|E_j) = \frac{\mu(A \cap E_j)}{\mu(E_j)}. \tag{20.2}$$

Conversely, the assignment of any kernel (20.1) allows "lifting" any probability measure μ_0 on $\mathcal{F}_0$ to a probability measure, denoted $\mathcal{E}^*\mu_0$, on $\mathcal{F}$ by the formula

$$(\mathcal{E}^*\mu_0)(A) = \sum_{j=1}^n \mu_0(E_j)P(A|E_j), \quad A \subseteq \Omega; A \in \mathcal{F}. \tag{20.3}$$

This lifting $\mathcal{E}^*$ is called a "nondemolition for μ_0" when the restriction of the lifted measure $\mathcal{E}^*\mu_0$ to $\mathcal{F}_0$ is μ_0 itself

$$\mu_0(E_k) = (\mathcal{E}^*\mu_0)(E_k) = \sum_{j=1}^n \mu_0(E_j)P(E_k|E_j), \quad \forall k=1,\dots,n.$$

If $\mathcal{E}^*$ is a nondemolition for every initial measure μ_0 on $\mathcal{F}_0$, then it is called a "nondemolition lifting".

If the kernel $P(\cdot|E_j)$ does not depend on the measure μ_0, then the lifting $\mathcal{E}^*$ is "linear". It is clear that for any linear map

$$\mathcal{E}^* : \mathcal{P}(\mathcal{F}_0) \to \mathcal{P}(\mathcal{F}), \tag{20.4}$$

where $\mathcal{P}(\mathcal{F})$ is the set of all probability measures on $\mathcal{F}$, there exists a kernel $P(\cdot|E_j)$, in the sense of Chap. 6, of the form (20.3) such that $\mathcal{E}^*$ has the form (20.2).

In general, a map as in (20.3) shall be called a "lifting" (linear lifting if $\mathcal{E}^*$ is linear) as introduced by Accardi and Ohya [19].

20.1.2 Conditional Probability

Thus the notion of a "lifting" is a natural generalization of that of "conditional probability", whose special dual linear map is the transition expectation of Accardi [4] and is linked to the conditional expectation by Umegaki [761] and Takesaki [748]. The definitions of these expectations are given and their relations are discussed below.

Notice that the integral defines a natural duality between a (probability) measure μ and a bounded measurable function f as

$$\langle \mu, f\rangle \bigl(= \mu(f)\bigr) = \int_\Omega f(\omega)\mu(d\omega) = \int_\Omega f\,d\mu.$$

This duality can be used to define the "dual of a linear lifting" of (20.4).

Definition 20.1 Denote $L^\infty(\Omega, \mathcal{F})$ (resp., $L^\infty(\Omega, \mathcal{F}_0)$) the algebra (by the pointwise operations) of complex-valued bounded measurable functions on $(\Omega, \mathcal{F})$ (resp., $(\Omega, \mathcal{F}_0)$). A linear map

$$\mathcal{E} : L^\infty(\Omega, \mathcal{F}) \to L^\infty(\Omega, \mathcal{F}_0) \tag{20.5}$$

shall be called a "transition expectation" if there exists a linear lifting $\mathcal{E}^*$ of the form (20.4) such that for any probability measure μ_0 on $\mathcal{F}_0$ and any bounded measurable function f on $(\Omega, \mathcal{F})$ one has

$$\langle \mathcal{E}^* \mu_0, f \rangle = \big\langle \mu_0, \mathcal{E}(f) \big\rangle. \tag{20.6}$$

Proposition 20.2 *The following statements are equivalent*:

(i) $\mathcal{E}$ *is a transition expectation of the form* (20.5).
(ii) $\mathcal{E}$ *is positive and normalized, i.e.*,

$$f \ge 0 \quad \Longrightarrow \quad \mathcal{E}(f) \ge 0, \tag{20.7}$$

$$\mathcal{E}(1) = 1. \tag{20.8}$$

Proof (i)$\Rightarrow$(ii) Using (20.6), we see that if $f \ge 0$ then for any probability measure μ_0 on $\mathcal{F}_0$ one has

$$\big\langle \mu_0, \mathcal{E}(f) \big\rangle = \langle \mathcal{E}^* \mu_0, f \rangle \ge 0,$$

and this implies (20.7). If $f = 1$, then for any μ_0

$$\big\langle \mu_0, \mathcal{E}(1) \big\rangle = \langle \mathcal{E}^* \mu_0, 1 \rangle = 1,$$

and this implies (20.8).

(ii)$\Rightarrow$(i) If (ii) is satisfied, then for any probability measure μ_0 on $\mathcal{F}_0$ the map

$$A \subseteq \Omega, \quad A \in \mathcal{F} \mapsto (\mathcal{E}^* \mu_0)(A) \equiv \mu_0\big(\mathcal{E}(1_A)\big)$$

is a probability measure on $\mathcal{F}$. It is clear that $\mathcal{E}^* \mu_0$ linearly depends on μ_0. Hence $\mathcal{E}^*$ is a linear lifting of (20.5). Notice that, by construction, condition (20.6) is satisfied. □

It is now natural to ask the following question: Which liftings come from a conditional probability?

This is answered by the following theorem:

Theorem 20.3 *Let* $\mathcal{E}^*$ *be a lifting of the form* (20.4). *Then the following statements are equivalent*:

(i) $\mathcal{E}^*$ *is linear and nondemolition.*
(ii) $\mathcal{E}^*$ *is linear, and the associated transition expectation* $\mathcal{E}^*$ *satisfies the condition*

$$\mathcal{E}(f_0) = f_0, \quad \forall f_0 \in L^\infty(\Omega, \mathcal{F}_0). \tag{20.9}$$

(iii) $\mathcal{E}^*$ *is linear, and the kernel associated to it by* (20.3) *satisfies the condition*

$$P(E_k|E_j) = \delta_{k,j}, \quad k, j = 1, \dots, n. \tag{20.10}$$

(iv) $\mathcal{E}^*$ *is linear, and there exists a probability measure* μ *on* $\mathcal{F}$ *such that the kernel associated to* $\mathcal{E}^*$ *by* (20.3) *satisfies the Bayes formula* (20.2).

Proof (iv)⇒(iii) is obvious by choosing $A = E_k$ in (20.2).

(iii)⇒(ii) For any $E_k \in \mathcal{F}_0$ and any probability measure μ_0 on $\mathcal{F}_0$, one has

$$\begin{aligned}
\big\langle \mu_0, \mathcal{E}(1_{E_k})\big\rangle &= \langle \mathcal{E}^*\mu_0, 1_{E_k}\rangle \\
&= (\mathcal{E}^*\mu_0)(E_k) = \sum_{j=1}^{n} \mu_0(E_j) P(E_k|E_j) \\
&= \mu_0(E_k) = \langle \mu_0, 1_{E_k}\rangle.
\end{aligned}$$

Since this is true for any probability measure μ_0 on $\mathcal{F}_0$, it follows that for any $k = 1, \dots, n$

$$\mathcal{E}(1_{E_k}) = 1_{E_k},$$

and this implies (20.9).

(ii)⇒(i) For any $f_0 \in L^\infty(\Omega, \mathcal{F}_0)$, one has

$$\langle \mathcal{E}^*\mu_0, f_0\rangle = \big\langle \mu_0, \mathcal{E}(f)\big\rangle = \langle \mu_0, f_0\rangle.$$

Since $f_0 \in L^\infty(\Omega, \mathcal{F}_0)$ is arbitrary, this implies

$$\mathcal{E}^*\mu_0 = \mu_0, \quad \forall \mu_0 \in \mathcal{P}(\mathcal{F}_0).$$

So $\mathcal{E}^*$ is a nondemolition.

(i)⇒(iv) If $\mathcal{E}^*$ is linear, then it has an associated kernel $P(\cdot|E_j)$ (cf. (20.2)). If $\mathcal{E}^*$ is a nondemolition, it follows that for any probability measures μ_0 on $\mathcal{F}_0$ one has, for any $k = 1, \dots, n$,

$$\mu_0(E_k) = (\mathcal{E}^*\mu_0)(E_k) = \sum_{j=1}^{n} \mu_0(E_j) P(E_k|E_j).$$

Since μ_0 is arbitrary, it follows that (20.10) must hold. Now fix $\mu_0 \in \mathcal{P}(\mathcal{F}_0)$ arbitrarily and define

$$\mu \equiv \mathcal{E}^*\mu_0.$$

Then for any $j = 1, \dots, n$ and $A \subset \Omega$, $A \in \mathcal{F}$,

$$\frac{\mu(A \cap E_j)}{\mu(E_j)} = \frac{1}{(\mathcal{E}^*\mu_0)(E_j)} (\mathcal{E}^*\mu_0)(A \cap E_j) \tag{20.11}$$

by the nondemolition assumption and the definition of $P(\cdot|E_k)$. The right hand side of (20.11) is equal to

$$\frac{1}{\mu_0(E_j)}\sum_{k=1}^{n}\mu_0(E_k)P(A\cap E_j|E_k). \tag{20.12}$$

But because of (20.10), it follows that $P(A\cap E_j|E_k)=0$, if $k\neq j$. Therefore, (20.12) is equal to

$$\frac{1}{\mu_0(E_j)}\mu_0(E_j)P(A\cap E_j|E_j)=P(A\cap E_j|E_j).$$

Finally, again by (20.10), one has

$$P(A\cap E_j|E_j)=P(A|E_j),$$

and therefore (20.2) holds. □

20.1.3 Various Liftings

The passage from classical to quantum probability is accomplished by replacing the algebra $L^\infty(\Omega,\mathcal{F})$ by a general $*$-algebra $\mathcal{A}$ and the algebra $L^\infty(\Omega,\mathcal{F}_0)$ by a $*$-subalgebra $\mathcal{B}\subset\mathcal{A}$. Correspondingly, the probability measures on $(\Omega,\mathcal{F})$ (resp., on $(\Omega,\mathcal{F}_0)$) are replaced by the "states" on $\mathcal{A}$ (resp., $\mathcal{B}$). An important case is obtained when

$$\mathcal{A}=\mathcal{B}\otimes\mathcal{C}$$

for some other $*$-algebra $\mathcal{C}$.

Definition 20.4

(i) In the case when $\mathcal{A}=\mathcal{B}\otimes\mathcal{C}$, a map $\mathcal{E}^*\colon\mathfrak{S}(\mathcal{B})\to\mathfrak{S}(\mathcal{A})$ is called a lifting, where $\mathfrak{S}(\mathcal{B})$ is the set of all states on $\mathcal{B}$.
(ii) In the case when $\mathcal{A}=\mathcal{B}\otimes\mathcal{C}$, a completely positive map $\mathcal{E}:\mathcal{A}\to\mathcal{B}, \mathcal{E}(1)=1$ is called a transition expectation.
(iii) Let $\mathcal{B}\subset\mathcal{A}$. A lifting $\mathcal{E}^*:\mathfrak{S}(\mathcal{B})\to\mathfrak{S}(\mathcal{A})$ is called a nondemolition for $\varphi\in\mathfrak{S}(\mathcal{B})$ if

$$\mathcal{E}^*\varphi\restriction\mathcal{B}=\varphi. \tag{20.13}$$

(iv) $\mathcal{E}^*$ is called a nondemolition if (20.13) holds for any $\varphi\in\mathfrak{S}(\mathcal{B})$. When $\mathcal{E}^*$ is a linear nondemolition, its dual $\mathcal{E}:\mathcal{A}\to\mathcal{B}$ is called a conditional expectation in the sense of Umegaki.

The analogue of the previous theorem holds.

Theorem 20.5 *Let $\mathcal{B} \subset \mathcal{A}$ and consider a linear lifting $\mathcal{E}^* : \mathfrak{S}(\mathcal{B}) \to \mathfrak{S}(\mathcal{A})$. Then for the transition expectation $\mathcal{E} : \mathcal{A} \to \mathcal{B}$ the following statements are equivalent*:

(i) *$\mathcal{E}$ is a conditional expectation.*
(ii) $\mathcal{E}(B) = B, \forall B \in \mathcal{B}$.
(iii) $\mathcal{E}(AB) = \mathcal{E}(A)B, \forall A \in \mathcal{A}, \forall B \in \mathcal{B}$.

Proof The equivalence (i)⇔(ii) is clear. The equivalence (ii)⇔(iii) is a Tomiyama's theorem [748]. □

Example 20.6 (The trivial liftings) Let $\mathcal{A} = \mathcal{B} \otimes \mathcal{C}$. A class of (trivial) liftings is obtained as follows: Fix a state $\psi \in \mathfrak{S}(\mathcal{C})$ and define

$$\mathcal{E}^* : \varphi \in \mathfrak{S}(\mathcal{B}) \to \mathcal{E}^*\varphi \equiv \varphi \otimes \psi.$$

Then $\mathcal{E}^*$ is a lifting, called a "product (or trivial) lifting".

The notion of a nondemolition linear lifting is too strong for the applications to a model in the quantum theory of measurement. This is shown by the following proposition.

Proposition 20.7 *Let $\mathcal{B} = M(n; \mathbb{C})$ be set of the $n \times n$ complex matrices. Then any linear nondemolition lifting $\mathcal{E}^* : \mathfrak{S}(\mathcal{B}) \to \mathfrak{S}(\mathcal{B} \otimes \mathcal{B})$ is trivial.*

Proof By Theorem 20.5, the dual of $\mathcal{E}^*$ is a conditional expectation $\mathcal{E} : \mathcal{B} \otimes \mathcal{B} \to \mathcal{B} \sim \mathcal{B} \otimes I$. Let $B, B' \in \mathcal{B}$. Then

$$\begin{aligned}(B \otimes I)\mathcal{E}(I \otimes B') &= \mathcal{E}(B \otimes B') \\ &= \mathcal{E}\big((I \otimes B')(B \otimes I)\big) \\ &= \mathcal{E}(I \otimes B')(B \otimes I),\end{aligned}$$

i.e., $\mathcal{E}(I \otimes \mathcal{B}) \subset I \otimes \mathcal{B}$.

By definition, one also has $\mathcal{E}(I \otimes \mathcal{B}) \subset \mathcal{B} \otimes I$. Therefore,

$$\mathcal{E}(I \otimes \mathcal{B}) \subset (\mathcal{B} \otimes I) \cap (I \otimes \mathcal{B}) = \mathbb{C}(I \otimes I).$$

So there exists a state $\psi \in \mathfrak{S}(\mathcal{B})$ such that

$$\mathcal{E}(I \otimes B) = \psi(B) I \otimes I, \quad \forall B \in \mathcal{B}.$$

It follows that, for any $B, B' \in \mathcal{B}$, one has

$$\begin{aligned}(\mathcal{E}^*\varphi)(B' \otimes B) &= \varphi\big(\mathcal{E}(B' \otimes B)\big) \\ &= \varphi\big(B'\mathcal{E}(I \otimes B)\big) = \varphi(B')\psi(B).\end{aligned}$$

This means that

$$\mathcal{E}^*\varphi = \varphi \otimes \psi,$$

i.e., $\mathcal{E}^*$ is a trivial lifting. □

For this reason, it was proposed in [12] that the appropriate generalization of the notion of conditional expectation is not that of a universally nondemolition lifting, but more generally of a "lifting", connection with POM, operations effects, etc. (see Chap. 5).

Let $\mathcal{E}^* : \mathfrak{S}(\mathcal{B}) \to \mathfrak{S}(\mathcal{B} \otimes \mathcal{C})$ be a linear lifting and let $\mathcal{E} : \mathcal{B} \otimes \mathcal{C} \to \mathcal{B}$ be the associated transition expectation.

We suppose that for some separable Hilbert spaces $\mathcal{H}$ and $\mathcal{K}$ one has

$$\mathcal{B} = \mathbf{B}(\mathcal{H}), \qquad \mathcal{C} = \mathbf{B}(\mathcal{K}).$$

If $\mathcal{C}_0$ is an abelian von Neumann subalgebra of $\mathcal{C}$, we can realize $\mathcal{C}_0$ as the algebra $L^\infty(\Omega, \mathcal{F}, \mu)$ for some probability space $(\Omega, \mathcal{F}, \mu)$. Then, if $E \subseteq \Omega$, $E \in \mathcal{F}$, $1_E \in L^\infty(\Omega, \mathcal{F}, \mu) \subset \mathcal{C}$, and we can consider the map

$$M(E) : B \in \mathcal{F} \in \mathcal{B} \to M(E)(B) \equiv \mathcal{E}(B \otimes 1_E).$$

If $B \in \mathcal{B}^+$ (the set of all positive operators of $\mathcal{B}$) then the map

$$E \subset \Omega, \quad E \in \mathcal{F} \to M(E)(B)$$

is a positive operator-valued measure.

Example 20.8 Let $\{F_i\}_{i \in Z}$ be a partition of the identity by orthogonal projections in $\mathcal{B}$ and define $\mathcal{B}_0$ the (atomic) algebra generated by the $\{F_i\}$ and

$$\mathcal{E} : \mathcal{A} \to \mathcal{B}_0{}'(= \text{commutant of } \mathcal{B}_0)$$

by

$$\mathcal{E}(A) = \sum_{i \in Z} F_i A F_i, \quad A \in \mathcal{A}.$$

Then $\mathcal{E}$ is a conditional expectation, and we can define, for each $A \in \mathcal{A}^+$, the POM

$$M(\Delta)(A) = F_\Delta \mathcal{E}(A) = \sum_{i \in \Delta} F_i A F_i, \quad \Delta \subset Z$$

where

$$F_\Delta \equiv \sum_{j \in \Delta} F_j.$$

This is an example of Theorem 20.5.

Example 20.9 $\mathcal{E}(B \otimes 1_E)$ is the conditional expectation of $B \in \mathcal{B}$ given that the result of a measurement of A is a number in the interval E.

Before closing this section, we mention a generalized Markov chain by Accardi and Frigerio [6, 8].

For two C^*-algebras $\mathcal{A}, \mathcal{B}$ and a transition expectation $\mathcal{E}$ from $\mathcal{A} \otimes \mathcal{B}$ to $\mathcal{A}$, when a state φ on $\mathcal{A}$ satisfies $\varphi \circ \mathcal{E}(A \otimes I) = \varphi(A)$, $A \in \mathcal{N}$, $(\varphi, \mathcal{E})$ is called a Markov pair. If $\{\mathcal{E}_n\}$ is a sequence of transition expectations from $\mathcal{A} \otimes \mathcal{B}$ to $\mathcal{A}$, then there exists a unique complete positive unital map $E_{0]} : \bigotimes_N \mathcal{A} \to \mathcal{B}$ such that

$$E_{n]}(A_0 \otimes A_1 \otimes \cdots \otimes A_n \otimes I \otimes I \otimes \cdots) = \mathcal{E}_0\big(A_0 \otimes \mathcal{E}_1\big(A_1 \otimes \cdots \otimes \mathcal{E}_n(A_n \otimes I)\big)\big).$$

Let φ_0 be a state on $\mathcal{A}$ and take $\varphi \equiv \varphi_0 \circ E_{n]}$. Then φ satisfies

$$\varphi(A_0 \otimes A_1 \otimes \cdots \otimes A_n \otimes I \otimes \cdots) = \varphi_0\big(\mathcal{E}_0\big(A_0 \otimes \mathcal{E}_1\big(A_1 \otimes \cdots \otimes \mathcal{E}_n(A \otimes I)\big)\big)\big).$$

This state φ is called a generalized Markov chain associated to the pair $(\varphi_0, \{\mathcal{E}_n\})$.

20.2 Various Topics on Entropies

20.2.1 Boltzmann's Entropy

Entropy in thermodynamics, introduced by Clausius, is a measure of the unavailability of a system's energy to do work due to inevitable loss of some energy in the form of irretrievable heat.

Boltzmann gave a definition of the entropy of a macroscopic system in terms of its macro and microstates. If Γ is the phase space of the system and Γ_m is the region in Γ corresponding to the value m of a macroscopic observables M then the Boltzmann entropy is

$$S_B(m) = k \log |\Gamma_m|,$$

where $|\Gamma_m|$ is the volume of Γ_m and k is the Boltzmann constant. The macroscopic observable M is a function on the phase space Γ which gives a crude, coarse grained description of Γ. There are many microstates $X \in \Gamma$ which correspond to the same value of the macroscopic observable, $M(X) = m$. One says that all such X form the macrostate m. For example, the average number of particles, or the average energy or momentum in a given cell of the system provide macroscopic observables.

20.2.2 Entropy Increase

If the evolution of microstates is given by the curve X_t in Γ, then one gets the induced evolution of the macroscopic observables $M(X_t) = m_t$. One assumes the Hamiltonian dynamics with a Hamiltonian function $H(X)$, so one considers the time evolution on the energy surface $\Omega_E = \{X \in \Gamma | H(X) = E\}$ corresponding to the value E of the energy of the system. There is a special value m^{eq} of the macroscopic observable M on the energy surface Ω_E, which is called the equilibrium state. It is characterized, in particular, by the property $|\Gamma_{m^{\mathrm{eq}}}| \simeq |\Omega_E|$.

Boltzmann has argued that if one starts in a nonequilibrium microstate X then, typically, the entropy $S_B(m_t)$ will increase, in accordance with the second law of thermodynamics, up to some time shorter than the Poincaré recurrence time. The argument is the following. Since Ω_E has many more points from the equilibrium state $\Gamma_{m^{\mathrm{eq}}}$ than from nonequilibrium microstates, it is expected that for a generic dynamics with a very high probability the evolution X_t arising from a general nonequilibrium microstate X should rather quickly carry out X_t into $\Gamma_{m^{\mathrm{eq}}}$ and keep it there for a very long time. In this argument, the initial conditions play an important role.

Example 20.10 Consider a gas consisting of a large number N of identical particles in a box V with a volume $|V|$ with positions $r_i \in V$ and momenta $p_i \in \mathbb{R}^3, i = 1, 2, \ldots, N$. The microstate of the system is given by a point $X = (r_1, \ldots, r_N, p_1, \ldots, p_N)$ in the phase space $\Gamma = V^N \times \mathbb{R}^{3N}$. In this case,

$$|\Gamma_m| = \frac{1}{N!h^{3N}} \int_{\Gamma_m} dX,$$

where $dX = \prod_{i=1}^{N} dr_i\, dp_i$ and h is the Planck constant.

As a famous example consider the gas that initially (at time $t = t_0$) is compressed by a piston in the left half of a box; then at time t_1 the piston is released so that the gas expands into the whole box. Let the macrostate of the gas be given by $M(X) = N_L/N$, the fraction of particles in the left half of V (within a given accuracy). The macrostate at time t_0 is $m_0 = 1$. The phase space region, available to the gas for $t > t_1$, contains many microstates, corresponding to various fractions of particles in the left half of the box, with phase space volumes being very large when compared to the initial phase space volume. Therefore, it is expected that as a generic phase point X evolves, it will with a very high probability enter new macrostates m_t for which $|\Gamma_{m_t}|$ is increasing, until it is contained in $\Gamma_{m^{\mathrm{eq}}}$. Here m^{eq} is an equilibrium macroscopic state, in this case $m^{\mathrm{eq}} = 1/2$.

20.2.3 *Gibbs Entropy*

The Gibbs entropy S_G is defined for the ensemble density $\rho(X)$ by

$$S_G(\rho) = -k \int \rho(X) \log \rho(X)\, dX.$$

One can establish the relation with the Boltzmann entropy as follows. If one takes $\rho(X)$ to be the microcanonical ensemble associated with a macrostate m,

$$\rho_m(X) = \begin{cases} |\Gamma_m|^{-1}, & \text{if } X \in \Gamma_m, \\ 0, & \text{otherwise} \end{cases}$$

then

$$S_G(\rho_m) = k \log |\Gamma_m| = S_B(m).$$

There is an important difference between the Bolztmann and the Gibbs entropies: the Gibbs entropy is an integral of motion, so it is not immediately related with the second law of thermodynamics.

20.2.4 Boltzmann Entropy in Quantum Mechanics

In quantum mechanics, the entropy is a quantity which describes the density of levels in the energy spectrum of a macroscopic system.

Let us consider a macroscopic closed system in a box of volume V with the energy E, and let $\Delta\Gamma$ be the number of states corresponding to the interval ΔE of energy, where ΔE is equal in the order of magnitude to the mean fluctuation of the energy. The quantity $\Delta\Gamma$ is called the statistical weight of the macroscopical state of the system, characterized by E and V, and its logarithm

$$S = S(E, V) = k \log \Delta\Gamma$$

is called the (Boltzmann) entropy of the system.

Thermodynamic Quantities

If one has the entropy $S = S(E, V)$ then one derives the temperature T and the pressure p as follows:

$$\frac{1}{T} = \frac{\partial S}{\partial E}, \qquad p = -\frac{\partial S}{\partial V},$$

and one gets the important relation

$$dS = \frac{1}{T} dS + \frac{p}{T} dV.$$

One can express the thermodynamic quantities in terms of any two variables: (E, V), (T, V), (S, V), etc. The Helmholtz free energy $F = F(T, V)$ is defined by

$$F = E - TS.$$

One gets

$$dF = -S\,dT - p\,dV$$

in a Hilbert space. We can compute the free energy by using the formula

$$F = -kT \log \operatorname{tr} e^{-H/kT}$$

where H is the Hamiltonian operator.

One has the following inequalities:

$$S \geq 0, \quad C_v = T\left(\frac{\partial S}{\partial T}\right)_V > 0, \qquad \left(\frac{\partial p}{\partial V}\right)_T < 0,$$

and the Nernst's theorem: $S|_{T=0} = 0$.

20.2.5 von Neumann Microscopic and Macroscopic Entropies

As we extensively discussed in Chap. 7, von Neumann introduced the well known microscopic entropy for the density operator ρ in a Hilbert space $\mathcal{H}$:

$$S(\rho) = -k \operatorname{tr} \rho \log \rho.$$

This entropy, like the classical Gibbs entropy, does not change in time under the unitary evolution. von Neumann also introduced what he calls the macroscopic entropy of the system. Let $\hat{M}$ be an operator in the Hilbert space $\mathcal{H}$ describing a macroscopic observable with the projection operators $\{E_i\}$ and eigenvalues m_i. Let $\mathcal{H}_i = E_i\mathcal{H}$. The von Neumann macroscopic entropy of a system is

$$S_{\text{mac}}\big(\rho, \{E_i\}\big) = k\sum_i p_i \log \dim \mathcal{H}_i - k\sum_i p_i \log p_i = -k \operatorname{tr} \rho_E \log \rho_E.$$

Here

$$p_i = \operatorname{tr}(E_i \rho)$$

and

$$\rho_E = \sum_i \frac{p_i}{\dim \mathcal{H}_i} E_i$$

is the density operator which is macroscopically indistinguishable from ρ.

The quantum analogue of the Boltzmann entropy $S_B(m)$ in this situation would be

$$S_{B,q}(m_i) = k \log \dim \mathcal{H}_i.$$

20.2.6 Entropy Change for Open Systems

One thinks that the entropy increases in a dissipative system. Such dissipative systems are often described by dynamical semigroups. The Gorini–Kossakowski–Sudarshan–Lindblad master equation is one of these, which comes from the complete positivity of the dynamics. One expects that the GKSL master equation gives us an increase of entropy. However, let us consider an example of a solution of GKSL master equation when entropy decreases.

The GKSL master equation for the density operator ρ in the Hilbert space $\mathcal{H}$ has the form

$$\frac{d}{dt}\rho = -\mathrm{i}[H, \rho] + \sum_{\alpha}\left(L_\alpha \rho L_\alpha^* - \frac{1}{2}(L_\alpha^* L_\alpha \rho + \rho L_\alpha^* L_\alpha)\right).$$

Here H is a self-adjoint operator, and L_α are bounded operators in $\mathcal{H}$.

Here as an example, we discuss the one-qubit case when $\mathcal{H} = \mathbb{C}^2$ and there is only one operator

$$L = \sqrt{\gamma}|0\rangle\langle 1|, \quad \gamma > 0,$$

where the set of vectors $\{|0\rangle, |1\rangle\}$ is an orthonormal basis in $\mathbb{C}^2$. Moreover, we take $H = 0$. Then the GKSL master equation gets the form

$$\frac{d}{dt}\rho = \gamma\left(|0\rangle\langle 1|\rho|1\rangle\langle 0| - \frac{1}{2}|1\rangle\langle 1|\rho - \frac{1}{2}\rho|1\rangle\langle 1|\right).$$

Such a master equation can be obtained by the method of the stochastic limit, for example, for the two-level atom interacting with the electromagnetic field or for the spin–boson model. We denote

$$\rho_{ij} = \langle i|\rho|j\rangle,$$

then we obtain the following system of equations:

$$\dot{\rho}_{11} = -\gamma\rho_{11}, \qquad \dot{\rho}_{00} = \gamma\rho_{11},$$
$$\dot{\rho}_{01} = -\frac{\gamma}{2}\rho_{01}, \qquad \dot{\rho}_{10} = -\frac{\gamma}{2}\rho_{10}.$$

We look for the solution subject to

$$\rho_{11}(t) = p(t), \qquad \rho_{00}(t) = 1 - p(t),$$
$$\rho_{01}(0) = \rho_{10}(0) = 0,$$

where

$$p(t) = xe^{-\gamma t}, \quad 0 \le x \le 1, t \ge 0.$$

Then the density matrix which solves the GKSL equation reads

$$\rho(t) = \big(1 - p(t)\big)|0\rangle\langle 0| + p(t)|1\rangle\langle 1|.$$

The von Neumann entropy for this density matrix is

$$S\big(\rho(t)\big) = -\operatorname{tr}\rho(t)\log\rho(t) = -p(t)\log p(t) - \big(1 - p(t)\big)\log\big(1 - p(t)\big).$$

The time derivative of the entropy is

$$\frac{d}{dt}S\big(\rho(t)\big)=\dot{p}(t)\log\left(\frac{1}{p(t)}-1\right)=-x\gamma e^{-\gamma t}\log\left(\frac{e^{\gamma t}}{x}-1\right).$$

Let us take $0<x<1/2$. Then we get

$$\frac{d}{dt}S\big(\rho(t)\big)<0,\quad t>0.$$

So, we have proved the theorem:

Theorem 20.11 *If $\rho(t)$ is the above described solution of the GKSL master equation then the von Neumann entropy monotonically decreases from*

$$S\big(\rho(0)\big)=-x\log x-(1-x)\log(1-x)$$

to

$$S\big(\rho(\infty)\big)=0$$

when time increases from $t=0$ to infinity. Note that one has $S(\rho(0))=\log 2$ for $x=1/2$.

Let us write the solution of the GKSL equation in the form of the Kraus–Sudarshan representation for the linear completely positive map.

Let us define two operators in $\mathcal{H}=\mathbb{C}^2$:

$$V_1(t)=|0\rangle\langle 0|+e^{-\gamma t/2}|1\rangle\langle 1|,\qquad V_2(t)=\sqrt{1-e^{-\gamma t}}|0\rangle\langle 1|,\quad t\geq 0.$$

Note that $V_1^*(t)=V_1(t)$, $V_2^*(t)=\sqrt{1-e^{-\gamma t}}|1\rangle\langle 0|$, and there is a relation:

$$\sum_{i=1}^{2}V_i^*(t)V_i(t)=I.$$

The channel Λ_t^* is defined as the dual map of the following Λ_t

$$\Lambda_t(A)=\sum_{i=1}^{2}V_i(t)AV_i^*(t),$$

for an arbitrary operator A in $\mathcal{H}=\mathbb{C}^2$.

Λ_t is a linear completely positive map from $\mathbf{B}(\mathbb{C}^2)$ to $\mathbf{B}(\mathbb{C}^2)$ which has the following properties:

1. The channel preserves the trace for any $t\geq 0$ and $A\in\mathbf{B}(\mathbb{C}^2)$:

$$\operatorname{tr}\Lambda_t(A)=\operatorname{tr}A,$$

but is not unital:

$$\Lambda_t(I) = \left(2 - e^{-\gamma t}\right)|0\rangle\langle 0| + e^{-\gamma t}|1\rangle\langle 1|.$$

2. It satisfies

$$\Lambda_t(A)|_{t=0} = A, \qquad \lim_{t\to\infty} \Lambda_t(A) = (\operatorname{tr} A)|0\rangle\langle 0|, \quad \text{and}$$
$$\Lambda_t(|0\rangle\langle 0|) = |0\rangle\langle 0|, \quad t \geq 0.$$

3. The von Neumann entropy has the following properties for any density operator ρ in $\mathcal{H} = \mathbb{C}^2$:

$$S\left(\Lambda_t^*(\rho)\right)\big|_{t=0} = S(\rho) \quad \text{and} \quad \lim_{t\to\infty} S\left(\Lambda_t^*(\rho)\right) = 0.$$

The results in the previous subsection can be written as:

Theorem 20.12 *Let us take the initial density operator ρ in $\mathcal{H} = \mathbb{C}^2$ in the form*

$$\rho = (1-x)|0\rangle\langle 0| + x|1\rangle\langle 1|, \quad 0 \leq x \leq 1.$$

Then for the above introduced channel one has

$$\Lambda_t^*(\rho) = \left(1 - p(t)\right)|0\rangle\langle 0| + p(t)|1\rangle\langle 1|, \quad p(t) = xe^{-\gamma t},$$

and the entropy monotonically decreases:

$$S\left(\Lambda_t^*(\rho)\right) < S\left(\Lambda_\tau^*(\rho)\right), \quad 0 \leq \tau < t.$$

Proof We have the relations:

$$V_1(t)|0\rangle\langle 0|V_1^*(t) = |0\rangle\langle 0|,$$
$$V_2(t)|0\rangle\langle 0|V_2^*(t) = 0,$$
$$V_1(t)|1\rangle\langle 1|V_1^*(t) = e^{-\gamma t}|1\rangle\langle 1|,$$
$$V_2(t)|1\rangle\langle 1|V_2^*(t) = \left(1 - e^{-\gamma t}\right)|0\rangle\langle 0|.$$

Then the last statement follows. Other properties are obvious. □

20.2.7 Reversible and Irreversible Processes

The irreversibility problem can be formulated in the following way: Why do the microscopic dynamical equations (of Newton, Schrödinger, Maxwell, etc.) are time-symmetric while macroscopic dynamical equations (the diffusion equation, Navier–Stokes, Boltzmann, second law of thermodynamics of entropy increasing, etc.) are time-asymmetric? What is a relation between the micro- and macroscopic equations? Can we derive the macroscopic equations from the microscopic? There are lots of discussions on irreversible and nonequilibrium processes in terms of entropy, ergodicity, open system dynamics, etc. [131, 190, 222, 223, 557, 611].

In the approach to the irreversibility problem going back to Boltzmann, the following points are stressed:

- The great disparity between microscopic and macroscopic scales; this includes applications of macroscopic or coarse-grained variables, in particular the Boltzmann entropy.
- The role of special (or, better to say, generic) initial conditions; in particular, a low entropy state of the early universe.
- Probabilistic and information theory approach to dynamical problems.

We remind that the Gibbs entropy S_{G} and the von Neumann quantum entropy S_{vN} are constant under the microscopic evolution. It is outlined above that one expects that the Boltzmann entropy $S_{\mathrm{B}}(m_t)$ should increase in accordance with the second law of thermodynamics. The derivation of the kinetic Boltzmann equation form the microscopic Newton–Hamilton equations was performed under the additional assumptions by Bogoliubov.

The master equations in quantum theory were derived from the quantum field theory equations by Van Hove and by Accardi, Lu and Volovich by using the stochastic limit method. As it was demonstrated in the previous section, the master equation in the general case does not lead to increasing of the von Neumann entropy.

There is an important general property of monotonicity of the relative entropy: $S(\Lambda^*\rho, \Lambda^*\sigma) \le S(\rho, \sigma)$ which is valid for any channel Λ^* and any density operators ρ and σ (see Chap. 7). This does not means that one has the entropy increase in the general case, but for the unital channel in the finite-dimensional Hilbert space one gets an increase of the von Neumann entropy: $S_{\mathrm{vN}}(\Lambda^*\rho) \ge S_{\mathrm{vN}}(\rho)$ (see Chap. 7). For the trace-preserving channels under additional restrictions, one also has the entropy increase, [570].

The problem of reversible and irreversible classical and quantum computations is discussed in Chaps. 2 and 14.

20.2.8 Entropy Production

Before closing this chapter we briefly discuss the entropy production.

As is seen in the preceding section, there are many approaches to explain the nonequilibrium irreversible processes.

The entropy production is one of such approaches. Mathematical formulation of this concept was started by Spohn [730], although there were many other studies of the entropy production.

Here we briefly explain Spohn's formulation [730] and Ojima's more recent formulations [609, 612] of the entropy production.

In nonequilibrium balance equations, the balance equation for the entropy is

$$\frac{\partial S}{\partial t} = -\operatorname{div} \mathbf{J}_S + \sigma$$

where S is local entropy, $\mathbf{J}_S$ is the vector of the entropy flow per unit area and unit time, and σ is the (local) entropy production associated with an irreversible change of the system ($\sigma = 0$ for a reversible process).

In quantum statistical mechanics, Spohn discussed the entropy production as follows, [730]: Let Λ_t^* ($t \in \mathbb{R}^+$) be a dynamical semigroup for the set of all trace operators $\mathfrak{S}(\mathcal{H})$ to $\mathfrak{S}(\mathcal{H})$, namely, Λ_t^* is a CP (completely positive) channel satisfying $\Lambda_{t+s}^* = \Lambda_t^* \Lambda_s^*$ for any $t, s \in \mathbb{R}^+$ and $\lim_{t\to 0} \|\Lambda_t^* \rho - \rho\| = 0$ for all ρ (strong continuity). The strong continuity of $\{\Lambda_t^*; t \in \mathbb{R}^+\}$ implies the existence of the generator L [482] such that $\lim_{t\to 0} \|L\rho - \frac{1}{t}(\Lambda_t^* \rho - \rho)\|_1 = 0$. If one assumes

$$S(\rho) = -\operatorname{tr} \rho \log \rho,$$

$$\operatorname{div} \mathbf{J}_S = \frac{d}{dt} \operatorname{tr} \Lambda_t^* \rho \log \rho_\beta \bigg|_{t=0},$$

where β is the inverse temperature of the system, then the entropy production is

$$\begin{aligned}\sigma(\rho) &= \frac{d}{dt}\{\operatorname{tr} \Lambda_t^* \rho \log \rho_\beta - \operatorname{tr} \Lambda_t^* \rho \log \Lambda_t^* \rho\}\bigg|_{t=0} \\ &= -\frac{d}{dt} S(\Lambda_t^* \rho, \rho_\beta)\bigg|_{t=0}.\end{aligned}$$

Spohn defined the entropy production $\sigma(\rho)$ in relation to the Λ_t^*-invariant state ρ_0 by

$$\sigma(\rho) = -\frac{d}{dt} S(\Lambda_t^* \rho, \rho_0)\bigg|_{t=0}.$$

It is easily shown that $S(\Lambda_t^* \rho, \rho_0)$ is decreasing and continuous in t from the right.

Theorem 20.13 *Let* $\dim \mathcal{H} < \infty$ *and the range* $\operatorname{ran} \rho_0 = \mathcal{H}$. *Then the entropy production* σ *is defined on* $\mathfrak{S}(\mathcal{H})$ *and is given by*

$$\sigma(\rho) = \operatorname{tr}\bigl\{L(\rho)(\log \rho_0 - \log \rho)\bigr\}, \quad \rho \in \mathfrak{S}(\mathcal{H}),$$

where $\Lambda_t^* = e^{tL}$, $t \in \mathbb{R}^+$. *Moreover,* σ *is convex and it takes values in* $[0, \infty]$.

Theorem 20.14

1. $\{\rho \in \mathfrak{S}(\mathcal{H}); L(\rho) = 0\} \subset \{\rho \in \mathfrak{S}(\mathcal{H}); \sigma(\rho) = 0\}$. *If* $\sigma(\rho) = 0$ *only for* $\rho = \rho_0$, *then* $\lim_{t\to 0} \Lambda_t \rho = \rho_0$ *for all* $\rho \in \mathfrak{S}(\mathcal{H})$.
2. $\sigma = 0$ *iff* $L = -\mathrm{i}[H, \cdot]$.

Ojima discussed the entropy production in terms of Kubo's linear response theory.

Let $\mathcal{A}$ be a C^*-algebra, α_t $(t \in \mathbb{R})$ be a one-parameter group of automorphisms of $\mathcal{A}$ and let ω_β be a KMS state with respect to α_t and inverse temperature β. The KMS state ω_β is an equilibrium state prepared at $t = -\infty$ and an external perturbation $\mathbf{V}(t) = (V_1(t), \ldots, V_n(t))$ is adiabatically applied to the system. The interaction energy of this perturbation with the system is expressed by an observable vector $\mathbf{Q} = (Q_1, \ldots, Q_n)$, $Q_k \in \mathcal{A}$ as follows

$$H_{\text{int}}(t) = -\mathbf{Q} \cdot \mathbf{V}(t) = -\sum_{i=1}^{n} Q_i V_i(t).$$

Using the derivation $\delta(\cdot)$ [428, 671] obtained by

$$\frac{d\alpha_t(A)}{dt} = \alpha_t\big(\delta(A)\big), \quad A \in \mathcal{A},$$

the time evolution $\{\alpha_{t,s;\mathbf{V}}; t, s \in \mathbb{R}\}$ with the perturbation $\mathbf{V}$ is determined by

$$\frac{d}{dt}\alpha_{t,s;\mathbf{V}}(A) = \alpha_{t,s;V}\big(\delta(A) + [H_{\text{int}}(t), A]\big)$$

and

$$\alpha_{t=s;\mathbf{V}}(A) = A, \quad A \in \mathcal{A}.$$

Put $U(t,s;\mathbf{V}) = T\exp\{\mathrm{i}\int_s^t \alpha_\tau(\mathbf{Q}) \cdot \mathbf{V}(\tau)\,d\tau\}$ with the time ordering operator T. Then

$$\alpha_{t,s;\mathbf{V}}(A) = \alpha_{-s}\big[U(t,s;\mathbf{V})^*\alpha_t(A)U(t,s;\mathbf{V})\big], \quad A \in \mathcal{A}.$$

When the input state ω_β $(= \varphi_{t=t_0=-\infty})$ changes to $\varphi_t \equiv \omega_\beta \circ \alpha_{t_0,t;\mathbf{V}}$ at time t, the relative entropy between ω_β and φ_t is evaluated as

$$S(\varphi_t, \varphi_0 = \omega_\beta) = \beta \int_{t_0}^{t} \varphi_s\big(\delta(\mathbf{Q})\big)\mathbf{V}(s)\,ds \geq 0.$$

The current operator $\mathbf{J}$ conjugated to the external force $\mathbf{V}(t)$ is defined by

$$\mathbf{J} = \delta(\mathbf{Q}).$$

Then the time-dependent entropy production becomes

$$\sigma(t, t_0; \mathbf{V}) = \frac{d}{dt}S(\varphi_t, \varphi_0 = \omega_\beta) = \beta\varphi_t\big(\delta(\mathbf{Q})\big)\mathbf{V}(t)$$

which is not always positive because ω_β is not $\alpha_{t,t_0;\mathbf{V}}$-invariant, so that Ojima, Hasegawa and Ichiyanagi [612] considered the mean entropy production as

$$\begin{aligned}\bar{\alpha} &= \lim_{T\to\infty}\frac{1}{T}\int_0^T dt \lim_{T_0\to\infty}\frac{1}{T_0}\int_{-T_0}^0 dt_0\sigma(t+t_0,t_0;\mathbf{V}) \\ &= \lim_{T\to\infty,T_0\to\infty}\frac{1}{TT_0}\int_{-T_0}^0 dt_0 S(\varphi_{T+t_0};\varphi_{t_0}=\omega_\beta)\geq 0. \end{aligned} \tag{20.14}$$

In order for these two limiting procedures to be meaningful, Ojima gave certain characterization to the external force $\mathbf{V}(t)$, namely, he assumed that the continuous function $\mathbf{V}(t)$ is almost periodic, i.e., it is uniformly approximated by a linear combination of periodic functions. Here "uniformly" means the convergence in the uniform topology with the norm $\|\mathbf{V}\| = \sup\{|\mathbf{V}(t)|; t\in\mathbb{R}\}$.

By Brochner's theorem, this is equivalent to the condition that the orbit $\{|\mathbf{V}(t)|; t\in\mathbb{R}\}$ is precompact in the uniform topology whose completion $M_{\mathbf{V}} = \overline{\{\mathbf{V}(t-\lambda);\lambda\in\mathbb{R}\}}$ becomes an abelian compact group. Therefore, the external force $\mathbf{V}(t)$ can be expressed as an element in the set $C(M_{\mathbf{V}})$ of all continuous functions on $M_{\mathbf{V}}$, that is,

$$\exists\widetilde{\mathbf{V}}\in C(M_{\mathbf{V}}),\xi_0\in M_{\mathbf{V}}\quad\text{such that}\quad \mathbf{V}(t)=\widetilde{\mathbf{V}}\big(\lambda_t(\xi_0)\big),$$

where λ_t is the time flow on $M_{\mathbf{V}}$ defined by

$$\lambda_t\xi=\xi_t,\quad \xi\in M_{\mathbf{V}}.$$

We construct a compound system of $(\mathcal{A},\mathbb{R},\alpha)$ and $(C(M_{\mathbf{V}}),\mathbb{R},\lambda)$, which is denoted by $(\mathcal{B},\mathbb{R},\gamma)$:

$$\begin{aligned}\mathcal{B}&=\mathcal{A}\otimes C(M_{\mathbf{V}})=C(M_{\mathbf{V}},\mathcal{A}),\\ \gamma_t(\tilde{B})(\xi)&=\alpha_{0,t;\xi}\big((\tilde{B})(\lambda_{-t}\xi)\big),\quad \tilde{B}\in\mathcal{B}.\end{aligned}$$

Since the Haar measure μ on $M_{\mathbf{V}}$ is ergodic, we have

$$\begin{aligned}(\varphi\otimes\mu)\big(\gamma_t(\tilde{B})\big)&=\int_{M_{\mathbf{V}}}\varphi\big(\gamma_t(\tilde{B})(\xi)\big)\,d\mu(\xi)\\ &=\lim_{T_0\to\infty}\frac{1}{T_0}\int_{-T_0}^0\varphi\big(\alpha_{t_0t+t_0;\xi}\big(\tilde{B}(\lambda_{-t-t_0}\xi)\big)\big)\,dt_0\end{aligned}$$

for any state φ on $\mathcal{A}$. Then the relative entropy and the mean entropy production becomes

$$\begin{aligned}S\big(\gamma_t^*(\omega_\beta\otimes\mu),\omega_\beta\otimes\mu\big)&=\lim_{T_0\to\infty}\frac{1}{T_0}\int_{-T_0}^0 S(\varphi_{t+t_0},\varphi_{t_0}=\omega_\beta)\,dt_0,\\ \bar{\sigma}&=\lim_{n\to\infty}\frac{1}{T_n}\int_0^{T_n}S\big(\gamma_t^*(\omega_\beta\otimes\mu),\omega_\beta\otimes\mu\big)\,dt,\end{aligned}$$

where

$$\tilde{\varphi} \equiv \lim_{n\to\infty} \frac{1}{T_n} \int_0^{T_n} \gamma_t^*(\omega_\beta \otimes \mu)\, dt$$

with the proper sequence $\{T_n\}$.

Ojima discussed the relation between the above expression (20.14) and the linear response theory of Kubo [612].

20.3 Quantum Dynamical Entropy

The Kolmogorov–Sinai dynamical entropy is used in the theory of classical dynamical systems to compute the mean information and the degree of chaos for dynamical systems.

Here we consider some of the above quantum generalizations of the dynamical entropy.

20.3.1 Formulation by CNT

Let $\mathcal{A}$ be a unital C^*-algebra, θ be an automorphism of $\mathcal{A}$, and φ be a stationary state over $\mathcal{A}$ with respect to θ, i.e., $\varphi \circ \theta = \varphi$. Let $\mathcal{B}$ be a finite-dimensional C^*-subalgebra of $\mathcal{A}$.

The CNT entropy [176] for a subalgebra $\mathcal{B}$ is given by

$$H_\varphi(\mathcal{B}) = \sup\left\{\sum_k \lambda_k S(\omega_k|\mathcal{B}, \varphi|\mathcal{B});\ \varphi = \sum_k \lambda_k \omega_k \text{ finite decomposition of } \varphi\right\},$$

where $\varphi|\mathcal{B}$ is the restriction of the state φ to $\mathcal{B}$ and $S(\cdot,\cdot)$ is the relative entropy for C^*-algebra [63, 760, 763].

The CNT dynamical entropy with respect to θ and $\mathcal{B}$ is given by

$$\tilde{H}_\varphi(\theta, \mathcal{B}) = \limsup_{N\to\infty} \frac{1}{N} H_\varphi\big(\mathcal{B} \vee \theta\mathcal{B} \vee \cdots \vee \theta^{N-1}\mathcal{B}\big),$$

and the dynamical entropy for θ is defined by

$$\tilde{H}_\varphi(\theta) = \sup_{\mathcal{B}} \tilde{H}_\varphi(\theta, \mathcal{B}).$$

20.3.2 Kolmogorov–Sinai Type Complexities

Let $\mathcal{A}$ and $\bar{\mathcal{A}}$ be $*$-algebras and let φ be a state on $\mathcal{A}$. Let θ (resp., $\bar{\theta}$) be an automorphism of the algebra $\mathcal{A}$ (resp., $\bar{\mathcal{A}}$) such that

$$\varphi \circ \theta = \varphi$$

and Λ be a covariant (i.e., $\Lambda \circ \bar{\theta} = \theta \circ \Lambda$) CP map from $\bar{\mathcal{A}}$ to $\mathcal{A}$. Take a finite subalgebra $\mathcal{A}_k$, $k \in \mathbb{N}$ (resp., $\bar{\mathcal{A}}_k$) of $\mathcal{A}$ (resp., $\bar{\mathcal{A}}$) and a unital map α_k (resp., $\bar{\alpha}_k$) from $\mathcal{A}_k$ (resp., $\bar{\mathcal{A}}_k$) to $\mathcal{A}$ (resp., $\bar{\mathcal{A}}$). Put $\alpha^M \equiv (\alpha_1, \alpha_2, \ldots, \alpha_M)$, $\bar{\alpha}_\Lambda^N \equiv (\Lambda \circ \bar{\alpha}_1, \Lambda \circ \bar{\alpha}_2, \ldots, \Lambda \circ \bar{\alpha}_N)$. Let $\mathcal{S}$ be a weak *-convex subset of the state space $\mathfrak{S}(\mathcal{A})$ as was introduced in Chap. 7. Then two generalized compound states for α^M and $\bar{\alpha}_\Lambda^N$ w.r.t. a measure $\mu \in M_\varphi(\mathcal{S})$ decomposing a state $\varphi \in \mathcal{S}$ such that $\varphi(A) = \int_{\mathcal{S}} \omega(A)\, d\mu(\omega), \forall A \in \mathcal{A}$ are defined by

$$\Phi_\mu^{\mathcal{S}}(\alpha^M) = \int_{\mathcal{S}} \bigotimes_{m=1}^{M} \alpha_m^* \omega\, d\mu,$$

$$\Phi_\mu^{\mathcal{S}}(\alpha^M \cup \bar{\alpha}_\Lambda^N) = \int_{\mathcal{S}} \bigotimes_{m=1}^{M} \alpha_m^* \omega \bigotimes_{n=1}^{N} \bar{\alpha}_n^* \Lambda^* \omega\, d\mu.$$

Define the transmitted complexities as follows: Using the notations given in Chap. 10, the complexity, the mutual entropy, the quasi-mutual entropy respectively are

$$\begin{aligned} &T^{\mathcal{S}}(\varphi;\, \alpha^M, \bar{\alpha}_\Lambda^N) \\ &\quad \equiv \sup \int_{\mathcal{S}} S\left(\bigotimes_{m=1}^{M} \alpha_m^* \omega \bigotimes_{n=1}^{N} \bar{\alpha}_n^* \Lambda^* \omega,\, \Phi_\mu^{\mathcal{S}}(\alpha^M) \otimes \Phi_\mu^{\mathcal{S}}(\bar{\alpha}_\Lambda^N) \right) d\mu, \quad \mu \in M_\varphi(S); \\ &I^{\mathcal{S}}(\varphi;\, \alpha^M, \bar{\alpha}_\Lambda^N) \\ &\quad \equiv \sup\left\{ S\left(\Phi_\mu^{\mathcal{S}}(\alpha^M \cup \bar{\alpha}_\Lambda^N), \Phi_\mu^{\mathcal{S}}(\alpha^M) \otimes \Phi_\mu^{\mathcal{S}}(\bar{\alpha}_\Lambda^N)\right), \mu \in M_\varphi(\mathcal{S}) \right\}; \\ &J^{\mathcal{S}}(\varphi;\, \alpha^M, \bar{\alpha}_\Lambda^N) \\ &\quad \equiv \sup \int_{\mathcal{S}} S\left(\bigotimes_{m=1}^{M} \alpha_m^* \omega \bigotimes_{n=1}^{N} \bar{\alpha}_n^* \Lambda^* \omega,\, \Phi_\mu^{\mathcal{S}}(\alpha^M) \otimes \Phi_\mu^{\mathcal{S}}(\bar{\alpha}_\Lambda^N) \right) d\mu_f, \quad \mu_f \in F_\varphi(\mathcal{S}). \end{aligned}$$

In the case of $\mathfrak{S}(\mathcal{A}) = \mathcal{S}$, we denote $T^{\mathcal{S}}$ by T for simplicity, and similarly with I, J. When $\mathcal{A}_k = \mathcal{A}_0 = \bar{\mathcal{A}}_k$, $\mathcal{A} = \bar{\mathcal{A}}$, $\theta = \bar{\theta}$, $\alpha_k = \theta^{k-1} \circ \alpha = \bar{\alpha}_k$ ($\alpha : \mathcal{A}_0 \to \mathcal{A}$; unital CP),

$$\tilde{T}_\varphi^{\mathcal{S}}(\theta, \alpha, \Lambda^*) \equiv \lim \sup_{N\to\infty} \frac{1}{N} T^{\mathcal{S}}(\varphi;\, \alpha^N, \bar{\alpha}_\Lambda^N),$$

$$\tilde{T}_\varphi^{\mathcal{S}}(\theta, \Lambda^*) \equiv \sup_\alpha \tilde{T}_\varphi^{\mathcal{S}}(\theta, \alpha, \Lambda^*).$$

$\tilde{T}_\varphi^{\mathcal{S}}(\theta, \Lambda^*)$ is the mean transmitted complexity w.r.t. θ and Λ^*. We similarly define $\tilde{I}_\varphi^{\mathcal{S}}$, $\tilde{J}_\varphi^{\mathcal{S}}$. Then the Connes–Narnhofer–Thirring type theorem holds for these complexities.

Theorem 20.15 *If there exists* $\alpha'_m : \mathcal{A} \to \mathcal{A}_m$ *such that* $\alpha_m \circ \alpha'_m = \mathrm{id}$, $\bar{\alpha}_m \circ \bar{\alpha}'_m = \mathrm{id}$ *as* $m \to \infty$, *then* $\tilde{T}^{\mathcal{S}}_\varphi(\theta, \Lambda^*) = \lim_{m\to\infty} \tilde{T}^{\mathcal{S}}_\varphi(\theta, \alpha_m, \Lambda^*)$. *The same holds for* $\tilde{I}^{\mathcal{S}}_\varphi$, $\tilde{J}^{\mathcal{S}}_\varphi$.

See [536] for the proof.

Our complexities generalize the usual dynamical entropy in the following senses:

1. $T^{\mathcal{S}}(\varphi;\ \mathrm{id}, \Lambda^*) = T^{\mathcal{S}}(\varphi, \Lambda^*)$, where $\mathrm{id} : \mathcal{A} \to \mathcal{A}$.
2. When $\mathcal{A}_n, \mathcal{A}$ are abelian C^*-algebras and α_k is an embedding,

$$T\left(\mu; \alpha^M\right) = S_\mu^{\text{classical}}\left(\bigvee_{m=1}^{M} \tilde{A}_m\right),$$

$$I\left(\mu; \alpha^M, \bar{\alpha}^N\right) = I_\mu^{\text{classical}}\left(\bigvee_{n=1}^{M} \tilde{A}_n, \bigvee_{n=1}^{N} \tilde{B}_n\right)$$

for any finite partitions $\tilde{A}_n, \tilde{B}_n$ on the probability space $(\Omega = \mathrm{spec}(\mathcal{A}), \mathcal{F}, \mu)$.
3. When Λ is the restriction of $\mathcal{A}$ to a subalgebra $\mathcal{M}$ of $\mathcal{A}$, $\Lambda = |_{\mathcal{M}}$ (see Chaps. 7 and 10),

$$J(\varphi; |_{\mathcal{M}}) = J(\varphi;\ \mathrm{id}; |_{\mathcal{M}}) = H_\varphi(\mathcal{M}) = \text{CNT-entropy}.$$

The CNT-entropy $\tilde{H}_\varphi$ is equivalent to our complexity J: Let $\mathcal{M} \subset \mathcal{A}_0, \mathcal{A} = \otimes^N \mathcal{A}_0$, $\theta \in \mathrm{Aut}(\mathcal{A})$, $\alpha^N \equiv (\alpha, \theta \circ \alpha, \ldots, \theta^{N-1} \circ \alpha)$, $\alpha = \bar{\alpha}$; $\mathcal{A}_0 \to \mathcal{A}$ (embedding) and $\mathcal{M}_N \equiv \bigotimes_1^N \mathcal{M}$. Then we have

$$\tilde{J}_\varphi(\theta; \mathcal{M}) = \limsup_{N\to\infty} \frac{1}{N} J\left(\varphi;\ \alpha^N; |_{\mathcal{M}_N}\right)$$

and

$$\tilde{J}_\varphi(\theta) \equiv \sup_{\mathcal{M}} \left\{\tilde{J}_\varphi(\theta; \mathcal{M});\ \mathcal{M} \subset \mathcal{A}_0\right\}.$$

$\tilde{T}^{\mathcal{S}}_\varphi(\theta), \tilde{I}^{\mathcal{S}}_\varphi(\theta)$ are computed similarly.

20.3.3 *Model Computation*

Numerical computation of the dynamical entropy can be used to see which state (or modulated state) is most effective for optical fiber communication. Let $\mathcal{H}_0$ and $\overline{\mathcal{H}}_0$ be input and output Hilbert spaces, respectively. In order to send a state carrying information to the output system, we might need to modulate the state in a proper way. A modulation M is a channel, denoted by $\Gamma^*_{(M)}$, from $\mathfrak{S}(\mathcal{H}_0)$ to a certain state space $\mathfrak{S}(\mathcal{H}_{(M)})$ on a proper Hilbert space $\mathcal{H}_{(M)}$. Take

$$\mathcal{A} \equiv \bigotimes_{-\infty}^{\infty} \mathbf{B}(\mathcal{H}_0), \qquad \overline{\mathcal{A}} \equiv \bigotimes_{-\infty}^{\infty} \mathbf{B}(\overline{\mathcal{H}}_M).$$

Let $\mathfrak{S}(\mathcal{A})$ (resp., $\overline{\mathfrak{S}(\mathcal{A})}$) be the set of all density operators in $\mathcal{A}$ (resp., $\overline{\mathcal{A}}$). Let θ (resp., $\overline{\theta}$) be a shift on $\mathcal{A}$ (resp., $\overline{\mathcal{A}}$) and α (resp., $\overline{\alpha}$) be an embedding map from $\mathbf{B}(\mathcal{H}_0)$ to $\mathcal{A}$ (resp., $\mathbf{B}(\overline{\mathcal{H}}_0)$ to $\overline{\mathcal{A}}$) as before. Let Λ^* be an attenuation channel, i.e., $\Lambda_{\sqrt{\eta},\sqrt{1-\eta}}$ with the transmission rate η discussed in Chap. 7. Put $\tilde{\Lambda}^* \equiv \bigotimes_{-\infty}^{\infty} \Lambda^*$ and $\tilde{\Gamma}^*_{(M)} \equiv \bigotimes_{-\infty}^{\infty} \Gamma^*_{(M)}$ and define

$$\alpha^N_{(M)} \equiv \left(\alpha \circ \tilde{\Gamma}_{(M)}, \dots, \theta^{N-1} \circ \alpha \circ \tilde{\Gamma}_{(M)}\right),$$

$$\overline{\alpha}^N_{(M)\tilde{\Lambda}^*} \equiv \left(\tilde{\Gamma}_{(M)} \circ \tilde{\Lambda}^* \circ \overline{\alpha}, \dots, \tilde{\Gamma}_{(M)} \circ \tilde{\Lambda}^* \circ \overline{\theta}^{N-1} \circ \overline{\alpha}\right).$$

We only consider two modulations here: the PAM (pulse amplitude modulation) and the PPM (pulse position modulation). The modulations $\Gamma_{(\mathrm{PAM})}$ for PAM and $\Gamma_{(\mathrm{PPM})}$ for PPM are written as

$$\Gamma_{(\mathrm{PAM})}(E_n) = |n\rangle\langle n|,$$

$$\Gamma_{(\mathrm{PPM})}(E_n) = \underbrace{|0\rangle\langle 0| \otimes \cdots \otimes |0\rangle\langle 0| \otimes \overbrace{|d\rangle\langle d|}^{n\text{-th}} \otimes |0\rangle\langle 0| \otimes \cdots \otimes |0\rangle\langle 0|}_{M},$$

where E_n is a pure state in $\mathfrak{S}(\mathcal{H}_0)$ coding the n-th symbol and $|k\rangle\langle k|$ is a k-photon number state.

For a stationary initial state $\rho = \sum_m \mu_m (\bigotimes_{-\infty}^{\infty} \rho_m^{(i)}) \in \mathfrak{S}(\mathcal{A})$ with $\rho_m^{(i)} = \sum_{n_i} \lambda_{n_i}^{(m)} E_{n_i}$, $E_{n_i} \in \operatorname{ex} \mathfrak{S}(\mathcal{H}_0)$ and the attenuation channel Λ^*, the transmitted complexities of mutual entropy type for two modulations PAM and PPM are calculated as

$$\begin{aligned} I&\left(\rho;\ \alpha^N_{(\mathrm{PAM})}, \overline{\alpha}^N_{(\mathrm{PAM})\tilde{\Lambda}^*}\right) \\ &= \sum_{j_0=0}^{M} \cdots \sum_{j_{N-1}=0}^{M} \sum_{n_0=J_0}^{M} \cdots \sum_{n_{N-1}=J_{N-1}}^{M} \left(\sum_m \mu_m \prod_{k=0}^{N-1} \lambda_{n_k}^{(m)}\right)\left(\prod_{k'=0}^{N-1} |C_{j_{k'}}^{n_{k'}}|^2\right) \\ &\quad \times \left\{\log \prod_{k=0}^{N-1} |C_{j_k}^{n_k}|^2 - \log\left(\sum_{n'_0=J_0}^{M} \cdots \sum_{n'_{N-1}=J_{N-1}}^{M} \left(\sum_{m'} \mu_{m'} \prod_{k'=0}^{N-1} \lambda_{n'_{k'}}^{(m')}\right)\right.\right. \\ &\quad \left.\left. \times \left(\prod_{k''=0}^{N-1} |C_{j_{k''}}^{n'_{k''}}|^2\right)\right)\right\}, \end{aligned}$$

$$I\left(\rho;\ \alpha^N_{(\mathrm{PPM})}, \overline{\alpha}^N_{(\mathrm{PPM})\tilde{\Lambda}^*}\right)$$

$$= -\sum_{n_0=1}^{M}\cdots\sum_{n_{N-1}=1}^{M}\left(\sum_m \mu_m \prod_{k=0}^{N-1}\lambda_{n_k}^{(m)}\right)\left\{\sum_{p=1}^{N}\sum_{\{q_1,\ldots,q_p\}\subset\{1,2,\ldots,N\}}\right.$$

$$\left.\times\left(\sum_{\ell_1=1}^{d}\cdots\sum_{\ell_p=1}^{d}\left|C^d_{\ell_1}\right|^2\cdots\left|C^d_{\ell_p}\right|^2(1-\eta)^{N-p}\eta^p\log\left(\sum_{m'}\mu_{m'}\prod_{k'=0}^{p}\lambda_{n_{q_{k'}}}^{(m')}\right)\right)\right\},$$

where

$$\left|C^{n_i}_{j_i}\right|^2 = \frac{n_i!}{j_i!(n_i-j_i)!}\eta^{j_i}(1-\eta)^{(n_i-j_i)}.$$

Then we have

Theorem 20.16 *If $\mathcal{A} = \overline{\mathcal{A}}$, $\theta = \overline{\theta}$, $\alpha = \overline{\alpha}$ and $d \geq N$, then*

$$\tilde{I}_\rho(\theta, \alpha_{(\mathrm{PPM})}, \Lambda) \geq \tilde{I}_\rho(\theta, \alpha_{(\mathrm{PAM})}, \Lambda^*).$$

This theorem means that the PPM sends more information than the PAM.

20.3.4 Various Quantum Dynamical Entropies

We will discuss several attempts to define the quantum dynamical entropy.

(1) *Formulation by AF* (Alicki–Fannes [53]). Let $\mathcal{A}$ be a C^*-algebra, θ be an automorphism on $\mathcal{A}$, φ be a stationary state with respect to θ, and let $\mathcal{B}$ be a unital $*$-subalgebra of $\mathcal{A}$. A set $\gamma = \{\gamma_1, \gamma_2, \ldots, \gamma_k\}$ of elements of $\mathcal{B}$ is called a finite operational partition of unity of size k if γ satisfies the following condition:

$$\sum_{i=1}^{k}\gamma_i^*\gamma_i = I.$$

The operation $\circ$ is defined by

$$\gamma \circ \xi \equiv \{\gamma_i\xi_j;\ i = 1, 2, \ldots, k,\ j = 1, 2, \ldots, l\}$$

for any partitions $\gamma = \{\gamma_1, \gamma_2, \ldots, \gamma_k\}$ and $\xi = \{\xi_1, \xi_2, \ldots, \xi_l\}$. For any partition γ of size k, a $k \times k$ density matrix $\rho[\gamma] = (\rho[\gamma]_{i,j})$ is given by

$$\rho[\gamma]_{i,j} = \varphi(\gamma_j^*\gamma_i).$$

Then the dynamical entropy $\tilde{H}_\varphi(\theta, \mathcal{B}, \gamma)$ with respect to the partition γ and shift θ is defined by von Neumann entropy $S(\cdot)$ as

$$\tilde{H}_\varphi(\theta, \mathcal{B}, \gamma) = \limsup_{n\to\infty}\frac{1}{n}S\left(\rho\left[\theta^{n-1}(\gamma)\circ\cdots\circ\theta(\gamma)\circ\gamma\right]\right).$$

The dynamical entropy $\tilde{H}_\varphi(\theta, \mathcal{B})$ is given by taking the supremum over operational partition of unity in $\mathcal{B}$ as

$$\tilde{H}_\varphi(\theta, \mathcal{B}) = \sup\bigl\{\tilde{H}_\varphi(\theta, \mathcal{B}, \gamma); \gamma \subset \mathcal{B}\bigr\}.$$

(2) *Formulation by AOW*. A construction of dynamical entropy is done by using a quantum Markov chain in [17].

Let $\mathcal{A}$ be a von Neumann algebra acting on a Hilbert space $\mathcal{H}$, let φ be a state on $\mathcal{A}$, and set $\mathcal{A}_0 = M_d$ ($d \times d$ matrix algebra). Take the transition expectation $\mathcal{E}_\gamma : \mathcal{A}_0 \otimes \mathcal{A} \to \mathcal{A}$ of Accardi such that

$$\mathcal{E}_\gamma(\tilde{A}) = \sum_i \gamma_i A_{ii} \gamma_i,$$

where $\tilde{A} = \sum_{i,j} e_{ij} \otimes A_{ij} \in \mathcal{A}_0 \otimes \mathcal{A}$ and $\gamma = \{\gamma_j\}$ is a finite partition of unity $I \in \mathcal{A}$. A quantum Markov chain is defined by $\psi \equiv \{\varphi, \mathcal{E}_{\gamma,\theta}\} \in \mathfrak{S}(\bigotimes_1^\infty \mathcal{A}_0)$ such that

$$\psi\bigl(j_1(A_1)\cdots j_n(A_n)\bigr) \equiv \varphi\bigl(\mathcal{E}_{\gamma,\theta}\bigl(A_1 \otimes \mathcal{E}_{\gamma,\theta}\bigl(A_2 \otimes \cdots \otimes A_{n-1}\mathcal{E}_{\gamma,\theta}(A_n \otimes I)\bigr)\bigr)\bigr),$$

where $\mathcal{E}_{\gamma,\theta} = \theta \circ \mathcal{E}_\gamma$, $\theta \in \mathrm{Aut}(\mathcal{A})$, and j_k is an embedding of $\mathcal{A}_0$ into $\bigotimes_1^\infty \mathcal{A}_0$ such that $j_k(A) = I \otimes \cdots \otimes I \otimes \underbrace{A}_{k\text{-th}} \otimes I \cdots$.

Suppose that for φ there exists a unique density operator ρ such that $\varphi(A) = \operatorname{tr} \rho A$ for any $A \in \mathcal{A}$. Let us define a state ψ_n on $\bigotimes_1^n \mathcal{A}_0$ expressed as

$$\psi_n(A_1 \otimes \cdots \otimes A_n) = \psi\bigl(j_1(A_1)\cdots j_n(A_n)\bigr).$$

The density operator ξ_n for ψ_n is given by

$$\xi_n \equiv \sum_{i_1} \cdots \sum_{i_n} \operatorname{tr}_{\mathcal{A}}\bigl(\theta^n(\gamma_{i_n})\cdots \gamma_{i_1}\rho\gamma_{i_1}\cdots\theta^n(\gamma_{i_n})\bigr) e_{i_1 i_1} \otimes \cdots \otimes e_{i_n i_n}.$$

Put

$$P_{i_n \ldots i_1} = \operatorname{tr}_{\mathcal{A}}\bigl(\theta^n(\gamma_{i_n})\cdots \gamma_{i_1}\rho\gamma_{i_1}\cdots\theta^n(\gamma_{i_n})\bigr).$$

The dynamical entropy through QMC is defined by

$$\tilde{S}_\varphi(\theta; \gamma) \equiv \limsup_{n\to\infty} \frac{1}{n}(-\operatorname{tr} \xi_n \log \xi_n) = \limsup_{n\to\infty} \frac{1}{n}\Bigl(-\sum_{i_1,\ldots,i_n} P_{i_n \cdots i_1} \log P_{i_n \cdots i_1}\Bigr).$$

If $P_{i_n \cdots i_1}$ satisfies the Markov property, then the above equality can be written as

$$\tilde{S}_\varphi(\theta; \gamma) = -\sum_{i_1,i_2} P(i_2|i_1)P(i_1)\log P(i_2|i_1).$$

The dynamical entropy through QMC with respect to θ and a von Neumann subalgebra $\mathcal{B}$ of $\mathcal{A}$ is given by

$$\tilde{S}_\varphi(\theta; \mathcal{B}) \equiv \sup\bigl\{\tilde{S}_\varphi(\theta; \gamma); \gamma \subset \mathcal{B}\bigr\}.$$

(3) *Quantum dynamical entropy for CP map* (KOW dynamical entropy [441]). Let $\mathbf{B}(\mathcal{K})$ (resp., $\mathbf{B}(\mathcal{H})$) be the set of all bounded linear operators on a separable Hilbert space $\mathcal{K}$ (resp., $\mathcal{H}$). We denote the set of all density operators on $\mathcal{K}$ (resp., $\mathcal{H}$) by $\mathfrak{S}(\mathcal{K})$ (resp., $\mathfrak{S}(\mathcal{H})$). Let

$$\Gamma : \mathbf{B}(\mathcal{K}) \otimes \mathbf{B}(\mathcal{H}) \to \mathbf{B}(\mathcal{K}) \otimes \mathbf{B}(\mathcal{H}) \tag{20.15}$$

be a normal, unital CP linear map, that is, Γ satisfies

$$\tilde{B}_\alpha \uparrow \tilde{B} \quad \Longrightarrow \quad \Gamma(\tilde{B}_\alpha) \uparrow \Gamma(\tilde{B}),$$
$$\Gamma(I_\mathcal{K} \otimes I_\mathcal{H}) = I_\mathcal{K} \otimes I_\mathcal{H} \quad \bigl(I_\mathcal{H} \text{ (resp., } I_\mathcal{K}\text{) is the unity in } \mathcal{H} \text{ (resp., } \mathcal{K}\text{)}\bigr)$$

for any increasing net $\{\tilde{B}_\alpha\} \subset \mathbf{B}(\mathcal{K}) \otimes \mathbf{B}(\mathcal{H})$ converging to $\tilde{B} \in \mathbf{B}(\mathcal{K}) \otimes \mathbf{B}(\mathcal{H})$ and the inequalities

$$\sum_{i,j=1}^{n} \tilde{B}_j^* \Gamma(\tilde{A}_j^* \tilde{A}_i) \tilde{B}_i \geq 0$$

hold for any $n \in \mathbb{N}$ and any $\tilde{A}_i, \tilde{B}_j \in \mathbf{B}(\mathcal{K}) \otimes \mathbf{B}(\mathcal{H})$. For a normal state ω on $\mathbf{B}(\mathcal{K})$, there exists a density operator $\tilde{\omega} \in \mathfrak{S}(\mathcal{K})$ associated to ω (i.e., $\omega(A) = \operatorname{tr} \tilde{\omega} A$, $\forall A \in \mathbf{B}(\mathcal{K})$). Then the map

$$\mathcal{E}^{\Gamma,\omega} : \mathbf{B}(\mathcal{K}) \otimes \mathbf{B}(\mathcal{H}) \to \mathbf{B}(\mathcal{H})$$

defined as

$$\mathcal{E}^{\Gamma,\omega}(\tilde{A}) = \omega\bigl(\Gamma(\tilde{A})\bigr) = \operatorname{tr}_\mathcal{K} \tilde{\omega} \Gamma(\tilde{A}), \quad \forall \tilde{A} \in \mathbf{B}(\mathcal{K}) \otimes \mathbf{B}(\mathcal{H})$$

is a transition expectation in the sense of [14] (i.e., $\mathcal{E}^{\Gamma,\omega}$ is a linear unital CP map from $\mathbf{B}(\mathcal{K}) \otimes \mathbf{B}(\mathcal{H})$ to $\mathbf{B}(\mathcal{H})$), whose dual is the map

$$\mathcal{E}^{*\Gamma,\omega}(\rho) : \mathfrak{S}(\mathcal{H}) \to \mathfrak{S}(\mathcal{K} \otimes \mathcal{H})$$

given by

$$\mathcal{E}^{*\Gamma,\omega}(\rho) = \Gamma^*(\tilde{\omega} \otimes \rho).$$

The dual map $\mathcal{E}^{*\Gamma,\omega}$ is a lifting in the sense of [19]; that is, it is a continuous map from $\mathfrak{S}(\mathcal{H})$ to $\mathfrak{S}(\mathcal{K} \otimes \mathcal{H})$.

For a normal, unital CP map $\Lambda : \mathbf{B}(\mathcal{H}) \to \mathbf{B}(\mathcal{H})$, $\mathrm{id} \otimes \Lambda : \mathbf{B}(\mathcal{K}) \otimes \mathbf{B}(\mathcal{H}) \to \mathbf{B}(\mathcal{K}) \otimes \mathbf{B}(\mathcal{H})$ is a normal, unital CP map, where id is the identity map on $\mathbf{B}(\mathcal{K})$. Then one defines the transition expectation

$$\mathcal{E}_\Lambda^{\Gamma,\omega}(\tilde{A}) = \omega\bigl((\mathrm{id} \otimes \Lambda)\Gamma(\tilde{A})\bigr), \quad \forall \tilde{A} \in \mathbf{B}(\mathcal{K}) \otimes \mathbf{B}(\mathcal{H}) \tag{20.16}$$

and the lifting

$$\mathcal{E}_\Lambda^{*\Gamma,\omega}(\rho) = \Gamma^*\bigl(\tilde{\omega} \otimes \Lambda^*(\rho)\bigr), \quad \forall \rho \in \mathfrak{S}(\mathcal{H}). \tag{20.17}$$

The above Λ^* is a quantum channel from $\mathfrak{S}(\mathcal{H})$ to $\mathfrak{S}(\mathcal{H})$, in which ρ is regarded as an input signal state and $\tilde{\omega}$ is as a noise state.

The equality

$$\begin{aligned}&\operatorname{tr}_{(\bigotimes_1^n \mathcal{K})\otimes\mathcal{H}} \Phi_{\Lambda,n}^{\Gamma,\omega}(\rho)(A_1\otimes\cdots\otimes A_n\otimes B)\\ &\quad\equiv \operatorname{tr}_{\mathcal{H}}\rho\big(\mathcal{E}_\Lambda^{\Gamma,\omega}\big(A_1\otimes\mathcal{E}_\Lambda^{\Gamma,\omega}\big(A_2\otimes\cdots A_{n-1}\otimes\mathcal{E}_\Lambda^{\Gamma,\omega}(A_n\otimes B)\big)\big)\big)\end{aligned}$$

for all $A_1, A_2, \ldots, A_n \in \mathbf{B}(\mathcal{K})$, $B \in \mathbf{B}(\mathcal{H})$ and any $\rho \in \mathfrak{S}(\mathcal{H})$ defines:

1. A lifting

$$\Phi_{\Lambda,n}^{\Gamma,\omega} : \mathfrak{S}(\mathcal{H}) \to \mathfrak{S}\left(\left(\bigotimes_1^n \mathcal{K}\right)\otimes\mathcal{H}\right).$$

2. Marginal states

$$\rho_{\Lambda,n}^{\Gamma,\omega} \equiv \operatorname{tr}_{\mathcal{H}} \Phi_{\Lambda,n}^{\Gamma,\omega}(\rho) \in \mathfrak{S}\left(\bigotimes_1^n \mathcal{K}\right),$$

$$\bar{\rho}_{\Lambda,n}^{\Gamma,\omega} \equiv \operatorname{tr}_{\bigotimes_1^n \mathcal{K}} \Phi_{\Lambda,n}^{\Gamma,\omega}(\rho) \in \mathfrak{S}(\mathcal{H}).$$

Here, the state

$$\Phi_{\Lambda,n}^{\Gamma,\omega}(\rho) \in \mathfrak{S}\left(\left(\bigotimes_1^n \mathcal{K}\right)\otimes\mathcal{H}\right)$$

is a compound state for $\bar{\rho}_{\Lambda,n}^{\Gamma,\omega}$ and $\rho_{\Lambda,n}^{\Gamma,\omega}$ in the sense of [560]. Note that generally $\bar{\rho}_{\Lambda,n}^{\Gamma,\omega}$ is not equal to ρ.

Definition 20.17 The quantum dynamical entropy with respect to Λ^*, ρ, Γ^* and ω is defined by

$$\tilde{S}(\Lambda^*; \rho, \Gamma^*, \omega) \equiv \limsup_{n\to\infty} \frac{1}{n} S\big(\rho_{\Lambda,n}^{\Gamma,\omega}\big), \tag{20.18}$$

where $S(\cdot)$ is the von Neumann entropy [805], that is, $S(\sigma) \equiv -\operatorname{tr}\sigma\log\sigma, \sigma \in \mathfrak{S}(\bigotimes_1^n \mathcal{K})$. The dynamical entropy with respect to Λ^* and ρ is defined as

$$\tilde{S}(\Lambda^*; \rho) \equiv \sup\big\{\tilde{S}(\Lambda^*; \rho, \Gamma^*, \omega); \Gamma^*, \omega\big\}.$$

Let us discuss the transition expectation $\mathcal{E}_\Lambda^{*\Gamma,\omega}$ associated to the CP maps Γ, Λ, and a state ω.

For a complete orthonormal system (CONS) $\{e_i\}$ in $\mathcal{K}$, put $E_{ij} = |e_i\rangle\langle e_j|$. There exist operators $v_\alpha \in \mathbf{B}(\mathcal{H})$ and complex numbers $\lambda_{kl\alpha,mn\beta} \in \mathbb{C}$ such that the unital CP map Γ of (20.15) can be written in the form (see p. 145 of [578])

$$\Gamma(\tilde{A}) = \sum_{k,l,m,n}\sum_{\alpha,\beta} \lambda_{kl\alpha,mn\beta}(E_{kl}^*\otimes v_\alpha^*)\tilde{A}(E_{mn}\otimes v_\beta), \quad \tilde{A} \in \mathbf{B}(\mathcal{K})\otimes\mathbf{B}(\mathcal{H}). \tag{20.19}$$

Let $\tilde{A} \in \mathbf{B}(\mathcal{K}) \otimes \mathbf{B}(\mathcal{H})$ be

$$\tilde{A} = \sum_{i,j} E_{ij} \otimes A_{ij}, \quad A_{ij} \in \mathbf{B}(\mathcal{H}). \tag{20.20}$$

From (20.19) and (20.20), we have

$$\begin{aligned}
\Gamma(\tilde{A}) &= \sum_{k,l,m,n} \sum_{\alpha,\beta} \lambda_{kl\alpha,mn\beta} (E_{kl}^* \otimes v_\alpha^*) \Big(\sum_{i,j} E_{ij} \otimes A_{ij} \Big) (E_{mn} \otimes v_\beta) \\
&= \sum_{k,l,m,n} \sum_{\alpha,\beta} \sum_{i,j} \lambda_{kl\alpha,mn\beta} (E_{kl}^* \otimes v_\alpha^*)(E_{ij} \otimes A_{ij})(E_{mn} \otimes v_\beta) \\
&= \sum_{k,l,m,n} \sum_{\alpha,\beta} \sum_{i,j} \lambda_{kl\alpha,mn\beta} (E_{kl}^* E_{ij} E_{mn} \otimes v_\alpha^* A_{ij} v_\beta) \\
&= \sum_{k,l,m,n} \sum_{\alpha,\beta} \sum_{i,j} \lambda_{kl\alpha,mn\beta} \big(|e_l\rangle \langle e_k, e_i\rangle \langle e_j, e_m\rangle \langle e_n| \otimes v_\alpha^* A_{ij} v_\beta \big) \\
&= \sum_{k,l,m,n} \sum_{\alpha,\beta} \lambda_{kl\alpha,mn\beta} (E_{ln} \otimes v_\alpha^* A_{km} v_\beta).
\end{aligned}$$

The equality

$$\Gamma(I_{\mathcal{K}} \otimes I_{\mathcal{H}}) = I_{\mathcal{K}} \otimes I_{\mathcal{H}}$$

implies

$$\sum_k \sum_{\alpha,\beta} \lambda_{kl\alpha,kn\beta} v_\alpha^* v_\beta = \delta_{ln} I_{\mathcal{H}}.$$

As the matrix $\sigma \equiv (\lambda_{kl\alpha,mn\beta})$ is positive definite, taking a proper basis of $\{v_\alpha\}$, σ can be written as

$$\sigma = \Big(\sum_q \tau_q \xi_{kl\alpha}^{*(q)} \xi_{mn\beta}^{(q)} \Big),$$

with some proper numbers $\tau_q, \xi_{kl\alpha}^{(q)}$, where τ_q is a positive number and $\xi_{kl\alpha}^{(q)}$ is a normalized orthogonal vector associated to τ_q. From the positivity of $\tilde{\omega}$, we have

$$\begin{aligned}
\omega(E_{ln}) &= \operatorname{tr}\big(\tilde{\omega}(E_{ln})\big) = \operatorname{tr}\big(\tilde{\omega} |e_l\rangle\langle e_n|\big) = \langle e_n, \tilde{\omega} e_l\rangle \\
&= \sum_p \big\langle \sqrt{\tilde{\omega}} e_n, f_p \big\rangle \big\langle f_p, \sqrt{\tilde{\omega}} e_l \big\rangle,
\end{aligned}$$

where $\{f_p\}$ is a CONS of $\mathcal{K}$. From the above equality, the transition expectation $\mathcal{E}^{*\Gamma,\omega}$ is expressed as (20.21) below:

$$\begin{aligned}
&\mathcal{E}^{*\Gamma,\omega}\Bigl(\sum_{i,j} E_{ij}\otimes A_{ij}\Bigr)\\
&\quad=\omega\Bigl(\Gamma\Bigl(\sum_{i,j} E_{ij}\otimes A_{ij}\Bigr)\Bigr)=\sum_{k,l,m,n}\sum_{\alpha,\beta}\lambda_{kl\alpha,mn\beta}\,\omega(E_{ln})v_\alpha^* A_{km}v_\beta\\
&\quad=\sum_{k,l,m,n}\sum_{\alpha,\beta}\Bigl(\sum_q \tau_q \tilde{\xi}^{*(q)}_{kl\alpha}\tilde{\xi}^{(q)}_{mn\beta}\Bigr)\sum_p \langle\sqrt{\bar{\omega}}e_n, f_p\rangle\langle f_p,\sqrt{\bar{\omega}}e_l\rangle v_\alpha^* A_{km}v_\beta\\
&\quad=\sum_{k,m}\Bigl(\sum_p\sum_q \tau_q\Bigl(\sum_{l,\alpha}\tilde{\xi}^{*(q)}_{kl\alpha}\overline{\langle\sqrt{\bar{\omega}}e_l, f_p\rangle}v_\alpha^*\Bigr)A_{km}\Bigl(\sum_{n,\beta}\tilde{\xi}^{(q)}_{mn\beta}\langle\sqrt{\bar{\omega}}e_n, f_p\rangle v_\beta\Bigr)\Bigr)\\
&\quad=\sum_{k,m}\Bigl(\sum_p\sum_q\Bigl(\sqrt{\tau_q}\sum_{l,\alpha}\tilde{\xi}^{*(q)}_{kl\alpha}\overline{\langle\sqrt{\bar{\omega}}e_l, f_p\rangle}v_\alpha^*\Bigr)\\
&\qquad\times A_{km}\Bigl(\sqrt{\tau_q}\sum_{n,\beta}\tilde{\xi}^{(q)}_{mn\beta}\langle\sqrt{\bar{\omega}}e_n, f_p\rangle v_\beta\Bigr)\Bigr).
\end{aligned}\tag{20.21}$$

Putting $u_{pqm}=\sqrt{\tau_q}\sum_{n,\beta}\tilde{\xi}^{(q)}_{mn\beta}\langle\sqrt{\bar{\omega}}e_n, f_p\rangle v_\beta$,

$$E^{\Gamma,\omega}\Bigl(\sum_{i,j} E_{ij}\otimes A_{ij}\Bigr)=\sum_{k,m}\sum_p\sum_q u^*_{pqk}A_{km}u_{pqm},$$

so that we can rewrite Γ,ω by u, and the dual of $E^{\Gamma,\omega}$ $(=E^u)$ as

$$E^{*u}(\rho)=\sum_{k,m}\sum_p\sum_q E_{km}\otimes u_{pqk}\rho u^*_{pqm}.$$

Since $\Gamma(I_{\mathcal{K}}\otimes I_{\mathcal{H}})=I_{\mathcal{K}}\otimes I_{\mathcal{H}}$ holds, one obtains

$$\sum_{p,q,k} u^*_{pqk}u_{pqk}=I_{\mathcal{H}}.$$

For a channel $\Lambda^*:\mathfrak{S}(\mathcal{H})\to\mathfrak{S}(\mathcal{H})$, the transition expectation (20.16) and the lifting (20.17) are expressed as follows:

$$\mathcal{E}^{*u}_{\Lambda}\Bigl(\sum_{i,j} E_{ij}\otimes A_{ij}\Bigr)=\sum_{k,m}\sum_p\sum_q \Lambda^*(u^*_{pqk}A_{km}u_{pqm})$$

and

$$\mathcal{E}^{*u}_{\Lambda}(\rho)=\sum_{k,m}\sum_p\sum_q E_{km}\otimes u_{pqk}\Lambda^*(\rho)u^*_{pqm}.\tag{20.22}$$

We generalize both the AF (Alicki–Fannes) entropy and the AOW (Accardi–Ohya–Watanabe) entropy. Then we compare the generalized AF entropy with the generalized AOW entropy.

Let θ be an automorphism of $\mathbf{B}(\mathcal{H})$, ρ be a density operator on $\mathcal{H}$ and $\mathcal{E}_\theta^{*u}$ be the transition expectation on $\mathbf{B}(\mathcal{K}) \otimes \mathbf{B}(\mathcal{H})$ with $\Lambda = \theta$ defined in Sect. 20.2.

One introduces the transition expectation $\mathcal{E}_\theta^{*u}$ from $\mathbf{B}(\mathcal{K}) \otimes \mathbf{B}(\mathcal{H})$ to $\mathbf{B}(\mathcal{H})$ as

$$\mathcal{E}_\theta^u\left(\sum_{i,j} E_{ij} \otimes A_{ij}\right) \equiv \sum_{k,m,p,q} \theta(u_{pqk}^* A_{km} u_{pqm}) = \sum_{k,m,p,q} \theta(u_{pqk}^*)\theta(A_{km})\theta(u_{pqm}).$$

The quantum Markov state $\{\rho_{\theta,n}^u\}$ on $\bigotimes_1^n \mathbf{B}(\mathcal{K})$ is defined through this transition expectation E_θ^u by

$$\begin{aligned}
&\operatorname{tr}_{\bigotimes_1^n \mathcal{K}}\left[\rho_{\theta,n}^u(A_1 \otimes \cdots \otimes A_n)\right] \\
&\quad \equiv \operatorname{tr}_{\mathcal{H}}\left[\rho \mathcal{E}_\theta^u\left(A_1 \otimes \mathcal{E}_\theta^u\left(A_2 \otimes \cdots \otimes A_{n-1} \otimes \mathcal{E}_\theta^{*u}(A_n \otimes I)\right)\right)\right]
\end{aligned}$$

for all $A_1, \ldots, A_n \in \mathbf{B}(\mathcal{K})$ and any $\rho \in \mathfrak{S}(\mathcal{H})$.

Let us consider another transition expectation $\hat{\mathcal{E}}_\theta^u$ such that

$$\hat{\mathcal{E}}_\theta^u\left(\sum_{i,j} E_{ij} \otimes A_{ij}\right) \equiv \sum_{k,l,p,q} \theta(u_{pqk}^*) A_{kl} \theta(u_{pql}).$$

One can define the quantum Markov state $\{\tilde{\rho}_{\theta,n}^u\}$ in terms of $\hat{\mathcal{E}}_\theta^u$

$$\begin{aligned}
&\operatorname{tr}_{\bigotimes_1^n \mathcal{K}}\left[\tilde{\rho}_{\theta,n}^u(A_1 \otimes \cdots \otimes A_n)\right] \\
&\quad \equiv \operatorname{tr}_{\mathcal{H}}\left[\rho \hat{\mathcal{E}}_\theta^u\left(A_1 \otimes \hat{\mathcal{E}}_{\theta^2}^u\left(A_2 \otimes \cdots \otimes A_{n-1} \otimes \hat{\mathcal{E}}_{\theta^n}^u(A_n \otimes I)\right)\right)\right] \qquad (20.23)
\end{aligned}$$

for all $A_1, \ldots, A_n \in \mathbf{B}(\mathcal{K})$ and any $\rho \in \mathfrak{S}(\mathcal{H})$. Then we have the following theorem.

Theorem 20.18 $\rho_{\theta,n}^u = \tilde{\rho}_{\theta,n}^u$.

Proof It is easily seen that

$$\hat{\mathcal{E}}_{\theta^i}^u(A \otimes B) = \theta^i \hat{\mathcal{E}}^u\left(A \otimes \theta^{-i}(B)\right) \qquad (20.24)$$

for any $A \in \mathbf{B}(\mathcal{K})$ and any $B \in \mathbf{B}(\mathcal{H})$, where $\hat{\mathcal{E}}^u = \hat{\mathcal{E}}_{\theta=\text{identity}}^u$. From Equality (20.24), we have

$$\hat{\mathcal{E}}_\theta^u(A_1 \otimes I) = \theta \hat{\mathcal{E}}^u(A_1 \otimes I) = \mathcal{E}_\theta^u(A_1 \otimes I)$$

and

$$\begin{aligned}
\hat{\mathcal{E}}_\theta^u\left(A_1 \otimes \hat{\mathcal{E}}_{\theta^2}^u(A_2 \otimes I)\right) &= \hat{\mathcal{E}}_\theta^u\left(A_1 \otimes \theta^2 \hat{\mathcal{E}}^u(A_2 \otimes I)\right) \\
&= \theta \hat{\mathcal{E}}^u\left(A_1 \otimes \theta^{-1} \circ \theta^2 \hat{\mathcal{E}}^u(A_2 \otimes I)\right)
\end{aligned}$$

$$= \theta\hat{\mathcal{E}}^u\big(A_1 \otimes \theta\hat{\mathcal{E}}^u(A_2 \otimes I)\big)$$
$$= \mathcal{E}_\theta^u\big(A_1 \otimes \mathcal{E}_\theta^u(A_2 \otimes I)\big).$$

Generally, we observe

$$\hat{\mathcal{E}}_\theta^u\big(A_1 \otimes \hat{\mathcal{E}}_{\theta^2}^u\big(A_2 \otimes \cdots \otimes \hat{\mathcal{E}}_{\theta^n}^u(A_n \otimes I)\big)\big)$$
$$= \mathcal{E}_\theta^u\big(A_1 \otimes \mathcal{E}_\theta^u\big(A_2 \otimes \cdots \otimes \mathcal{E}_\theta^u(A_n \otimes I)\big)\big)$$

for any $n \geq 1$. Hence $\rho_{\theta,n}^u = \tilde{\rho}_{\theta,n}^u$ holds. □

Let $\mathcal{B}_0$ be a subalgebra of $\mathbf{B}(\mathcal{K})$. Taking the restriction of a transition expectation $\mathcal{E} : \mathbf{B}(\mathcal{K}) \otimes \mathbf{B}(\mathcal{H}) \to \mathbf{B}(\mathcal{H})$ to $\mathcal{B}_0 \otimes \mathbf{B}(\mathcal{H})$, i.e., $\mathcal{E}_0 = \mathcal{E}|_{\mathcal{B}_0 \otimes \mathbf{B}(\mathcal{H})}$, $\mathcal{E}_0$ is the transition expectation from $\mathcal{B}_0 \otimes \mathbf{B}(\mathcal{H})$ to $\mathbf{B}(\mathcal{H})$. The QMC (quantum Markov chain) defines the state $\rho_{\theta,n}^{u(0)}$ on $\bigotimes_1^n \mathcal{B}_0$ through (20.23), which is

$$\rho_{\theta,n}^{u(0)} = \rho_{\theta,n}^u|_{\bigotimes_1^n \mathcal{B}_0}.$$

The subalgebra $\mathcal{B}_0$ of $\mathbf{B}(\mathcal{K})$ can be constructed as follows: Let $P_1, \ldots P_m$ be projection operators on mutually orthogonal subspaces of $\mathcal{K}$ such that $\sum_{i=1}^m P_i = I_{\mathcal{K}}$. Putting $\mathcal{K}_i = P_i\mathcal{K}$, the subalgebra $\mathcal{B}_0$ is generated by

$$\sum_{i=1}^m P_i A P_i, \quad A \in \mathbf{B}(\mathcal{K}).$$

One observes that in the case of $n = 1$

$$\rho_{\theta,1}^{u(0)} \equiv \rho_{\theta,1}^u|_{\mathcal{B}_0} = \sum_{i=1}^m P_i \rho_{\theta,1}^u P_i,$$

and one has for any $n \in \mathbb{N}$

$$\rho_{\theta,n}^{u(0)} = \sum_{i_1,\ldots,i_n} (P_{i_1} \otimes \cdots \otimes P_{i_n}) \rho_{\theta,n}^u (P_{i_1} \otimes \cdots \otimes P_{i_n})$$

from which the following theorem is proved.

Theorem 20.19

$$S\big(\rho_{\theta,n}^u\big) \leq S\big(\rho_{\theta,n}^{u(0)}\big).$$

Proof Let $\{\tilde{P}_{i_1,\ldots,i_n}\}$ be a PVM (projection-valued measure) on $\bigotimes_1^n \mathcal{K}$. If $\mathcal{E}_{\tilde{P}}(\tilde{A})$ is a map defined by

$$\mathcal{E}_{\tilde{P}}(\tilde{A}) \equiv \sum_{i_1,\ldots,i_n} \tilde{P}_{i_1,\ldots,i_n} \tilde{A} \tilde{P}_{i_1,\ldots,i_n}, \quad \forall \tilde{A} \in \bigotimes_1^n \mathbf{B}(\mathcal{K}),$$

then $\mathcal{E}_{\tilde{P}}$ satisfies: (i) $\mathcal{E}_{\tilde{P}}(I_{\bigotimes_1^n \mathcal{K}}) = I \in \bigotimes_1^n \mathcal{B}_0$, (ii) $\mathcal{E}_{\tilde{P}} \circ \mathcal{E}_{\tilde{P}}(\tilde{A}) = \mathcal{E}_{\tilde{P}}(\tilde{A}), \forall \tilde{A} \in \bigotimes_1^n \mathbf{B}(\mathcal{K})$, (iii) $\operatorname{tr} \mathcal{E}_{\tilde{P}}(\tilde{A})\tilde{B} = \operatorname{tr} \tilde{A}\tilde{B}$, $\forall \tilde{A} \in \bigotimes_1^n \mathbf{B}(\mathcal{K}), \forall \tilde{B} \in \bigotimes_1^n \mathcal{B}_0$. Therefore, $\mathcal{E}_{\tilde{P}}$ is a conditional expectation from $\bigotimes_1^n \mathbf{B}(\mathcal{K})$ to $\bigotimes_1^n \mathcal{B}_0$, so that

$$\begin{aligned} S(\rho_{\theta,n}^{u(0)}) &= S(\mathcal{E}_{\tilde{P}}^*(\rho_{\theta,n}^u)) = \operatorname{tr} \eta(\mathcal{E}_{\tilde{P}}^*(\rho_{\theta,n}^u)) \\ &\geq \operatorname{tr} \mathcal{E}_{\tilde{P}}^*(\eta(\rho_{\theta,n}^u)) = \operatorname{tr} \eta(\rho_{\theta,n}^u) = S(\rho_{\theta,n}^u), \end{aligned}$$

where $\eta(t) \equiv -t \log t\,(t \geq 0)$. We used the concavity:

$$\eta(\mathcal{E}_{\tilde{P}}^*(\rho_{\theta,n}^u)) \geq \mathcal{E}_{\tilde{P}}^*(\eta(\rho_{\theta,n}^u)).$$

□

Taking into account the construction of the subalgebra $\mathcal{B}_0$ of $\mathbf{B}(\mathcal{K})$, one can construct a transition expectation in the case that $\mathbf{B}(\mathcal{K})$ is a finite subalgebra of $\mathbf{B}(\mathcal{H})$.

Let $\mathbf{B}(\mathcal{K})$ be the $d \times d$ matrix algebra M_d $(d \leq \dim \mathcal{H})$ in $\mathbf{B}(\mathcal{H})$ and $E_{ij} = |e_i\rangle\langle e_j|$ with normalized vectors $e_i \in \mathcal{H}$ $(i = 1, 2, \ldots, d)$. Let $\gamma_1, \ldots, \gamma_d \in \mathbf{B}(\mathcal{H})$ be a finite operational partition of unity, that is, $\sum_{i=1}^d \gamma_i^* \gamma_i = I$, then the transition expectation

$$\mathcal{E}^\gamma : M_d \otimes \mathbf{B}(\mathcal{H}) \to \mathbf{B}(\mathcal{H})$$

is defined by

$$\mathcal{E}^\gamma\left(\sum_{i,j=1}^d E_{ij} \otimes A_{ij}\right) \equiv \sum_{i,j=1}^d \gamma_i^* A_{ij} \gamma_j.$$

Let M_d^0 be a subalgebra of M_d consisting of diagonal elements of M_d. Since an element of M_d^0 has the form $\sum_{i=1}^d b_i E_{ii}$ $(b_i \in \mathbb{C})$, one can see that the restriction $\mathcal{E}^{\gamma(0)}$ of $\mathcal{E}^\gamma$ to M_d^0 is defined as

$$\mathcal{E}^{\gamma(0)}\left(\sum_{i,j=1}^d E_{ij} \otimes A_{ij}\right) \equiv \sum_{i=1}^d \gamma_i^* A_{ii} \gamma_i.$$

When $\Lambda : \mathbf{B}(\mathcal{H}) \to \mathbf{B}(\mathcal{H})$ is a normal unital CP map, the transition expectations $\mathcal{E}_\Lambda^\gamma$ and $\mathcal{E}_\Lambda^{\gamma(0)}$ are defined by

$$\mathcal{E}_\Lambda^\gamma\left(\sum_{i,j=1}^d E_{ij} \otimes A_{ij}\right) \equiv \sum_{i,j=1}^d \Lambda(\gamma_i^* A_{ij} \gamma_j),$$

$$\mathcal{E}_\Lambda^{\gamma(0)}\left(\sum_{i,j=1}^d E_{ij} \otimes A_{ij}\right) \equiv \sum_{i=1}^d \Lambda(\gamma_i^* A_{ii} \gamma_i).$$

Then one obtains the quantum Markov states $\{\rho_{\Lambda,n}^{\gamma}\}$ and $\{\rho_{\Lambda,n}^{\gamma(0)}\}$

$$\rho_{\Lambda,n}^{\gamma} = \sum_{i_1,\dots,i_n=1}^{d} \sum_{j_1,\dots,j_n=1}^{d} \operatorname{tr}_{\mathcal{H}} \rho \Lambda\big(W_{j_1 i_1}\big(\Lambda\big(W_{j_2 i_2}\big(\cdots \Lambda\big(W_{j_n i_n}(I_{\mathcal{H}})\big)\big)\big)\big)\big) \times E_{i_1 j_1} \otimes \cdots \otimes E_{i_n j_n} \tag{20.25}$$

and

$$\begin{aligned}\rho_{\Lambda,n}^{\gamma(0)} &= \sum_{i_1,\dots,i_n=1}^{d} \operatorname{tr}_{\mathcal{H}} \rho \Lambda\big(W_{i_1 i_1}\big(\Lambda\big(W_{i_2 i_2}\big(\cdots \Lambda\big(W_{i_n i_n}(I_{\mathcal{H}})\big)\big)\big)\big)\big) E_{i_1 i_1} \otimes \cdots \otimes E_{i_n i_n} \\ &= \sum_{i_1,\dots,i_n=1}^{d} p_{i_1,\dots,i_n} E_{i_1 i_1} \otimes \cdots \otimes E_{i_n i_n},\end{aligned} \tag{20.26}$$

where we put

$$\begin{aligned}W_{ij}(A) &\equiv \gamma_i^* A \gamma_j, \quad A \in \mathbf{B}(\mathcal{H}),\\ W_{ij}^*(\rho) &\equiv \gamma_j \rho \gamma_i^*, \quad \rho \in \mathfrak{S}(\mathcal{H}),\\ p_{i_1,\dots,i_n} &\equiv \operatorname{tr}_{\mathcal{H}} \rho \Lambda\big(W_{i_1 i_1}\big(\Lambda\big(W_{i_2 i_2}\big(\cdots \Lambda\big(W_{i_n i_n}(I_{\mathcal{H}})\big)\big)\big)\big)\big)\\ &= \operatorname{tr}_{\mathcal{H}} W_{i_n i_n}^*\big(\Lambda^* \cdots \Lambda^*\big(W_{i_2 i_2}^*\big(\Lambda^*\big(W_{i_1 i_1}^*\big(\Lambda^*(\rho)\big)\big)\big)\big)\big).\end{aligned}$$

The above $\rho_{\Lambda,n}^{\gamma}, \rho_{\Lambda,n}^{\gamma(0)}$ become the special cases of $\rho_{\Lambda,n}^{\Gamma,\omega}$ defined in Sect. 20.2 by taking Γ and ω in (20.22) as

$$\begin{aligned}\Gamma\left(\sum_{i,j=1}^{d} E_{ij} \otimes A_{ij}\right) &\equiv \sum_{i,j=1}^{d} E_{ij} \otimes \gamma_i A_{ij} \gamma_j,\\ \omega\left(\sum_{i,j=1}^{d} E_{ij} \otimes \gamma_i A_{ij} \gamma_j\right) &\equiv \sum_{i,j=1}^{d} \operatorname{tr}_{\mathcal{H}}(\rho \gamma_i A_{ij} \gamma_j) E_{ij}\end{aligned}$$

and

$$\begin{aligned}\Gamma\left(\sum_{i,j=1}^{d} E_{ij} \otimes A_{ij}\right) &\equiv \sum_{i=1}^{d} E_{ii} \otimes \gamma_i A_{ii} \gamma_i,\\ \omega\left(\sum_{i=1}^{d} E_{ii} \otimes \gamma_i A_{ii} \gamma_i\right) &\equiv \sum_{i=1}^{d} \operatorname{tr}_{\mathcal{H}}(\rho \gamma_i A_{ii} \gamma_i) E_{ii}.\end{aligned}$$

Therefore, the dynamical entropy (20.18) becomes

$$\tilde{S}\big(\Lambda^*; \rho, \{\gamma_i\}\big) \equiv \limsup_{n\to\infty} \frac{1}{n} S\big(\rho_{\Lambda^*,n}^{\gamma}\big),$$

$$\tilde{S}^{(0)}\big(\Lambda^*; \rho, \{\gamma_i\}\big) \equiv \limsup_{n\to\infty} \frac{1}{n} S\big(\rho_{\Lambda^*,n}^{\gamma(0)}\big).$$

The dynamical entropies of Λ^* with respect to a finite-dimensional subalgebra $\mathcal{B} \subset \mathbf{B}(\mathcal{H})$ and the transition expectations $E_{\Lambda^*}^{\gamma}$ and $E_{\Lambda^*}^{\gamma(0)}$ are given by

$$\tilde{S}_{\mathcal{B}}(\Lambda^*; \rho) \equiv \sup\big\{\tilde{S}\big(\Lambda^*; \rho, \{\gamma_i\}\big), \{\gamma_i\} \subset \mathcal{B}\big\}, \tag{20.27}$$

$$\tilde{S}_{\mathcal{B}}^{(0)}(\Lambda^*; \rho) \equiv \sup\big\{\tilde{S}^{(0)}\big(\Lambda^*; \rho, \{\gamma_i\}\big), \{\gamma_i\} \subset \mathcal{B}\big\}. \tag{20.28}$$

We call (20.27) and (20.28) a generalized AF entropy and a generalized AOW entropy [17], respectively. When $\{\gamma_i\}$ is a PVM (projection-valued measure) and Λ^* is an automorphism θ, $\tilde{S}_{\mathcal{B}}^{(0)}(\theta; \rho)$ is equal to the AOW entropy. When $\{\gamma_i^* \gamma_i\}$ is a POV (positive-operator valued measure) and $\Lambda^* = \theta$, $\tilde{S}_{\mathcal{B}}(\theta; \rho)$ is equal to the AF entropy [53].

From Theorem 20.19, one obtains an inequality

Theorem 20.20

$$\tilde{S}_{\mathcal{B}}(\Lambda^*; \rho) \le \tilde{S}_{\mathcal{B}}^{(0)}(\Lambda^*; \rho).$$

That is, the generalized AOW entropy is greater than the generalized AF entropy. Moreover, the dynamical entropy $\tilde{S}_{\mathcal{B}}(\Lambda^*; \rho)$ is rather difficult to compute because there exist off-diagonal parts in (20.25), so we mainly consider the dynamical entropy $\tilde{S}_{\mathcal{B}}^{(0)}(\Lambda^*; \rho)$ in the following.

Here, we note that the dynamical entropy defined in terms of $\rho_{\theta,n}^{u}$ on $\bigotimes_1^n \mathbf{B}(\mathcal{K})$ is related to that of flows by Emch [223], which was defined in terms of the conditional expectation, provided $\mathbf{B}(\mathcal{K})$ is a subalgebra of $\mathbf{B}(\mathcal{H})$.

20.3.5 Some Models

We numerically compute the KOW dynamical entropy for several models.

Let $\gamma_i = \gamma_i^*$ be projection operators on one-dimensional mutually orthogonal subspaces of $\mathcal{H}$ such that $\sum_i \gamma_i = I_{\mathcal{H}}$ holds. For unitary operators U_i and V on $\mathcal{H}$, $\widetilde{\gamma}_i \equiv U_i \gamma_i V$ satisfies

$$\sum_i \widetilde{\gamma}_i^* \widetilde{\gamma}_i = \sum_i V^* \gamma_i^* U_i^* U_i \gamma_i V = I_{\mathcal{H}}.$$

Let us consider the transition expectation

$$\mathcal{E}^{\gamma(0)} : \mathbf{A}(\mathcal{H}) \otimes \mathbf{B}(\mathcal{H}) \to \mathbf{B}(\mathcal{H})$$

defined by

$$\mathcal{E}^{\gamma(0)}\bigg(\sum_{i,j=1} E_{ij} \otimes A_{ij}\bigg) \equiv \sum_i \tilde{\gamma}_i A_{ii} \tilde{\gamma}_i, \quad A_{ii} \in \mathbf{B}(\mathcal{H}),$$

where $\mathbf{A}(\mathcal{H})$ is an abelian subalgebra of $\mathbf{B}(\mathcal{H})$ generated by $\sum_i b_i E_{ii} (b_i \in \mathbb{C})$. Let $\Lambda : \mathbf{B}(\mathcal{H}) \to \mathbf{B}(\mathcal{H})$ be the normal unital CP map given by

$$\Lambda(A) \equiv \sum_{k,l,m,n} \lambda_{kl,mn} E_{kl}^{*} A E_{mn} \tag{20.29}$$

satisfying

$$\sum_k \lambda_{kl,kn} = \delta_{ln} \quad (\Longleftrightarrow \Lambda(I_{\mathcal{H}}) = I_{\mathcal{H}}),$$

where $\lambda_{kl,mn}$ are complex numbers.

We will compute the dynamical entropy of Λ based on $\mathcal{E}^{\gamma(0)}$ above.

Theorem 20.21 *When $\gamma_i = E_{ii}$, the quantum dynamical entropy with respect to Λ, ρ, and $\{\gamma_i = E_{ii}\}$ is*

$$\tilde{S}^{(0)}\big(\Lambda; \rho, \{\gamma_i = E_{ii}\}\big) = -\sum_{i,j}\sum_{l,k} \lambda_{ij,ij}\lambda_{jl,jk} \operatorname{tr}_{\mathcal{H}}(\rho E_{lk}) \log \lambda_{ij,ij},$$

where $\{\lambda_{ij,ij}\}$ are the coefficients of (20.29) *associated to the CP map Λ.*

Proof From (20.26), it follows that one obtains

$$\rho_{\Lambda,n}^{\gamma(0)} = \sum_{j_1,\dots,j_n=1} p_{j_1,\dots,j_n} E_{j_1 j_1} \otimes \cdots \otimes E_{j_n j_n},$$

where

$$\begin{aligned}
p_{j_1,\dots,j_n} &\equiv \operatorname{tr}_{\mathcal{H}}\big(\rho\Lambda\big(W_{j_1 j_1}\big(\Lambda\big(W_{j_2 j_2}\big(\cdots\Lambda\big(W_{j_n j_n}(I_{\mathcal{H}})\big)\big)\big)\big)\big)\big)\\
&= \operatorname{tr}_{\mathcal{H}}\big(\rho\Lambda\big(W_{j_1 j_1}\big(\Lambda\big(W_{j_2 j_2}\big(\cdots W_{j_{n-1} j_{n-1}}\big(\Lambda(E_{j_n j_n})\big)\big)\big)\big)\big)\big)\\
&= \operatorname{tr}_{\mathcal{H}}\left(\rho\Lambda\left(W_{j_1 j_1}\left(\Lambda\left(W_{j_2 j_2}\left(\cdots W_{j_{n-1} j_{n-1}}\left(\sum_{l,k} \lambda_{j_n l, j_n k} E_{lk}\right)\right)\right)\right)\right)\right)\\
&= \lambda_{j_n j_{n-1}, j_n j_{n-1}} \operatorname{tr}_{\mathcal{H}}\big(\rho\Lambda\big(W_{j_1 j_1}\big(\Lambda\big(W_{j_2 j_2}\big(\cdots\Lambda(E_{j_{n-1} j_{n-1}})\big)\big)\big)\big)\big)\\
&= \lambda_{j_n j_{n-1}, j_n j_{n-1}} \cdots \lambda_{j_2 j_1, j_2 j_1} \operatorname{tr}_{\mathcal{H}}\big(\rho\Lambda(E_{j_1 j_1})\big)\\
&= \prod_{k=2}^{n} \lambda_{j_k j_{k-1}, j_k j_{k-1}} \sum_{l,k} \lambda_{j_1 l, j_1 k} \operatorname{tr}_{\mathcal{H}}(\rho E_{lk}).
\end{aligned}$$

From the complete positivity of Λ and $\Lambda(I_{\mathcal{H}}) = I_{\mathcal{H}}$, it follows that

$$\lambda_{kl,kl} \geq 0$$

and

$$\sum_k \lambda_{kl,kl} = 1,$$

which implies that $P_n = \{p_{j_1,\dots,j_n}\}$ has the Markov property. Therefore, the quantum dynamical entropy with respect to Λ, ρ and $\{\gamma_i = E_{ii}\}$ is computed as

$$\tilde{S}^{(0)}\big(\Lambda; \rho, \{\gamma_i = E_{ii}\}\big) = -\sum_{i,j}\sum_{l,n} \lambda_{ij,ij}\lambda_{jl,jn}\,\mathrm{tr}_{\mathcal{H}}(\rho E_{ln}) \log \lambda_{ij,ij}.$$

□

Theorem 20.22 *When $\gamma_i = U^* E_{ii} U (= |f_i\rangle\langle f_i| = F_{ii})$ for a unitary operator U on $\mathcal{H}$, the quantum dynamical entropy with respect to Λ, ρ, and $\{\gamma_i = F_{ii}\}$ is*

$$\tilde{S}^{(0)}\big(\Lambda; \rho, \{\gamma_i = F_{ii}\}\big) = -\sum_{i,j}\sum_{l,k} \lambda^{(F)}_{ij,ij}\lambda_{jl,jk}\,\mathrm{tr}_{\mathcal{H}}(\rho U^* E_{lk} U) \log \lambda^{(F)}_{ij,ij},$$

where $\{\lambda_{ij,ij}\}$ are the coefficients of (20.29) associated to the CP map Λ and $\lambda^{(F)}_{ij,ij} = \sum_{p,q,r,s} \lambda_{pq,rs}\,\mathrm{tr}_{\mathcal{H}}\, E_{qp} U^* E_{ii} U E_{rs} U^* E_{jj} U$.

Proof Let $F_{ij} \equiv U^* E_{ij} U$. Then

$$|f_i\rangle = U^*|e_i\rangle, \qquad F_{ij} = |f_i\rangle\langle f_j|, \quad e_i, f_i \in \mathcal{H},$$

and one has

$$\begin{aligned}
E_{kl}x &= U F_{kl} U^* x = U|f_k\rangle\langle f_l| U^* x = U f_k \langle f_l, U^* x\rangle = U f_k \langle U f_l, x\rangle \\
&= \sum_{m,n} |f_m\rangle\langle f_m, U f_k\rangle\langle U f_l, f_n\rangle\langle f_n, x\rangle \\
&= \sum_{m,n} \langle f_m, U f_k\rangle\langle U f_l, f_n\rangle |f_m\rangle\langle f_n| x \\
&= \sum_{m,n} \langle f_m, U f_k\rangle\langle U f_l, f_n\rangle F_{mn} x, \quad \forall x \in \mathcal{H},
\end{aligned}$$

which implies

$$E_{kl} = \sum_{m,n} u_{kl,mn} F_{mn}, \quad u_{kl,mn} \equiv \overline{\langle U f_k, f_m\rangle}\langle U f_l, f_n\rangle. \tag{20.30}$$

Inserting (20.30) into (20.29), one obtains

$$\begin{aligned}
\Lambda(A) &= \sum_{p,q,r,s} \lambda_{pq,rs} E^*_{pq} A E_{rs} \\
&= \sum_{p,q,r,s} \lambda_{pq,rs}\Big(\sum_{k,l} \bar{u}_{pq,kl} F^*_{kl}\Big) A \Big(\sum_{m,n} u_{rs,mn} F_{mn}\Big)
\end{aligned}$$

$$= \sum_{k,l,m,n} \left(\sum_{p,q,r,s} \lambda_{pq,rs} \bar{u}_{pq,kl} u_{rs,mn} \right) F_{kl}^* A F_{mn}$$

$$= \sum_{k,l,m,n} \lambda_{kl,mn}^{(F)} F_{kl}^* A F_{mn},$$

where

$$\lambda_{kl,mn}^{(F)} \equiv \sum_{p,q,r,s} \lambda_{pq,rs} \bar{u}_{pq,kl} u_{rs,mn}$$

$$= \sum_{p,q,r,s} \lambda_{pq,rs} \operatorname{tr}_{\mathcal{H}} E_{qp} U^* E_{ii} U E_{rs} U^* E_{jj} U.$$

Proceeding as before, it is shown that

$$p_{j_1,\dots,j_n}^{(F)} = \lambda_{j_n j_{n-1}; j_n j_{n-1}}^{(F)} \cdots \lambda_{j_2 j_1; j_2 j_1}^{(F)} \sum_{l,k} \lambda_{j_1 l, j_1 k} \operatorname{tr}_{\mathcal{H}} (\rho F_{lk})$$

$$= \lambda_{j_n j_{n-1}; j_n j_{n-1}}^{(F)} \cdots \lambda_{j_2 j_1; j_2 j_1}^{(F)} \sum_{l,n} \lambda_{j_1 l, j_1 k} \operatorname{tr}_{\mathcal{H}} (\rho U^* E_{lk} U)$$

because of $\gamma_i = F_{ii}$. Since $P_n^{(F)} = \{p_{j_1,\dots,j_n}^{(F)}\}$ has the Markov property, we obtain the quantum dynamical entropy with respect to Λ, ρ, and $\{\gamma_i = F_{ii}\}$ as

$$\tilde{S}^{(0)}\big(\Lambda; \rho, \{\gamma_i = F_{ii}\}\big) = -\sum_{i,j} \sum_{l,k} \lambda_{ij,ij}^{(F)} \lambda_{jl,jk} \operatorname{tr}_{\mathcal{H}} (\rho U^* E_{lk} U) \log \lambda_{ij,ij}^{(F)}.$$

□

Theorem 20.23 *Take $\Lambda_U(A) \equiv U^* A U$ as a unitary operator U on $\mathcal{H}$. When U has a simple point spectrum $\{e^{i\varphi_k}\}$ and its eigenvector f_k, the dynamical entropy with respect to Λ_U, ρ, and $\{\gamma_k = |f_k\rangle\langle f_k|\}$ is*

$$\tilde{S}^{(0)}\big(\Lambda_U; \rho, \{\gamma_k = |f_k\rangle\langle f_k|\}\big) = 0.$$

Proof For a CONS $e = \{e_m\}$ of $\mathcal{H}$, U is written by

$$U = \sum_{m,n} \lambda_{mn}^{(e)} E_{mn},$$

where

$$\lambda_{mn}^{(e)} \equiv \langle e_m, U e_n \rangle,$$

$$E_{mn} \equiv |e_m\rangle\langle e_n|.$$

From the definition of Λ_U, one obtains

$$\Lambda_U(A) = \sum_{k,l,m,n} \lambda_{kl,mn}^{(e)} E_{lk} A E_{mn},$$

where $\lambda^{(e)}_{kl,mn} \equiv \overline{\lambda}^{(e)}_{kl} \lambda^{(e)}_{mn}$. Therefore, one has

$$p^{(e)}_{j_1,\dots,j_n} \equiv \lambda^{(e)}_{j_n j_{n-1}, j_n j_{n-1}} \cdots \lambda^{(e)}_{j_2 j_1, j_2 j_1} \sum_{l,k} \lambda^{(e)}_{j_1 l, j_1 k} \operatorname{tr}_{\mathcal{H}}(\rho E_{lk}),$$

$$\lambda^{(e)}_{mn,mn} \equiv \left|\langle e_m, U e_n\rangle\right|^2.$$

$P_n = \{p_{j_1,\dots,j_n}\}$ has the Markov property. Moreover, for the eigenvectors $\{f_k\}$ of the simple point spectrum $\{e^{\mathrm{i}\varphi_k}\}$ of U such that $U f_k = e^{\mathrm{i}\varphi_k} f_k$, since

$$\lambda^{(f)}_{mn,mn} = \left|\langle f_m, U f_n\rangle\right|^2 = \delta_{mn},$$

$$p^{(f)}_{j_1} \equiv \sum_{l,k} \lambda^{(f)}_{j_1 l, j_1 k} \operatorname{tr}_{\mathcal{H}}\bigl(\rho |f_l\rangle\langle f_k|\bigr) = \langle f_{j_1}, \rho f_{j_1}\rangle,$$

the dynamical entropy with respect to Λ_U, ρ, and $\{\gamma_k = |f_k\rangle\langle f_k|\}$ is

$$\tilde{S}^{(0)}\bigl(\Lambda^*_U; \rho, \{\gamma_k = |f_k\rangle\langle f_k|\}\bigr) = \lim_{n\to\infty} \frac{1}{n}\left(-\sum_{j_1=1}^{d} \langle f_{j_1}, \rho f_{j_1}\rangle \log\langle f_{j_1}, \rho f_{j_1}\rangle\right) = 0.$$ □

We remark here that for another choice of base $\{g_i\} \subset \mathcal{H}$, one has

$$\tilde{S}^{(0)}\bigl(\Lambda^*_U; \rho, \{\gamma_k = |g_k\rangle\langle g_k|\}\bigr) > 0.$$

Now we study the dynamical entropy for a quantum communication process, in particular, the attenuation process. That is, Λ^* is the attenuation channel [559] defined as follows: Let $\mathcal{H} = \mathcal{L}^2(\mathbb{R})$, $|\alpha\rangle$ be a coherent state vector in $\mathcal{H}$, and take $\gamma_\alpha \equiv \frac{1}{\sqrt{\pi}}|\alpha\rangle\langle\alpha|$ ($\alpha \in \mathbb{C}$). The attenuation channel Λ^* with a transmission rate η is defined by

$$\Lambda^*\bigl(|\alpha\rangle\langle\alpha|\bigr) \equiv \bigl|\sqrt{\eta}\alpha\bigr\rangle\bigl\langle\sqrt{\eta}\alpha\bigr|.$$

Theorem 20.24 *When ρ is a superposition of coherent vectors $|\xi_m\rangle$ given by $\rho = \sum_{m=1}^{N} \lambda_m |\xi_m\rangle\langle\xi_m|$ with $\sum_m \lambda_m = 1$ and Λ^* is the attenuation channel with a transmission rate η, the quantum dynamical entropy with respect to Λ^*, ρ, and $\{\gamma_\alpha\}$ is given by*

$$\tilde{S}^{(0)}\bigl(\Lambda^*; \rho, \{\gamma_\alpha\}\bigr)$$

$$= -\int_{\mathbb{R}^2} \frac{1}{\pi} e^{-|\alpha_2 - \sqrt{\eta}\alpha_1|^2} \sum_{m=1}^{N} \lambda_m e^{-|\alpha_1 - \sqrt{\eta}\xi_m|^2} \log \frac{1}{\pi} e^{-|\alpha_2 - \sqrt{\eta}\alpha_1|^2}\, d^2\alpha_1\, d^2\alpha_2.$$

Proof Formula (20.26) can be rewritten in the form (for $n \ge 3$).

$$\rho^{\gamma(0)}_{\Lambda,n} = \int_{\mathbb{R}^n} d^2\alpha_1 \cdots d^2\alpha_n P_\rho(\alpha_1,\dots,\alpha_n) \bigotimes_{i=1}^{n} |\alpha_i\rangle\langle\alpha_i|$$

where

$$P_\rho(\alpha_1, \ldots, \alpha_n) \equiv \mathrm{tr}_{\mathcal{H}}\, W^*_{\alpha_n}\big(\Lambda^*(\cdots \Lambda^*\big(W^*_{\alpha_2}\big(\Lambda^*\big(W^*_{\alpha_1}\big(\Lambda^*(\rho)\big)\big)\big)\big)\big)\big),$$

$$W^*_\alpha(A) \equiv \gamma^*_\alpha A \gamma_\alpha = \frac{1}{\pi}|\alpha\rangle\langle\alpha|A|\alpha\rangle\langle\alpha|.$$

When ρ is given by

$$\rho = \sum_{m=1}^{N} \lambda_m |\xi_m\rangle\langle\xi_m| \quad \text{with} \sum_m \lambda_m = 1,$$

one has

$$\begin{aligned} P_\rho(\alpha_1, \ldots, \alpha_n) &= \sum_{m=1}^{N} \lambda_m P_{\xi_m}(\alpha_1, \ldots, \alpha_n) \\ &= \frac{1}{\pi^n} \prod_{k=2}^{n} e^{-|\alpha_k - \sqrt{\eta}\alpha_{k-1}|^2} \sum_{m=1}^{N} \lambda_m e^{-|\alpha_1 - \sqrt{\eta}\xi_m|^2}. \end{aligned}$$

From $\{P_\rho(\alpha_1, \ldots, \alpha_n)\}$, we obtain the dynamical entropy with respect to Λ^*, ρ, and $\{\gamma_\alpha\}$ as

$$\begin{aligned} &\tilde{S}^{(0)}\big(\Lambda^*; \rho, \{\gamma_\alpha\}\big) \\ &\quad = -\int_{\mathbb{R}^2} \frac{1}{\pi} e^{-|\alpha_2 - \sqrt{\eta}\alpha_1|^2} \sum_{m=1}^{N} \lambda_m e^{-|\alpha_1 - \sqrt{\eta}\xi_m|^2} \log \frac{1}{\pi} e^{-|\alpha_2 - \sqrt{\eta}\alpha_1|^2} d^2\alpha_1\, d^2\alpha_2. \end{aligned}$$ □

Note that the details of the above theorem are discussed in [525].

20.4 Fractal Dimension of State

Usual fractal theory mostly treats geometrical sets. It is desirable to extend the fractal theory so as to be applicable to some other objects. For this purpose, we introduced fractal dimensions for general states. First, we recall two usual fractal dimensions of geometrical sets.

Scaling Dimension. We observe a complex set F built from a fundamental pattern. If the number of the patterns observed is $N(1)$ when the scale is very rough, say 1, and the number is $N(r)$ when the scale is r, then we call the dimension defined through

$$d_s(F) = \frac{\log(N(r)/N(1))}{\log(1/r)}$$

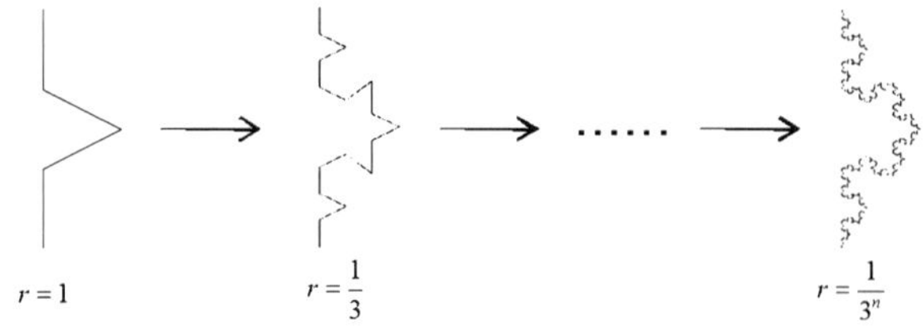

Fig. 20.1 The construction of Koch's curve

the scaling dimension of the set F. Let us explain the scale r in Koch's curve as an example (the case n goes to infinity in Fig. 20.1).

The scaling dimension of this curve is easily computed: $d_s = \frac{\log 4}{\log 3}$ because $N(\frac{1}{3^n}) = 4^n$.

Capacity Dimension. Let us cover a set F in the n-dimension Euclidean space $\mathbb{R}^n$ by a family of convex sets (e.g., balls) whose diameter is ε.

If the smallest number of the convex sets needed to cover the set F is $N(\varepsilon)$, then we call the dimension given by

$$d_c(F) = \lim_{\varepsilon \to 0} \frac{\log N(\varepsilon)}{\log(1/\varepsilon)}$$

the capacity dimension (or the ε-entropy dimension) of the set F.

These two fractal dimensions become equal for almost all sets in which we can compute these dimensions.

The ε-entropy was extensively studied by Kolmogorov and his ε-entropy is defined for a probability measure, which gives us an idea to define the ε-entropy for a general quantum state.

Kolmogorov introduced the notion of ε-entropy in a probability space $(\Omega, \mathcal{F}, \mu)$. His formulation is as follows: For two random variables $f, g \in M(\Omega)$, the mutual entropy $I(f, g)$ is defined by the joint probability measure $\mu_{f,g}$ and the direct product measure $\mu_f \otimes \mu_g$ as

$$I(f, g) = S(\mu_{f,g}, \mu_f \otimes \mu_g),$$

where $S(\cdot, \cdot)$ is the relative entropy.

The ε-entropy of Kolmogorov for a random variable f with values on a separable metric space (X, d) is given by

$$S_{\mathrm{K}}(f, \varepsilon) \equiv \inf\bigl\{I(f, g); g \in M_d(f, \varepsilon)\bigr\},$$

where

$$M_d(f, \varepsilon) \equiv \bigl\{g \in M(\Omega);\ d(f, g) \leq \varepsilon\bigr\}$$

with

$$d(f, g) \equiv \sqrt{\int_{X \times X} d(x, y)^2 \, d\mu_{fg}(x, y)},$$

where μ_{fg} is the joint probability measure on $\Omega \times \Omega$ induced from f and g. For a general probability measure μ on $(\Omega, \mathcal{F})$, the Kolmogorov ε-entropy $S_{\mathrm{K}}(\mu; \varepsilon)$ is given by

$$S_{\mathrm{K}}(\mu; \varepsilon) = \inf\bigl\{S(\mu_{co}, \mu \otimes \bar{\mu}); \bar{\mu} \in P_0(\Omega)\bigr\},$$

where μ_{co} is the joint (compound) probability measure of μ and $\bar{\mu}$ and $P_0(\Omega)$ is the set of all probability measures $\bar{\mu}$ satisfying $\|\mu - \bar{\mu}\| \le \varepsilon$.

We introduced the ε-entropy of a general quantum state φ and the fractal dimensions of the state φ. Let $\mathcal{C}$ be the set of all channels physically interesting, and define two sets

$$\mathcal{C}_1(\Lambda^*; \varphi) = \{\Gamma^* \in \mathcal{C}; \Gamma^*\varphi = \Lambda^*\varphi\},$$

$$\mathcal{C}_2(\varphi; \varepsilon) = \bigl\{\Gamma^* \in \mathcal{C}; \|\varphi - \Gamma^*\varphi\| \le \varepsilon\bigr\}.$$

Then the ε-entropy of a state of φ w.r.t. $\mathcal{S}$ is defined by means of the transmitted complexity (see Chap. 10) $T^{\mathcal{S}}(\varphi;\ \Lambda^*)$ as

$$S^{\mathcal{S}}_{\mathcal{C},T}(\varphi; \varepsilon) = \inf\bigl\{T^{\mathcal{S}}_{\max,\mathcal{C}}(\varphi;\ \Lambda^*);\ \Lambda^* \in \mathcal{C}_2(\varphi; \varepsilon)\bigr\},$$

where

$$T^{\mathcal{S}}_{\max,\mathcal{C}}(\varphi;\ \Lambda^*) = \sup\bigl\{T^{\mathcal{S}}(\varphi;\ \Gamma^*);\ \Gamma^* \in \mathcal{C}_1(\Lambda^*; \varphi)\bigr\}.$$

When $\mathcal{S} = \mathfrak{S}(\mathcal{A})$ and $\mathcal{C}$ is the set of all channels on $\mathfrak{S}(\mathcal{A})$, our ε-entropy is simply denoted by $S_{O,T}(\varphi; \varepsilon)$.

The capacity dimension of a state φ w.r.t. $\mathcal{S}$ and $\mathcal{C}$ is defined by

$$d^{\mathcal{S}}_{\mathcal{C},T}(\varphi) \equiv \lim_{\varepsilon \to 0} d^{\mathcal{S}}_{\mathcal{C},T}(\varphi; \varepsilon),$$

where

$$d^{\mathcal{S}}_{\mathcal{C},T}(\varphi; \varepsilon) = \frac{S^{\mathcal{S}}_{\mathcal{C},T}(\varphi; \varepsilon)}{\log(1/\varepsilon)}.$$

The above $d^{\mathcal{S}}_{\mathcal{C},T}(\varphi; \varepsilon)$ is called the capacity dimension of ε-order. The information dimension of a state φ for $\mathcal{S}$ and $\mathcal{C}$ of ε order is defined by

$$d^{\mathcal{S}}_{I,\mathcal{C},T}(\varphi; \varepsilon) = \frac{S^{\mathcal{S}}_{\mathcal{C},T}(\varphi; \varepsilon)}{S^{\mathcal{S}}(\varphi)}.$$

20.4.1 Applications

These ε-entropy and fractal dimensions are applied to several physical phenomena and mathematical objects. For instance, we can classify the shapes of the seas, of the moon, and of the rivers, and we can consider a symmetry breaking in the Ising system. Here we state the main results concerning the Gaussian measures. For a random variable $f=(f_1,\dots,f_n)$ from Ω to $\mathbb{R}^n$, the random variable norm $\|\cdot\|_{\mathrm{R.V.}}$ of the measure μ_f associated with f is defined by

$$\|\mu_f\|_{\mathrm{R.V.}}=\sqrt{\frac{1}{n}\sum_{i=1}^{n}\int_{\Omega}|f_i|^2\,d\mu.}$$

Theorem 20.25 *If the distance of two states is defined through the above random variable norm on $\Omega=\mathbb{R}^n$ and the transmitted complexity T is the mutual entropy in CDS, then*

1. $S_{\mathcal{C},I}(\mu_f;\varepsilon)=S_K(f;\varepsilon)=\frac{1}{2}\sum_{i=1}^{n}\log\max(\frac{\lambda_i}{\theta^2},1)$, *where $\lambda_1,\dots,\lambda_n$ are the eigenvalues of the covariance operator R for μ_f and θ^2 is a constant uniquely determined by the equation $\sum_{i=1}^{n}\min(\lambda_i,\theta^2)=\varepsilon^2$.*
2. $d_{I,\mathcal{C},I}(\mu_f)=d_K(\mu_f)=n$.

According to this theorem the Kolmogorov ε-entropy coincides with our ε-entropy when the norm of states is given by the random variable norm on $\mathbb{R}^n$. The difference between $S_{\mathcal{C},I}$ and S_K comes from the norm for states taken. For example, when we take the norm of states as the total variation norm, namely,

$$\|\mu\|=|\mu|(\Omega).$$

Let $\Omega=\mathbb{R}$ for simplicity. An input state μ is described by the mean 0 and the covariance σ^2, and we take the set $\mathcal{C}$ of the channels Λ^* sending a Gaussian measure to a Gaussian measure with a noise expressed by the one-dimensional Gaussian measure $\mu_0=[0,\sigma_0^2]$ so that the output state $\Lambda^*\mu$ is represented by $[0,\alpha^2\sigma^2+\sigma_0^2]$ with a certain constant α. Since the channel Λ^* depends on α and σ_0^2, we put $\Lambda^*=\Lambda^*_{(\bar{\alpha},\sigma_0^2)}$.

Theorem 20.26 *When $\|\mu-\Lambda^*_{(\bar{\alpha},\sigma_0^2)}\mu\|=\delta$, we denote $\alpha^2\sigma^2+\sigma_0^2$ by C_δ. If α^2satisfies $\alpha^2\le\frac{C_\delta-\delta}{\sigma^2}$, then we have*

$$S_{\mathcal{C},I}(\mu;\varepsilon)=\frac{1}{2}\log\frac{1}{\varepsilon}+\frac{1}{2}\log\frac{\sigma^2}{(1+\frac{\sqrt{2\pi}}{4}(\varepsilon+o(\varepsilon)))}>S_K(\mu;\varepsilon)=0,$$

and $d_{\mathcal{C},I}(\mu)=\frac{1}{2}$, where $o(\varepsilon)$ is the order of ε, that is, $\lim_{\varepsilon\to 0}o(\varepsilon)=0$.

This theorem tells the difference between the Kolmogorov ε-entropy and our ε-entropy. It concludes that our fractal dimension enables the classification of the Gaussian measures.

Remark 20.27 The proofs of all theorems above are given in [368].

20.5 Entanglement in Janes–Cummings Model

The exactly solvable Jaynes–Cummings model (JCM) has been studied by many researchers from various points of view. One of the most interesting features of this model is the entanglement developed between an atom and a field during the process of interaction. There have been several approaches to analyze the time evolution of this model, for instance, by von Neumann entropy and by atomic inversion. Here we will apply the DEN (degree of entanglement) discussed in Chap. 8 to analyze the entanglement of JCM. The details of this study can be seen in [266].

The resonant JCM Hamiltonian can be expressed by a rotating wave approximation in the following form:

$$H = H_A + H_F + H_I,$$

$$H_A = \frac{1}{2}\hbar\omega_0\sigma_z, \qquad H_F = \hbar\omega_0 a^* a,$$

$$H_I = \hbar g\bigl(a \otimes \sigma^+ + a^* \otimes \sigma^-\bigr)$$

where g is a coupling constant, $\sigma^\pm$ are the pseudo-spin operators of two-level atom, $\sigma^\pm = \frac{1}{2}(\sigma_x \pm \sigma_y)$ with Pauli spin operators $\sigma_x, \sigma_y, \sigma_z$, a (resp., a^*) is the annihilation (resp., creation) operator of a photon, $[a, a^*] = 1$, and $\hbar, \omega_0, g$ are real constants. Put $\hbar = 1$ in the sequel. Suppose that the initial state of the atom is a superposition of the upper and lower levels:

$$\rho = \lambda_0 E_0 + \lambda_1 E_1 \in \mathfrak{S}_A$$

where $E_0 = |1\rangle\langle 1|$ (resp., $E_1 = |2\rangle\langle 2|$) is the state in the upper level (resp., lower level) and $\lambda_0 + \lambda_1 = 1$. Remark that $\sigma^+|2\rangle = |1\rangle, \sigma^-|1\rangle = |2\rangle$.

We also assume that the field is in a coherent state:

$$\omega = |\theta\rangle\langle\theta| \in \mathfrak{S}_F, \quad |\theta\rangle = \exp\left[-\frac{1}{2}|\theta|^2\right]\sum_l \frac{\theta^l}{\sqrt{l!}}|l\rangle, \ \theta \in \mathbb{C}$$

where $|l\rangle$ is the photon number state; $l = 0, 1, 2 \ldots$. Note that $|l\rangle = a^{*l}|0\rangle/\sqrt{l!}, a|0\rangle = 0$.

Solving the eigenequation

$$H\Phi = E\Phi,$$

we have the eigenvalue

$$E_{n,j} = \omega_0\left(n + \frac{1}{2}\right) + (-1)^j \Omega_n, \quad n = 0, 1, \ldots, j = 0, 1,$$

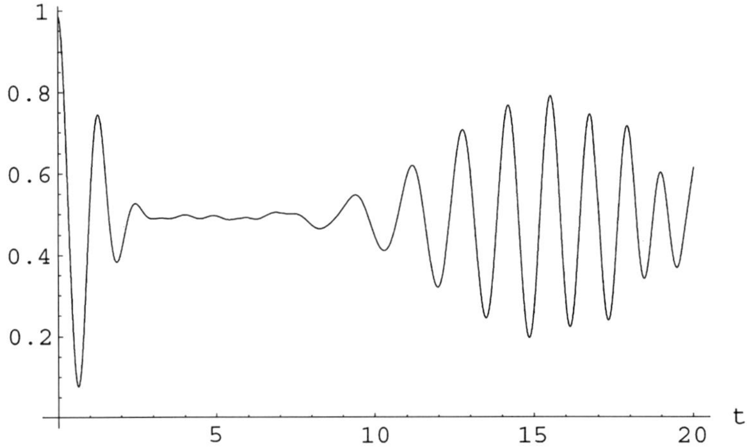

Fig. 20.2 Transition probability as a function of time t

where Ω_n is the Rabi frequency $\Omega_n = g\sqrt{n+1}$. Denote the corresponding eigenvector by $|\Phi_j^{(n)}\rangle$.

The continuous lifting $\mathcal{E}^* : \mathfrak{S}_A \to \mathfrak{S}_A \otimes \mathfrak{S}_F$ describing the time evolution between the atom and the field for the JCM is defined by the unitary operator generated by H such that

$$\mathcal{E}^* \rho = U_t(\rho \otimes \omega)U_t^*, \quad U_t = \exp(-\mathrm{i}tH/\hbar). \tag{20.31}$$

This unitary operator U_t is written as

$$U_t = \exp(-\mathrm{i}tH/\hbar) = \sum_{n=0}^{\infty}\sum_{j=0}^{1} \exp(-\mathrm{i}tE_{n,j}/\hbar)\big|\Phi_j^{(n)}\big\rangle\big\langle\Phi_j^{(n)}\big|. \tag{20.32}$$

The transition probability that the atom is initially prepared in the upper state and stays at the upper state at time t is given by

$$\begin{aligned} c(t) &= \big|\langle n,2|U_t|n,2\rangle\big|^2 \\ &= \exp\big(-|\theta|^2\big)\sum_n \frac{|\theta|^{2n}}{n!}\cos^2[\Omega_n t]. \end{aligned}$$

Figure 20.2 shows the transition probability for a coupling constant $g = 1$ and a mean photon number $|\theta|^2 = 5$. This model yields the *dephase* around the time $t_c \approx 1/g$ and shows the damped oscillation with a Gaussian envelope. This damping is caused by the difference and interference of each Rabi frequency Ω_n. This damping phenomenon is called the *Cummings collapse* ($t \approx 3 \sim 7$). Later the system shows *revival* around $t_r \approx 2\pi|\theta|/g$. The reason of *revival* is considered as a *rephase*, and its revival periodically appears at each $T_k = kt_r$ ($k = 1, 2, 3, \ldots$).

From (20.31) and (20.32), the entangled state between the atom and the field is

$$\begin{aligned}\mathcal{E}^*\rho &= e_1(t)|n,2\rangle\langle n,2| + e_2(t)|n,2\rangle\langle n+1,1| \\ &\quad + e_3(t)|n+1,1\rangle\langle n,2| + e_4(t)|n+1,1\rangle\langle n+1,1|,\end{aligned}$$

where

$$\begin{aligned}e_1(t) &= \lambda_0 s(t) + \lambda_1 c(t), \\ e_2(t) &= \frac{\mathrm{i}}{2}\exp(-|\theta|^2)(\lambda_1 - \lambda_0)\sum_n \frac{|\theta|^{2n}}{n!}\sin 2\Omega_n t, \\ e_3(t) &= \frac{-\mathrm{i}}{2}\exp(-|\theta|^2)(\lambda_1 - \lambda_0)\sum_n \frac{|\theta|^{2n}}{n!}\sin 2\Omega_n t, \\ e_4(t) &= \lambda_0 c(t) + \lambda_1 s(t), \\ s(t) &= \exp[-|\theta|^2]\sum_n \frac{|\theta|^{2n}}{n!}\sin^2 \Omega_n t.\end{aligned}$$

Taking the partial trace of $\mathcal{E}^*\rho$, we have the reduced density operators ρ_t^A and ρ_t^F for the atom and the field, respectively:

$$\begin{aligned}\rho_t^A &= \mathrm{tr}_F\, \mathcal{E}^*\rho = e_1(t)|2\rangle\langle 2| + e_4(t)|1\rangle\langle 1|, \\ \rho_t^F &= \mathrm{tr}_A\, \mathcal{E}^*\rho = e_1(t)|n\rangle\langle n| + e_4(t)|n+1\rangle\langle n+1|.\end{aligned}$$

Thus, the degree of the entangled state $\mathcal{E}^*\rho$ for the Jaynes–Cummings model can be computed as

$$\begin{aligned}I_{\mathcal{E}^*\rho}\big(\rho_t^A, \rho_t^F\big) &= \mathrm{tr}\,\mathcal{E}^*\rho\big(\log \mathcal{E}^*\rho - \log \rho_t^A \otimes \rho_t^F\big) \\ &= S\big(\rho_t^A\big) + S\big(\rho_t^F\big) - S(\mathcal{E}^*\rho) \\ &= -e_1(t)\log e_1(t) - e_4(t)\log e_4(t) \\ &\quad + e_2(t)\log e_2(t) + e_3(t)\log e_3(t).\end{aligned}$$

The numerical results of this measure showed the time development of DEN for the Jaynes–Cummings model with the mean photon number $|\theta|^2 = 5$, the coupling constant $g = 1$, and the parameter $(\lambda_0, \lambda_1) = (0.7, 0.3)$.

From Fig. 20.3, the time-development of DEN for JCM periodically shows the Rabi oscillation, which is an important feature of this model. Moreover, the entanglement degree takes the highest value around the collapse interval ($t \approx 3 \sim 7$) indicated in Fig. 20.3.

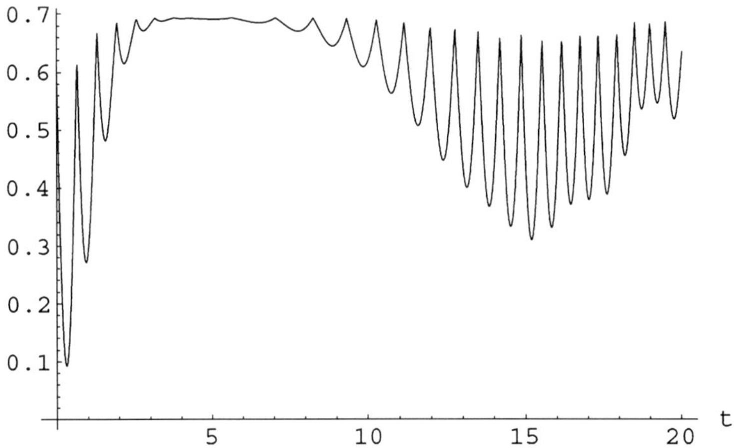

Fig. 20.3 DEM as a function of time t

20.6 Decoherence in Quantum Systems: Stochastic Limit

There already exist small quantum computers working with a few qubits. However, it is not yet clear whether large quantum computers suitable for practical use are possible to build. One reason that quantum computer is difficult to build is decoherence of quantum superpositions due to the interaction with the environment. Decoherence can be defined as the decay of the off-diagonal matrix elements of the density operator in the computational basis. By using simple models like a spin system coupling with a quantum field, it was found that decoherence is quite large. Different methods were discussed to reduce the decoherence in quantum computers, still it is considered to be one of the main open problems in this area.

Here we consider an approach to the problem of reducing decoherence in quantum memory by using a control on the parameters of the system. The Hida white noise calculus and the stochastic limit method are used, and it is discussed that one can choose the parameters of the system in such a way that decoherence is drastically reduced. This is based on the analysis of the spin–boson Hamiltonian performed in the stochastic limit approximation. Considerations of the decoherence problem in quantum computers performed and used a special case of the spin–boson interaction when no spin-flip transitions (or tunneling, in another interpretation) were present. Here we consider the complete spin–boson model, including the spin-flip term, and show that this term plays, in fact, a crucial role in reducing decoherence. Recently, Viola and Lloyd have proposed using the spin-flip transitions for the dynamical suppression of decoherence. The control procedure in their scheme is implemented as a sequence of radio-frequency pulses that repetitively flip the state of the system.

20.6.1 Reducing of Decoherence in Quantum Computers

The maintenance of quantum coherence is a crucial requirement for the ability of quantum computers to be more efficient in certain problems than classical computers. One can simulate the environment as a classical or quantum white noise. In a simple model of spin coupling with the massless quantum field, it was found that for quantum computations not only the coupling with the environment must be small, but also the time taken in the quantum calculation must be less than the thermal time scale $1/T$ where T is the temperature of the initial state of the environment.

The tape (memory) cells are taken to be two-level systems (qubits), with each of the levels having the same energy, and the two states are taken to be the eigenstates of the spin operator σ_z.

Any (pure or mixed) state of a quantum circuit S_l can be described by a density operator of the form

$$\rho_S(t) = \sum_{a,b=0}^{2^l-1} \rho_{ab}(t)|a\rangle\langle b|$$

where $\{|a\rangle\}$ is the computational basis in H^l. The degree of the quantum coherence is described by the off-diagonal elements $\rho_{ab}, a \neq b$ of the density operator. The decoherence is the decay of the off-diagonal elements of the density operator in the computational basis.

In the simple model of the quantum computer interacting with the environment (reservoir) represented by a quantum field (a family of harmonic oscillators), one assumes that the total Hilbert space is $H^l \otimes \mathcal{F}$ where H^l is the l-qubit space while $\mathcal{F}$ is the Bosonic Fock space. If $\rho_{\mathrm{Tot}}(t)$ is the density operator of the total system then to get the density operator $\rho_S(t)$ of the quantum computer one has to take the partial trace over the reservoir space

$$\rho_S(t) = \mathrm{tr}_{\mathcal{F}}\big(\rho_{\mathrm{Tot}}(t)\big).$$

The Hamiltonian describing the coupling of qubit with the environment (the spin–boson Hamiltonian) has the form

$$H_\lambda = -\frac{1}{2}\Delta\sigma_x + \frac{1}{2}\epsilon\sigma_z + \int dk\omega(k)a^*(k)a(k) + \lambda\sigma_z\big(A(g^*) + A^*(g)\big) \quad (20.33)$$

where σ_x and σ_z are Pauli matrices, ϵ and Δ are real parameters interpreted respectively as the energy of the spin and the spin-flip parameter. Here

$$A^*(g) = \int a^*(k)g(k)\,dk, \qquad A(g^*) = \int a(k)g^*(k)\,dk,$$

where $a(k)$ and $a^*(k)$ are the bosonic annihilation and creation operators

$$\big[a(k), a^*(k')\big] = \delta(k - k')$$

which describe the environment.

The one-particle energy of the environment is denoted $\omega(k)$, and we assume $\omega(k) \geq 0$. The function $g(k)$ is a form factor describing the interaction of the system with the environment, λ is the coupling constant. It is well known that, in times of order t/λ^2, the interaction produces effects of order t. Thus λ provides a natural time scale for the observable effects of the system–environment interaction.

Leggett et al. have found a very rich behavior of the dynamics of the Hamiltonian (20.33) ranging from undamped oscillations, to exponential relaxation, to power-law types of behavior, and to total localization. Main qualitative features of the system dynamics can be described in terms of the temperature (i.e., the initial state of the environment) and of the behavior, for low frequencies ω, of the spectral function

$$J(\omega) = \int dk \left| g(k) \right|^2 \delta\left(\omega(k) - \omega\right). \tag{20.34}$$

As for the dynamics of the Hamiltonian (20.33) in the so-called *stochastic approximation*, it was found that the pure oscillating regime, when no damping and no decoherence are present, is described by the simple equation

$$J(\nu \Delta) = 0, \tag{20.35}$$

where

$$\nu = \sqrt{1 + \left(\frac{\epsilon}{\Delta}\right)^2}.$$

20.6.2 Stochastic Limit

The basic idea of the stochastic approximation is the following. If one has a Hamiltonian of the form

$$H_\lambda = H_0 + \lambda V \tag{20.36}$$

then the stochastic limit of the evolution operator

$$U^{(\lambda)}(t) = e^{\mathrm{i}tH_0} e^{-\mathrm{i}tH_\lambda}$$

is the following limit (when it exists in the sense of the convergence of matrix elements):

$$U(t) = \lim_{\lambda \to \infty} U^{(\lambda)}\left(\frac{t}{\lambda^2}\right).$$

The limiting evolution operator $U(t)$ describes the behavior of the model in the time scale t/λ^2.

In order to apply the stochastic approximation to the Hamiltonian (20.33), we write (20.33) in the form (20.36) where

$$H_0 = H_S + H_R.$$

The system Hamiltonian H_S is

$$H_S = -\frac{1}{2}\Delta\sigma_x + \frac{1}{2}\epsilon\sigma_z,$$

and the reservoir Hamiltonian H_R is

$$H_R = \int dk\omega(k)a^*(k)a(k).$$

The evolution operator $U^{(\lambda)}(t)$ satisfies the equation

$$\frac{dU^{(\lambda)}(t)}{dt} = -\mathrm{i}\lambda V(t)U^{(\lambda)}(t)$$

where $V(t) = e^{\mathrm{i}tH_0}Ve^{-\mathrm{i}tH_0}$ has the form

$$V(t) = \sigma_z(t)\big(A\big(e^{-\mathrm{i}t\omega}g^*\big) + A^*\big(e^{\mathrm{i}t\omega}g\big)\big)$$

and

$$\sigma_z(t) = e^{\mathrm{i}tH_S}\sigma_z e^{-\mathrm{i}tH_S}. \tag{20.37}$$

Let us compute (20.37). The eigenvalues of the Hamiltonian H_S are

$$H_S|e_\pm\rangle = \lambda_\pm|e_\pm\rangle$$

where

$$\lambda_\pm = \pm\frac{1}{2}\Delta\nu, \qquad |e_\pm\rangle = \frac{1}{\sqrt{1+\mu_\mp^2}}\begin{pmatrix}1\\ \mu_\mp\end{pmatrix},$$

and

$$\mu_\pm = \frac{\epsilon}{\Delta} \pm \nu, \quad \nu = \sqrt{1 + \left(\frac{\epsilon}{\Delta}\right)^2}.$$

Notice that

$$\langle e_\pm|\sigma_z|e_\pm\rangle = \frac{1-\mu_\mp^2}{1+\mu_\mp^2}, \qquad \langle e_+|\sigma_z|e_-\rangle = \langle e_-|\sigma_z|e_+\rangle = 1/\nu.$$

Therefore,

$$\sigma_z(t) = \frac{1-\mu_-^2}{1+\mu_-^2}DD^+ + \frac{1-\mu_+^2}{1+\mu_+^2}D^+D + \nu^{-1}e^{\mathrm{i}t\nu\Delta}D + \nu^{-1}e^{-\mathrm{i}t\nu\Delta}D^+$$

where

$$D = |e_+\rangle\langle e_-|.$$

The interaction Hamiltonian can now be written in the form:

$$V(t) = \sum_{\alpha=1}^{3} \left(D_\alpha^+ \otimes A\left(e^{-\mathrm{i}t\omega_\alpha} g^*\right) + h.c.\right)$$

where the three spectral frequencies correspond respectively to the down-, zero-, and up-transitions of the two-level system, i.e.,

$$\omega_1(k) = \omega(k) - \nu\Delta, \qquad \omega_2(k) = \omega(k), \qquad \omega_3(k) = \omega(k) + \nu\Delta,$$

$$D_1 = \nu^{-1} D^+, \qquad D_2 = \frac{1-\mu_-^2}{1+\mu_-^2} DD^+ + \frac{1-\mu_+^2}{1+\mu_+^2} D^+ D, \qquad D_3 = \nu^{-1} D^+.$$

The spectral frequencies $\omega_2(k)$ and $\omega_3(k)$ are positive. Therefore, in the stochastic limit one gets only one white noise field. Still a remnant of the interaction remains because, after taking the limit, the system evolves with a new Hamiltonian, equal to the old one plus a shift term depending on the interaction and on the initial state of the field. This was called a *Cheshire Cat effect.*

The limiting evolution equation can then be written as

$$\frac{dU(t)}{dt} = Db^+(t)U(t) - D^+U(t)b(t) - (\gamma + \mathrm{i}\kappa)D^+DU(t) - \mathrm{i}\zeta U(t) \tag{20.38}$$

where

$$\gamma = \nu^{-2}\pi J(\nu\Delta),$$

$$\kappa = \nu^{-2}\big(I(-\nu\Delta) - I(\nu\Delta)\big) + \left(\left(\frac{1-\mu_-^2}{1+\mu_-^2}\right)^2 - \left(\frac{1-\mu_+^2}{1+\mu_+^2}\right)^2\right) I(0),$$

$$\zeta = \nu^{-2} I(-\nu\Delta) + \left(\frac{1-\mu_-^2}{1+\mu_-^2}\right)^2 I(0),$$

and we denote

$$J(\omega) = \int dk \left|g(k)\right|^2 \delta\big(\omega(k) - \omega\big), \qquad I(\omega) = \mathrm{P}\int_0^\infty \frac{d\omega' J(\omega')}{\omega' - \omega},$$

where P means the principal part of the integral. The operators $b(t)$, $b^*(t)$ satisfy the quantum white noise relations

$$\left[b(t), b^*(t')\right] = \gamma\delta(t - t').$$

In the notations of quantum stochastic equations, (20.38) reads

$$dU(t) = \big(D\,dB_t^* - D^+\,dB_t - (\gamma + \mathrm{i}\kappa)D^+D - \mathrm{i}\zeta\big)U(t).$$

Notice that all parameters γ, κ and ζ in the evolution equation (20.40) are expressed in terms of the spectral density $J(\omega)$ and parameters Δ and ϵ of the original Hamiltonian.

For zero temperature, the stochastic approximation to the vacuum expectation value of the Heisenberg evolution of σ_z is given by

$$P(t) = \langle U^*(t)\sigma_z(t)U(t)\rangle.$$

From (20.38), one gets the Langevin equation for $P(t)$ the solution of which is

$$P(t) = \nu^{-1}e^{-\gamma t}\left(D^+ e^{\mathrm{i}(\sigma-\nu\Delta)t} + De^{-\mathrm{i}(\sigma-\nu\Delta)t}\right)$$
$$+ D^+D\left(\frac{1-\mu_+^2}{1+\mu_+^2} - \frac{1-\mu_-^2}{1+\mu_-^2}\right)e^{-2\gamma t} + \frac{1-\mu_-^2}{1+\mu_-^2}. \quad (20.39)$$

We obtain the pure oscillating behavior and no decoherence if

$$\gamma = \nu^{-2}\pi J(\nu\Delta) = 0. \quad (20.40)$$

For a non-zero temperature, we get a stochastic evolution equation of the same form as before only with new constants γ, κ, and ζ. More precisely,

$$\gamma = \nu^{-2}\pi J(\nu\Delta)\coth\frac{\beta\nu\Delta}{2},$$
$$\kappa = \left[\left(\frac{1-\mu_+^2}{1+\mu_+^2}\right)^2 - \left(\frac{1-\mu_-^2}{1+\mu_-^2}\right)^2\right]\left(I_+(0) + I_-(0)\right)$$
$$+ \nu^{-2}\left(I_+(-\nu\Delta) - I_+(\nu\Delta) + I_-(-\nu\Delta) - I_-(\nu\Delta)\right),$$

where spectral densities are

$$J_+(\omega) = \frac{J(\omega)}{1-e^{-\beta\omega}}, \qquad J_-(\omega) = \frac{J(\omega)e^{-\beta\omega}}{1-e^{-\beta\omega}}.$$

Here $J(\omega)$ is the spectral density (20.34) and β is the inverse temperature. The functions $I_\pm(\omega)$ are defined by

$$I_\pm(\omega) = \mathrm{P}\int \frac{d\omega' J_\pm(\omega')}{\omega'-\omega}.$$

One has the same as for the zero–temperature expression (20.39) for $P(t)$ but now with new constants γ and κ depending on temperature. The condition for the reduction of coherence is still the same (20.40). It seems this condition is a rather weak requirement on the parameters of the interaction between a quantum computer and the environment, so one can hope to use it to reduce decoherence.

20.7 Properties of the Quantum Baker's Map

In this section, we consider the semiclassical properties and chaos degree for the quantum Baker's map [366].

The study of chaotic behavior in classical dynamical systems is dating back to Lobachevsky and Hadamard who have been studying the exponential instability property of geodesics on manifolds of negative curvature, and to Poincaré who initiated the inquiry into the stability of the solar system. One believes now that the main features of chaotic behavior in the classical dynamical systems are rather well understood; see, for example, [59, 718]. However, the status of "quantum chaos" is much less clear, although significant progress has been made on this front.

Sometimes one says that an approach to quantum chaos, which attempts to generalize the classical notion of sensitivity to initial conditions, fails for two reasons: first, there is no quantum analogue of the classical phase space trajectories, and second, the unitarity of linear Schrödinger equation precludes sensitivity to initial conditions in the quantum dynamics of a state vector. Let us remind, however, that, in fact, there exists a quantum analogue of the classical phase space trajectories. It is the quantum evolution of expectation values of appropriate observables in suitable states. Also let us remind that the dynamics of a classical system can be described either by the Hamilton equations or by the liner Liouville equations. In quantum theory, the linear Schrödinger equation is the counterpart of the Liouville equation while the quantum counterpart of the classical Hamilton's equation is the Heisenberg equation. Therefore, the study of the quantum expectation values should reveal the chaotic behavior of quantum systems. In this section, we demonstrate this fact for the quantum Baker's map.

If one has the classical Hamilton's equations

$$dq/dt = p, \qquad dp/dt = -V'(q),$$

then the corresponding quantum Heisenberg equations have the same form

$$dq_h/dt = p_h, \qquad dp_h/dt = -V'(q_h),$$

where q_h and p_h are the quantum canonical operators of position and momentum. For the expectation values one gets the Ehrenfest equations

$$d\langle q_h\rangle/dt = \langle p_h\rangle, \qquad d\langle p_h\rangle/dt = -\big\langle V'(q_h)\big\rangle.$$

Note that the Ehrenfest equations are classical equations but for a nonlinear $V'(q_h)$ they are neither Hamilton's equations nor even differential equations because one cannot write $\langle V'(q_h)\rangle$ as a function of $\langle q_h\rangle$ and $\langle p_h\rangle$. However, these equations are very convenient for the consideration of the semiclassical properties of a quantum system. The expectation values $\langle q_h\rangle$ and $\langle p_h\rangle$ are functions of time and initial data. They also depend on the quantum states. One of important problems is to study the dependence of the expectation values on the initial data. In this section, we will study this problem for the quantum Baker's map.

The main objective of “quantum chaos” is to study the correspondence between classical chaotic systems and their quantum counterparts in the semiclassical limit [150, 303]. The quantum–classical correspondence for dynamical systems has been studied for many years; see, for example, [68, 224, 322, 327, 837, 838] and reference therein. Significant progress in understanding this correspondence has been achieved in the WKB approach when one considers the Planck constant h as a small variable parameter. Then it is well known that in the limit $h \to 0$ the quantum theory is reduced to the classical one [509]. However, in physics the Planck constant is a fixed constant, although it is very small. Therefore, it is important to study the relation between classical and quantum evolutions when the Planck constant is fixed. There is a conjecture [113, 116, 832, 838] that a characteristic timescale τ appears in the quantum evolution of chaotic dynamical systems. For time less then τ, there is a correspondence between quantum and classical expectation values, while for times greater that τ the predictions of the classical and quantum dynamics no longer coincide. The important problem is to estimate the dependence of τ on the Planck constant h. Probably, a universal formula expressing τ in terms of h does not exist, and every model should be studied case by case. It is expected that certain quantum and classical expectation values diverge on a timescale inversely proportional to some power of h [78]. Other authors suggest that a breakdown may be anticipated on a much smaller logarithmic timescale [204, 398, 550, 625, 627, 675, 680, 681]. The characteristic time τ associated with the hyperbolic fixed points of the classical motion is expected to be of the logarithmic form $\tau = \frac{1}{\lambda} \ln \frac{C}{h}$ where λ is the Lyapunov exponent and C is a constant which can be taken to be the classical action. Such a logarithmic timescale has been found in the numerical simulations of some dynamical models [837]. It was shown also that the discrepancy between quantum and classical evolutions is decreased even by a small coupling with the environment, which in the quantum case leads to decoherence [837].

The chaotic behavior of the classical dynamical systems is often investigated by computing the Lyapunov exponents. An alternative quantity measuring chaos in dynamical systems which is called the chaos degree has been suggested in [590], in the general framework of information dynamics [359]. The chaos degree was applied to various models in [360]. An advantage of the chaos degree is that it can be applied not only to classical systems but also to quantum systems.

In this section, we study the chaotic behavior and the quantum-classical correspondence for the Baker's map [78, 674]. The quantum Baker's map is a simple model invented for the theoretical study of quantum chaos. Its mathematical properties have been studied in numerical works. In particular, its semiclassical properties have been considered [204, 398, 550, 625, 627, 675, 680, 681], quantum computing and optical realizations have been proposed [139, 626, 679], various quantization procedures have been discussed [79, 461, 675, 682], a symbolic dynamics representation has been given [682].

It is well known that for the consideration of the semiclassical limit in quantum mechanics it is very useful to use coherent states. We define an analogue of the coherent states for the quantum Baker's map. We study the quantum Baker's map by using the correlation functions of the special form which corresponds to the expectation values of Weyl operators, translated in time by the unitary evolution operator and taken in the coherent states.

To explain our formalism, we first discuss the classical limit for correlation functions in ordinary quantum mechanics. Correspondence between quantum and classical expectation values for the Baker's map will be investigated and it will be numerically shown that it is lost at the logarithmic timescale. The chaos degree for the quantum Baker's map will be computed, and it will be demonstrated that it describes the chaotic features of the model. The dependence of the chaos degree on the Planck constant will be studied, and the correspondence between classical and quantum chaos degrees will be established.

20.7.1 Quantum vs. Classical Dynamics

In this section, we discuss an approach to the semiclassical limit in quantum mechanics by using the coherent states, see [327]. Then in the next section, an extension of this approach to the quantum Baker's map will be given.

Consider the canonical system with the Hamiltonian function

$$H = \frac{p^2}{2} + V(x) \tag{20.41}$$

in the plane $(p, x) \in \mathbb{R}^2$. We assume that the canonical equations

$$\dot{x}(t) = p(t), \qquad \dot{p}(t) = -V'\big(x(t)\big) \tag{20.42}$$

have a unique solution $(x(t), p(t))$ for times $|t| < T$ with the initial data

$$x(0) = x_0, \qquad p(0) = v_0. \tag{20.43}$$

This is equivalent to the solution of the Newton equation

$$\ddot{x}(t) = -V'\big(x(t)\big) \tag{20.44}$$

with the initial data

$$x(0) = x_0, \qquad \dot{x}(0) = v_0. \tag{20.45}$$

We denote

$$\alpha = \frac{1}{\sqrt{2}}(x_0 + iv_0). \tag{20.46}$$

The quantum Hamiltonian operator has the form

$$H_h = \frac{p_h^2}{2} + V(q_h)$$

where p_h and q_h satisfy the commutation relations

$$[p_h, q_h] = -\mathrm{i}h.$$

The Heisenberg evolution of the canonical variables is defined as

$$p_h(t) = U(t) p_h U(t)^*, \qquad q_h(t) = U(t) q_h U(t)^*$$

where

$$U(t) = \exp(-\mathrm{i}t H_h / h).$$

For the consideration of the classical limit, we take the following representation

$$p_h = -\mathrm{i}h^{1/2}\partial/\partial x, \qquad q_h = h^{1/2} x$$

acting on functions of the variable $x \in \mathbb{R}$. We also set

$$a = \frac{1}{\sqrt{2}h^{1/2}}(q_h + \mathrm{i}p_h) = \frac{1}{\sqrt{2}}\left(x + \frac{\partial}{\partial x}\right),$$
$$a^* = \frac{1}{\sqrt{2}h^{1/2}}(q_h - \mathrm{i}p_h) = \frac{1}{\sqrt{2}}\left(x - \frac{\partial}{\partial x}\right),$$

then

$$[a, a^*] = 1.$$

The coherent state $|\alpha\rangle$ is defined as

$$|\alpha\rangle = W(\alpha)|0\rangle \tag{20.47}$$

where α is a complex number, $W(\alpha) = \exp(\alpha a^* - a\alpha^*)$, and $|0\rangle$ is the vacuum vector, $a|0\rangle = 0$. The vacuum vector is the solution of the equation

$$(q_h + \mathrm{i}p_h)|0\rangle = 0. \tag{20.48}$$

In the x-representation, one has

$$|0\rangle = \exp\bigl(-x^2/2\bigr)\big/\sqrt{2\pi}. \tag{20.49}$$

The operator $W(\alpha)$ can also be written in the form

$$W(\alpha) = C e^{\mathrm{i}q_h v_0/h^{1/2}} e^{-\mathrm{i}p_h x_0/h^{1/2}} \tag{20.50}$$

where $C = \exp(-v_0 x_0/2h)$.

The mean value of the position operator with respect to the coherent vectors is the real valued function

$$q(t, \alpha, h) = \bigl\langle h^{-1/2}\alpha \big| q_h(t) \big| h^{-1/2}\alpha \bigr\rangle. \tag{20.51}$$

Now one can present the following basic formula describing the semiclassical limit

$$\lim_{h\to 0} q(t,\alpha,h) = x(t,\alpha). \tag{20.52}$$

Here $x(t,\alpha)$ is the solution of (20.44) with the initial data (20.45), and α is given by (20.46).

Let us notice that for time $t=0$ the quantum expectation value $q(t,\alpha,h)$ is equal to the classical one:

$$q(0,\alpha,h) = x(0,\alpha) = x_0 \tag{20.53}$$

for any h. We are going to compare the time dependence of two real functions $q(t,\alpha,h)$ and $x(t,\alpha)$. For small t, these functions are approximately equal. The important problem is to estimate for which t the large difference between them will appear. It is expected that certain quantum and classical expectation values diverge on a timescale inversely proportional to some power of h [327]. Other authors suggest that a breakdown may be anticipated on a much smaller logarithmic timescale [204, 398, 550, 625, 627, 675, 680, 681]. One of very interesting examples [68] of classical systems with chaotic behavior is described by the Hamiltonian function

$$H = \frac{p_1^2}{2} + \frac{p_2^2}{2} + \lambda x_1^2 x_2^2.$$

The consideration of this classical and quantum model within the described framework will be presented elsewhere.

20.7.2 Coherent States for the Quantum Baker's Map

The classical Baker's transformation maps the unit square $0 \le q, p \le 1$ onto itself according to

$$(q,p) \to \begin{cases} (2q, p/2), & \text{if } 0 \le q \le 1/2, \\ (2q-1, (p+1)/2), & \text{if } 1/2 < q \le 1. \end{cases}$$

This corresponds to compressing the unit square in the p-direction and stretching it in the q-direction, while preserving the area, then cutting it vertically and stacking the right part on top of the left part.

The classical Baker's map has a simple description in terms of its symbolic dynamics [52]. Each point (q,p) is represented by a symbolic string

$$\xi = \cdots \xi_{-2}\xi_{-1}\xi_0 \bullet \xi_1\xi_2 \cdots, \tag{20.54}$$

where $\xi_k \in \{0,1\}$, and

$$q = \sum_{k=1}^{\infty} \xi_k 2^{-k}, \qquad p = \sum_{k=0}^{\infty} \xi_{-k} 2^{-k-1}.$$

The action of the Baker's map on a symbolic string s is given by the shift map (Bernoulli shift) U defined by $U\xi = \xi'$, where $\xi'_k = \xi_{k+1}$. This means that, at each time step, the dot $\bullet$ is shifted one place to the right while the entire string remains fixed. After n steps the q coordinate becomes

$$q_n = \sum_{k=1}^{\infty} \xi_{n+k} 2^{-k}. \tag{20.55}$$

This relation defines the classical trajectory with the initial data

$$q = q_0 = \sum_{k=1}^{\infty} \xi_k 2^{-k}. \tag{20.56}$$

Quantum Baker's maps are defined on the D-dimensional Hilbert space of the quantized unit square. To quantize the unite square one defines the Weyl unitary displacement operators $\hat{U}$ and $\hat{V}$ in a D-dimensional Hilbert space, which produces displacements in the momentum and position directions, respectively, and the following commutation relation is obeyed

$$\hat{U}\hat{V} = \epsilon \hat{V}\hat{U},$$

where $\epsilon = \exp(2\pi \mathrm{i}/D)$. We choose $D = 2^N$, so that our Hilbert space will be the N-qubit space $\mathbb{C}^{\otimes N}$. The constant $h = 1/D = 2^{-N}$ can be regarded as the Planck constant. The space $\mathbb{C}^2$ has a basis

$$|0\rangle = \begin{pmatrix} 0 \\ 1 \end{pmatrix}, \qquad |1\rangle = \begin{pmatrix} 1 \\ 0 \end{pmatrix}.$$

The basis in $\mathbb{C}^{\otimes N}$ is

$$|\xi_1\rangle \otimes |\xi_2\rangle \otimes \cdots \otimes |\xi_N\rangle, \quad \xi_k = 0, 1.$$

We write

$$\xi = \sum_{k=1}^{N} \xi_k 2^{N-k}$$

when $\xi = 0, 1, \ldots, 2^N - 1$, and denote

$$|\xi\rangle = |\xi_1 \xi_2 \cdots \xi_N\rangle = |\xi_1\rangle \otimes |\xi_2\rangle \otimes \cdots \otimes |\xi_N\rangle.$$

For this basis, we will also use the notations $\{|\eta\rangle = |\eta_1 \eta_2 \cdots \eta_N\rangle, \eta_k = 0, 1\}$ and $\{|j\rangle = |j_1 j_2 \cdots j_N\rangle, j_k = 0, 1\}$.

The operators $\hat{U}$ and $\hat{V}$ can be written as

$$\hat{U} = e^{2\pi \mathrm{i} \hat{q}}, \qquad \hat{V} = e^{2\pi \mathrm{i} \hat{p}}$$

where the position and momentum operators $\hat{q}$ and $\hat{p}$ are operators in $\mathbb{C}^{\otimes N}$ which are defined as follows. The position operator is

$$\hat{q} = \sum_{j=0}^{2^N-1} q_j |j\rangle\langle j| = \sum_{j_1,\dots,j_N} q_j |j_N \cdots j_1\rangle\langle j_1 \cdots j_N|$$

where

$$|j\rangle = |j_1 j_2 \cdots j_N\rangle, \quad j_k = 0, 1$$

is the basis in $\mathbb{C}^{\otimes N}$,

$$j = \sum_{k=1}^{N} j_k 2^{N-k}$$

and

$$q_j = \frac{j + 1/2}{2^N}, \quad j = 0, 1, \dots, 2^N - 1.$$

The momentum operator is defined as

$$\hat{p} = F_N \hat{q} F_N^*$$

where F_N is the quantum Fourier transform acting on the basis vectors as

$$F_N |j\rangle = \frac{1}{\sqrt{D}} \sum_{\xi=0}^{D-1} e^{2\pi i \xi j / D} |\xi\rangle,$$

here $D = 2^N$.

The symbolic representation of quantum Baker's map T was introduced by Schack and Caves [682] and studied in [683, 729]. Let us explain the symbolic representation of quantum Baker's map as a special case [682]: By applying a partial quantum Fourier transform $G_m = \overbrace{I \otimes \cdots \otimes I}^{m} \otimes F_{N-m}$ to the position eigenstates, one obtains the following quantum Baker's map T:

$$T|_\bullet \xi_1 \cdots \xi_N\rangle \equiv |\xi_{1\bullet} \xi_2 \cdots \xi_N\rangle,$$

where

$$T = G_{N-1} \circ G_N^{-1}$$

and

$$\begin{aligned} &|\xi_1 \cdots \xi_{N-m\bullet} \xi_{N-m+1} \cdots \xi_N\rangle \\ &\equiv G_m |\xi_{N-m+1} \cdots \xi_N \xi_{N-m} \cdots \xi_1\rangle \\ &= |\xi_{N-m+1}\rangle \otimes \cdots \otimes |\xi_N\rangle \otimes F_{N-m} |\xi_{N-m}\rangle \otimes \cdots \otimes |\xi_1\rangle. \end{aligned}$$

The quantum Baker's map T is the unitary operator in $\mathbb{C}^{\otimes N}$ with the following matrix elements

$$\langle \xi | T | \eta \rangle = \frac{1-\mathrm{i}}{2} \exp\left(\frac{\pi}{2} \mathrm{i} |\xi_1 - \eta_N| \right) \prod_{k=2}^{N} \delta(\xi_k - \eta_{k-1}), \tag{20.57}$$

where $|\xi\rangle = |\xi_1 \xi_2 \cdots \xi_N\rangle$, $|\eta\rangle = |\eta_1 \eta_2 \cdots \eta_N\rangle$, and $\delta(x)$ is the Kronecker symbol, $\delta(0) = 1$; $\delta(x) = 0, x \neq 0$.

We define the coherent states by

$$|\alpha\rangle = C e^{2\pi \mathrm{i} \hat{q} v} e^{-2\pi \mathrm{i} \hat{p} x} |\psi_0\rangle. \tag{20.58}$$

Here $\alpha = x + \mathrm{i}v$, x and v are integers, C is the normalization constant, and $|\psi_0\rangle$ is the vacuum vector. This definition should be compared with (20.50). The vacuum vector can be defined as the solution of the equation

$$(q_h + \mathrm{i} p_h)|\psi_0\rangle = 0$$

(compare with (20.48)). We will use a simpler definition which in the position representation is

$$\langle q_j | \psi_0 \rangle = C \exp\left(-q_j^2/2\right)$$

(compare with (20.49)). Here C is a normalization constant.

The classical chaos degree was applied to several dynamical maps such as the logistic map, Baker's transformation, and Tinkerbel map in Chap. 10. We will apply the chaos degree to the quantum Baker's transformation here.

20.7.3 Expectation Values and Chaos Degree

In this section, we show a general representation of the mean value of the position operator $\hat{q}$ for the time evolution, which is constructed by the quantum Baker's map. Then we give an algorithm to compute the chaos degree for the quantum Baker's map.

To study the time evolution and the classical limit $h \to 0$ which corresponds to $N \to \infty$ of the quantum Baker's map T, we introduce the following the mean value of the position operator $\hat{q}$ for time $n \in \mathbb{N}$ with respect to a single basis $|\xi\rangle$:

$$r_n^{(N)} = \langle \xi | T^n \hat{q} T^{-n} | \xi \rangle, \tag{20.59}$$

where $|\xi\rangle = |\xi_1 \xi_2 \cdots \xi_N\rangle$.

From (20.57), the following formula of the matrix elements of T^n for any $n \in \mathbb{N}$ is easily obtained:

$$\langle \xi | T_0^n | \zeta \rangle = \begin{cases} (\frac{1-i}{2})^n (\prod_{k=1}^{N-n} \delta(\xi_{n+k} - \zeta_k))(\prod_{l=1}^{n} A_{\xi_l \zeta_{N-n+l}}), & \text{if } n < N, \\ (\frac{1-i}{2})^n (\prod_{k=1}^{n} A_{\xi_k \zeta_k}), & \text{if } n = N, \\ (\frac{1-i}{2})^n (\prod_{k=1}^{p} (A^{m+1})_{\xi_k \zeta_{N-p+k}}) & \\ \quad \times (\prod_{l=1}^{N-p} (A^m)_{\xi_{p+l} \zeta_l}), & \text{if } n = mN + p, \\ (\frac{1-i}{2})^n \prod_{k=1}^{N} (A^m)_{\xi_k \zeta_k}, & \text{if } n = mN, \end{cases} \tag{20.60}$$

where A is the 2×2 matrix with the element $A_{x_1 x_2} = \exp(\frac{\pi}{2} \mathrm{i} |x_1 - x_2|)$ for $x_1, x_2 = 0, 1$, $p = 1, \ldots, N-1$ and $m \in \mathbb{N}$.

Using this formula, the following theorems are obtained and their proofs are given

Theorem 20.28

$$r_n^{(N)} = \begin{cases} \sum_{k=1}^{N-n} \xi_{n+k} 2^{-k} + \frac{2^n}{2^{N+1}}, & \text{if } n < N, \\ \frac{1}{2}, & \text{if } n = N, \\ \frac{1}{2^n} \sum_{j=0}^{2^N-1} \frac{j+1/2}{2^N} \prod_{k=1}^{p} |(A^{m+1})_{\xi_k j_{N-p+k}}|^2 & \\ \quad \times \prod_{l=1}^{N-p} |(A^m)_{\xi_{p+l} j_l}|^2, & \text{if } n = mN + p, \\ \frac{1}{2^n} \sum_{j=0}^{2^N-1} \frac{j+1/2}{2^N} \prod_{k=1}^{N} |(A^m)_{\xi_k j_k}|^2, & \text{if } n = mN, \end{cases} \tag{20.61}$$

where A is the 2×2 matrix with the element $A_{x_1 x_2} = \exp(\frac{\pi}{2} \mathrm{i} |x_1 - x_2|)$ for $x_1, x_2 = 0, 1$, $p = 1, \ldots, N-1$ and $m \in \mathbb{N}$.

Proof By a direct calculation, we obtain

$$\begin{aligned} r_n^{(N)} &= \langle \xi | T^n \hat{q} T^{-n} | \xi \rangle = \langle \xi | T^n \left(\sum_{j=0}^{2^N-1} \frac{j+1/2}{2^N} |j\rangle\langle j| \right) T^{-n} | \xi \rangle \\ &= \sum_{j=0}^{2^N-1} \frac{j+1/2}{2^N} \langle \xi | T^n | j \rangle \langle j | T^{-n} | \xi \rangle = \sum_{j=0}^{2^N-1} \frac{j+1/2}{2^N} \langle \xi | T^n | j \rangle \langle j | T^{*n} | \xi \rangle \\ &= \sum_{j=0}^{2^N-1} \frac{j+1/2}{2^N} \langle \xi | T^n | j \rangle \overline{\langle \xi | T^n | j \rangle} = \sum_{j=0}^{2^N-1} \frac{j+1/2}{2^N} \left| \langle \xi | T^n | j \rangle \right|^2. \end{aligned}$$

Using (20.60), the mean value $r_n^{(N)}$ in the case $n < N$ can be expressed as

$$r_n^{(N)} = \sum_{j=0}^{2^N-1} \frac{j+1/2}{2^N} \left| \langle \xi | T^n | j \rangle \right|^2$$

$$= \sum_{j=0}^{2^N-1} \frac{j+1/2}{2^N} \left| \left(\frac{1-\mathrm{i}}{2}\right)^n \left(\prod_{k=1}^{N-n} \delta(\xi_{n+k} - j_k)\right) \left(\prod_{l=1}^{n} A_{\xi_l j_{N-n+l}}\right) \right|^2$$

$$= \sum_{j=0}^{2^N-1} \frac{j+1/2}{2^N} \left(\frac{1-\mathrm{i}}{2}\right)^n \left(\frac{1+\mathrm{i}}{2}\right)^n \left(\prod_{k=1}^{N-n} \delta(\xi_{n+k} - j_k)\right) \left(\prod_{l=1}^{n} |A_{\xi_l j_{N-n+l}}|^2\right)$$

$$= \sum_{j=0}^{2^N-1} \frac{j+1/2}{2^N} \left(\frac{1-\mathrm{i}}{2}\right)^n \left(\frac{1+\mathrm{i}}{2}\right)^n \left(\prod_{k=1}^{N-n} \delta(\xi_{n+k} - j_k)\right).$$

We write the sum over j as follows by using $j_k = 0, 1 (k = 1, \dots, N)$

$$r_n^{(N)} = \frac{1}{2^{N+n}} \sum_{j_1,\dots,j_N} \left\{ \left(\sum_{k=1}^{N} j_k 2^{N-k}\right) + \frac{1}{2} \right\} \left(\prod_{k=1}^{N-n} \delta(\xi_{n+k} - j_k)\right)$$

$$= \frac{1}{2^{N+n}} \sum_{j_1,\dots,j_N} \left(\sum_{k=1}^{N} j_k 2^{N-k}\right) \left(\prod_{k=1}^{N-n} \delta(\xi_{n+k} - j_k)\right)$$

$$+ \frac{1}{2^{N+n+1}} \sum_{j_1,\dots,j_N} \left(\prod_{k=1}^{N-n} \delta(\xi_{n+k} - j_k)\right)$$

$$= \frac{1}{2^{N+n}} \sum_{j_{N-n+1},\dots,j_N} \left(\sum_{l=1}^{N-n} \xi_{n+l} 2^{N-l} + \sum_{l=N-n+1}^{N} j_l 2^{N-l}\right)$$

$$+ \frac{1}{2^{N+n+1}} \left(\sum_{j_{N-n+1},\dots,j_N} 1\right)$$

$$= \frac{1}{2^{N+n}} \left(\sum_{l=1}^{N-n} \xi_{n+l} 2^{N-l}\right) \left(\sum_{j_{N-n+1},\dots,j_N} 1\right)$$

$$+ \frac{1}{2^{N+n}} \sum_{j_{N-n+1},\dots,j_N} \left(\sum_{l=N-n+1}^{N} j_l 2^{N-l}\right)$$

$$+ \frac{1}{2^{N+n+1}} \left(\sum_{j_{N-n+1},\dots,j_N} 1\right).$$

Since $\sum_{j_{N-n+1},\dots,j_N} 1 = 2^{N-p}$, we get

$$r_n^{(N)} = \frac{1}{2^N} \left(\sum_{l=1}^{N-n} \xi_{n+l} 2^{N-l}\right) + \frac{1}{2^{N+n}} \sum_{j_{N-n+1},\dots,j_N} \left(\sum_{l=1}^{n} j_{N-n+l} 2^{n-l}\right) + \frac{1}{2^{N+1}}$$

$$= \frac{1}{2^N}\left(\sum_{l=1}^{N-n} \xi_{n+l} 2^{N-l}\right) + \frac{1}{2^{N+n}} \frac{1}{2}(2^n - 1)2^n + \frac{1}{2^{N+1}}$$

$$= \frac{1}{2^N}\left(\sum_{l=1}^{N-n} \xi_{n+l} 2^{N-l}\right) + \frac{2^n}{2^{N+1}}.$$

For the case $n = N$, we similarly obtain

$$r_n^{(N)} = \sum_{j=0}^{2^N-1} \frac{j+1/2}{2^N} |\langle \xi | T^n | j \rangle|^2$$

$$= \sum_{j=0}^{2^N-1} \frac{j+1/2}{2^N} \left| \left(\frac{1-\mathrm{i}}{2}\right)^N \left(\prod_{k=1}^{N} A_{\xi_k j_k}\right) \right|^2$$

$$= \sum_{j=0}^{2^N-1} \frac{j+1/2}{2^N} \left| \left(\frac{1-\mathrm{i}}{2}\right)^N \right|^2 \prod_{k=1}^{N} |A_{\xi_k j_k}|^2$$

$$= \frac{1}{2^{2N}} \sum_{j=0}^{2^N-1} (j+1/2) = \frac{1}{2^{2N}} \frac{1}{2}(2^N - 1)2^N + \frac{1}{2^{N+1}} = \frac{1}{2}.$$

For $n = mN + p$, $p = 1, 2, \ldots, N-1$, $m \in \mathbb{N}$,

$$r_n^{(N)} = \sum_{j=0}^{2^N-1} \frac{j+1/2}{2^N} |\langle \xi | T^n | j \rangle|^2$$

$$= \sum_{j=0}^{2^N-1} \frac{j+1/2}{2^N} \left| \left(\frac{1-\mathrm{i}}{2}\right)^n \left(\prod_{k=1}^{p} (A^{m+1})_{\xi_k j_{N-p+k}}\right) \left(\prod_{l=1}^{N-p} (A^m)_{\xi_{p+l} j_l}\right) \right|^2$$

$$= \frac{1}{2^n} \sum_{j=0}^{2^N-1} \frac{j+1/2}{2^N} \prod_{k=1}^{p} |(A^{m+1})_{\xi_k j_{N-p+k}}|^2 \prod_{l=1}^{N-p} |(A^m)_{\xi_{p+l} j_l}|^2,$$

and for $n = mN$, $m \in \mathbb{N}$,

$$r_N^{(n)} = \sum_{j=0}^{2^N-1} \frac{j+1/2}{2^N} |\langle \xi | T^n | j \rangle|^2 = \sum_{j=0}^{2^N-1} \frac{j+1/2}{2^N} \left| \left(\frac{1-\mathrm{i}}{2}\right)^n \prod_{k=1}^{N} (A^m)_{\xi_k j_k} \right|^2$$

$$= \frac{1}{2^n} \sum_{j=0}^{2^N-1} \frac{j+1/2}{2^N} \prod_{k=1}^{N} |(A^m)_{\xi_k j_k}|^2.$$

□

By diagonalizing the matrix A, we obtain the following formula of the absolute square of the matrix elements of A^n for any $n \in \mathbb{N}$.

Lemma 20.29 *For any $n \in \mathbb{N}$, we have*

$$\left|(A^n)_{kj}\right|^2 = \begin{cases} 2^n \cos^2(\frac{n\pi}{4}), & \text{if } k = j, \\ 2^n \sin^2(\frac{n\pi}{4}), & \text{if } k \neq j. \end{cases}$$

Proof By a direct calculation, the matrix A is diagonalized as follows:

$$A = FDF^*, \tag{20.62}$$

where

$$F = \frac{1}{\sqrt{2}}\begin{pmatrix} 1 & -1 \\ 1 & 1 \end{pmatrix}, \qquad D = \begin{pmatrix} 1+\mathrm{i} & 0 \\ 0 & 1-\mathrm{i} \end{pmatrix}.$$

From (20.62), we have

$$\begin{aligned} A^n &= FD^nF^* \\ &= \frac{1}{\sqrt{2}}\begin{pmatrix} 1 & -1 \\ 1 & 1 \end{pmatrix}\begin{pmatrix} (1+\mathrm{i})^n & 0 \\ 0 & (1-\mathrm{i})^n \end{pmatrix}\frac{1}{\sqrt{2}}\begin{pmatrix} 1 & 1 \\ -1 & 1 \end{pmatrix} \\ &= \frac{1}{2}\begin{pmatrix} (1+\mathrm{i})^n + (1-\mathrm{i})^n & (1+\mathrm{i})^n - (1-\mathrm{i})^n \\ (1+\mathrm{i})^n - (1-\mathrm{i})^n & (1+\mathrm{i})^n + (1-\mathrm{i})^n \end{pmatrix}. \end{aligned} \tag{20.63}$$

Using (20.63), it follows that for any $k = j, k = 1, 2$,

$$\begin{aligned} \left|(A^n)_{kj}\right|^2 &= \frac{1}{2}\{(1+\mathrm{i})^n + (1-\mathrm{i})^n\}\overline{\frac{1}{2}\{(1+\mathrm{i})^n + (1-\mathrm{i})^n\}} \\ &= \frac{1}{4}\{(1+\mathrm{i})^n + (1-\mathrm{i})^n\}^2 \\ &= \frac{1}{4}\left\{\left(\sqrt{2}\frac{1+\mathrm{i}}{\sqrt{2}}\right)^n + \left(\sqrt{2}\frac{1-\mathrm{i}}{\sqrt{2}}\right)^n\right\}^2 \\ &= \frac{2^n}{4}\left\{\left(\exp\left(\frac{\pi}{4}\mathrm{i}\right)\right)^n + \left(\exp\left(-\frac{\pi}{4}\mathrm{i}\right)\right)^n\right\}^2 \\ &= \frac{2^n}{4}\left[\left\{\cos\left(\frac{n\pi}{4}\right) + \mathrm{i}\sin\left(\frac{n\pi}{4}\right)\right\} + \left\{\cos\left(\frac{n\pi}{4}\right) - \mathrm{i}\sin\left(\frac{n\pi}{4}\right)\right\}\right]^2 \\ &= \frac{2^n}{4}\left\{2\cos\left(\frac{n\pi}{4}\right)\right\}^2 = 2^n\cos^2\left(\frac{n\pi}{4}\right), \end{aligned}$$

and for any $k \neq j, k = 1, 2$,

$$\begin{aligned} \left|(A^n)_{kj}\right|^2 &= \frac{1}{2}\{(1+\mathrm{i})^n - (1-\mathrm{i})^n\}\overline{\frac{1}{2}\{(1+\mathrm{i})^n - (1-\mathrm{i})^n\}} \\ &= -\frac{1}{4}\left\{(\sqrt{2})^n\left(\frac{1+\mathrm{i}}{\sqrt{2}}\right)^n - (\sqrt{2})^n\left(\frac{1-\mathrm{i}}{\sqrt{2}}\right)^n\right\}^2 \end{aligned}$$

$$= -\frac{2^n}{4}\left\{\left(\exp\left(\frac{\pi}{4}\mathrm{i}\right)\right)^n - \left(\exp\left(-\frac{\pi}{4}\mathrm{i}\right)\right)^n\right\}^2$$

$$= -\frac{2^n}{4}\left\{2\mathrm{i}\sin\left(\frac{n\pi}{4}\right)\right\}^2 = 2^n \sin^2\left(\frac{n\pi}{4}\right). \qquad \square$$

Combining the above theorem and lemma, we obtain the following two theorems with respect to the mean value $r_n^{(N)}$ of the position operator.

Theorem 20.30 *For the case $n = mN + p$, $p = 1, 2, \ldots, N-1$ and $m \in \mathbb{N}$, we have*

$$r_n^{(N)} = \begin{cases} \sum_{k=1}^{N-p} \xi_{p+k} 2^{-k} + \frac{2^p}{2^{N+1}}, & \text{if } m = 0 \ (\mathrm{mod}\, 4), \\ \sum_{k=N-p+1}^{N} \eta_{k-(N-p)} 2^{-k} + \frac{2^N - 2^p + 1}{2^{N+1}}, & \text{if } m = 1 \ (\mathrm{mod}\, 4), \\ \sum_{k=1}^{N-p} \eta_{p+k} 2^{-k} + \frac{2^p}{2^{N+1}}, & \text{if } m = 2 \ (\mathrm{mod}\, 4), \\ \sum_{k=N-p+1}^{N} \xi_{k-(N-p)} 2^{-k} + \frac{2^N - 2^p + 1}{2^{N+1}}, & \text{if } m = 3 \ (\mathrm{mod}\, 4), \end{cases} \tag{20.64}$$

where $\eta_k = \xi_k + 1 (\mathrm{mod}\, 2)$, $k = 1, \ldots, N$.

Proof For the case $n = mN + p$, $p = 1, \ldots, N-1$ and $m \in \mathbb{N}$,

$$r_n^{(N)} = \frac{1}{2^n} \sum_{j=0}^{2^N-1} \frac{j + 1/2}{2^N} \prod_{k=1}^{p} \left|\left(A^{m+1}\right)_{\xi_k j_{N-p+k}}\right|^2 \prod_{l=1}^{N-p} \left|\left(A^m\right)_{\xi_{p+l} j_l}\right|^2.$$

By a direct calculation, we obtain

$$r_n^{(N)} = \frac{1}{2^n} \sum_{j=0}^{2^N-1} \frac{j + 1/2}{2^N} \prod_{k=1}^{p} \left|\left(A^{m+1}\right)_{\xi_k j_{N-p+k}}\right|^2 \prod_{l=1}^{N-p} \left|\left(A^m\right)_{\xi_{p+l} j_l}\right|^2$$

$$= \frac{1}{2^n} \sum_{j=0}^{2^N-1} \frac{j + 1/2}{2^N} \prod_{l=1}^{N-p} \left|\left(A^m\right)_{\xi_{p+l} j_l}\right|^2 \prod_{k=1}^{p} \left|\left(A^{m+1}\right)_{\xi_k j_{N-p+k}}\right|^2$$

$$= \frac{1}{2^{n+N}} \sum_{j=0}^{2^N-1} \left(j + \frac{1}{2}\right) \prod_{l=1}^{N-p} \left|\left(A^m\right)_{\xi_{p+l} j_l}\right|^2 \prod_{k=1}^{p} \left|\left(A^{m+1}\right)_{\xi_k j_{N-p+k}}\right|^2$$

$$= \frac{1}{2^{n+N}} \sum_{j_1, \ldots, j_N} \left\{\left(\sum_{k=1}^{N} j_k 2^{N-k}\right) + \frac{1}{2}\right\} \prod_{l=1}^{N-p} \left|\left(A^m\right)_{\xi_{p+l} j_l}\right|^2$$

$$\times \prod_{k=1}^{p} \left|\left(A^{m+1}\right)_{\xi_k j_{N-p+k}}\right|^2$$

$$
\begin{aligned}
&= \frac{1}{2^{n+N}} \sum_{j_1,\dots,j_N} \left\{ \left(\sum_{k=1}^{N-p} j_k 2^{N-k} \right) + \left(\sum_{k=N-p+1}^{N} j_k 2^{N-k} \right) + \frac{1}{2} \right\} \\
&\quad \times \prod_{l=1}^{N-p} \left| (A^m)_{\xi_{p+l} j_l} \right|^2 \prod_{k=1}^{p} \left| (A^{m+1})_{\xi_k j_{N-p+k}} \right|^2 \\
&= \frac{1}{2^{(m+1)N+p}} \sum_{j_1,\dots,j_N} \left\{ \left(\sum_{k=1}^{N-p} j_k 2^{N-k} \right) + \left(\sum_{k=N-p+1}^{N} j_k 2^{N-k} \right) + \frac{1}{2} \right\} \\
&\quad \times \prod_{l=1}^{N-p} \left| (A^m)_{\xi_{p+l} j_l} \right|^2 \prod_{k=N-p+1}^{N} \left| (A^{m+1})_{\xi_{k-(N-p)} j_k} \right|^2 .
\end{aligned} \tag{20.65}
$$

Case (i) $m = 0 \pmod 4$. From the above lemma, we have

$$
\left| (A^m)_{\xi_{p+l} j_l} \right|^2 = \begin{cases} 2^m, & \text{if } j_l = \xi_{p+l}, \\ 0, & \text{if } j_l \neq j_{p+l}, \end{cases}
$$

$$
\left| (A^{m+1})_{\xi_{k-(N-p)} j_k} \right|^2 = 2^m
$$

for any $l = 1, \dots, N-p$ and $k = N-p+1, \dots, N$. Using this formula, the product of absolute squares can be expressed as

$$
\begin{aligned}
&\prod_{l=1}^{N-p} \left| (A^m)_{\xi_{p+l} j_l} \right|^2 \prod_{k=N-p+1}^{N} \left| (A^{m+1})_{\xi_{k-(N-p)} j_k} \right| \\
&= \begin{cases} (2^m)^{N-p} (2^m)^p, & \text{if } j_l = \xi_{p+l} \text{ for all } l = 1, \dots, N-p, \\ 0, & \text{otherwise,} \end{cases} \\
&= \begin{cases} 2^{mN}, & \text{if } j_l = \xi_{p+l} \text{ forall } l = 1, \dots, N-p, \\ 0, & \text{otherwise.} \end{cases}
\end{aligned}
$$

Equation (20.65) can be rewritten as

$$
\begin{aligned}
r_n^{(N)} &= \frac{1}{2^{(m+1)N+p}} \sum_{j_1,\dots,j_N} \left\{ \left(\sum_{k=1}^{N-p} j_k 2^{N-k} \right) + \left(\sum_{k=N-p+1}^{N} j_k 2^{N-k} \right) + \frac{1}{2} \right\} \\
&\quad \times \prod_{l=1}^{N-p} \left| (A^m)_{\xi_{p+l} j_l} \right|^2 \prod_{k=N-p+1}^{N} \left| (A^{m+1})_{\xi_{k-(N-p)} j_k} \right|^2 \\
&= \frac{2^{mN}}{2^{(m+1)N+p}} \sum_{j_{N-p+1},\dots,j_N} \left\{ \left(\sum_{k=1}^{N-p} \xi_{p+k} 2^{N-k} \right) + \left(\sum_{k=N-p+1}^{N} j_k 2^{N-k} \right) + \frac{1}{2} \right\}
\end{aligned}
$$

$$= \frac{1}{2^{N+p}} \left(\sum_{k=1}^{N-p} \xi_{p+k} 2^{N-k} \right) \left(\sum_{j_{N-p+1},\ldots,j_N} 1 \right)$$

$$+ \frac{1}{2^{N+p}} \sum_{j_{N-p+1},\ldots,j_N} \left(\sum_{k=N-p+1}^{N} j_k 2^{N-k} \right) + \frac{1}{2^{N+p}} \frac{1}{2} \left(\sum_{j_{N-p+1},\ldots,j_N} 1 \right)$$

$$= \frac{1}{2^N} \sum_{k=1}^{N-p} \xi_{p+k} 2^{N-k} + \frac{1}{2^{N+p}} \sum_{j_{N-p+1},\ldots,j_N} \left(\sum_{k=N-p+1}^{N} j_k 2^{N-k} \right) + \frac{1}{2^{N+1}}$$

$$= \frac{1}{2^N} \sum_{k=1}^{N-p} \xi_{p+k} 2^{N-k} + \frac{1}{2^{N+p}} \sum_{j_{N-p+1},\ldots,j_N} \left(\sum_{k=1}^{p} j_{N-p+k} 2^{p-k} \right) + \frac{1}{2^{N+1}}$$

$$= \frac{1}{2^N} \sum_{k=1}^{N-p} \xi_{p+k} 2^{N-k} + \frac{1}{2^{N+p}} \sum_{k=0}^{2^p-1} k + \frac{1}{2^{N+1}}$$

$$= \frac{1}{2^N} \sum_{k=1}^{N-p} \xi_{p+k} 2^{N-k} + \frac{1}{2^{N+p}} \frac{1}{2} (2^p - 1) 2^p + \frac{1}{2^{N+1}}$$

$$= \sum_{k=1}^{N-p} \xi_{p+k} 2^{-k} + \frac{2^p}{2^{N+1}}. \tag{20.66}$$

Case (ii) $m = 1 \pmod 4$. From the above lemma, we have

$$\left|(A^m)_{\xi_{p+l} j_l}\right|^2 = 2^{m-1}, \qquad \left|(A^{m+1})_{\xi_{k-(N-p)} j_k}\right|^2 = \begin{cases} 2^{m+1}, & \text{if } j_k \neq \xi_{k-(N-p)}, \\ 0, & \text{if } j_k = \xi_{k-(N-p)} \end{cases}$$

for any $l = 1, \ldots, N-p$ and $k = N-p+1, \ldots, N$. Using this formula, the product of absolute squares can be expressed as

$$\prod_{l=1}^{N-p} \left|(A^m)_{\xi_{p+l} j_l}\right|^2 \prod_{k=N-p+1}^{N} \left|(A^{m+1})_{\xi_{k-(N-p)} j_k}\right|$$

$$= \begin{cases} (2^{m-1})^{N-p} (2^{m+1})^p, & \text{if } j_k \neq \xi_{k-(N-p)} \text{ for all } k = N-p+1, \ldots, N, \\ 0, & \text{otherwise,} \end{cases}$$

$$= \begin{cases} 2^{(m-1)N+2p}, & \text{if } j_k \neq \xi_{k-(N-p)} \text{ for all } k = N-p+1, \ldots, N, \\ 0, & \text{otherwise.} \end{cases}$$

Let $\eta_{k-(N-p)} = \xi_{k-(N-p)} + 1 \pmod 2$, $k = N-p+1, \ldots, N$. It follows that

$$r_n^{(N)} = \frac{1}{2^{(m+1)N+p}} \sum_{j_1,\ldots,j_N} \left\{ \left(\sum_{k=1}^{N-p} j_k 2^{N-k} \right) + \left(\sum_{k=N-p+1}^{N} j_k 2^{N-k} \right) + \frac{1}{2} \right\}$$

$$\times \prod_{l=1}^{N-p} \left|(A^m)_{\xi_{p+l} j_l}\right|^2 \prod_{k=N-p+1}^{N} \left|(A^{m+1})_{\xi_{k-(N-p)} j_k}\right|^2$$

$$= \frac{2^{(m-1)N+2p}}{2^{(m+1)N+p}} \sum_{j_1,\dots,j_{N-p}} \left\{ \left(\sum_{k=1}^{N-p} j_k 2^{N-k} \right) \right.$$

$$\left. + \left(\sum_{k=N-p+1}^{N} \eta_{k-(N-p)} 2^{N-k} \right) + \frac{1}{2} \right\}$$

$$= \frac{1}{2^{2N-p}} \sum_{j_1,\dots,j_{N-p}} \left\{ \left(\sum_{k=1}^{N-p} j_k 2^{N-k} \right) + \left(\sum_{k=N-p+1}^{N} \eta_{k-(N-p)} 2^{N-k} \right) + \frac{1}{2} \right\}$$

$$= \frac{1}{2^{2N-p}} \sum_{j_1,\dots,j_{N-p}} \left(\sum_{k=1}^{N-p} j_k 2^{N-k} \right) + \frac{1}{2^{2N-p}} \left(\sum_{k=N-p+1}^{N} \eta_{k-(N-p)} 2^{N-k} \right)$$

$$\times \left(\sum_{j_1,\dots,j_{N-p}} 1 \right) + \frac{1}{2^{2N-p}} \frac{1}{2} \left(\sum_{j_1,\dots,j_{N-p}} 1 \right). \tag{20.67}$$

Since $\sum_{j_{N-n+1},\dots,j_N} 1 = 2^{N-p}$, we get

$$r_n^{(N)} = \frac{1}{2^{2N-p}} \sum_{j_1,\dots,j_{N-p}} \left(\sum_{k=1}^{N-p} j_k 2^{N-k} \right)$$

$$+ \frac{2^{N-p}}{2^{2N-p}} \left(\sum_{k=N-p+1}^{N} \eta_{k-(N-p)} 2^{N-k} \right) + \frac{2^{N-p}}{2^{2N-p}} \frac{1}{2}$$

$$= \frac{1}{2^{2N-p}} \sum_{j_1,\dots,j_{N-p}} \left(\sum_{k=1}^{N-p} j_k 2^{N-k} \right) + \sum_{k=N-p+1}^{N} \eta_{k-(N-p)} 2^{-k} + \frac{1}{2^{N+1}}$$

$$= \frac{2^p}{2^{2N-p}} \sum_{k=0}^{2^{N-p}-1} k + \sum_{k=N-p+1} N \eta_{k-(N-p)} 2^{-k} + \frac{1}{2^{N+1}}$$

$$= \frac{2^p}{2^{2N-p}} \frac{1}{2} (2^{N-p} - 1) 2^{N-p} + \sum_{k=N-p+1}^{N} \eta_{k-(N-p)} 2^{-k} + \frac{1}{2^{N+1}}$$

$$= \sum_{k=N-p+1}^{N} \eta_{k-(N-p)} 2^{-k} + \frac{2^N - 2^p + 1}{2^{N+1}}.$$

Case (iii) $m = 2 \pmod 4$. From the above lemma, we have

$$\left|\left(A^m\right)_{\xi_{p+l} j_l}\right|^2 = \begin{cases} 2^m, & \text{if } j_l \neq \xi_{p+l}, \\ 0, & \text{if } j_l = \xi_{p+l}, \end{cases} \qquad \left|\left(A^{m+1}\right)_{\xi_{k-(N-p)} j_k}\right|^2 = 2^m$$

for any $l = 1, \ldots, N-p$ and $k = N-p+1, \ldots, N$. Using this formula, the product of absolute squares can be expressed as

$$\begin{aligned} &\prod_{l=1}^{N-p} \left|\left(A^m\right)_{\xi_{p+l} j_l}\right|^2 \prod_{k=N-p+1}^{N} \left|\left(A^{m+1}\right)_{\xi_{k-(N-p)} j_k}\right| \\ &\quad = \begin{cases} (2^m)^{N-p}(2^m)^p, & \text{if } j_l \neq \xi_{p+l} \text{ for all } l = 1, \ldots, N-p, \\ 0, & \text{otherwise,} \end{cases} \\ &\quad = \begin{cases} 2^{mN}, & \text{if } j_l \neq \xi_{p+l} \text{ for all } l = 1, \ldots, N-p, \\ 0, & \text{otherwise.} \end{cases} \end{aligned}$$

Let $\eta_{p+l} = \xi_{p+l} + 1 \pmod 2$, $l = 1, \ldots, N-p$. It follows that

$$\begin{aligned} r_n^{(N)} &= \frac{1}{2^{(m+1)N+p}} \sum_{j_1, \ldots, j_N} \left\{ \left(\sum_{k=1}^{N-p} j_k 2^{N-k} \right) + \left(\sum_{k=N-p+1}^{N} j_k 2^{N-k} \right) + \frac{1}{2} \right\} \\ &\quad \times \prod_{l=1}^{N-p} \left|\left(A^m\right)_{\xi_{p+l} j_l}\right|^2 \prod_{k=N-p+1}^{N} \left|\left(A^{m+1}\right)_{\xi_{k-(N-p)} j_k}\right|^2 \\ &= \frac{2^{mN}}{2^{(m+1)N+p}} \sum_{j_{N-p+1}, \ldots, j_N} \left\{ \left(\sum_{k=1}^{N-p} \eta_{p+k} 2^{N-k} \right) + \left(\sum_{k=N-p+1}^{N} j_k 2^{N-k} \right) + \frac{1}{2} \right\}. \end{aligned}$$

Substituting η_{p+k} for ξ_{p+k} in (20.67), we get

$$r_n^{(N)} = \sum_{k=1}^{N-p} \eta_{p+k} 2^{-k} + \frac{2^p}{2^{N+1}}.$$

Case (iv) $m = 3 \pmod 4$. From the above lemma, we have

$$\left|\left(A^m\right)_{\xi_{p+l} j_l}\right|^2 = 2^{m-1}, \qquad \left|\left(A^{m+1}\right)_{\xi_{k-(N-p)} j_k}\right|^2 = \begin{cases} 2^{m+1}, & \text{if } j_k = \xi_{k-(N-p)}, \\ 0, & \text{if } j_k \neq \xi_{k-(N-p)} \end{cases}$$

for any $l = 1, \ldots, N-p$ and $k = N-p+1, \ldots, N$. Using this formula, the product of absolute squares can be expressed as

$$\prod_{l=1}^{N-p} \left|\left(A^m\right)_{\xi_{p+l} j_l}\right|^2 \prod_{k=N-p+1}^{N} \left|\left(A^{m+1}\right)_{\xi_{k-(N-p)} j_k}\right|$$

$$= \begin{cases} (2^{m-1})^{N-p}(2^{m+1})^{p}, & \text{if } j = \xi_{k-(N-p)} \text{ for all } k = N-p+1, \ldots, N, \\ 0, & \text{otherwise,} \end{cases}$$

$$= \begin{cases} 2^{(m-1)N+2p}, & \text{if } j_k \neq \xi_{k-(N-p)} \text{ for all } k = N-p+1, \ldots, N, \\ 0, & \text{otherwise.} \end{cases}$$

Equation (20.65) can be rewritten as

$$r_n^{(N)} = \frac{1}{2^{(m+1)N+p}} \sum_{j_1,\ldots,j_N} \left\{ \left(\sum_{k=1}^{N-p} j_k 2^{N-k} \right) + \left(\sum_{k=N-p+1}^{N} j_k 2^{N-k} \right) + \frac{1}{2} \right\}$$

$$\times \prod_{l=1}^{N-p} \left| (A^m)_{\xi_{p+l} j_l} \right|^2 \prod_{k=N-p+1}^{N} \left| (A^{m+1})_{\xi_{k-(N-p)} j_k} \right|^2$$

$$= \frac{2^{(m-1)N+2p}}{2^{(m+1)N+p}} \sum_{j_{N-p+1},\ldots,j_N} \left\{ \left(\sum_{k=1}^{N-p} j_k 2^{N-k} \right) \right.$$

$$\left. + \left(\sum_{k=N-p+1}^{N} \xi_{k-(N-p)} 2^{N-k} \right) + \frac{1}{2} \right\}.$$

Substituting $\xi_{k-(N-p)}$ for $\eta_{k-(N-p)}$ in (20.67), we get

$$r_n^{(N)} = \sum_{k=N-p-1}^{N-p} \xi_{k-(N-p)} 2^{-k} + \frac{2^N - 2^p + 1}{2^{N+1}}.$$

□

Theorem 20.31 *For the case $n = mN$, $m \in \mathbb{N}$, we have*

$$r_N^{(n)} = \begin{cases} \sum_{k=1}^{N} \xi_k 2^{-k} + \frac{1}{2^{N+1}}, & \text{if } m = 0 \pmod 4, \\ \frac{1}{2}, & \text{if } m = 1, 3 \pmod 4, \\ \sum_{k=1}^{N} \eta_k 2^{-k} + \frac{1}{2^{N+1}}, & \text{if } m = 2 \pmod 4. \end{cases} \tag{20.68}$$

Proof For any $n = mN$, $m \in \mathbb{N}$,

$$r_n^{(N)} = \frac{1}{2^n} \sum_{j=0}^{2^N-1} \frac{j+1/2}{2^N} \prod_{k=1}^{N} \left| (A^m)_{\xi_k j_k} \right|^2.$$

By a direct calculation, we obtain

$$r_n^{(N)} = \frac{1}{2^n} \sum_{j=0}^{2^N-1} \frac{j+1/2}{2^N} \prod_{k=1}^{N} \left| (A^m)_{\xi_k j_k} \right|^2$$

$$= \frac{1}{2^{n+N}} \sum_{j=0}^{2^N-1} (j+1/2) \prod_{k=1}^{N} \left| \left(A^m\right)_{\xi_k j_k} \right|^2$$

$$= \frac{1}{2^{(m+1)N}} \sum_{j_1,\dots,j_N} \left\{ \left(\sum_{k=1}^{N} j_k 2^{N-k} \right) + \frac{1}{2} \right\} \prod_{k=1}^{N} \left| \left(A^m\right)_{\xi_k j_k} \right|^2.$$

Case (i) $m = 0 \pmod 4$. From the above lemma, we have

$$\left| \left(A^m\right)_{\xi_k j_k} \right|^2 = \begin{cases} 2^m, & \text{if } j_k = \xi_k, \\ 0, & \text{if } j_k \neq \xi_k \end{cases}$$

for any $k = 1, \dots, N$. Using this formula, the product of absolute squares can be expressed as

$$\prod_{l=1}^{N} \left| \left(A^m\right)_{\xi_k j_k} \right|^2 = \begin{cases} 2^{mN}, & \text{if } j_k = \xi_k \text{ for all } k = 1, \dots, N, \\ 0, & \text{otherwise.} \end{cases}$$

Using this formula, the mean value $r_n^{(N)}$ of the position operator can be expressed as

$$r_n^{(N)} = \frac{1}{2^{(m+1)N}} \sum_{j_1,\dots,j_N} \left\{ \left(\sum_{k=1}^{N} j_k 2^{N-k} \right) + \frac{1}{2} \right\} \prod_{k=1}^{N} \left| \left(A^m\right)_{\xi_k j_k} \right|^2$$

$$= \frac{2^{mN}}{2^{(m+1)N}} \left\{ \left(\sum_{k=1}^{N} \xi_k 2^{N-k} \right) + \frac{1}{2} \right\}$$

$$= \sum_{k=1}^{N} \xi_k 2^{-k} + \frac{1}{2^{N+1}}.$$

Case (ii) $m = 1, 3 \pmod 4$. From the above lemma, we have

$$\left| \left(A^m\right)_{\xi_k j_k} \right|^2 = 2^{m-1}$$

for any $k = 1, \dots, N$. Note that

$$\prod_{l=1}^{N} \left| \left(A^m\right)_{\xi_k j_k} \right|^2 = 2^{(m-1)N}.$$

Using this formula, the mean value $r_n^{(N)}$ of the position operator can be expressed as

$$\begin{aligned}
r_n^{(N)} &= \frac{1}{2^{(m+1)N}} \sum_{j_1,\ldots,j_N} \left\{ \left(\sum_{k=1}^{N} j_k 2^{N-k} \right) + \frac{1}{2} \right\} \prod_{k=1}^{N} \left| (A^m)_{\xi_k j_k} \right|^2 \\
&= \frac{2^{(m-1)N}}{2^{(m+1)N}} \sum_{j_1,\ldots,j_N} \left\{ \left(\sum_{k=1}^{N} j_k 2^{N-k} \right) + \frac{1}{2} \right\} \\
&= \frac{1}{2^{2N}} \left\{ \sum_{j_1,\ldots,j_N} \left(\sum_{k=1}^{N} j_k 2^{N-k} \right) + \frac{1}{2} \left(\sum_{j_1,\ldots,j_N} 1 \right) \right\} \\
&= \frac{1}{2^{2N}} \sum_{k=0}^{2^N-1} k + \frac{1}{2^{N+1}} = \frac{1}{2}.
\end{aligned}$$

Case (iii) $m = 2 \pmod 4$. From the above lemma, we have

$$\left| (A^m)_{\xi_k j_k} \right|^2 = \begin{cases} 2^m, & \text{if } j_k \neq \xi_k, \\ 0, & \text{if } j_k = \xi_k \end{cases}$$

for any $k = 1, \ldots, N$. Using this formula, the product of absolute squares can be expressed as

$$\prod_{l=1}^{N} \left| (A^m)_{\xi_k j_k} \right|^2 = \begin{cases} 2^{mN}, & \text{if } j_k \neq \xi_k \text{ for all } k = 1, \ldots, N, \\ 0, & \text{otherwise.} \end{cases}$$

Let $\eta_k = \xi_k + 1 \pmod 2$, $k = 1, \ldots, N$. It follows that

$$\begin{aligned}
r_n^{(N)} &= \frac{1}{2^{(m+1)N}} \sum_{j_1,\ldots,j_N} \left\{ \left(\sum_{k=1}^{N} j_k 2^{N-k} \right) + \frac{1}{2} \right\} \prod_{k=1}^{N} \left| (A^m)_{\xi_k j_k} \right|^2 \\
&= \frac{2^{mN}}{2^{(m+1)N}} \left\{ \left(\sum_{k=1}^{N} \eta_k 2^{N-k} \right) + \frac{1}{2} \right\} = \sum_{k=1}^{N} \eta_k 2^{-k} + \frac{1}{2^{N+1}}.
\end{aligned}$$

□

Using Formulas (20.61), (20.64), and (20.68), the probability distribution $(p_i^{(n)})$ of the orbit of the mean value $r_n^{(N)}$ of the position operator $\hat{q}$ for the time evolution, which is constructed by the quantum Baker's map, is given by

$$p_i^{(n)} \equiv \frac{1}{m+1} \sum_{k=n}^{m+n} 1_{B_i}\left(r_n^{(N)}\right)$$

for an initial value $r_0^{(N)} \in [0,1]$ and the characteristic function 1_A. The joint distribution $(p_{ij}^{(n,n+1)})$ between the time n and $n+1$ is given by

$$p_{ij}^{(n,n+1)} \equiv \frac{1}{m+1} \sum_{k=n}^{m+n} 1_{B_i}\left(r_k^{(N)}\right) 1_{B_j}\left(r_{k+1}^{(N)}\right).$$

Thus the chaos degree for the quantum Baker's map is calculated by

$$D_q\left(p^{(n)}; \Lambda_n^*\right) = \sum_{i,j} p_{ij}^{(n,n+1)} \log \frac{p_i^{(n)}}{p_{ij}^{(n,n+1)}}, \tag{20.69}$$

whose numerical value is shown in the next section.

20.7.4 Numerical Simulation of the Chaos Degree and Classical–Quantum Correspondence

We compare the dynamics of the mean value $r_n^{(N)}$ of the position operator $\hat{q}$ with that of the classical value q_n in the q-direction. We take an initial value of the mean value as

$$r_0^{(N)} = \sum_{l=1}^{N} \xi_l 2^{-l} + 1/2^{N+1} = 0.\xi_1 \xi_2 \cdots \xi_N 1,$$

where ξ_i is a pseudo-random number taking values 0 or 1. At the time $n = 0$, we assume that the classical value q_0 in the q-direction takes the same value as the mean value $r_0^{(N)}$ of position operator $\hat{q}$. The distribution of $r_n^{(N)}$ for the case $N = 500$ is shown in Fig. 20.4 up to the time $n = 1000$. The distribution of the classical value q_n for the case $N = 500$ in the q-direction is shown in Fig. 20.5 up to the time $n = 1000$.

Figure 20.6 presents the change of the chaos degree for the cases $N = 100, 300, 500, 700$ up to the time $n = 1000$.

The correspondence between the chaos degree D_q for the quantum Baker's map and the chaos degree D_c for the classical Baker's map for some fixed Ns (100, 300, 500, 700 here) is shown for the time less than $T = \log_2 \frac{1}{h} = \log_2 2^N = N$, and it is lost at the logarithmic time scale T. Here we took a finite partition $\{B_k\}$ of $I = [0,1]$ such as $B_k = [\frac{k}{100}, \frac{k+1}{100})$ $(k = 0, 1, \ldots, 98)$ and $B_{99} = [\frac{99}{100}, 1]$ to compute the chaos degree numerically.

The difference of the chaos degrees between the chaos degree D_q for the quantum Baker's map and the chaos degree D_c for the classical Baker's map for a fixed time n (1000, here) is displayed w.r.t. N in Fig. 20.7.

Thus we conclude that the dynamics of the mean value q_n reduces the classical dynamics q_n in the q-direction in the classical limit $N \to \infty$ $(h \to 0)$.

The appearance of the logarithmic time scale has been proved rigorously in [364], see the next subsection.

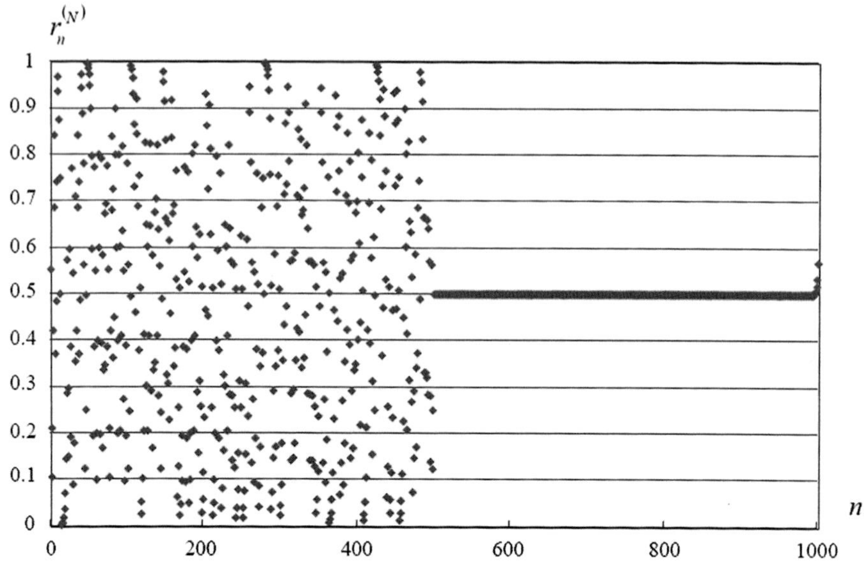

Fig. 20.4 The distribution of $r_N^{(n)}$ for the case $N = 500$

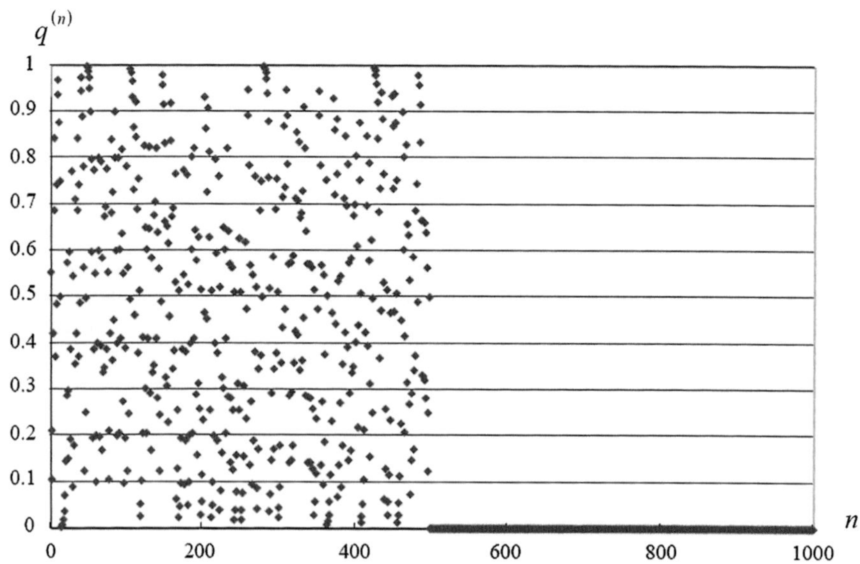

Fig. 20.5 The distribution of the classical value $q^{(n)}$ for the case $N = 500$

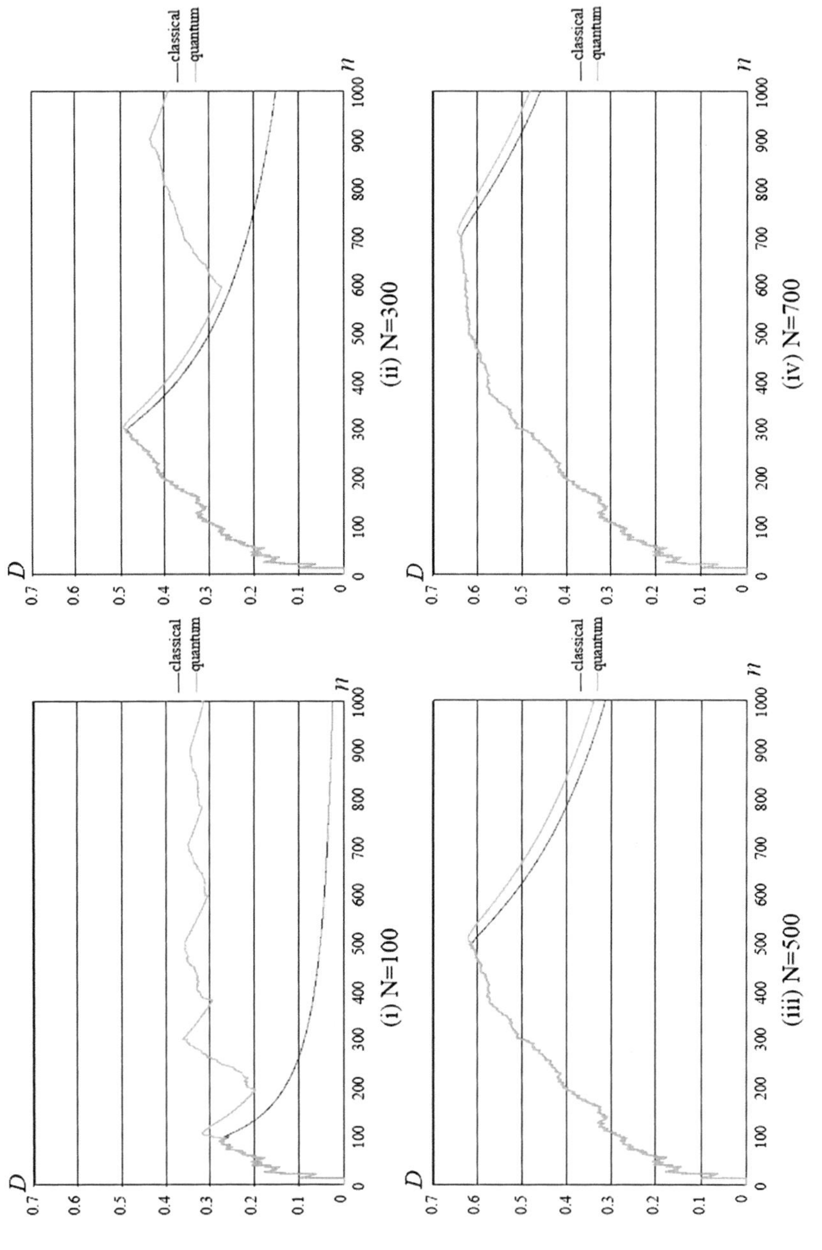

Fig. 20.6 The change of the chaos degree for several Ns up to time $n = 1000$

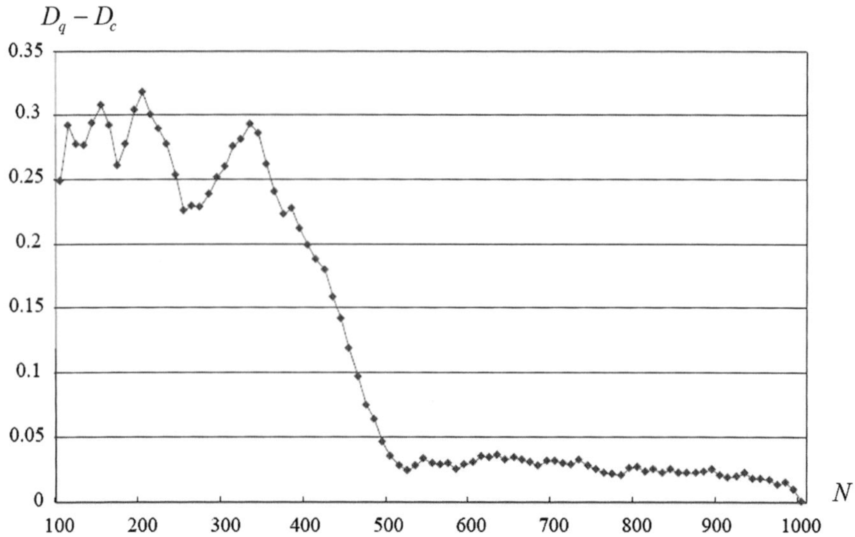

Fig. 20.7 The difference of the chaos degree between quantum and classical for the case $n = 1000$

20.7.5 Quantum–Classical Correspondence for Quantum Baker's Map

The quantum–classical correspondence for dynamical systems has been studied for many years; see, for example, [289, 838] and reference therein. Significant progress in understanding this correspondence has been achieved in a mathematical approach when one considers the Planck constant $\hbar$ as a small variable parameter. It is well known that in the limit $\hbar \to 0$ the quantum theory is reduced to the classical one [327, 509].

However, in physics the Planck constant is a fixed constant, although it is very small. Therefore, it is important to study the relation between classical and quantum evolutions when the Planck constant is fixed. There is a conjecture [113, 115, 636] that a characteristic time scale $t_\hbar$ appears in the quantum evolution of chaotic dynamical systems. For a time less then $t_\hbar$, there is a correspondence between quantum and classical expectation values, while for times greater than $t_\hbar$ the predictions of the classical and quantum dynamics no longer coincide.

An important problem is to estimate the dependence $t_\hbar$ on the Planck constant $\hbar$. Probably a universal formula expressing $t_\hbar$ in terms of $\hbar$ does not exist, and every model should be studied case by case. It is expected that certain quantum and classical expectation values diverge on a time scale inversely proportional to some power of $\hbar$. Other authors suggest that for chaotic systems a breakdown may be anticipated on a much smaller logarithmic time scale (see [400, 838] for a discussion). Numerous works are devoted to the analytical and numerical study of classical and quantum chaotic systems [52, 59, 68, 78, 116, 150, 204, 225, 311, 359, 360, 365, 398, 508, 530, 590, 625, 627, 675, 683, 723, 729, 832].

Most results concerning various time scales are obtained numerically. In this section, we will present some exact results on a quantum chaos model. We compute explicitly the expectation value for the quantum Baker's map and prove rigorously the appearance of the logarithmic time scale.

The quantum Baker's map is a model invented to study the chaotic behavior [78]. The model has been studied in [680, 682].

In this section, quantum dynamics of the position operator for the quantum Baker's map is considered. We use a simple symbolic description of the quantum Baker's map proposed by Schack and Caves [681]. We find an exact expression for the expectation value of the time dependent position operator. In this sense, the quantum Baker's map is an exactly solvable model though a stochastic one. A relation between quantum and classical trajectories is investigated. For some matrix elements the breakdown of the quantum–classical correspondence at the logarithmic time scale is established.

Here we would like to note that, in fact, the notion of the time scale is not uniquely defined. Actually, we will obtain the formula

$$\left|\langle\hat{q}_m\rangle - q_m\right| \le \hbar 2^{m-1}$$

where $\hat{q}_m$ and q_m are respectively the quantum and classical positions at time m. This formula will be interpreted as the derivation of the logarithmic time scale. The result of this section is presented in Proposition 20.33.

We consider the following mean value of the position operator $\hat{q}$ for time $m = 0, 1, \ldots,$ with respect to a vector $|\xi\rangle$:

$$r_m^{(N)} = \langle\xi|T^m \hat{q} T^{-m}|\xi\rangle, \tag{20.70}$$

where $|\xi\rangle = |\xi_1\xi_2\cdots\xi_N\rangle$. First, we show that there is an explicit formula for the expectation value $r_m^{(N)}$. In this sense, the quantum Baker's map is an explicitly solvable model. Then we compare the dynamics of the mean value $r_m^{(N)}$ of position operator $\hat{q}$ with that of the classical value q_m, (20.55). We will establish a logarithmic time scale for the breakdown of the quantum–classical correspondence for the quantum Baker's map.

From (20.57), one gets for $m = 0, 1, \ldots, N-1$

$$\langle\xi|T^m|\eta\rangle = \left(\frac{1-\mathrm{i}}{2}\right)^m \left(\prod_{k=1}^{N-m} \delta(\xi_{m+k} - \eta_k)\right)\left(\prod_{l=1}^{m} \exp\left(\frac{\pi}{2}\mathrm{i}|\xi_l - \eta_{N-m+l}|\right)\right), \tag{20.71}$$

and for $m = N$

$$\langle\xi|T^N|\eta\rangle = \left(\frac{1-\mathrm{i}}{2}\right)^N \left(\prod_{l=1}^{N} \exp\left(\frac{\pi}{2}\mathrm{i}|\xi_l - \eta_l|\right)\right). \tag{20.72}$$

Using this formula and Theorem 20.28, we have

$$r_m^{(N)} = \langle \xi | T^m \hat{q} T^{-m} | \xi \rangle = \sum_{k=1}^{N-m} \frac{\xi_{m+k}}{2^k} + \frac{1}{2^{N-m+1}} \quad (0 \le m < N) \tag{20.73}$$

and

$$r_N^{(N)} = \frac{1}{2}. \tag{20.74}$$

We consider here the quantum–classical correspondence for the quantum Baker's map. First, let us mention that $2^N = 1/\hbar$ and the limit $\hbar \to 0$ corresponds to the limit $N \to \infty$. Therefore, from Proposition 20.33 and (20.57), one has the mathematical correspondence between quantum and classical trajectories as $\hbar \to 0$:

$$\lim_{N\to\infty} r_m^{(N)} = q_m, \quad m = 0, 1, \ldots.$$

Now let us fix the Planck constant $\hbar = 2^{-N}$ and investigate on which time scale the quantum and classical expectation values start to differ from each other. From Proposition 20.33 and (20.57), we obtain the following.

Proposition 20.32 *Let $r_m^{(N)}$ be the mean value of position operator $\hat{q}$ at the time m and q_m is the classical trajectory* (20.57). *Then we have*

$$q_m - r_m^{(N)} = \sum_{j=N-m+1}^{\infty} \xi_{m+j} 2^{-j} - \frac{1}{2^{N-m+1}} \tag{20.75}$$

for any $0 \le m \le N$.

Let us estimate the difference between the quantum and classical trajectories.

Proposition 20.33 *Let q_m and $r_m^{(N)}$ be the same as in the above proposition. Then we have*

$$\left| r_m^{(N)} - q_m \right| \le \frac{1}{2^{N-m+1}} \tag{20.76}$$

for any string $\xi = \xi_1 \xi_2 \cdots$ and any time $0 \le m \le N$.

Proof Note that

$$\begin{aligned} 0 &\le \sum_{j=N-m+1}^{\infty} \xi_{m+j} 2^{-j} \\ &\le \frac{1}{2^{N-m+1}} \left(1 + \frac{1}{2} + \left(\frac{1}{2}\right)^2 + \cdots \right) = \frac{1}{2^{N-m}}. \end{aligned}$$

Using the above inequality, one gets from (20.75)

$$-\frac{1}{2^{N-m+1}} \le q_m - r_m^{(N)} \le \frac{1}{2^{N-m}} - \frac{1}{2^{N-m+1}} = \frac{1}{2^{N-m+1}}.$$

This means that we have

$$\left| r_m^{(N)} - q_m \right| \le \frac{1}{2^{N-m+1}}$$

for any $0 \le m \le N$. □

This proposition shows an exact correspondence between quantum and classical expectation value for the Baker's map. We can write the relation (20.76) in the form

$$\left| r_m^{(N)} - q_m \right| \le \frac{1}{2^{N-m+1}} = \hbar 2^{m-1} \tag{20.77}$$

since the Planck constant $\hbar = 2^{-N}$. In particular, for $m = 0$ we have

$$\left| r_0^{(N)} - q_0 \right| \le \frac{\hbar}{2} \tag{20.78}$$

for any $\xi = \xi_1 \xi_2 \cdots$.

Now let us estimate at what time $m = t_\hbar$ an essential difference between the classical trajectory and quantum expectation value appears. From (20.77), we can expect that the time $m = t_\hbar$ corresponds to the maximum of the function $2^m/2^{N-1}$ for $0 \le m \le N$, i.e.,

$$t_\hbar = N = \log_2 \frac{1}{\hbar}. \tag{20.79}$$

For time $0 \le m < t_\hbar$, the difference between classical and quantum trajectories in (20.77) is bounded by $1/4$ since

$$\hbar 2^{m-1} = \frac{1}{2^{N-m+1}} \le \frac{1}{4}.$$

One can see that the bound is saturated. Indeed, let us take a string ξ with arbitrary $\xi_1, \dots, \xi_N$ but with $\xi_{N+1} = 0, \xi_{N+2} = 0, \dots$. Then one has

$$r_m^{(N)} - q_m = \hbar 2^{m-1}, \quad m = 0, 1, \dots, N.$$

Therefore, we have established the logarithmic dependence of the time scale on the Planck constant $\hbar$.

Here we have computed the expectation values for the position operator in the quantum Baker's map. Breakdown of the quantum–classical correspondence at the logarithmic time scale is rigorously established. For better understanding of the quantum–classical correspondence and the decoherence process, it is important to perform similar computations for more general matrix elements which include also the momentum operators and coherent vectors. Some further properties of the quantum Baker's map and the entropic chaos degree are considered in the next section.

20.8 Combined Quantum Baker's Map

Chaos shows complicated and difficult to predict behavior since it has a property of exponential sensibility to initial conditions. The property means that the divergence of infinitesimally nearby initial conditions grows exponentially. Dynamics showing such a chaotic behavior is called a chaotic classical dynamics. A quantum–classical correspondence for a chaotic dynamics has been studied in the field of study of quantum chaos.

A chaotic classical dynamics is often studied by computing the Lyapunov exponents of the dynamics. A chaos degree has been introduced to measure degree of chaos in dynamics in the framework of Information Dynamics in Chap. 10.

In this section, an expectation value for a combined quantum Baker's map is given as

$$(1-a)\langle\hat{q}_m\rangle + \frac{a}{2}\big(\langle\hat{q}_{m-1}\rangle + \langle\hat{q}_{m+1}\rangle\big)$$

for a satisfying $0 \le a \le 1$ where $\hat{q}_m$ is the position operator quantizing a classical value q_m at the time m in the q-direction. Semiclassical properties for the combined quantum Baker's map are considered by the entropic chaos degree.

20.8.1 A Combined Classical Baker's Map and Its Entropic Chaos Degree

In this subsection, we combine several dynamics of position operators which are produced by the quantum Baker's maps. Chaos of the combined dynamics is studied by the entropic chaos degree.

We consider the following combination of classical orbit,

$$(c_q)_m = (1-a)q_m + \frac{a}{2}(q_{m-1} + q_{m+1}), \quad a \in [0,1] \tag{20.80}$$

as a combined classical Baker's transformation. In order to compute the entropic chaos degree D_c of the combined classical Baker's transformation, we divide the interval $[0,1]$ into a finite equi-partition $\{B_i\}$. The following Figs. 20.8, 20.9 and 20.10 show the entropic chaos degree D_c of the combined classical Baker's transformation versus a for equi-partitions having 100 to 300 intervals, respectively.

Zooming-in into these figures, one finds that $D_c(a)$ has an almost self-similar structure on a concave curve for the equi-partition with 100 intervals.

The next Fig. 20.11 shows the number of the half-sized concave curve, i.e., fundamental figure in the fractal theory. For the equi-partition with 100 intervals, this number is one, that for the 200 interval equi-partition is two, and then that for the 300 interval equi-partition is three.

The above property indicates the fractal structure of Baker's map. Moreover, we can roughly estimate chaotic behavior of the combined classical Baker's transformation by the entropic chaos degree set even only for the case of 100 interval equi-partition.

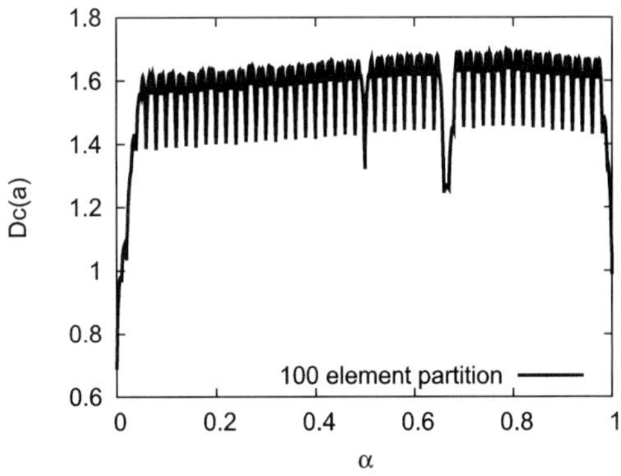

Fig. 20.8 $D_c(a)$ versus a for the equi-partition with 100 intervals

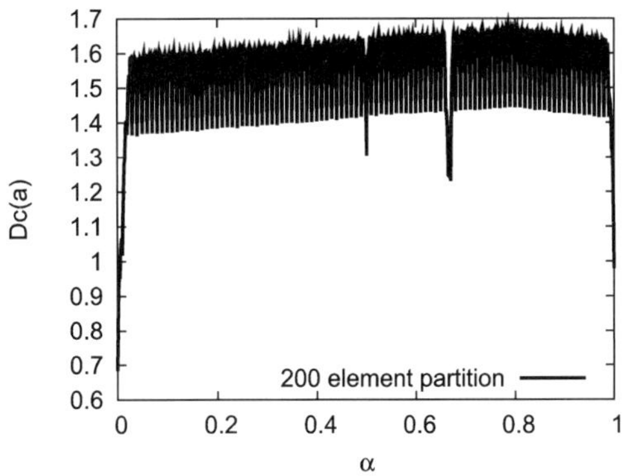

Fig. 20.9 $D_c(a)$ versus a for the equi-partition having 200 intervals

20.8.2 A Combined Quantum Baker's Map and Its Entropic Chaos Degree

We consider an expectation value of a combined position operator,

$$\begin{aligned}(c_r)_m^{(N)} &= \langle \xi | (1-a)T^m \hat{q} T^{-m} + \frac{a}{2}\big(T^{(m-1)}\hat{q}T^{-(m-1)} + T^{(m+1)}\hat{q}T^{-(m+1)}\big)|\xi\rangle \\ &= (1-a)r_m^{(N)} + \frac{a}{2}\big(r_{m-1}^{(N)} + r_{m+1}^{(N)}\big), \quad a \in [0,1] \end{aligned} \tag{20.81}$$

as an orbit value of a combined quantum Baker's map. The expectation value $(c_r)_m^{(N)}$ gets the classical value $(c_q)_m$ at the classical limit as $N \to \infty$.

Then the probability distribution $(p_i^{(m)})$ at the time m and the joint probability distribution $(p_{ij}^{(m,m+1)})$ between the time m and $m+1$ are given as

$$p_i^{(m)} \equiv \frac{1}{M+1} \sum_{k=m}^{M+m} 1_{B_i}\big((c_r)_k^{(N)}\big),$$

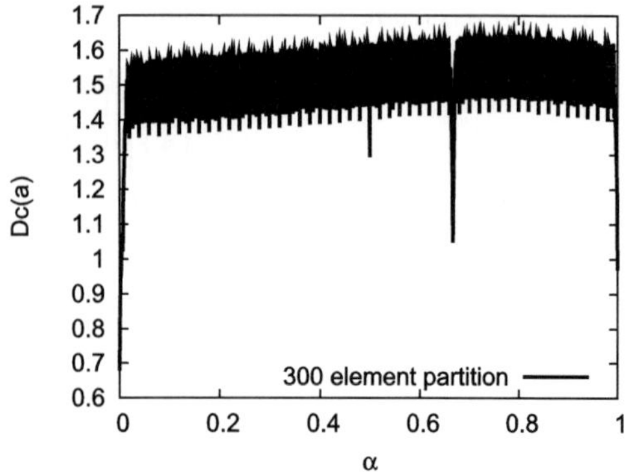

Fig. 20.10 $D_c(a)$ versus a for the equi-partition having 300 intervals

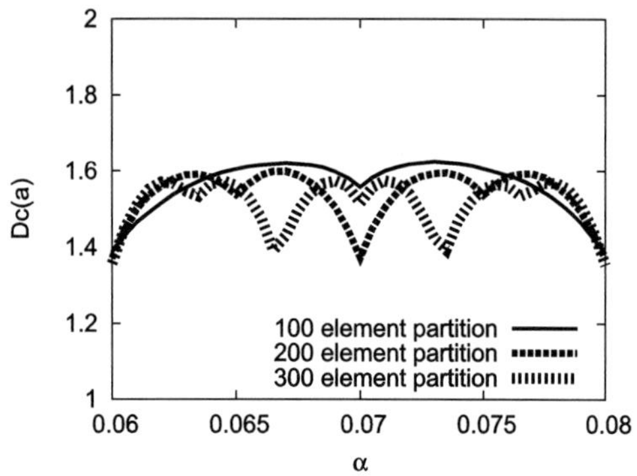

Fig. 20.11 $D_c(a)$ versus a on $[0.06, 0.08]$

$$p_{ij}^{(m,m+1)} \equiv \frac{1}{M+1} \sum_{k=m}^{M+m} 1_{B_i}\left((c_r)_k^{(N)}\right) 1_{B_j}\left((c_r)_{k+1}^{(N)}\right)$$

for an initial value $r_0^{(N)} \in [0, 1]$, respectively. Then the entropic chaos degree of the combined quantum Baker's map is computed as

$$D_q\left(p^{(m)}; \Lambda_m^*\right) = \sum_{i,j} p_{ij}^{(m,m+1)} \log \frac{p_i^{(m)}}{p_{ij}^{(m,m+1)}}.$$

In the sequel, a quantum and classical initial values $r_0^{(N)}$ and q_0 of the Baker's map are set as

$$r_0^{(N)} = \sum_{l=1}^{N} \xi_l 2^{-l} + 1/2^{N+1} = 0.\xi_1\xi_2 \cdots \xi_N 1,$$

$$q_0 = \sum_{l=1}^{\infty} \xi_l 2^{-l} = 0.\xi_1\xi_2 \cdots$$

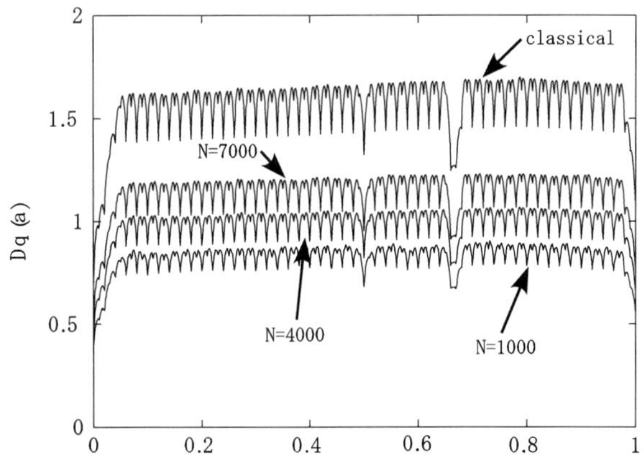

Fig. 20.12 $D_q(a)$ versus a on $[0, 1]$

where ξ_i is a pseudo-number 0 or 1, respectively. Then the quantum and classical initial values $(c_r)_1^{(N)}$ and $(c_q)_1$ of the combined Baker's map become

$$(c_r)_1^{(N)} = (1-a)r_1^{(N)} + \frac{a}{2}\left(r_0^{(N)} + r_2^{(N)}\right),$$

$$(c_q)_1 = (1-a)q_1 + \frac{a}{2}(q_0 + q_2).$$

According to the classical setting, we divide the interval $[0, 1]$ into 100 interval equi-partition $\{B_i\}$, and we set that $m = 1$ and the number M of orbit points $(c_r)_k$ is 10000. Then we can write the entropic chaos degree D_q as $D_q(N, a)$ since D_q depends on two parameters N and a.

20.8.3 Dependence of the Entropic Chaos Degree on the Combination Parameter a

We study the dependence of the entropic chaos degree on the combination parameter a for some fixed numbers of N. Then the entropic chaos degree is denoted by a one-parameter function $D_q(a)$ of a. The following Fig. 20.12 shows the entropic chaos degree $D_q(a)$ versus a on $[0, 1]$ for some fixed numbers of N.

To study the dependence of the entropic chaos degree on a for further details, we divide the domain $[0, 1]$ of a into five intervals. The following figures show that, in principle, the entropic chaos degree $D_q(a)$ firstly oscillates and increases (Fig. 20.13), secondly becomes periodic (Figs. 20.13, 20.14, 20.15, 20.16 and 20.17), and finally oscillates and decreases (Fig. 20.17). However, there exist several domains where $D_q(a)$ rapidly gets small. One is the small interval around the point $a = 0.5$ at which it holds $a : 1 - a = 1 : 1$ (Fig. 20.15). The other is the interval $[0.66, 0.67]$ which is an interval around the point such that $a : 1 - a = 2 : 1$ (Fig. 20.16).

Fig. 20.13 $D_q(a)$ versus a on $[0, 0.2)$

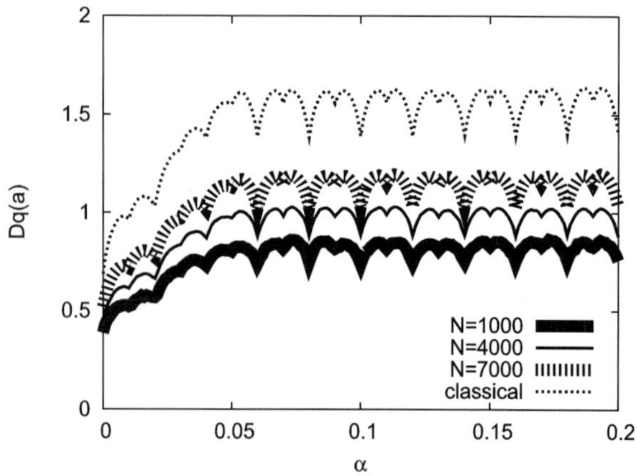

Fig. 20.14 $D_q(a)$ versus a on $[0.2, 0.4)$

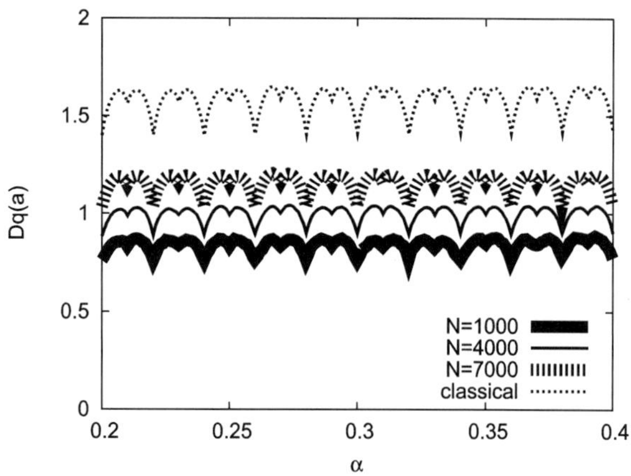

Fig. 20.15 $D_q(a)$ versus a on $[0.4, 0.6)$

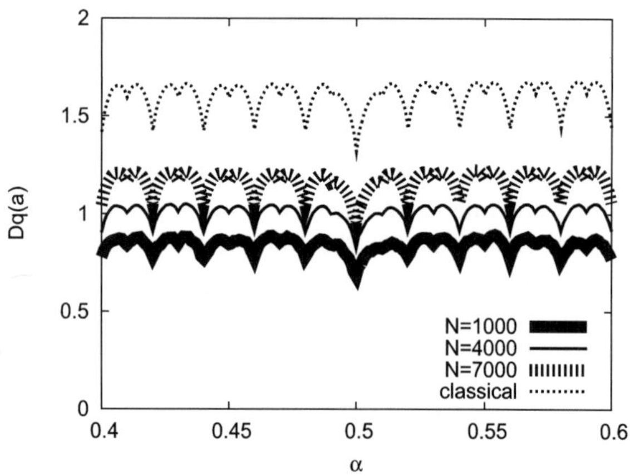

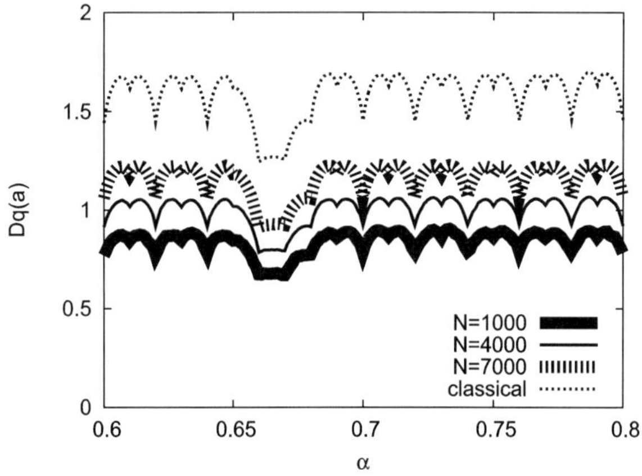

Fig. 20.16 $D_q(a)$ versus a on [0.6, 0.8)

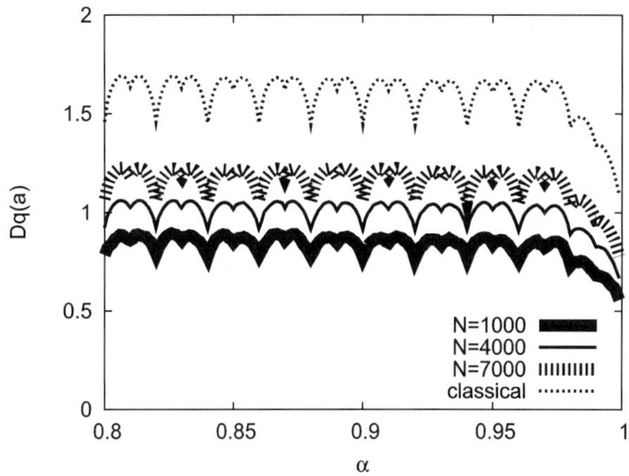

Fig. 20.17 $D_q(a)$ versus a on [0.8, 1.0)

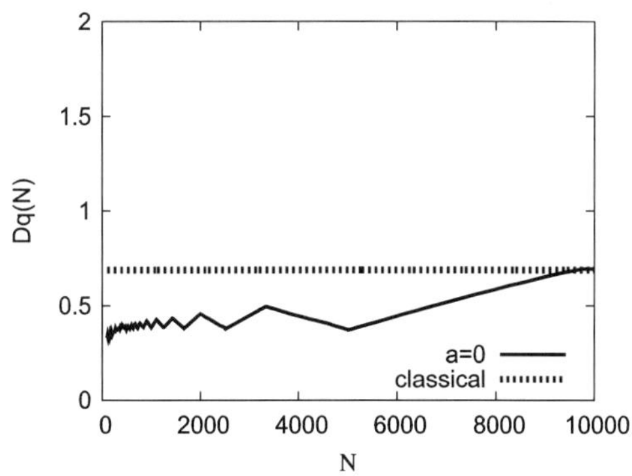

Fig. 20.18 $D_q(N)$ versus N for $a = 0$

One finds that the entropic chaos degree for the combined quantum Baker's map takes a smaller value than that of the entropic chaos degree for the combined classical Baker's transformation.

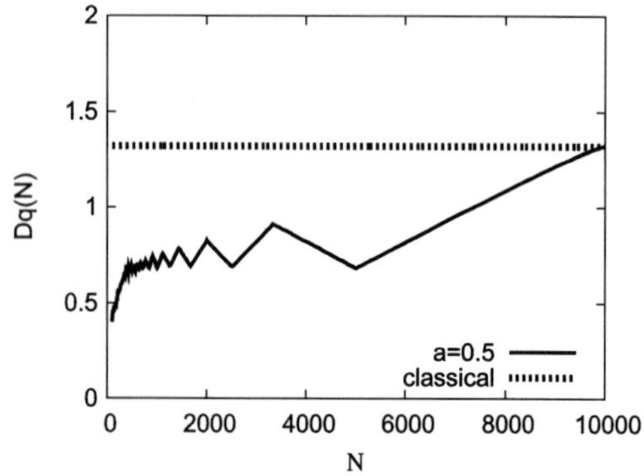

Fig. 20.19 $D_q(N)$ versus N for $a = 0.5$

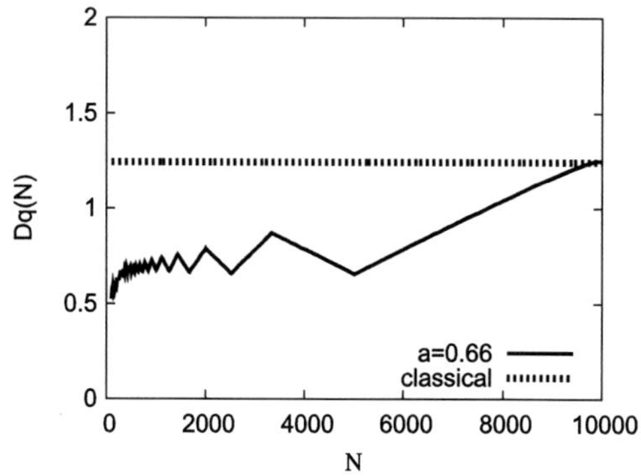

Fig. 20.20 $D_q(N)$ versus N for $a = 0.66$

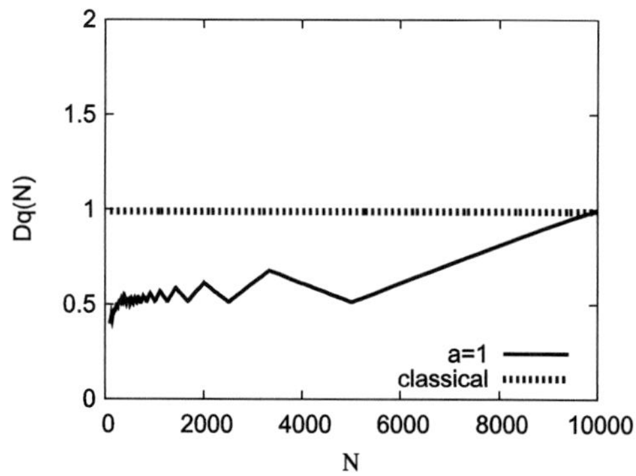

Fig. 20.21 $D_q(N)$ versus N for $a = 1$

20.8.4 Dependence of the Entropic Chaos Degree on N

According to (20.81), one finds that there is some correspondence between the quantum expectation value $r_k^{(N)}$ and the classical value q_k for Baker's map for $0 \le k \le N$, while there is no correspondence between their values for $N + 1 \le k \le M$ where M is the number of orbit points. Therefore if we assume N such that $N = M$, then the quantum entropic chaos degree D_q should take the same value as that of the classical entropic chaos degree D_c. In the sequel, we study the dependence of the chaos degree on N for some fixed numbers of a. Then the chaos degree is denoted by a one-parameter function $D_q(N)$ of N. Figures 20.18, 20.19, 20.20, 20.21 show that $D_q(N)$ versus N for some fixed a's.

We have chosen a from several domains where the entropic chaos degree shows some typical behavior. The results as shown in the figures above indicate difference between classical case and quantum case, which are not dependent on the value of a.

One finds that the entropic chaos degree $D_q(N)$ oscillates and increases for $1 \leq N < 5000$, while $D_q(N)$ monotonously increases for $5000 \leq N \leq M$ converging to the entropic chaos degree for the classical combined Baker's transformation.

Chaos of the combined dynamics for the quantum Baker's map was studied by the entropic chaos degree. We first studied the dependence of the entropic chaos degree on the combination parameter a for some fixed numbers of N where the Planck constant $\hbar$ was set to satisfy $2\pi\hbar = 2^{-N}$. The entropic chaos degree for the combined quantum Baker's map shows some typical behavior such as an oscillating and increasing motion, a periodic motion, an oscillating and decreasing motion, and so on, which just appear in the combined classical Baker's transformation. However, the entropic chaos degree for the combined quantum Baker's map takes a smaller value than that of the entropic chaos degree for the combined classical Baker's transformation. This means that the entropic chaos degree is reduced by quantizing the classical Baker's transformation. Next we studied the dependence of the chaos degree on N for some fixed numbers of a. Then we chose a from several domains where the entropic chaos degree showed some typical behavior. One finds that after the entropic chaos degree for the quantum combined Baker's map oscillates and increases, it monotonously increases and converges to the entropic chaos degree for the classical combined Baker's transformation at $N = M$.

20.9 Notes

For the definition of the conditional expectation in the sense of Umegaki, we refer to [761], and Tomiyama's theorem is written in [748]. The concept of a quantum Markov chain was introduced by Accardi [3]. In the theory of classical dynamical systems, Boltzmann entropy was introduced by Boltzmann [127].

For a more recent discussion of the Boltzmann and Gibbs entropy and the irreversibility problem, see [70, 121, 178, 193, 239, 295, 447, 449, 465, 469–471, 473, 616, 643, 667, 731]. Irriversible processes from the point of view of quantum information theory are considered by Ohya in [570]. The role of various entropies in the black hole information paradox is considered in [548] where a derivation of the kinetic equations in quantum gravity is proposed. The Gorini–Kossakowsky–Sudarshan–Lindblad master equation is considered in [300, 483]. Violation of the entropy increase for classical gases is considered in [386] and for quantum master equation in [376, 430].

Kolmogorov–Sinai entropy and its role in the homeomorphism problem related to symbolic dynamics is stated in Billingsley [118]. As for the dynamical entropy of operator algebras, which is called Connes–Narnhofer–Thirring entropy, we refer to [176], and CNT type convergence theorem with respect to Kolmogorov–Sinai type complexity is proved in Muraki and Ohya's paper [536]. Moreover, quantum theoretic generalization of Kolmogorov–Sinai type entropy is presented in Accardi, Ohya and Watanabe's paper [14]. For the inequality showing that the transmitted complexity of mutual entropy of PPM is greater than that of PAM, we refer to Ohya and Watanabe's paper [566]. The foundation of fractal geometry has been laid by Mandelbrot [506]. Kolmogorov has introduced the concept of ε-entropy of random variables in [436], and Kullback–Leibler entropy, which is used in the definition of ε-entropy of random variables in this book, plays an important role in the theory of statistical sufficiency [455]. Moreover, quantum theoretic generalization of Kolmogorov's ε-entropy has been given by Ohya [572, 582]. For some numerical examples illustrating the values of quantum states and their applications to Ising models, we refer to Matsuoka and Ohya's paper [511]. An approach to the problem reducing decoherence in quantum memory is stated in Volovich's paper [786] and the method to make the decoherence drastically reduced, which is based on stochastic limits, is given in Accardi, Kozyrev and Volovich's paper [18]. The decoherence problem in quantum computers is discussed in [628, 769]. Especially, the spin-flip transition for the dynamical suppression of decoherence is treated in [780]. For a recent consideration of quantum control see [638]. As for the white noise calculus, we refer to the fundamental book written by Hida, Kuo, Potthoff, and Streit [436]. As for stochastic approximation theoretic aspects of the dynamics of the Hamiltonian, we refer to Accardi, Kozyrev and Volovich's paper [18]. Leggett, Chakravarty, Garg and Chang have given the behavior of the Hamiltonian such as undamped oscillation and exponential relaxation [475]. Many results related to quantum field theory represented in quantum probability scheme are given in Accardi, Lu and Volovich [34].

Quantum dynamical entropy was first introduced by Emch [223] and Connes-Störmer [175] around 1975. In 1987, Connes, Narnhoffer and Thirring [176] defined a dynamical entropy (CNT entropy) in C^*-dynamical systems. Park computed the CNT entropy for several models [629]. In 1994, Alicki and Fannes [8] defined a quantum dynamical entropy (AF entropy) by means of a finite operational partition of unity. In 1994, Hudetz [352] discussed the dynamical entropy in terms of topological entropy. In 1995, Ohya [582] introduced a quantum dynamical entropy and a quantum dynamical mutual entropy based on the C^*-mixing entropy [570] and the complexity in Information Dynamics [359]. In 1995, Voiculescu [784] introduced a dynamical approximation entropy for C^*- and W^*-algebra automorphisms based on a general approximation approach [332]. In 1997, Accardi, Ohya and Watanabe [17] defined a quantum dynamical entropy (AOW entropy) through quantum Markov chain. Ojima considered the entropy production in [609, 612]. The relations among some definitions of the dynamical entropy were discussed in [17, 97, 441]. Some computations of the dynamical entropy are done in several papers such as [161, 578]. Semiclassical properties of quantum Baker's map are considered by Inoue, Ohya and Volovich in [364–366]. The discrete time dynamics and energy levels in classical theory are considered in [418, 421].

Chapter 21
Applications to Life Science

In this chapter, we discuss some applications of mathematics in the quantum theory and information science to the biosystems and life sciences. In particular, the following topics are discussed: characteristics of biological systems, the sequence alignment by quantum algorithm in molecular biology, quantum-like models for brain and cognitive psychology, applications of information dynamics to study of the HIV and the influenza A viruses, code structure of HIV-1, p-adic modeling of genome and genetic code, mathematical model of drug delivery system, the folding problem and molecular dynamics, quantum photosynthesis.

There is a long tradition of applications of physical and mathematical methods to various biological systems. Up to now, most of the presently known mathematical models have been inspired by physics problems. In the recent years, the fascinating challenge of creating a similar fruitful interaction between mathematics and biology is gaining the attention of more and more scientists.

These models will surely play an important role, but it would be naive to believe that interesting biological models may be built simply by cleverly combining known physics and mathematics. We now understand that the laws of nature are distributed over several levels of scales and of complexity and that each of these levels introduces new qualitative features which, in their turn, require new types of models and substantial new ideas.

We are not going to discuss here the numerous results in biophysics, biochemistry, and mathematical biology. We only consider in this chapter some ideas and results for possible applications of quantum theory and quantum information to certain biological and life science phenomena which emerged mainly in our own trials on the Quantum Bio-Informatics in 2005–2010. We give also an overview of some ideas and some recent experimental results on applications of quantum mechanics to life sciences.

M. Ohya, I. Volovich, *Mathematical Foundations of Quantum Information and Computation and Its Applications to Nano- and Bio-systems*, Theoretical and Mathematical Physics,
DOI 10.1007/978-94-007-0171-7_21,

21.1 Introduction

The main topics in this chapter will be: a discussion of possible non-trivial applications of quantum theory in life sciences; the quantum algorithm of multiple alignment; the quantum brain; a study of the HIV and influenza A viruses in the domain of Information Dynamics; p-adic modeling of the genome and the genetic code; drug delivery systems; the quantum algorithm for molecular dynamics; quantum photosynthesis; quantum-like models for cognitive psychology.

21.1.1 Quantum Theory and Molecular Biology

Chemistry and molecular biology are based ultimately on quantum mechanics since the quantum laws govern the structure and dynamics of a molecule. However, there is an important question of which specific quantum phenomena can be seen in biology. There are several such phenomena, for instance, we shall discuss recent remarkable experimental discoveries of quantum effects in photosynthesis.

The founders of quantum theory Niels Bohr and Erwin Schrödinger introduced an important trend in biology and life sciences. Bohr extended the principle of complementarity to the question of life versus mechanism in 1932 in a lecture entitled "Light and Life." He pointed out that an exhaustive investigation of the basic units of life was impossible because those life units would most likely be destroyed by the particles needed for their observation. According to Bohr, the units of life represented irreducible entities similar to the quantum of energy, the "essential non-analyzability of atomic stability in mechanical terms presents a close analogy to the impossibility of a physical or chemical explanation of the peculiar functions characteristic of life."

The principle of complementarity recognized that since the observer and the observed represented a continuous interaction, the rigid line of separation between the subjective and the objective needed some modification. This meant a radical modification of the physicist's concept of the external world. The enunciation of the principle of complementarity by Bohr produced a block for a mechanistic or reductionist explanation of reality as it is conceived and experienced by a man.

Schrödinger, in his famous book "What is life?", suggested that an aperiodic crystal could be the "storing device" for biological information. This and also Delbruck's influence stimulated the search leading to the discovery of the structure of the DNA molecule by Watson and Crick.

After the discovery of the structure of DNA, George Gamow attempted to solve the problem of how the order of the four different kinds of bases (adenine, cytosine, thymine, and guanine) in DNA chains could control the synthesis of proteins from amino acids. This led later to the discovery of genetic code.

Since the discovery of DNA, molecular biology has attempted to describe the properties of living systems in terms of molecules. The goal is to trace down chemical processes from the level of cells to that of the composing molecules and to understand the functioning of each molecule in biological processes, e.g., heredity, metabolism, motility, and so on.

21.1.2 On Quantum Mind and Brain

Researchers in quantum physics such as Planck, Bohr, Heisenberg, Schrödinger, von Neumann, Pauli, Wigner, and others emphasized the possible role of quantum theory in reconsidering the well known conflict between physical determinism and conscious free will.

Bohr suggested a hypothesis that thought involves such small amounts of energy that quantum-theoretical limitations play an essential role in determining its character.

These ideas have been developed by Bohm who proposed that a holistic paradigm should take the place of reductionist quantum physics. "Information contributes fundamentally to the qualities of substance... Wave function, which operates through form, is closer to life and mind... The electron has a mindlike quality." In his theory of "wholeness and the implicate order", Bohm proposed a new model of reality. In this model, as in a hologram, any element contains enfolded within itself the totality of its universe including both matter and mind. "One can view consciousness as a self-organizing process on the border between the quantum and the classical worlds, relating biological systems and physical systems." Several modes of existence should be considered in some mathematical frames, which is recently proposed in [608] as discussed in Chap. 10.

Psychological, or cognitive, description of the mental activity can be given as follows. A mental system can be in many different conscious, intentional mental states. In a state space, a sequence of such states forms a process representing the stream of consciousness. Since different subsets of the state space are typically associated with different stability properties, a mental state can be assumed to be more or less stable, depending on its position in the state space. Stable states are distinguished by a residence time longer than that of metastable or unstable states. If a mental state is stable with respect to perturbations, it "activates" a mental representation encoding a content that is consciously perceived.

Moving from this purely psychological, or cognitive, description to its neurophysiological counterpart leads us to the question: What is the neural correlate of a mental representation? A quantum like model of recognition shall be described in this chapter. We note also that there is a p-adic approach to the description of the mental space proposed by Khrennikov.

Penrose proposed controversial ideas on the connection between fundamental physics and human consciousness. In the book "The Emperor's New Mind" (1989) he argues that known laws of physics are inadequate to explain the phenomenon of consciousness. He claims that the present computer is unable to have intelligence because it is an algorithmically deterministic system. He argues against the viewpoint that the rational processes of the mind are completely algorithmic and can thus be duplicated by a sufficiently complex computer. This is based on claims that consciousness transcends formal logic systems because there are such things as the insolubility of the halting problem and Godel's incompleteness theorem. Penrose believes that a deterministic non-algorithmic processes may come into play in the quantum mechanical wave function reduction, and may be harnessed by the brain. These claims were originally made by the philosopher John Lucas.

A new approach to the problem of quantum mind was suggested by the authors in [600, 601, 608] by going beyond the usual paradigm of quantum computation based on a quantum Turing machine. It is proposed that one can make a new type of device combining the quantum computer with the chaotic amplifier, see Chaps. 10, 11, and 14.

There are many discussions of the quantum theory in relation to consciousness that adapt ideas of quantum theory in a purely metaphorical manner. Terms such as observation, complementarity, superposition, collapse, duality, entanglement, and others are used without specific reference to how they are applicable to specific situations. For instance, conscious acts are just suggested to be considered somehow similarly to acts of measurement, or correlations in psychological systems are interpreted somehow analogously to quantum entanglement. Such speculations could be interesting but it is not enough to claim that we really have some understanding of connections between quantum theory and consciousness.

21.1.3 Characteristics of Biological Systems

Below we have listed, with no pretense of completeness, the 12 characteristic of biological systems and, for each of these characteristics, we have tried to identify some mathematical, physical, or information-theoretical technique which seems particularly suitable to deal with this characteristics.

1. Biological systems are *open systems* (interaction with environment is fundamental). Therefore, we expect that the theory of open systems will play a relevant role in their description.
2. Biological systems are *multi-component systems*: even the simplest unicellular organism, even a single ion channel, from the point of view of physics is made up of a huge number of molecules with different structural and functional roles. Therefore, we expect a connection with the theory of complex systems, chaos theory, etc.
3. Biological subsystems are *strongly interacting*. So the models of interacting particle systems widely studied in probability and in physics will be a good starting point to introduce additional biological features. In the quantum case, we expect a role for entanglement.
4. Biological *interactions have a local nature*. The probabilistic counterpart of the locality of interactions is the Markov property.
5. In biological systems, the *three-dimensional structure* is essential: *Flatland* is rare among biological systems and *Lineland* is even rarer. This means that random fields rather than random processes will be needed.
6. In biological systems, *spatially periodic*, *translation invariant structures* are typically absent: biological systems are rarely like crystals. The distribution patterns of neurons, animal cells, etc. look more like complex nonhomogeneous graphs

than lattices. Therefore, we will need random fields on graphs rather than random fields on regular lattices.

7. In biological systems, there is an *interplay between geometry and interaction*; think, for example, of protein folding. This fits with the intuition, coming from general relativity, that a free structure on a complex geometry may be equivalent to an interacting structure on a simple geometry.
8. Biological systems *elaborate information*: they *code* information, they *store* information, they have mechanisms to *activate* the stored information: memory, consciousness, reproduction, etc.

   Where is information stored in the biological beings? Surely not all information in biosystems is of genetical origin, for example, only a fraction of the information on a human being is stored in the human genome. Probably a great deal of information is stored in some *collective mode* of organization among say neurons, or cells, etc.
9. Biological systems use information to transform *matter into energy, new information, new matter and structure*: this is the food, growth, and reproduction chain.
10. Biological systems use information to *communicate*. The fine structure of biological communication is largely unknown. For example, cells communicate among themselves through various types of channels. These channels transport ions, so they are called ionic channels. As far as we understand, the role of these ions is not that of being some kind of *food* for the cells: their role is much more similar to bits of information. Could we extrapolate from the theory of ionic channels the notion of an *ion bit*? Could we put to use what we have learned from quantum information, quantum computer, based on the notion of qubit, to develop an analogue theory of bio-information, based on the notion of an *ion bit*.
11. Biological systems are *adaptive*. This means that Newtonian determinism must be replaced by adaptive determinism. Adaptive systems lie between the Newtonian determinism and the theory of feedback and control. Furthermore, we now understand that there are two types of adaptedness:

    (i) Observable adaptive, e.g., from environment to individual: *if one meets a certain type of environment, then one will react so and so.*
    (ii) State adaptive, e.g., from individual to environment: *if, when an interaction with the environment begins, one will be in this state, then one will react so and so.*

    The details of adaptive dynamics is discussed in Chap. 10.
12. Biological systems should be described in an *infinite-dimensional Hilbert space* as far as a certain self-consistency is required.

The present mathematical models of biological structures are far from meeting all these requirements. However, we feel that to lay them down explicitly may be useful as a benchmark for checking future models.

Let us start to review the fundamentals of the genome.

21.2 Genome and Proteins

The goal of theoretical molecular approaches to biology is to describe living systems in terms of physics, chemistry, and mathematics. In this section, the basic notions of molecular biology and bio-informatics are briefly described, and some topics in molecular dynamics approach to molecular modeling and simulation are reviewed.

The major achievements in molecular biology are the Watson and Crick discovery (1953) of the DNA double helix structure and the completion in 2003 of the Human Genome Project which goal was to identify all the approximately 25 000 genes in a human DNA and determine the sequences of the 3 billion chemical base pairs that make up a human DNA.

The flow of biological information is described by Crick's Central dogma of molecular biology.

Common activities in bio-informatics include mapping and analyzing DNA and protein sequences, aligning different DNA and protein sequences to compare them and creating and viewing three-dimensional models of protein structures.

21.2.1 Cell

All organisms are composed of one or more *cells*. Cells are the building bricks of life, they are the smallest structural and functional units of an organism that is considered as living. Some organisms, such as most bacteria, consist of a single cell. Humans have an estimated 10^{14} cells. A typical cell size is 10^4 nanometers ($1 \text{ nm} = 10^{-9}$ m); a typical cell mass is 1 nanogram.

All cells come from preexisting cells. Vital functions of an organism occur within cells, and all cells contain the hereditary information necessary for regulating cell functions and for transmitting information to the next generation of cells.

There are two types of cells: *eukaryotic* and *prokaryotic*. Bacteria and archaea are prokaryotes, while protists, fungi, plants, and animals are eukaryotes.

Cell Structure

All cells have a *membrane* that envelops the cell. The cell membrane is the interface between the cellular machinery inside the cell and the fluid outside. Most of the cell volume inside the membrane takes a salty cytoplasm. Within the cytoplasm there are *cytoskeleton* and *organelles*. Cytoskeleton is a dynamic structure that maintains cell's shape, enables cellular motion, and plays a role in the intracellular transport and cellular division. Organelles are specialized subunits which are usually separately enclosed within their own lipid membranes. They are *nuclei*, *mitochondria*, *microsomes*, *lysosomes*, *the Golgi apparatus*.

Cell nucleus is a eukaryotic cell's information center. The genetic material is contained in the biomolecules DNA and RNA, see below. Eukaryotic organisms store their DNA inside the cell nucleus, while in prokaryotes it is found in the cell's cytoplasm. A human cell has genetic material in the nucleus (the nuclear genome) and in the mitochondria (the mitochondrial genome). In humans, the nuclear genome is divided into 23 pairs of linear DNA molecules called *chromosomes*. Within the chromosomes, chromatin proteins such as histones compact and organize DNA.

Mitochondria are "cellular power plants", they generate the cell energy by the process of oxidative phosphorylation, utilizing oxygen to release energy stored in cellular nutrients to generate adenosine triphosphate (ATP) which is used as a source of chemical energy.

Cell Functions

Cell functions include *cell movement*, *cell growth*, *cell division*, *and metabolism*. Cell division is a process by which a cell divides into two cells. This leads to growth in multicellular organisms. Metabolism is the set of chemical reactions that occur in living organisms in order to maintain life. These processes allow organisms to grow and reproduce, maintain their structures, and respond to their environments. Metabolism uses energy to construct components of cells such as proteins, nucleic acids and other biomolecules.

21.2.2 Biomolecules

A biomolecule is any organic molecule (i.e., it contains carbon) that is produced by a living organism. A diverse range of biomolecules exists, including polymers, monomers, and small molecules. Polymers include *peptides*, *proteins*, *polysaccharides* (including cellulose), *lignin*, *nucleic acids*, i.e., deoxyribonucleic acid (DNA) and ribonucleic acid (RNA). Monomers include *amino acids*, *nucleotides*, *monosaccharides*. There are also small biomolecules such as lipids, sterols, vitamins, hormones, sugars.

Biomolecules consist primarily of carbon and also hydrogen, nitrogen, and oxygen, as well as phosphorus, sulfur, and other elements.

Proteins

Proteins are biomolecules made of amino acids arranged in a linear chain. An *amino acid* is a molecule containing both amine and carboxyl functional groups. Each amino acid consists of a central tetrahedral carbon (the alpha carbon C^{α}) attached to four units: a hydrogen atom, a protonated amino group (NH_3^+), a dissociated carboxyl group (COO^-), and a distinguished side chain, or R group.

Amino acids in a protein are joined together by CO–NH peptide bonds between the carboxyl and amino groups of adjacent amino acid residues. Each protein is

Table 21.1 List of the 20 standard amino acids with the abbreviations

Amino acid	3-letter	1-letter
Alanine	Ala	A
Arginine	Arg	R
Asparagine	Asn	N
Aspartic acid	Asp	D
Cysteine	Cys	C
Glutamine	Gln	E
Glutamic acid	Glu	Q
Glycine	Gly	G
Histidine	His	H
Isoleucine	Ile	I
Leucine	Leu	L
Lysine	Lys	K
Methionine	Met	M
Phenylalanine	Phe	F
Proline	Pro	P
Serine	Ser	S
Threonine	Thr	T
Tryptophan	Trp	W
Tyrosine	Tyr	Y
Valine	Val	V

composed of 20 different kinds of amino acids. The sequence of amino acids in a protein is defined by the sequence of a gene, which is encoded in the genetic code. In general, the genetic code specifies 20 standard amino acids.

Table 21.1 is the list of the 20 standard amino acids with the abbreviations.

21.2.3 Nucleic Acids (DNA and RNA)

Nucleic acids molecules are linear, unbranched polymers of simple units called nucleotides. The most common nucleic acids are deoxyribonucleic acid (DNA) and ribonucleic acid (RNA). DNA molecules made of two long polymers of nucleotides.

Nucleotides consist of three parts:

1. *A five-carbon sugar* (hence a pentose). Two kinds are found:
   (i) Deoxyribose, which has a hydrogen atom attached to its #2 carbon atom (designated 2′).
   (ii) Ribose, which has a hydroxyl group atom there.

   Deoxyribose-containing nucleotides, the deoxyribonucleotides, are the monomers of DNA.

Ribose-containing nucleotides, the ribonucleotides, are the monomers of RNA.

2. A nitrogen-containing ring structure called a *base*. The base is attached to the 1′ carbon atom of the pentose. In DNA, four different bases are found:

   (i) Two purines, called adenine (A) and guanine (G).
   (ii) Two pyrimidines, called thymine (T) and cytosine (C).

   RNA contains:

   The same purines, adenine (A) and guanine (G).

   RNA also uses the pyrimidine cytosine (C), but instead of thymine, it uses the pyrimidine uracil (U).

   The combination of a base and a pentose is called a *nucleoside*.

3. One, two, or three *phosphate groups*. These are attached to the 5′ carbon atom of the pentose.

   Both DNA and RNA are assembled from nucleoside triphosphates.

   For DNA, these are dATP, dCTP, dGTP, and dTTP.

   For RNA, these are ATP, CTP, GTP, and UTP.

   In both cases, as each nucleotide is attached, the second and third phosphates are removed.

The nucleic acids, both DNA and RNA, consist of polymers of nucleotides. The nucleotides are linked covalently between the 3′ carbon atom of the pentose and the phosphate group attached to the 5′ carbon of the adjacent pentose.

In the DNA double helix described by Watson and Crick, a flexible ladder-like structure is formed with the polymer wrapped around an imaginary central axis. The two rails of the ladder consist of alternating sugar (deoxyribose) and phosphate units; the rungs of the ladder consist of nitrogenous base pairs held together by hydrogen bonds.

RNA molecules, while single-stranded, usually contain regions where two portions of the strand twist around each other to form helical regions.

The two strands of DNA and the helical regions of RNA are held together by base pairing.

21.2.4 Genetic Code

As we just discussed, DNA consists of two long polymers made of nucleotides, with backbones made of sugars and phosphate groups joined by ester bonds. These two strands run in opposite directions to each other. Attached to each sugar is one of four types (they are denoted A, G, T, and C) of molecules called bases. It is the sequence of these four bases along the backbone that encodes information.

More than 100 000 proteins in our bodies are produced from a set of only 20 building blocks, amino acids.

There is a mapping from a set of sequences of nucleotides in DNA to a set of sequences of amino acids in proteins. One considers the set of *triplets* of nucleotide

Table 21.2 The standard (Watson–Crick) table of the vertebral mitochondrial code. Ter denotes the terminal (stop) signal

UUU Phe	ACC Thr	UCC Ser	GCC Ala
UUC Phe	ACA Thr	UCA Ser	GCA Ala
UUA Leu	ACU Thr	UCU Ser	GCU Ala
UUG Leu	ACG Thr	UCG Ser	GCG Ala
CUU Leu	AAC Asn	UAC Tyr	GAC Asp
CUC Leu	AAA Lys	UAA Ter	GAA Glu
CUA Leu	AAU Asn	UAU Tyr	GAU Asp
CUG Leu	AAG Lys	UAG Ter	GAG Glu
AUU Ile	AUC Ile	UUC Phe	GUC Val
AUC Ile	AUA Met	UUA Leu	GUA Val
AUA Met	AUU Ile	UUU Phe	GUU Val
AUG Met	AUG Met	UUG Leu	GUG Val
GUU Val	AGC Ser	UGC Cys	GGC Gly
GUC Val	AGA Ter	UGA Trp	GGA Gly
GUA Val	AGU Ser	UGU Cys	GGU Gly
GUG Val	AGG Ter	UGG Trp	GGG Gly

bases which are cytosine, guanine, adenine, and thymine abbreviated as C, G, A, and T, respectively. So, we consider the set of triplets of these letters. It is supposed that a triplet can contain a letter more than once and the letters appear in a certain order. For example, the triplet TTA is different from the triplet TAT. There are $4^3 = 64$ such triplets. This mapping is provided by the genetic code.

The genetic code is a map from the set of 64 triplets of nucleotides in RNA to the set of 20 amino acids used in the synthesis of proteins. With three exceptions, each codon encodes for one of the 20 amino acids. That produces some redundancy in the code: most of the amino acids being encoded by more than one codon. The information of DNA is converted to the RNA, the sequence of four bases A, G, C, U instead of T. A triple of these four bases is called a *codon*. Table 21.2 represents the genetic code.

For example, UUU Phe means the triplet of thymine is mapped to phenylalanine, i.e., phenylalanine is encoded by UUU.

The genetic code specifies the sequence of the amino acids within proteins during the process of the synthesis of proteins. The code is read by copying stretches of DNA into the related nucleic acid RNA, in a process called *transcription*. The stop codons are signals for the termination of the process.

Flow of Biological Information

The so-called *Central dogma of molecular biology*, stated by Crick, is a framework for understanding the transfer of sequence information between sequential information-carrying biopolymers. It says: Information cannot be transferred back from protein to either protein or nucleic acid. The general transfers describe the normal *flow of biological information*: DNA can be copied to DNA (DNA replication), DNA information can be copied into messenger ribonucleic acid mRNA (transcription), and proteins can be synthesized using the information in mRNA as a template through translation (DNA→RNA→protein).

21.2.5 Human Genome

The basic unit of heredity in a living organism is a *gene*. Genes hold the information to build and maintain their cells and pass genetic traits to offspring. In molecular biology, a gene is a segment of nucleic acid that, taken as a whole, specifies a trait or phenotype.

A *phenotype* is any observable characteristic or trait of an organism: such as its morphology, development, behavior, biochemical or physiological properties. Phenotypes result from the expression of an organism's genes as well as the influence of environmental factors and possible interactions between the two. The *genotype* of an organism is the inherited instructions it carries within its genetic code.

The *genome* of an organism is its hereditary information encoded in DNA. In classical genetics, the genome of a diploid organism including eukarya refers is a full set of chromosomes or genes in a gamete.

The Human Genome Project was organized in 1990 to map and to sequence the human genome. The Human Genome Project (HGP) was an international scientific research project with a goal to determine the sequence of chemical base pairs which make up DNA and to identify and map the approximately 20 000–25 000 genes of the human genome from both a physical and functional standpoint. A complete draft of the genome was released in 2003, with further analysis still being published.

The HGP aimed to map the nucleotides contained in a haploid reference human genome (more than three billion).

The "genome" of any given individual (except for identical twins and cloned organisms) is unique; mapping "the human genome" involves sequencing multiple variations of each gene. The project did not study the entire DNA found in human cells; some heterochromatic areas (about 8% of the total) remain un-sequenced.

The process of identifying the boundaries between genes and other features in raw DNA sequence is called *genome annotation* and is the domain of bioinformatics.

21.3 Sequence Alignment by Quantum Algorithm

When we analyze life on a gene level, we examine the homology of genome or amino acid sequences to compare these sequences, for which we have to align the sequences. The alignment of two sequences is called the pairwise alignment, and that for more than three sequences is called the multiple alignment. To align the sequences, we insert a gap (indel) "$*$" into the position of a sequence where a base or an amino acid is considered to deviate. Such alignment should be first done to analyze genome sequences or amino acid sequences, so that it is one of the fundamental operations for the study of life.

At the present stage, the algorithms of the pairwise alignment are done by applying dynamic programming [319, 543, 577, 702]. The most accurate algorithm is recently made by introducing a sort of entanglement between closed residues in the different sequences [319]. However, it is rather difficult to use a similar algorithm for the multiple alignment because the computational complexity of the N sequences with their length L by dynamic programming becomes $O(L^N)$, whose alignment will be very difficult as N increases. Therefore, the various methods have been considered to reduce the computational complexity. Among those, the Simulated Annealing [1] has been used in [528, 551]. The simulated annealing is one of the methods solving some combinatorics optimization problems such as the traveling salesman problem. Even if the simulated annealing works effectively, it is difficult to demonstrate the multiple alignment in polynomial time of N, so that the multiple alignment is considered as one of the NP-problems. In this section, we discuss a quantum algorithm for the multiple alignment, that is, how we can construct a quantum gate solving alignment by the simulated annealing.

21.3.1 Outline of Alignment

Let

$$\begin{array}{llllllllllll}
\mathcal{A}_1: & \mathrm{M} & \mathrm{N} & \mathrm{P} & \mathrm{W} & \mathrm{Y} & \mathrm{S} & \mathrm{T} & \mathrm{W} & \mathrm{Q} & \mathrm{Y} & \mathrm{T} \\
\mathcal{A}_2: & \mathrm{M} & \mathrm{N} & \mathrm{P} & \mathrm{Q} & \mathrm{Y} & \mathrm{T} & \mathrm{V} & \mathrm{W} & \mathrm{P} & \mathrm{Y} & \\
\mathcal{A}_3: & \mathrm{M} & \mathrm{N} & \mathrm{W} & \mathrm{Y} & \mathrm{S} & \mathrm{T} & \mathrm{Q} & \mathrm{P} & \mathrm{Y} & \mathrm{V} &
\end{array}$$

be the amino acid sequences of three organisms or identical proteins.

These sequences $\mathcal{A}_1$, $\mathcal{A}_2$, and $\mathcal{A}_3$ look not so close each other. It is considered that some amino acids are changed, deleted, or inserted during the course of the biological evolution from a common origin of $\mathcal{A}_1$, $\mathcal{A}_2$, and $\mathcal{A}_3$. Therefore, it is important to align the sequences $\mathcal{A}_1$, $\mathcal{A}_2$, and $\mathcal{A}_3$ to study the similarity or difference of organisms properly. After the alignment, they become

$$\begin{array}{lllllllllllll}
\mathcal{A}_1: & \mathrm{M} & \mathrm{N} & \mathrm{P} & \mathrm{W} & \mathrm{Y} & \mathrm{S} & \mathrm{T} & * & \mathrm{W} & \mathrm{Q} & \mathrm{Y} & \mathrm{T} \\
\mathcal{A}_2: & \mathrm{M} & \mathrm{N} & \mathrm{P} & \mathrm{Q} & \mathrm{Y} & * & \mathrm{T} & \mathrm{V} & \mathrm{W} & \mathrm{P} & \mathrm{Y} & * \\
\mathcal{A}_3: & \mathrm{M} & \mathrm{N} & * & \mathrm{W} & \mathrm{Y} & \mathrm{S} & \mathrm{T} & * & \mathrm{Q} & \mathrm{P} & \mathrm{Y} & \mathrm{V}
\end{array}$$

by which we can see the similarity of $\mathcal{A}_1$, $\mathcal{A}_2$, and $\mathcal{A}_3$.

21.3.2 Quantum Algorithm of Multiple Alignment

Here we construct a quantum algorithm of the multiple alignment from the classical algorithm using simulated annealing. Let us consider N amino acid sequences $\mathcal{A}_1, \mathcal{A}_2, \dots, \mathcal{A}_N$:

$$\begin{array}{lllll} \mathcal{A}_1: & a_1^1 & a_2^1 & \cdots & a_{m_1}^1 \\ \mathcal{A}_2: & a_1^2 & a_2^2 & \cdots & a_{m_2}^2 \\ & \cdots & & & \\ \mathcal{A}_N: & a_1^N & a_2^N & \cdots & a_{m_N}^N, \end{array}$$

where m_i $(i = 1, \dots, N)$ is the number of amino acids for each sequence.

In order to make the quantum algorithm of the multiple alignment, we need to modify the input data for the alignment. Let us explain this process for the following three sequences:

$$\begin{array}{lllllll} G & G & I & P & G & & \\ G & G & Q & P & I & G & A \\ G & I & P & Q & I & G. & \end{array}$$

First, we add some gaps at the end of the amino acid sequences to make all sequences have the same length such as

$$\begin{array}{lllllllll} G & G & I & P & G & * & * & * & * \\ G & G & Q & P & I & G & A & * & * \\ G & I & P & Q & I & G & * & * & *. \end{array}$$

Here, the maximum number of the gaps needed for the multiple alignment is due to the rule of simulated annealing. Then let L be the length of the arranged amino acid sequences, so that all amino acid sequences can be written as

$$\begin{array}{lllll} \mathcal{A}_1: & a_1^1 & a_2^1 & \cdots & a_L^1 \\ \mathcal{A}_2: & a_1^2 & a_2^2 & \cdots & a_L^2 \\ & \cdots & & & \\ \mathcal{A}_N: & a_1^N & a_2^N & \cdots & a_L^N. \end{array}$$

Further, we set the total sequence

$$\mathcal{A} \equiv \overbrace{a_1^1 \cdots a_L^1}^{\mathcal{A}_1} \overbrace{a_1^2 \cdots a_L^2}^{\mathcal{A}_2} \cdots \overbrace{a_1^N \cdots a_L^N}^{\mathcal{A}_N},$$

and define the objective function $f(\mathcal{A})$ to apply the simulated annealing by

$$f(\mathcal{A}) = \sum_{k=1}^{L} \sum_{i=1}^{N-1} \sum_{j=i+1}^{N} \frac{d(a_k^i, a_k^j)}{{}_N C_2},$$

where

$$d\left(a_k^i, a_k^j\right) = \begin{cases} 0, & a_k^i = a_k^j, \\ 1, & a_k^i \neq a_k^j \text{ and } (a_k^i \neq * \text{ and } a_k^j \neq *), \\ w, & a_k^i \neq a_k^j \text{ and } (a_k^i = * \text{ or } a_k^j = *). \end{cases}$$

In the above definition, w is called the weight having a value in $(0, 2]$, we take $w = 2$ in the sequel. This objective function f is an averaged difference among all sequences.

In order to compute the minimum value of $f(\mathcal{A})$ by the simulated annealing [528], we have to replace any amino acid with 1 and the gap with 0, and the resulting sequence of 0 and 1 is called the labeled sequence.

For example,

$$\begin{array}{llllllll} \mathcal{A}_1: & G & G & I & P & * & * & G \\ \mathcal{A}_2: & G & G & P & Q & I & G & * \\ \mathcal{A}_3: & G & I & * & P & Q & I & G \end{array} \Longrightarrow \begin{array}{l} \mathcal{B}_1: 1111001 \\ \mathcal{B}_2: 1111110 \\ \mathcal{B}_3: 1101111. \end{array}$$

A perturbation in the simulated annealing here means exchanging some 0 and 1 in the labeled sequences, for instance,

$$\begin{array}{l} \mathcal{B}_1 \\ \mathcal{B}_2 \\ \mathcal{B}_3 \end{array} \Longrightarrow \begin{array}{l} \mathcal{B}_1': 1111001 \\ \mathcal{B}_2': 1101111 \\ \mathcal{B}_3': 1011111 \end{array} \Longrightarrow \begin{array}{llllllll} \mathcal{A}_1': & G & G & I & P & * & * & G \\ \mathcal{A}_2': & G & G & * & P & Q & I & G \\ \mathcal{A}_3': & G & * & I & P & Q & I & G. \end{array}$$

Note that the order of the amino acids in the sequences should not be changed by this perturbation, and the perturbed sequence is denoted by $\mathcal{A}'$.

When we apply the quantum algorithm to the multiple alignment, the symbols 0 and 1 are considered as the vectors $|0\rangle$ and $|1\rangle$ of qubits. Therefore, an initial state vector for the quantum algorithm is

$$|v_{\text{in}}\rangle \equiv \underbrace{\underbrace{\otimes_1^{k_1} |1\rangle \otimes_1^{g_1 = L - k_1} |0\rangle}_{L} \otimes_1^{k_2} |1\rangle \otimes_1^{g_2 = L - k_2} |0\rangle \otimes \cdots \otimes_1^{k_N} |1\rangle \otimes_1^{g_N = L - k_N} |0\rangle}_{NL}$$

$$\otimes_1^X |0\rangle \otimes |\text{a certain constant}\rangle \underbrace{\otimes |0\rangle}_{\text{check}},$$

where g_j is the number of gaps in $\mathcal{A}_j$, and the last qubit $|0\rangle$ expresses the so-called check bit indicating whether a perturbation is accepted or not. Moreover, $\otimes_1^X |0\rangle$ are the bits to represent the value of the objective function.

The algorithm involves several steps:

Step 1. This step is composed of the following three:

S(1.1) Construct the unitary operator U_P for the perturbation as follows: Make a pair of $|0\rangle$ and $|1\rangle$ in $\mathcal{A}$ by choosing randomly. Then, repeat this operation until all $|0\rangle$ make a pair with a certain $|1\rangle$ in $\mathcal{A}$. Let m be the position of the $|1\rangle$ and n be the position of $|0\rangle$. The Controlled-NOT gate $U_{\text{C-NOT}}(m,n)$ (see Chap. 14) on $\otimes_1^L \mathbb{C}^2$ attached to the positions m and n is given by

$$\begin{aligned} U_{\text{C-NOT}}(m,n) \equiv {} & \otimes_1^{n-1} I \otimes |0\rangle\langle 0| \otimes_1^{L-n} I + \otimes_1^{m-1} I \\ & \otimes \big(|0\rangle\langle 1| + |1\rangle\langle 0|\big) \\ & \otimes_1^{n-m-1} I \otimes |1\rangle\langle 1| \otimes_1^{L-n} I, \end{aligned}$$

where we put $U_{\text{C-NOT}}(m,n) = U_{m,n}$. Then construct U_P as a combination of the above unitary operators. We will see this procedure in an example below.

Example 21.1 Let $|v_{\text{in}}\rangle$ be an initial state such that

$$|v_{\text{in}}\rangle \equiv \overbrace{|1\rangle \otimes |1\rangle \otimes |1\rangle \otimes |1\rangle \otimes |1\rangle}^{\text{amino acid}} \otimes \overbrace{|0\rangle \otimes |0\rangle \otimes |0\rangle}^{\text{gap}},$$

and make pairs of $|0\rangle$ and $|1\rangle$ as

$$\begin{aligned} \text{Pair1} &\equiv (2,7), \\ \text{Pair2} &\equiv (5,6), \\ \text{Pair3} &\equiv (3,8), \end{aligned}$$

where the first number of the pair denotes the position of $|1\rangle$, and the second number denotes the position of $|0\rangle$.

Then U_P has the form as

$$U_P \equiv U_{2,7} \cdot U_{5,6} \cdot U_{3,8}.$$

S(1.2) Apply the Hadamard transformation H to $|0\rangle$ (the vector of a gap). Let the unitary operator U_F be defined as

$$U_H \equiv \overbrace{I \otimes I \otimes \cdots \otimes \underbrace{H}_{\text{gap}} \otimes \cdots \otimes I}^{\text{sequence length}},$$

$$H|0\rangle \equiv \frac{1}{\sqrt{2}}\big(|0\rangle + |1\rangle\big).$$

In the example, we have

$$\begin{aligned} U_H|v_{\text{in}}\rangle &= U_H \cdot \big(|1\rangle \otimes |1\rangle \otimes |1\rangle \otimes |1\rangle \otimes |1\rangle \otimes |0\rangle \otimes |0\rangle \otimes |0\rangle\big) \\ &= \big(\otimes_1^5 I \otimes_1^3 H\big) \otimes \big(|1\rangle \otimes |1\rangle \otimes |1\rangle \otimes |1\rangle \otimes |1\rangle \otimes |0\rangle \otimes |0\rangle \otimes |0\rangle\big) \\ &= \frac{1}{\sqrt{2^3}}\big(|1\rangle \otimes |1\rangle \otimes |1\rangle \otimes |1\rangle \otimes |1\rangle \otimes |0\rangle \otimes |0\rangle \otimes |0\rangle \end{aligned}$$

$$+ |1\rangle \otimes |1\rangle \otimes |1\rangle \otimes |1\rangle \otimes |1\rangle \otimes |0\rangle \otimes |0\rangle \otimes |1\rangle$$
$$+ |1\rangle \otimes |1\rangle \otimes |1\rangle \otimes |1\rangle \otimes |1\rangle \otimes |0\rangle \otimes |1\rangle \otimes |0\rangle$$
$$+ |1\rangle \otimes |1\rangle \otimes |1\rangle \otimes |1\rangle \otimes |1\rangle \otimes |0\rangle \otimes |1\rangle \otimes |1\rangle$$
$$+ |1\rangle \otimes |1\rangle \otimes |1\rangle \otimes |1\rangle \otimes |1\rangle \otimes |1\rangle \otimes |0\rangle \otimes |0\rangle$$
$$+ |1\rangle \otimes |1\rangle \otimes |1\rangle \otimes |1\rangle \otimes |1\rangle \otimes |1\rangle \otimes |0\rangle \otimes |1\rangle$$
$$+ |1\rangle \otimes |1\rangle \otimes |1\rangle \otimes |1\rangle \otimes |1\rangle \otimes |1\rangle \otimes |1\rangle \otimes |0\rangle$$
$$+ |1\rangle \otimes |1\rangle \otimes |1\rangle \otimes |1\rangle \otimes |1\rangle \otimes |1\rangle \otimes |1\rangle \otimes |1\rangle\Big).$$

S(1.3) The third step is to apply U_P to the state $U_H|v_{\text{in}}\rangle$. The resulting state $U_P U_H|v_{\text{in}}\rangle$ is called the perturbed state of the sequence.
For the example, it holds that

$$U_P U_H|v_{\text{in}}\rangle = \frac{1}{\sqrt{2^3}}\Big(|1\rangle \otimes |1\rangle \otimes |1\rangle \otimes |1\rangle \otimes |1\rangle \otimes |0\rangle \otimes |0\rangle \otimes |0\rangle$$
$$+ |1\rangle \otimes |1\rangle \otimes |\mathbf{0}\rangle \otimes |1\rangle \otimes |1\rangle \otimes |0\rangle \otimes |0\rangle \otimes |\mathbf{1}\rangle$$
$$+ |1\rangle \otimes |\mathbf{0}\rangle \otimes |1\rangle \otimes |1\rangle \otimes |1\rangle \otimes |0\rangle \otimes |\mathbf{1}\rangle \otimes |0\rangle$$
$$+ |1\rangle \otimes |\mathbf{0}\rangle \otimes |\mathbf{0}\rangle \otimes |1\rangle \otimes |1\rangle \otimes |0\rangle \otimes |\mathbf{1}\rangle \otimes |\mathbf{1}\rangle$$
$$+ |1\rangle \otimes |\mathbf{0}\rangle \otimes |\mathbf{0}\rangle \otimes |1\rangle \otimes |1\rangle \otimes |0\rangle \otimes |\mathbf{1}\rangle \otimes |\mathbf{1}\rangle$$
$$+ |1\rangle \otimes |1\rangle \otimes |\mathbf{0}\rangle \otimes |1\rangle \otimes |\mathbf{0}\rangle \otimes |\mathbf{1}\rangle \otimes |0\rangle \otimes |\mathbf{1}\rangle$$
$$+ |1\rangle \otimes |\mathbf{0}\rangle \otimes |1\rangle \otimes |1\rangle \otimes |\mathbf{0}\rangle \otimes |\mathbf{1}\rangle \otimes |\mathbf{1}\rangle \otimes |0\rangle$$
$$+ |1\rangle \otimes |\mathbf{0}\rangle \otimes |\mathbf{0}\rangle \otimes |1\rangle \otimes |\mathbf{0}\rangle \otimes |\mathbf{1}\rangle \otimes |\mathbf{1}\rangle \otimes |\mathbf{1}\rangle\Big),$$

where the bold letters mean that an amino acid changes due to the computation U_P.

Let us do the multiple alignment of N sequences with their common sequence length L.

S(1.1) Taking a certain value (i.e., a threshold) λ, we compute the objective function $f(\mathcal{A})$ and we look for the minimum value of this function less than the given value λ, which gives us the desired alignment. Let $|v_{\text{in}}\rangle$ be an initial state,

$$|v_{\text{in}}\rangle \equiv \otimes_1^{k_1} |1\rangle \otimes_1^{g_1=L-k_1} |0\rangle \otimes_1^{k_2} |1\rangle \otimes_1^{g_2=L-k_2} |0\rangle \otimes \cdots \otimes_1^{k_n} |1\rangle$$
$$\otimes_1^{g_n=L-k_n} |0\rangle \otimes_1^{X} |0\rangle \otimes |\lambda\rangle \otimes |0\rangle.$$

Make pairs of $|0\rangle$ and $|1\rangle$ in each sequence randomly, and let the unitary operator U_P be defined as

$$U_P \equiv \prod_{i=1}^{N} U_{P_i},$$

$$U_{P_i} \equiv U_{m_1,n_1} \cdot U_{m_2,n_2} \cdots U_{m_{g_i},n_{g_i}}$$

where $m_i \le k_i$ and $g_i = L - k_i \le n_i \le L$.

S(1.2) Apply the discrete Fourier transformation U_H to the initial state:

$$U_H \equiv \prod_{i=1}^{N} U_{H_i},$$

where

$$U_{H_i} \equiv \overbrace{I \otimes I \otimes \cdots \otimes \underbrace{H}_{\text{gap}} \otimes \cdots \otimes I}^{L}.$$

Then we have

$$\begin{aligned}
U_H|v_{\text{in}}\rangle &= \frac{1}{\sqrt{2^{g_1+\cdots+g_N}}} \otimes_1^{k_1} |1\rangle \otimes_1^{g_1} \big(|0\rangle + |1\rangle\big) \otimes \cdots \otimes_1^{k_N} |1\rangle \\
&\otimes_1^{g_N} \big(|0\rangle + |1\rangle\big) \otimes_1^{N} |0\rangle \otimes |\lambda\rangle \otimes |0\rangle \\
&= \frac{1}{\sqrt{2^{g_1+\cdots+g_N}}} \big(\otimes_1^{k_1} |1\rangle \otimes_1^{g_1} |0\rangle \otimes \cdots \otimes_1^{k_N} |1\rangle \otimes_1^{g_N} |0\rangle \otimes_1^{N} |0\rangle \\
&\otimes |\lambda\rangle \otimes |0\rangle \\
&+ \otimes_1^{k_1} |1\rangle \otimes |1\rangle \otimes_1^{g_1-1} |0\rangle \otimes \cdots \otimes_1^{k_N} |1\rangle \otimes_1^{g_N} |0\rangle \otimes_1^{N} |0\rangle \\
&\otimes |\lambda\rangle \otimes |0\rangle \\
&+ \otimes_1^{k_1} |1\rangle \otimes |0\rangle \otimes |1\rangle \otimes_1^{g_1-2} |0\rangle \otimes \cdots \otimes_1^{k_N} |1\rangle \otimes_1^{g_N} |0\rangle \otimes_1^{N} |0\rangle \\
&\otimes |\lambda\rangle \otimes |0\rangle \\
&+ \otimes_1^{k_1} |1\rangle \otimes |1\rangle \otimes |1\rangle \otimes_1^{g_1-2} |0\rangle \otimes \cdots \otimes_1^{k_N} |1\rangle \otimes_1^{g_N} |0\rangle \otimes_1^{N} |0\rangle \\
&\otimes |\lambda\rangle \otimes |0\rangle \\
&\cdots \\
&+ \otimes_1^{k_1} |1\rangle \otimes_1^{g_1} |1\rangle \otimes \cdots \otimes_1^{k_N} |1\rangle \otimes_1^{g_N} |1\rangle \otimes_1^{N} |0\rangle \otimes |\lambda\rangle \otimes |0\rangle\big).
\end{aligned}$$

S(1.3) Applying U_P to the state $U_H|v_{\text{in}}\rangle$, we obtain the superposition of $2^{g_1} + \cdots + 2^{g_N}$ sequences as follows:

$$\begin{aligned}
&U_P U_H |v_{\text{in}}\rangle \\
&= \frac{1}{\sqrt{2^{g_1+\cdots+g_N}}} \big(\otimes_1^{k_1} |1\rangle \otimes_1^{g_1} |0\rangle \otimes \cdots \otimes_1^{k_N} |1\rangle \otimes_1^{g_N} |0\rangle \otimes_1^{N} |0\rangle \\
&\otimes |\lambda\rangle \otimes |0\rangle \\
&+ |0\rangle \otimes_1^{k_1-1} |1\rangle \otimes |1\rangle \otimes_1^{g_1-1} |0\rangle \otimes \cdots \otimes_1^{k_N} |1\rangle \otimes_1^{g_N} |0\rangle \otimes_1^{N} |0\rangle
\end{aligned}$$

$$\otimes |\lambda\rangle \otimes |0\rangle$$
$$+ |1\rangle \otimes |0\rangle \otimes_1^{k_1-2} |1\rangle \otimes |0\rangle \otimes |1\rangle \otimes_1^{g_1-2} |0\rangle \otimes \cdots \otimes_1^{k_N} |1\rangle \otimes_1^{g_N} |0\rangle$$
$$\otimes_1^N |0\rangle \otimes |\lambda\rangle \otimes |0\rangle$$
$$+ \otimes_1^2 |0\rangle \otimes_1^{k_1-2} |1\rangle \otimes |1\rangle \otimes |1\rangle \otimes_1^{g_1-2} |0\rangle \otimes \cdots \otimes_1^{k_N} |1\rangle \otimes_1^{g_N} |0\rangle$$
$$\otimes_1^N |0\rangle \otimes |\lambda\rangle \otimes |0\rangle$$
$$\cdots$$
$$+ \otimes_1^{g_1} |0\rangle \otimes_1^{k_1-g_1} |1\rangle \otimes_1^{g_1} |1\rangle \otimes \cdots \otimes_1^{g_n} |0\rangle \otimes_1^{k_N-g_N} |1\rangle \otimes_1^{g_N} |1\rangle$$
$$\otimes_1^N |0\rangle \otimes |\lambda\rangle \otimes |0\rangle\big).$$

Step 2. The computation step of the algorithm involves the following two:

S(2.1) Compute the objective function (distance among the sequences). The result is represented in $|f(\mathcal{A}'_i)\rangle$ $(i = 1, \ldots, 2^{g_1+\cdots+g_N})$.

$$U_C U_P U_H |v_{\text{in}}\rangle$$
$$= \frac{1}{\sqrt{2^{g_1+\cdots+g_N}}} \big(\otimes_1^{k_1} |1\rangle \otimes_1^{g_1} |0\rangle \otimes \cdots \otimes_1^{k_N} |1\rangle \otimes_1^{g_N} |0\rangle$$
$$\otimes \big|f(\mathcal{A}'_1)\big\rangle \otimes |\lambda\rangle \otimes |0\rangle$$
$$+ |0\rangle \otimes_1^{k_1-1} |1\rangle \otimes |1\rangle \otimes_1^{g_1-1} |0\rangle \otimes \cdots \otimes_1^{k_N} |1\rangle \otimes_1^{g_N} |0\rangle$$
$$\otimes \big|f(\mathcal{A}'_2)\big\rangle \otimes |\lambda\rangle \otimes |0\rangle$$
$$+ |1\rangle \otimes |0\rangle \otimes_1^{k_1-2} |1\rangle \otimes |0\rangle \otimes |1\rangle \otimes_1^{g_1-2} |0\rangle \otimes \cdots \otimes_1^{k_N} |1\rangle \otimes_1^{g_N} |0\rangle$$
$$\otimes \big|f(\mathcal{A}'_3)\big\rangle \otimes |\lambda\rangle \otimes |0\rangle$$
$$+ \otimes_1^2 |0\rangle \otimes_1^{k_1-2} |1\rangle \otimes |1\rangle \otimes \big|1\big\rangle \otimes_1^{g_1-2} \big|0\big\rangle \otimes \cdots \otimes_1^{k_N} |1\rangle \otimes_1^{g_N} |0\rangle$$
$$\otimes \big|f(\mathcal{A}'_4)\big\rangle \otimes |\lambda\rangle \otimes |0\rangle$$
$$\cdots$$
$$+ \otimes_1^{g_1} |0\rangle \otimes_1^{k_1-g_1} |1\rangle \otimes_1^{g_1} |1\rangle \otimes \cdots \otimes_1^{g_N} |0\rangle \otimes_1^{k_N-g_N} |1\rangle \otimes_1^{g_N} |1\rangle$$
$$\otimes \big|f(\mathcal{A}'_{2^{g_1+\cdots+g_n}})\big\rangle \otimes |\lambda\rangle \otimes |0\rangle\big),$$

where U_C is the unitary operator computing the objective function f.

S(2.2) Define the unitary operator U_A as the following

$$U_A \otimes_1^{LN} I \otimes \big|f(\mathcal{A}'_i)\big\rangle \otimes |\lambda\rangle \otimes |0\rangle$$
$$\equiv \otimes_1^{LN} I \otimes \big|f(\mathcal{A}'_i)\big\rangle \otimes |\lambda\rangle \otimes \begin{cases} |1\rangle & f(\mathcal{A}'_i) \le \lambda, \\ |0\rangle & f(\mathcal{A}'_i) > \lambda. \end{cases}$$

Apply U_A to the state $U_C U_P U_H |v_{\text{in}}\rangle$, the check bit of the sequences satisfying $\lambda \geq f(\mathcal{A}_i')$, namely the sequences accepted, becomes $|1\rangle$.

Step 3. The observation step proceeds as follows:

If the acceptance probability is very small (say, less than $\frac{1}{2^{LN}}$), we use the chaotic dynamics to amplify the probability as discussed in Chap. 14. If the acceptance probability is not so small, calculate again with the same input state and smaller λ. About 100 rounds of calculation is adequate. Therefore, we can determine whether there exist alignments such that $\lambda \geq f(\mathcal{A}_i')$ holds with a probability $1/2$.

21.3.3 Computational Complexity of Quantum Algorithm

We discuss the difference of the computational complexity between the quantum algorithm and the classical algorithm for the multiple alignment. Table 21.3 shows the computational complexity of the classical algorithm and that of the quantum algorithm. This quantum algorithm is finished in polynomial time of the size N of input sequences and the sequence length L.

21.4 A Brain Model Describing the Recognition Process

Specialists in modern brain research are convinced that signals in the brain should be coded by populations of excited neurons (cf. [320, 649, 719, 722]). When considering models based on classical probability theory, the states of signals should be identified with probability distributions of certain random point fields located inside the volume G of the brain. However, researchers as, for instance, Singer [722] and Stapp [732] expressed doubts whether classical models could give satisfactory answers to fundamental problems of modern brain research, and pleaded for using quantum models. The paper [252] represents a first attempt to explain the process of recognition in terms of quantum statistics. Here (pure) states of signals are described by complex functions Ψ of point configurations inside of G where $|\Psi|^2$ is the probability density of a random point field. Thus a probability distribution of a random point field called the position distribution corresponding to the quantum state is obtained. On the other hand, the probability distribution of each random point field can

Table 21.3 Computational complexity

	Classical algorithm	Quantum algorithm
DFT	$2^{N\times L}$	$N\times L$
Perturbation	$2^{N\times L}$	$N\times L$
Calculation	$2^{N\times L}\times L\times_N C_2$	$L\times_N C_2$
Total	$(2+L\times_N C_2)\times 2^{N\times L}$	$(2N+_N C_2)\times L$

be identified with the position distribution corresponding to a certain quantum state. In this sense, quantum models are more general than classical models, i.e., the use of quantum theory gives rise to a more detailed description of reality.

Here there is no intention to create a physiological model of recognition taking into account exact biological, chemical and/or physical reactions. We are searching for a mathematical model capable of explaining certain aspects of the recognition process.

In Sect. 21.4.3, we enumerate 10 postulates [252] any model of brain activities has to fulfill. In the subsequent subsections, we present some aspects of the general outline of the model and we check its compliance with the postulates.

In this section, we will discuss how to make a quantum-like brain model to explain its activities by means of the ideas of quantum entanglement and teleportation processes of Chaps. 8 and 18. It will be shown that the mathematical model reproduces the basic properties of the recognition of signals. That is, we confine ourselves to a simplified description of the basic spaces and operators used in this approach. Details of the brain model are given in [252–259]. In these papers, the memory space, the procedures of creation of signals from the memory, amplification, accumulation and transformation of input signals, and measurements like EEG (Electroencephalogram) or MEG (Magnetroencephalogram) are treated in detail.

21.4.1 A Quantum-Like Model of Brain

It is the processing speed that we take as a particular character of the brain, so that the high speed of processing in the brain is here supposed to come from the coherent effects of substances in the brain like quantum computer, as was pointed out by Penrose. Having this in our mind, we propose a model of a brain describing its function as follows:

The brain system *BS* is supposed to be described by a triple $(\mathbf{B}(\mathcal{H}), \mathfrak{S}(\mathcal{H}), \Lambda^*)$ on a certain Hilbert space $\mathcal{H}$ where $\mathbf{B}(\mathcal{H})$ is the set of all bounded operators on $\mathcal{H}$, $\mathfrak{S}(\mathcal{H})$ is the set of all states, and Λ^* is a channel giving a state change.

Further we assume the following:

1. *BS* is described by a quantum state, and the brain itself is divided into several parts, each of which corresponds to a Hilbert space so that $\mathcal{H} = \bigoplus_k \mathcal{H}_k$ and $\varphi = \bigoplus_k \varphi_k$, $\varphi_k \in \mathfrak{S}(\mathcal{H}_k)$. However, in the sequel we simply assume that the brain is in one Hilbert space $\mathcal{H}$ because we only consider the basic mechanism of recognition.
2. The function (action) of the brain is described by a channel $\Lambda^* = \bigoplus_k \Lambda_k^*$. Here as in assumption 1 we take only one channel Λ^*.
3. *BS* is composed of mainly two parts: information processing part "*P*" and other parts "*O*" (consciousness, memory, recognition) so that $\mathcal{H} = \mathcal{H}_P \otimes \mathcal{H}_O$.

In our model, the whole brain may be considered as a quantum computer [603], but we here explain the function of the brain like a process in a quantum computer,

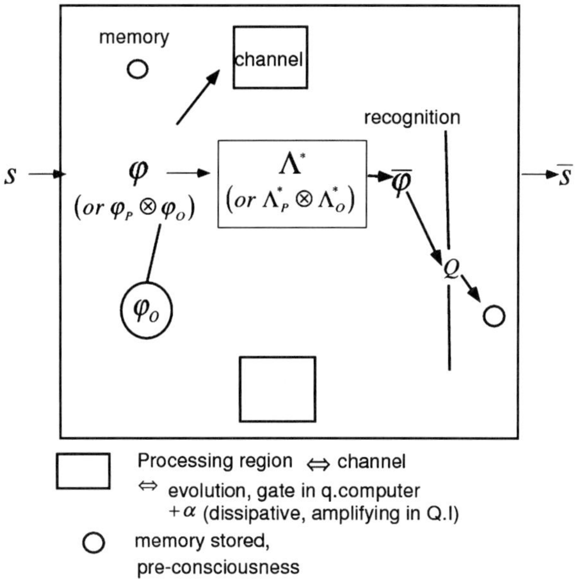

Fig. 21.1 Scheme of the brain model

more precisely, like a quantum communication process with entanglements as in a quantum teleportation process. We will explain the mathematical structure of our model.

Let $s = \{s^1, s^2, \dots, s^n\}$ be a given (input) signal (perception), and let $\overline{s} = \{\overline{s}^1, \overline{s}^2, \dots, \overline{s}^n\}$ be the output signal. After the signal s enters the brain, each element s^j of s is coded into a proper quantum state $\rho^j \in \mathfrak{S}(\mathcal{H}_P)$ so that the state corresponding to the signal s is $\rho = \bigotimes_j \rho^j$. This state may be regarded as a state processed by the brain, and it is coupled to a state ρ_O stored as a memory (preconsciousness) in brain. The processing in the brain is expressed by a properly chosen quantum channel Λ^* (or $\Lambda^*_P \otimes \Lambda^*_O$). The channel is determined in the form of a network of neurons and some other biochemical actions, and its function is like that of a (quantum) gate in a quantum computer. The outcome state $\overline{\rho}$ comes in contact with an operator F describing the work as noema of consciousness (Husserl's noema), after the contact a certain reduction of state occurs, which may correspond to the noesis (Husserl's) of consciousness. A part of the reduced state is stored in brain as a memory. The scheme of our model is represented in Fig. 21.1.

21.4.2 Value of Information in Brain

The complex system responds to the information and has a particular role to choose the information (value of information). Brain selects some information (inputs) from a huge flow of information (inputs). It will be important to find a rule or rules of such a selection mechanism, which will be discussed in the framework of adaptive dynamics. In the above model of the brain, an output signal s (information) is somehow coded into a quantum state φ_P, then it runs in the brain with a certain processing effect Λ^* and a memory stored by a state φ_O, and it changes its own figure. Thus we have two standpoints to catch the value of information in the brain. Suppose that we have a fixed "*purpose (intention) of the brain*" described by a certain operator Q, then *one view of the value of information is whether the signal s is important for the purpose Q and the processing Λ^*, and another is whether the processing Λ^* chosen in the brain is effective for s and Q*. From these views, the value should be estimated by a function of $\varphi_P \otimes \varphi_O$, Λ^*, and Q so that it is important to find a proper measure $V(\varphi_P \otimes \varphi_O, \Lambda^*, Q)$ estimating the effect of a signal and a function of the brain. The following properties should hold:

Definition 21.2 (Value of information)

1. $s = \{s^1, s^2, \dots, s^n\}$ is more valuable than $s' = \{s'^1, s'^2, \dots, s'^n\}$ for Λ^* and Q iff

$$V(\varphi_P \otimes \varphi_O, \Lambda^*, Q) \geqq V(\varphi'_P \otimes \varphi_O, \Lambda^*, Q).$$

2. Λ^* is more valuable than Λ'^* for given $s = \{s^1, s^2, \dots, s^n\}$ and Q iff

$$V(\varphi_P \otimes \varphi_O, \Lambda^*, Q) \geqq V(\varphi_P \otimes \varphi_O, \Lambda'^*, Q).$$

An example of this estimator is discussed in [253]. There exist some relations between the information of value and the complexity or the chaos degree discussed in Chap. 10 with the properly chosen complexity C and transmitted complexity T, for which we conjecture

$$\begin{aligned} &D(\varphi_P \otimes \varphi_O, \Lambda^*; Q) \le D(\varphi_P \otimes \varphi_O, \Lambda'^*; Q) \quad \Longleftrightarrow \\ &V(\varphi_P \otimes \varphi_O, \Lambda^*, Q) \ge V(\varphi_P \otimes \varphi_O, \Lambda'^*, Q). \end{aligned}$$

The detailed study of this value in the brain is very subtle and more evaluation is needed.

21.4.3 Postulates Concerning the Recognition of Signals

Singer, a modern brain researcher, summarizes in [722] some common experiences discovered mainly in the last decade. Following the main ideas in [722], we extract some postulates any mathematical model of the procedure of recognition of signals should fulfill:

(P1) *The brain acts discrete in time.*
(P2) *Signals are represented by populations of excited neurons.*
(P3) *Signals can be decomposed into parts in compliance with the fact that there are different regions of the brain being responsible for different tasks.*
(P4) *The brain acts in parallel with respect to the different regions.*

(P5) *Signals stored in the brain are superpositions of finitely many elementary signals.*
(P6) *The brain permanently creates complex signals representing a hypothesis concerning an "expected view of the world".*
(P7) *Recognition of a signal produced by our senses is a random event which can occur as a consequence of the interaction of that signal and a signal created by the brain.*
(P8) *Recognition causes a loss of excited neurons in some regions of the brain.*
(P9) *Recognition changes the state of the signal coming from our senses. One will be aware of that changed signal.*
(P10) *Changes in some region of the brain have immediate consequences concerning the other regions.*

In what follows, we will introduce some basic components of the model—the basic spaces for the signals and the memory, and the basic operators describing the process of recognition. We will check whether this approach is in accordance with the above formulated postulates. The main components of the model are

- a Hilbert space $\mathcal{H}$ representing the space of *signals*,
- a Hilbert space $\overline{\mathcal{H}}$ representing the *memory*,
- an isometry $\mathcal{D}: \overline{\mathcal{H}} \to \overline{\mathcal{H}} \otimes \mathcal{H}$ describing the *creation of signals from the memory*,
- a unitary operator $\mathcal{V}$ on $\mathcal{H} \otimes \mathcal{H}$ describing the *interaction of two signals*.

21.4.4 The Space of Signals

Our quantum-like model of signals in the brain is based on the so-called bosonic Fock space which was discussed in Chap. 18. The use of that Fock space has several advantages. For instance, as it was pointed out in [719, 722], the space G representing the whole volume of the brain can be divided into disjoint regions $G_1, \dots, G_n$ responsible for different aspects of the signals. More precisely, corresponding to that decomposition of the space the signal should be decomposed into different parts representing a special type of information contained in the signal. This can be reflected in the fact that the bosonic Fock space corresponding to G can be identified with the tensor product of the bosonic Fock spaces corresponding to the subsets $G_1, \dots, G_n$. Moreover, we need the double Fock space, namely, the Fock space over the Fock space of neurons, the use of such Fock spaces is very essential for the study of brain in the following reasons: (i) The first (basic) Fock space is needed for the description of neurons as the number of the excited neurons is not fixed and it will be from 0 up to a huge number (which can be considered as ∞). (ii) A set of excited neurons is described by a state of the first Fock space. We call the set of all such states the EN-set. The second (upstairs) Fock space is needed for the study of the memory and recognition because every combination of some EN-sets will be based on the brain function, and the number of the EN-sets used for the combination will not be fixed as in the case of neurons. Thus we need the double Fock space to describe the function of the brain.

Our brain contains about 100 billions of neurons. Thus a lot of signals could be described in terms of a classical probabilistic model according to postulate (P2). But there are many well-known facts being in contradiction to the classical models. In a classical model, it is difficult to explain that activities in one region of the brain have immediate consequences concerning the other regions (this is the so-called *binding problem*). However, because of nonlocality of quantum theory a quantum model may be in accordance with postulate (P10). We need a quantum-like model of the brain to explain its activities being dependent on nonlocality and correlation existing in the brain.

The starting point for our quantum model is a suitable compact subset $G \subset \mathbb{R}^3$ representing the physical volume of the brain. The Hilbert space $L^2(G)$ of square integrable complex-valued functions on G describes a quantum particle inside of G. Now, a neuron is excited if a certain amount of energy (electric potential) is stored in it. For that reason, we identify *an excited neuron* with a *quantum particle localized inside the neuron*, i.e., in our model the quantum particle represents that amount of energy which makes the difference between an excited and a nonexcited neuron. In this sense, the bosonic Fock space

$$\Gamma\big(L^2(G)\big) = \bigoplus_{n=0}^{\infty} S_+ L^2(G)^{\otimes n}$$

represents *populations of excited neurons* inside the volume G of the brain, where S_+ is defined in Chap. 18. In accordance with postulate (P2), we will use (a subspace of) $\mathcal{H} \equiv \Gamma(L^2(G))$ as our basic *space of signals*. Incoming signals and signals taken from the memory are states on the Fock space $\mathcal{H}$. The use of the Fock space has several advantages. The most striking one is that the Fock space over G can be identified with the tensor product of the bosonic Fock spaces corresponding to the subsets $G_1, \dots, G_n$ for an arbitrary disjoint partition of the space G:

$$\Gamma\big(L^2(G)\big) = \Gamma\big(L^2(G_1)\big) \otimes \cdots \otimes \Gamma\big(L^2(G_n)\big). \tag{21.1}$$

The property (21.1) is conformed to postulate (P3). Postulate (P5) states that there is a finite number of elementary signals generating all signals. Due to the exceptional properties of *exponential vectors* being in accordance with our postulates, we restrict ourselves to signals given by *exponential vectors*. Besides the above decomposition property, the Fock space has the property that *exponential vectors* from $\mathcal{H}$ decompose in the same way into *independent partial signals* from $\Gamma(L^2(G_r))$, $r \in \{1, \dots, n\}$, and the partial signals are again given by exponential vectors. More precisely, an exponential vector $\exp(f) \in \mathcal{H}$ corresponding to $f \in L^2(G)$ is given by

$$\exp(f) = \big(\exp(f)_n\big)_{n=0}^{\infty} \in \Gamma\big(L^2(G)\big)$$

with

$$\exp(f)_0 = 1, \qquad \exp(f)_n(x_1, \dots, x_n) = \prod_{k=1}^{n} f(x_k) \quad (n \geq 1,\ x_1, \dots, x_n \in G).$$

For $f \in L^2(G)$ and for a disjoint decomposition $G_1, \dots, G_n$ of G one has

$$\exp(f) = \exp(f_{|1}) \otimes \cdots \otimes \exp(f_{|n}) \tag{21.2}$$

where $f_{|k}$ is the restriction of f to G_k, $k \in \{1, \dots, n\}$. This property is in accordance with postulate (P3). Moreover, this property of the exponential vectors enables us to reflect in a clear sense the fundamental experimental experience (P4) that the brain is acting in parallel in the disjoint regions.

Now, for a partition $G_1, \dots, G_n$ of G we fix positive numbers $\lambda_1, \dots, \lambda_n$ representing the *intensity* of the signals in the single regions, and we fix finite orthonormal systems

$$f_1^{(k)}, \dots, f_{r_k}^{(k)} \in \Gamma\big(L^2(G_k)\big) \quad \big(f^{(k)} \in L^2(G_k),\ n_k \in \mathbb{N},\ k \in \{1, \dots, n\}\big).$$

The exponential vectors $\exp(\lambda_k f_1^{(k)}), \dots, \exp(\lambda_k f_{r_k}^{(k)}) \in \Gamma(L^2(G_k))$ span finite-dimensional subspaces $\mathcal{H}_k \subset \Gamma(L^2(G_k))$ representing the *space of signals in the region* G_k. Since $|\lambda_k|^2$ is the expectation of the number operator with respect to the pure (coherent) state given by $\lambda_k f_j^{(k)}$ the intensity of the signals in the region G_k is represented by $|\lambda_k f_j^{(k)}|^2$. By choosing the above-mentioned orthonormal systems, we pass from the infinite-dimensional spaces $\Gamma(L^2(G_k))$ to the finite-dimensional subspaces $\mathcal{H}_k$, and on G we get a finite-dimensional space

$$\mathcal{H}_{\text{sig}} \equiv \mathcal{H}_1 \otimes \cdots \otimes \mathcal{H}_n$$

of *regular* signals. The functions $f_j^{(k)}$ represent the type of signals, and the number λ_k the intensity of the signals in the region G_k.

21.4.5 The Memory Space

Now we want to describe the (long-term) memory. We have used the Hilbert space $\mathcal{H} = \Gamma(L^2(G))$ in order to describe signals (coded by sets of excited neurons). The memory contains information concerning sets of signals. For that reason, it seems to be reasonable to describe the memory using a space consisting of sets of configurations. Mathematically, this can be done by constructing the (symmetric) Fock space $\overline{\mathcal{H}}$ over the Hilbert space $\mathcal{H}$, i.e.,

$$\overline{\mathcal{H}} \equiv \Gamma(\mathcal{H}) = \bigoplus_{n=0}^{\infty} S_+ \mathcal{H}^{\otimes n}.$$

We will call this space *compound Fock space over* G. Let us remark that $\mathcal{H}$ itself can be represented as an L^2-space over a space of configurations M. Consequently, the compound Fock space $\overline{\mathcal{H}}$ is a Fock space of the same kind as $\mathcal{H}$. In an analogous way to (21.2) we can define exponential vectors $\mathrm{EXP}(\Psi)$ in the memory space $\overline{\mathcal{H}}$ (one only has to replace points from G by configurations of points from M (see Sect. 18.4.1)):

$$\mathrm{EXP}(\Psi) = \bigl(\bigl(\mathrm{EXP}(\Psi)\bigr)_n\bigr)_{n=0}^{\infty} \quad \bigl(\mathrm{EXP}(\Psi)\bigr)_0 = 1,$$

$$\bigl(\mathrm{EXP}(\Psi)\bigr)_n(\varphi_1, \ldots, \varphi_n) = \prod_{k=1}^{n} \Psi(\varphi_k) \quad (\Psi \in \mathcal{H}, \varphi_k \in M, n \in \mathbb{N}).$$

It is well known [722] that there are finite sets of certain elementary signals corresponding to the different regions of the brain, and the only signals which can be stored in the memory are superpositions of such elementary signals. That's why we first passed over from $\mathcal{H}$ to the finite-dimensional subspace $\mathcal{H}_{\mathrm{sig}}$ of regular signals. Now, the same arguments lead to a restriction of $\overline{\mathcal{H}}$ to $\mathcal{H}_{\mathrm{mem}} \subset \overline{\mathcal{H}}$. This subspace we will call the *memory space*.

Now, according to a postulate, recognition is based on a comparison of incoming signals with signals coming from the memory. For that reason, according to postulate (P6), we have to define an operator describing the removal of single signals from the sets of signals stored in the memory. This will be done with the aid of an isometry being a bounded variant of the *Hida–Malliavin derivative*.

Let $\mathcal{D} : \mathcal{H}_{\mathrm{mem}} \to \mathcal{H}_{\mathrm{mem}} \otimes \mathcal{H}_{\mathrm{sig}}$ be the isometry given on the exponential vectors $\mathrm{EXP}(R)$ by

$$\mathcal{D}\mathrm{EXP}(R) = \mathrm{EXP}(R) \otimes \frac{1}{\sqrt{N+1}} R \quad (R \in \mathcal{H}_{\mathrm{sig}})$$

where N denotes the number operator in $\mathcal{H}$.

More precisely, if φ is a set of excited neurons $\varphi \in M$, and Φ a family of such sets of neurons we have

$$\bigl(\mathcal{D}\mathrm{EXP}(R)\bigr)(\Phi, \varphi) = \frac{1}{\sqrt{|\Phi| + 1}} \cdot \mathrm{EXP}(R)(\Phi) \cdot R(\varphi).$$

21.4.6 The Exchange Operator

Now let us consider a unitary operator $\mathcal{V}$ on $\mathcal{H} \otimes \mathcal{H}$ describing the interaction of two signals, one coming from our senses and one created by the memory with the aid of the operator $\mathcal{D}$. We fix a decomposition of G into subsets $G_1, \ldots, G_n$ in compliance with the fact that there are different parts of the brain being responsible for different tasks. Because of the decomposition (21.1) of the Fock space, we can make the following identification

$$\mathcal{H}\otimes\mathcal{H}=\big(\Gamma\big(L^2(G_1)\big)\otimes\Gamma\big(L^2(G_1)\big)\big)\otimes\cdots$$
$$\otimes\big(\Gamma\big(L^2(G_n)\big)\otimes\Gamma\big(L^2(G_n)\big)\big). \tag{21.3}$$

Postulate (P4) states that the brain acts in parallel with respect to the different regions. For that reason, the unitary operator $\mathcal{V}$ should decompose in the same way as the Fock space does. Experiments show that the regions can increase or decrease [541]. Further, they can change their location on the surface of the brain [735]. On the other hand, the physical structure of the brain does not change, i.e., there are always the same neurons, irrespective of the task they are involved.

Consequently, for each possible finite disjoint decomposition $G_1,\dots,G_n$ of G the operator $\mathcal{V}:\mathcal{H}\otimes\mathcal{H}\to\mathcal{H}\otimes\mathcal{H}$ has to have the representation

$$\mathcal{V}=\mathcal{V}_{G_1}\otimes\cdots\otimes\mathcal{V}_{G_n} \tag{21.4}$$

where $\mathcal{V}_{G_r}$ is a unitary operator on $\Gamma(L^2(G_r))\otimes(L^2(G_r))$, $r\in\{1,\dots,n\}$. Condition (21.4) is a very strong one. It was shown in [258] that only operators $\mathcal{V}:\mathcal{H}\otimes\mathcal{H}\to\mathcal{H}\otimes\mathcal{H}$ acting on exponential vectors of the type

$$\mathcal{V}\exp(f)\otimes\exp(g)=\exp(b_1f+b_2g)\otimes\exp(b_3f+b_4g)\quad\big(f,g\in L^2(G)\big)$$

fulfill condition (21.4). Hereby, $(b_1,b_2,b_3,b_4):G\to\mathbb{C}^4$ are functions satisfying

$$\big|b_1(x)\big|^2+\big|b_2(x)\big|^2=1=\big|b_3(x)\big|^2+\big|b_4(x)\big|^2\quad(x\in G),$$
$$b_1(x)\overline{b}_3(x)+b_2(x)\overline{b}_4(x)=0\quad(x\in G).$$

Moreover, taking into account that neurons are "nonspecialized" (cf. [722]) the functions b_k should be constant. Adding a symmetry condition, we finally infer

$$b_4=-\frac{1}{\sqrt{2}},\qquad b_1=b_2=b_3=\frac{1}{\sqrt{2}}.$$

Summarizing we can conclude that the interaction of two signals should be described by the *symmetric beam splitter* $\mathcal{V}$ on $\mathcal{H}\otimes\mathcal{H}$ given on exponential vectors $f,g\in L^2(G)$ by

$$\mathcal{V}\exp(f)\otimes\exp(g)=\exp\bigg(\frac{1}{\sqrt{2}}(f+g)\bigg)\otimes\exp\bigg(\frac{1}{\sqrt{2}}(f-g)\bigg). \tag{21.5}$$

Observe that the operator $\mathcal{V}$ is unitary and self-adjoint which implies $\mathcal{V}\mathcal{V}=I_{\mathcal{H}\otimes\mathcal{H}}$ (I denotes the identity operator). Further, $\mathcal{V}$ is a so-called *exchange operator*, i.e., $\mathcal{V}$ preserves the number of excited neurons supporting both of the signals in each part of G. For that reason, the use of the unitary operator $\mathcal{V}$ reflects the hypothesis that signals permanently exchange their supports without any loss of information [722]. A detailed investigation of the class of exchange operators was given in [250].

21.4.7 Processing of the Signals

Processing is described on the space

$$\boxed{\tilde{\mathcal{H}} \equiv \mathcal{H}_{\text{sig}} \otimes \mathcal{H}_{\text{sig}} \otimes \mathcal{H}_{\text{mem}}}$$

$\uparrow$	$\uparrow$	$\nwarrow$
signal chosen from memory	incoming signal	memory before and after processing

With the aid of $\mathcal{D}$, a signal is created from memory. Then this signal and the incoming signal interact via the operator $\mathcal{V}$, and two new signals arise. The operator $\mathcal{D}^*$ adjoint to $\mathcal{D}$ reimplants the changed signal to the memory. This is, of course, only the processing in a very simplified way where only one step is taken into account. In a more realistic model, one has to incorporate also repeated interactions, amplification processes, etc. Schematically, one step of the processing could be illustrated by the following picture

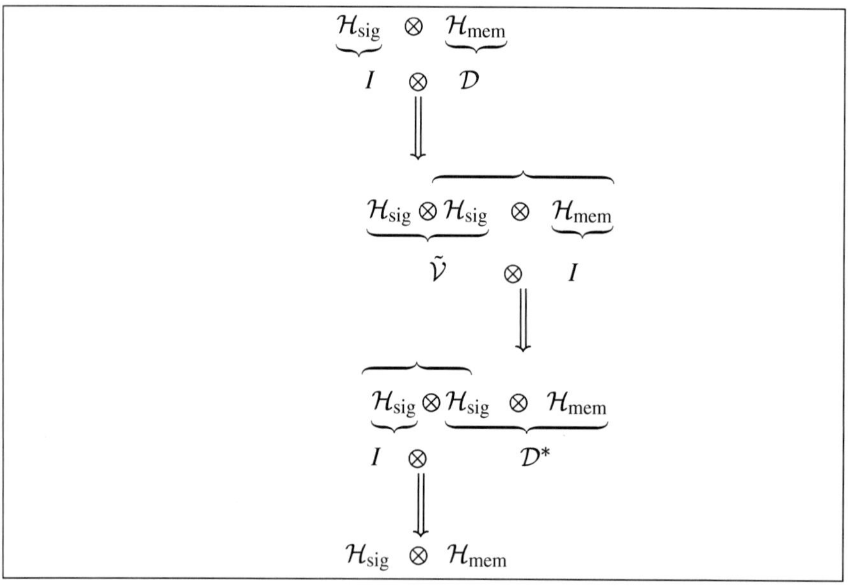

21.4.8 Recognition of Signals

In order to describe the basic idea of recognition, we consider the special case of coherent states of the input-signal given by the exponential vectors $\Psi^{\text{in}} = \exp(g)$

and coherent signals $\Psi^{\text{mem}} = \exp(f)$ chosen from the memory. The interaction with the beam splitter $\mathcal{V}$ of the signals gives again a pair of coherent states

$$\mathcal{V}\big(\exp(f) \otimes \exp(g)\big) = \exp\left(\frac{1}{\sqrt{2}}(f+g)\right) \otimes \exp\left(\frac{1}{\sqrt{2}}(f-g)\right).$$

If we have $f = g$ then $\exp(\frac{1}{\sqrt{2}}(f-g)) = \exp(0)$ represents the vacuum state. Even in the case $f \neq g$, the vacuum can occur with probability $e^{-\frac{1}{2}\|f-g\|^2}$. In the case $f = g$, the next step gives the pair of states

$$\mathcal{V}\left(\exp\left(\frac{1}{\sqrt{2}}(f+g)\right) \otimes \exp(0)\right) = \exp\left(\frac{1}{2}(f+g)\right) \otimes \exp\left(\frac{1}{2}(f+g)\right),$$

i.e., the pair of created signal and signal arising from the senses coincide after that two steps of exchange if the above mentioned event occurs. For that reason, we will interpret this event as "full recognition" (postulate (P7)). If nothing happens, the second step of exchange reconstructs the original pair of states

$$\mathcal{V}\big(\Psi^{\text{mem}} \otimes \Psi^{\text{in}}\big) = \Psi^{\text{mem}} \otimes \Psi^{\text{in}},$$

and the procedure can start again. Of course, in a real brain we would have only an approximate recognition of the function g.

Observe that the event of recognition causes a change of the signals. Furthermore, the repetitions of the procedure after the first recognition will not cause further changes of the pair of states. The recognition causes a certain loss of excited neurons, with reflects postulate (P8). In the case of our example, the expectation of that loss is

$$\|f\|^2 + \|g\|^2 - \frac{1}{2}\|f+g\|^2 = \frac{1}{2}\|f-g\|^2.$$

Furthermore, the model reflects another well known experience, namely that in the case of signals being "unexpected" from the point of view of the memory, the probability of recognition is small but the measured activity, i.e., the loss of excited neurons, is large.

Normally, full recognition does not occur—only some kind of partial recognition. We fix a decomposition $G_1, \dots, G_n$ of G. In a first step, in one of the regions G_k recognition can occur. All further applications of V_{G_k} will not cause any change. However, occasionally after several applications of $\mathcal{V}$ in further regions we may get additional partial recognition. Such a procedure is in accordance with postulate (P1) claiming that recognition is a process discrete in time. That means the recognition of the signal will be improved step by step up to the maximal level. Such a model agrees with experimental experiences. Following [735], one step of this process lasts 10^{-13}–10^{-10} seconds, and the whole procedure lasts 10^{-3}–10^{-1} seconds.

Further mathematical study of the recognition is given in the papers [256, 259].

21.4.9 Measurements of the EEG-Type

Let u be a function on G representing the electric potential of one excited neuron measured by an electrode placed on the surface of the scalp. If φ denotes a finite subset of G representing the positions of a configuration of excited neurons then the electric potential of that configuration measured by the electrode is given by

$$U(\varphi) \equiv \left(\sum_{x \in \varphi} u(x) \right).$$

We mentioned already that the bosonic Fock space $\mathcal{H}$ can be identified with a certain L^2-space of functions of finite point configurations, i.e., functions of finite subsets φ of G [243], and the Fock space $\overline{\mathcal{H}}$ can be identified with a certain L^2-space of functions of finite families Φ of point configurations from G. For that reason, the elements of $\overline{\mathcal{H}} \otimes \mathcal{H} \otimes \mathcal{H}$ are functions $\Psi(\Phi, \varphi_1, \varphi_2)$ where φ_1 and φ_2 are finite subsets of G, and Φ denotes a finite family of point configurations from G. One may interpret

φ_1	as the "support" of the created signal,
φ_2	as the "support" of the arriving signal,
$\bigcup_{\varphi \in \Phi} \varphi$	as the "support" of all signals in the memory.

Then

$$Z(\Phi, \varphi_1, \varphi_2) \equiv \varphi_1 \cup \varphi_2 \cup \bigcup_{\varphi \in \Phi} \varphi$$

represents the positions of all excited neurons in the brain. Consequently, the operator of multiplication corresponding to the function $U(Z(\varphi_1, \varphi_2, \Phi))$ represents the measurement of the electric potential of the brain corresponding to one electrode.

Now, in the case of an EEG-device one uses a sequence of potentials $(u_k)_{k=1}^r$ of the type $u_k(x) \equiv u(x - y_k)$ $(x \in G)$. Hereby, y_k represents the position of the kth electrode placed on the surface of the scalp. The corresponding operators of multiplication commute [255], i.e., one is able to perform the corresponding measurements simultaneously.

In this section, we briefly discussed the fundamentals of mathematical description of the brain and its activities, whose details are in [256, 259, 603].

21.5 Evolution Tree and Study of the HIV and the Influenza A Viruses: As Applications of Information Dynamics

In this section, we study the evolution of species and the characterization of the HIV-1 and the Influenza A viruses by applying various entropies and the chaos degree in dynamical processes.

21.5.1 Genetic Difference and Evolution Tree

It is interesting to study the evolution of species only from the data of the genome or sequences of amino acids. We discuss the evolution tree by introducing a measure defined by the Shannon's entropies.

Almost all genetic differences are based on counting or predicting the number of substitutions during the biological evolution of each organism. There are several problems with this type of genetic difference when one considers multiple changes in one site. In such a case, entropy will play a rather nice role. Moreover, a measure by means of entropy allows us not only to give a difference for the substitution rate in two sequences but also to enable including information transmitted between two sequences.

Let X, Y be the aligned amino acid sequences of two organisms. X and Y are composed of 20 kinds of amino acids and the gap (indel) "$*$". Then we have three complete event systems:

$$\begin{pmatrix} X \\ p \end{pmatrix} = \begin{pmatrix} a_0 & a_1 & \cdots & a_{20} \\ p_0 & p_1 & \cdots & p_{20} \end{pmatrix},$$

$$\begin{pmatrix} Y \\ q \end{pmatrix} = \begin{pmatrix} a_0 & a_1 & \cdots & a_{20} \\ q_0 & q_1 & \cdots & q_{20} \end{pmatrix},$$

$$\begin{pmatrix} X \times Y \\ r \end{pmatrix} = \begin{pmatrix} ** & *a_1 & \cdots & a_{20}a_{20} \\ r_{00} & r_{01} & \cdots & r_{2020} \end{pmatrix}.$$

That is, the complete event system (X, p) of X is determined by the occurrence probability p_i $(0 \le i \le 20)$ of each amino acid and the gap $*$, which is the probability distribution $p = (p_i)_{i=0}^{20}$. Similarly, the complete event system (Y, q) of Y is determined by the probability distribution $q = (q_j)_{j=0}^{20}$. In addition, the compound event system $(X \times Y, r)$ of X and Y is determined by the joint probability distribution $r = (r_{ij})_{i,j=0}^{20}$, that is, $r_{ij} \equiv p(x_i \in X, y_j \in Y)$, which satisfies the conditions $\sum_j r_{ij} = p_i$ and $\sum_i r_{ij} = q_j$.

These event systems define various entropies, among which the following two are important:

1. Shannon entropy

$$S(X) = -\sum_i p_i \log p_i,$$

which expresses the amount of information carried by (X, p).

2. Mutual entropy

$$I(X, Y) = \sum_{i,j} r_{ij} \log \frac{r_{ij}}{p_i q_j},$$

which expresses the amount of information transmitted from (X, p) to (Y, q).

Using the entropy and the mutual entropy, a quantity measuring the similarity between X and Y was introduced as

$$r(X,Y)=\frac{1}{2}\left\{\frac{I(X,Y)}{S(X)}+\frac{I(X,Y)}{S(Y)}\right\},$$

which is called the symmetrized entropy ratio, and it takes values in the domain [0, 1]. We can also define the following similarity

$$r'(X,Y)=\frac{I(X,Y)}{S(X)+S(Y)-I(X,Y)}.$$

As the similarity between X and Y becomes larger, the values of the above $r(X,Y)$ and $r'(X,Y)$ get larger. Using one of these rates, a measure, called the *entropy evolution rate* (EER) and indicating the difference between X and Y, was introduced as follows [571]:

$$\rho(X,Y)=1-r(X,Y)\quad \bigl(\text{or } 1-r'(X,Y)\bigr),\quad 0\le\rho(X,Y)\le 1.$$

As the difference between X and Y becomes larger, the value of the entropy evolution rate gets larger.

When one is interested in studying the evolution tree for n-species, X_k $(k=1,\dots,n)$, one can compute the EER $\rho_{ij}\equiv\rho(X_i,X_j)$ for the ith and jth species, which define the so-called genetic matrix $G\equiv(\rho_{ij})$.

21.5.2 A Method Constructing Phylogenetic Trees

In this subsection, we briefly review the UPG (unweighted pair group clustering) method and the NJ (neighbor-joining) method writing a phylogenetic tree.

The UPG method was initiated by Sneath and Sokai and by Nei [728]; it is now understood as a way to divide organisms into several groups. The pair having the smallest difference makes the first group. Then we try to find the next group (pair or triple) giving the second smallest difference calculated for any pair out of organisms and the first group. Moreover, we consider the difference between two groups, that is, the averaged difference of all pairs of organisms in two groups. We repeat this procedure and make a final group as well as a final relation among all organisms. Let us show this procedure by an example.

Let the difference between an organism $\mathcal{A}$ and an organism $\mathcal{B}$ be denoted by $\rho(\mathcal{A},\mathcal{B})$, which is an element of the properly defined genetic (distance) matrix given in advance. The averaged difference between an organism $\mathcal{A}$ and a group $(\mathcal{B},\mathcal{C})$ is denoted by $\rho(\mathcal{A},(\mathcal{B},\mathcal{C}))$, and it is computed by

$$\rho\bigl(\mathcal{A},(\mathcal{B},\mathcal{C})\bigr)=\frac{\rho(\mathcal{A},\mathcal{B})+\rho(\mathcal{A},\mathcal{C})}{2}.$$

Table 21.4 Genetic distance matrix

	$\mathcal{A}$	$\mathcal{B}$	$\mathcal{C}$	$\mathcal{D}$	$\mathcal{E}$
$\mathcal{A}$	0	3	5	7	8
$\mathcal{B}$	3	0	6	8	6
$\mathcal{C}$	5	6	0	9	10
$\mathcal{D}$	7	8	9	0	7
$\mathcal{E}$	8	8	10	7	0

The averaged difference between a group $(\mathcal{A}, (\mathcal{B}, \mathcal{C}))$ and a group $(\mathcal{D}, \mathcal{E})$ is computed as

$$\rho\big(\mathcal{A}, (\mathcal{B}, \mathcal{C}), (\mathcal{D}, \mathcal{E})\big) = \frac{\rho(\mathcal{A}, \mathcal{D}) + \rho(\mathcal{A}, \mathcal{E}) + \rho(\mathcal{B}, \mathcal{D}) + \rho(\mathcal{B}, \mathcal{E}) + \rho(\mathcal{C}, \mathcal{D}) + \rho(\mathcal{C}, \mathcal{E})}{6}.$$

Now, suppose that the genetic distance matrix is given as in Table 21.4. Then an organism $\mathcal{A}$ and an organism $\mathcal{B}$ are first combined together because the difference between $\mathcal{A}$ and $\mathcal{B}$ is the smallest. Secondly, we compute the differences for a group $(\mathcal{A}, \mathcal{B})$ and one of three organisms $\mathcal{C}$, $\mathcal{D}$, and $\mathcal{E}$:

$$\begin{aligned}
\rho\big((\mathcal{A}, \mathcal{B}), \mathcal{C}\big) &= \frac{\rho(\mathcal{A}, \mathcal{C}) + \rho(\mathcal{B}, \mathcal{C})}{2} = \frac{5+6}{2} = 5.5, \\
\rho\big((\mathcal{A}, \mathcal{B}), \mathcal{D}\big) &= \frac{\rho(\mathcal{A}, \mathcal{D}) + \rho(\mathcal{B}, \mathcal{D})}{2} = \frac{7+8}{2} = 7.5, \\
\rho\big((\mathcal{A}, \mathcal{B}), \mathcal{E}\big) &= \frac{\rho(\mathcal{A}, \mathcal{E}) + \rho(\mathcal{B}, \mathcal{E})}{2} = \frac{8+8}{2} = 8, \\
&\big(\rho(\mathcal{C}, \mathcal{D}) = 9,\ \rho(\mathcal{C}, \mathcal{E}) = 10,\ \text{and}\ \rho(\mathcal{D}, \mathcal{E}) = 7\big).
\end{aligned}$$

Since the pair having the smallest difference forms a group, the group $(\mathcal{A}, \mathcal{B})$ and the organism $\mathcal{C}$ are combined.

We next compute the following three differences:

$$\begin{aligned}
\rho\big(((\mathcal{A}, \mathcal{B}), \mathcal{C}), \mathcal{D}\big) &= \frac{\rho(\mathcal{A}, \mathcal{D}) + \rho(\mathcal{B}, \mathcal{D}) + \rho(\mathcal{C}, \mathcal{D})}{3} \\
&= \frac{7+8+9}{3} = 8, \\
\rho\big(((\mathcal{A}, \mathcal{B}), \mathcal{C}), \mathcal{E}\big) &= \frac{\rho(\mathcal{A}, \mathcal{E}) + \rho(\mathcal{B}, \mathcal{E}) + \rho(\mathcal{C}, \mathcal{E})}{3} \\
&= \frac{8+8+10}{3} = 8.3, \\
&\big(\rho(\mathcal{D}, \mathcal{E}) = 7\big).
\end{aligned}$$

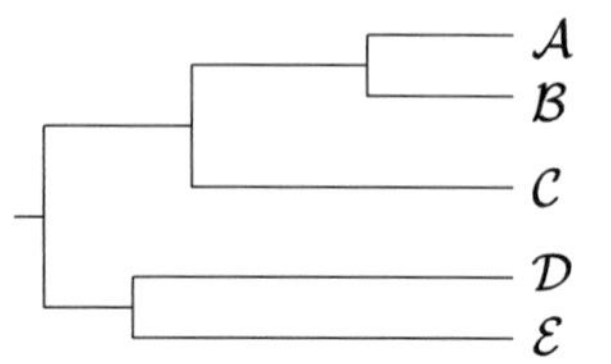

Fig. 21.2 Example of phylogenetic tree by UPG

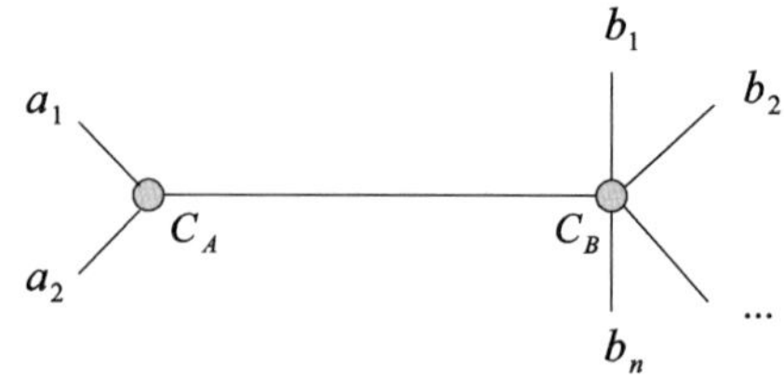

Fig. 21.3 Network of organisms

Accordingly, we have a group $(\mathcal{D}, \mathcal{E})$. Finally,

$$\rho\big(((\mathcal{A}, \mathcal{B}), \mathcal{C}), (\mathcal{D}, \mathcal{E})\big) = 8.3.$$

On the basis of the above results, we can write a phylogenetic tree of these organisms as in Fig. 21.2.

The NJ method was initiated by Saito and Nei [670]. In this method, the branch length is reflected from the genetic difference.

Let us assume that there exist n organisms, and call a node by a virtual point through which some organisms are connected. We call the two organisms connected by one node the "neighborhood", and the set of neighborhoods makes one group. The NJ method is an iterative algorithm combining two neighborhoods (i.e., an organism and an organism, an organism and a group, a group and a group) to make a new group until all organisms make one group.

First, we have to find the group of just two organisms, a "neighborhood". In order to find such a neighborhood, let us consider the following network as in Fig. 21.3. In this network, two organisms a_1 and a_2 is supposed to be a neighborhood with a node C_A and the other organisms are connected through another node C_B.

Put $A = \{a_1, a_2\}$ and $B = \{b_1, b_2, \ldots, b_n\}$. The sum of all branch lengths $T_{A,B}$ in the above network is defined as

$$\begin{aligned} T_{A,B} &= T_{(a_1,a_2),(b_1,b_2,\ldots,b_n)} \\ &= \frac{1}{2n}\sum_{i=1}^{2}\sum_{j=1}^{n} D_{a_i b_j} + \frac{1}{2} D_{a_1 a_2} + \frac{1}{n}\sum_{i=1}^{n-1}\sum_{j=i+1}^{n} D_{b_i b_j}. \end{aligned}$$

We look for the pair giving the minimum value of the above $T_{A,B}$. The pair of organisms $\{a_1, a_2\}$ which gives the minimum value makes a neighborhood.

Once certain two organisms are selected as a neighborhood, we can calculate each branch length as follows. Let D_{a_1,C_A} be the length between an organism a_1

and a node C_A, and let $D_{a_2C_A}$ be the length between an organism a_2 and a node C_A defined as

$$D_{a_1C_A} = \frac{1}{2}D_{a_1a_2} + \frac{1}{2n}\sum_{i=1}^{n}(D_{a_1b_i} - D_{a_2b_i}),$$

$$D_{a_2C_A} = \frac{1}{2}D_{a_1a_2} + \frac{1}{2n}\sum_{i=1}^{n}(D_{a_2b_i} - D_{a_1b_i}).$$

Then, we also define a genetic difference between a neighborhood of organisms in the pair $A = \{a_1, a_2\}$ and another organism c as

$$\begin{aligned} D_{A,\{c\}} &= D_{(a_1,a_2),c} \\ &= \frac{1}{2}(D_{a_1c} + D_{a_2c}). \end{aligned}$$

After we find the pair $\{a_1, a_2\}$, we regard this pair as one organism and repeat the above procedures.

Let us explain the above procedure by an example having the genetic differences as in Table 21.4.

When $A = \{a_1, a_2\}$, we denote the sum of all the branches $T_{A,B}$ as

$$T_{A,B} = T_{(a_1,a_2),(b_1,b_2,\ldots,b_n)} = T_{a_1a_2}.$$

Then we calculate a new distance matrix (T_{ij}). For example, $T_{\mathcal{AB}}$ is calculated as

$$\begin{aligned} T_{\mathcal{AB}} &= T_{(\mathcal{A},\mathcal{B}),(\mathcal{C},\mathcal{D},\mathcal{E})} \\ &= \frac{1}{2\cdot 3}\sum_{i=1}^{2}\sum_{j=1}^{3} D_{a_ib_j} + \frac{1}{2}D_{a_1a_2} + \frac{1}{3}\sum_{i=1}^{2}\sum_{j=i+1}^{3} D_{b_ib_j} \\ &= \frac{D_{\mathcal{AC}} + D_{\mathcal{AD}} + D_{\mathcal{AE}} + D_{\mathcal{BC}} + D_{\mathcal{BD}} + D_{\mathcal{BE}} + 3D_{\mathcal{AB}} + 2(D_{\mathcal{CD}} + D_{\mathcal{CE}} + D_{\mathcal{DE}})}{6} \\ &= \frac{5+7+8+6+8+8+3\cdot 3+2(9+10+7)}{6} \\ &= 17.16. \end{aligned}$$

Table 21.5 shows the distances T_{ij} calculated as above. From this table, the organisms D and E are paired as they give a minimum value. Here, the length between D and C_A, and the length between E and C_A are

$$D_{\mathcal{D}C_A} = 3.17,$$

$$D_{\mathcal{E}C_A} = 3.83.$$

Now, the tree is like in Fig. 21.4.

The distance matrix is rewritten as Table 21.6.

Table 21.5 Genetic distance matrix

T_{ij}	$\mathcal{A}$	$\mathcal{B}$	$\mathcal{C}$	$\mathcal{D}$	$\mathcal{E}$
$\mathcal{A}$	0	17.2	17.3	18.2	18.3
$\mathcal{B}$	17.2	0	17.5	18.3	18.0
$\mathcal{C}$	17.3	17.5	0	18.0	18.2
$\mathcal{D}$	18.2	18.3	18.0	0	16.5
$\mathcal{E}$	18.3	18.0	18.2	16.5	0

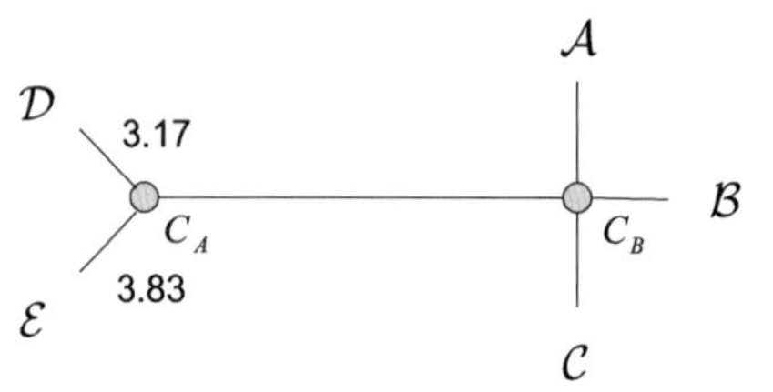

Fig. 21.4 The network based on the above data

Table 21.6 Refined distance matrix

	$\mathcal{A}$	$\mathcal{B}$	$\mathcal{C}$	$\mathcal{D}, \mathcal{E}$
$\mathcal{A}$	0	3	5	7.5
$\mathcal{B}$	3	0	6	8
$\mathcal{C}$	5	6	0	9.5
$\mathcal{D}, \mathcal{E}$	7.5	8	9.5	0

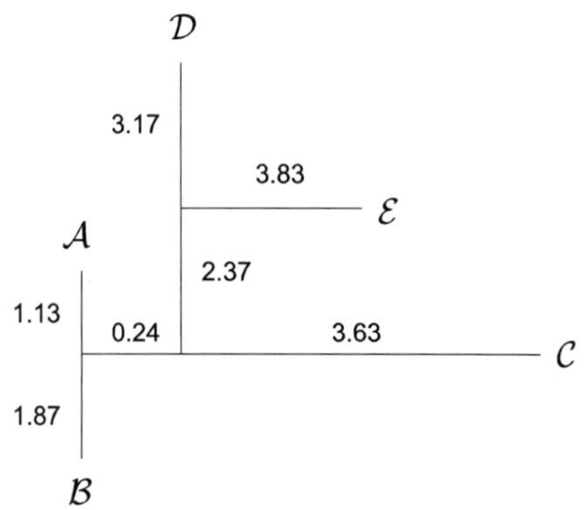

Fig. 21.5 Unrooted tree by the NJ method

We repeat this procedure until all organisms are connected into one group. Finally, the difference between two groups is calculated by selecting organisms from each group. The final network is in Fig. 21.5. Taking the middle point of the longest path as a location of a common ancestor, we get the final phylogenetic tree (Fig. 21.6).

Fig. 21.6 Rooted tree by the NJ method

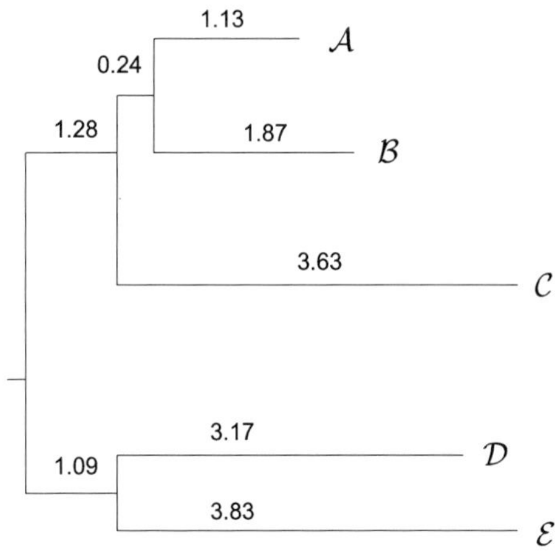

In the next subsection, we write phylogenetic trees by using these methods with the genetic matrix constructed from our entropy evolution rate $\rho(\mathcal{A},\mathcal{B})$ and the substitution rate.

21.5.3 Phylogenetic Tree for Hemoglobin α

In this subsection, we write phylogenetic trees for hemoglobin α. We here consider the following species: Monodelphia (human, horse), Marsupialia (gray kangaroo), Monotremata (platypus), Aves (ducks, greylag goose), Crocodilia (alligator, nile crocodile), viper, bullfrog tadpole, Osteinthyes (carp, goldfish), Port Jackson shark. All data are taken from the PIR [282]. Let the degree of difference between two organisms $\mathcal{A}$ and $\mathcal{B}$ be given as

$$\rho_1(\mathcal{A},\mathcal{B}) = 1 - r(\mathcal{A},\mathcal{B})$$

or

$$\rho_2(\mathcal{A},\mathcal{B}) = \frac{n}{N},$$

where n is the number of replaced amino acids between the sequences $\mathcal{A}$ and $\mathcal{B}$ and N be the number of amino acids in $\mathcal{A}$ or $\mathcal{B}$ both after the alignment. Then we can make the genetic matrix ρ_k from $\rho_k(\mathcal{A},\mathcal{B})$ $(k = 1, 2)$ such that $\rho_k = (\rho_k(\mathcal{A},\mathcal{B}))$, by which we can construct the phylogenetic tree for the above species. The difference ρ_1 is new, but the difference ρ_2 (or its modification) is used on several occasions.

The phylogenetic trees written by ρ_1 and ρ_2 are shown in Fig. 21.7 and Fig. 21.8, respectively.

Figure 21.9 is a result estimated from the fossils of species. At first glance, Fig. 21.7 is closer to Fig. 21.9 than Fig. 21.8.

For more scientific judgment among the resulted phylogenetic trees, Robinson and Foulds [662] considered a certain operation which expresses the movement of a

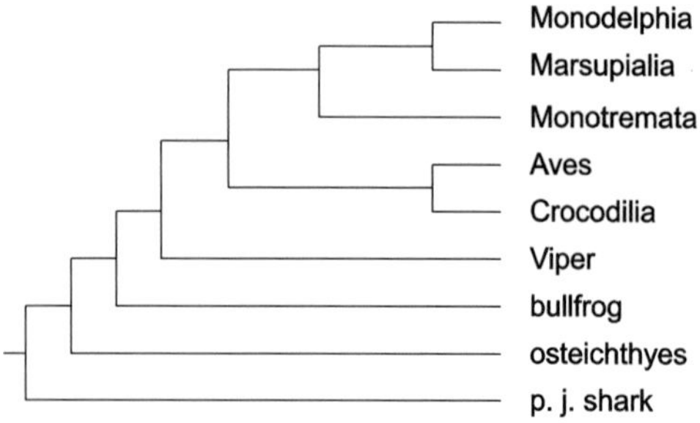

Fig. 21.7 Tree constructed by entropy evolution ratio

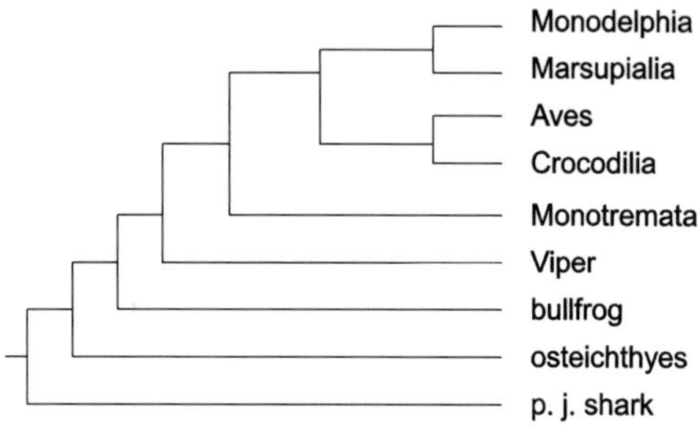

Fig. 21.8 Tree constructed by substitution rate

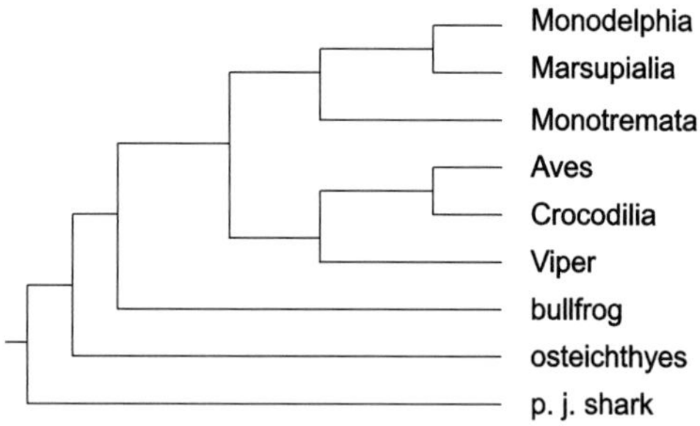

Fig. 21.9 Tree constructed by fossils

branch between two phylogenetic trees. For two phylogenetic trees, say X and Y, if we can overlap X with Y by moving n branches in X, then the difference between X and Y is said to be n. There are two such operations α and β: α is the operation adding a branch in a tree, and β is that eliminating a branch, as shown in Fig. 21.10.

The difference of three phylogenetic trees in Fig. 21.7, Fig. 21.8, and Fig. 21.9 are shown in Table 21.7. Therefore, if we believe the phylogenetic tree written by the data of fossils and if the UPG method is a plausible way to write the phylogenetic tree, then the genetic matrix ρ_1 constructed by the entropy ratio will be better than the genetic matrix ρ_2. Even so, we have a little difference between Fig. 21.7 and Fig. 21.9, so that we might need to refine both the UPG method and the genetic matrix in order to write more accurate phylogenetic trees.

Fig. 21.10 RF operations

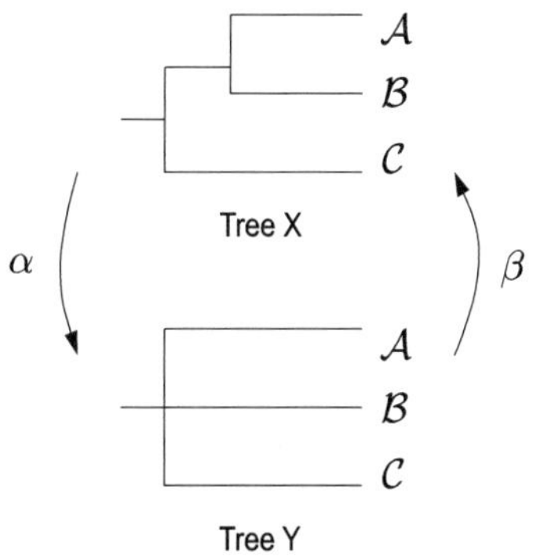

Table 21.7 RF-distance between trees

	Figure 21.7	Figure 21.8	Figure 21.9
Figure 21.7		2	2
Figure 21.8	2		4

21.5.4 Evolution and Its Tree of HIV by EER

We used the sequence data of V3 region of the *env* gene obtained from 25 HIV-1 infected patients (P1–P25) reported in [170, 269, 356, 705, 819] and those of the protease (PR) region of the *pol* gene from 13 different patients (P13–P25) reported in [269, 356] (Table 21.8). For each infected patient, data were obtained at several points in time after HIV-1 infection. Some patients had progressed to AIDS and died of AIDS-related complications, while others have been asymptomatic during the period of the follow-up.

To estimate prognosis in patients with human immunodeficiency virus type 1 (HIV-1) infection, we analyzed sequences of V3 region obtained at several points in time following seroconversion of each patient by means of a genetic measure called entropy evolution rate. Our results indicated that the entropy evolution rate can be one of surrogate markers to estimate the stage of disease.

Data used in our analysis are amino acid and nucleotide sequences of the V3 region for virus clones that were isolated at several time points from each of HIV-1-infected patients. Here, 25 patients were designated as patients P1–P25. Table 21.8 shows the summary of data for patients used in this study.

In order to examine the relation between sequence variability in V3 and disease progression, we constructed phylogenetic trees using the neighbor-joining (NJ) method with genetic matrix made using the entropy evolution rate. The denotation of viral sequences in phylogenetic analysis means that the first part is the time following seroconversion (year (y), month (m)) or the date (e.g., 11/'87). The second part indicated in parentheses is the number of identical sequences (e.g., (9)). [M] and [T] denote the macrophage-tropic virus and T-cell-tropic virus, respectively. The filled circles indicate SI (syncytium inducing) phenotype, the others indicate NSI (non-syncytium inducing) phenotype.

Table 21.8 HIV-1 infected patients

Patients	AIDS diagnosis	Death diagnosis	Molecular type	Region of HIV-1 gene
P1	×	o109 months (9.1 years) after seroconversion	viral RNA of plasma and viral DNA of PBMC	V3
P2	×	×	viral RNA of plasma and viral DNA of PBMC	V3
P3	o	o109 months (9.1 years) after seroconversion	viral RNA of plasma and viral DNA of PBMC	V3
P4	o	o 97 months (8.1 years) after seroconversion	viral RNA of plasma and viral DNA of PBMC	V3
P5	o	o 85 months (7.1 years) after seroconversion	viral RNA of plasma and viral DNA of PBMC	V3
P6	o	o109 months (9.1 years) after seroconversion	viral RNA of plasma and viral DNA of PBMC	V3
P7	×	×	viral RNA of plasma and viral DNA of PBMC	V3
P8	×	×	viral RNA of plasma and viral DNA of PBMC	V3
P9	o	o around 10 years after sero-conversion	viral RNA of serum	V3
P10	o	o around 14 years after sero-conversion	viral RNA of serum	V3
P11	o	×	viral RNA of serum	V3
P12	×	×	viral RNA of serum	V3
P13	×	×	viral DNA of PBMC	V3, PR
P14	×	×	viral DNA of PBMC	V3, PR
P15	×	×	viral DNA of PBMC	V3, PR
P16	×	×	viral DNA of PBMC	V3, PR
P17	×	×	viral DNA of PBMC	V3, PR
P18	×	×	viral DNA of PBMC	V3, PR
P19	×	×	viral DNA of PBMC	V3, PR
P20	×	×	viral DNA of PBMC	V3, PR
P21	×	×	viral DNA of PBMC	V3, PR
P22	×	×	viral DNA of PBMC	V3, PR
P23	×	×	viral DNA of PBMC	V3, PR
P24	×	×	viral DNA of PBMC	V3, PR
P25	×	×	viral DNA of PBMC	V3, PR

PBMCs; Peripheral blood mononuclear cells

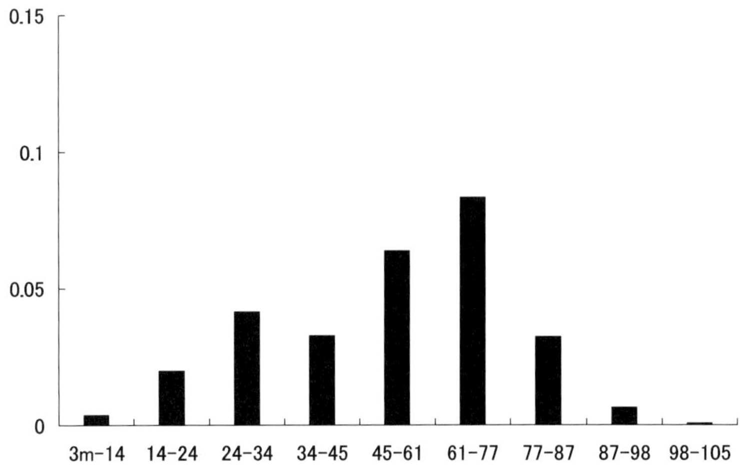

Fig. 21.11 Months post seroconversion

We analyzed sequences of the V3 region obtained at several time points after seroconversion of each patient by means of entropy evolution rate.

We here explain the evolution tree of two persons P1 and P12. Person P1 is a patient who had a decline of CD4 + T cells to less than 200 cells/μl in approximately six years (diagnosis as AIDS) after seroconversion, and died in 9.1 years. Person P12 is a patient who has been asymptomatic during the more than five years of follow-up. The analysis of other patients has been done in [596, 677]. The change of entropy evolution rate for Patient P1 is given in Fig. 21.11 and the evolution tree for P1 is given in Fig. 21.12. Those for patient P12 are given in Figs. 21.13 and 21.14, respectively.

21.5.5 Use of the Entropic Chaos Degree to Study Influenza A Viruses and HIV-1

Influenza pandemics threaten our lives. The twentieth century saw three influenza pandemics: the 1918 ("Spanish influenza") H1N1 virus, the 1957 ("Asian influenza") H2N2 virus, and the 1968 ("Hong Kong influenza") H3N2 virus. According to the World Health Organization (WHO), the deaths of the 1918, 1957, and 1968 pandemics are estimated at 40–50 million people, 2 million people, and 1 million people worldwide, respectively. And in the spring of 2009, a new H1N1 influenza virus emerged and caused another pandemic. The new 2009 H1N1 virus still remains as a very high transmission among humans and has spread worldwide. As of January 31, 2010, WHO reported that at least 15 174 people worldwide have been killed by the 2009 H1N1 infections.

Moreover, though it is not considered a pandemic, an H1N1 strain which appeared in 1977 (called "Russian influenza") caused severe epidemic in children

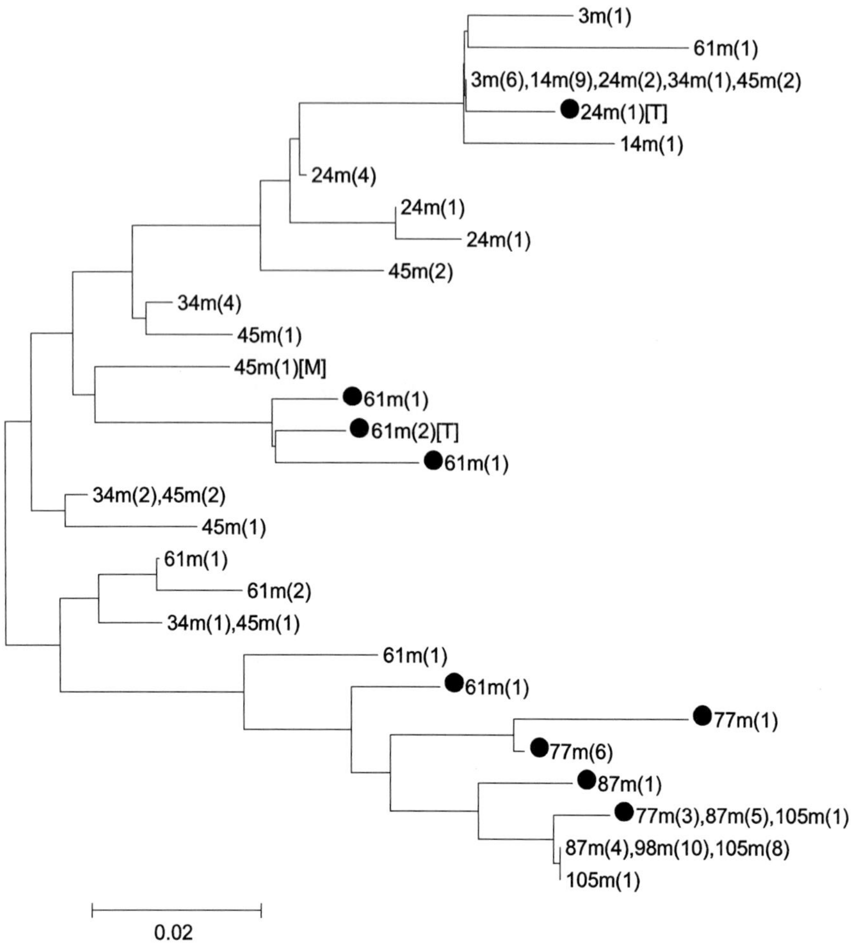

Fig. 21.12 The evolution tree for P1

and young people worldwide. Like the 1977 epidemic, a large mortality was also recorded for the epidemics that occurred in 1928–1929 (H1N1), 1932–1933 (H1N1), 1951–1953 (H1N1), and 1997–1999 (H3N2) [291, 531, 545]. By contrast, the epidemic H1N1 virus which emerged in 1947 was globally distributed like a pandemic virus, but the mortality was relatively low like a yearly influenza epidemic [531, 545, 708].

It is very interesting to find the dynamics causing the changes of viruses from one type to another. If one could get a rule to describe the dynamics such as Schrödinger equation of motion and trace the changes of the viruses, it would be a great step in studying the genome of the influenza A virus, even more in studying the disease due to that virus. However, it is also very difficult to find such an equation of motion and more difficult to solve it because the micro-dynamics of the virus change will be one of the multi-body problems with very complicated interactions. In any case, we have

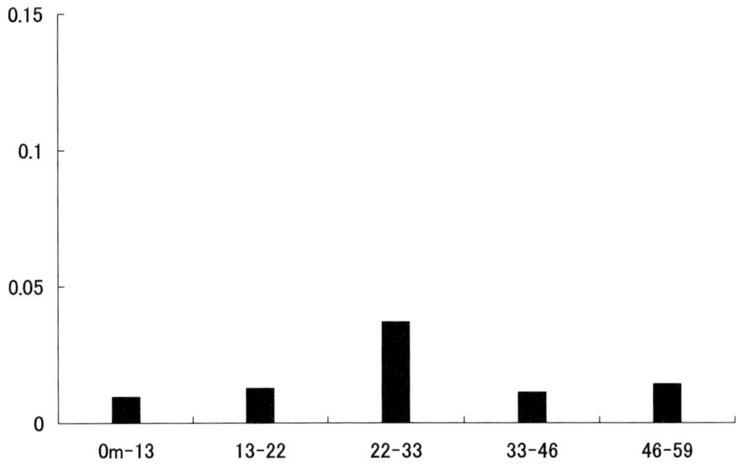

Fig. 21.13 Months post seroconversion

to look for some rule representing the dynamics even in not complete micro-level, that is, in a certain macro-level. Most of biologists discuss the changes of viruses by looking at each site of genome or counting the substitution rate of the sites. We use the probability theory and the chaos degree to study such changes, which provide us with a few more rules than merely counting.

Influenza A viruses have eight pieces of segmented RNA that encode 11 proteins [614]. The antigenic properties in the two viral surface proteins, hemagglutinin (HA) and neuraminidase (NA), are used to classify influenza A viruses into different subtypes. Influenza A viruses representing 16 HA (H1 to H16) and 9 NA (N1 to N9) antigens have been detected in wild birds and poultry all over the world [453, 614]. Wild aquatic birds are considered to be natural reservoirs for all subtypes of Influenza A viruses [263, 544].

Avian influenza A viruses are classified as either highly pathogenic avian influenza (HPAI) or low pathogenic avian influenza (LPAI) viruses, based on a polybasic amino acid cleavage site within the HA and the severity of disease. To date, only subtypes containing H5 or H7 have been found in the highly pathogenic form. Since 2003, the HPAI H5N1 viruses have spread throughout Asia, Europe, Middle East, North and West Africa with outbreaks in poultry and continuing cases of human infection [214, 633]. Despite widespread exposures to poultry infected with the H5N1 viruses, the H5N1 viruses have never circulated widely among humans [2]. According to the WHO, almost all human infections can be linked to contact with infected poultry.

Currently, influenza A viruses circulating among humans are H1N1, H1N2, and H3N2 subtypes. Influenza A viruses continually change by the accumulation of point mutation in the gene that encodes the two HA and NA proteins. This continual change is called "antigenic drift", and is associated with seasonal influenza. Besides antigenic drift, influenza A viruses suddenly change to form a new influenza A subtype or a virus with HA segment or HA and NA segments emerging by re-

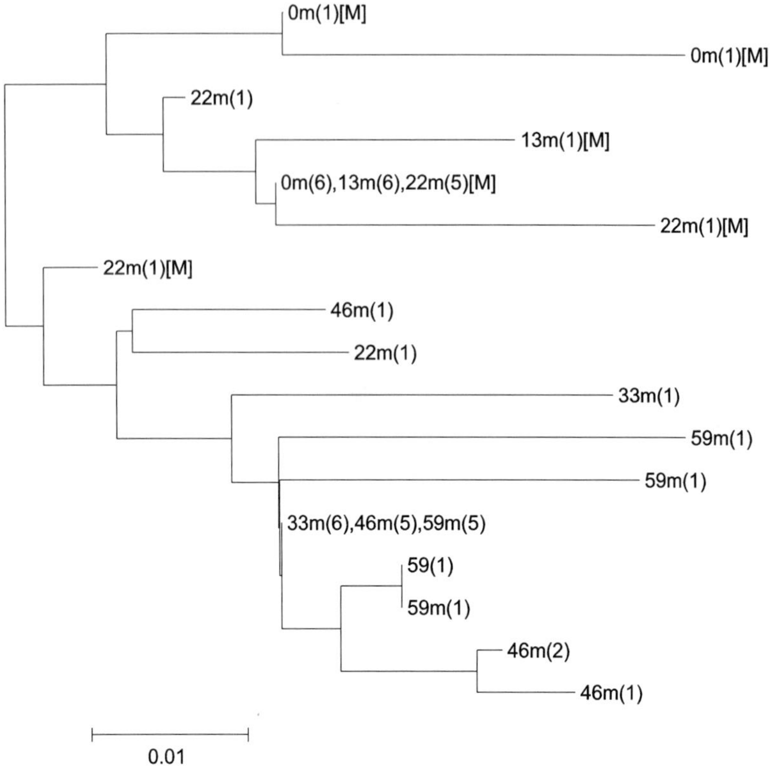

Fig. 21.14 The evolution tree for P12

assortment when two more influenza virus strains of the same or different subtypes co-infect a single host cell and whereby exchanging RNA segments [544]. This sudden change is called "antigenic shift", and is associated with a pandemic. The cause of the 1957, 1968, and 2009 pandemics is considered as antigenic shift [708, 726], though the prerequisite condition for pandemic emergence is unknown [744, 755]. Moreover, the cause of the 1947, 1951–1953, and 1997–1999 epidemics is considered as intrasubtypic antigenic shift [531, 744]. Other severe epidemic viruses with high mortality are caused by antigenic drift, such as seasonal influenza viruses [531].

Given this, we have the following questions: (i) Why the H5N1 viruses have not initiated sustained human-to-human transmission? (ii) What evolutional processes determine the severity of influenza A viruses? (iii) What are the conditions leading to differences between severe epidemic influenza viruses, pandemic influenza viruses, and seasonal influenza viruses? (iv) If the reason for the differences is antigenic change in the HA segment or HA and NA segments, we would like to know how much change HA and NA proteins or proteins other than HA protein need to acquire severe epidemic or pandemic capability.

Although these questions may be difficult to answer, we would like to find an answer from an information-theoretical point of view. Entropic Chaos Degree that has been introduced in Information Dynamics can explore a change in the amount of information that the whole sequence holds. We applied the entropic chaos degree to the course of sequence changes in 10 proteins (PB1, PB2, PA, HA, NP, NA, M1, M2, NS1, and NS2) of influenza A viruses that were sampled between 1918 and 2009.

21.5.6 Entropic Chaos Degree

Entropic chaos degree (ECD for short) has been used to characterize the chaotic aspects of the dynamics leading sequence changes [573, 590]. We will remind the ECD that was discussed in Chap. 10. The ECD for the amino acid sequences is given as follows [599]. Take two influenza A viruses X and Y. We would like to find a rule how X changes to Y if X is supposed to be ahead of Y. As we explained above, it is almost impossible to write down the equation of motion in the very micro-level, say the level of quantum mechanics. However, it is true that there exists some micro-dynamics causing the change from X to Y even though we cannot state the exact form of the dynamics. We denote this dynamics by Λ^*_{micro} and its extension to a macro-scale properly considered by Λ^*. Even when the micro-dynamics Λ^*_{micro} is hidden, we can somehow find the macro-dynamics Λ^*. The macro-dynamics we consider here is one in the usual discrete probability theory, in which we can apply the usual Shannon's information theory. So the macro-dynamics Λ^* is nothing but a channel from the probability space of X to the probability space of Y. After two amino acid sequences are aligned, they are symbolized as X and Y. The complete event system of X is determined by the occurrence probability p_i of each amino acid a_i and p_0 of the gap $*$;

$$(X, p) = \begin{pmatrix} * & a_1 & \cdots & a_{20} \\ p_0 & p_1 & \cdots & p_{20} \end{pmatrix}.$$

In the same way, the complete event system of Y is denoted by

$$(Y, \overline{p}) = \begin{pmatrix} * & a_1 & \cdots & a_{20} \\ \overline{p_0} & \overline{p_1} & \cdots & \overline{p_{20}} \end{pmatrix}.$$

The compound event system is denoted by

$$(X \times Y, r) = \begin{pmatrix} ** & *a_1 & \cdots & a_{20}a_{20} \\ r_{00} & r_{01} & \cdots & r_{2020} \end{pmatrix}$$

$$\left(\sum_{j=0}^{20} r_{ij} = p_i, \sum_{i=0}^{20} r_{ij} = \overline{p_j} \right),$$

and it represents the joint probability for the event of X and the event of Y. From the hidden dynamics for a sequence change, a mapping Λ^* coming from this dynamics can be considered to describe the change of the sequence from X to Y, and it is a channel sending the probability distribution p to $\overline{p} \equiv \Lambda^* p$. As stated above, it is difficult to know the details of the hidden dynamics in the course of sequence changes. The ECD can be used to measure the complexity without knowing the hidden dynamics.

The ECD for the amino acid sequences is given by the following formula:

$$\mathrm{ECD}(X, Y) \equiv \sum_i p_i S(\Lambda^* \delta_i),$$

where $S(\cdot)$ is the Shannon entropy and

$$p = \sum_i p_i \delta_i, \quad \delta_i(j) = \begin{cases} 1 & (i = j), \\ 0 & (i \neq j). \end{cases}$$

Note that the $\mathrm{ECD}(X, Y)$ is written as $\mathrm{ECD}(p, \Lambda^*)$ to indicate when p and Λ^* are desired. By a simple computation, when $p, \overline{p}\ (= \Lambda^* p)$ and r are obtained by some proper means, the ECD above is represented as

$$\mathrm{ECD} = \sum_{i,j} r_{i,j} \log \frac{p_i}{r_{ij}}.$$

This chaos degree was originally considered to find how much chaos is produced by the dynamics Λ^* [608]. As discussed in Chap. 10, one considers that

1. Λ^* produces chaos if $\mathrm{ECD} > 0$
2. Λ^* does not produce chaos if $\mathrm{ECD} = 0$.

Moreover, the chaos degree $\mathrm{ECD}(X, Y)$ provides a certain difference between X and Y through a change from X to Y, so that the chaos degree characterizes the dynamics changing X to Y.

21.5.7 Evolution Analysis of Influenza A Viruses

We apply the rate of entropic chaos degree (RECD for short)

$$\mathrm{RECD}(X, Y) \equiv \frac{\mathrm{ECD}(X, Y)}{S(Y)}$$

to characterize influenza A virus strains that were sampled between 1918 and 2009.

We calculate the RECD between an influenza A virus and another one. Amino acid sequences of 10 proteins (i.e., PB2, PB1, PA, HA, NP, NA, M1, M2, NS1, and NS2, except PB1-F2) from all 8 genomic segments that were sampled between 1918 and 2009 were downloaded from Influenza Virus Resources at the National

Center for Biotechnology Information (NCBI) and Influenza Research Database at the Biodefense and Public Health Database (BioHealthBase) Bio-informatics Resource Center (BRC). In this study, we used the 10 encoded protein sequences from available influenza A virus strains where all 8 genomic segments were completely sequenced. All strain names are listed in Fig. 21.15.

For each of the 10 proteins, we aligned the amino acid sequences with some alignment methods [319, 527] and evaluated the RECD between the sequences. The RECD's $\mathrm{RECD}_{\mathrm{HA}}(X,Y)$, $\mathrm{RECD}_{\mathrm{NA}}(X,Y)$, and $\mathrm{RECD}_{\mathrm{inter}}(X,Y)$ represent the RECD of HA protein, NA protein, and internal viral protein (PB2, PB1, PA, NS, M1, M2, NS1, and NS2) for a strain X with another strain Y, respectively. $\mathrm{RECD}_{\mathrm{inter}}(X,Y)$ was defined as

$$\mathrm{RECD}_{\mathrm{inter}}(X,Y)=\frac{1}{8}\left\{\begin{array}{l}\mathrm{RECD}_{\mathrm{PB2}}(X,Y)+\mathrm{RECD}_{\mathrm{PB1}}(X,Y)\\ \quad+\mathrm{RECD}_{\mathrm{PA}}(X,Y)+\mathrm{RECD}_{\mathrm{NP}}(X,Y)\\ \quad+\mathrm{RECD}_{\mathrm{M1}}(X,Y)+\mathrm{RECD}_{\mathrm{M2}}(X,Y)\\ \quad+\mathrm{RECD}_{\mathrm{NS1}}(X,Y)+\mathrm{RECD}_{\mathrm{NS2}}(X,Y)\end{array}\right\}.$$

To classify influenza A viruses according to the value of the RECD, phylogenetic tree was constructed with the Neighbor-joining method of PHYLIP version 3.69 (Felsenstein J., PHYLIP: Phylogeny Inference Package) [235, 236] using the difference measured by the RECD. We carried out phylogenetic analysis of the HA, NA, and internal virus protein sequences, and examined the evolutionary process of influenza A viruses to find a factor causing various degree of influenza activity.

Results and Discussion

Phylogenetic analysis of the internal viral protein gave the following results as shown in Figs. 21.16, 21.17 and 21.18:

1. H5N1 viruses isolated from humans in Eurasia between 2003 and 2007 were closely related to H5N1 viruses isolated from birds in Eurasia after 2001.
2. Influenza A viruses adapting to humans and transmitting among humans were distinguished from swine lineage and avian lineage, which is one of the most important results.
3. Human influenza A viruses of human lineage fell into separate clusters (1918, 1933–1936, 1940–2008, 1968–1982, 1977, 1999–2008, and 2009).

We insist that influenza A viruses were clearly classified into three lineages (i.e., Avian lineage, Human lineage, and Swine lineage) for the phylogenetic analysis by the RECD, which gives better classification compared with the tree by another measure in common use, called the substitution rate.

Although we will not discuss the details here (they are given in [678]), our results indicate that the influenza strains that caused severe epidemics with high mortality or pandemics emerged under the following conditions:

Influenza A virus strain name used in this study			
A/Brevig Mission/1/1918(H1N1)	A/USSR/92/77(H1N1)	A/swine/Bakum/1832/2000(H1N1)	A/duck/Viet Nam/12/2005(H5N1)
A/swine/1931(H1N1)	A/swine/Tennessee/25/1977(H1N1)	A/duck/Guangxi/xa/2001(H5N1)	A/duck/Viet Nam/1/2005(N5N1)
A/Wilson-Smith/33(H1N1)	A/Hong Kong/117/77(H1N1)	A/swine/Spain/33601/2001(H3N2)	A/Indonesia/5/2005(H5N1)
A/Melbourne/35(H1N1)	A/Tientsin/78/1977(H1N1)	A/swine/Spain/39139/2002(H3N2)	A/chicken/Nigeria/641/2006(H5N1)
A/Phila/1935(H1N1)	A/mallard/Alberta/77/1977(H2N3)	A/New York/78/2002(H1N2)	A/Indonesia/CDC594/20068H5N1)
A/Henry/1936(H1N1)	A/California/10/1978(H1N1)	A/chicken/Hebei/108/02(H5N1)	A/China/GD01/2006(H5N1)
A/Hickox/1940(H1N1)	A/red-necked stint/Australia/4189/1980(H4N8)	A/chicken/Jilin/hd/2002(H5N1)	A/Cygnus cygnus/Iran/754/2006(H5N1)
A/Bellamy/1942(H1N1)	A/pheasant/MN/917/1980(H7N3)	A/duck/Zhejiang/bj/2002(H5N1)	A/duck/Italy/194659/2006(H3N2)
A/Cameron/1946(H1N1)	A/Nanjing/2/82(H3N2)	A/blue-winged teal/Ohio/908/2002(H3N2)	A/Shanghai/1/2006(H5N1)
A/Fort Monmouth/1/1947(H1N1)	A/chicken/Pennsylvania/1/1983(H5N2)	A/chiken/Shantou/3744/2003(H5N1)	A/grey heron/Hong Kong/3088/2007(H5N1)
A/duck/England/1/1956(H11N6)	A/laughing gull/New Jersey/75/1985(H2N9)	A/Beijing/01/2003(H5N1)	A/Jiangsu/1/2007(H5N1)
A/Japan/305/1957(H2N2)	A/Memphis/12/1986(H1N1)	A/duck/Guangxi/12/2003(H5N1)	A/turkey/Saudi Arabia/6732-6/2007(H5N1)
A/Denver/57(H1N1)	A/mallard/Ohio/48/1986(H3N2)	A/duck/Guangxi/27/2003(H5N1)	A/swine/Minnesota/SG-00239/2007(H1N2)
A/chicken/Scotland/1959(H5N1)	A/swine/Virginia/671/1987(H1N1)	A/goose/Jilin/hb/2003(H5N1)	A/Taiwan/70120/2008(H3N2)
A/tern/South Africa/1961(H5N3)	A/ruddy turnstone/DE/2378/1988(H7N7)	A/quail/Italy/4610/2003(H7N2)	A/Pennsylvania/PIT43/2008(H5N1)
A/equine/Sao Paulo/6/1963(H3N8)	A/chicken/New York/28263/1989(H6N3)	A/swine/Spain/51915/2003(H1N1)	A/whooper swan/Akita/1/2008(H5N1)
A Virus A/Albany/10/1968(H3N2)	A/mallard duck/ALB/155/1990(H6N3)	A/Swine/Spain/50047/2003(H1N1)	A/District of Columbia/WRAMC-1154047/2008(H1N1)
A Virus A/Beijing/1/68(H3N2)	A/blue-winged teal/Alberta/141/1992(1N1)	A/New York/32/2003(H3N2)	A/Taiwan/70167/2008(H1N1)
A/Hong Kong/1-1-MA-12/1968(H3N2)	A/chicken/Hidalgo/281590232/1994(H5N2)	A/New York/61A/2003(H3N2)	A/peregrine falcon/Hong Kong/2142/2008(H5N1)
A/equine/Sachiyama/1/197(H3N8)	A/chicken/New York/13828-3/1995(H2N2)	A/swine/Ontario/52156/03(H1N2)	A/whooper swan/Hokkaido/1/2008(H5N1)
A/Memphis/1/1971(H3N2)	A/Goose/Guangdong/1/96(H5N1)	A/chicken/Kyoto/3/2004(H5N1)	A/swine/Hong Kong/294/2009(H1N2)
A/Memphis/102/1972(H3N2)	A/chicken/Hubei/wj/1997(H1N1)	A/crow/Kyoto/53/2004(H5N1)	A/California/04/2009(H1N1)
A/equine/Kentucky/1a/1975(H7N7)	A/Hong Kong/487/97(H5N1)	A/swine/Henan/wy/2004(H5N1)	A/New York/3002/2009(H1N1)
A/duck/Hong Kong/7/1975(H3N2)	A/chicken/Chis/15224/1997(H5N2)	A/Thailand/16/2004(H5N1)	A/Mexico/47N/2009(H1N1)
A/pintail duck/ALB/86/1976(H3N2)	A/blue wing teal/Ohio/31/1999(H3N2)	A/Viet Nam/1203/2004(H5N1)	A/New York/3227/2009(H1N1)
A/swine/Colorado/1/77(H3N2)	A Virus A/New South Wales/8/1999(H3N2)	A/swine/Spain/53207/2004(H1N1)	A/Italy/05/2009(H1N1)
A/mallard duck/ALB/127/1977(H1N1)	A/Hong Kong/1774/99(H3N2)	A/turkey/Italy/4479/2004(H7N3)	A/Shanghai/37T/2009(H1N1)
A/swine/Arizona/148/1977(H1N1)	A/New York/146/2000(H1N1)	A/chukar/New York/11653-1/2005(H7N2)	A/Guangdong/02/2009(H1N1)

Fig. 21.15 Name of influenza A virus strain

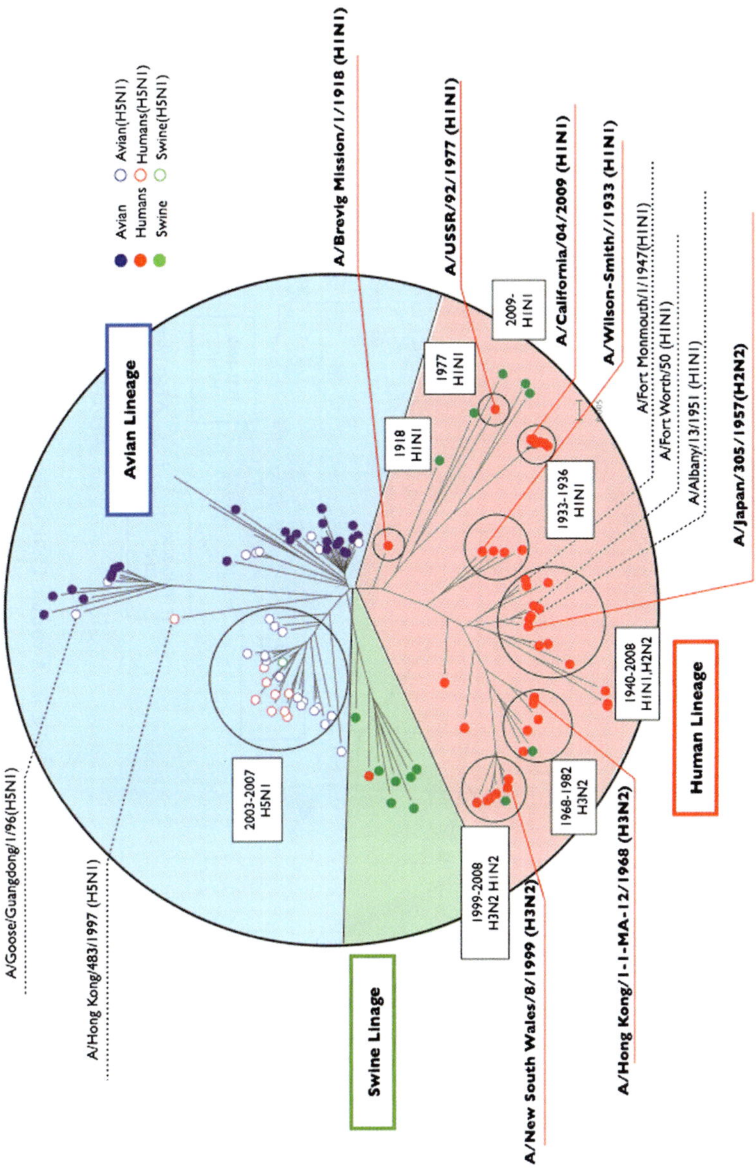

Fig. 21.16 Phylogenetic classification of the internal protein of influenza A viruses. Avian, human, and swine influenza A virus strains are represented by *blue*, *red*, and *green circles*, respectively. Influenza A virus strains that caused pandemics or severe epidemics with high mortality are indicated by boldface with *red line*

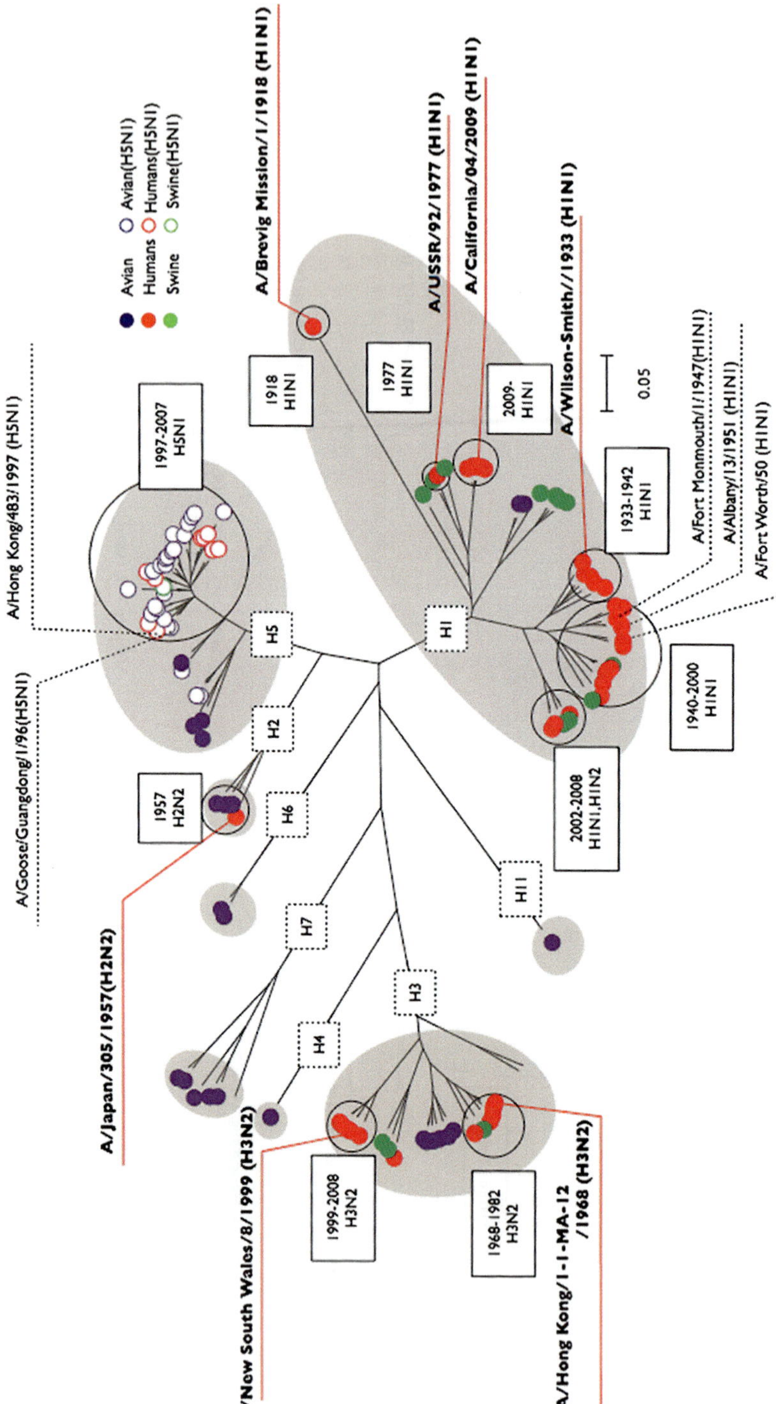

Fig. 21.17 Phylogenetic classification of the HA protein of influenza A viruses. Color scheme is the same as that used in Fig. 21.16

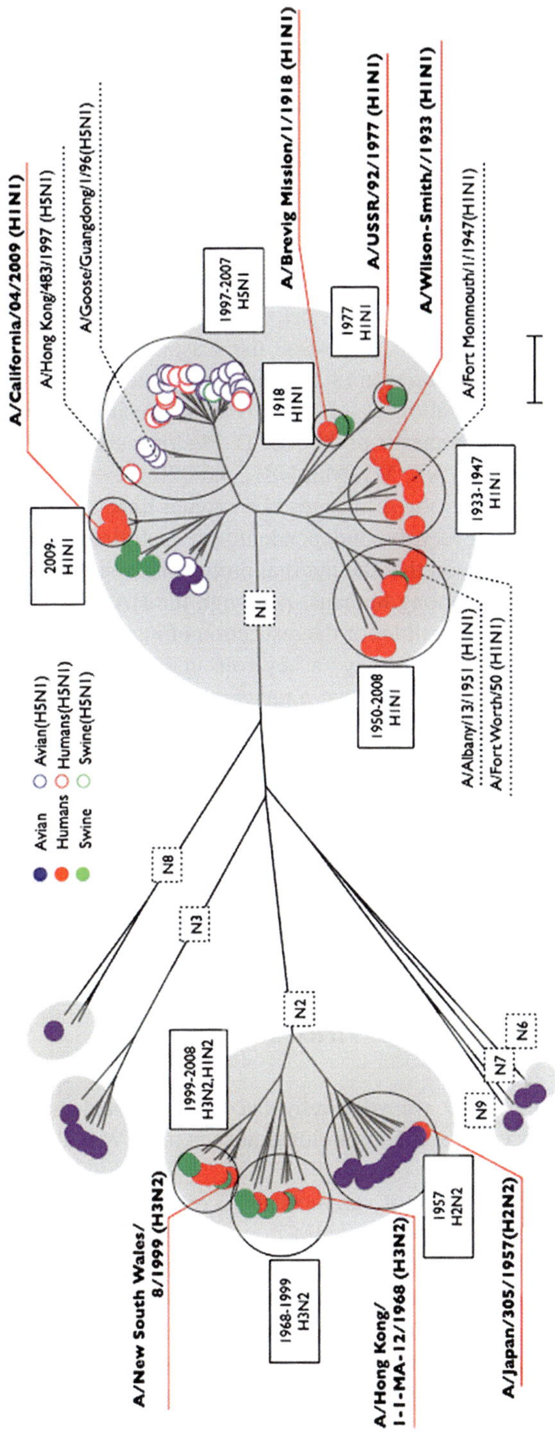

Fig. 21.18 Phylogenetic classification of the NA protein of influenza A viruses. Color scheme is the same as that used in Fig. 21.16

1. The HA and NA proteins of such epidemic or pandemic strain necessarily have to be different from those of human strains that had already appeared prior to that strain, namely new HA protein and new NA protein.
2. As for the internal protein (PB2, PB1, PA, NS, M1, M2, NS1, and NS2), the influenza strain that caused such an epidemic or pandemic necessarily has to be different from previous severe epidemic and pandemic strains.
3. A prerequisite for human–human transmission is that the internal protein is located in the human lineage.

In contrast to the severe epidemic or pandemic strains, each of the seasonal influenza strains was not the origin of strains belonging to a cluster for the HA protein and/or the NA protein. In addition, almost all the seasonal influenza strains were derived from a preexisting human strain for the internal protein.

We performed the classification of influenza A viruses using the difference between sequences measured by means of the RECD. Phylogenetic analysis of the internal protein (PB2, PB1, PA, NS, M1, M2, NS1, and NS2) revealed that influenza A viruses can be divided into three lineages (i.e., Avian lineage, Human lineage, and Swine lineage), and they evolve independently in each lineage. In this study, we have come to the conclusion that the internal protein has a significant impact on the ability for transmission among humans. Although the HA protein of influenza A viruses is known to be responsible for the restriction of interspecies transmission, we found that the internal protein plays a key role in species transmission. Furthermore, our present results indicate that a pandemic strain or a severe epidemic strain emerges in a combination of new HA, new NA, and new internal proteins that are phylogenetically distinct from those of previous pandemic and severe epidemic strains.

Based on this study, we are convinced that entropic chaos degree describes the dynamics hidden in the evolution of the influenza A virus and can be a useful measure for understanding the classification and severity of an isolated strain of influenza A virus.

21.5.8 Evolution of the HIV-1 Viruses by ECD

As another application of ECD, we will briefly discuss evolutional change of HIV-1. We used the V3 sequences for some virus clones at each time. We calculate the ECD and take its average.

Figure 21.19 shows the course of variation of V3 region studied by the entropic chaos degree in two HIV-1-infected patients.

P1 has died of AIDS-related complications at 109 months post-seroconversion. In contrast, P12 has been asymptomatic during the follow-up period. The vertical axis shows the value of the entropic chaos degree. x–y at the horizontal axis means x months and y months after seroconversion.

Since the ECD describes the stage of the variation, it can be considered that the state of the disease progression is characterized by this degree. P1 has progressed to

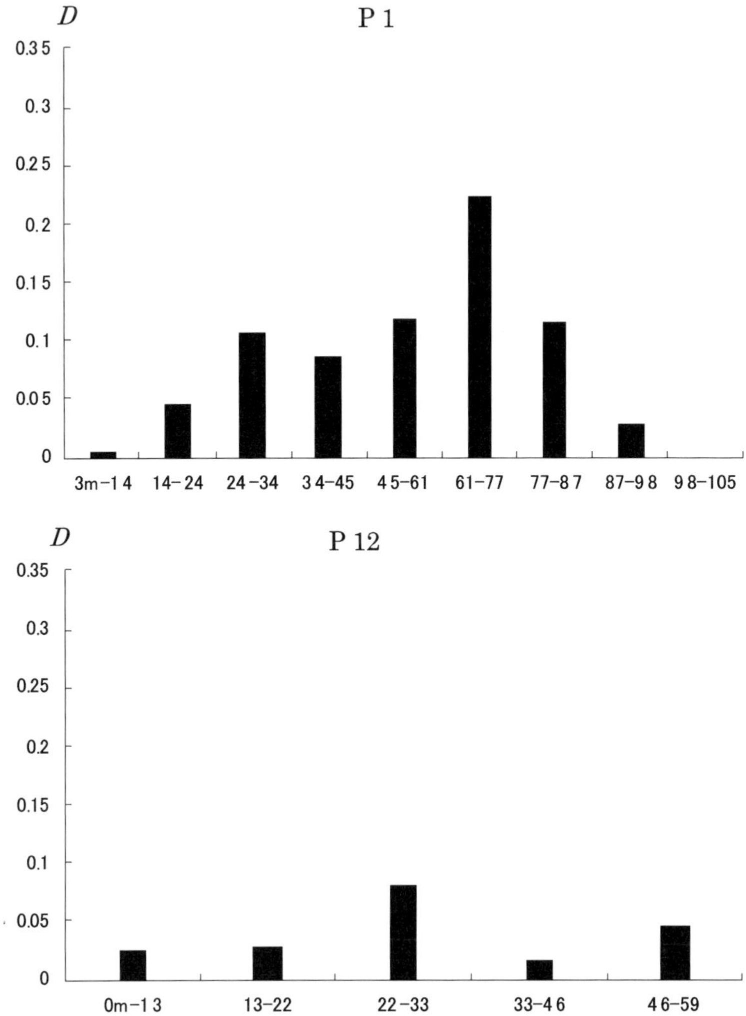

Fig. 21.19 The course of variation of V3 region by ECD

immunologic AIDS (about six years post-seroconversion 200 CD4 + T cells per μl) and has died of AIDS-related complications at 109 months post-seroconversion. The value of the ECD gradually increases from the primary HIV-1 infection to around the time of AIDS diagnosis, and then continues to decrease up to death for duration of AIDS. For almost all patients who died of AIDS-related complications, the entropic chaos degree showed the same variation patterns.

In contrast, P12 has been asymptomatic for 59 months after infection. The value of the ECD for asymptomatic patients like P12 has stayed stable and low. Our results indicate that the value is comparatively low at the early stage of HIV-1 infection, and then gradually increases from asymptomatic HIV-1 infection to around the time of

AIDS diagnosis, then decreases up to death. That is, the variation pattern of the ECD indicated that the degree is useful to infer patient's stage of disease progression after HIV-1 infection.

21.6 Code Structure of HIV-1

The purpose of this study is to find the similarity between the code structure of nucleotide sequences in HIV-1 genes and the code structure of an artificial code, and to examine whether a nucleotide sequence can be explained by an artificial code.

Information of life is stored as a sequence of nucleotides and the sequence which is composed of four bases seems to be a sort of code. Hence we can consider that the DNA or gene in each organism is a code showing its inherent structure. Thus we ask what kind of structure each code has. More precisely, we ask what roles the code structure has for the emergence of life and how it is concerned with the changes of the living body. On such questions, we explore the code structure of different genetic sequences of HIV-1 (Human Immunodeficiency Virus Type 1) and then we study the characteristics of HIV-1. Therefore, as the first step, we search for the similarity between the nucleotide sequences of HIV-1 genes and the sequences encoded by various artificial codes in information transmission and examine whether a nucleotide sequence can be explained by a certain artificial code [596].

21.6.1 Coding Theory

In a transmission system, errors that occur in communication processes are mainly divided into two groups: burst error and random error. It is the coding theory that teaches us how to symbolize the information and how to add the redundancy detecting and correcting the errors (see Chaps. 15, 17 for the related topics). The encoder that transforms the information sequence into the code sequence usually encodes each regular block, which is a segment, dividing the information sequence into several blocks. Each block of the code is called a code-word. The set of all code-words is called a code. Suppose that the information block $\mathbf{i} = (i_1, i_2, \ldots, i_k)$ consisting of numbers of k information symbols, $i_1, i_2, \ldots, i_k$, is encoded, and the code-word $\mathbf{x} = (i_1, i_2, \ldots, i_k, p_1, p_2, \ldots, p_{n-k})$ is obtained. The check symbols $p_1, p_2, \ldots, p_{n-k}$ added to the information block are dependent on $\mathbf{i}$ and form a redundancy to detect and correct errors. The code structured as above is called a systematic code, and the systematic code with the code length n, the k information symbols and the $n - k$ check symbols is called a code with information speed $R = \frac{k}{n}$, or a (n, k) code. Moreover, if the relation between the information symbols and the check symbols is linear, then the systematic code is called linear. A cyclic code and a convolutional code that we describe below are linear codes.

Cyclic Code

A cyclic code in a Galois field GF(q) is a (n,k) linear code for which any cyclic permutation of elements of any code-word is again a code-word.

In order to encode information with q-element (n,k) cyclic code, we use a polynomial $G(t)$ of degree $n-k$ on GF(q) such that the coefficient of the highest power is 1, which is called a generator polynomial. Every generator polynomial of a cyclic code is always a factor of t^n-1.

Let k information symbols be represented by a $k-1$ degree polynomial:

$$I(t) = i_1 + i_2 t + \cdots + i_k t^{k-1}.$$

Multiplying $I(t)$ by t^{n-k} and then dividing by $G(t)$, the remainder $R(t)$ is obtained:

$$R(t) = p_1 + p_2 t + \cdots + p_{n-k} t^{n-k-1}.$$

Put $X(t)$ as follows:

$$X(t) = I(t)t^{n-k} + R(t),$$

which represents the information symbols in the coefficients of degree from $n-k$ to $n-1$ and the check symbols in the coefficients from 0 to $n-k-1$. $X(t)$ is called a code polynomial.

Convolutional Code

With the cyclic code, the information is encoded for each regular block. However, convolutional codes have a property that past blocks also affect the present block.

In the convolutional code, the information sequence is divided into information blocks of length k, and the k information symbols on GF(q) go into the encoder and are encoded per unit time. The information sequence $I^j(D)$ $(j=1,2,\ldots,k)$ is expressed by a polynomial:

$$I^j(D) = i_0^j + i_1^j D + i_0^j D^2 + \cdots \quad (j=1,2,\ldots,k),$$

where i_t^j is the jth information symbol of the information block at time t. D represents the delay per unit time in the encoder and is called the delay operator. The code sequence $X^i(D)$ $(i=1,2,\ldots,n)$ is also expressed as

$$X^i(D) = x_0^i + x_1^i D + x_0^i D^2 + \cdots \quad (i=1,2,\ldots,n),$$

where $x_t^i = i_t^j$ $(i=j=1,2,\ldots,k)$ is the ith symbol of the code block with length n at time t.

The convolutional code is also a systematic code. Therefore, the following equation holds:

$$X^i(D) = I^j(D) \quad (i=j=1,2,\ldots,k).$$

The other $X^i(D)$ $(i = k+1, \ldots, n)$ is a check sequence represented by a proper polynomial $G_i^j(D)$ $(j = 1, \ldots, k, i = k+1, \ldots, n)$ with the delay operator D:

$$X^i(D) = \sum_{j=1}^{k} G_i^j(D) I^j(D).$$

This $G_i^j(D)$ is called the generator polynomial. The information speed R of the convolutional code is $R = \frac{k}{n}$. When the number of delay operators is m, the number of code blocks on which the information symbol has direct influence is $m + 1$. In addition, since each code block has length n, the influence of the information symbol extends to $n(m + 1)$ symbols, which is called the constraint length.

21.6.2 Application of Artificial Codes to Genome

How to Encode Genes

Information can be expressed by the elements of GF(q), the Galois field. Since DNA is composed of four different nucleotide bases, A, G, T, and C, it is natural to take $q = 4$. Let α be a root of the algebraic equation over GF(2) taken to be $t^2 + t + 1 = 0$, i.e., α satisfies $\alpha^2 + \alpha + 1 = 0$. The set of elements of GF(4) can be denoted by $\{0, 1, \alpha, \alpha^2\}$ and any power of α can be expressed by these four elements. In other words, the set GF(4) $= \{0, 1, \alpha, \alpha^2\}$ is closed under addition $+$ and multiplication $\cdot$.

As it was mentioned, a gene is an ordered sequence of nucleotide bases, and it is considered to be a sort of symbol sequence (code). Therefore, to find out the code structure of genes or DNA as our ultimate goal, we first investigate the similarity between the sequences encoded by artificial codes and nucleotide sequences of DNA, and we examine in which sense those nucleotide sequences can be explained by an artificial code.

Three consecutive nucleotides correspond to one amino acid, and these three nucleotides are called a codon. The total number of a three-nucleotide codes is $4^3 = 64$, which means we have 64 codons. However, only 20 amino acids exist in nature. Moreover, it is considered that the third nucleotide for a codon will not play an essential role in making an amino acid. This shows that the gene or DNA has redundancies to correct errors to a certain extent, that is, a similar structure as an error-correcting code. Based on the above consideration, in order to study the code structure of a genome sequence, we examine the similarity between an artificial error-correcting code and the nucleotide sequence of the genome. First, we select artificial correcting codes satisfying the hypothesis below, and then we investigate which artificial code characterizes the nucleotide sequence of the gene. The hypothesis we make is the following:

Each codon determines an amino acid and the third nucleotide of the codon will not have much influence on the amino acid, so that the third nucleotide is supposed to play a role of a check symbol in an error-correcting code. That is, an error-correcting code that a genome has is considered to be a code which has the code length that does not destroy the codon unit and changes the third nucleotide.

Under this hypothesis, we consider how the code structure of a gene is analyzed. Since GF(4) consists of four elements, 0, 1, α, and α^2, the four bases can be expressed as

$$\mathrm{A} \to 0, \qquad \mathrm{T} \to 1, \qquad \mathrm{C} \to \alpha, \qquad \mathrm{G} \to \alpha^2.$$

We rewrite an important part of the sequence in a gene by that of these four elements, and we make the error-correcting code by using an artificial code. The total length of such a code is a multiple of 3, and the length of the information symbols is a multiple of 2.

Artificial error-correcting codes used for our study are BCH (Bose–Chandra–Hocquenghem) codes, self-orthogonal codes, and Iwadare-codes, which are briefly explained below.

BCH Code

Let α be a primitive element of GF(q). The minimal polynomial of α is the nonzero polynomial $f(t)$, having the smallest degree such that $f(\alpha)=0$. When α is a primitive element of GF(q^m) (extended field of degree m over GF(q)), a BCH code is a cyclic code whose generator is the minimal polynomial such that $\alpha, \alpha^2, \ldots, \alpha^{2s}$ are the roots of the polynomial, where s is the error correcting capacity. Let $M_i(t)$ be the minimal polynomial of an element β^i of GF(q^m), then the q-element (n,k)-BCH codes are characterized by

code length n	: $n = q^m - 1$,
the number of information symbols k	: $k \geq n - 2ms$,
the number of check symbols $n-k$	: $n - k \leq 2ms$,
minimal distance $d_{\min}$	: $d_{\min} \geq 2s + 1$,
generator polynomial $G(t)$	: $G(t) = \mathrm{LCM}[M_1(t), M_2(t), \ldots, M_{2s}(t)]$.

Let us encode a certain nucleotide sequence by a $(12,8)$-BCH code. The $(12,8)$-BCH code is obtained from the $(15,11)$-BCH code with $q=4$, $m=2$, $s=1$.

Let α be a root of the irreducible polynomial t^2+t+1 over GF(2) and β be a root of the irreducible polynomial $t^2+t+\alpha$ over GF(4). Then the generator polynomial $G(t)$ is given by

$$G(t) = \mathrm{LCM}\big[M_1(t), M_2(t)\big] = \big(t^2+t+\alpha\big)\big(t^2+t+\alpha^2\big) = t^4+t+1.$$

Example 21.3 Let us encode the nucleotide sequence $\mathcal{A}$: GCAAGGCTACCTTCCAGGAATGCT by the $(12,8)$-BCH code of error-correcting capability $s=1$ with the generator polynomial $G(t)=t^4+t+1$. We proceed as follows:

1. Transform bases into the elements of the Galois field as follows:

$$\alpha^2\alpha\underline{0}0\alpha^2\underline{\alpha}^2\alpha 1\underline{0}\alpha\alpha\underline{1}1\alpha\underline{\alpha}0\alpha^2\underline{\alpha}^2 0 0\underline{1}\alpha^2\alpha\underline{1},$$

   where the underlines represent the third base of each codon, corresponding to the check symbols.
2. Remove the check symbols and divide the remainder (information symbols) into 8 blocks:

$$\left(\alpha^2\alpha 0\alpha^2\alpha 1\alpha\alpha\right)\left(1\alpha 0\alpha^2 00\alpha^2\alpha\right).$$

3. Calculate the check symbols from the information symbols in each block by means of the code rule:

$$\begin{aligned} R_1(t) &= I_1(t)t^4 \mod G(t) \\ &= \left(\alpha^2 t^4 + \alpha t^5 + \alpha^2 t^7 + \alpha t^8 + t^9 + \alpha t^{10} + \alpha t^{11}\right) \mod G(t) \\ &= \alpha^2 t, \\ R_2(t) &= I_2(t)t^4 \mod G(t) \\ &= \left(t^4 + \alpha t^5 + \alpha^2 t^7 + \alpha^2 t^{10} + \alpha t^{11}\right) \mod G(t) \\ &= 1 + t + \alpha^2 t^2 + t^3, \\ & \left(\underline{0\alpha^2 00}\alpha^2\alpha 0\alpha^2\alpha 1\alpha\alpha\right)\left(\underline{11\alpha^2 1}1\alpha 0\alpha^2 00\alpha^2\alpha\right). \end{aligned}$$

4. Put the calculated check symbols of each block back into the corresponding positions of the third nucleotide of a codon:

$$\left(\alpha^2\alpha\underline{0}0\alpha^2\underline{\alpha}^2\alpha 1\underline{0}\alpha\alpha\underline{0}\right)\left(1\alpha\underline{1}0\alpha^2\underline{1}00\underline{\alpha}^2\alpha^2\alpha\underline{1}\right).$$

5. Finally, rewrite the encoded symbol sequence back into the encoded nucleotide sequence, and obtain the coded sequence $\mathcal{A}^C$ by way of the (12, 8)-BCH code:

$$\mathcal{A}^C : \text{GCAAGGCTACCATCTAGTAAGGCT}.$$

Self-orthogonal Code

Let us take a set of integers $M = \{d_j : j = 1, 2, \ldots, J\}$ such that all $d_i - d_j$ $(i \neq j)$ are different. This set is called a complete simple difference set of multiplicative order J. For such a set M, a self-orthogonal convolutional code with information speed $R = \frac{1}{2}$ and constraint length $L_c = 2(d_J + 1)$ is obtained if we use a generator polynomial such as $G_2^1(D) = D^{d_1} + D^{d_2} + \cdots + D^{d_J}$. This code can correct $\frac{J}{2}$ random errors.

For example, a set $M = \{0, 2, 5, 6\}$ is such a set of $J = 4$. The self-orthogonal code with $R = \frac{1}{2}$ and $L_C = 14$ is made by the generator polynomial $G_2^1(D) = 1 + D^2 + D^5 + D^6$, whose error-correcting capacity is 2. It is easy to make the self-orthogonal code with $R = \frac{2}{3}$ from the self-orthogonal code with $R = \frac{1}{2}$. The difference set, M, is divided into two sets $\{0, 2\}, \{5, 6\} \approx \{0, 1\} \pmod 5$, by which we can obtain the generator polynomial of the self-orthogonal code with $R = \frac{2}{3}$ as

$$G_3^1(D) = 1 + D^2,$$

$$G_3^2(D) = 1 + D.$$

In this case, the constraint length is 9 and the error-correcting capacity is 1.

Iwadare Code

The burst-correcting convolutional code with information speed $R = \frac{1}{2}$ is called an Iwadare code if the generator polynomial satisfies either of the following two:

$$\text{(a)} \quad G_n^j(D) = D^{n-2+2b(n-j)+\sum_{i=1}^{n-2-j} i}\left(1 + D^{b+n-j-1}\right),$$

$$\text{(b)} \quad G_n^j(D) = D^{\beta(n-j)+n-j-1}\left(1 + D^{\beta n+n+j-2}\right),$$

where $j = 1, 2, \ldots, n-1$. This code can correct burst errors of length bn.

21.6.3 Code Structure of HIV-1

Table 21.9 shows the codes used in our study. We encoded the n nucleotide sequences of each patient at a specific point in time, and then got the encoded amino acid sequences X_j^C $(j = 1, 2, \ldots, n)$ by code C. Here, a degree to measure the similarity between an artificial code of X_j^C and the code of amino acid sequences X_j $(j = 1, 2, \ldots, n)$ before coding is defined by

$$D_C(j) = \rho\left(X_j, X_j^C\right).$$

When the code structure of X_j gets closer to C, the value of $D_C(j)$ gets closer to 0. Moreover, to look for a common code of a group of several V3 sequences or PR sequences, we use the degree of code difference D_C defined by

$$D_C = \frac{\sum_{j=1}^{n} \rho(X_j, X_j^C)}{n}.$$

By calculating D_C for various C codes, we can find a common code structure of V3 sequences (or PR sequences) obtained at each point in time after HIV-1 infection if D_C for a code is smaller than that of any other codes. Then we can infer that the group has the property that code C owns.

To study the code structure of the nucleotide sequences of HIV-1, we take the sequences from the V3 region in the *env* gene and the PR region in the *pol* gene.

Table 21.9 Cyclic and convolutional codes used

Cyclic codes	Designation	Code length	Error-correcting capacity	Generator polynomial
(3, 2)-cyclic code	$J(3,2)$	3	1 (random error)*	$t+1$
(12, 8)-BCH code	$J(12,8)$	12	1 (random error)	t^4+t^3+1
(15, 10)-BCH code	$J(15,10)$	15	1 (random error)	t^5+t^3+1
(27, 18)-BCH code	$J(27,18)$	27	2 (random error)	$t^9+\alpha^2t^8+\alpha t^7+t^6 +t^5+t^3+\alpha t^2+1$
(36, 24)-BCH code	$J(36,24)$	36	3 (random error)	$t^{12}+t^{11}+\alpha t^{10}+\alpha^2t^9 +t^7+\alpha t^6+\alpha t^5+\alpha^2t^3 +\alpha^2t^2+\alpha^2t+1$
(99, 66)-BCH code	$J(99,66)$	99	7 (random error)	$t^{33}+t^{32}+t^{31}+\alpha^2t^{30} +\alpha t^{29}+t^{28}+\alpha^2t^{27} +\alpha^2t^{25}+\alpha^2t^{24}+\alpha^2t^{23} +\alpha^2t^{22}+\alpha^2t^{21}+t^{20} +\alpha t^{19}+t^{18}+\alpha^2t^{15} +t^{13}+\alpha t^{11}+\alpha^2t^{10} +t^8+\alpha t^7+\alpha t^6+\alpha^2t^5 +\alpha t^4+\alpha^2t^3+t+1$
(105, 70)-BCH code	$J(105,70)$	105	5 (random error)	$t^{35}+t^{34}+\alpha^2t^{33}+\alpha t^{32} +\alpha t^{30}+t^{29}+\alpha t^{26} +\alpha^2t^{25}+\alpha t^{24}+\alpha^2t^{21} +\alpha^2t^{19}+\alpha^2t^{18}+\alpha^2t^{17} +\alpha t^{16}+t^{14}+\alpha^2t^{12} +t^{11}+\alpha t^9+t^7+t^5 +t^4+t^2+\alpha^2t+\alpha$
Convolutional codes	**Designation**	**Code lengths**	**Error-correcting capacity**	**Generator polynomial**
Self-orthogonal code 1	$T(j1)$	9	1 (random error)	$D+1,\ D^2+1$
Self-orthogonal code 2	$T(j2)$	42	2 (random error)	$D^{12}+D^9+D^8+1,\ D^{13}+D^{11}+D^6+1$
Self-orthogonal code 3	$T(j3)$	120	3 (random error)	$D^{39}+D^{36}+D^{23} +D^{21}+D^{14}+1,\ D^{38}+D^{33}+D^{28} +D^{27}+D^8+1$
Iwadare code 1	$T(b1)$	27	3 (burst error)	$D^8+D^3,\ D^7+D$
Iwadare code 2	$T(b2)$	24	3 (burst error)	$D^7+D^5,\ D^4+D^3$

*Error-detecting capacity

Then we encode, by means of various C codes, these nucleotide sequences obtained from infected patients, and then calculate the index D_C mentioned above.

As a result, at all points in time observed for all patients, the codes for both V3 region and PR region are close to the (3, 2)-cyclic code, which means that both regions do not have the error-correcting capacity. We think that this result is deeply

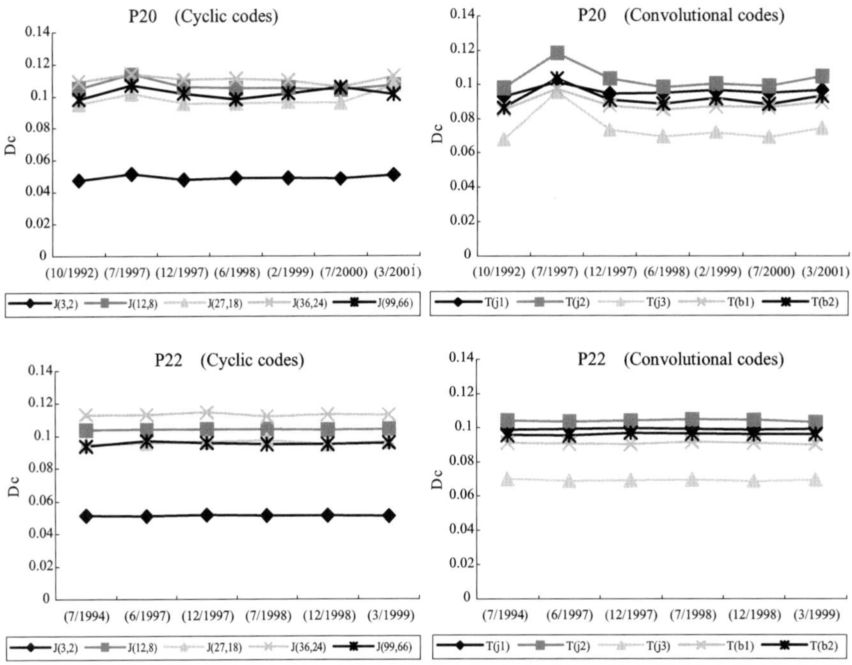

Fig. 21.20 The value of D_C by various C codes for the PR region of the *pol* gene obtained at each point in time from two patients (P20 and P22). We also got the same result for other patients as discussed above

related to a high mutation of the HIV-1 gene. In the PR region, the value of D_C by the $(3,2)$-cyclic code is almost a constant 0.05, the lowest among values of all codes at all points in time for all patients. In other words, the code of the PR region is closest to the $(3,2)$-cyclic code. Moreover, we observe that the PR region is second closest to the code structure of the self-orthogonal code 3 which has the random error-correcting capacity $s=3$. Accordingly, we can infer that the third base of the check symbol in a certain block affects widely and randomly the first and second bases of the information symbols in many other blocks due to the long constraint length of the self-orthogonal code 3 when correcting mutations in the nucleotide sequence of the PR region.

On the other hand, the V3 region has a similar structure not only for the $(3,2)$-cyclic code but also for the $(15,10)$-BCH code with the error-correcting capacity $s=1$, the self-orthogonal code 1 with $s=1$ and the self-orthogonal code 3 with $s=3$. For almost all patients, the V3 region is closest to the self-orthogonal code 1 among all error-correcting codes. On the other hand, the value D_C of the PR region for all 13 patients was not low for the self-orthogonal code 1. Therefore, the code structure of the V3 region is characterized by this self-orthogonal code 1 with $s=1$ whose constraint length is 9 so that the effect range of every base to other bases will be narrow.

However, the fact that the code structure of V3 region is close to various artificial codes, such as the $(3,2)$-cyclic code, the $(15,10)$-BCH code, the self-orthogonal code 1, and the self-orthogonal code 3, indicates the variety of its structure. In addition, the code structure of the V3 region for patients whose condition remains rather good during the latency period without any indication of AIDS-related illnesses is completely different from the structure of the $(12,8)$-BCH code with $s=1$ and that of the $(27,18)$-BCH code with $s=2$, where the value of D_C is high when compared with patients who died of AIDS-related illnesses.

We tried to check whether it is possible to see the different functions of several HIV-1 genes through the code structure. First, we encoded the nucleotide sequences of the V3 region and that of the PR region obtained at each point in time throughout longitudinal follow-ups of all patients, and then we plotted the value of D_C by three different codes, the self-orthogonal codes with $s=1$, $s=2$, and $s=3$, on x-axis, y-axis, and z-axis, respectively (Fig. 21.22). This 3D graph shows that two respective genes (in V3 and in PR) clearly form two different clusters. This implies that the self-orthogonal codes will be useful to observe the differences of HIV-1 genes.

21.7 p-adic Modeling of the Genome and the Genetic Code

In this section, we consider the p-adic modeling in genomics. p-adic numbers are an alternative to the real numbers. For an introduction to p-adic analysis and its applications, see [210, 404, 783]. We follow the work of Dragovich and Dragovich [209]. There is also another approach to p-adic modeling in genomics developed by Khrennikov and Kozyrev [423].

Considering nucleotides, codons, DNA and RNA sequences, amino acids, and proteins as information systems, the corresponding p-adic formalisms for their investigations will be formulated. Each of these systems has its characteristic prime number p used for construction of the related information space. Relevance of this approach is illustrated by some examples. In particular, it is shown that degeneration of the genetic code is a p-adic phenomenon. A hypothesis is also put forward on the evolution of the genetic code assuming that the primitive code was based on single nucleotides and chronologically first four amino acids. This formalism of p-adic genomic information systems can be implemented in computer programs and applied to various concrete cases.

A p-adic diffusion equation was suggested and studied in [783]. It was discovered in [76] that the equation can be used to describe the protein dynamics. It was shown that with the help of p-adic methods obtained results on protein dynamics coincide with the data of spectroscopic experiments for Mb–CO rebinding. One of the most important applications of the dynamics on energy landscapes and interbasin kinetics is the application to conformational dynamics of proteins. In this case, an ultrametric parameter describes the conformational coordinate for the protein [77].

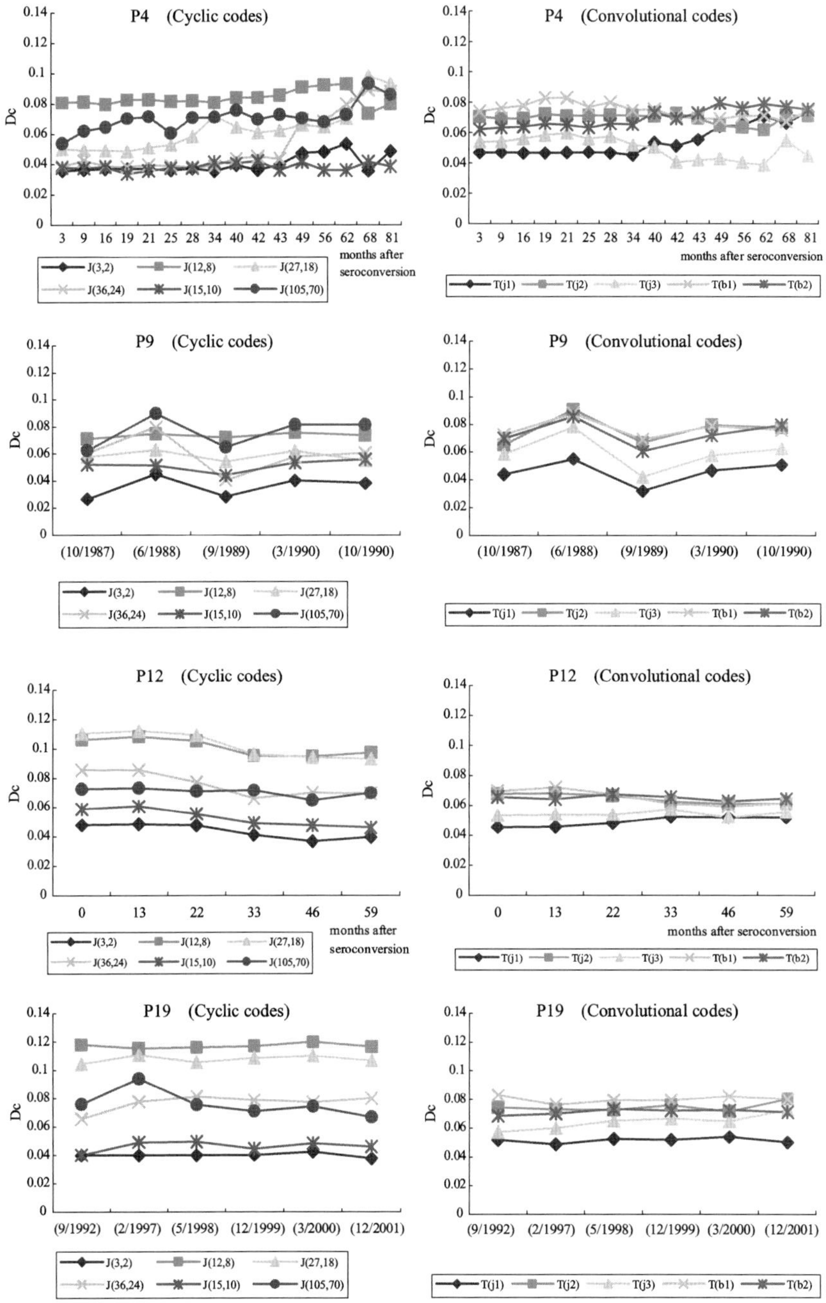

Fig. 21.21 The value of D_C by various C codes for the V3 region of *env* gene obtained from four patients as representatives of 25 patients: two patients (P4 and P9) who died of AIDS-related illnesses during the follow-up period and two patients (P12 and P19) who have been asymptomatic during the follow-up period

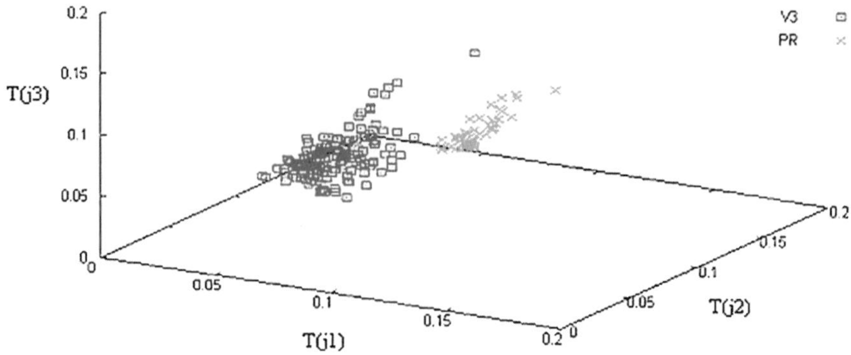

Fig. 21.22 3-D graph of D_C with the self-orthogonal code 1 ($T(j1)$), the self-orthogonal code 2 ($T(j2)$), and the self-orthogonal code 3 ($T(j3)$) for the nucleotide sequences of the V3 region and that of the PR region, where □ denotes the sequences in the V3 region and × denotes those in the RT region

21.7.1 Genetic Code

For a comprehensive information on molecular biology aspects of DNA, RNA, and the genetic code one can use [809]. To have a self-contained exposition, we shall briefly review some necessary basic properties of genomics.

The DNA is a macromolecule composed of two polynucleotide chains with a double-helical structure. Nucleotides consist of a base, a sugar, and a phosphate group. Helical backbone is a result of the sugar and phosphate groups. There are four bases and they are the building blocks of the genetic information. They are called adenine (A), guanine (G), cytosine (C), and thymine (T). Adenine and guanine are derived from purine, while cytosine and thymine from pyrimidine. In the sense of information, the nucleotide and its base present the same object. Nucleotides are arranged along chains of double helix through base pairs A–T and C–G bonded by 2 and 3 hydrogen bonds, respectively. As a consequence of this pairing, there is an equal number of cytosine and guanine as well as the equal number of adenine and thymine bases. DNA is packaged in chromosomes which are localized in the nucleus of the eukaryotic cells.

The main role of DNA is to store genetic information, and there are two main processes to exploit this information. The first one is replication in which DNA duplicates giving two new DNA containing the same information as the original one. This is possible owing to the fact that each of two chains contains complementary bases of the other one. The second process is related to the gene expression, i.e., the passage of DNA gene information to proteins. It is performed by the messenger ribonucleic acid (mRNA) which is usually a single polynucleotide chain. The mRNA is synthesized during the first part of this process, known as transcription, when nucleotides C, A, T, G from DNA are respectively transcribed into their complements G, U, A, C in mRNA, where T is replaced by U (U is the uracil, which is a pyrimidine). The next step in gene expression is translation, when the information coded by codons in the mRNA is translated into proteins. In this process, the transfer tRNA and ribosomal rRNA also participate.

Codons are ordered trinucleotides composed of C, A, U (T), and G. Each of them presents information which controls the use of one of the 20 standard amino acids or a stop signal in the synthesis of proteins.

Protein synthesis in all eukaryotic cells is performed in the ribosomes of the cytoplasm. Proteins [260] are organic macromolecules composed of amino acids arranged in a linear chain. Amino acids are molecules that consist of amino-, carboxyl- and R- (side chain) groups. Depending on the R-group, there are 20 standard amino acids. These amino acids are joined together by a peptide bond. The sequence of amino acids in a protein is determined by sequence of codons contained in DNA genes. The relation between codons and amino acids is known as the *genetic code*. Although there are at least 16 codes (see, e.g., [267]), two of them are the most important: the standard (eukaryotic) code and the vertebral mitochondrial code.

In the sequel, we shall mainly have in mind the vertebral mitochondrial code because it is a simple one, and the others may be regarded as its slight modifications. It is obvious that there are $4 \times 4 \times 4 = 64$ codons. However (in the vertebral mitochondrial code), 60 of them are distributed on the 20 different amino acids and 4 make stop codons which serve as termination signals. According to experimental observations, two amino acids are coded by six codons, six amino acids by four codons, and 12 amino acids by two codons. This property that some amino acids are coded by more than one codon is known as *genetic code degeneracy*. This degeneracy is a very important property of the genetic code and gives an efficient way to minimize errors caused by mutations.

Since there is, in principle, a huge number (between 10^{71} and 10^{84} [347]) of all possible assignments between codons and amino acids, and only a very small number of them is represented in living cells, it has been a persistent theoretical challenge to find an appropriate model explaining contemporary genetic codes. An interesting model based on the quantum algebra $\mathcal{U}_q(\mathrm{sl}(2) \oplus \mathrm{sl}(2))$ in the $q \to 0$ limit was proposed as a symmetry algebra for the genetic code (see [267] and references therein). In a sense, this approach mimics quark model of baryons. To describe the correspondence between codons and amino-acids, an operator was constructed which acts on the space of codons, and its eigenvalues are related to amino acids. Despite some successes of this approach, there is a problem with rather many parameters in the operator. There is still no generally accepted explanation of the genetic code.

Ultrametric and p-adic methods seem to be very promising tools in further investigation of life.

Modeling of the genome, the genetic code, and proteins is a challenge as well as an opportunity for applications of p-adic mathematical physics. Recently [208], a p-adic approach to DNA and RNA sequences and to the genetic code was introduced. The central point of the approach is an appropriate identification of four nucleotides with digits 1, 2, 3, 4 of 5-adic integer expansions and an application of p-adic distances between obtained numbers. 5-adic numbers with three digits form 64 integers which correspond to 64 codons. In [209], p-adic degeneracy of the genetic code was analyzed. As one of the main results, an explanation of the structure of the genetic code degeneracy using p-adic distance between codons was obtained. A similar approach to the genetic code was considered on diadic plane [423].

Let us mention that p-adic models in mathematical physics have been actively considered since 1987 [785] (see [137, 210, 404, 783] for reviews). It is worth noting that p-adic models with pseudo-differential operators have been successfully applied to interbasin kinetics of proteins [76]. Some p-adic aspects of cognitive, psychological, and social phenomena have been also considered [416]. The recent application of p-adic numbers in physics and related branches of sciences is reflected in the proceedings of the second International Conference on p-adic Mathematical Physics [422].

21.7.2 p-adic Genome

We have already presented a brief review of the genome and the genetic code, as well as some motivations for their p-adic theoretical investigations. To consider p-adic properties of the genome and the genetic code in a self-contained way, we shall also recall some mathematical preliminaries.

Some Mathematical Preliminaries and p-adic Codon Space

As a new tool to study the Diophantine equations, p-adic numbers were introduced by a German mathematician Kurt Hensel in 1897. They are involved in many branches of modern mathematics, either as rapidly developing topics or as suitable applications. An introduction to p-adic numbers can be found in [783]. However, for our purposes we will use here only a small portion of p-adics, mainly some finite sets of integers and ultrametric distances between them.

Let us introduce the set of natural numbers

$$\mathcal{C}_5[64] = \{n_0 + n_1 5 + n_2 5^2 \; : \; n_i = 1, 2, 3, 4\}, \tag{21.6}$$

where n_i are digits related to nucleotides by the following assignments: C (cytosine) = 1, A (adenine) = 2, T (thymine) = U (uracil) = 3, G (guanine) = 4. This is a finite expansion to the base 5. It is obvious that 5 is a prime number and that the set $\mathcal{C}_5[64]$ contains 64 numbers between 31 and 124 in the usual base 10. In the sequel, we shall often denote the elements of $\mathcal{C}_5[64]$ by their digits to the base 5 in the following way: $n_0 + n_1 5 + n_2 5^2 \equiv n_0 n_1 n_2$. Note that here ordering of digits is the same as in the expansion, i.e., this ordering is the opposite to the usual one. There is now evident one-to-one correspondence between codons in three-letter notation and the number $n_0 n_1 n_2$ representation.

It is also often important to know a distance between numbers. Distance can be defined by a norm. On the set $\mathbb{Z}$ of integers, there are two kinds of nontrivial norm: the usual absolute value $|\cdot|_\infty$ and p-adic absolute value $|\cdot|_p$, where p is any prime number. The usual absolute value is well known from elementary mathematics and the corresponding distance between two numbers x and y is $d_\infty(x, y) = |x - y|_\infty$.

The p-adic absolute value is related to the divisibility of integers by prime numbers. The difference of two integers is again an integer. The p-adic distance between two integers can be understood as a measure of divisibility by p of their difference (the more divisible, the shorter). By definition, the p-adic norm of an integer $m \in \mathbb{Z}$, is $|m|_p = p^{-k}$, where $k \in \mathbb{N} \cup \{0\}$ is the degree of divisibility of m by the prime number p (i.e., $m = p^k, m', p \nmid m'$) and $|0|_p = 0$. This norm is a mapping from $\mathbb{Z}$ into non-negative rational numbers and has the following properties:

(i) $|x|_p \geq 0$, $|x|_p = 0$ if and only if $x = 0$,
(ii) $|xy|_p = |x|_p|y|_p$,
(iii) $|x + y|_p \leq \max\{|x|_p, |y|_p\} \leq |x|_p + |y|_p$ for all $x, y \in \mathbb{Z}$.

Because of the strong triangle inequality $|x + y|_p \leq \max\{|x|_p, |y|_p\}$, p-adic absolute value belongs to the class of non-Archimedean (ultrametric) norms. One can easily conclude that $0 \leq |m|_p \leq 1$ for any $m \in \mathbb{Z}$ and any prime p.

p-adic distance between two integers x and y is

$$d_p(x, y) = |x - y|_p. \tag{21.7}$$

Since p-adic absolute value is ultrametric, the p-adic distance (21.7) is also ultrametric, i.e., it satisfies

$$d_p(x, y) \leq \max\{d_p(x, z), d_p(z, y)\} \leq d_p(x, z) + d_p(z, y), \tag{21.8}$$

where x, y and z are any three integers.

The above introduced set $\mathcal{C}_5[64]$ endowed with the p-adic distance we shall call the *p-adic codon space*, i.e., the elements of $\mathcal{C}_5[64]$ are codons denoted by $n_0 n_1 n_2$. The 5-adic distance between two codons $a, b \in \mathcal{C}_5[64]$ is

$$d_5(a, b) = \left|a_0 + a_1 5 + a_2 5^2 - b_0 - b_1 5 - b_2 5^2\right|_5, \tag{21.9}$$

where $a_i, b_i \in \{1, 2, 3, 4\}$. When $a \neq b$, $d_5(a, b)$ may have three different values:

- $d_5(a, b) = 1$ if $a_0 \neq b_0$,
- $d_5(a, b) = 1/5$ if $a_0 = b_0$ and $a_1 \neq b_1$,
- $d_5(a, b) = 1/5^2$ if $a_0 = b_0$, $a_1 = b_1$ and $a_2 \neq b_2$.

We see that the largest 5-adic distance between codons is 1 and it is the maximum p-adic distance on $\mathbb{Z}$. The smallest 5-adic distance on the codon space is 5^{-2}. Let us also note that 5-adic distance depends only on the first two nucleotides of different codons.

If we apply the real (standard) distance $d_\infty(a, b) = |a_0 + a_1 5 + a_2 5^2 - b_0 - b_1 5 - b_2 5^2|_\infty$, then the third nucleotides a_2 and b_2 would play a more important role than those at the second position (i.e., a_1 and b_1), and nucleotides a_0 and b_0 are of the smallest importance. In the real $\mathcal{C}_5[64]$ space distances are also discrete, but take values $1, 2, \ldots, 93$. The smallest real and the largest 5-adic distances are equal to 1.

While the real distance describes a metric of the ordinary physical space, this p-adic one serves to describe ultrametricity of the codon space.

It is worth emphasizing that the metric role of digits depends on their position in the number expansion, and it is quite opposite in real and p-adic cases. We shall see later, when we consider the genetic code, that the first two nucleotides in a codon are more important than the third one and that p-adic distance between codons is a natural one in describing their information content (the closer, the more similar).

p-adic Genomic and Bioinformation Spaces

Appropriateness of the p-adic codon space $\mathcal{C}_5[64]$ to the genetic code is already shown in [209] and will be reconsidered in Sect. 21.7.3. Now we want to extend the $\mathcal{C}_5[64]$ space approach to more general genetic and bioinformation spaces.

Let us recall that four nucleotides are related to the prime number 5 by their correspondence to the four nonzero digits $(1, 2, 3, 4)$ of $p = 5$. It is inappropriate to use the digit 0 for a nucleotide because it leads to non-uniqueness in the representation of the codons by natural numbers. For example, $123 = 123000$ as numbers, but 123 represents one and 123000 represents two codons. This is also a reason why we do not use 4-adic representation for codons, since it would contain a nucleotide represented by the digit 0. One can use 0 as a digit to denote the absence of any nucleotide.

Let us note also that we have used on $\mathcal{C}_5[64]$, in [208] and [209], not only the 5-adic but also the 2-adic distance.

Definition 21.4 We shall call *(p, q)-adic genomic space* a pair $(\Gamma_p[(p-1)^m], d_q)$ where

$$\Gamma_p\big[(p-1)^m\big] = \{n_0 + n_1 p + \cdots + n_{m-1} p^{m-1} : \\ n_i = 1, 2, \ldots, p-1, m \in \mathbb{N}\} \tag{21.10}$$

is the set of natural numbers, d_q is the corresponding q-adic distance on $\Gamma_p[(p-1)^m]$ and nonzero digits n_i are related to some $p-1$ basic constituents of a genomic system (or to any other biological information system) in a unique way. The index q is a prime number.

Here m can be called also *multiplicity* of space elements with respect to their constituents. In addition to d_p, there can be a few other d_q useful distances on $\Gamma_p[(p-1)^m]$.

For simplicity, we shall often call $\Gamma_p[(p-1)^m]$ *p-adic genomic space* and use the notation

$$n_0 + n_1 p + \cdots + n_{m-1} p^{m-1} \equiv n_0 n_1 \cdots n_{m-1}, \tag{21.11}$$

where the ordering of digits is in the opposite direction to the standard one and seems here more natural. Earlier introduced codon space $\mathcal{C}_5[64]$ can be regarded as a significant example of the p-adic genomic spaces, i.e., $\mathcal{C}_5[64] = \Gamma_5[(5-1)^3]$ as the space of trinucleotides. Two other examples, which will be used later, are: $\Gamma_5[4]$—the space of nucleotides and $\Gamma_5[4^2]$—the space of dinucleotides.

Now we can introduce a p-adic bioinformation space as a space composed of some p-adic genomic spaces.

Definition 21.5 Let $\mathcal{B}_p[N]$ be a p-adic space composed of N natural numbers. We shall call $\mathcal{B}_p[N]$ *p-adic bioinformation space* when it can be represented as

$$\mathcal{B}_p[N] \subset \prod_{m=m_1}^{m_2} \Gamma_p\big[(p-1)^m\big],$$

where m_1 and m_2 are positive integers ($m_1 \le m_2$), which determine the range of multiplicity between m_1 and m_2. In the sequel, we shall present some concrete examples of the $\mathcal{B}_p[N]$ spaces.

DNA and RNA Spaces

DNA sequences can be considered as a union of coding and non-coding segments. Coding parts are composed of codons included into genes, which are rather complex systems. In coding segments, information is stored, which through a series of complex processes is translated into proteins. The space of coding DNA sequences (cDNA) can be represented as

$$\mathrm{cDNA}[N] \subset \prod_{m=m_1}^{m_2} \Gamma_{61}\big[60^m\big],$$

where $p = 61$ because there are 60 codons coding amino acids (in the vertebral mitochondrial code). Thus the cDNA space can be regarded as a set of N coded sequences as well as a set of N discrete points (a lattice) of the $\prod_{m=m_1}^{m_2} \Gamma_{61}[60^m]$ space. While in $\mathcal{C}_5[64]$ codons are space elements, in cDNA[N] they are building units.

The structure and function of non-coding sequences is still highly unknown. They include information on various regulatory processes in a cell. We assume that the space of non-coding DNA sequences (ncDNA) is a subspace

$$\mathrm{ncDNA} \subset \prod_{m=m_1}^{m_2} \Gamma_5\big[4^m\big],$$

where m_1 and m_2 are the minimum and maximum values of the size of non-coding segments.

In a similar way, one can construct a space of all RNA sequences in a cell.

Table 21.10 List of 20 standard amino acids used in proteins by living cells. 3-Letter and 1-letter abbreviations, and chemical structure of their side chains are presented

Amino acid	Abbr.	Side chain (R)
Alanine	Ala, A	$-CH_3$
Cysteine	Cys, C	$-CH_2SH$
Aspartate	Asp, D	$-CH_2COOH$
Glutamate	Glu, E	$-(CH_2)_2COOH$
Phenynalanine	Phe, F	$-CH_2C_6H_5$
Glycine	Gly, G	$-H$
Histidine	His, H	$-CH_2-C_3H_3N_2$
Isoleucine	Ile, I	$-CH(CH_3)CH_2CH_3$
Lysine	Lys, K	$-(CH_2)_4NH_2$
Leucine	Leu, L	$-CH_2CH(CH_3)_2$
Methionine	Met, M	$-(CH_2)_2SCH_3$
Asparagine	Asn, N	$-CH_2CONH_2$
Proline	Pro, P	$-(CH_2)_3$
Glutamine	Gln, Q	$-(CH_2)_2CONH_2$
Arginine	Arg, R	$-(CH_2)_3NHC(NH)NH_2$
Serine	Ser, S	$-CH_2OH$
Threonine	Thr, T	$-CH(OH)CH_3$
Valine	Val, V	$-CH(CH_3)_2$
Tryptophan	Trp, W	$-CH_2C_8H_6N$
Tyrosine	Tyr, Y	$-CH_2-C_6H_4OH$

Protein Space

We mentioned some basic properties of proteins in Introduction. Recall also that the functional properties of proteins depend on their three-dimensional structure. There are four distinct levels of protein structure (primary, secondary, tertiary, and quaternary) [260]. The primary structure is determined by the amino acid sequence and the other ones depend on side chains of amino acids (see Table 21.10). In addition to the 20 standard amino acids, presented in Table 21.10, there are also 2 special non-standard amino acids: selenocysteine and pyrrolysine [816]. They are also coded by codons, but are very rare in proteins. Thus there are 22 amino acids encoded in the genetic code. According to Jukes [395], a non-freezing code may contain 28 amino acids.

The 20 standard (canonical) amino acids employed by the genetic code in proteins of the living cells are listed in Table 21.10. Some their important chemical properties are presented in Table 21.11.

Table 21.11 Some chemical properties of 20 standard amino acids. $p+n$ is the number of nucleons

Amino acids	$p+n$	Polar	Hydro-phobic	In proteins %
Ala, A	89	no	yes	7.8
Cys, C	121	no	yes	1.9
Asp, D	133	yes	no	5.3
Glu, E	147	yes	no	6.3
Phe, F	165	no	yes	3.9
Gly, G	75	no	yes	7.2
His, H	155	yes	no	2.3
Ile, I	131	no	yes	5.3
Lys, K	146	yes	no	5.9
Leu, L	131	no	yes	9.1
Met, M	149	no	yes	2.3
Asn, N	132	yes	no	4.3
Pro, P	115	no	yes	5.2
Gln, Q	146	yes	no	4.2
Arg, R	174	yes	no	5.1
Ser, S	105	yes	no	6.8
Thr, T	119	yes	no	5.9
Val, V	117	no	yes	6.6
Trp, W	204	no	yes	1.4
Tyr, Y	181	yes	yes	3.2

Now we want to construct an appropriate space whose elements are proteins. We propose the protein space $\mathcal{P}_p$ to be a subspace of product of genomic spaces

$$\mathcal{P}_p[N] \subset \prod_{m=m_1}^{m_2} \Gamma_p\big[(p-1)^m\big],$$

where the building units are amino acids. Thus $\mathcal{P}_p[N]$ is a space of N proteins with size measured by the number of amino acids between m_1 and m_2 ($m_1 \sim 10$ and $m_2 \sim 10^4$).

In (20.50), the prime number p is related to the number of amino acids by the relation: $p-1 =$ *number of different amino acids* used as building blocks in proteins. At present, there are 22 amino acids (20 standard and 2 special), and consequently $p = 23$. One can argue that not all 22 amino acids have been there from the very beginning of life, and that there has been an evolution of amino acids. Namely, using 60 different criteria for temporal order of appearance of the 20 standard amino acids, the obtained result [752] is presented in Table 21.12. The first four amino acids (Gly, Ala, Asp, and Val) have the highest production rate in Miller's experi-

Table 21.12 Temporal appearance of the 20 standard amino acids

(1) Gly	(2) Ala	(3) Asp	(4) Val
(5) Pro	(6) Ser	(7) Glu	(8) Leu
(9) Thr	(10) Arg	(11) Ile	(12) Gln
(13) Asn	(14) His	(15) Lys	(16) Cys
(17) Phe	(18) Tyr	(19) Met	(20) Trp

ment of an imitation of the atmosphere of the early Earth. This could correspond to $p = 5$ and single nucleotide codons in a primitive code. In the case of dinucleotide code, there are 16 codons, and the maximum amino acids that can be coded is 16, i.e., $p = 17$. As we already mentioned, according to Jukes [395], it is possible to code 28 amino acids by trinucleotide code, and it gives the corresponding $p = 29$.

21.7.3 p-adic Genetic Code

An intensive study of the connection between the ordering of nucleotides in DNA (and RNA) and the ordering of amino acids in proteins led to the experimental deciphering of the standard genetic code in the mid-1960s. The genetic code is understood as a dictionary for translation of information from DNA (through RNA) to synthesis of proteins by amino acids. The information on amino acids is contained in codons: each codon codes either an amino acid or the termination signal (see, e.g., Table 21.12 as a standard table of the vertebral mitochondrial code). To the sequence of codons in RNA there corresponds a quite definite sequence of amino acids in a protein, and this sequence of amino acids determines the primary structure of the protein. The genetic code is comma-free and non-overlapping. At the time of deciphering, it was mainly believed that the standard code was unique, result of a chance, and fixed a long time ago. Crick [181] expressed such belief in his "frozen accident" hypothesis which has not been supported by later observations. Moreover, so far at least 16 different codes have been discovered and some general regularities have been found. At first glance, the genetic code looks rather arbitrary, but it is not. Namely, mutations between synonymous codons give the same amino acid. When a mutation alters an amino acid, it is like a substitution of the original by a similar one. In this respect, the code is almost optimal.

Despite the remarkable experimental successes, there is no simple and generally accepted theoretical understanding of the genetic code. There are many papers in this direction (in addition to already cited, see also, e.g., [668] and [739]), scattered in various journals, with theoretical approaches based more or less on chemical, biological, and mathematical aspects of the genetic code. Even before deciphering of the code, there have been very attractive theoretical inventions (of Gamow and Crick), but the genetic code happened to be quite different (for a review on the early inventions around the genetic code, see [323]). However, the foundation of biological coding is still an open problem. In particular, it is not clear why the genetic

Table 21.13 Our table of the vertebral mitochondrial code in the usual notation

CCC Pro	UCU Ser	UAU Tyr	UGU Cys
CCA Pro	UCC Ser	UAC Tyr	UGC Cys
CCU Pro	UCA Ser	UAA Ter	UGA Trp
CCG Pro	UCG Ser	UAG Ter	UGG Trp
CAC His	CCU Pro	CAU His	CGU Arg
CAA Gln	CCC Pro	CAC His	CGC Arg
CAU His	CCA Pro	CAA Gln	CGA Arg
CAG Gln	CCG Pro	CAG Gln	CGG Arg
CUC Leu	ACU Thr	AAU Asn	AGU Ser
CUA Leu	ACC Thr	AAC Asn	AGC Ser
CUU Leu	ACA Thr	AAA Lys	AGA Ter
CUG Leu	ACG Thr	AAG Lys	AGG Ter
CGC Arg	GCU Ala	GAU Asp	GGU Gly
CGA Arg	GCC Ala	GAC Asp	GGC Gly
CGU Arg	GCA Ala	GAA Glu	GGA Gly
CGG Arg	GCG Ala	GAG Glu	GGG Gly

code exists just in a few known ways and not in many other possible ones. What is a principle (or principles) employed in establishment of a basic (mitochondrial) code? What are properties of codons connecting them into definite multiplets which code the same amino acid or termination signal? Answers to these and some other questions should lead us to a discovery of an appropriate theoretical model of the genetic code.

Let us now turn to Table 21.2. We observe that this table can be regarded as a big rectangle divided into 16 equal smaller rectangles: 8 of them are quadruplets which are in one-to-one correspondence to the 8 amino acids, and other 8 rectangles are divided into 16 doublets coding 14 amino acids and termination (stop) signal (by two doublets at different places). However, there is no manifested symmetry in the distribution of these quadruplets and doublets.

In order to get a symmetry, we have rewritten this standard table into a new one by rearranging 16 rectangles. As a result, we obtained Table 21.13 which exhibits a symmetry with respect to the distribution of codon quadruplets and codon doublets. Namely, in our table quadruplets and doublets form separately two figures, which are symmetric with respect to the mid-vertical line (a left-right symmetry), i.e., they are invariant under interchange $\mathrm{C} \leftrightarrow \mathrm{G}$ and $\mathrm{A} \leftrightarrow \mathrm{U}$ at the first position in codons at all horizontal lines. Recall that DNA is also symmetric under simultaneous interchange of complementary nucleotides $\mathrm{C} \leftrightarrow \mathrm{G}$ and $\mathrm{A} \leftrightarrow \mathrm{T}$ between its strands. All doublets in this table form a nice figure which looks like letter $\mathbb{T}$.

Table 21.14 contains the same distribution of amino acids as Table 21.13, but codons are now presented by 5-adic numbers $n_0 n_1 n_2$ instead of capital letters (recall: $\mathrm{C} = 1$, $\mathrm{A} = 2$, $\mathrm{U} = 3$, $\mathrm{G} = 4$). This new table can be also regarded as a representation of the $\mathcal{C}_5[64]$ codon space with gradual increasing of integers from 111

Table 21.14 Our 5-adic table of the vertebral mitochondrial code, which is a representation of the $\mathcal{C}_5[64]$ codon space

111 Pro	211 Thr	311 Ser	411 Ala
112 Pro	212 Thr	312 Ser	412 Ala
113 Pro	213 Thr	313 Ser	413 Ala
114 Pro	214 Thr	314 Ser	414 Ala
121 His	221 Asn	321 Ter	421 Asp
122 Gln	222 Lys	322 Ter	422 Glu
123 His	223 Asn	323 Tyr	423 Asp
124 Gln	224 Lys	324 Ter	424 Glu
131 Leu	231 Ile	331 Phe	431 Val
132 Leu	232 Met	332 Leu	432 Val
133 Leu	233 Ile	333 Phe	433 Val
134 Leu	234 Met	334 Leu	434 Val
141 Arg	241 Ser	341 Cys	441 Gly
142 Arg	242 Ter	342 Trp	442 Gly
143 Arg	243 Ser	343 Cys	443 Gly
144 Arg	244 Ter	344 Trp	444 Gly

to 444. The observed left–right symmetry is now invariant under the corresponding transformations $1 \leftrightarrow 4$ and $2 \leftrightarrow 3$. In other words, at each horizontal line one can perform *doublet* $\leftrightarrow$ *doublet* and *quadruplet* $\leftrightarrow$ *quadruplet* interchange around the vertical midline.

It is worth noting that the above invariance leaves also unchanged the polarity and hydrophobicity of the corresponding amino acids in all but three cases: Asn $\leftrightarrow$ Tyr, Arg $\leftrightarrow$ Gly, and Ser $\leftrightarrow$ Cys.

Degeneracy of the Genetic Code

Let us now explore distances between codons and their role in the formation of the genetic code degeneration.

To this end, let us again turn to Table 21.14 as a representation of the $\mathcal{C}_5[64]$ codon space. Namely, we observe that there are 16 quadruplets such that each of them has the same first two digits. Hence the 5-adic distance between any two different codons within a quadruplet is

$$\begin{aligned} d_5(a,b) &= \left|a_0 + a_1 5 + a_2 5^2 - a_0 - a_1 5 - b_2 5^2\right|_5 \\ &= \left|(a_2 - b_2)5^2\right|_5 = \left|(a_2 - b_2)\right|_5 \left|5^2\right|_5 = 5^{-2}, \end{aligned} \tag{21.12}$$

because $a_0 = b_0$, $a_1 = b_1$ and $|a_2 - b_2|_5 = 1$. According to (21.12), nucleotides within every quadruplet are at the smallest distance apart, i.e., they are closest compared to all other nucleotides.

Since codons are composed of three arranged nucleotides, each of which is either a purine or a pyrimidine, it is natural to try to quantify the similarity inside purines and pyrimidines, as well as the distinction between elements from these two groups of nucleotides. Fortunately, there is a tool, which is again related to the p-adics, and now it is the 2-adic distance. One can easily see that 2-adic distance between pyrimidines C and U is $d_2(1,3) = |3-1|_2 = 1/2$ as the distance between purines A and G, namely $d_2(2,4) = |4-2|_2 = 1/2$. However, the 2-adic distance between C and A or G, as well as the distance between U and A or G is 1 (i.e., maximum).

With respect to the 2-adic distance, the above quadruplets may be regarded as composed of two doublets: $a = a_0a_1 1$ and $b = a_0a_1 3$ make the first doublet, and $c = a_0a_1 2$ and $d = a_0a_1 4$ form the second one. The 2-adic distance between codons within each of these doublets is $\frac{1}{2}$, i.e.,

$$d_2(a,b) = \left|(3-1)5^2\right|_2 = \frac{1}{2}, \qquad d_2(c,d) = \left|(4-2)5^2\right|_2 = \frac{1}{2}, \tag{21.13}$$

because $3-1 = 4-2 = 2$.

One can now look at Table 21.14 as a system of 32 doublets. Thus 64 codons are clustered by a very regular way into 32 doublets. Each of 21 subjects (20 amino acids and one termination signal) is coded by one, two, or three doublets. In fact, there are two, six, and 12 amino acids coded by three, two, and one doublet, respectively. Residual two doublets code the termination signal.

Note that two of 16 doublets code two amino acids (Ser and Leu) which are already coded by two quadruplets, thus amino acids Serine and Leucine are coded by six codons (three doublets).

To have a more complete picture on the genetic code, it is useful to consider possible distances between codons of different quadruplets as well as between different doublets. Also, we introduce a distance between quadruplets and between doublets, especially when distances between their codons have the same value. Thus the 5-adic distance between any two quadruplets in the same column is $1/5$, while such distance between other quadruplets is 1. The 5-adic distance between doublets coincides with distance between quadruplets, and this distance is $\frac{1}{5^2}$ when doublets are within the same quadruplet.

The 2-adic distances between codons, doublets, and quadruplets are more complex. There are three basic cases:

- Codons differ only in one digit
- Codons differ in two digits
- Codons differ in all three digits.

In the first case, the 2-adic distance can be $\frac{1}{2}$ or 1 depending on whether the difference between digits is 2 or not, respectively.

Let us now look at the 2-adic distances between doublets coding leucine and also between doublets coding serine. These are two cases of amino acids coded by three doublets. One has the following distances:

- $d_2(332, 334) = d_2(132, 134) = \frac{1}{2}$ for leucine
- $d_2(311, 241) = d_2(313, 243) = \frac{1}{2}$ for serine.

If we use the usual distance between codons, instead of the p-adic one, then we would observe that two synonymous codons are very far (at least 25 units apart), and that those which are close code different amino acids. Thus we conclude that not the usual metric but the ultrametric is inherent to codons.

How is degeneracy of the genetic code connected with the p-adic distances between codons? The answer is in the following p-adic degeneracy principle: *Two codons have the same meaning with respect to amino acids if they are at smallest* 5-adic and 0.5, 2-*adic distance*. Here the p-adic distance plays a role of similarity: the closer, the more similar. Taking into account all known codes (see the next subsection) there is a slight violation of this principle. Now it is worth noting that in modern particle physics just breaking the fundamental gauge symmetry gives its standard model. It makes sense to introduce a new principle (let us call it reality principle): *Reality is a realization of some broken fundamental principles*. It seems that this principle is valid not only in physics but also in all sciences. In this context, modern genetic code is an evolutionary breaking the above p-adic degeneracy principle.

Evolution of the Genetic Code

The origin and early evolution of the genetic code are among the most interesting and important investigations related to the origin and whole evolution of life. However, since there are no concrete facts from that early period, it gives rise to many speculations. Nevertheless, one can hope that some of the hypotheses may be tested looking for their traces in the contemporary genomes.

It seems natural to consider biological evolution as an adaptive development of simpler living systems to more complex ones. Namely, living organisms are open systems in permanent interaction with environment. Thus the evolution can be modeled by a system with given initial conditions and guided by some internal rules taking into account environmental factors.

We are going now to conjecture on the evolution of the genetic code using our p-adic approach to the genomic space, and assuming that preceding codes used simpler codons and older amino acids.

Recall that the p-adic genomic space $\Gamma_p[(p-1)^m]$ has two parameters: p—related to $p-1$ building blocks, and m—multiplicity of the building blocks in space elements.

- The case $\Gamma_2[1]$ is a trivial one and useless for a primitive code.
- The case $\Gamma_3[2^m]$ with $m = 1, 2, 3$, does not seem to be realistic.
- The case $\Gamma_5[4^m]$ with $m = 1, 2, 3$, offers a possible pattern to consider evolution of the genetic code. Namely, the codon space could evolve in the following way: $\Gamma_5[4] \to \Gamma_5[4^2] \to \Gamma_5[4^3] = \mathcal{C}_5[64]$ (see also Table 21.15).

Table 21.15 The 5-adic system including digit 0 and containing single nucleotide, dinucleotide and trinucleotide codons. If one ignores numbers which contain digit 0 in front of any 1, 2, 3, or 4, then one has a one-to-one correspondence between 1-digit, 2-digits, 3-digits numbers, and single nucleotides, dinucleotides, trinucleotides, respectively. It seems that evolution of codons has followed transitions: single nucleotides → dinucleotides → trinucleotides

000	100 C	200 A	300 U	400 G
010	110 CC	210 AC	310 UC	410 GC
020	120 CA	220 AA	320 UA	420 GA
030	130 CU	230 AU	330 UU	430 GU
040	140 CG	240 AG	340 UG	440 GG
001	101	201	301	401
011	111 CCC	211 ACC	311 UCC	411 GCC
021	121 CAC	221 AAC	321 UAC	421 GAC
031	131 CUC	231 AUC	331 UUC	431 GUC
041	141 CGC	241 AGC	341 UGC	441 GGC
002	102	202	302	402
012	112 CCA	212 ACA	312 UCA	412 GCA
022	122 CUA	222 AAA	322 UAA	422 GAA
032	132 CGA	232 AUA	332 UUA	432 GUA
042	142 CGA	242 AGA	342 UGA	442 GGA
003	103	203	303	403
013	113 CCU	213 ACU	313 UCU	413 GCU
023	123 CAU	223 AAU	323 UAU	423 GAU
033	133 CUU	233 AUU	333 UUU	433 GUU
043	143 CGU	243 AGU	343 UGU	443 GGU
004	104	204	304	404
014	114 CCG	214 ACG	314 UCG	414 GCG
024	124 CAG	224 AAG	324 UAG	424 GAG
034	134 CUG	234 AUG	334 UUG	434 GUG
044	144 CGG	244 AGG	344 UGG	444 GGG

According to Table 21.12, this primary code containing codons in the single nucleotide form (C, A, U, G) encoded the first four amino acids: Gly, Ala, Asp, and Val. From the last column of Table 21.14, we conclude that the connection between digits and amino acids is: 1 = Ala, 2 = Asp, 3 = Val, 4 = Gly. In the primary code, these digits occupied the first position in the 5-adic expansion (Table 21.15), and at the next step, i.e., $\Gamma_5[4] \to \Gamma_5[4^2]$, they moved to the second position adding digits $1, 2, 3, 4$ in front of each of them.

In $\Gamma_5[4^2]$, one has 16 dinucleotide codons which can code up to 16 new amino acids. The addition of the digit 4 in front of already existing codons $1, 2, 3, 4$ leaves their meaning unchanged, i.e., 41 = Ala, 42 = Asp, 43 = Val, and 44 = Gly. Adding

Table 21.16 The dinucleotide genetic code based on the p-adic genomic space $\Gamma_5[4^2]$. Note that it encodes 15 amino acids without the stop codon, but encodes serine twice

11 Pro	21 Thr	31 Ser	41 Ala
12 His	22 Asn	32 Tyr	42 Asp
13 Leu	23 Ile	33 Phe	43 Val
14 Arg	24 Ser	34 Cys	44 Gly

digits $3, 2, 1$ in front of the primary $1, 2, 3, 4$ codons, one obtains 12 possibilities for coding some new amino acids. To decide which amino acid was encoded by which of 12 dinucleotide codons, we use as a criterion their immutability in the trinucleotide coding on the $\Gamma_5[4^3]$ space. This criterion assumes that amino acids encoded earlier are more fixed than those encoded later. According to this criterion, we decide in favor of the first row in each rectangle of Table 21.14, and the result is presented in Table 21.16.

Transition from dinucleotide to trinucleotide codons occurred by attaching nucleotides $1, 2, 3, 4$ at the third position, i.e., behind each dinucleotide. By this way, one obtains a new codon space $\Gamma_5[4^3] = \mathcal{C}_5[64]$ which is significantly enlarged and provides a pattern to generate known genetic codes. This codon space $\mathcal{C}_5[64]$ gives a possibility to realize at least three general properties of the modern code:

(i) Encoding of more than 16 amino acids
(ii) Diversity of codes
(iii) Stability of the gene expression.

Let us give some relevant clarifications:

(i) For functioning of contemporary living organisms it is necessary to code at least 20 standard (Table 21.10) and 2 non-standard amino acids (selenocysteine and pyrrolysine). Probably these 22 amino acids are also sufficient building units for biosynthesis of all necessary contemporary proteins. While $\Gamma_5[4^2]$ is insufficient, the genomic space $\Gamma_5[4^3]$ offers approximately three codons per one amino acid.
(ii) The eukaryotic (often called standard or universal) code is established around 1966 and was thought to be universal, i.e., common to all organisms. When the vertebral mitochondrial code was discovered in 1979, it gave rise to a belief that the code is not frozen and that there are also some other codes which are mutually different. According to later evidences, one can say that there are at least 16 slightly different mitochondrial and nuclear codes (for a review, see [433, 618] and references therein). Different codes have some codons with different meaning. So, in the standard code there are the following changes in Table 21.14:

   - 232 (AUA): Met $\to$ Ile
   - 242 (AGA) and 244 (AGG): Ter $\to$ Arg
   - 342 (UGA): Trp $\to$ Ter.

Modifications in these 16 codes are not homogeneously distributed on 16 rectangles of Table 21.14. For instance, in all 16 codes codons $41i$ ($i = 1, 2, 3, 4$) have the same meaning.

(iii) Each of the 16 codes is degenerate, and degeneration provides their stability against possible mutations. In other words, degeneration helps to minimize codon errors.

Genetic codes based on single nucleotide and dinucleotide codons were mainly directed to code amino acids with rather different properties. This may be the reason why amino acids Glu and Gln are not coded in dinucleotide code (Table 21.16), since they are similar to Asp and Asn, respectively. However, to become almost optimal, trinucleotide codes have taken into account structural and functional similarities of amino acids.

We presented here a hypothesis on the genetic code evolution taking into account possible codon evolution, from 1-nucleotide to 3-nucleotide, and amino acids temporal appearance. This scenario may be extended to the cell evolution, which probably should be considered as a coevolution of all its main ingredients (for an early idea of the coevolution, see [820]).

21.7.4 Remarks

There are two aspects of the genetic code related to:

(i) Multiplicity of codons which code the same amino acid,
(ii) Concrete assignment of codon multiplets to particular amino acids.

The above presented p-adic approach gives a quite satisfactory description of the aspect (i). Ultrametric behavior of p-adic distances between elements of the $\mathcal{C}_5[64]$ codon space radically differs from the usual ones. Quadruplets and doublets of codons have natural explanation within 5-adic and 2-adic nearness. Degeneracy of the genetic code in the form of doublets, quadruplets, and sextuplets is a direct consequence of p-adic ultrametricity between codons. The p-adic $\mathcal{C}_5[64]$ codon space is our theoretical pattern to consider all variants of the genetic code: some codes are a direct representation of $\mathcal{C}_5[64]$ and the others are its slight evolutional modifications.

Aspect (ii) is related to the question: Which amino acid corresponds to which multiplet of codons? An answer to this question should be expected from connections between physicochemical properties of amino acids and anticodons. Namely, the enzyme aminoacyl-tRNA synthetase links a specific tRNA anticodon and a related amino acid. Thus there is no direct interaction between amino acids and codons, as it was believed for some time in the past.

Note that there are in general 4! ways to assign digits $1, 2, 3, 4$ to nucleotides C, A, U, G. After an analysis of all 24 possibilities, we have taken C $= 1$, A $= 2$, U $=$ T $= 3$, G $= 4$ as a quite appropriate choice. In addition to various properties already presented in this chapter, also the complementarity of nucleotides in the DNA double helix is exhibited by the relation $C+G=A+T=5$.

It would be useful to find an analogous connection between 22 amino acids and digits $1, 2, \ldots, 22$ in $p=23$ representation. Now there are 22! possibilities and to explore all of them seems to be a hard task. However, a computer analysis may help to find a satisfactory solution.

One can express many above considerations of the p-adic information theory in linguistic terms and investigate possible linguistic applications.

In this chapter, we have employed the p-adic distances to measure similarity between codons, which have been used to describe degeneracy of the genetic code. It is worth noting that in other contexts the p-adic distances can be interpreted in quite different meanings. For example, the 3-adic distance between cytosine and guanine is $d_3(1,4)=\frac{1}{3}$, and between adenine and thymine $d_3(2,3)=1$. This 3-adic distance seems to be natural to relate to hydrogen bonds between complements in the DNA double helix: the smaller the distance, the stronger the hydrogen bond. Recall that C–G and A–T are bonded by 3 and 2 hydrogen bonds, respectively.

The translation of codon sequences into proteins is highly an information-processing phenomenon. The p-adic information modeling presented in this section [208–210] offers a new approach to systematic investigation of ultrametric aspects of DNA and RNA sequences, the genetic code and the world of proteins. It can be embedded in computer programs to explore the p-adic side of the genome and related subjects.

The above considerations and obtained results may be regarded as contributions mainly towards foundations of (i) p-adic theory of information and (ii) p-adic theory of the genetic code.

Summarizing, contributions to

(i) p-adic theory of information contain:

- Formulation of p-adic genomic space (whose examples are spaces of nucleotides, dinucleotides and trinucleotides).
- Formulation of p-adic bioinformation space (whose examples are DNA, RNA and protein spaces).
- Relation between building blocks of information spaces and some prime numbers.

(ii) p-adic theory of the genetic code include:

- Description of codon quadruplets and doublets by 5-adic and 2-adic distances.
- Observation of a symmetry between quadruplets as well as between doublets at our table of codons.
- Formulation of degeneracy principle.
- Formulation of hypothesis on codon evolution.

Many problems remain to be explored in the future on the above p-adic approach to genomics. Among the most attractive and important themes are:

- Elaboration of the p-adic theory of information towards genomics.

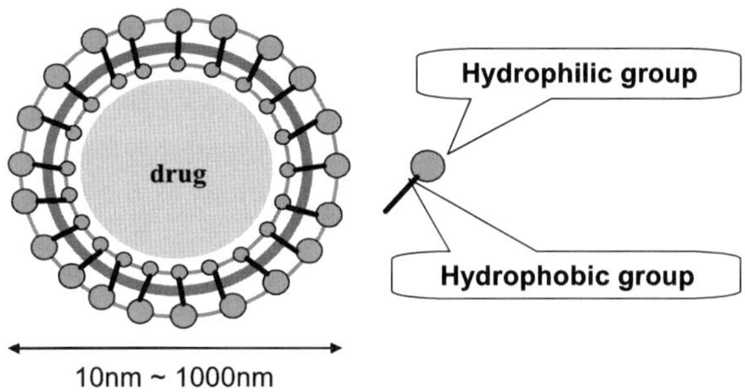

Fig. 21.23 Liposomal drug

- Evolution of the genome and the genetic code.
- Structure and function of non-coding DNA.
- Ultrametric aspects of proteins.
- Creation of the corresponding computer programs.

21.8 Mathematical Model of Drug Delivery System

In this section, we apply stochastic analysis discussed in Chaps. 3 and 4 to the drug delivery systems (DDS) following the paper of Hara, Iriyama, Makino, Terada, and Ohya [318].

Standard chemotherapy for many tumors includes intravenous injection of anticancer drugs, which in most cases shows insufficient therapeutic effects accompanied by severe side effects. The main problems associated with systemic drug administration are: the lack of drugs having specific affinity toward a pathological site; the necessity of a large total dose to achieve high local concentration; non-specific toxicity and other adverse side-effects due to high drug doses. Drug-targeting is expected to resolve most of these problems. That is, minimally required amount of drug should be delivered to the target site when needed. Nanoparticles, such as liposomes, polymer micelles, and biodegradable nanospheres, have been studied as drug carriers for several decades, and the features that make them attractive drug carriers are well known. Nanoparticles encompass a variety of submicron (<1 μm) colloidal nanosystems, and one of their major advantages is their small size which allows them to pass through certain biological barriers.

A typical nanoscale drug, liposomal drug, is shown in Fig. 21.23.

Surface functionalities can sometimes be incorporated into nanoparticles since the challenge includes finding a means to make drug carrier systems avoid the immunogenic and nonspecific interactions that efficiently clear foreign materials from the body. The most noteworthy surface modification of the nanoparticles is the incorporation of polyethylene glycol (PEGylation) which serves as a barrier, preventing interactions with plasma proteins, and thus retarding recognition by the reticuloendothelial system (RES) and enhancing the circulation time of the drug carriers [321].

Passive delivery is targeted to solid tumor. Aggressive tumors inherently develop leaky vasculature with 100–1000 nm pores due to rapid formation of vessels that must serve the fast-growing tumor. This defect in vasculature coupled with poor lymphatic drainage serves to enhance the permeation and retention of nanoparticles with the tumor region. This is called the EPR (enhanced permeability and retention) effect of a tumor, and is a representative form of passive targeting [496–499]. The basis for increased tumor specificity is the differential accumulation of drug-loaded nanoparticles in tumor tissue versus normal cells, which results from particle size rather than binding. Normal tissues contain capillaries with tight junctions that are less permeable to nanosized particles. Passive targeting can therefore result in increases of drug concentrations in solid tumors of several-fold relative to those attained with free drugs.

For the treatment of solid tumors, both liposomal and solid nanoparticle formulations have received clinical approval for delivery of some anticancer drugs. Examples of liposomal formulations include doxorubicin (Doxil/Caelyx and Myocet) and daunorubicin (Daunosome) [270, 666]. The mechanism of drug release may depend on diffusion of the drug from the carrier into the tumor interstitium. This is followed by the subsequent uptake of the released drugs by tumor cells. Also, the Food and Drug Administration (FDA) approved Abraxane, an albumin-bound paclitaxel nanoparticles in an injectable suspension for the treatment of metastatic breast cancer [353, 354]. Other solid nanoparticle-based cancer therapies have been approved for clinical trials. Thus, nanoparticles are potentially useful as carriers of active drugs for making the drugs as efficiently as possible. In this section, the effects of nanoparticle size on the efficiency of passive targeting to the pores existing in tumor vessels will be discussed.

21.8.1 Mathematical Model

In this subsection, we estimate the configuration of a drug which is most effective as a remedy for tumors variously situated in a living body, following the steps below.

1. We start by constructing a dynamical model of nanoparticles inside the body, based on mathematical physics.

Taking various interactions, the dynamical model should be one of the many-body problems. This dynamical model is difficult to solve analytically, so that we treat it under the following conditions by:

2. Taking some approximations to describe the interactions.
3. Analyzing the movement of drug particles in the blood.

In addition, we consider the movement of drug particles as a stochastic process, and we approximate the blood plasma as a Newtonian fluid and the force working on the drug particle as a mean field.

Based on above conditions, we

4. Calculate the probability that a drug particle is in the target tumor after a fixed time.

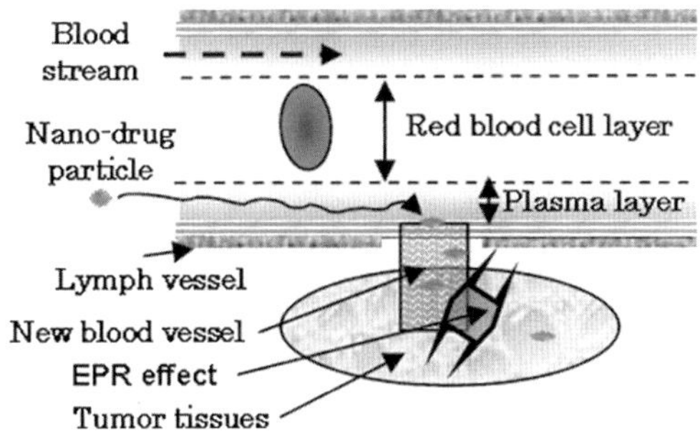

Fig. 21.24 The relation between terminal capillary and tumor tissues

5. Estimate the shape of a drug particle giving the maximum probability that the injected drug stays in the tumor.

We will discuss our model of the drug movement below with the particle circulation and the EPR effect of tumor.

Drug Particle Circulation and the Probability of Staying in a Tumor

The injected drug particles are circulating inside the body in the blood flow. The diameter of a terminal capillary is measured in microns, and it is known that red blood cells concentrate in the center of a capillary and a layer of blood plasma forms near the vessel wall, which is called the sigma effect. Our target drug and other substances such as water and proteins exist in this layer. Near a tumor, new blood vessels come out from a normal vessel and connect to the tumor (Fig. 21.24). Therefore, the drug particles circulating over the body pass close to the vessel near the tumor, and some of them get into the new blood vessel and will be retained by the tumor due to the EPR effect of the tumor.

Based on the above conditions, we construct a dynamical model of drug particles in the body. We assume that (i) the stream in a blood vessel is divided into two layers, red blood cell layer and plasma layer; (ii) the particles distribute uniformly in blood. Let $L_{\text{once}} \equiv L_{\text{capillary}} + L_{\text{body}}$ be the length for one circulation of a drug particle, where $L_{\text{capillary}}$ is the length of the capillary near the target tumor, L_{body} is the length getting back to that place after passing through the entrance of the new blood vessel. In addition, let $T_{\text{once}} \equiv T_{\text{capillary}} + T_{\text{body}}$ be the time for a single body circulation, where $T_{\text{capillary}}$ is the time in $L_{\text{capillary}}$ and T_{body} is the time in L_{body}. Then, let the places where drug particles stay as follows: B_1 is blood, B_2 is tumor, and B_3 is RES, and let $p_i^{(n)}$ be the probability of a drug staying in B_i after n times of circulation. One has

$$p_i^{(n)} = \sum_k t_{ik}^{(n)} p_k^{(n-1)}, \quad n > 0,$$

where $t_{ij}^{(n)}$ is the transition probability from B_j to B_i at the nth body circulation. The initial condition $p_i^{(0)}$ at time $n = 0$ should be

$$p_1^{(0)} = 1, \qquad p_2^{(0)} = 0, \qquad p_3^{(0)} = 0.$$

Let $T_{\max}$ be the lifetime of a drug particle given by the drug maker, and let $n_{\max}$ be the maximum integer n such that $\sum_{k=1}^{n} T_{\text{once}} < T_{\max}$. The probability of a drug staying in B_i at time $T_{\max}$ is denoted by $p_i^{(n_{\max})}$. We will calculate the value "$p_i^{(n_{\max})}$" with respect to several parameters characterizing the drug and the tumor.

The transition probabilities $t_{ij}^{(n)}$ should depend on the mass m of a drug so that $t_{ij}^{(n)} = g_{ij}^{(n)}(m)$, where g is a proper function related to the characters of a drug particle and the personal data of a patient. We have to look for this function, here we take the first approximation such that $t_{31}^{(n)} = t = c_1 m + c_2$, where c_1, c_2 are certain constants determined by the drug and the personal data of a patient. From these considerations, we took $t_{ij}^{(n)}$ as

$$\begin{aligned}
t_{11}^{(n)} &= (1 - P)(1 - t), \\
t_{21}^{(n)} &= P(1 - t), \\
t_{31}^{(n)} &= t, \\
t_{22}^{(n)} &= t_{33}^{(n)} = 1, \\
t_{ij}^{(n)} &= 0 \quad \text{(otherwise)}.
\end{aligned}$$

Here P is the probability that a drug particle gets into the tumor. This probability can be written as

$$P \equiv P_{\text{EPR}} P_{\text{plasma}}.$$

Here $P_{\text{EPR}} = \overline{h(v)}$ which is the rate that a drug particle near the tumor is captured due to the EPR effect of the tumor. We will describe $\overline{h(v)}$ in the following subsection. Then P_{plasma} is the probability that a drug particle is in the plasma layer. Since we assumed that all particles are independent and identically distributed, we set P_{plasma} as

$$P_{\text{plasma}} \equiv \left[1 - \left(\frac{d}{2} - d_{\text{eff}}\right)^2 \Big/ \left(\frac{d}{2}\right)^2\right] \times \frac{\varepsilon}{\pi d},$$

where d is the diameter of a capillary, d_{eff} is the thickness of the plasma layer, and ε is the size of the new blood vessel.

Drug Particle Movement Model

We discuss in this section how we construct a mathematical model to compute $\overline{h(v)}$ above. Let X be a set of drug particles and Y be a set of proteins in blood. The potential energy of a particle $i \in X$ located at $x_i \in \mathbb{R}^3$ is denoted by

$$U_i = \sum_{j \neq i} I(x_i, x_j) + \sum_{k \in Y} I'(x_i, y_k) + I''(x_i),$$

where I describes the interactions among drug particles: I' denotes the interaction between a drug particle and a protein in the blood at location y, and I'' that due to the blood flow. These interactions come from physical considerations of particles, which are of various types; one example of the potential energy which we used in our simulation is:

$$\begin{aligned} I(x, x') &= I_0 e^{-\gamma |x - x'|}, \\ I'(x, y) &= \gamma' \theta(y - x), \\ I''(x) &= \alpha \cdot x, \end{aligned}$$

where I_0, γ, γ' are constants, α is a constant vector $\alpha = (\alpha_1, 0, 0)$ and θ is the step function, $\theta(s) \equiv \theta(s_1)\theta(s_2)\theta(s_3)$ for $s = (s_1, s_2, s_3)$, and $\theta(\tau) = 1$ for $\tau > 0$, $\theta(\tau) = 0$ for $\tau \leq 0$.

In addition, we might need to consider a noise effect because the size of drug particles is measured in nanometers. We suppose this noise is a white noise because it is the most common due to Brownian motion. It is denoted by $W_i(t)$ for the particle i. Let μ be the strength of the noise, ζ be friction from viscosity of the blood stream, and m be the mass of a drug particle. We assume that the drug particle in the blood plasma layer follows the Langevin stochastic differential equation

$$m\, dv_i = f_i\, dt - \zeta v_i\, dt + \mu\, dW_i, \tag{21.14}$$

where v_i is the three dimensional vector of the velocity of particle i, $m = 4/3\pi\rho a^3$ (ρ is the mass density of a drug particle, and a is the radius of the particle), $f_i = -\operatorname{grad} U_i$, $\zeta = 6\pi\eta a$ (η is the viscosity of blood plasma), and $\mu = \sqrt{2\zeta k_B \theta_B}$ (k_B is Boltzmann's constant and θ_B is blood temperature).

It is very difficult to solve this stochastic differential equation, so that we need to take some approximations. We consider the various forces affecting the drug particles in the blood by a mean field. We suppose that the injected drug particles first diverge and will be almost static near the tumor; therefore, we might consider all particles as identical and independent. Under this assumption, we take a mean field approximation for the potential energy such as

$$\begin{aligned} \sum_{j \neq i} I(x_i, x_j) &\fallingdotseq \widetilde{I}(x_i), \\ \sum_{\alpha \in Y} I'(x_i, y_\alpha) &\fallingdotseq \widetilde{I'}(x_i). \end{aligned}$$

Therefore, the potential energy U_i is rewritten as

$$U_i = \widetilde{I}(x_i) + \widetilde{I'}(x_i) + I''(x_i).$$

Difference Equation for Numerical Calculation

According to the property of Brownian motion, the difference between each instant of time $dW_i(t)$ follows a Gaussian distribution. Thus the differences

$$\Delta W_i(k) = W_i(t_{k+1}) - W_i(t_k)$$

are independent and identically distributed Gaussian random variables with mean 0 and variance $t_{k+1} - t_k$, for a given discretization

$$t_0 = 0 < t_1 < \cdots < t_n = T$$

of the time interval $[0, T]$. We can use Gaussian pseudo-random numbers generated by a computer.

In order to calculate the above processes, we need to replace the stochastic differential equation (21.14) by the difference equation. Using Euler–Maruyama method [432], the stochastic differential equation becomes the following stochastic difference equation

$$v_i(t_{n+1}) = v_i(t_n) + \frac{f_i(t_n) - \zeta v_i(t_n)}{m}(t_{n+1} - t_n) + \frac{\mu}{m}\Delta W_i(n).$$

Based on the above formula, we use a computer simulation to obtain the velocity v_i.

EPR Effect of Tumor and Its Mathematical Treatment

The Enhanced Permeability and Retention (EPR) effect of a tumor was proposed by Maeda [496–499], and it has the following properties.

1. (Enhanced Permeability) The new blood vessels appearing near the tumor have high permeability even for large molecules, compared with normal blood vessels.
2. (Retention) Since the retrieval mechanism of lymph vessel is not complete, the large molecules tend to accumulate in tumor tissues.

A mathematical model of this effect was suggested by Hara, Iriyama, Makino, Terada, and Ohya [318]. This mathematical model was formulated by taking various parameters to characterize a tumor and the diverse spreading of the new blood vessels. They focus on the velocity of a drug particle at a proper place (called the "entrance") of the new blood vessels. Here, we define the absorbing rate for drug particles by the EPR effect of the tumor as follows. First, let L be the length from a fixed place to the entrance of the new blood vessel associated with the tumor, and v be the velocity of the drug particle arriving at the entrance. In addition, λ is a parameter characterizing the EPR effect of the tumor. Since the EPR effect of the tumor is applicable to the drug particle getting into the entrance, we have

$$\left|v(t)\right| > 0,\ \exists T \quad \text{such that} \quad \int_0^T v\,dt \geq L.$$

Here, T_L is the time satisfying

$$\int_0^{T_L} v\,dt = L.$$

Put

$$h\bigl(v(t)\bigr) = \begin{cases} e^{-\lambda v(t)} & (t = T_L), \\ 0 & (t < T_L). \end{cases}$$

Then we calculate the mean value of $h(v)$ by using a computer: Let N denote the total number of drug particles, then $h(v)$ is obtained as

$$\overline{h(v)} \cong \frac{1}{N} \sum_i h\bigl(v_i(T_L)\bigr).$$

21.8.2 Results and Discussion

In this section, we show the results of calculating the probability that a drug is at each place B_i (B_1 blood, B_2 tumor, B_3 RES) after a proper fixed time. First of all, we should find and set the suitable values of parameters for the target condition. We assign the parameters for a body in Table 21.17 and those for a drug in Table 21.18, taking the conditions of the EPC liposome in [758] in our mind. The values of the parameters d, D_{eff}, $L_{\mathrm{capillary}}$, η, α, θ_B, and T_{body} are taken from [196, 500], those of the parameters $\lambda, \varepsilon, c_1, c_2$ are determined by sensitivity analysis and optimization to fit with experimental data. We plot the probabilities $p_i^{(n_{\max})}$ in Fig. 21.25 with respect to the diameter of the drug. There is a peak of the probability $p_2^{(n_{\max})}$ near 100 nm and the probability $p_3^{(n_{\max})}$ has the opposite curve. These results well match the results of in-vivo study [758]. The results imply that our mathematical model can simulate the drug movement well. Note that in our computations, we took the simplest approximation for U_i, that is, the linear approximation of $\tilde{I}(x_i)$ with respect to x_i.

We show some results of sensitivity analysis for the parameters $\varepsilon, \lambda, a, m, T_{\mathrm{body}}$, and $T_{\max}$.

(i) We take three $\varepsilon = 0.5 \times 10^{-6}$, 1.0×10^{-6}, 1.5×10^{-6} meter and plot the probability $p_3^{(n_{\max})}$ in RES for each ε in Fig. 21.26.

(ii) We take three $\lambda = 1.0 \times 10^2, 1.5 \times 10^2, 2.0 \times 10^2$ s/m and plot the probability for each λ in Fig. 21.27. The other parameters are the same as above. The parameters ε and λ, which are the parameters characterizing a tumor, affect not only $p_2^{(n_{\max})}$ in the tumor but also $p_3^{(n_{\max})}$ in RES; however, the change of ε, λ does not affect $p_3^{(n_{\max})}$ when the diameter of the drug particle is less than about 10 nm. From these results, we may understand the total disposition of a drug in body.

Table 21.17 Parameters for the body

Parameter	Physical meaning	Value
d	Diameter of blood capillary	10.0 μm
d_{eff}	Height of blood plasma layer	1.0 μm
$L_{\mathrm{capillary}}$	Length of the capillary near the target tumor	1.0 cm
η	Viscosity of blood plasma	1.3×10^{-3} Pa s
ε	Size of new blood vessel	1.57×10^{-5} m
λ	Constant corresponding to Tumor and its EPR effect of tumor	2.0×10^{2} s/m
α_1	Intensity of blood flow	3.77×10^{-12} J/m
θ_B	Blood temperature	310 K

Table 21.18 Parameters for the drug

Parameter	Physical meaning	Value
ρ	Mass density of a drug particle	1.0×10^{3} kg/m^3 $= 1.0$ g/cm^3
t	$= c_1 m + c_2$; transition rate	$c_1 = 9.69 \times 10^{15}$ kg^{-1}
	from blood to RES	$c_2 = 5.0 \times 10^{-3}$
T_{circ}	Circulation time	5 min
T_{max}	Lifetime of a drug particle	24 hour
a	Radius of a drug particle	Variable

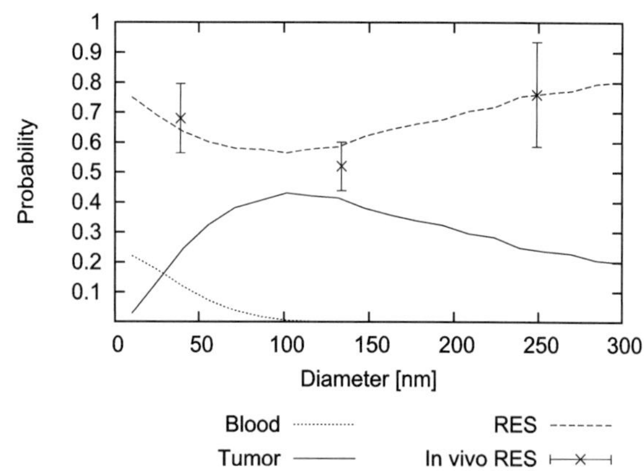

Fig. 21.25 Our simulations (*curve*) and experiments (*bar*): This figure describes the probabilities $p_1^{(n_{\max})}$, $p_2^{(n_{\max})}$, $p_3^{(n_{\max})}$ and the experiment data about the RES uptake of EPC liposomes in [758]. In the figure, Blood, Tumor, RES mean $p_1^{(n_{\max})}$, $p_2^{(n_{\max})}$, $p_3^{(n_{\max})}$ and "In-vivo RES" means the experiment result (average+/−SD)

(iii) We take two $T_{\mathrm{circ}} = 5$, 15 min and plot the probability $p_3^{(n_{\max})}$ for each T_{body} in Fig. 21.28. In the same figure, we also plot the RES uptake of EPC liposomes and that of HEPC liposomes (i.e., whose membranes are less fluid than those of EPC liposome) given in [758]. Comparing our result with their result, we understand that HEPC liposome has a bigger circulation time T_{body} than that

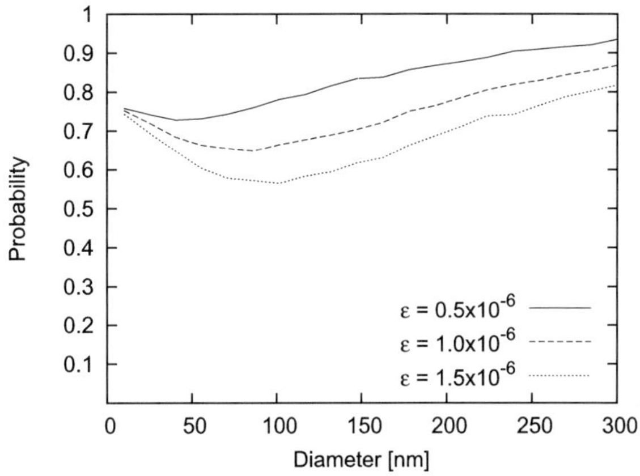

Fig. 21.26 Sensitivity of the parameter ε to the probability in RES: The *lines* indicate the probability $p_3^{(n_{\max})}$ for fixed three $\varepsilon = 0.5 \times 10^{-6}$, 1.0×10^{-6}, 1.5×10^{-6}

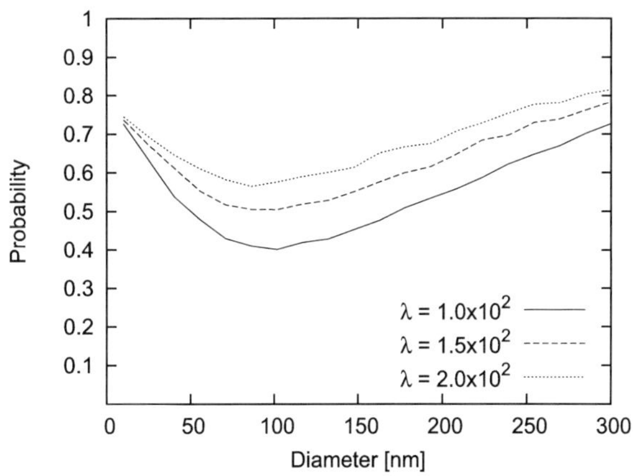

Fig. 21.27 Sensitivity of the parameter λ to the probability in RES: The *lines* indicate the probability $p_3^{(n_{\max})}$ for three fixed $\lambda = 1.0 \times 10^2$, 1.5×10^2, 2.0×10^2

of EPC liposome. That is, our simulation can explain the fluid ability of a liposome through the parameter T_{body}.

(iv) We take $T_{\max} = 6$ hour and plot (a) the probability $p_2^{(n_{\max})}$ in tumor in Fig. 21.29 with respect to the diameter of the drug and its mass density; (b) the probability $p_2^{(n_{\max})}$ with respect to the diameter for three fixed $\rho = 0.5$, 1.0, 2.0 g/cm^3 in Fig. 21.30; and (c) the probability $p_1^{(n_{\max})}$ in blood in Fig. 21.31. These are the sensitivity analysis of parameter a and ρ. The sensitivity analysis suggests that (i) the size of a drug maximizing the probability $p_2^{(n_{\max})}$ depends on its mass density; and (ii) the probability $p_1^{(n_{\max})}$ decreases as the mass density increase. These results (i) and (ii) imply that we can find the most effective size of a drug for DDS with respect to the mass density ρ which is designed.

(v) We take three $T_{\max} = 1, 3, 6$ hour and plot the probability $p_2^{(n_{\max})}$ for each $T_{\max}$ in Fig. 21.32. From the simulation, we conclude that the most effective size of a drug depends on its survival time. The size of a drug particle should be determined with respect to the survival time $T_{\max}$ which is designed.

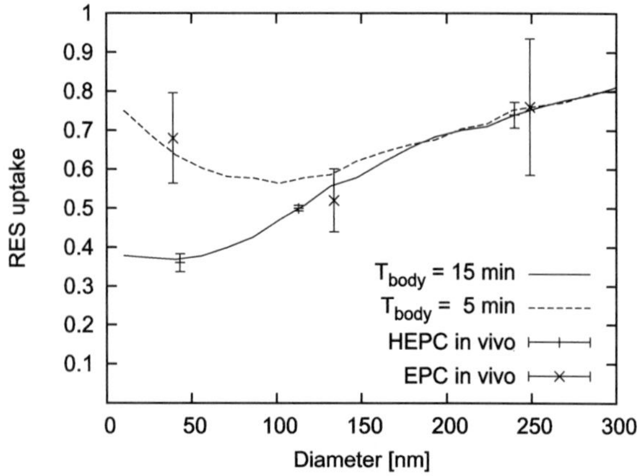

Fig. 21.28 Simulations (*curve*) and experiments (*bar*) for RES uptake: This figure describes the probability $p_3^{(n_{\max})}$ in RES for two fixed $T_{\text{body}} = 5,\ 15$ min, and the RES uptake of EPC liposomes and those of HEPC liposomes in [758] (average+/−SD)

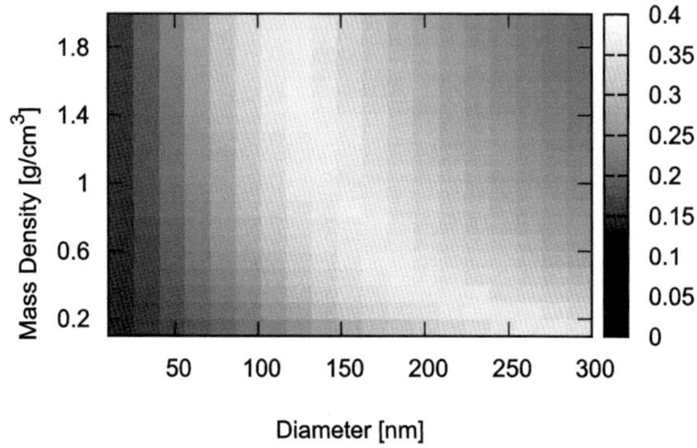

Fig. 21.29 Sensitivity analysis of parameter a and ρ: This figure describes the probability $p_2^{(n_{\max})}$ in tumor after the time $T_{\max} = 6$ hour with respect to the diameter and the mass density

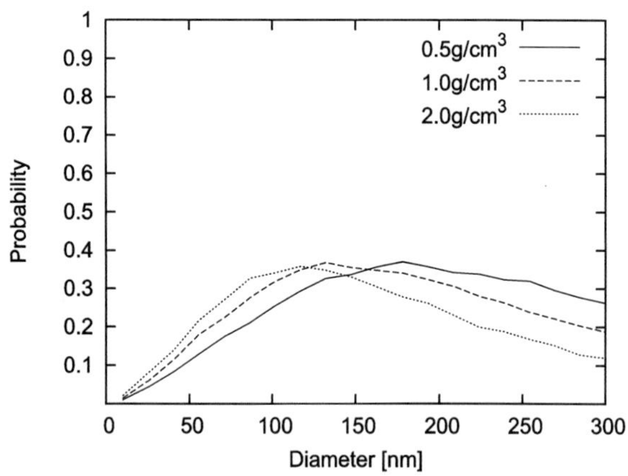

Fig. 21.30 Sensitivity of the parameter ρ to the probability in tumor: The *lines* indicate the probability $p_2^{(n_{\max})}$ for three fixed $\rho = 0.5$, 1.0, 2.0 g/cm^3

From (i)–(v), the model makes clear a relation between the changing characters of drug and its effect, therefore the simulation will indicate the way how to design a drug.

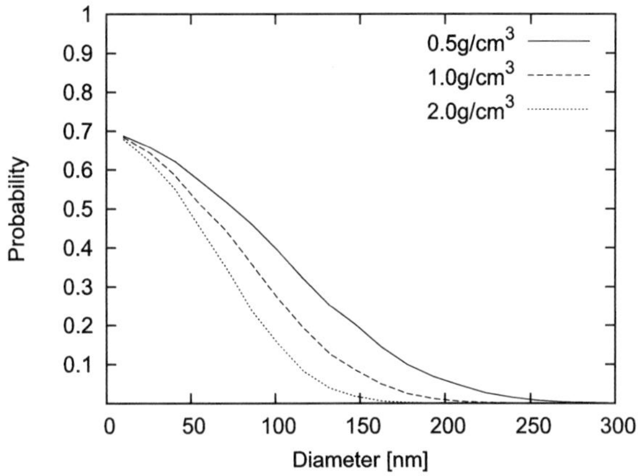

Fig. 21.31 Sensitivity of the parameter ρ to the probability in blood: The *lines* indicate the plobability $p_1^{(n_{\max})}$ for three fixed $\rho = 0.5, 1.0, 2.0\ \mathrm{g/cm^3}$

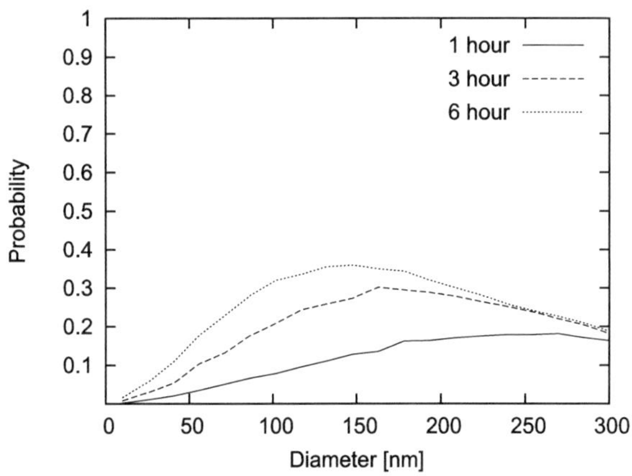

Fig. 21.32 Sensitivity of the parameter $T_{\max}$ to the probability in tumor: The *lines* indicate the probability $p_2^{(n_{\max})}$ for three fixed $T_{\max} = 1, 3, 6$ hour

In this section, we explained some results of experiments by a mathematical model and showed an ideal design of a drug for the drug delivery system. It was found that the optimal diameter of a nanoparticle to be absorbed by the tumor is about 100 nm. The model will help make drug delivery effective. Moreover, this research could be used to understand which surface-modification of a drug particle or some others such as ligand-receptor is effective for a drug delivery system.

21.9 An Application of Quantum Information to a Biosystem: Biomolecular Modeling

21.9.1 Folding Problem

The sequence of amino acids in a protein defines its primary structure. As discussed in Sect. 21.1, the blueprint for each amino acid is laid down by sets of three letters, known as base triplets, that are found in the coding regions of genes. These base triplets are recognized by ribosomes, the protein building sites of the cell, which create and successively join the amino acids together. This is a quick process: a protein of 300 amino acids will be made in little more than a minute.

The result is a linear chain of amino acids, but this only becomes a functional protein when it *folds* into its three-dimensional (tertiary structure) form which is called the *native state*. Protein folding involves physical timescales—microseconds to seconds. This occurs through an intermediate form, known as secondary structure, the most common of which are the rod-like α-helix and the plate-like β-pleated sheet. These secondary structures are formed by a small number of amino acids that are close together, which then, in turn, interact, fold and coil to produce the tertiary structure that contains its functional regions.

Although it is possible to deduce the primary structure of a protein from a gene's sequence, its tertiary structure (i.e., its shape) currently cannot be determined theoretically. It can only be determined by complex experimental analyses and, at present, this information is only known for about 10% of proteins. It is therefore not yet known how an amino-acid chain folds into its three-dimensional structure in the short time scale (fractions of a second) that occurs in the cell.

Can the three-dimensional structure of a protein be predicted given only its amino acid sequence? What is the pathway from the unfolded to the folded state for any given protein? What is the physical basis for the stability of the folded conformation? These questions constitute the "protein folding problem".

Analogous folding problems are relevant also to DNA and RNA molecules. Another fundamental problem is the interaction of nucleic acids and proteins. One hopes to make progress in these problems by using molecular dynamics simulation and other methods.

We here refer to the book by Schlick [689] for molecular dynamics and to the recent reviews by Ando and Yamato [57] and [686] for folding problem.

Note that there exists a p-adic approach to study the protein dynamics, see also Sect. 21.7. It was shown that the results on protein dynamics obtained with the help of p-adic methods coincide with the data of spectroscopic experiments for Mb–CO rebinding. A p-adic diffusion equation which was suggested and studied in [783] was used in [76] to describe the protein dynamics. It was shown that p-adic methods are relevant for the dynamics on energy landscapes, interbasin kinetics and conformational dynamics of proteins. In this case, an ultrametric parameter describes the conformational coordinate for the protein [77].

21.9.2 Molecular Dynamics

The aim of the molecular dynamics simulations of the behavior of biomolecules is not at describing of individual trajectories of a given molecule but rather at gaining structural insights and predicting statistical properties of complex biomolecules with the goal of relating structure and dynamics of biomolecules to biological functions.

Quantum Mechanics and Molecular Dynamics

The structure and dynamics of any molecule, including the biomolecules, are described in quantum mechanics by the Schrödinger equation

$$i\hbar \frac{\partial \psi}{\partial t} = H\psi.$$

Here the wave function ψ and the Hamiltonian $H = H_0 + V$ depend on the atomic coordinates. In principle, one could try to calculate the electronic and nuclear structures of atoms and molecules from the Schrödinger equation. Though this fundamental approach is impractical yet, quantum mechanics is used to study various effects on biological macromolecules. Usually, one uses the Hartree–Fock and Born–Oppenheimer approximations.

Molecular Mechanics

In the Born–Oppenheimer approximation to the Schrödinger equation, the nuclei remain fixed on the timescale of electronic motion. A molecule is considered as built up from just two types of particles, nuclei with negligible size and electrons, the motion of which can be separated. In this approximation, a molecule is described as a system of N point masses satisfying the laws of classical mechanics and moving in an effective potential field V. *A molecule is modeled as a mechanical body*. It is considered as a collection of masses centered at the nuclei (atoms) connected by springs (bonds); the molecule can stretch, bend and rotate about those bonds.

For a molecule consisting from N atoms, let $\mathbf{x}_i = (x_{i1}, x_{i2}, x_{i3})$, $i = 1, \dots, N$, denote the position vector of atom i in Cartesian coordinates, and let $\mathbf{r}_{ij} = \mathbf{x}_i - \mathbf{x}_j$ denote the distance vector from atom i to j. The magnitude of the vector $\mathbf{r}_{ij}$ is denoted by r_{ij}.

The potential energy V of a molecular model depends on all Cartesian variables of the atoms: $V = V(\mathbf{x}_1, \dots, \mathbf{x}_N)$. The classical Hamiltonian of a molecule has the form

$$H = H_0 + V$$

where

$$H_0 = \sum_{i=1}^{N} \frac{\mathbf{p}_i^2}{2m_i}$$

is the kinetic energy, $\mathbf{p}_i$ are momenta, and m_i are masses of atoms.

The Newton equations for this Hamiltonian read

$$m_i \frac{d^2\mathbf{x}_i}{dt^2} = -\frac{\partial V}{\partial \mathbf{x}_i}, \quad i = 1, \dots, N.$$

The potential energy V is constructed as the sum of contributions from the following five types of terms: bond length and bond strain terms (V_{bond} and V_{bang}), a torsional potential (V_{tor}), a Lennard–Jones potential (V_{LJ}) and a Coulomb potential (V_{Coul}):

$$V = V_{\text{bond}} + V_{\text{bang}} + V_{\text{tor}} + V_{\text{LJ}} + V_{\text{Coul}}.$$

The typical form of the potentials is:

$$V_{\text{bond}} = \sum_{i,j \in M_{\text{B}}} B_{ij}(r_{ij} - \overline{r}_{ij})^2,$$

$$V_{\text{bang}} = \sum_{i,j,k \in M_{\text{BA}}} K_{ijk}(\theta_{ijk} - \overline{\theta}_{ijk})^2,$$

$$V_{\text{tor}} = \sum_{i,j,k,l \in M_{\text{DA}}} \sum_{n} V_{i,j,k,l}^{(n)}\left[1 \pm \cos(n\tau_{ijkl})\right],$$

$$V_{\text{LJ}} = \sum_{i,j \in M_{\text{NB}}} \left(-\frac{C_{ij}}{r_{ij}^6} + \frac{D_{ij}}{r_{ij}^{12}}\right),$$

$$V_{\text{Coul}} = \sum_{i,j \in M_{\text{NB}}} \frac{q_i q_j}{\epsilon r_{ij}}.$$

The symbols M_{B}, M_{BA}, and M_{DA} denote the sets of all bonds, bond angles, and dihedral angles. The nonbonded set M_{NB} includes the (i, j) atom pairs separated by three bonds or more. B_{ij}, K_{ijk}, ... are constant parameters, ϵ is the dielectric constant. Bond and angle variables capped by bar symbols denote reference values associated with these quantities. Typical values of n are 1, 2, 3, or 4.

A bond angle θ_{ijk} formed by a bonded triplet of atoms $i - j - k$ can be expressed as

$$\cos\theta_{ijk} = \frac{\langle \mathbf{r}_{kj}, \mathbf{r}_{ij} \rangle}{r_{kj} r_{ij}}.$$

A dihedral angle τ_{ijkl}, defining the rotation of bond $i - j$ around $j - k$ with respect to $k - l$, is expressed as

$$\cos\tau_{ijkl} = \langle \mathbf{n}_{ab}, \mathbf{n}_{bc} \rangle.$$

Here vectors $\mathbf{n}_{ab}$ and $\mathbf{n}_{bc}$ denote unit normals to planes spanned by the vectors $\{\mathbf{a}, \mathbf{b}\}$ and $\{\mathbf{b}, \mathbf{c}\}$, respectively, where $\mathbf{a} = \mathbf{r}_{ij}$, $\mathbf{b} = \mathbf{r}_{jk}$ and $\mathbf{c} = \mathbf{r}_{kl}$.

Stochastic Dynamics

There are various mathematical models to represent the effects of the environment (water or other stochastic dynamics method in which the influence of solvent particle of the solute is incorporated through additional frictional and random terms. The

Newton equation is replaced by the stochastic Langevin equation (on the mathematical discussion of stochastic differential equations see Sect. 21.5.5 on Hida white noise analysis): When there exists the density of the noise,

$$m_i \frac{d^2\mathbf{x}_i}{dt^2} = -\frac{\partial V}{\partial \mathbf{x}_i} - \gamma \frac{d\mathbf{x}_i}{dt} + \mathbf{R}_i(t), \quad i = 1, \ldots, N$$

where γ is the damping constant and $\mathbf{R}_i(t) = (R_{i1}(t), R_{i2}(t), R_{i3}(t))$, the density of the noise, with mean $E(\mathbf{R}_i(t)) = 0$ and the covariance

$$E\big(R_{i\alpha}(t)R_{j\beta}(s)\big) = 2\gamma k_B T \delta_{ij}\delta_{\alpha\beta}\delta(t-s), \quad i, j = 1, \ldots, N, \ \alpha, \beta = 1, 2, 3.$$

Here k_B is Boltzmann's constant, and T is the temperature. The damping constant γ controls both the magnitude of the frictional force and the variance of the random forces; it follows from the fluctuation/dissipation theorem. According to the Stokes law for the frictional resistance of a spherical particle in a solution, one can take $\gamma = 6\pi\eta a$ where η is the solvent viscosity, and a is the hydrodynamic radius of atom.

The Brownian Limit

In the limit when the solvent damping is large and the velocity relaxation time is much more rapid than position relaxation time, one can take the following approximation to the Langevin equation:

$$\gamma \frac{d\mathbf{x}_i}{dt} = -\frac{\partial V}{\partial \mathbf{x}_i} + \mathbf{R}_i(t), \quad i = 1, \ldots, N$$

which is called the Brownian dynamics.

21.9.3 Molecular Dynamics for Harmonic Oscillator

Let us consider the method of molecular dynamics for studying the equation of motion for the simple harmonic oscillator to see at which scale it could give a good approximation. One of widely used algorithms for integrating the equations of motion in molecular dynamics is the velocity Verlet algorithm below. The method is symplectic and time-reversable.

If we have to integrate numerically an equation

$$\ddot{x}(t) = F(t)$$

then the velocity Verlet algorithm is often used, which reads:

$$x(t+h) = x(t) + hv(t) + \frac{h^2}{2}F(t),$$

$$v(t+h) = v(t) + \frac{h}{2}\big[F(t+h) + F(t)\big].$$

Here $h > 0$ is the time step. We denote

$$x_n = x(nh), \qquad v_n = v(nh), \quad n = 0, 1, 2, \ldots.$$

As an example, we take a harmonic oscillator which is one of the terms in the potential energy for biomolecules,

$$\ddot{x}(t) = -\omega^2 x(t),$$

we have

$$\begin{pmatrix} x_{n+1} \\ v_{n+1} \end{pmatrix} = A \begin{pmatrix} x_n \\ v_n \end{pmatrix}$$

where

$$A = \begin{pmatrix} 1 - \frac{h^2\omega^2}{2} & h \\ -h\omega^2 + \frac{h^3\omega^4}{4} & 1 - \frac{h^2\omega^2}{2} \end{pmatrix}.$$

The solutions of the characteristic equation

$$\det(A - \lambda I) = 0$$

are

$$\lambda_{1,2} = 1 - \frac{h^2\omega^2}{2} \pm \mathrm{i}\sqrt{h^2\omega^2 - \frac{h^4\omega^4}{4}} = e^{\pm \mathrm{i}\varphi}.$$

We represent the matrix A in the diagonal form:

$$A = QDQ^{-1},$$

where

$$D = \begin{pmatrix} \lambda_1 & 0 \\ 0 & \lambda_2 \end{pmatrix}.$$

For simplicity let us set $\omega^2 = 2$. Then

$$Q = \begin{pmatrix} h & h \\ \lambda_1 - 1 + h^2 & \lambda_2 - 1 + h^2 \end{pmatrix},$$

$$Q^{-1} = \frac{1}{h(\lambda_1 - \lambda_2)} \begin{pmatrix} -\lambda_2 + 1 - h^2 & h \\ \lambda_1 - 1 + h^2 & -h \end{pmatrix}.$$

The solution of our discrete equation is

$$\begin{pmatrix} x_n \\ v_n \end{pmatrix} = QD^n Q^{-1} \begin{pmatrix} x_0 \\ v_0 \end{pmatrix} = Q \begin{pmatrix} e^{\mathrm{i}n\varphi} & 0 \\ 0 & e^{-\mathrm{i}n\varphi} \end{pmatrix} Q^{-1} \begin{pmatrix} x_0 \\ v_0 \end{pmatrix}.$$

In particular, for x_n we get

$$x_n = \frac{1}{\sin\varphi}\big[x_0\big(\sin\varphi\cos n\varphi + \big(1 - h^2 - \cos\varphi\big)\sin n\varphi\big) + h v_0 \sin n\varphi\big]$$

where

$$\sin\varphi = h\sqrt{2-h^2}.$$

It would be interesting to compare the behavior of the obtained x_n with $v(nh)$ where $v(nh)$ is the known solution of the differential equation

$$x(t) = x_0\cos(\omega t) + \frac{v_0}{\omega}\sin(\omega t)$$

and to see for which n and h the values x_n will lead to a good approximation to $v(nh)$.

21.9.4 Examples of Molecular Dynamics Simulations

Long molecular dynamics computations are mathematically ill-defined and generate cumulative errors in numerical integration that can be minimized with appropriate selection of algorithms and parameters, but not eliminated entirely. Furthermore, known potential functions are not sufficiently accurate to reproduce the dynamics of molecular systems.

Better results, in principle, can be obtained from first principles by using a quantum mechanical method known as Ab Initio Molecular Dynamics (AIMD), such as the theory of the density functional. Due to the cost of treating the electronic degrees of freedom, the computational cost of this simulation is higher than classical molecular dynamics. This implies that AIMD is limited to smaller systems and shorter periods of time. Classical molecular dynamics techniques allow a rather detailed time and space insight into dynamical behavior for some systems.

In particular, in 2006 folding simulations of the Villin Headpiece with 20 000 atoms were performed; Simulation time: 500 μs, Program: "folding@home". This simulation was run in 200 000 central processing units (CPU) of participating personal computers around the world with a distributed computing effort coordinated by Vijay Pande at Stanford University. The kinetic properties of the Villin Headpiece protein were probed by using many independent, short trajectories run by CPUs without continuous real-time communication. One technique employed was the Pfold value analysis, which measures the probability of folding before unfolding of a specific starting conformation. Pfold gives information about transition state structures and an ordering of conformations along the folding pathway.

21.9.5 Quantum Algorithm of Protein Folding

Molecular dynamics (MD) is an invaluable tool with which to study protein folding in silico. Although just a few years ago the dynamic behavior of a protein molecule could be simulated only in the neighborhood of the experimental conformation (or protein unfolding could be simulated at high temperature), the advent of distributed computing, new techniques such as replica-exchange MD, new approaches (based on, e.g., the stochastic difference equation), and physics-based reduced models of proteins now make it possible to study protein-folding pathways from completely unfolded structures. In this review, we present algorithms for MD and their extensions and applications to protein-folding studies, using all-atom models with explicit and implicit solvent as well as reduced models of polypeptide chains.

We can write the quantum algorithm to calculate the position x_i of all moleculars in the Hilbert space $\otimes\mathbb{C}^2$ in the following steps:

1. Prepare an input state of the tensor products of initial positions of molecules in a protein.
2. Construct unitary operator which represents a dynamics of molecules.
3. Apply it into an initial state several times.
4. Obtain the final state (the positions of all molecules) for each specified time and check whether the protein finally folds.

The details of this algorithm is discussed in [375].

21.10 Quantum Photosynthesis

In this section, we describe some results on photosynthesis. Recently, it was discovered [172] that quantum mechanics might be involved in the process of photosynthesis in some marine algae at the room temperature. This experimental result is very important for establishing quantum biology as an experimental science. We will describe applications to theoretical investigation of photosynthesis of some results from quantum theory and quantum information discussed in this book.

We describe the relation between the chaotic amplifier in generalized quantum algorithm discussed in Chaps. 10 and 14, and the efficient excitation transport in the photosynthetic antenna. We also discuss a possible role of the entropy decreasing mentioned in Chap. 20 for the process of photosynthesis.

Photosynthesis is vital for life on Earth. Photosynthesis changes the energy from the sun into chemical energy and splits water to liberate oxygen and convert carbon dioxide into organic compounds, especially sugars. Energy from sunlight is used to convert carbon dioxide and water into organic materials to be used in cellular functions such as biosynthesis and respiration. Photosynthesis occurs in plants, algae, and many species of bacteria. In plants, algae, and cyanobacteria photosynthesis uses carbon dioxide and water, releasing oxygen as a waste product and maintains the normal level of oxygen in the atmosphere. Nearly all life either depends on it directly as a source of energy, or indirectly as the ultimate source of the energy in their food. The amount of energy trapped by photosynthesis, is about six times larger than the power consumption of human civilization. Photosynthesis is also the source of the carbon in all the organic compounds within organisms' bodies.

21.10.1 Photosystems

Photosystems (or Reaction Centers) are protein complexes in cells involved in photosynthesis. They are found in the thylakoid membranes of cells in plants and algae where these are located in the chloroplasts, or in the cytoplasmic membrane of photosynthetic bacteria. Chloroplasts are organelles found in plant cells and other eukaryotic organisms that conduct photosynthesis. The thylakoid membrane is the site of the light-dependent reactions of photosynthesis with the photosynthetic pigments embedded in the membrane.

The thylakoid membranes of higher plants are composed primarily of phospholipids and galactolipids that are asymmetrically arranged along and across the membranes.

In and around photosystems, there are chlorophyll molecules. The chlorophyll molecule is the active part that absorbs the sunlight, it is attached to the backbone of a complicated protein.

Chlorophyll is a green pigment which absorbs light most strongly in the red, a bit in the blue, but poorly in the green portions of the electromagnetic spectrum, hence the green color of chlorophyll-containing tissues such as plant leaves. The basic structure of a chlorophyll molecule is a porphyrin ring, coordinated to a central atom. There are actually two main types of chlorophyll, named *a* and *b*. They differ only slightly, in the composition of a sidechain; in *a* it is CH_3, in *b* it is CHO. Both of these two chlorophylls are effective photoreceptors because they contain a network of alternating single and double bonds, and the orbitals can delocalize stabilizing the structure. Such delocalized polyenes have very strong absorption bands in the visible regions of the spectrum, allowing the plant to absorb the energy from sunlight.

Chlorophyll serves two primary functions. The function of the vast majority of chlorophyll (up to several hundred molecules per photosystem) is to absorb light and transfer that light energy by resonance energy transfer to a specific chlorophyll pair in the reaction center of the photosystems. Because of chlorophyll's selectivity regarding the wavelength of light it absorbs, areas of a leaf containing the molecule will appear green.

The two currently accepted photosystem units are Photosystem II and Photosystem I, which have their own distinct reaction center chlorophylls, named P680 and P700, respectively. These pigments are named after the wavelength (in nanometers) of their red-peak absorption maximum.

The function of the reaction center chlorophyll is to use the energy absorbed by and transferred to it from the other chlorophyll pigments in the photosystems to undergo a charge separation, a specific redox reaction in which the chlorophyll donates an electron into a series of molecular intermediates called an electron transport chain. The charged reaction center chlorophyll (P680+) is then reduced back to its ground state by accepting an electron. In Photosystem II, the electron which reduces P680+ ultimately comes from the oxidation of water into O_2 and H^+ through several intermediates. This reaction is how photosynthetic organisms like plants produce O_2 gas, and is the source for practically all the O_2 in Earth's atmosphere. Photosystem I typically works in series with Photosystem II, thus the P700+ of Photosystem I is usually reduced, via many intermediates in the thylakoid membrane, by electrons ultimately from Photosystem II. Electron transfer reactions in the thylakoid membranes are complex, however, and the source of electrons used to reduce P700+ can vary.

The electron flow produced by the reaction center chlorophyll pigments is used to shuttle H^+ ions across the thylakoid membrane, setting up a chemiosmotic potential mainly used to produce ATP chemical energy, and those electrons ultimately reduce NADP+ to NADPH a universal reductant used to reduce CO_2 into sugars as well as for other biosynthetic reductions.

Reaction center chlorophyll-protein complexes are capable of directly absorbing light and performing charge separation events without other chlorophyll pigments, but the absorption cross section (the likelihood of absorbing a photon under a given light intensity) is small. Thus, the remaining chlorophylls in the photosystem and antenna pigment protein complexes associated with the photosystems all cooperatively absorb and funnel light energy to the reaction center. Besides chlorophyll, there are other pigments, called accessory pigments, which occur in these pigment–protein antenna complexes.

A photosystem is an enzyme which uses light to reduce molecules. The membrane protein complex is made of several subunits and contains numerous cofactors. In the photosynthetic membranes, reaction centers provide the driving force for the bioenergetic electron and proton transfer chain. When light is absorbed by a reaction center (either directly or passed by neighboring pigment-antennae), a series of oxido-reduction reactions is initiated, leading to the reduction of a terminal acceptor. Two families of photosystems exist: type I reaction centers (like photosystem I (P700) in chloroplasts and in green-sulphur bacteria) and type II reaction centers (like photosystem II (P680) in chloroplasts and in non-sulphur purple bacteria). Each photosystem can be identified by the wavelength of light to which it is most reactive (700 and 680 nanometers, respectively, for PSI and PSII in chloroplasts), and the type of terminal electron acceptor. Type I photosystems use ferredoxin-like iron–sulfur cluster proteins as terminal electron acceptors, while type II photosystems ultimately shuttle electrons to a quinone terminal electron acceptor.

21.10.2 Biophysics of Photosynthesis

The process of photosynthesis begins when energy from light is absorbed by proteins of photosynthetic reaction centers that contain chlorophylls. In plants, these proteins are held inside organelles called chloroplasts, while in bacteria they are embedded in the plasma membrane. Some of the light energy gathered by chlorophylls is stored in the form of adenosine triphosphate (ATP). The rest of the energy is used to remove electrons from a substance such as water. These electrons are then used in the reactions that turn carbon dioxide into organic compounds. In plants, algae and cyanobacteria this is done by a sequence of reactions called the Calvin cycle.

In photosynthesis carbon dioxide and water are converted into glucose and oxygen. The photosynthesis equation is

$$6CO_2+12H_2O+\text{light}\rightarrow C_6H_{12}0_6+60_2+6H_20.$$

Photosynthesis occurs in two stages. In the first stage, light reactions capture the energy of light and use it to make the energy-storage molecules. During the second stage, the light-independent reactions use these products to capture and reduce carbon dioxide. Most organisms use visible light to produce oxygen.

The proteins that gather light for photosynthesis are embedded within cell membranes. In the light reactions, one molecule of the pigment chlorophyll absorbs one photon and loses one electron. This electron is passed to a modified form of chlorophyll called pheophytin, which passes the electron to a quinone molecule, allowing the start of a flow of electrons down an electron transport chain.

21.10.3 Quantum Mechanics in Photosynthesis

It was discovered [172] that quantum mechanics might be involved in the process of photosynthesis in some marine algae at room temperature. Previously the role of quantum effects in the photosynthesis at room temperature was ruled out because of the quantum decoherence.

The evidence comes from a study of how energy travels across the light-harvesting molecules involved in photosynthesis. The work by Scholes et al. [172] demonstrated that the light-harvesting molecules involved in photosynthesis in a marine algae may exploit quantum processes at room temperature to transfer energy almost without loss.

The antenna proteins absorb light and transmit the resultant excitation energy between molecules to a reaction center. The efficiency of these electronic energy transfers was investigated in many works on antenna proteins isolated from photosynthetic organisms to uncover the basic mechanisms at play. Recent work has documented that light-absorbing molecules in some photosynthetic proteins capture and transfer energy according to quantum-mechanical probability laws instead of classical laws at temperatures up to 180 K. Photosynthesis starts with the absorption of a photon of sunlight by one of the light-harvesting pigments, followed by transfer of the energy to the reaction center, where the primary electron transfer reactions convert the solar energy into an electrochemical gradient. The transfer of this excitation energy towards the reaction center occurs with a near unity quantum yield. The Fenna–Matthews–Olson (FMO) pigment–protein complex is found in low light-adapted green sulfur bacteria. Under physiological conditions, this complex is situated between the so-called base-plate protein of the large peripheral chlorosome antenna and the reaction center complex, and it is transporting sunlight energy harvested in the chlorosome to the reaction center pigments. The complex is a trimer made of identical subunits, each of which contains seven bacteriochlorophyll molecules.

In [377], the spatial and temporal dynamics of excitation energy transfer through the FMO complex at physiological temperature are investigated. The numerical results demonstrate that quantum wave-like motion persists for several hundred femtoseconds even at physiological temperature, and suggest that the FMO complex may work as a rectifier for unidirectional energy flow from the peripheral light-harvesting antenna to the reaction center complex by taking advantage of quantum coherence and the energy landscape of pigments tuned by the protein scaffold. The observation of long-lasting and robust quantum coherence prompts the speculation that quantum effects may play a significant role in achieving the remarkable efficiency of photosynthetic excitation energy transfer. In [227], it is proposed that the FMO complex performs a quantum search algorithm that is more efficient than a classical random walk suggested by the hopping mechanism. Quantum coherence enables the excitation to rapidly and reversibly sample multiple pathways to search for bacteriochlorophyll molecules that connects to the reaction center.

This contrasts with the long-held view that long-range quantum coherence between molecules cannot be sustained in complex biological systems, even at low temperatures. In [172], two-dimensional photon echo spectroscopy measurements on two evolutionarily related light-harvesting proteins isolated from marine cryptophyte algae are presented, which reveal exceptionally long-lasting excitation oscillations with distinct correlations and anti-correlations even at ambient temperature.

These observations provide compelling evidence for quantum coherent sharing of electronic excitation across the 5-nm-wide proteins under biologically relevant conditions, suggesting that distant molecules within the photosynthetic proteins are 'wired' together by quantum coherence for more efficient light-harvesting in cryptophyte marine algae.

For the experiments [172], the proteins were isolated from the algae and suspended at low concentration in aqueous buffer at ambient temperature (294 K). The femtosecond laser pulse (25-fs duration) excites a coherent superposition of the antenna protein's electronic vibrational eigenstates (absorption bands). The initial state of the system is thus prepared in a non-stationary state, where electronic excitation is localized to a greater or lesser degree compared to the eigenstates. The time-dependent solution to quantum dynamics for electronically coupled molecules with this initial condition predicts that excitation subsequently oscillates among the molecules under the influence of the system Hamiltonian until the natural eigenstates are restored owing to interactions with the environment.

21.10.4 Quantum Network Model

Electron transport in organic molecules, such as proteins and polymers, may be described by quantum graphs [112, 292, 446, 529, 635]. Indeed, it follows one-dimensional pathways (the bonds) changing from one path to other due to scattering centers (the vertices). Charge transport in solids is also well described by quantum graphs. A simplified version of a quantum graph is given by a quantum network. Here we describe some recent works on application of quantum networks to photosynthesis.

Light-harvesting complexes are typically constituted of multiple chromophores which transform photons into excitons and transport them to a reaction center. Experimental studies of the exciton dynamics in such systems reveal rich transport dynamics consisting of short-time coherent quantum dynamics which evolve, in the presence of noise, into an incoherent population transport which irreversibly transfers excitations to the reaction center. In order to elucidate the basic phenomena clearly without overburdening the description with detail, we consider the relevant complexes as systems composed of several distinct sites, one of which is connected to the chromosomes while another is connected to the reaction center. This complex effective dynamics will then be modeled by a combination of simple Hamiltonian dynamics which describe the coherent exchange of excitations between sites, and local Lindblad terms that take into account the dephasing and dissipation caused by the external environment.

The pigment–protein complex will be considered as a network composed of distinct sites, one of which receives a single initial excitation, while another is connected to the reaction center.

A network of N sites will be described by the Hamiltonian [151]

$$H = \sum_{j=1}^{N} \hbar\omega_j \sigma_j^+ \sigma_j^- + \sum_{j \neq k} \hbar v_{jk} \left(\sigma_j^- \sigma_j^+ + \sigma_j^+ \sigma_k^-\right)$$

where $\sigma_j^+ = |j\rangle\langle 0|$ and $\sigma_j^- = |0\rangle\langle j|$ are raising and lowering operators for site j, the state $|j\rangle$ denotes one excitation in site j and $|0\rangle$ is the zero exciton state. The local site energies are ω_j, and v_{jk} is the coherent tunneling amplitude between the sites j and k.

The dynamics of the network's density matrix $\rho(t)$ is described by a Markovian master equation of the form

$$\frac{d}{dt}\rho(t) = -\mathrm{i}\big[H, \rho(t)\big] + L_{\mathrm{diss}}\big(\rho(t)\big) + L_{\mathrm{deph}}\big(\rho(t)\big)$$

where the local dissipative and pure dephasing terms are described respectively by GKS-L super-operators

$$L_{\mathrm{diss}}(\rho) = \sum_{j=1}^{N} \Gamma_j \left(-\{\sigma_j^+ \sigma_j^-, \rho\} + 2\sigma_j^- \rho \sigma_j^+\right),$$

$$L_{\text{deph}}(\rho) = \sum_{j=1}^{N} \gamma_j \big(-\{\sigma_j^+ \sigma_j^-, \rho\} + 2\sigma_j^+ \sigma_j^- \rho \sigma_j^+ \sigma_j^- \big).$$

The total transfer of excitation is measured by the population in the 'sink', numbered $N+1$, which is populated by an irreversible decay process with rate Γ_{N+1} from a chosen site k as described by the GKS-L super-operator

$$L_{\text{sink}}(\rho) = \Gamma_{N+1} \big(2\sigma_{N+1}^+ \sigma_k^- \rho \sigma_k^+ \sigma_{N+1}^- - \{\sigma_k^+ \sigma_{N+1}^- \sigma_{N+1}^+ \sigma_k^-, \rho\} \big).$$

The initial state of the network at $t = 0$ is assumed to be a single excitation in site 1 (i.e., state $|1\rangle$). The model is completed by introducing the quantity by which we measure. The efficiency of network's transport properties will be measured by the population transferred to the sink $p_{\text{sink}}(t)$, which is given by

$$p_{\text{sink}}(t) = 2\Gamma_{N+1} \int_0^t \rho_{kk}(\tau)\, d\tau.$$

For a fully connected uniform network, when $\hbar v_{jk} = J$ for any $j \neq k$, and moreover, when ω_j, Γ_j and γ_j are the same on every site, i.e., $\omega_j = \omega$, $\Gamma_j = \Gamma$ and $\gamma_j = \gamma$, an exact analytical solution is found in [151] for the density matrix. It follows that for $\Gamma = 0$ (no dissipation) one gets different behavior of the network for the cases $\gamma = 0$ and $\gamma \neq 0$. If $\gamma = 0$ then

$$\lim_{t\to\infty} p_{\text{sink}}(t) = \frac{1}{N-1}.$$

If $\gamma \neq 0$ then

$$\lim_{t\to\infty} p_{\text{sink}}(t) = 1. \tag{21.15}$$

The result $p_{\text{sink}}(\infty) = 1$ means that there is the complete excitation transfer. Therefore, it is shown that the dephasing noise leads to the enhancement of the transport of excitons in this quantum network modeling the photosynthetic complexes.

It is found in [152] that the quantum and classical capacities for a family of quantum channels in the complex network dynamics can be enhanced by introducing dephasing noise.

21.10.5 Chaotic Amplifier and Decreasing of Entropy

Note that a constructive role of chaos in quantum computations was investigated in [369, 600, 602], see Chap. 14.

The phenomenon of the enhancement of the transport of excitons might be related with the decreasing of entropy discussed in Chap. 20.

21.10.6 Channel Representation of Photosynthesis

Here, we introduce the complete positive trace-preserving quantum channel which describes the photosynthesis. Let $\mathcal{H} = (\mathbb{C}^2)^{\otimes N}$ be the Hilbert space of the total system where N is the number of vertices, $\rho_A \in \mathfrak{S}(\mathbb{C}^2)$ the initial state of the site 1, and $\rho_E \in \mathfrak{S}((\mathbb{C}^2)^{\otimes N-1})$ the state of the environment. The output state at time t_{out} is given by

$$\rho_B(t_{\text{out}}) = \Lambda^*(\rho_A) = \operatorname{tr}_E U(t_{\text{out}})\rho_A \otimes \rho_E U^*(t_{\text{out}})$$

where $U(t_{\text{out}})$ of the unitary evolution of system and environment for time t_{out}. Let us see the change of von Neumann entropy of the state. Let $\rho(0)$ be the state at time $t = 0$ as

$$\rho(0) = \begin{pmatrix} p & \gamma \\ \gamma^* & 1-p \end{pmatrix}.$$

According to [152], the output state $\rho_B(t) = \rho(t)$ at time t can be written by

$$\rho(t) = \begin{pmatrix} \eta(t)p & \sqrt{\eta(t)s(t)}\gamma \\ \sqrt{\eta(t)s(t)}\gamma^* & 1-\eta(t)p \end{pmatrix}.$$

Now we represent $\rho(t)$ in the form given in Sect. 20.2.6 such that

$$\rho_{00}(t) = \eta(t)p.$$

Then we solve the GKSL equation. One has

$$\eta(t)p = \frac{r_{01}}{r_{01}+r_{10}} - \frac{r_{01}}{r_{01}+r_{10}} e^{-(r_{01}+r_{10})t} + pe^{-(r_{01}+r_{10})t}.$$

Since $\eta(t)$ should not depend on p, we take $r_{01} = 0$, then we have

$$\eta(t) = e^{-r_{10}t}. \tag{21.16}$$

Similarly, $\rho_{01}(t)$ is written as

$$\begin{aligned} \rho_{01}(t) &= \sqrt{\eta(t)s(t)}\gamma \\ &= \gamma e^{-\frac{1}{2}(r_{00}+r_{01}+r_{10}+r_{11})}. \end{aligned}$$

According to (21.16), $s(t)$ becomes

$$s(t) = e^{-(r_{00}+r_{11})t}.$$

If the von Neumann entropy of $\rho(t)$ increases in time t, p should be

$$p > \frac{r_{01}}{r_{01}+r_{10}} + \frac{r_{10}-r_{01}}{2(r_{01}+r_{10})} e^{(r_{01}+r_{10})t}.$$

In the case of $r_{01} = 0$, this condition becomes

$$p > \frac{r_{10}}{2r_{10}} e^{r_{10}t} = \frac{1}{2} e^{r_{10}t},$$

or equivalently,

$$\eta(t) > \frac{1}{2p}.$$

This condition might correspond to a condition in [151] for obtaining (21.15). The general version of the above discussion will be given in [374].

21.11 Quantum-Like Models for Cognitive Psychology

A few authors pointed to a possibility to apply the mathematical formalism of quantum mechanics to cognitive psychology, in particular, to games of the Prisoners Dilemma (PD) type [44, 73, 144–146, 177, 256, 415–417, 420, 424–426]. It was found that statistical data obtained in some experiments of cognitive psychology [182, 339, 340, 703, 704] cannot be described either by a classical probability model (Kolmogorov's model) or by the conventional quantum mechanics [420, 426]. These experiments play an important role in behavioral economics; these are tests for rationality of agents acting in the market (including the financial market). Note though that in [44] statistical data from one of experiments (Tversky–Shafir [704]) was described by a quantum Markov chain. Khrennikov suggested that the decision-making process in PD-type games could be described by a "quantum-like model", which has a mathematical structure different from the conventional quantum mechanics. Recently, a quantum-like model of decision making for two-players games was proposed [73]. The quantum-like decision-making in this model is described by a simple system of differential equations for the quantum state. Our final aim is to describe the decision making process in the PD-type game intuitively more plausible, for which quantum-like representation should be essential. For instance, we are looking for a new interpretation of equilibrium such as Nash's.

21.11.1 Quantum-Like Model for Decision-Making in Two-Player Game

Pay-off Table of Two-Player Game

Let us consider a two-player game with two strategies. The two strategies players A and B choose are denoted by "0" and "1". Table 21.19 shows pay-offs assigned to possible four consequences of "$0_A 0_B$", "$0_A 1_B$", "$1_A 0_B$", and "$1_A 1_B$".

In Table 21.19 a, b, c and d are the values of pay-offs.

Table 21.19 The values of pay-off

$A\backslash B$	0_B	1_B
0_A	$b\backslash b$	$d\backslash a$
1_A	$a\backslash d$	$c\backslash c$

We have the relation $a > b > c > d$ for the game of prisoner's dilemma (PD) type. *The goal of this game is to maximize the pay-off for each player independently.* For the player A, his pay-off will be a or c if the player B chooses "0" and b or d if the player B chooses "1". In both cases, from the relations of $a > b$ and $c > d$, he can obtain larger pay-offs if he chooses 1. The condition for the player B is same as that for the player A. We conclude that, in a PD game, a "rational" player always chooses "1".

However, the above discussion does not completely explain the process of decision-making in real player's mind. Actually, as seen in statistical data in some experiments, real players frequently behave "irrationally". The model proposed in [73] is an attempt to explain such real player's behaviors in "a quantum-like model".

21.11.2 Decision-Making Process in Player's Mind

Let us focus on player A's mind. In principle, the player A is not informed about the action the player B chooses. The player A will be conscious of two possibilities of B's action. This indeterminacy of the player A is described by means of the following quantum superposition:

$$|\phi_B\rangle = \alpha|0_B\rangle + \beta|1_B\rangle \in \mathbb{C}^2. \tag{21.17}$$

The values of α and β relate to the degrees of consciousness to B's actions. This vector is called a prediction state vector. In accordance with the formalism of quantum mechanics, $|\alpha|^2 + |\beta|^2 = 1$ is assumed.

When the player A decides to choose the action "0", he will be conscious of two consequences of "$0_A 0_B$" and "$0_A 1_B$" with the weights α and β, respectively. This situation is described by the vector on $\mathbb{C}^2 \otimes \mathbb{C}^2$ such as

$$\begin{aligned} |\Phi_{0_A}\rangle &= \alpha|0_A\rangle \otimes |0_B\rangle + \beta|0_A\rangle \otimes |1_B\rangle \\ &= |0_A\rangle \otimes |\phi_B\rangle. \end{aligned} \tag{21.18}$$

Similarly,

$$|\Phi_{1_A}\rangle = |1_A\rangle \otimes |\phi_B\rangle \tag{21.19}$$

is given for the situation when A decides to choose "1". By using these state vectors $|\Phi_{0_A}\rangle$ and $|\Phi_{1_A}\rangle$, we define the following vector for the player A:

$$|\Psi_A\rangle = x|\Phi_{0_A}\rangle + y|\Phi_{1_A}\rangle \in \mathbb{C}^2 \otimes \mathbb{C}^2, \tag{21.20}$$

$(|x|^2+|y|^2=1)$, which is called a mental state vector. The player A with this mental state chooses his own action probabilistically. His decision is described as "quantum measurement" of $|\Phi_{0_A}\rangle$ or $|\Phi_{1_A}\rangle$ in the state $|\Psi_A\rangle$. The probabilities of "0" and "1" are given by $P(0_A)=P_{0_A}=|x|^2$ and $P(1_A)=P_{1_A}=|y|^2$.

The decision-making process is described as a dynamics changing $|x|^2$ and $|y|^2$, and its dynamics has a stability solution. Such a stabilization of mental state explains the following psychological activity in the player A's mind: The player A has two psychological tendencies, one is to choose 0 and the other to choose 1. The two tendencies change in his mind, and they become stable balance. We do not consider here quantum dynamics of the mental state vector and focus only on the diagonal part of the density matrix. As a most simple classical dynamics, the equations like chemical equilibration are assumed as

$$\begin{aligned}\frac{d}{dt}P_{0_A}&=-kP_{0_A}+\tilde{k}P_{1_A},\\ \frac{d}{dt}P_{1_A}&=kP_{0_A}-\tilde{k}P_{1_A}.\end{aligned} \tag{21.21}$$

The parameter of k $(\tilde{k})$ corresponds to the velocity of psychological reaction from 0_A to 1_A (from 1_A to 0_A). The stable state of the probabilities P_{0_A} and P_{1_A} is given as

$$P^E_{0_A}=\frac{\tilde{k}}{k+\tilde{k}},\qquad P^E_{1_A}=\frac{k}{k+\tilde{k}}, \tag{21.22}$$

where E means equilibrium. From the above discussions, the player's tendency to choose 1 or 0 is proportional to the velocity k or $\tilde{k}$, and these parameters decide the stability solution of (21.22). It is assumed that $\tilde{k}$ is not zero in the case of PD, which explains the irrational behavior of the player A. The next question is how we can derive the velocities k and $\tilde{k}$ from fundamental considerations.

21.11.3 Definition of Velocities of k and $\tilde{k}$

In this model, the values k and $\tilde{k}$ are given by the "comparison" of possible consequences, 0_A0_B, 0_A1_B, 1_A0_B, and 1_A1_B. The player will consider the following four kinds of comparisons:

$$\begin{aligned} 0_A0_B &\underset{\tilde{k}_1}{\overset{k_1}{\rightleftharpoons}} 1_A0_B, &\quad 0_A1_B &\underset{\tilde{k}_2}{\overset{k_2}{\rightleftharpoons}} 1_A1_B,\\ 0_A1_B &\underset{\tilde{k}_3}{\overset{k_3}{\rightleftharpoons}} 1_A0_B, &\quad 0_A0_B &\underset{\tilde{k}_4}{\overset{k_4}{\rightleftharpoons}} 1_A1_B. \end{aligned} \tag{21.23}$$

These comparisons are represented again by chemical-like equilibrations, each of which is specified by reaction velocities, k_i and $\tilde{k}_i$. We note, in the next subsection, the relation between k_i ($\tilde{k}_i$), and the pay-off table of game is explained. Here, let us consider four maps from $\mathcal{B}(\mathbb{C}^2 \otimes \mathbb{C}^2)$ to $\mathcal{B}(\mathbb{C}^2 \otimes \mathbb{C}^2)$ with forms of

$$\mathcal{V}_i(\cdot) \equiv V_i \cdot V_i^*, \qquad \tilde{\mathcal{V}}_i(\cdot) \equiv V_i^* \cdot V_i \quad (i = 1, 2, 3, 4), \tag{21.24}$$

where $\{V_i\}$ are the transition operators defined by

$$\begin{aligned} V_1 &= |1_A 0_B\rangle\langle 0_A 0_B|, & V_2 &= |1_A 1_B\rangle\langle 0_A 1_B|, \\ V_3 &= |1_A 0_B\rangle\langle 0_A 1_B|, & V_4 &= |1_A 1_B\rangle\langle 0_A 0_B|. \end{aligned} \tag{21.25}$$

Each $\mathcal{V}_i$ gives state transitions in the four comparisons of (21.23). Furthermore, the map providing state transitions between $|\Phi_{0A}\rangle\langle\Phi_{0A}|$ and $|\Phi_{1A}\rangle\langle\Phi_{1A}|$ is considered

$$\begin{aligned} \mathcal{V}(\cdot) &\equiv V \cdot V^*, \\ \tilde{\mathcal{V}}(\cdot) &\equiv V^* \cdot V, \end{aligned} \tag{21.26}$$

where

$$V = |\Phi_{1A}\rangle\langle\Phi_{0A}|. \tag{21.27}$$

Using $\mathcal{V}$ and $\tilde{\mathcal{V}}$, the differential equations of (21.21) are rewritten as

$$\begin{aligned} \frac{d}{dt}\operatorname{tr}\bigl(\mathcal{V}(\Theta)\bigr) &= -k \operatorname{tr}\bigl(\mathcal{V}(\Theta)\bigr) + \tilde{k} \operatorname{tr}\bigl(\tilde{\mathcal{V}}(\Theta)\bigr), \\ \frac{d}{dt}\operatorname{tr}\bigl(\tilde{\mathcal{V}}(\Theta)\bigr) &= k \operatorname{tr}\bigl(\mathcal{V}(\Theta)\bigr) - \tilde{k} \operatorname{tr}\bigl(\tilde{\mathcal{V}}(\Theta)\bigr), \end{aligned} \tag{21.28}$$

where Θ is the mental state $|\Psi_A\rangle\langle\Psi_A|$ of (21.20). The maps $\mathcal{V}$ and $\tilde{\mathcal{V}}$ specify the "comparison" in the player's mind. Thus, depending on the forms of $\mathcal{V}$ and $\tilde{\mathcal{V}}$, the values of k and $\tilde{k}$ are decided. $\mathcal{V}$ and $\tilde{\mathcal{V}}$ have forms depending on α and β, which are the coefficients of the prediction state vector $|\phi_B\rangle$ of (21.17). In general, $\mathcal{V} \neq \mathcal{V}_i$, if α or $\beta \neq 0$ or 1. Actually, from the definition of $|\Phi_{0_A,1_A}\rangle$ of (21.18) and (21.19), one can easily check that

$$\begin{aligned} V &= |\alpha|^2 |1_A 0_B\rangle\langle 0_A 0_B| + |\beta|^2 |1_A 1_B\rangle\langle 0_A 1_B| \\ &\quad + \alpha\beta^* |1_A 0_B\rangle\langle 0_A 1_B| + \alpha^*\beta |1_A 1_B\rangle\langle 0_A 0_B| \\ &= \sum_{i=1,2,3,4} c_i V_i, \end{aligned} \tag{21.29}$$

where $\{c_i\} = \{|\alpha|^2, |\beta|^2, \alpha\beta^*, \alpha^*\beta\}$. Equation (21.27) is rewritten as

$$\begin{aligned} \mathcal{V}(\cdot) &= \sum_{i=1,2,3,4} |c_i|^2 \mathcal{V}_i(\cdot) + \sum_{i \neq j} c_i c_j^* V_i \cdot V_j^*, \\ \tilde{\mathcal{V}}(\cdot) &= \sum_{i=1,2,3,4} |c_i|^2 \tilde{\mathcal{V}}_i(\cdot) + \sum_{i \neq j} c_i^* c_j V_i^* \cdot V_j. \end{aligned} \tag{21.30}$$

The term of $\sum_i |c_i|^2 \mathcal{V}_i(\cdot)$ (or $\sum_i |c_i|^2 \tilde{\mathcal{V}}_i(\cdot)$) indicates that the four kinds of comparisons affect the player's tendency to choose 0 (or 1) simultaneously. The terms of $\sum_{i \neq j} c_i c_j^* V_i \cdot V_j^*$ and $\sum_{i \neq j} c_i^* c_j V_i^* \cdot V_j$ indicate that the four comparisons are not done independently in the player A's mind. The player A in our model holds indeterminacy about B's action, so his concerns about the consequence of "$0_A 0_B$" and "$0_A 1_B$" (or "$1_A 0_B$" and "$1_A 1_B$") are always "fluctuated". Under such a situation, the player A cannot do four comparisons independently. The "correlation" terms of $\sum_{i \neq j} c_i c_j^* V_i \cdot V_j^*$ and $\sum_{i \neq j} c_i^* c_j V_i^* \cdot V_j$ represent psychological influences from the situation where the player A is not informed of B's action. The velocities of k and $\tilde{k}$ should have the forms reflecting effects of the four comparisons and the correlations between them. Such the k and $\tilde{k}$ are defined as

$$k = |\mu|^2, \qquad \tilde{k} = |\tilde{\mu}|^2, \tag{21.31}$$

where

$$\begin{aligned} \mu &\equiv |\alpha|^2 \mu_1 + |\beta|^2 \mu_2 + \alpha\beta^* \mu_3 + \alpha^* \beta \mu_4 = \sum_i c_i \mu_i, \\ \tilde{\mu} &\equiv |\alpha|^2 \tilde{\mu}_1 + |\beta|^2 \tilde{\mu}_2 + \alpha^* \beta \tilde{\mu}_3 + \alpha\beta^* \tilde{\mu}_4 = \sum_i c_i^* \tilde{\mu}_i. \end{aligned} \tag{21.32}$$

Here $\mu_{i=1,2,3,4}$ and $\tilde{\mu}_{i=1,2,3,4}$ are complex numbers satisfying $|\mu_i|^2 = k_i$, $|\tilde{\mu}_i|^2 = \tilde{k}_i$ for given k_i and $\tilde{k}_i$, then k and $\tilde{k}$ are given as

$$\begin{aligned} k &= \sum_{i=1,2,3,4} |c_i|^2 k_i + \sum_{i \neq j} c_i c_j^* \mu_i \mu_j^*, \\ \tilde{k} &= \sum_{i=1,2,3,4} |c_i|^2 \tilde{k}_i + \sum_{i \neq j} c_i^* c_j \tilde{\mu}_i \tilde{\mu}_j^*. \end{aligned} \tag{21.33}$$

These are similar forms as $\mathcal{V}$ and $\tilde{\mathcal{V}}$ represented in (21.30). In (21.33), the most important point is to present the complex numbers μ_i and $\tilde{\mu}_i$. These parameters will have no meanings as physical quantities, rather, they have meanings similar to "amplitudes" introduced in the quantum mechanical sense. The "correlation terms" as $\sum_{i \neq j} c_i c_j^* \mu_i \mu_j^*$ will give the effect similar as "quantum interference" to the value of k.

One can simplify the representation of k and $\tilde{k}$, by defining the operator called "comparison operator":

$$\begin{aligned} \mathcal{T}_\sigma &\equiv |\Phi_{1_A}\rangle\langle\Phi_{1A}|T|\Phi_{0_A}\rangle\langle\Phi_{0A}| + |\Phi_{0_A}\rangle\langle\Phi_{0A}|T|\Phi_{1_A}\rangle\langle\Phi_{1A}| \\ &= \mu|\Phi_{1_A}\rangle\langle\Phi_{0A}| + \tilde{\mu}|\Phi_{0_A}\rangle\langle\Phi_{1A}|, \end{aligned} \tag{21.34}$$

where T is a matrix given in the form of

$$T = \begin{pmatrix} 0 & 0 & \tilde{\mu}_1 & \tilde{\mu}_3 \\ 0 & 0 & \tilde{\mu}_4 & \tilde{\mu}_2 \\ \mu_1 & \mu_4 & 0 & 0 \\ \mu_3 & \mu_2 & 0 & 0 \end{pmatrix}, \tag{21.35}$$

and the index σ of $\mathcal{T}_\sigma$ denotes the prediction state $\sigma \equiv |\phi_B\rangle\langle\phi_B|$. By $\mathcal{T}_\sigma$, k and $\tilde{k}$ are represented as

$$\begin{aligned} k &= \langle \Phi_{0A}|\mathcal{T}_\sigma^* \mathcal{T}_\sigma|\Phi_{0A}\rangle \equiv \langle \Phi_{0A}|K_\sigma|\Phi_{0A}\rangle, \\ \tilde{k} &= \langle \Phi_{1A}|\mathcal{T}_\sigma^* \mathcal{T}_\sigma|\Phi_{1A}\rangle \equiv \langle \Phi_{1A}|K_\sigma|\Phi_{1A}\rangle. \end{aligned} \tag{21.36}$$

Here, $K_\sigma = \mathcal{T}_\sigma^* \mathcal{T}_\sigma$ is Hermitian because it represents an operator of the velocity. This operator is different from conventional operators of physical quantities defined in quantum mechanics since its form is determined depending on the prediction state σ. This property indicates that the dynamics in our model has the "state adaptivity" which is an important concept in the adaptive dynamics theory proposed in [608].

21.11.4 Decision-Making in PD-Type Game and Irrational Choice

The parameters k_i and $\tilde{k}_i$ introduced in the previous subsection specify the player's four kinds of comparisons, see (21.23). It is natural that these comparisons depend on a given game, namely its pay-off table is Table 21.19.

The most simple relation of the pay-offs and the parameters k_i, $\tilde{k}_i$ is determined by the magnitude relation between the pay-offs. In the case of prisoner's dilemma (PD) type game, the relation of pay-offs is $a > b > c > d$, and then, k_i and $\tilde{k}_i$ are given as

$$\begin{aligned} &k_1 = 1, \quad k_2 = 1, \quad k_3 = 1, \quad k_4 = 0, \\ &\tilde{k}_1 = 0, \quad \tilde{k}_2 = 0, \quad \tilde{k}_3 = 0, \quad \tilde{k}_4 = 1. \end{aligned} \tag{21.37}$$

Such a setting is very simple, but not realistic. The real player's decision-making will depend on differences between the pay-offs, not only magnitude relations. That is, the following setting will be more realistic:

$$\begin{aligned} &k_1 = f_1\big(|a-c|\big), \quad k_2 = f_2\big(|b-d|\big), \quad k_3 = f_3\big(|b-c|\big), \quad k_4 = 0, \\ &\tilde{k}_1 = 0, \quad \tilde{k}_2 = 0, \quad \tilde{k}_3 = 0, \quad \tilde{k}_4 = \tilde{f}_4\big(|a-d|\big). \end{aligned} \tag{21.38}$$

The functions $f_i(x)$ are assumed to be monotone increasing functions.

Under the settings of k_i and $\tilde{k}_i$ of (21.37) or (21.38), the probability P_{0A}^E of (21.22) is not zero as a result. Thus, our model explains that the player A generally has the possibility to choose the "irrational" choice of 0 in a PD game. The reason of this result is that the parameter of $\tilde{k}_4$ is not zero. $\tilde{k}_4$ represents the degree of tendency to choose 0 which occurs from the comparison between consequences of $0_A 0_B$ and $1_A 1_B$. It should be noted that such a comparison is not considered in the classical game theory.

21.11.5 Non-Kolmogorovian Structure

This model is a non-Kolmogorovian model. Let us consider the following probabilities relating with the player A's decision.

$P(m_A)$: Probability that the player A chooses the action m_A ($m_A = 0$ or 1).
$P(n_B)$: Probability that the player A decides "the player B will choose n_B ($n_B = 0$ or 1)" in a definitive way.
$P(m_A|n_B)$: Conditional probability that the player A chooses m under the condition that he decided "the player B will choose n".

The player A in this model has a prediction state of σ. The probability of $P(m_A)$ corresponds to P_{m_A} of (21.22), whose form depends on the prediction state σ. The probability of $P(n_B)$ is assumed to be $\mathrm{tr}(\sigma|n_B\rangle\langle n_B|)$. When A decides that B's choice will be n_B, A's prediction is changed from σ to $|n_B\rangle\langle n_B|$. One can check that

$$P(0_A|n_B) + P(1_A|n_B) = 1,$$
$$P(m_A|0_B) + P(m_A|1_B) \neq 1,$$

which can be seen in the classical probability theory. However, one can find that

$$P(m_A) \neq P(m_A|0_B)P(0_B) + P(m_A|1_B)P(1_B),$$

in general. This is a violation of the total probability law, so that the PD has given us one example of non-Kolmogorovian models.

In the vein of the above researches, the "subjectivity" will play very important role, and mathematical study to describe decision making process with the subjectivity should be done.

References

1. Aarts, E., Korst, J.: Simulated Annealing and Boltzmann Machines. Wiley, New York (1989)
2. Abdel-Ghafar, A.N., Chotpitayasunondh, T., Gao, Z., Hayden, F.G., Nguyen, D.H., de Jong, M.D., Naghdaliyev, A., Peiris, J.S., Shindo, N., Soeroso, S., Uyeki, T.M.: Update on avian influenza A (H5N1) virus infection in humans. N. Engl. J. Med. **358**(3), 261–273 (2008)
3. Accardi, L.: Noncommutative Markov chain. In: International School of Mathematical Physics, Camerino, pp. 268–295 (1974)
4. Accardi, L.: Topics in quantum probability. Phys. Rep. **71**, 169–192 (1981)
5. Accardi, L., Fedullo, A.: On the statistical meaning of complex numbers in quantum theory. Lett. Nuovo Cimento **34**, 161–172 (1982). University of Salerno, Preprint, May 1981
6. Accardi, L., Frigerio, A., Lewis, J.: Quantum stochastic processes. Publ. Res. Inst. Math. Sci. **18**, 97–133 (1982)
7. Accardi, L.: The probabilistic roots of the quantum mechanical paradoxes. In: Diner, S., Lochak, G., Selleri, F. (eds.) The Wave–Particle Dualism, pp. 297–330. Reidel, Dordrecht (1984)
8. Accardi, L., Frigerio, A.: Markov cocycles. Proc. R. Ir. Acad. Sci. A **83**, 251–269 (1983)
9. Accardi, L.: Quantum Probability and Applications II: Proceedings Workshop Held Heidelberg. Springer, Berlin (1985)
10. Accardi, L.: An outline of quantum probability. Unpublished manuscript (1990)
11. Accardi, L., Ohya, M., Suyari, H.: Computation of mutual entropy in quantum Markov chains. Open Syst. Inf. Dyn. **2**, 337–354 (1994)
12. Accardi, L., Ohya, M., Suyari, H.: An application of lifting theory to optical communication processes. Rep. Math. Phys. **36**, 403–420 (1995)
13. Accardi, L., Ohya, M., Suyari, H.: Mutual entropy in quantum Markov chains. In: Quantum Communications and Measurement, vol. 2, pp. 351–358. Plenum, New York (1995)
14. Accardi, L., Ohya, M., Watanabe, N.: Note on quantum dynamical entropies. Rep. Math. Phys. **38**(3), 457–469 (1996)
15. Accardi, L.: Urne e Camaleoni: Dialogo sulla realta, le leggi del caso e la teoria quantistica. Il Saggiatore, Rome (1997)
16. Accardi, L.: An open system approach to quantum computers. In: Quantum Communication, Computing and Measurement, vol. 3, pp. 387–393 (1997)
17. Accardi, L., Ohya, M., Watanabe, N.: Dynamical entropy through quantum Markov chains. Open Syst. Inf. Dyn. **4**, 71–87 (1997)
18. Accardi, L., Kozyrev, S.V., Volovich, I.V.: Dynamics of dissipative two-state systems in the stochastic approximation. Phys. Rev. A **56**(3), 2557–2562 (1997)
19. Accardi, L., Ohya, M.: Compound channels, transition expectations, and liftings. Appl. Math. Optim. **39**, 33–59 (1999)
20. Accardi, L., Ohya, M.: Teleportation of general quantum states. In: Hida, T., Saito, K. (eds.) Quantum Information, pp. 59–70. World Scientific, Singapore (1999)

M. Ohya, I. Volovich, *Mathematical Foundations of Quantum Information and Computation and Its Applications to Nano- and Bio-systems*, Theoretical and Mathematical Physics, DOI 10.1007/978-94-007-0171-7, © Springer Science+Business Media B.V. 2011

21. Accardi, L., Kozyrev, S.V., Volovich, I.V.: Non-exponential decay for polaron model. quant-ph/9904084 (1999)
22. Accardi, L., Regoli, M.: Locality and Bell's inequality. Volterra Center Preprint (2000)
23. Accardi, L., Regoli, M.: Non-locality and quantum theory: new experimental evidence. Volterra Center Preprint (2000)
24. Accardi, L., Aref'eva, I.Ya, Volovich, I.V.: Non-equilibrium quantum field theory and entangled commutation relations. Proc. Steklov Inst. Math. **228**, 116–136 (2000)
25. Accardi, L., Aref'eva, I.Ya, Volovich, I.V.: Stochastic limit and interacting commutation relations. To be published in the volume dedicated to L. Streit
26. Accardi, L.: Locality and Bell's inequality. Volterra Center Preprint (2000)
27. Accardi, L.: Quantum filtering as a dynamical covariance condition. In: Proceedings International Conference: Quantum Information II, Meijo University, Nagoya, 1–5 Mar 1999, pp. 1–15. World Scientific, Singapore (2000)
28. Accardi, L.: Teleportation of general quantum states, squeezing and the decoherence problem, in quantum computers. In: Kumar, P., D'Ariano, G.M., Hirota, O. (eds.) Quantum Communication, Computing and Measurement, vol. 2, pp. 229–239. Kluwer Academic, Dordrecht (2000)
29. Accardi, L.: Quantum algorithms. In: Volterra–CIRM International School: Quantum Computer and Quantum Information, Levico Terme, 25–31 July 2001
30. Accardi, L., Sabbadini, R.: On the Ohya–Masuda quantum SAT Algorithm. In: Antoniou, I., Calude, C.S., Dinneen, M. (eds.) Proceedings International Conference "Unconventional Models of Computations". Springer, Berlin (2001)
31. Accardi, L., Imafuku, K., Regoli, M.: On the physical meaning of the EPR-chameleon experiment. quant-ph/0112067 (2001)
32. Accardi, L., Imafuku, K., Regoli, M.: On the physical meaning of the EPR–chameleon experiment. Infin. Dimens. Anal. Quantum Probab. Relat. Top. **5**(1), 1–20 (2002). Volterra Center Preprint (2002)
33. Accardi, L., Sabbadini, R.: A generalization of Grover's algorithm. In: Proceedings International Conference: Quantum Information III. World Scientific, Singapore (2002). quant-ph/0012143
34. Accardi, L., Lu, Y.G., Volovich, I.V.: Quantum Theory and Its Stochastic Limit. Springer, Berlin (2002)
35. Accardi, L.: Urne e camaleonti: Dialogo sulla realta, le leggi del caso e la teoria quantistica. Il Saggiatore (1997). English ed. World Scientific (2002); Japanese ed. Makino (2002), Russian ed. Regular and Chaotic Dynamics (2002)
36. Accardi, L., Fagnola, F.: Quantum interacting particle systems. In: Quantum Probability and White Noise Analysis, vol. 14. World Scientific, Singapore (2002)
37. Accardi, L., Ohya, M.: A stochastic limit approach to the SAT problem. Open Syst. Inf. Dyn. **11**, 1–16 (2004)
38. Accardi, L., Fidaleo, F.: Entangled Markov chains. Ann. Mat. Pura Appl. (2004)
39. Accardi, L., Matsuoka, T., Ohya, M.: Entangled Markov chains are indeed entangled. Infin. Dimens. Anal. Quantum Probab. Relat. Top. **9**(3), 379–390 (2006)
40. Accardi, L., Matsuoka, T., Ohya, M.: Entangled quantum Markov chains satisfying the PPT condition. TUS preprint
41. Accardi, L., Freudenberg, W., Ohya, M.: Quantum bio-informatics. In: QP-PQ Quatum Probability and White Noise Analysis, vol. 21. World Scientific, Singapore (2007)
42. Accardi, L., Freudenberg, W., Ohya, M.: Quantum bio-informatics II. In: QP-PQ Quatum Probability and White Noise Analysis, vol. 24. World Scientific, Singapore (2008)
43. Accardi, L., Freudenberg, W., Ohya, M.: Quantum bio-informatics III. In: QP-PQ Quatum Probability and White Noise Analysis, vol. 26. World Scientific, Singapore (2009)
44. Accardi, L., Khrennikov, A., Ohya, M.: Quantum Markov model for data from Shafir-Tversky experiments in cognitive psychology. Open Syst. Inf. Dyn. **16**, 371–385 (2009)
45. Accardi, L.: Foundation of quantum probability. To be published
46. Achiezer, A.I., Berestecky, V.B.: Quantum Electrodynamics. Nauka, Moscow (1969)

47. Adleman, L.M., DeMarrais, J., Huang, M.A.: Quantum computability. SIAM J. Comput. **26**, 1524–1540 (1997)
48. Afriat, A., Selleri, F.: The Einstein, Podolsky, and Rosen Paradox in Atomic, Nuclear, and Particle Physics. Plenum, New York (1999)
49. Aharonov, D., Ben-Or, M.: Fault-tolerant quantum computation with constant error. In: Proc. 29th Ann. ACM Symp. on Theory of Computing, pp. 176–188. ACM, New York (1997). arXiv:quant-ph/9611025
50. Aharonov, D., Ben-Or, M.: Fault-tolerant quantum computation with constant error rate. arXiv:quant-ph/9906129 (1999)
51. Akashi, S.: Superposition representability problems of quantum information channels. Open Syst. Inf. Dyn. **4**(1), 45–52 (1997)
52. Alekseev, V.M., Yakobson, M.N.: Symbolic dynamics and hyperbolic dynamic systems. Phys. Rep. **75**, 287–325 (1981)
53. Alicki, R., Fannes, M.: Defining quantum dynamical entropy. Lett. Math. Phys. **32**, 75–82 (1994)
54. Alicki, R., Horodecki, M., Horodecki, P., Horodecki, R.: Dynamical description of quantum computing: generic nonlocality of quantum noise. Phys. Rev. A **65**, 062101 (2002). quant-ph/0105115
55. Alicki, R., Lidar, D.A., Zanardi, P.: Are the assumptions of fault-tolerant quantum error correction internally consistent? quant-ph/0506201 (2005)
56. Aliferis, P., Gottesman D. Preskill, J.: Quantum accuracy threshold for concatenated distance-3 codes. quant-ph/0504218 (2005)
57. Ando, T., Yamato, I: Basics of molecular simulation and its application to biomolecules. In: Quantum Bio-Informatics II, pp. 228–241. World Scientific, Singapore (2009)
58. Anosov, D.V.: Geodesic flows on closed Riemann manifolds with negative curvature. Proc. Steklov Inst. Math. **90**, 1–235 (1960)
59. Anosov, D.V., Arnold, V.I. (eds.): Dynamical Systems. VINITI, Moscow (1996)
60. Araki, H., Yanase, M.M.: Measurement of quantum mechanical operators. Phys. Rev. **120**, 622 (1960)
61. Araki, H., Lieb, E.: Entropy inequalities. Commun. Math. Phys. **18**, 160170 (1970)
62. Araki, H.: Some properties of modular conjugation operator of a von Neumann algebra and non-commutative Radon Nikodym theorem with a chain rule. Pac. J. Math. **50**, 309 (1974)
63. Araki, H.: Relative entropy of states of von Neumann algebras. Publ. Res. Inst. Math. Sci., Kyoto Univ. **11**, 809–833 (1976)
64. Araki, H.: Relative entropy for states of von Neumann algebras II. Publ. Res. Inst. Math. Sci., Kyoto Univ. **13**, 173–192 (1977)
65. Araki, H.: Mathematical Theory of Quantum Fields. Oxford University Press, London (1999)
66. Aref'eva, I.Ya., Volovich, I.V.: Quantum decoherence and higher order corrections to the large time exponential behaviour. Infin. Dimens. Anal. Quantum Probab. **3**, 453–482 (2000)
67. Aref'eva, I.Ya., Volovich, I.V.: The large time behaviour in quantum field theory and quantum chaos. quant-ph/9910109 (1999). In the volume dedicated to S. Albeverio
68. Aref'eva, I.Y., Medvedev, P.B., Rytchkov, O.A., Volovich, I.V.: Chaos Solitons Fractals **10**(2–3), 213 (1999)
69. Arndt, C.: Information Measures. Springer, Berlin (2001)
70. Arnold, V.I., Avez, A.: Ergodic Problems of Classical Mechanics. Benjamin, Elmsford (1968)
71. Arnold, V.I.: Mathematical Methods of Classical Mechanics. Springer, Berlin (1978)
72. Asano, M., Ohya, M., Togawa, Y.: Entropic chaos degree of rotations and log-linear dynamics. In: QP-PQ:Quantum Prob. White Noise Anal. Quantum Bio-Informatics, vol. 21, pp. 36–52. World Scientific, Singapore (2008)
73. Asano, M., Khrennikov, A., Ohya, M.: Quantum like model for decision making process in two players game. Found. Phys. (2010). doi:10.1007/s10701-010-9454-y
74. Ash, R., Information Theory. Wiley, New York (1965)

75. Atature, M., Di Giuseppe, G., Shaw, M.D., Sergienko, A.V., Saleh, B.E.A., Teich, M.C.: Multi-parameter entanglement in femtosecond parametric down-conversion. Phys. Rev. A **65**, 023808 (2002)
76. Avetisov, V.A., Bikulov, A.H., Kozyrev, S.V., Osipov, V.A.: p-adic models of ultra-metric diffusion constrained by hierarchical energy landscapes. J. Phys. A, Math. Gen. **35**(2), 177 (2002)
77. Avetisov, V.A., Bikulov, A.Kh.: Protein ultrametricity and spectral di usion in deeply frozen proteins. Biophys. Rev. Lett. **3**(3), 387–396 (2008)
78. Balazs, N.L., Voros, A.: The quantized baker's transformation. Ann. Phys. **190**, 1–31 (1989)
79. Balazs, N.L., Lakshminarayan, A.: The classical and quantum mechanics of lazy baker maps. Ann. Phys. **226**, 350 (1993)
80. Banaszek, K., Wodkiewicz, K.: Nonlocality of the Einstein-Podolsky-Rosen state in the Wigner representation. Phys. Rev. A **58**, 4345 (1998)
81. Barenco, A., Bennett, C.H., Cleve, R., DiVicenzo, D., Margolus, N., Shor, P., Sleator, T., Smolin, J., Weinfurter, H.: Elementary gates for quantum computation. Phys. Rev. A **52**, 3457–3467 (1995)
82. Barett, J., Collins, D., Hardy, L., Kent, A., Popescu, S.: Quantum nonlocality, Bell inequalities and the memory loophole. Phys. Rev. A **66**, 042111 (2002)
83. Barnum, H., Caves, C.M., Fuchs, C.A., Jozsa, R., Schumacher, B.: Noncommuting mixed states cannot be broadcast. Phys. Rev. Lett. **76**(15), 2818–2821 (1996)
84. Barnum, H., Nielsen, M.A., Schumacher, B.W.: Information transmission through a noisy quantum channel. Phys. Rev. A **57**(6), 4153–4175 (1998)
85. Barnum, H., Nielsen, M.A., Schumacher, B.: Information-theoretic approach to quantum error correction and reversible measurement. Proc. R. Soc. Lond. Ser. A, Math. Phys. Eng. Sci. **454**(1969), 277–304 (1998). Quantum Coherence and Decoherence (Santa Barbara, CA, 1996)
86. Barnum, H., Knill, E., Nielsen, M.A.: On quantum fidelities and channel capacities. IEEE Trans. Inf. Theory **46**(4), 1317–1329 (2000)
87. Bashford, J.D., Tsohantjis, I., Jarvis, P.D.: Codon and nucleotide assignments in a supersymmetric model of the genetic code. Phys. Lett. A **233**, 481–488 (1997)
88. Bechmann-Pasquinucci, H., Peres, A.: Quantum cryptography with 3-state systems. Phys. Rev. Lett. **85**, 3313 (2000)
89. Beige, A., Munro, W.J., Knight, P.L.: Bell's inequality test with entangled atoms. Phys. Rev. A **62**, 052102 (2000)
90. Belavkin, V.P., Stratonovich, P.L.: Optimization of processing of quantum signals according to an information criterion. Radio Eng. Electron. Phys. **9**, 1839–1844 (1973)
91. Belavkin, V.P.: Conditional entropy and capacity of quantum channels. In: Proc. of VIII-th Conference on Coding Theory and Information Transmission, Moscow-Kuibishev, pp. 15–19 (1982)
92. Belavkin, V.P.: Quantum stochastic calculus and quantum nonlinear filtering. J. Multivar. Anal. **42**, 171–201 (1992)
93. Belavkin, V.P., Ohya, M.: Quantum entropy and information in discrete entangled states. Infin. Dimens. Anal., Quantum Probab. Relat. Top. **4**(2), 137–160 (2001)
94. Belavkin, V.P., Ohya, M.: Entanglement, quantum entropy and mutual information. Proc. R. Soc. Lond. A **458**, 209–231 (2002)
95. Bell, J.S.: On the Einstein–Podolsky–Rosen paradox. Physics **1**, 195–200 (1964)
96. Bell, J.S.: Speakable and Unspeakable in Quantum Mechanics: Collected Papers on Quantum Mechanics. Cambridge University Press, Cambridge (1989)
97. Benatti, F.: Deterministic Chaos in Infinite Quantum Systems. Trieste Notes in Physics. Springer, Berlin (1993)
98. Benioff, P.: The computer as a physical system: a microscopic quantum mechanical Hamiltonian. J. Stat. Phys. **22**(5), 563–591 (1980)
99. Bennett, C.H.: Logical reversibility of computation. IBM J. Res. Dev. **17**, 525–532 (1973)

100. Bennett, C.H., Brassard, G., Breidbart, S., Wiesner, S.: Quantum cryptography, or unforgeable subway tokens. In: Chaum, D., Rivest, R.L., Sherman, A.T. (eds.) Advances in Cryptology: Proceedings of Crypto, vol. 82, pp. 267–275. Plenum, New York (1982)
101. Bennett, C.H.: The thermodynamics of computation-a review. Int. J. Theor. Phys. **21**(12), 905–940 (1982)
102. Bennett, C.H., Brassard, G.: Quantum cryptography: public key distribution and coin tossing. In: Proceedings of IEEE International Conference on Computers, Systems and Signal Processing, pp. 175–179. IEEE Press, New York (1984)
103. Bennett, C.H.: Quantum cryptography using any two nonorthogonal states. Phys. Rev. Lett. **68**(21), 3121 (1992)
104. Bennett, C.H., Brassard, G., Mermin, N.D.: Quantum cryptography without Bell's theorem. Phys. Rev. Lett. **68**(5), 557–559 (1992)
105. Bennett, C.H., Brassard, G., Crepeau, C., Jozsa, R., Peres, A., Wootters, W.K.: Teleporting an unknown quantum state via Dual Classical and Einstein-Podolsky-Rosen channels. Phys. Rev. Lett. **70**, 1895–1899 (1993)
106. Bennett, C.H.: Quantum and classical information: transmission and computation. Phys. Today **48**(10), 24–30 (1995)
107. Bennett, C.H., Brassard, G., Popescu, S., Schumacher, B., Smolin, J.A., Wootters, W.K.: Purification of noisy entanglement and faithful teleportation via noisy channels. Phys. Rev. Lett. **76**, 722–725 (1996)
108. Bennett, C.H., DiVincenzo, D.P., Smolin, J.A., Wootters, W.K.: Mixed-state entanglement and quantum error correction. Phys. Rev. A **54**, 3824 (1996)
109. Bennett, C.H., Bernstein, E., Brassard, G., Vazirani, U.: Strengths and weaknesses of quantum computing. SIAM J. Comput. **26**, 1510–1523 (1997)
110. Bennett, C.H., Shor, P.W., Smolin, J.A., Thapliyalz, A.V.: Entanglement-assisted capacity of a quantum channel and the reverse Shannon theorem. quant-ph/0106052 (2001)
111. Berestetzky, V.B., Livshiz, E.M., Pitaevsky, L.P.: Quantum Electro-Dynamics. Nauka, Moscow (1980) (in Russian)
112. Beratran, D.N., Betts, J.N., Onuchic, J.N.: Protein electron transfer rates are predicted to be set by the bridging secondary and tertiary structure. Science **252**, 1285 (1991)
113. Berman, G.P., Zaslavsky, G.M.: Condition of stochasticity in quantum non-linear systems. Physica A **91**, 450–460 (1978)
114. Bernstein, E., Vazirani, U.: Quantum complexity theory. In: Proc. of the 25th Annual ACM Symposium on Theory of Computing, pp. 11–22. ACM, New York (1993). SIAM J. Comput. **26**, 1411 (1997)
115. Berry, M.V.: Evolution of semiclassical quantum states in phase space. J. Phys. A **12**, 625–642 (1979)
116. Berry, M.V.: Some quantum to classical asymptotics. In: Giannoni, M.J., Voros, A., Justi, Z. (eds.) Les Houches Summer School "Chaos and Quantum Physics". North-Holland, Amsterdam (1991)
117. Biham, E., Boyer, M., Boykin, P.O., Mor, T., Roychowdhury, V: A proof of the security of quantum key distribution. quant-ph/9912053 (1999)
118. Billingsley, P.: Ergodic Theory and Information. Wiley, New York (1968)
119. Blohinzev, D.I.: Principal Questions of Quantum Mechanics. Nauka, Moscow (1986) (in Russian)
120. Bogolyubov, N.N.: Problems of Dynamic Theory in Statistical Physics. OGIZ, Moscow (1946) (in Russian)
121. Bogoliubov, N.N.: Problems of a Dynamical Theory in Statistical Physics (1959). Translated by E. Gora, Providence College, Providence
122. Bogoliubov, N.N., Schirkov, D.V.: Introduction to Theory of Quantized Fields. Nauka, Moscow (1985)
123. Bogoliubov, N.N., Logunov, A.A., Oksak, A.I., Todorov, I.T.: General Principles of Quantum Field Theory. Nauka, Moscow (1987)
124. Bohm, D.: Quantum Theory. Prentice-Hall, New York (1951)

125. Boltzmann, L.: Lectures on Gas Theory. University of California Press, Berkeley (1964)
126. Boltzmann, L.: Ober die Beziehung eines allgemeine mechanischen Satzes zum zweiten Hauptsatze der Warmetheorie. Sitzungsber. Akad. Wiss., Wien. Part II **75**, 67–73 (1977). English transl: Stephen Brush, Kinetic Theory **2**, 188 (1977)
127. Boltzmann, L.: Vorlesungen Ëuber Gastheorie, vol. 2. Barth, Leipzig (1896/1898). This book has been translated into English by S.G. Brush, Lectures on Gas Theory. Cambridge University Press, London (1964), reprinted Dover (1995)
128. Bourennane, M., Karlsson, A., Bjork, G., Gisin, N., Cerf, N.: Quantum key distribution using multilevel encoding: security analysis. J. Phys. A **35**(47), 10065–10076 (2002)
129. Bouwmeester, D., Pan, J.W., Mattle, K., Eibl, M., Weinfurter, H., Zeilinger, A.: Experimental quantum teleportation. Nature **390**, 575–579 (1997)
130. Bratteli, O., Robinson, D.W.: Operator Algebras and Quantum Statistical Mechanics, vol. I. Springer, New York (1979)
131. Bratteli, O., Robinson, D.W.: Operator Algebras and Quantum Statistical Mechanics, vol. II. Springer, New York (1981)
132. Brassard, G., Lutkenhaus, N., Mor, T., Sanders, B.C.: Security aspects of practical quantum cryptography. Phys. Rev. Lett. **85**, 1330–1333 (1999)
133. Braunstein, S.L., Mann, A., Revzen, M.: Maximal violation of Bell inequalities for mixed states. Phys. Rev. Lett. **68**, 3259 (1992)
134. Braunstein, S.L., Kimble, H.J.: Teleportation of continuous quantum variables. Phys. Rev. Lett. **80**, 869 (1998)
135. Braunstein, S.L., Fuchs, C.A., Kimble, H.J., Loock, van P.: Quantum versus classical domains for teleportation with continuous variables. Phys. Rev. A **64**, 022321 (2001)
136. Breiman, L.: On achieving channel capacity in finite-memory channels. Ill. J. Math. **4**, 246–252 (1960)
137. Brekke, L., Freund, P.G.O.: p-adic numbers in physics. Phys. Rep. **233**, 1–66 (1993)
138. Bricmont, J.: Science of chaos or chaos in science? In: The Flight from Science and Reason, Annals of the N.Y. Academy of Sciences, vol. 775, pp. 131–182 (1996)
139. Brun, T., Schack, R.: Realizing the quantum baker's map on a NMR quantum computer. Phys. Rev. A **59**, 2649 (1999)
140. Brun, T.A., Carteret, H.A., Ambainis, A.: Quantum walks driven by many coins. Phys. Rev. A **67**, 052317 (2003), 17 pp.
141. Bruss, D.: Optimal eavesdropping in quantum cryptography with six states. Phys. Rev. Lett. **81**, 3018 (1998)
142. Bulinsky, A.V., Shiryaev, A.N.: Theory of Random Processes. Fizmatlit, Moscow (2003)
143. Burkard, G., Loss, D., DiVincenzo, D.P.: Coupled quantum dots as quantum gates. Phys. Rev. B **59**, 2070 (1998)
144. Busemeyer, J.R., Matthews, M., Wang, Z.: A quantum information processing explanation of disjunction effects. In: Sun, R., Myake, N. (eds.) The 29th Annual Conference of the Cognitive Science Society and the 5th International Conference of Cognitive Science, pp. 131–135. Erlbaum, Mahwah (2006)
145. Busemeyer, J.B., Wang, Z., Townsend, J.T.: Quantum dynamics of human decision making. J. Math. Psychol. **50**, 220–241 (2006)
146. Busemeyer, J.R., Santuy, E., Lambert-Mogiliansky, A.: Comparison of Markov and quantum models of decision making. In: Bruza, P., Lawless, W., van Rijsbergen, K., Sofge, D.A., Coeke, B., Clark, S. (eds.) Quantum Interaction: Proceedings of the Second Quantum Interaction Symposium, pp. 68–74. College Publications, London (2008)
147. Bush, P., Lahti, P., Mittelstaedt, P.: The Quantum Theory of Measurement. Springer, Berlin (1996)
148. Calderbank, A.R., Shor, P.W.: Good quantum error-correcting code exists. Phys. Rev. A **54**, 1098–1105 (1996)
149. Calderbank, A.R., Rains, E.M., Shor, P.W., Sloane, N.J.A.: Quantum error correction and orthogonal geometry. Phys. Rev. Lett. **78**, 405–408 (1997)
150. Casati, G., Chirikov, B.V. (eds.): Quantum Chaos: Between Order and Disorder. Cambridge University Press, Cambridge (1995)

151. Caruso, F., Chin, A.W., Datta, A., Huelga, S.F., Plenio, M.B.: Highly efficient energy excitation transfer in light-harvesting complexes: The fundamental role of noise-assisted transport. J. Chem. Phys. **131**, 105106 (2009)
152. Caruso, F., Huelga, S.F., Plenio, M.B.: Noise-enhanced classical and quantum capacities in communication networks. arXiv:1003.5877 (2010)
153. Caves, C.M., Drummond, P.D.: Quantum limits on bosonic communication rates. Rev. Mod. Phys. **66**, 481–537 (1994)
154. Cerf, N.J., Bourennane, M., Karlsson, A., Gisin, N.: Security of quantum key distribution using d-level systems. Phys. Rev. Lett. **88**, 127902 (2002)
155. Cerf, N.J., Iblisdir, S., van Assche, G.: Cloning and cryptography with quantum continuous variables. Eur. Phys. J. D At. Mol. Opt. Phys. **18**(2), 211–218 (2002)
156. Chaitin, G.J.: On the length of programs for computing finite binary sequences: statistical considerations. J. ACM **16**, 145–159 (1969)
157. Chen, Z.-B, Pan, J.-W., Hou, G., Zhang, Y.-D.: Maximal violation of Bell's inequalities for continuous variable systems. Phys. Rev. Lett. **88**, 040406 (2002)
158. Chester, G.V.: The theory of irreversible processes. Rep. Prog. Phys. **26**(1), 411–472 (1963)
159. Childs, A.M., Farhi, E., Gutmann, S.: An example the difference between quantum and classical random walks. Quantum Inf. Process. **1**, 35–43 (2002)
160. Chin, A.W., Datta, A., Caruso, F., Huelga, S.F., Plenio, M.B.: Noise-assisted energy transfer in quantum networks and light-harvesting complexes. arXiv:0910.4153 (2009)
161. Choda, M.: Entropy for extensions of Bernoulli shifts. Ergod. Theory Dyn. Syst. **16**(6), 1197–1206 (1996)
162. Choquet, G.: Lecture Analysis, vols. I, II, III. Benjamin, Elmsford (1969)
163. Chruściński, D., Kossakowski, A.: A new class of states with positive partial transposition. Phys. Rev. A **74**, 022308 (2006)
164. Chruscinski, D., Hirota, Y., Matsuoka, T., Ohya, M.: Remarks on the degree of entanglement. In: Quantum Bio-Informatics, vol. IV. World Scientific, Singapore (2011, to appear)
165. Chuang, I.L., Vandersypen, L.M.K., Xinlan, Z., Leung, D.W., Lloyd, S.: Experimental realization of a quantum algorithm. Nature **393**, 143 (1998)
166. Chuang, I.L., Gershenfeld, N., Kubinec, M.: Experimental implementation of fast quantum searching. Phys. Rev. Lett. **18**(15), 3408–3411 (1998)
167. Cirac, J.I., Zoller, P.: Quantum computations with cold trapped ions. Phys. Rev. Lett. **74**, 4091 (1995)
168. Clauser, J.F., Horne, M.A., Shimony, A., Holt, R.A.: Proposed experiment to test local hidden-variable theories. Phys. Rev. Lett. **23**, 880 (1969)
169. Clauser, J.F., Shimony, A.: Bell's theorem: experimental tests and implications. Rep. Prog. Phys. **41**, 1881–1927 (1978)
170. Cleland, A., Watson, H.G., Robertson, P., Ludlam, C.A., Brown, A.J.: Evolution of zidovudine resistance-associated genotypes in human immunodeficiency virus type 1-infected patients. J. Acquir. Immune Defic. Syndr. Hum. Retrovirol. **12**(1), 6 (1996)
171. Cleve, R.: An introduction to quantum complexity theory. In: Macchiavello, C., Palma, G.M., Zeilinger, A. (eds.) Collected Papers on Quantum Computation and Quantum Information Theory. World Scientific, Singapore (1999)
172. Collini, E., Wong, C.Y., Wilk, K.E., Curmi, P.M.G., Brumer, P., Scholes, G.D.: Coherently wired light-harvesting in photosynthetic marinealgae at ambient temperature. Nature **463**, 644–647 (2010)
173. Collins, D., Gisin, N., Linden, N., Massar, S., Popescu, S.: Bell inequalities for arbitrarily high-dimensional systems. Phys. Rev. Lett. **88**, 040404 (2002)
174. Connes, A.: Characterization des espaces vectoriels ordonn és sous-jacents aux algébres de von Neumann. Ann. Inst. Fourier **24**, 121 (1974)
175. Connes, A., Störmer, E.: Entropy for automorphisms of II von Neumann algebras. Acta Math. **134**, 289–306 (1975)
176. Connes, A., Narnhofer, H., Thirring, W.: Dynamical entropy of C^*-algebras and von Neumann algebras. Commun. Math. Phys. **112**, 691–719 (1987)

177. Conte, E., Khrennikov, A., Todarello, O., Federici, A., Zbilut, J.P.: Mental states follow quantum mechanics during perception and cognition of ambiguous figures. Open Syst. Inf. Dyn. **16**, 1–17 (2009)
178. Cornfeld, I., Fomin, S., Sinai, Ya.G.: Ergodic Theory. Springer, New York (1982)
179. Cory, D.G., Price, M.D., Havel, T.F.: Nuclear magnetic resonance spectroscopy: an experimentally accessible paradigm for quantum computing. quant-ph/9709001 (1997)
180. Cover, T.M., Thomas, J.A.: Elements of Information Theory. Wiley, New York (1991)
181. Crick, F.: The origin of the genetic code. J. Mol. Biol. **38**, 367–379 (1968)
182. Croson, R.: The disjunction effect and reasoning-based choice in games. Organ. Behav. Hum. Decis. Process. **80**, 118–133 (1999)
183. Chruściński, D., Kossakowski, A.: Circulant states with positive partial transpose. Phys. Rev. A **76**(3), 032308 (2007), 14 pp.
184. Csiszár, I.: I-divergence geometry of probability distributions and minimization problems. Ann. Probab. **3**, 146–158 (1975)
185. Csiszar, I., Korner, J.: Broadcast channels with confidential messages. IEEE Trans. Inf. Theory **24**(3), 339–348 (1978)
186. Cuculescu, I.: Some remarks on tensor products of standard forms of von Neumann algebras. Boll. Unione Mat. Ital. **7**(b), 907 (1993)
187. Czachor, M.: Notes on nonlinear quantum algorithm. Acta Phys. Slovaca **48**, 157 (1998)
188. Davis, M.: Computability and Unsolvability. Dover, New York (1958)
189. Davies, E.B., Lewis, J.T.: An operational approach to quantum probability. Commun. Math. Phys. **17**, 239–260 (1970)
190. Davies, E.B.: Quantum Theory of Open System. Academic Press, San Diego (1976)
191. Delerue, C., Lannoo, M.: Nanostructures: Theory and Modelling. Springer, Berlin (2004)
192. DeLuca, A., Termini, S.: A definition of a non-probabilistic entropy in the setting of fuzzy sets theory. Inf. Control **20**, 301–312 (1972)
193. De Masi, A., Presutti, E.: Mathematical Methods for Hydrodynamic Limits. Lecture Notes in Math., vol. 1501. Springer, New York (1991)
194. Demoen, B., Vanheuverzwijn, P., Verbeure, A.: Completely positive maps on the CCR-algebra. Lett. Math. Phys. **2**, 161–166 (1977)
195. DeMuynck, W.M., DeBaere, W., Martens, H.: Interpretations of quantum mechanics, joint measurement of incompatible observables, and counterfactual definiteness. Found. Phys. **24**, 1589 (1994)
196. Derrickson, B.H., Tortora, G.J.: Principles of Anatomy and Physiology, 11th edn. Wiley, New York (2005)
197. Deutsch, D.: Quantum theory, the Church-Turing principle and the universal quantum computer. Proc. R. Soc. Lond., Ser. A **400**, 96–117 (1985)
198. Deutsch, D.: Quantum computational networks. Proc. R. Soc. Lond. Ser. A **425**, 73–90 (1989)
199. Deutsch, D., Jozsa, R.: Rapid solution of problems by quantum computation. Proc. R. Soc. Lond. Ser. A **439**, 553–558 (1992)
200. Devetak, I., Shor, P.W.: The capacity of a quantum channel for simultaneous transmission of classical and quantum information. quant-ph/0311131 (2004)
201. Di Giuseppe, G., Atature, M., Shaw, M.D., Sergienko, A.V., Saleh, B.E.A., Teich, M.C.: Entangled-photon generation from parametric down-conversion in media with inhomogeneous nonlinearity. quant-ph/0112140 (2001)
202. Diffie, W., Hellman, M.E.: New directions in cryptography. IEEE Trans. Inf. Theory **22**, 644–654 (1976)
203. Dirac, P.A.M.: The Principles of Quantum Mechanics, 4th edn. Oxford University Press, London (1958)
204. Dittes, F.M., Doron, E., Smilansky, U.: Long-time behavior of the semiclassical baker's map. Phys. Rev. E **49**, R963–R966 (1994)
205. DiVincenzo, D.P., Shor, P.W.: Fault-tolerant error correction with efficient quantum codes. Phys. Rev. Lett. **77**, 3260–3265 (1996)

206. Donald, M.J.: On the relative entropy. Commun. Math. Phys. **105**, 13–34 (1985)
207. Doob, J.L.: Stochastic Processes. Wiley, New York (1953)
208. Dragovich, B., Dragovich, A.: A p-adic model of DNA sequence and genetic code. arXiv:q-bio/0607018 (2006)
209. Dragovich, B., Dragovich, A.: p-adic modelling of the genome and the genetic code. arXiv:0707.3043 (2007)
210. Dragovich, B., Khrennikov, A.Yu., Kozyrev, S.V., Volovich, I.V.: On p-adic mathematical physics. p-Adic Numb. Ultrametric Anal. Appl. **1**(1), 1–17 (2009)
211. Drexler, K.E.: Nanosystems: Molecular Machinery, Manufacturing, and Computation. Wiley, New York (1992)
212. Duan, L.M., Guo, G.C.: Preserving coherence in quantum computation by pairing quantum bits. Phys. Rev. Lett. **79**, 1953–1958 (1997)
213. Dubrovin, B.A., Fomenko, A.T., Novikov, S.P.: Modern Geometry. Methods and Applications. GTM, vol. 93, Part 1. Springer, Berlin (1984)
214. Dugan, V.G., Chen, R., Spiro, D.J., Sengamalay, N., Zaborsky, J., Ghedin, E., Nolting, J., Swayne, D.E., Runstadler, J.A., Happ, G.M., Senne, D.A., Wang, R., Slemons, R.D., Holmes, E.C., Taubenberger, J.K.: The evolutionary genetics and emergence of avian influenza viruses in wild birds. PLoS Pathog **4**(5), e1000076 (2008)
215. Dusek, M., Bradler, K.: The effect of multi-pair signal states in quantum cryptography with entangled photons. quant-ph/0011007 (2000)
216. Dynin, E.A.: Statistical moments in quantum tunneling. Sov. Phys. Semicond. **24**(44), 480–481 (1990)
217. Ebanks, B.R.: On measures of fuzziness and their representations. J. Math. Anal. Appl. **94**, 24–37 (1983)
218. Eistein, A., Podolsky, B. Rosen N.: Can quantum-mechanical description of physical reality be considered complete? Phys. Rev. **47**, 777 (1935)
219. Ekert, A.K.: Quantum cryptography based on Bell's theorem. Phys. Rev. Lett. **67**, 661–663 (1991)
220. Ekert, A., Jozsa, R.: Quantum computation and Shor's factoring algorithm. Rev. Mod. Phys. **68**, 733–753 (1996)
221. Ekimov, A.I., Onushchenko, A.A.: Quantum size effect in three-dimensional microscopic semiconductor crystals. JETP Lett. **34**, 345–349 (1981)
222. Emch, G.G.: Algebraic Methods in Statistical Mechanics and Quantum Field Theory. Wiley, New York (1972)
223. Emch, G.G.: Positivity of the K-entropy on non-abelian K-flows. Z. Wahrscheinlichkeitstheor. Verw. Geb. **29**, 241–252 (1974)
224. Emch, G.G., Narnhofer, H., Thirring, W., Sewell, G.L.: Anosov actions on noncommutative algebras. J. Math. Phys. **35**(11), 5582–5599 (1994)
225. Emerson, J., Ballentine, L.E.: Quantum-classical correspondence for the equilibrium distributions of two interacting spins. Phys. Rev. E **64**, 026217 (2001)
226. Engel, A.K., Singer, W.: Temporal binding and the neural correlates of sensory awarenes. Trends Cogn. Sci. **5**(1), 16–25 (2001)
227. Engel, G.S., et al.: Evidence for wavelike energy transfer through quantum coherence in photosynthetic systems. Nature **446**, 782–786 (2007)
228. Faddeev, L.D., Volkov, A.Yu.: Algebraic quantization of integrable models in discrete spacetime. In: Discrete Integrable Geometry and Physics. Oxford Lecture Ser. Math. Appl., vol. 16, pp. 301–319, Vienna, 1996. Oxford University Press, London (1999)
229. Faddeev, L.D., Yakubovsky, O.A.: Lectures on Quantum Mechanics for Mathematical Students, 2nd edn. NITS, Moscow (2001)
230. Fagnola, F., Rebolledo, R.: On the existence of stationary states for quantum dynamical semigroups. J. Math. Phys. **42**(3), 1296–1308 (2001)
231. Fannes, M., Nachtergaele, B., Werner, R.F.: Fintiely correlated pure states. J. Funct. Anal. **120**, 511–534 (1994)
232. Feinstein, A.: Foundations of Information Theory. McGraw-Hill, New York (1958)

233. Feinstein, A.: On the coding theorem and its converse for finite memory channels. Inf. Control **2**, 25–44 (1959)
234. Feller, W.: An Introduction to Probability Theory and Its Applications, vol. I. Wiley, New York (1950)
235. Felsenstein, J.: PHYLIP phylogeny inference package. http://evolution.genetics.washington.edu/phylip/doc/main.html
236. Felsenstein, J.: Neighbor–Neighbor-joining and UPGMA methods. http://evolution.genetics.washington.edu/phylip/doc/neighbor.html
237. Feynman, R.P.: The Feynman Lectures on Physics. Quantum Mechanics, vol. 3. Addison-Wesley, Reading (1965)
238. Feynman, R.: The Character of Physical Law. A Series of Lectures Recorded by the BBC at Cornell University USA. Cox and Wyman, London (1965)
239. Feynman, R.P.: The Character of Physical Law. MIT Press, Cambridge (1967). Chap. 5
240. Feynman, R.: Simulating physics with computers. Int. J. Theor. Phys. **21**(6/7), 467–488 (1982)
241. Feynman, R.: Quantum mechanical computers. Found. Phys. **16**, 11–20 (1986)
242. Fichtner, K.-H., Freudenberg, W.: Point processes and the position. distribution of infinite boson systems. J. Stat. Phys. **47**, 959 (1987)
243. Fichtner, K.-H., Freudenberg, W.: Characterization of states of infinite Boson systems I. On the construction of states. Commun. Math. Phys. **137**, 315–357 (1991)
244. Fichtner, K.-H., Freudenberg, W., Liebscher, V.: Time evolution and invariance of Boson systems given by beam splittings. Infin. Dimens. Anal. Quantum Probab. Relat. Top. **1**(4), 511–531 (1998)
245. Fichtner, K.-H., Freudenberg, W., Liebscher, V.: Non-independent splittings and Gibbs states. Math. Notes **64**(3–4), 518–523 (1998)
246. Fichtner, K.-H., Ohya, M.: Quantum teleportation with entangled states given by beam splittings. Commun. Math. Phys. **222**, 229–247 (2001)
247. Fichtner, K.-H., Miyadera, T., Ohya, M.: Fidelity of quantum teleportation by beam splittings. In: Proc. of International Symposium on Quantum Computing, vol. 3 (2001)
248. Fichtner, K.-H., Ohya, M.: Quantum teleportation and beam splitting. Commun. Math. Phys. **225**, 67–89 (2002)
249. Fichtner, K.-H., Freudenberg, W., Ohya, M.: Recognition and teleportation. In: Freudenberg, W. (ed.) Quantum Probability and Infinite Dimensional Analysis. QP–PQ: Quantum Probability and White Noise Analysis, vol. XV, pp. 85–105. Singapore, World Scientific (2003)
250. Fichtner, K.-H., Freudenberg, W., Liebscher, V.: On exchange mechanisms for bosons. Random Oper. Stoch. Equ. **12**(4), 331–348 (2004)
251. Fichtner, K.-H., Freudenberg, W., Ohya, W.: Teleportation scheme in infinite dimensional Hilbert space. J. Math. Phys. **46**(10), 102103 (2005), 14 pp.
252. Fichtner, K.-H., Fichtner, L.: Bosons and a quantum model of the brain. Jenaer Schriften zur Mathematik und Informatik Math/Inf/08/05, FSU Jena, Faculty of Mathematics and Informatics, Jena (2005), 27 pp.
253. Fichtner, K.-H., Freudenberg, W., Ohya, M.: Recognition and teleportation. In: Quantum Information, vol. V, pp. 1–17 (2006)
254. Fichtner, K.-H., Fichtner, L., Freudenberg, W., Ohya, M.: On a mathematical model of brain activities. In: Quantum Theory, Reconsideration of Foundations, vol. 4. AIP Conference Proceedings, vol. 962, pp. 85–90, Melville, New York. Am. Inst. Phys., New York (2007)
255. Fichtner, K.-H., Fichtner, L.: Quantum Markov chains and the process of recognition. Jenaer Schriften zur Mathematik und Informatik Math/Inf/02/07, FSU Jena, Faculty of Mathematics and Informatics, Jena (2007), 24 pp.
256. Fichtner, K.-H., Fichtner, L., Freudenberg, W. Ohya, M.: On a quantum model of the recognition process. In: QP-PQ: Quantum Prob. White Noise Anal. Quantum Bio-Informatics, vol. 21, pp. 64–84. World Scientific, Singapore (2008)

257. Fichtner, K.-H., Freudenberg, W.: The compound Fock space and its application to brain models. In: Accardi, L., Freudenberg, W., Ohya, M. (eds.) Quantum Bio-Informatics II. QP–PQ: Quantum Probability and White Noise Analysis, vol. XXIV, pp. 55–67. Singapore, World Scientific (2009)
258. Fichtner, L., Gäbler, M.: Characterisation of beam splitters. In: Accardi, L., Freudenberg, W., Ohya, M. (eds.) Quantum Bio-Informatics II. QP–PQ: Quantum Probability and White Noise Analysis, vol. XXIV, pp. 68–80. World Scientific, Singapore (2009)
259. Fichtner, K.-H., Fichtner, L., Freudenberg, W., Ohya, M.: Quantum models of the recognition process—on a convergence theorem. Open Syst. Inf. Dyn. **17**(2), 161–187 (2010)
260. Finkelshtein, A.V., Ptitsyn, O.B.: Physics of Proteins. Academic Press, San Diego (2002)
261. Flugge, S.: Practical Quantum Mechanics. Springer, Berlin (1994)
262. Forger, M., Sachse, S.: Lie superalgebras and the multiplet structure of the genetic code I: codon representations. J. Math. Phys. **41**(8), 5407–5422 (2000). arXiv:math-ph/9808001
263. Fouchier, R.A.M., Munster, V., Wallensten, A., Bestebroer, T.M., Herfst, S., Smith, D., Rimmelzwaan, G.F., Olsen, B., Osterhaus, A.D.M.E.: Characterization of a novel influenza a virus hemagglutinin subtype (H16) obtained from black-headed gulls. J. Virol. **79**(5), 2814–2822 (2005)
264. Fujiwara, A., Nagaoka, H.: Capacity of memoryless quantum communication channels. Math. Eng. Tech. Rep. 94–22, University of Tokyo (1994)
265. Furuichi, S., Ohya, M.: Quantum mutual entropy for Jaynes-Cummings model. Rep. Math. Phys. **44**, 81–86 (1999)
266. Furuichi, S., Ohya, M.: Entanglement degree in the time development of the Jaynes-Cummings model. Lett. Math. Phys. **49**, 279–285 (1999)
267. Frappat, L., Sciarrino, A., Sorba, P.: Crystalizing the genetic code. J. Biol. Phys. **27**, 1–38 (2001). arXiv:physics/0003037
268. Fredkin, E., Toffoli, T.: Conservative logic. Int. J. Theor. Phys. **21**, 219–253 (1982)
269. Frenkel, L.M., Wang, Y., Learn, G.H., McKernan, J.L., Ellis, G.M., Mohan, K.M., Holte, S.E., De Vange, S.M., Pawluk, D.M., Melvin, A.J., Lewis, P.F., Heath, L.M., Beck, I.A., Mahalanabis, M., Naugler, W.E., Tobin, N.H., Mullins, J.I.: Multiple viral genetic analyses detect low-level human immunodeficiency virus type 1 replication during effective highly active antiretroviral therapy. J. Virol. **77**(10), 5721 (2003)
270. Frenkel, V., Etherington, A., Greene, M., Quijiano, J., Xie, J., Hunter, F., Dromi, S., Li, K.C.P.: Delivery of liposomal doxorubicin(Doxil) in a breast cancer tumor model: investigation of potential enhancement by pulsed high intensity focused ultrasound exposure. Acad. Radiol. **13**, 469–479 (2006)
271. Freudenberg, W., Ohya, M., Watanabe, N.: Generalized Fock space approach to Fredkin-Toffoli-Milburn gate. TUS preprint (2001)
272. Freudenberg, W., Ohya, M., Watanabe, N.: On quantum logical gates on a general Fock space. In: QP-PQ: Quantum Probability and White Noise Analysis, vol. 18, pp. 252–268 (2005)
273. Friedman, A.: Uber die Krummung des Raumes. Z. Phys. **10**(1), 377–386 (1922)
274. Fuchs, C.A., Gisin, N., Griffiths, R.B., Niu, C.-S., Peres, A.: Optimal eavesdropping in quantum cryptography. I. Information bound and optimal strategy. Phys. Rev. A **56**, 1163 (1997)
275. Fuchs, C.A.: Quantum foundations in the light of quantum information. In: Gonis, A. (ed.) Proceedings of the NATO Advanced Research Workshop on Decoherence and Its Implications in Quantum Computation and Information Transfer (2001)
276. Gallavotti, G.: Fluctuation relation, fluctuation theorem, thermostats and entropy creation in nonequilibrium statistical physics. arXiv:cond-mat/0612061 (2007)
277. Garg, A., Mermin, N.D.: Detector inefficiencies in the Einstein-Podolsky-Rosen experiment. Phys. Rev. D **35**, 3831–3835 (1987)
278. Garey, M., Johnson, D.: Computers and Intractability—A Guide to the Theory of NP-Completeness. Freeman, New York (1979)
279. Gelfand, I.M., Yaglom, A.M.: Calculation of the amount of information about a random function contained in another such function. Am. Math. Soc. Transl. **12**, 199–246 (1959)

280. Gelfand, I.M., Vilenkin, N.Ya.: Generalized Functions. Applications of Harmonic Analysis. Academic Press, San Diego (1977)
281. Gershenfeld, N.A., Chuang, I.L.: Bulk spin-resonance quantum computation. Science **275**, 350 (1997)
282. George, D.G., Barker, W.C., Hunt, L.T.: The protein identification resource (PIR). Nucleic Acids Res. **14**(1), 11–15 (1986)
283. Gill, R.: On quantum statistical inference. J. R. Stat. Soc. B **65**, 1–31 (2003)
284. Gill, R.: Bell's inequality and the coincidence-time loophole. Europhys. Lett. **67**, 707–713 (2004)
285. Gilbert, G., Hamrick, M.: Practical quantum cryptography: a comprehensive analysis (part one). quant-ph/0009027 (2000)
286. Ginzburg, V.L.: Which problems in physics are most important and interesting in the beginning of XXI century? In: Ginzburg, V.L. (ed.) On Science, on Myself and Others, pp. 11–74. Fizmatlit, Moscow (2003) (in Russian)
287. Gisin, N., Massar, S.: Optimal quantum cloning machines. Phys. Rev. Lett. **79**, 2153 (1997)
288. Gisin, N., Ribordy, G., Tittel, W., Zbinden, H.: Quantum cryptography. Rev. Mod. Phys. **74**(1), 145–195 (2002)
289. Giulini, D., Joos, E., Kiefer, C., Kupsch, J., Stamatescu, I.O., Zeh, H.D.: Decoherence and the Appearance of a Classical World in Quantum Theory. Springer, Berlin (1996)
290. Gleason, A.M.: Measures on closed subspaces of a Hilbert space. Math. Mech. **6**, 885 (1957)
291. Glezen, W.P.: Emerging infections: pandemic influenza. Epidemiol. Rev. **18**, 64–76 (1996)
292. Gnutzman, S., Smilansky, U.: Quantum graphs: applications to quantum chaos and universal spectral statistics. Adv. Phys. **55**, 527–625 (2006)
293. Gödel, K.: Undecidable Propositions of Formal Mathematical Systems. Lecture Notes in Princeton Institute for Advanced Study. Princeton Sci. Publ., Princeton (1934)
294. Goldstein, S.: Boltzmann's approach to statistical mechanics. In: Bricmont, J. et al. (eds.) Chance in Physics: Foundations and Perspectives. Lecture Notes in Physics, vol. 574, pp. 39–68. Springer, Berlin (2001)
295. Goldstein, S., Lebowitz, J.L.: On the (Boltzmann) entropy of nonequilibrium systems. Physica D **193**, 53–66 (2004)
296. Gordon, J.P.: Noise at optical frequencies; information theory. In: Miles, P.A. (ed.) Quantum Electronics and Coherent Light; Proceedings of the International School of Physics "Enrico Fermi," Course XXXI, pp. 156–181. Academic Press, San Diego (1964)
297. Gottesman, D.: Stabilizer codes and quantum error correction. Caltech Ph.D. Thesis (1997). arXiv:quant-ph/9705052
298. Gottesman, D.: A theory of fault-tolerant quantum computation. Phys. Rev. A **57**, 127–137 (1998)
299. Gottesman, D., Lo, H.-K.: Proof of security of quantum key distribution with two-way classical communications. IEEE Trans. Inf. Theory **49**(2), 457–475 (2003)
300. Gorini, V., Kossakowski, A., Sudarshan, E.C.G.: Completely positive semigroups of N-level systems. J. Math. Phys. **17**, 821 (1976)
301. Gu, J., Purdom, P.W., Franco, J., Wah, B.W.: Algorithms for the satisfiability (SAT) problem: a survey. DIMACS Ser. Discret. Math. Theor. Comput. Sci. **35**, 19–151 (1996)
302. Gudder, S., Marchand, J.P.: Noncommutative probability on von Neumann algebras. J. Math. Phys. **13**, 799–806 (1972)
303. Gutzwiller, M.C.: Chaos in Classical and Quantum Mechanics. Springer, Berlin (1990)
304. Grover, L.K.: A fast quantum mechanical algorithm for database search. In: STOC, pp. 212–219 (1996)
305. Grover, L.K.: Quantum mechanics helps in searching for a needle in a haystack. Phys. Rev. Lett. **78**(2), 325 (1997)
306. Grover, L.K.: Quantum computers can search rapidly by using almost any transformation. Phys. Rev. Lett. **80**(19), 4329–4332 (1998)
307. Gudder, S.P.: Probability manifolds. J. Math. Phys. **25**, 2397 (1984)
308. Guiasu, S.: Information Theory with Applications. McGraw-Hill, New York (1977)

309. Haag, R., Kastler, D.: An algebraic approach to quantum field theory. J. Math. Phys. **5**, 848 (1964)
310. Haag, R.: Local Quantum Physics. Fields, Particles, Algebras. Springer, Berlin (1996)
311. Habib, S., Shizume, K., Zurek, W.H.: Decoherence, chaos, and the correspondence principle. Phys. Rev. Lett. **80**, 4361–4365 (1998)
312. Haeffner, H., Roos, C.F., Blatt, R.: Quantum computing with trapped ions. Phys. Rep. **469**, 155–203 (2008)
313. Hall, M.J.W.: Information exclusion principle for complementary observables. Phys. Rev. Lett. **74**, 3307 (1995)
314. Halmos, P.R.: Introduction to Hilbert Space and the Theory of Spectral Multiplicity. Chelsea, New York (1951)
315. Halvorson, H.: The Einstein–Podolsky–Rosen state maximally violates Bell's inequalities. Lett. Math. Phys. **53**, 321 (2000)
316. Hameroff, S.: Quantum coherence in microtubules. A neural basis for emergent consciousness? J. Conscious. Stud. **1**, 91–118 (1994)
317. Hameroff, S.: Quantum computing in brain microtubules? The Penrose-Hameroff Orch or model of consciousness. Philos. Trans. R. Soc., Lond. A **356**(1743), 1869–1896 (1998)
318. Hara, T., Iriyama, S., Makino, K., Terada, H., Ohya, M.: Mathematical description of drug movement into tumor with EPR effect and estimation of its configuration for DDS. Colloids Surf. B, Biointerfaces **75**, 42–46 (2009)
319. Hara, T., Sato, K., Ohya, M.: MTRAP: Pairwise sequence alignment algorithm by a new measure based on transition probability between two consecutive pairs of residues. BMC Bioinform. **11**, 235 (2010)
320. Hari, R., Lounasmaa, O.V.: Neuromagnetism: tracking the dynamics of the brain. Phys. World **13**, 33–38 (2000)
321. Harris, J.M., Zalipsky, S.: Poly (ethylene glycol) Chemistry and Biological Application. ACS Symposium Series, vol. 680. Am. Chem. Soc., Washington (1997)
322. Hasegawa, H.: Dynamical formulation of quantum level statistics. Open Syst. Inf. Dyn. **4**, 359–377 (1997)
323. Hayes, B.: The invention of the genetic code. Am. Sci. **86**(1), 8–14 (1998)
324. Henderson, L., Vedral, V: Information, relative entropy of entanglement, and irreversibility. Phys. Rev. Lett. **84**, 2014 (2000)
325. Heisenberg, W.: Principle of indeterminacy. Z. Phys. **43**, 172 (1927)
326. Helstrom, C.W.: Quantum Detection and Estimation Theory. Mathematics in Science and Engineering, vol. 123. Academic Press, San Diego (1976)
327. Hepp, K.: The classical limit for quantum mechanical correlation functions. Commun. Math. Phys. **35**, 265–277 (1974)
328. Hiai, F., Ohya, M., Tsukada, M.: Sufficiency, KMS condition and relative entropy in von Neumann algebras. Pac. J. Math. **96**, 99–109 (1981)
329. Hiai, F., Ohya, M., Tsukada, M.: Sufficiency and relative entropy in *-algebras with applications to quantum systems. Pac. J. Math. **107**, 117–140 (1983)
330. Hiai, F., Ohya, M., Tsukada, M.: Introduction to Operator Algebras. Kyoritsu, Tokyo (1985)
331. Hiai, F., Petz, D.: The proper formula for relative entropy and its asymptotics in quantum probability. Commun. Math. Phys. **143**, 99–114 (1991)
332. Hiai, F., Petz, D.: Entropy density for algebraic states. J. Funct. Anal. **125**, 287–308 (1994)
333. Hiai, F., Ohya, M., Petz, D.: McMillan type convergence for quantum Gibbs states. Arch. Math. **64**, 154–158 (1995)
334. Hiai, F., Petz, D.: The Semicircle Law, Free Random Variables, and Entropy. Am. Math. Soc., Providence (2000)
335. Hida, T.: Brownian Motion. Springer, Berlin (1980)
336. Hida, T., Kuo, H.-H., Potthoff, J., Streit, L.: White Noise. An Infinite Dimensional Calculus. Kluwer Academic, Dordrecht (1993)
337. Hida, T., Si, S.: An innovation approach to random fields. In: Application of White Noise Theory. World Scientific, Singapore (2004)

338. Hirano, T., Konishi, T., Namiki, R.: Quantum cryptography using balanced homodyne detection. quant-ph/0008037 (2000)
339. Hofstader, D.R.: Dilemmas for superrational thinkers, leading up to a luring lottery. Sci. Am. **6** (1983)
340. Hofstader, D.R.: Metamagical Themes: Questing for the Essence of Mind and Pattern. Basic Books, New York (1985)
341. Holevo, A.S.: Some estimates for the amount of information transmittable by a quantum communication channel. Probl. Pereda. Inf. **9**, 3–11 (1973) (in Russian)
342. Holevo, A.S.: Statistical problems in quantum physics. In: Maruyama, G., Prokhorov, J.V. (eds.) Proceedings of the Second Japan–USSR Symposium on Probability Theory. Lecture Notes in Mathematics, vol. 330, pp. 104–119. Springer, Berlin (1973)
343. Holevo, A.S.: Probabilistic and Statistical Aspects of Quantum Theory. Nauka, Moscow (1980)
344. Holevo, A.S.: Probabilistic and Statistical Aspects of Quantum Theory. North-Holland Series in Statistics and Probability, vol. 1. North-Holland, Amsterdam (1982)
345. Holevo, A.S.: The capacity of quantum channel with general signal states. IEEE Trans. Inf. Theory **44**, 269–273 (1998)
346. Holevo, A.S.: Statistical Structure of Quantum Theory. Lect. Notes Phys., vol. 67. Springer, Berlin (2001)
347. Hornos, J.E.M., Hornos, Y.M.M.: Algebraic model for the evolution of the genetic code. Phys. Rev. Lett. **71**, 4401–4404 (1993)
348. Horodecki, R., Horodecki, M., Horodecki, P.: Teleportation, Bell's inequalities and inseparability. quant-ph/9606027 (1996)
349. Horodecki, M., Horodecki, P., Horodecki, R.: Separability of mixed states: necessary and sufficient conditions. Phys. Lett. A **223**, 1–8 (1996)
350. Horodecki, M., Horodecki, P., Horodecki, R.: Limits for entanglement measures. Phys. Rev. Lett. **84**, 2014 (2000)
351. Horodecki, M., Horodecki, P., Horodecki, R.: Mixed state entanglement and quantum condition. In: Quantum Information. Springer Tracts in Modern Physics, vol. 173, pp. 151–195 (2001)
352. Hudetz, T.: Topological entropy for appropriately approximated C^*-algebras. J. Math. Phys. **35**(8), 4303–4333 (1994)
353. Ibrahim, N.K., Desai, N., Legha, S., Shiong, P.S., Theriault, R.L., Rivera, E., Esmaeli, B., Ring, S.E., Bedikian, A., Hortobagyi, G.N., Ellerhorst, J.A.: Phase I and pharmacokinetic study of ABI-007, a cremophor-free protein-stabilized, nanoparticle formulation of paclitaxel. Clin. Cancer Res. **8**, 1038–1044 (2002)
354. Ibrahim, N.K., Samuels, B., Page, R., Doval, D., Patel, K.M., Rao, S.C., Naie, M.K., Bhar, P., Desai, N., Hortobagyi, G.N.: Multicenter phase II trial of ABI-007, an albumin-bound paclitaxel, in woman with metastatic breast cancer. J. Clin. Oncol. **23**, 6019–6025 (2005)
355. Ihara, S.: Stochastic Process and Entropy. Iwanami, Tokyo (1984) (in Japanese)
356. Imamichi, H., Crandall, K.A., Natarajan, V., Jiang, M.K., Dewar, R.L., Berg, S., Gaddam, A., Bosche, M., Metcalf, J.A., Davey, R.T. Jr., Lane, H.C.: Human immunodeficiency virus type 1 quasi species that rebound after discontinuation of highly active antiretroviral therapy are similar to the viral quasi species present before initiation of therapy. J. Infect. Dis. **183**(1), 36 (2001)
357. Ingarden, R.S.: Simplified axioms for information without probability. Pr. Mat. **9**, 273–282 (1965)
358. Ingarden, R.S.: Quantum information theory. Rep. Math. Phys. **10**, 43–73 (1976)
359. Ingarden, R.S., Kossakowski, A., Ohya, M.: Information Dynamics and Open Systems. Kluwer Academic, Dordrecht (1997)
360. Inoue, K., Ohya, M., Sato, K.: Application of chaos degree to some dynamical systems. Chaos Solitons Fractals **11**, 1377–1385 (2000)
361. Inoue, K., Waks, E., Yamamoto, Y.: Differential phase shift quantum key distribution. Phys. Rev. Lett. **89**(3), 037902 (2002)

362. Inoue, K., Waks, E., Yamamoto, Y.: Differential-phase-shift quantum key distribution using coherent light. Phys. Rev. A **68**(2), 022317 (2003)
363. Inoue, K., Ohya, M., Suyari, H.: Characterization of quantum teleportation by nonlinear quantum channel and quantum mutual entropy. Physica D **120**, 117–124 (1998)
364. Inoue, K., Ohya, M., Volovich, I.V.: On quantum-classical correspondence for baker's map. quant-ph/0108107 (2001)
365. Inoue, K., Ohya, M., Volovich, I.V.: Semiclassical properties and chaos degree for the quantum baker's map. J. Math. Phys. **43**(1), 734 (2002)
366. Inoue, K., Ohya, M., Volovich, I.V.: On quantum classical correspondence for baker's map. In: Quantum Probability and White Noise Analysis, vol. 17, pp. 177–187 (2003)
367. Inoue, K., Kossakowski, A., Ohya, M.: Description of quantum chaos by chaos degree. TUS preprint (2004)
368. Inoue, K., Matsuoka, T., Ohya, M.: New approach to ε-entropy and fractal dimension of a state for a Gaussian measure. Open Syst. Inf. Dyn. **7**(1), 41–53 (2000)
369. Iriyama, S., Ohya, M., Volovich, I.V.: Generalized quantum Turing machine and its application to the SAT chaos algorithm. In: QP-PQ: Quantum Prob. White Noise Anal., Quantum Information and Computing, vol. 19, pp 204–225. World Scientific, Singapore (2006)
370. Iriyama, S., Miyadera, T., Ohya, M.: Note on a universal quantum Turing machine. Phys. Lett. A **372**, 5120–5122 (2008)
371. Iriyama, S., Ohya, M.: Review on quantum chaos algorithm and generalized quantum Turing machine. In: QP-PQ: Quantum Prob. White Noise Anal., Quantum Bio-Informatics, vol. 21, pp. 126–141. World Scientific, Singapore (2008)
372. Iriyama, S., Ohya, M.: Rigorous estimation of the computational complexity for OMV SAT algorithm. Open Syst. Inf. Dyn. **15**(2), 173–187 (2008)
373. Iriyama, S., Ohya, M.: Language classes defined by generalized quantum Turing machine. Open Syst. Inf. Dyn. **15**(4), 383–396 (2008)
374. Iriyama, S., Ohya, M., Sato, K., Volovich, I.V.: Photosynthetic antenna and entropy decreasing. TUS preprint (2010)
375. Iriyama, S., Ohya, M., Yamato, I.: Quantum algorithm of protain folding. TUS preprint (2010)
376. Isar, A.: Density operator and entropy of the damped quantum harmonic oscillator. Helv. Phys. Acta **68**, 225–234 (1995)
377. Ishizakia, A., Fleming, G.F.: Theoretical examination of quantum coherence in a photosynthetic system at physiological temperature. Proc. Natl. Acad. Sci. USA **106**, 17255–17260 (2009)
378. Ito, K.: Probability Theory, vols. I, II, III. Iwanami, Tokyo (1976–1978) (in Japanese)
379. Jamiołkowski, A.: Linear transformation which preserve trace and positive semidefiteness of operators. Rep. Math. Phys. **3**, 275 (1972)
380. Jamiolkowski, A.: Minimal number of operators for observability. Int. J. Theor. Phys. **22**, 369 (1983)
381. Jamiolkowski, A.: On complete and incomplete sets of observables, the principle of maximal entropy-revisited. Rep. Math. Phys. **46**, 469 (2000)
382. Jamiolkowski, A.: A stroboscopic approach to quantum tomography. In: Accardi, L., Ohya, M., Watanabe, N. (eds.) Quantum Information and Computing. World Scientific, Singapore (2006)
383. Jamiołkowski, J., Matsuoka, T., Ohya, M.: Entangling operator and PPT condition. TUS preprint
384. Janszky, J., Domokos, P., Szabo, S., Adam, P.: Quantum-state engineering via discrete coherent-state superpositions. Phys. Rev. A **51**, 4191 (1995)
385. Jacak, L., Hawrylak, P., Wjs, A.: Quantum Dots. Springer, Berlin (1998)
386. Jaynes, E.T.: Violation of Boltzmann's H theorem in real gases. Phys. Rev. A **4**, 747–750 (1971)
387. Jeong, H., Son, W., Kim, M.S., Ahn, D., Brukner, C.: Quantum nonlocality test for continuous-variable states with dichotomic observables. Phys. Rev. A **67**, 012106 (2003)

388. Jumarie, G.: Relative Information. Springer, Berlin (1990)
389. Jurkowski, J., Chruściński, D., Rutkowski, A.: A class of bound entangled states of two qutits. Open Syst. Inf. Dyn. **16**, 235–242 (2009)
390. Johansen, L.M.: EPR correlations and EPW distributions revisited. Phys. Lett. A **236**, 173 (1997)
391. Jones, J.A., Mosca, M.: Implementation of a quantum algorithm on a nuclear magnetic resonance quantum computer. J. Chem. Phys. **109**, 1648 (1998)
392. Jozsa, R., Schumacher, B.: A new proof of the quantum noiseless coding theorem. J. Mod. Opt. **41**, 2343–2349 (1994)
393. Jozsa, R.: Fidelity for mixed quantum states. J. Mod. Opt. **41**, 2315–2323 (1994)
394. Jozsa, R.: Entanglement and quantum computation. In: Huggett, S. et al. (eds.) Geometric Issues in the Foundations of Science (1997)
395. Jukes, T.H.: Evolution of the amino acid code: inferences from mitochondrial codes. J. Mol. Evol. **19**, 219–225 (1983)
396. Kac, M.: Probability and Related Topics in Physical Sciences. Wiley, New York (1959)
397. Kadison, R.V., Ringrose, J.R.: Fundamentals of the Theory of Operator Algebras, vol. IV. Birkhäuser, Boston (1986)
398. Kaplan, L., Heller, E.J.: Overcoming the wall in the semiclassical baker's map. Phys. Rev. Lett. **76**, 1453–1456 (1996)
399. Karatsuba, A.A., Ofman, Y.: Multiplication of multidigit numbers on automata. Dokl. Akad. Nauk SSSR **145**, 293–294 (1962) (in Russian)
400. Karkuszewski, Z.P., Zakrzewski, J., Zurek, W.H.: Breakdown of correspondence in chaotic systems: Ehrenfest versus localization times. Phys. Rev. A **65**, 042113 (2002)
401. Katok, A., Hasselblatt, B.: Introduction to the Modern Theory of Dynamical Systems. Cambridge University Press, Cambridge (1997)
402. Keyl, M., Schlingemann, D., Werner, R.F.: Infinitely entangled states. quant-ph/0212014 (2002)
403. Khinchin, A.I.: Mathematical Foundations of Information Theory. Dover, New York (1958), English Translation
404. Khrennikov, A.: Non-Archimedean Analysis: Quantum Paradoxes, Dynamical Systems and Biological Models. Kluwer Academic, Dordrecht (1997)
405. Khrennikov, A.: Interpretations of Probability. Frontiers in Probability Theory. De Gruyter, Berlin (1999)
406. Khrennikov, A.: Superanalysis. Kluwer Academic, Dordrecht (1999)
407. Khrennikov, A.: Statistical measure of ensemble nonreproducibility and correction to Bell's inequality. Il Nuovo Cimento B **115**(2), 179–184 (1999)
408. Khrennikov, A.: Non-Kolmogorov probability and modified Bell's inequality. J. Math. Phys. **41**(4), 1768–1777 (2000). Reviewer: S. Gudder
409. Khrennikov, A.: Linear representations of probabilistic transformations induced by context transitions. J. Phys. A, Math. Gen. **34**, 9965 (2001)
410. Khrennikov, A.: Frequency analysis of the EPR-Bell argumentation. Found. Phys. **32**, 1159–1174 (2002)
411. Khrennikov, A. (ed.): Quantum Theory: Reconsideration of Foundations. Vaxjo University Press, Vaxjo (2002)
412. Khrennikov, A., Volovich, I.V.: Local realism, contextualism and loopholes in Bell's experiments. In: Foundations of Probability and Physics-2. Ser. Math. Model., vol. 5, pp. 325–344. Vaxjo University Press, Vaxjo (2002)
413. Khrennikov, A., Volovich, I.V.: Einstein, Podolsky and Rosen versus Bohm and Bell. quant-ph/0211078 (2002)
414. Khrennikov, A., Volovich, I.V.: Quantum nonlocality, EPR model, and Bell's theorem. In: Proceedings of the 3nd Sakharov Conference on Physics, vol. 2, pp. 269–276. World Scientific, Singapore (2003)
415. Khrennikov, A.: Contextual viewpoint to quantum statistics. J. Math. Phys. **44**(6), 2471–2478 (2003)

416. Khrennikov, A.: Information Dynamics in Cognitive, Psychological, Social and Anomalous Phenomena. Kluwer Academic, Dordrecht (2004)
417. Khrennikov, A.: On quantum-like probabilistic structure of mental information. Open Syst. Inf. Dyn. **11**(3), 267–275 (2004)
418. Khrennikov, A., Volovich, J.I.: Discrete time dynamical models and their quantum-like context-dependent properties. J. Mod. Opt. **51**, 113–114 (2004)
419. Khrennikov, A., Volovich, I.V.: Local realistic representation for correlations in the original EPR-model for position and momentum. Soft Comput. **10**, 521–529 (2005)
420. Khrennikov, A.: Quantum-like brain: Interference of minds. BioSystems **84**, 225–241 (2006)
421. Khrennikov, A., Volovich, Ya.I.: Energy levels of "Hydrogen atom" in discrete time dynamics. Open Syst. Inf. Dyn. **13**, 119–132 (2006)
422. Khrennikov, A.: p-Adic mathematical physics. In: Proceedings of the 2nd International Conference on p-Adic Mathematical Physics. AIP Conference Proceedings, vol. 826 (2006)
423. Khrennikov, A.Yu., Kozyrev, S.V.: Genetic code on the diadic plane. Phys. A, Stat. Mech. Appl. **381**, 265–272 (2007)
424. Khrennikov, A.: Contextual Approach to Quantum Formalism. Fundamental Theories of Physics. Springer, Berlin (2009)
425. Khrennikov, A., Haven, E.: Quantum mechanics and violations of the sure-thing principle: the use of probability interference and other concepts. J. Math. Psychol. **53**, 378–388 (2009)
426. Khrennikov, A.: Ubiquitous Quantum Structure: From Psychology to Finance. Springer, Berlin (2010)
427. Kim, M., Son, W., Buzek, V., Knight, P.: Entanglement by a beam splitter: nonclassicality as a prerequisite for entanglement. Phys. Rev. A **65**, 32323 (2002)
428. Kishimoto, A.: Dissipations and derivations. Commun. Math. Phys. **47**(1), 25–32 (1976)
429. Kitaev, A.Y.: Quantum computations: algorithms and error correction. Russ. Math. Surv. **52**, 1191–1249 (1997)
430. Klaus, D.: Decoherence by Lindblad motion. J. Phys. A, Math. Gen. **37**, 6143–6155 (2004)
431. Klimontovich, Yu.L.: Statistical Physics. Nauka, Moscow (1982)
432. Kloeden, P.E., Platen, E.: Numerical Solution of Stochastic Differential Equations. Springer, Berlin (1992)
433. Knight, R.D., Freeland, S.J., Landweber, L.F.: Rewiring the keyboard: evolvability of the genetic code. Nat. Rev. Genet. **2**, 49–58 (2001)
434. Knill, E., Laflamme, R., Viola, L.: Theory of quantum error correction for general noise. Phys. Rev. Lett. **84**, 2525–2528 (2000)
435. Knill, E., Laflamme, R., Zurek, W.H.: Resilient quantum computation: error models and thresholds. Proc. R. Soc. Lond., Ser. A **454**, 365 (1998). arXiv:quant-ph/9702058
436. Kolmogorov, A.N.: Theory of transmission of information. Am. Math. Soc. Transl., Ser. 2 **33**, 291–321 (1963)
437. Kolmogorov, A.N.: Three approaches to the quantitative definition of information. Probl. Inf. Transm. **1**, 1–7 (1965)
438. Kolmogorov, A.N.: Grundbegriffe der Wahrscheinlichkeitsrechnung, Erg. Mat., 1933. Springer, Berlin (1977)
439. Kolmogorov, A.N.: The general theory of dynamical systems and classical mechanics. In: Proceedings of the International Congress of Mathematicians, Amsterdam, 1954, vol. 1, pp. 315–333. North Holland, Amsterdam (1957) (in Russian). English translation as Appendix D in R.H. Abraham, Foundations of Mechanics, pp. 263–279. Benjamin (1967). Reprinted as Appendix in R.H. Abraham and J.E. Marsden, Foundations of Mechanics, 2nd edn., pp. 741–757. Benjamin/Cummings (1978)
440. Konno, N.: Quantum random walks in one dimensions. Quantum Inf. Process. **1**, 345–354 (2003)
441. Kossakowski, A., Ohya, M., Watanabe, N.: Quantum dynamical entropy for completely positive maps. Infin. Dimens. Anal. Quantum Probab. Relat. Top. **2**(2), 267–282 (1999)
442. Kossakowski, A., Ohya, M., Togawa, Y.: How can we observe and describe chaos? Open Syst. Inf. Dyn. **10**(3), 221–233 (2003)

443. Kossakowski, A., Ohya, M.: Can non-maximal entangled state achieve a complete quantum teleportation? In: Reconsideration of Foundation-3, vol. 810, pp. 211–216. Am. Inst. Phys., New York (2006)
444. Kossakowski, A., Ohya, M.: New scheme of quantum teleportation. Infin. Dimens. Anal. Quantum Probab. Relat. Top. **10**(3), 411–420 (2007)
445. Kimura, G., Kossakowski, A.: A note on positive maps and classification on states. Open Sys. Inf. Dyn. **12**, 1 (2005)
446. Kobrak, M.N., Bittner, E.R.: Quantum molecular dynamics study of polaron recombination in conjugated polymers. Phys. Rev. B **62**, 11473 (2000)
447. Kozlov, V.V.: Thermodynamic Equilibrium According to Gibbs and Poincare. Institute of Computer Science, Moscow-Ijevsk (2002) (in Russian)
448. Kozlov, V.V.: Temperature Equilibrium on Gibbs and Poincaré. Institute of Computer Science, Moscow-Ijevsk (2002) (in Russian)
449. Kozlov, V.V., Treshchev, D.V.: Fine-grained and coarse-grained entropy in problems of statistical mechanics. Theor. Math. Phys. **151**, 539–555 (2007)
450. Kozlov, V.V.: Gibbs Ensembles and Nonequilibrium Statistical Mechanics. Institute of Computer Science, Moscow-Ijevsk (2008) (in Russian)
451. Kraus, K.: States, Effects, and Operations: Fundamental Notions of Quantum Theory. Lecture Notes in Physics, vol. 190. Springer, Berlin (1983)
452. Kraus, K.: General state changes in quantum theory. Ann. Phys. **64**(2), 139–177 (1985)
453. Krauss, S., Obert, C.A., Franks, J., Walker, D., Jones, K., Seiler, P., Niles, L., Pryor, S.P., Obenauer, J.C., Naeve, C.W., Widjaja, L., Webby, R.J., Webster, R.G.: Influenza in migratory birds and evidence of limited intercontinental virus exchange. PLoS Pathog **3**(11), 1684–1693 (2007)
454. Krylov, N.S.: Works on the Foundations of Statistical Physics. Akad. Nauk SSSR, Leningrad (1950) (in Russian)
455. Kullback, S., Leibler, R.: On information and sufficiency. Ann. Math. Stat. **22**, 79–86 (1951)
456. Kunisawa, K., Umegaki, H.: Progress of Information Theory. Iwanami, Tokyo (1965) (in Japanese)
457. Kunisawa, K.: Probability Theory and Its Applications. Iwanami, Tokyo (1982) (in Japanese)
458. Kunisawa, K.: Information Theory, vol. I. Kyoritsu, Tokyo (1983) (in Japanese)
459. Kuzmich, A., Walmsley, I.A., Mandel, L.: Violation of Bell's inequality by a generalized Einstein-Podolsky-Rosen state using homodyne detection. Phys. Rev. Lett. **85**, 1349 (2000)
460. Labuschagne, L.E., Majewski, W.A., Marciniak, M.: On k-decomposable positive maps. Exp. Math. **24**, 103 (2006)
461. Lakshminarayan, A.: On the quantum Baker's map and its unusual traces. Ann. Phys. **239**, 272 (1995)
462. Landau, L.D., Lifshitz, E.M.: Quantum Mechanics Non-relativistic Theory. Course of Theoretical Physics, vol. 3. Pergamon, Elmsford (1958). Translated from the Russian by J.B. Sykes and J.S. Bell. Addison-Wesley Series in Advanced Physics. Pergamon, Elmsford and Addison-Wesley, Reading
463. Landau, L.D., Lifshiz, E.M.: Fluid Mechanics. Pergamon, Elmsford (1959)
464. Landau, L.D., Lifshitz, E.M.: Quantum Mechanics. Pergamon, Elmsford (1965)
465. Landau, L.D., Lifshiz, E.M.: Statistical Physics, Part 1. Pergamon, Elmsford (1980)
466. Landauer, R.: Irreversibility and heat generation in the computing process. IBM J. Res. Dev. **5**, 183 (1961)
467. Landauer, R.: Computation and physics: Wheeler's meaning circuit? Found. Phys. **16**(6), 551–564 (1986)
468. Landauer, R.: Information is physical. Phys. Today **44**(5), 22–29 (1991)
469. Landim, C., Kipnis, C.: Scaling Limits for Interacting Particle Systems. Springer, Berlin (1998)
470. Lanford, O.E.: Time evolution of large classical systems. In: Moser, J. (ed.) Lecture Notes in Physics, vol. 38, pp. 1–111. Springer, Berlin (1975)
471. Lanford, O.E.: Hard-sphere gas in the Boltzmann-Grad limit. Physica A **106**, 70–76 (1981)

472. Larsson, J.-A.: Modeling the singlet state with local variables. Phys. Lett. A **256**, 245 (1999)
473. Lebowitz, J.L.: From time-symmetric microscopic dynamics to time-asymmetric macroscopic behavior: an overview. arXiv:0709.0724 (2007)
474. Lee, H., Cheng, Y.-C., Fleming, G.R.: Coherence dynamics in photosynthesis: protein protection of excitonic coherence. Science **316**, 1462–1465 (2007)
475. Leggett, A.J., Chakravarty, S., Dorsey, A.T., Fisher, M.P.A., Garg, A., Zwerger, W.: Dynamics of the dissipative two-state system. Rev. Mod. Phys. **59**(1), 1–85 (1987)
476. Lenstra, A.K., Lenstra, H.W. Jr. (eds.): The Development of the Number Field Sieve. Lect. Notes in Math., vol. 1554. Springer, Berlin (1993)
477. Levitin, L.B.: On the quantum measure of the amount of information. In: Proceedings of the IV National Conference on Information Theory, Tashkent, pp. 111–115. Springer, Berlin (1969) (in Russian). Information theory for quantum systems. In: Blaqui'ere, A., Diner, S., Lochak, G. (eds.) Information, Complexity, and Control in Quantum Physics (1987)
478. Levy, P.: Processus Stochastiques et Mouvement Brownien. Gauthier-Villars, Paris (1948)
479. Levy, P.: Problems Concrets d'Analyse Fonctionnelle. Gauthier-Villars, Paris (1951)
480. Li, M., Vitànyi, P.: An Introduction to Kolmogorov Complexity and Its Applications. Springer, Berlin (1997)
481. Lindblad, G.: Entropy, information and quantum measurements. Commun. Math. Phys. **33**, 111–119 (1973)
482. Lindblad, G.: Completely positive maps and entropy inequalities. Commun. Math. Phys. **40**, 147–151 (1975)
483. Lindblad, G.: On the generators of quantum dynamical semigroups. Commun. Math. Phys. **48**, 119 (1976)
484. Linde, A.D.: Inflation and Quantum Cosmology. Academic Press, San Diego (1990)
485. Linden, N., Popescu, S.: The halting problem for quantum computers. quant-ph/9806054 (1998)
486. Lindsay, J.M.: Quantum and noncausal stochastic calculus. Probab. Theory Relat. Fields **97**, 65 (1993)
487. Lloyd, S.: Universe as quantum computer, how decoherence leads to complexity. Complexity **3**(1), 32–35 (1997)
488. Lo, H.-K., Chau, H.-F.: Unconditional security of quantum key distribution over arbitrarily long distances. Science **283**, 2050–2056 (1999)
489. Loeve, M.: Probability Theory. van Nostrand, Princeton (1955)
490. Loss, D., DiVincenzo, D.P.: Quantum computation with quantum dots. Phys. Rev. A **57**, 120–126 (1998)
491. Luis, A., Sanchez-Soto, L.: Phase-difference operator. Phys. Rev. A **48**, 4702 (1993)
492. Maassen, H.: Quantum Markov processes on Fock space described by integral kernels. In: Quantum Probability and Applications II. Lecture Notes in Mathematics, vol. 1136, pp. 361–374. Springer, Berlin (1985)
493. Maassen, H., Uffink, J.B.M.: Generalized entropic uncertainty relations. Phys. Rev. Lett. **60**, 1103 (1988)
494. Mackey, G.W.: Mathematical Foundations of Quantum Mechanics. Benjamin, Elmsford (1963)
495. Mackey, T.D., Bartlett, S.D., Stephanson, L.T., Sanders, B.C.: Quantum walks in higher dimensions. J. Phys. A **36**, 241–253 (2003)
496. Maeda, H.: SMANCS and polymer-conjugated macromolecular drugs: advantages in cancer chemotherapy. Adv. Drug Deliv. Rev. **6**, 181–202 (1991)
497. Maeda, H.: The tumor blood vessel as an ideal target for macromolecular anti cancer agents. J. Control. Release **19**, 315–324 (1992)
498. Maeda, H., Wu, J., Sawa, T., Matsumura, Y., Hori, K.: Tumor vascular permeability and the EPR effect in macromolecular therapeutics: a review. J. Control. Release **65**, 271–284 (2000)
499. Maeda, H., Sawa, T., Konno, T.: Mechanism of tumor-targeted delivery of macromolecular drugs, including the EPR effect in solid tumor and clinicaloverview of the prototype polymeric drug SMANCS. J. Control. Release **74**, 47–61 (2001)

500. Maeda, N., Mchedlishvili, G.: Blood flow structure related to red cell flow: a determinant of blood fluidity in narrow microvessels. Jpn. J. Physiol. **51**, 19–30 (2001)
501. Majewski, W.A.: On a characterization of PPT states. e-print arXiv:0708.3980
502. Majewski, W.A.: Separable and entangled state of composite quantum systems-rigorous description. Open Syst. Inf. Dyn. **6**, 79 (1999)
503. Majewski, W.A.: Measures of entanglement—a Hilbert space approach. In: Quantum Prob. and White Noise Analysis, vol. 24, p. 127 (2009)
504. Majewski, W.A., Matsuoka, T., Ohya, M.: Characterization of partial positive transposition states and measures of entanglement. J. Math. Phys. **50**, 113509 (2009)
505. Mandel, L., Wolf, E.: Optical Coherence and Quantum Optics. Cambridge University Press, Cambridge (1995)
506. Mandelbrot, B.B.: The Fractal Geometry of Nature. Freemann, New York (1982)
507. Manin, Yu.I.: Reflections on arithmetical physics. In: Manin, Y.I. (ed.) Mathematics as Metaphor: Selected Essays of Yuri I. Manin, pp. 149–155. Am. Math. Soc., Providence (2007)
508. Mariano, A., Facchi, P., Pascazio, S.: Decoherence and fluctuations in quantum interference experiments. Fortschr. Phys. **49**, 1033–1039 (2001)
509. Maslov, V.P.: Perturbation Theory and Asymptotic Methods. MGU, Moscow (1965)
510. Maslov, V.P., Fedoryuk, M.V.: Semi-Classical Approximation in Quantum Mechanics. Reidel, Dordrecht (1981)
511. Matsuoka, T., Ohya, M.: Fractal dimensions of states and its application to Ising model. Rep. Math. Phys. **36**, 365–379 (1995)
512. Matsuoka, T., Ohya, M.: A new measurement of time serial correlations in stock price movements and its application. In: Hida, T., Saito, K., Si, Si (eds.) Quantum Information and Complexity, pp. 341–361 (2004)
513. Matsuoka, T., Ohya, M.: Quantum entangled state and its characterization. In: Foundations of Probability and Physics-3. AIP, vol. 750, pp. 298–306 (2005)
514. Mayers, D., Yao, A.: Quantum cryptography with imperfect apparatus. quant-ph/9809039 (1998). To appear in FOCS
515. Mayers, D.: Unconditional security in quantum cryptography. J. ACM **48**(3), 351–406 (2001)
516. McEliece, R.J.: The Theory of Information Theory and Coding. Addison-Wesley, Reading (1977)
517. McMillan, B.: The basic theorem of information theory. Ann. Math. Stat. **24**, 196–219 (1953)
518. Mercer, I.P., et al.: Instantaneous mapping of coherently coupled electronic transitions and energy transfers in a photosynthetic complex using angleresolved coherent optical wavemixing. Phys. Rev. Lett. **102**, 057402 (2009)
519. Messian, A.: Mechanique Quantique. Dunod, Paris (1959)
520. Messiah, A.: Quantum Mechanics, vols. I, II. North-Holland, Amsterdam (1961). Translated from the French by G.M. Temmer
521. Meyer, D.A.: From quantum cellular automata to quantum lattice gases. J. Stat. Phys. **85**, 551–574 (1996)
522. Michalet, X., Pinaud, F.F., Bentolila, L.A., et al.: Quantum dots for live cells, in vivo imaging, and diagnostics. Science **307**(5709), 538–544 (2005)
523. Milburn, G.J.: Quantum optical Fredkin gate. Phys. Rev. Lett. **63**(18), 2124–2127 (1989)
524. Miyadera, T., Ohya, M.: On chameleon effect and Bell's inequality. TUS preprint (2004)
525. Miyadera, T., Ohya, M.: Quantum dynamical entropy of spin systems. Rep. Math. Phys. **56**(1), 1–10 (2005)
526. Miyadera, T., Ohya, M.: On halting process of quantum Turing machine. Open Syst. Inf. Dyn. **12**(3), 261–264 (2006)
527. Miyazaki, S., Ohshima, Y., Ohya, M.: A new method of alignment of amino acid sequences. Viva Orig. **17**, 139–151 (1989)
528. Miyazaki, S., Ogata, K., Ohya, M.: On multiple alignment of genome sequences. IEICE Trans. Commun. **E75-B**(6), 453–457 (1992)

529. Montroll, E.W.: Quantum theory on a network. I. A solvable model whose wavefunctions are elementary functions. J. Math. Phys. **11**, 635 (1970)
530. Monteoliva, D., Paz, J.P.: Decoherence in a classically chaotic quantum system: entropy production and quantum-classical correspondence. Phys. Rev. E **64**, 056238 (2001)
531. Morens, D.M., Taubenberger, J.K., Fauci, A.S.: The persistent legacy of the 1918 influenza virus. N. Engl. J. Med. **361**(3), 225–229 (2009)
532. Mosley, P.J., Christ, A., Eckstein, A., Silberhorn, C.: Direct measurement of the spatial-spectral structure of waveguided parametric down-conversion. Phys. Rev. Lett. **103**, 233901 (2009)
533. Mueller, M., Liang, L.-M., Lesanovsky, I., Zoller, P.: Trapped Rydberg ions: from spin chains to fast quantum gates. New J. Phys. **10**, 093009 (2008)
534. Munkres, J.R.: Topology. Prentice Hall, New York (1975)
535. Muraki, N., Ohya, M., Petz, D.: Note on entropy of general quantum systems. Open Syst. Inf. Dyn. **1**(1), 43–56 (1992)
536. Muraki, N., Ohya, M.: Entropy functionals of Kolmogorov–Sinai type and their limit theorems. Lett. Math. Phys. **36**, 327–335 (1996)
537. Murao, M., Plenio, M.B., Vedral, V.: Quantum information distribution via entanglement. Phys. Rev. A **60**, 032311 (2000)
538. Myers, J.M.: Can a universal quantum computer be fully quantum? Phys. Rev. Lett. **78**, 1823 (1997)
539. Naik, D.S., Peterson, C.G., White, A.G., Berglund, A.J., Kwiat, P.G.: Entangled state quantum cryptography: Eavesdropping on the Ekert protocol. quant-ph/9912105 (1999)
540. Naimark, M.A.: Normieren Rings. Nauka, Moscow (1968)
541. Nakamura, A., et al.: Somatosensory homunculus as drawn by meg. NeuroImage **7**(4), 377–386 (1998)
542. Nambu, Y., Hatanaka, T., Nakamura, K.: BB84 quantum key distribution system based on silica-based planar lightwave circuits. Jpn. J. Appl. Phys. B **43**(8), 1109–1110 (2004)
543. Needleman, S.B., Wunsch, C.D.: A general method applicable to search for similarities in the amino acid sequence of two proteins. J. Mol. Biol. **88**, 233–258 (1986)
544. Nelson, M.I., Holmes, E.C.: The evolution of epidemic influenza. Nat. Rev. Genet. **8**, 196–205 (2007)
545. Nelson, M.I., Viboud, C., Simonsen, L., Bennett, R.T., Griesemer, S.B., George, K.S., Taylor, J., Spiro, D.J., Sengamalay, N.A., Ghedin, E., Taubenberger, J.K., Holmes, E.C.: Multiple reassortment events in the evolutionary history of H1N1 influenza a virus since 1918. PLoS Pathog **4**(2), 1–12 (2008)
546. Neveu, J.: Bases Mathematiques du Calcul des Probabilites. Masson, Paris (1970)
547. Nielsen, M.A., Chuang, I.L.: Quantum Computation and Quantum Information. Cambridge University Press, Cambridge (2000)
548. Nieuwenhuizen, Th.M., Volovich, I.V.: Role of various entropies in the black hole information loss problem. In: Nieuwenhuizen, Th.M., et al. (eds.) Beyond the Quantum. World Scientific, Singapore (2007)
549. Noh, J., Fougeres, A., Mandel, L.: Measurement of the quantum phase by photon counting. Phys. Rev. Lett. **67**, 1426 (1991)
550. O'Connor, P.W., Tomsovic, S.: The usual nature of the quantum baker's transformation. Ann. Phys. **207**, 218 (1991)
551. Ogata, K., Ohya, M.: Multiple alignment for amino acid sequences by dynamic program. Electron. Commun. Jpn. Part 3 **81**(4), 12–20 (1998)
552. Ohya, M.: Dynamical processes in quantum statistical mechanics. UR Tech. Rep. **25**(2), 12–25 (1974)
553. Ohya, M.: Remarks on noncommutative ergodic properties. UR Tech. Rep. **26**(3), 30–38 (1975)
554. Ohya, M.: Stability of Weiss Ising model. J. Math. Phys. **19**(5), 967–971 (1978)
555. Ohya, M.: On linear response dynamics. L. Nuovo Cimento **21**(16), 573–576 (1978)
556. Ohya, M.: Dynamical process in linear response theory. Rep. Math. Phys. **16**(3), 305–315 (1979)

557. Ohya, M.: On open system dynamics—an operator algebraic study. Kodai Math. J. **3**, 287–294 (1979)
558. Ohya, M.: Quantum ergodic channels in operator algebras. J. Math. Anal. Appl. **84**(2), 318–327 (1981)
559. Ohya, M.: On compound state and mutual information in quantum information theory. IEEE Trans. Inf. Theory **29**, 770–777 (1983)
560. Ohya, M.: Note on quantum proability. L. Nuovo Cimento **38**(11), 203–206 (1983)
561. Ohya, M.: Entropy transmission in C^*-dynamical systems. J. Math. Anal. Appl. **100**(1), 222–235 (1984)
562. Ohya, M.: Construction and analysis of a mathematical model in quantum communication processes. Electron. Commun. Jpn. **68**(2), 29–34 (1985)
563. Ohya, M.: State Change and Entropies in Quantum Dynamical Systems. Springer Lecture Notes in Math., vol. 1136, pp. 397–408. Springer, Berlin (1985)
564. Ohya, M., Matsuoka, T.: Continuity of entropy and mutual entropy in C*-dynamical systems. J. Math. Phys. **27**(8), 2076–2079 (1986)
565. Ohya, M., Fujii, Y.: Entropy change in linear response dynamics. IL Nuovo Cimento B **91**(1), 25–30 (1986)
566. Ohya, M., Watanabe, N.: A new treatment of communication processes with Gaussian channels. Jpn. J. Appl. Math. **3**(1), 197–206 (1986)
567. Ohya, M., Tsukada, M., Umegaki, H.: A formulation of noncommutative McMillan theorem. Proc. Jpn. Acad. Ser. A **63**(3), 50–53 (1987)
568. Ohya, M.: Entropy operators and McMillan type convergence theorems in a noncommutative dynamical system. In: Watanabe, S., Prokhorov, Yu.V. (eds.) Probability Theory and Mathematical Statistics. Lecture Notes in Math., vol. 1299, pp. 384–390. Springer, Berlin (1988)
569. Ohya, M.: Fractal dimensions for general quantum states. Symp. Appl. Funct. Anal. **11**, 45–57 (1988)
570. Ohya, M.: Some aspects of quantum information theory and their applications to irreversible processes. Rep. Math. Phys. **27**, 19–47 (1989)
571. Ohya, M.: Information theoretical treatments of genes. Trans. IEICE **E72**(5), 556 (1989)
572. Ohya, M.: Fractal dimensions of states. In: Quantum Probability and Related Topics, vol. 6, pp. 359–369. World Scientific, Singapore (1991)
573. Ohya, M.: Information dynamics and its application to optical communication processes. In: Springer Lecture Note in Physics, vol. 378, pp. 81–92. Springer, Berlin (1991)
574. Ohya, M., Suyari, H.: Rigorous derivation of error probability in coherent optical communication. In: Springer Lecture Note in Physics, vol. 378, pp. 203–212. Springer, Berlin (1991)
575. Ohya, M.: On fuzzy relative entropy. Symp. Appl. Funct. Anal. **13**, 105–115 (1991)
576. Ohya, M., Matsunaga, S.: Coding and genes. Trans. IEICE A **J74**(7), 1075 (1991)
577. Ohya, M., Uesaka, Y.: Amino acid sequences and DP matching. Int. J. Inf. Sci. **63**, 139–151 (1992)
578. Ohya, M., Petz, D.: Quantum Entropy and Its Use. Springer, Berlin (1993)
579. Ohya, M.: Quantum entropies and their maximizations. In: Maximum Entropy and Bayesian Methods, vol. 12, pp. 189–194 (1993)
580. Ohya, M., Naritsuka, S.: On fuzzy relative entropy. Open Syst. Inf. Dyn. **1**(3), 397–408 (1993)
581. Ohya, M., Watanabe, N.: Information dynamics and its application to Gaussian communication processes. In: Maximum Entropy and Bayesian Methods, vol. 12, pp. 195–203 (1993)
582. Ohya, M.: State change, complexity and fractal in quantum systems. In: Quantum Communications and Measurement, vol. 2, pp. 309–320. Plenum, New York (1995)
583. Ohya, M., Watanabe, N.: A mathematical study of information transmission in quantum communication processes. In: Quantum Communications and Measurement, vol. 2, pp. 371–378. Plenum, New York (1995)
584. Ohya, M., Petz, D.: Notes on quantum entropy. Studia Sci. Math. Hung. **31**, 423–430 (1996)
585. Ohya, M., Watanabe, N.: Note on irreversible dynamics and quantum information. In: Contributions in Probability, Undine, Forum, pp. 205–220 (1996)

586. Ohya, M.: Complexity, fractal dimension for quantum states. Open Syst. Inf. Dyn. **4**, 141–157 (1997)
587. Ohya, M., Petz, D., Watanabe, N.: On capacities of quantum channels. Prob. Math. Stat. **17**, 179–196 (1997)
588. Ohya, M., Watanabe, N.: On mathematical treatment of Fredkin-Toffoli-Milburn gate. Physica D **120**, 206–213 (1998)
589. Ohya, M.: A mathematical foundation of quantum information and quantum computer—on quantum mutual entropy and entanglement. quant-ph/9808051 (1998)
590. Ohya, M.: Complexities and their applications to characterization of chaos. Int. J. Theor. Phys. **37**(1), 495–505 (1998)
591. Ohya, M., Petz, D., Watanabe, N.: Numerical computation of quantum capacity. Int. J. Theor. Phys. **38**(1), 507–510 (1998)
592. Ohya, M.: Foundation of entropy, complexity and fractal in quantum systems. In: Probability Towards the Year 2000, pp. 263–286 (1998)
593. Ohya, M.: Fundamentals of quantum mutual entropy and capacity. Open Syst. Inf. Dyn. **6**(1), 69–78 (1999)
594. Ohya, M.: Mathematical Foundation of Quantum Computer. Maruzen, Tokyo (1999)
595. Ohya, M., Masuda, N.: NP problem in quantum algorithm. Open Syst. Inf. Dyn. **7**(1), 33–39 (2000)
596. Ohya, M., Sato, K.: Use of information theory to study genome sequences. Rep. Math. Phys. **46**(3), 419–428 (2000)
597. Ohya, M., Volovich, I.V.: Quantum computing, NP-complete problems and chaotic dynamics. In: Hida, T., Saito, K. (eds.) Quantum Information, pp. 161–171. World Scientific, Singapore (2000)
598. Ohya, M.: Complexity in dynamics and computation. Acta Appl. Math. **63**, 293–306 (2000)
599. Ohya, M., Sato, K.: Analysis of the disease course of HIV-1 by entropic chaos degree. Amino Acids **20**, 155–162 (2001)
600. Ohya, M., Volovich, I.V.: Quantum computing and chaotic amplification. J. Opt. B **5**(6), 639–642 (2003)
601. Ohya, M., Volovich, I.V.: On quantum entropy and its bound. In: Infinite Dimensional Analysis and Quantum Probability, vol. 6, pp. 301–310 (2003)
602. Ohya, M., Volovich, I.V.: New quantum algorithm for studying NP-complete problems. Rep. Math. Phys. **52**(1), 25–33 (2003)
603. Ohya, M.: Information dynamics and its application to recognition process. In: A Garden of Quanta: Essays in Honor of Hiroshi Ezawa, pp. 445–462. World Scientific, Singapore (2003)
604. Ohya, M.: Foundation of chaos through observation. In: Hida, T., Saito, K., Si, Si (eds.) Quantum Information and Complexity, pp. 391–410. World Scientific, Singapore (2004)
605. Ohya, M.: On quantum information and algorithm. In: Mathematical Modelling in Physics, vol. 10, pp. 451–468. Vaxjo University Press, Vaxjo (2004)
606. Ohya, M.: Quantum algorithm for SAT problem and quantum mutual entropy. Rep. Math. Phys. **55**(1), 109–125 (2005)
607. Ohya, M., Matsuoka, T.: Quantum entangled state and its characterization. In: Foundation and Probability and Physics-3. AIP, vol. 750, pp. 298–306 (2005)
608. Ohya, M.: Adaptive dynamics and its applications to chaos and NPC problem. In: QP-PQ: Quantum Probability and White Noise Analysis, Quantum Bio-Informatics, vol. 21, pp. 181–216 (2007)
609. Ojima, I.: Entropy production and nonequilibrium stationarity in quantum dynamical systems. Physical meaning of Van Hove limit. J. Stat. Phys. **56**, 203–226 (1989)
610. Ojima, I.: A unified scheme for generalized sectors based on selection criteria—order parameters of symmetries and of thermality and physical meanings of adjunctions. Open Syst. Inf. Dyn. **10**, 235–279 (2003)
611. Ojima, I.: Micro-macro duality in quantum physics. In: Hida, T. (ed.) Stochastic Analysis: Classical and Quantum Perspectives of White Noise Theory, pp. 143–161. World Scientific, Singapore (2005)

612. Ojima, I., Hasegawa, H., Ichiyanagi, M.: Entropy production and its positivity in nonlinear response theory of quantum dynamical systems. J. Stat. Phys. **50**, 633–655 (1988)
613. Olmschenk, S., Matsukevich, D.N., Maunz, P., Hayes, D., Duan, L.-M., Monroe, C.: Quantum teleportation between distant matter qubits. Science **323**, 486–491 (2009)
614. Olsen, B., Munster, V.J., Wallensten, A., Waldenström, J., Osterhaus, A.D.M.E., Fouchier, R.A.M.: Global patterns of influenza a virus in wild birds. Science **312**(5772), 384–388 (2006)
615. Omnes, R.: The Interpretation of Quantum Mechanics. Princeton Series in Physics. Princeton University Press, Princeton (1994)
616. Onsager, L.: Thermodynamics and some molecular aspects of biology. In: Quarton, G.C., Nelnechuk, T., Schmitt, F.O. (eds.) The Neurosciences. A study program, pp. 75–79. Rockefeller University Press, New York (1967)
617. Oosterkamp, T.H., Fujisawa, T., van der Wiel, W.G., Ishibashi, K., Hijman, R.V., Tarucha, S., Kouwenhoven, L.P.: Microwave spectroscopy on a quantum-dot molecule. cond-mat/9809142 (1998)
618. Osawa, S., Jukes, T.H., Watanabe, K., Muto, A.: Recent evidence for evolution of the genetic code. Microbiol. Rev. **56**(1), 229–264 (1992)
619. Ospelkaus, C., Langer, C.E., Amini, J.M., Brown, K.R., Leibfried, D., Wineland, D.J.: Trapped-ion quantum logic gates based on oscillating magnetic fields. Phys. Rev. Lett. **101**, 090502 (2008)
620. Ott, E.: Chaos in Dynamical Systems. Cambridge University Press, Cambridge (1993)
621. Ozawa, M.: Measurability and computability, LANL e-print. quant-ph/9809048 (1998)
622. Ozawa, M.: Quantum nondemolition monitoring of universal quantum computers. Phys. Rev. Lett. **80**, 631 (1998)
623. Ozawa, M.: Quantum nondemolition monitoring of universal quantum computers. Theor. Inf. Appl. **34**, 379 (2000)
624. Ozawa, M.: Conservation laws, uncertainty relations, and quantum limits of measurements. Phys. Rev. Lett. **88**, 050402 (2002)
625. Ozorio de Almeida, A.M., Saraceno, M.: Periodic orbit theory for the quantized baker's map. Ann. Phys. **210**, 1–15 (1991)
626. Ozorio de Almeida, A.M., Hannay, J.H., Keating, J.P.: Optical realization of the quantum baker's transformation. Nonlinearity **7**, 1327 (1994)
627. Ozorio de Almeida, A.M., Luz, M.G.E.: Path integral for the quantum baker's map. Nonlinearity **8**, 43–64 (1995)
628. Palma, G.M., Suominen, K.-A., Ekert, A.K.: Quantum computers and dissipation. Proc. R. Soc. Lond. Ser. A **452**, 567–574 (1996)
629. Park, Y.M.: Dynamical entropy of generalized quantum Markov chains. Lett. Math. Phys. **32**, 63–74 (1994)
630. Parthasarathy, K.R.: On the integral representation of the rate of transmission of a stationary channel. Ill. J. Math. **5**, 299–305 (1961)
631. Parthasarathy, K.R.: Probability Measure on Metric Spaces. Academic Press, San Diego (1967)
632. Pasquinucci, H.B., Peres, A.: Quantum cryptography with 3-state systems. Phys. Rev. Lett. **85**, 3313 (2000)
633. Pattnaik, B., Pateriya, A.K., Khandia, R., Tosh, C., Nagarajan, S., Gounalan, S., Murugkar, H.V., Shankar, B.P., Shrivastava, N., Behera, P., Bhagat, S., Peiris, J.S.M., Pradhan, H.K.: Phylogenetic analysis revealed genetic similarity of the H5N1 avian influenza viruses isolated from HPAI outbreaks in chickens in Maharashtra, India with those isolated from swan in Italy and Iran in 2006. Curr. Sci. **96**(1), 77–81 (2006)
634. Pauli, W.: Theory of Relativity. Pergamon, Elmsford (1958)
635. Pauling, L.J.: The diamagnetic anisotropy of aromatic molecules. Chem. Phys. **4**, 673 (1936)
636. Paz, J.P., Zurek, W.H.: Decoherence, chaos, and the second law. Phys. Rev. Lett. **72**, 2508–2511 (1994)
637. Pearle, P.M.: Hidden-variable example based upon data rejection. Phys. Rev. D **2**, 1418–1825 (1970)

638. Pechen, A., Rabitz, H.: Unified analysis of terminal-time control in classical and quantum systems. Europhys. Lett. **91**, 60005 (2010). arXiv:1011.1269
639. Pechen, A.N., Volovich, I.V.: Quantum multipole noise and generalized quantum stochastic equations. Infin. Dimens. Anal. Quantum Probab. Rel. Top. **5**(4), 441–464 (2002)
640. Pegg, D., Barnett, S.: Phase properties of the quantized single-mode electromagnetic field. Phys. Rev. A **39**, 1665 (1989)
641. Penrose, R.: The Emperor's New Mind. Oxford University Press, London (1989)
642. Penrose, R.: Shadows of the Mind. Oxford University Press, London (1994)
643. Penrose, O.: Foundations of Statistical Mechanics. Pergamon, Elmsford (1970). Reprinted by Dover (2005)
644. Peres, A.: Quantum Theory: Concepts and Methods. Kluwer Academic, Dordrecht (1993)
645. Peres, A.: Separability criterion for density matrices. Phys. Rev. Lett. **77**, 1413 (1996)
646. Petz, D.: Sufficient subalgebras and the relative entropy of states of a von Neumann algebra. Commun. Math. Phys. **105**, 123–131 (1986)
647. Petz, D.: The Algebra of Canonical Communication Relation. Leuven University Press, Leuven (1990)
648. Petz, D., Mosonyi, M.: Stationary quantum source coding. J. Math. Phys. **42**, 4857–4864 (2001)
649. Philips, J.W., et al.: Imaging neural activity using MEG and EEG. IEEE Eng. Med. Biol. **16**, 34–42 (1997)
650. Poincaré, H.: Remarks on the Kinetic Theory of Gases. In: Selected Works, vol. 3. Nauka, Moscow (1974)
651. Poincaré, H.: L'etat actuel et l'avenir de la physique mathematique. Bull. Sci. Math. **28**(2), 302–324 (1904). English translation in Poincare, Henri (1904). The present and the future of mathematical physics, Bull. Am. Math. Soc. **37**, 25–38 (2000)
652. Poincaré, H.: Le mecanisme et l'experience. Re. Metaphys. Morale **1**, 534–537 (1893). English translation, Stephen Brush, Kinetic Theory, vol. 2, p. 203
653. Preskill, J.: Quantum computing: pro and con. Proc. R. Soc. Lond. A **454**, 469–486 (1998)
654. Preskill, J.: Reliable quantum computers. Proc. R. Soc. Lond. A **454**, 385–410 (1998)
655. Preskill, J.: Battling decoherence: the fault-tolerant quantum computer. Phys. Today **52**, 24–30 (1999)
656. Prigogine, I.: Les Lois du Chaos. Flammarion, Paris (1994)
657. Prokhorov, Yu.V., Rozanov, Yu.A.: Probability Theory. Springer, Berlin (1969)
658. Prokhorov, Yu.V., Rozanov, Yu.A.: Probability Theory. Nauka, Moscow (1973)
659. Reed, M., Simon, B.: Methods of Modern Mathematical Physics, vol. 2. Academic Press, San Diego (1975)
660. Reid, M.D.: Incompatibility of macroscopic local realism with quantum mechanics in measurements with macroscopic uncertainties. Phys. Rev. Lett. **84**, 2765 (2000)
661. Rivest, R.L., Shamir, A., Adleman, L.: A method for obtaining digital signatures and public key cryptography. Commun. ACM **21**, 120–126 (1978)
662. Robinson, D.F., Foulds, L.R.: Comparison of phylogenetic trees. Math. Biosci. **53**, 131–147 (1981)
663. Roschin, R.A., Volovich, I.V.: Relativistic corrections to spin correlators and Bell's theorem. In: Foundations of Probability and Physics, 2, Proc. Conf., Växjö, Sweden, Math. Modelling in Phys., Eng. and Cognitive Sci., vol. 5, pp. 533–545. Växjö University Press, Växjö (2002)
664. Roelfsema, P.R., Singer, W.: Detecting connectedness. Cereb. Cortex **8**, 385–396 (1998)
665. Roos, C., Zeiger, T., Rohde, H., Naegerl, H.C., Eschner, J., Leibfried, D., Schmidt-Kaler, F., Blatt, R.: Quantum state engineering on an optical transition and decoherence in a Paul trap. quant-ph/9909038 (1999)
666. Rose, P.G., Blessing, J.A., Lele, S., Abulafia, O.: Evaluation of pegylated liposomal doxorubicin (Doxil) as second-line chemotherapy of squamous cell carcinomaof the cervix: a phase II study of the Gynecologic Oncology Group. Gynecol. Oncol. **102**, 210–213 (2006)
667. Ruelle, D.: Chance and Chaos. Princeton University Press, Princeton (1991)
668. Rumer, Yu.B.: On systematization of codons in the genetic code. Dokl. Acad. Nauk USSR **167**(6), 1393–1394 (1966)

669. Ruskai, M.B., Szarek, S., Werner, L.: An analysis of completely-positive trace-preserving maps on 2 × 2 matrices. Linear Algebra Appl. **347**, 159–187 (2002)
670. Saitou, N., Nei, M.: The neighbor-joining method: a new method for reconstructing phylogenetic trees. Mol. Biol. Evol. **4**, 406–425 (1987)
671. Sakai, S.: C^*-algebras and W^*-algebras. Springer, Berlin (1971)
672. Sakurai, J.J.: Modern Quantum Mechanics. Addison-Wesley, Reading (1985)
673. Santos, E.: Unreliability of performed tests of Bell's inequality using parametric down-converted photons. Phys. Lett. A **212**, 10 (1996)
674. Saraceno, M.: Classical structures in the quantized baker transformation. Ann. Phys. **199**, 37–60 (1990)
675. Saraceno, M., Voros, A.: Towards a semiclassical theory of quantum baker's map. Physica D **79**, 206–268 (1994)
676. Sato, K., Ohya, M.: Can information measure be one of markers to estimate disease progression in HIV-1 infected patients? Open Syst. Inf. Dyn. **8**(2), 125–136 (2001)
677. Sato, K., Miyazaki, S., Ohya, M.: Analysis of HIV by entropy evolution rate. Amino Acids **14**, 343–352 (1998)
678. Sato, K., Tanabe, T., Ohya, M.: How can we classify Influenza A viruses and understand their severity? Open. Syst. Inf. Dyn. **17**(3), 297–310 (2010)
679. Schack, R.: Using a quantum computer to investigate quantum chaos. Phys. Rev. A **57**, 1634–1635 (1998)
680. Schack, R., Caves, C.M.: Hypersensitivity to perturbations in the quantum baker's map. Phys. Rev. Lett. **71**, 525–528 (1993)
681. Schack, R., Caves, C.M.: Information-theoretic characterization of quantum chaos. Phys. Rev. E **53**, 3257 (1996)
682. Schack, R., Caves, C.M.: Shifts on a finite qubit string: a class of quantum baker's maps. Appl. Algebra Eng. Commun. Comput. **10**, 305–310 (2000)
683. Schack, R., Soklakov, A.N.: Decoherence and linear entropy increase in the quantum baker's map. Phys. Rev. E **66**, 036212 (2002)
684. Schatten, R.: A Theory of Cross-Spaces. Ann. of Math. Studies, vol. 26. Princeton University Press, Princeton (1950)
685. Schatten, R.: Norm Ideals of Completely Continuous Operators. Springer, Berlin (1970)
686. Scheraga, H.A., Khalili, M., Liwo, A.: Protein-folding dynamics: overview of molecular simulation techniques. Ann. Rev. Phys. Chem. **58**, 57–83 (2007)
687. Schiff, L.I.: Quantum Mechanics. McGraw-Hill, New York (1949)
688. Schleich, W.P.: Quantum Optics in Phase Space. Wiley, New York (2001)
689. Schlick, T.: Molecular Modeling and Simulation. Springer, New York (2002)
690. Schmelcher, P., Cederbaum, L.S.: Magnetic field induced two-body phenomena in atoms. Comments At. Mol. Phys., Part D **2**, 123 (2000)
691. Schonhage, A., Strassen, V.: Schnelle Multiplication Grosser Zahlen. Computing **7**, 281–292 (1971)
692. Schrödinger, E.: Foundations of quantum mechanics. Naturwissenschaften **48**, 52 (1935)
693. Schrödinger, E.: Die gegenwärtige Situation in der Quantenmechanik. Naturwissenschaften **23**, 807–812 (1935), 823–828, 844–849
694. Schumacher, B.: Sending entanglement through noisy quantum channels. Phys. Rev. A **51**, 2614–2628 (1993)
695. Schumacher, B.: Quantum coding. Phys. Rev. A **51**, 2738–2747 (1995)
696. Schumacher, B., Nielsen, M.A.: Quantum data processing and error correction. quant-ph/9604022 (1996)
697. Schumacher, B., Westmoreland, M.D.: Sending classical information via noisy quantum channels. Phys. Rev. A **56**(1), 131–138 (1997)
698. Schweber, S.S.: An Introduction to Relativistic Quantum Field Theory. Dover, New York (2005)
699. Scully, M.O., Zubairy, M.: Quantum Optics. Cambridge University Press, Cambridge (1997)
700. Segal, I.B.: Postulates for general quantum mechanics. Ann. Math. **48**, 930 (1947)

701. Segal, I.B.: Mathematical Foundations of Quantum Field Theory. Benjamin, Elmsford (1964)
702. Sellers, P.H.: On the theory and computation of evolutionary distances. SIAM J. Appl. Math. **26**(4), 787–793 (1974)
703. Shafir, E., Tversky, A.: Thinking through uncertainty: nonconsequential reasoning and choice. Cogn. Psychol. **24**, 449–474 (1992)
704. Shafir, E., Tversky, A.: The disjunction effect in choice under uncertainty. Psychol. Sci. **3**, 305–309 (1992)
705. Shankarappa, R., Margolick, J.B., Gange, S.J., Rodrigo, A.G., Upchurch, D., Farzadegan, H., Gupta, P., Rinaldo, C.R., Learn, G.H., He, X., Huang, X.L., Mullins, J.I.: Consistent viral evolutionary changes associated with the progression of human immunodeficiency virus Type 1 infection. J. Virol. **73**(12), 10489 (1999)
706. Shannon, C.E.: A mathematical theory of communication. Bell Syst. Tech. J. **27**, 379–423 (1948), 623–656
707. Shannon, C.E., Weaver, W.: The Mathematical Theory of Communication. University of Illinois Press, Champaign (1949)
708. Shanta, M., Zimmer, M.D., Donald, S., Burke, M.D.: Historical perspective—emergence of influenza A (H1N1) viruses. N. Engl. J. Med. **361**(3), 279–285 (2009)
709. Shi, Y.: Remarks on universal quantum computer. Phys. Lett. A **293**, 277 (2002)
710. Shor, P.W.: Algorithms for quantum computations: discrete logarithms and factoring. In: Proc. of the 35th Annual Symposium on Foundations of Computer Science, pp. 124–134. IEEE Comput. Soc., Los Alamitos (1994)
711. Show, P.W.: Scheme for reducing decoherence in quantum computer memory. Phys. Rev. A **52**, R2493-6 (1995)
712. Shor, P.W.: Fault-tolerant quantum computation. In: Proc. 35th Annual Symposium on Fundamentals of Computer Science, pp. 56–65 (1996)
713. Shor, P.W.: Polynomial-time algorithms for prime factorization and discrete logarithms on a quantum computer. SIAM J. Comput. **26**, 1484 (1997)
714. Shor, P.W., Preskill, J.: Simple proof of security of the BB84 quantum key distribution protocol. Phys. Rev. Lett. **85**, 441–444 (2000)
715. Shor, P.W.: Quantum error correction. http://www.msri.org/publications/ln/msri/2002/quantumcrypto/shor/1/index.html
716. Shor, P.W.: The quantum channel capacity and coherent information. Lecture Notes, MSRI Workshop on Quantum Computation (2002)
717. Simon, D.R.: On the power of quantum computation. SIAM J. Comput. **26**(5), 1474–1483 (1997)
718. Sinai, Ya.G.: Introduction to Ergodic Theory. Princeton University Press, Princeton (1977)
719. Singer, W.: Consciousness and the structure of neuronal representations. Philos. Trans. R. Soc. Lond. B **353**, 1829–1840 (1998)
720. Singer, W.: Neuronal synchrony: a versatile code for the definition of relations? Neuron **24**, 49–65 (1999)
721. Singer, W.: Driving for coherence. News and views. Nature **397**, 391–393 (1999)
722. Singer, W.: Der Beobachter im Gehirn. Essays zur Hirnforschung. Suhrkamp, Frankfurt am Main (2002)
723. Sipser, M.: Introduction to the Theory of Computation. Brooks Cole, Boston (1996)
724. Sinai, Ya.G.: Introduction to Ergodic Theory. Fasis, Moscow (1996)
725. Slichter, C.P.: Principles of Magnetic Resonance. Springer, Berlin (1991)
726. Smith, G.J.D., Bahl, J., Vijaykrishna, D., Zhang, J., Poon, L.L.M., Chen, H., Webster, R.G., Peiris, J.S.M., Guan, Y.: Dating the emergence of pandemic influenza viruses. Proc. Natl. Acad. Sci. USA **106**(28), 11709–11712 (2009)
727. Sobelman, I.I.: Atomic Spectra and Radiative Transitions. Springer, Berlin (1991)
728. Sokal, R.R., Sneath, P.H.A.: Numerical Taxonomy. Freeman, New York (1973)
729. Soklakov, A.N., Schack, R.: Classical limit in terms of symbolic dynamics for the quantum baker's map. Phys. Rev. E **61**, 5108–5114 (2000)

730. Spohn, H.: Approach to equilibrium for completely positive dynamical semigroups of n-level systems. Rep. Math. Phys. **10**, 189–194 (1976)
731. Spohn, H.: Large Scale Dynamics of Interacting Particles. Springer, New York (1991)
732. Stapp, H.P.: Mind, Matter and Quantum Mechanics, 2nd edn. Springer, Berlin (2003)
733. Steane, A.M.: Error correcting codes in quantum theory. Phys. Rev. Lett. **77**, 793–797 (1996)
734. Steane, A.M.: Multiparticle interference and quantum error correction. Proc. R. Soc. Lond. A **452**, 2551–2577 (1996)
735. Sterr, A., et al.: Perceptual correlates of changes in cortical representation of fingers in blind multifinger braille readers. J. Neurosci. **18**(11), 4417 (1998)
736. Streater, R.F., Wightman, A.S.: PCT, Spin, Statistics and All That. Benjamin, New York (1964)
737. Stormer, E.: Contemp. Math. **62**, 345–356 (1987)
738. Sudarshan, E.C.G., Mathews, P.M., Rau, J.: Stochastic dynamics of quantum-mechanical systems. Phys. Rev. **121**, 920–924 (1961)
739. Swanson, R.: A unifying concept for the amino acid code. Bull. Math. Biol. **46**(2), 187–203 (1984)
740. Takesaki, M.: Theory of Operator Algebras, vol. I. Springer, Berlin (1979)
741. Takesaki, M.: Tomita's Theory of Modular Hilbert Algebras and Its Application. Lecture Note in Mathematics, vol. 128. Springer, Berlin (1970)
742. Taki, Y., Information Theory, vol. I. Iwanami, Tokyo (1978) (in Japanese)
743. Tanaka, Y., Asano, M., Ohya, M.: A physical realization of quantum teleportation for non-maximal entangled state. Phys. Rev. A **82**(2), 022308 (2010)
744. Taubenberger, J.K., Morens, D.M.: 1918 influenza: the mother of all pandemics. Emerg. Infect. Dis. **12**(1), 15–22 (2006)
745. Terhal, B.M.: Detecting quantum entanglement. J. Theor. Comput. Sci. **287**(1), 313–335 (2002)
746. Terhal, B.M., Burkard, G.: Fault-tolerant quantum computation for local non-Markovian noise. Phys. Rev. A **71**, 012336 (2005). arXiv:quant-ph/0402104
747. Ternov, I.N.M.: Introduction to Physics of Spin of Relativistic Particles. MSU press, Moscow (1997) (in Russian)
748. Tomiyama, J.: On the projection of norm one in W*-algebras. Proc. Jpn. Acad. **33**, 608–612 (1957)
749. Tomonaga, S.: Quantum Mechanics, vols. I, II. North-Holland, Amsterdam (1962)
750. Travaglione, B.C., Milburn, G.J.: Implementing the quantum random walk. Phys. Rev. A **65**, 032310 (2002), 5 pp.
751. Treiber, A., Poppe, A., Hentschel, M., Ferrini, D., Lorünser, T., Querasser, E., Matyus, T., Hübel, H., Zeilinger, A.: Fully automated entanglement-based quantum cryptography system for telecom fiber networks. New J. Phys. **11**, 045013 (2009)
752. Trifonov, E.N.: The triplet code from first principles. J. Biomol. Struct. Dyn. **22**(1), 1–11 (2004)
753. Trushechkin, A.S., Volovich, I.V.: Functional classical mechanics and rational numbers. P-Adic Numb. Ultrametric Anal. Appl. **1**(4), 365–371 (2009). arXiv:0910.1502
754. Trushechkin, A.S., Volovich, I.V.: On standards and specifications in quantum cryptography. Int. J. Quant. Inf. **6**(2), 347–367 (2008). arXiv:quant-ph/0508035
755. Tumpey, T.M., Belser, J.A.: Resurrected pandemic influenza viruses. Annu. Rev. Microbiol. **63**, 79–98 (2009)
756. Turchette, Q.A., Hood, C.J., Lange, W., Mabuchi, H., Kimble, H.J.: Measurement of conditional phase shifts for quantum logic. Phys. Rev. Lett. **75**, 4710 (1995)
757. Turing, A.M.: On computable numbers, with an application to the Entscheidungs problem. Proc. Lond. Math. Soc. Ser. 2 **42**, 230–265 (1936)
758. Uchiyama, K., Nagayasu, A., Yamagiwa, Y., Nishida, T., Harashima, H., Kiwada, H.: Effects of the size and fluidity of liposomes on their accumulation in tumors: a presumption of their interaction with tumors. Int. J. Pharm. **121**(2), 195–203 (1995)
759. Uhlmann, A.: The 'transition probability' in the state space of a*-algebra. Rep. Math. Phys. **9**, 273–279 (1976)

760. Uhlmann, A.: Relative entropy and the Wigner-Yanase-Dyson-Lieb concavity in interpolation theory. Commun. Math. Phys. **54**, 21–32 (1977)
761. Umegaki, H.: Conditional expectation in an operator algebra. Tohoku Math. J. **6**, 177–181 (1954)
762. Umegaki, H.: Conditional expectation in an operator algebra, III. Kodai Math. Semin. Rep. **11**, 51–64 (1959)
763. Umegaki, H.: Conditional expectations in an operator algebra IV (entropy and information). Kodai Math. Semin. Rep. **14**, 59–85 (1962)
764. Umegaki, H.: A functional method on amount of entropy. Kodai Math. Semin. Rep. **15**, 162–175 (1963)
765. Umegaki, H.: General treatment of alphabet message space and integral representation of entropy. Kodai Math. Semin. Rep. **16**, 8–26 (1964)
766. Umegaki, H., Ohya, M.: Probabilistic Entropy. Kyoritsu, Tokyo (1983) (in Japanese)
767. Umegaki, H., Ohya, M., Hiai, F.: Introduction to Operator Algebra. Kyoritsu, Tokyo (1984) (in Japanese)
768. Umegaki, H., Tsukada, M., Ohya, M.: Measure, Integral and Probability. Kyoritsu, Tokyo (1987) (in Japanese)
769. Unruh, W.: Maintaining coherence in quantum computers. Phys. Rev. A **51**, 992–997 (1995)
770. Urbanik, K.: Joint probability distribution of observables in quantum mechanics. Stud. Math. **21**, 317 (1961)
771. van Enk, S.J., Kimble, H.J.: On the classical character of control fields in quantum information processing. Quantum Inf. Comput. **2**(1), 1–13 (2002)
772. Vandersypen, L.M.K., Steffen, M., Breyta, G., Yannoni, C.S., Cleve, R., Chuang, I.L.: Experimental realization of an order-finding algorithm with an NMR quantum computer. Phys. Rev. Lett. **85**(25), 5452–5455 (2000)
773. Varadarajan, V.S.: Geometry of Quantum Theory. Springer, New York (1985)
774. Varadarajan, V.S.: Multipliers for the symmetry groups of p-adic spacetime. P-Adic Numbers. Ultrametric Anal. Appl. **1**(1), 69–78 (2009)
775. Varcoe, B.T.H., Sang, R., MacGillivray, W.R., Standage, M.C.: Quantum state reconstruction using atom optics. quant-ph/9910024 (1999)
776. Vedral, V., Plenio, M.B., Rippin, M.A., Knight, P.L.: Quantifying entanglement. Phys. Rev. Lett. **78**, 2275 (1997)
777. Viboud, C., Tam, T., Fleming, D., Handel, A., Miller, M.A., Simonsen, L.: Transmissibility and mortality impact of epidemic and pandemic influenza, with emphasis on the unusually deadly 1951 epidemic. Vaccine **24**(44–46), 6701–6707 (2006)
778. Viboud, C., Tam, T., Fleming, D., Miller, M.A., Simonsen, L.: 1951 Influenza epidemic, England and Wales, Canada, and the United States. Emerg. Infect. Dis. **12**(4), 661–668 (2006)
779. Vinogradov, I.M.: Basics of Number Theory. Nauka, Moscow (1963)
780. Viola, L., Lloyd, S.: Dynamical suppression of decoherence in two-state quantum systems. Phys. Rev. A **58**(4), 2733–2744 (1998)
781. Vladimirov, V.S.: Equations of Mathematical Physics. Dekker, New York (1971)
782. Vladimirov, V.S., Volovich, I.V.: Superanalysis. I. Differential calculus. Teor. Mat. Fiz. **59**, 3–27 (1984)
783. Vladimirov, V.S., Volovich, I.V., Zelenov, E.I.: p-Adic Analysis and Mathematical Physics. World Scientific, Singapore (1994)
784. Voiculescu, D.: Dynamical approximation entropies and topological entropy in operator algebras. Commun. Math. Phys. **170**, 249–281 (1995)
785. Volovich, I.V.: p-adic string. Class. Quantum Gravity **4**, L83–L87 (1987)
786. Volovich, I.V.: Quantum computers and neural networks. Invited talk at the International Conference on Quantum Information held at Meijo University, 4–8 Nov 1997
787. Volovich, I.V.: Models of quantum computers and decoherence problem. In: International Conference on Quantum Information, Meijo University, Nagoya, 4–8 Nov. Proc. of the Conference, pp. 211–224. World Scientific, Singapore (1999)
788. Volovich, I.V.: Atomic quantum computer. In: Physics of Elementary Particles and Atomic Nuclei, Dubna, vol. 31, pp. 133–136 (2000)

789. Volovich, I.V., Volovich, Y.I.: Bell's theorem and random variables. quant-ph/0009058 (2000)
790. Volovich, I.V.: Bell's theorem and locality in space. quant-ph/0012010 (2000)
791. Volovich, I.V.: An attack to quantum cryptography from space. quant-ph/0012054 (2000)
792. Volovich, I.V., Volovich, Y.I.: On classical and quantum cryptography. In: Lectures at the Volterra–CIRM International School, Quantum Computer and Quantum Information, Trento, Italy, 25–31 July 2001. quant-ph/0108133
793. Volovich, I.V.: Quantum cryptography in space and Bell's theorem. In: Khrennikov, A. (ed.) Foundations of Probability and Physics: Proceedings of the Conference, Vaxjo, Smoland, Sweden, 25 November–1 December 2000. QP-PQ: Quantum Probability and White Noise Analysis, vol. 13, pp. 364–373. World Scientific, Singapore (2002)
794. Volovich, I.V.: Quantum information in space and time. quant-ph/0108073 (2001)
795. Volovich, I.V.: Quantum computing and Shor's factoring algorithm. quant-ph/0109004 (2001)
796. Volovich, I.V.: In: Khrennikov, A. (ed.) Foundations of Probability and Physics, pp. 364–372. World Scientific, Singapore (2001)
797. Volovich, I.V.: Towards quantum information theory in space and time. In: Khrennikov, A. (ed.) Proceedings of the Conference, Quantum Theory: Reconsideration of Foundations, Vaxjo, Smaland, Sweden, 17–21 June 2001, pp. 423–441. Vaxjo University Press, Vaxjo (2002)
798. Volovich, I.V.: Seven principles of quantum mechanics. quant-ph/0212126 (2002)
799. Volovich, I.V.: Number theory as the ultimate physical theory. p-Adic Numbers, Ultrametric Anal. Appl. **2**, 77–87 (2010)
800. Volovich, I.V.: Randomness in classical mechanics and quantum mechanics. Found. Phys. (2010). doi:10.1007/s10701-010-9450-2. arXiv:0907.2445
801. Volovich, I.V.: Functional mechanics and time irreversibility problem. In: Accardi, L., Freudenberg, W., Ohya, M. (eds.) Quantum Bio-Informatics III, pp. 393–404. World Scientific, Singapore (2010)
802. Volovich, I.V.: Bogolyubov equations and functional mechanics. Theor. Math. Phys. **164**(3), 354–362 (2010)
803. Volovich, I.V., Trushechkin, A.S.: On a model of key distribution in quantum cryptography. Dokl. Akad. Nauk **404**(2), 169–172 (2005). arXiv:quant-ph/0504156 (Russian)
804. Volovich, I.V., Trushechkin, A.S.: On quantum compressed states on interval and uncertainty relation for nanoscopic systems. Proc. Steklov Math. Inst. **265**, 1–31 (2009)
805. von Neumann, J.: Die Mathematischen Grundlagen der Quantenmechanik. Springer, Berlin (1932)
806. von Neumann, J.: Mathematical Foundations of Quantum Mechanics. Princeton University Press, Princeton (1955)
807. von Mises, R.: The Mathematical Theory of Probability and Statistics. Academic Press, San Diego (1964)
808. Walls, D.F., Milburn, G.J.: Quantum Optics. Springer, Berlin (1994)
809. Watson, J.D., Baker, T.A., Bell, S.P., Gann, A., Levine, M., Losick, R.: Molecular Biology of the Gene. CSHL Press, Benjamin (2004)
810. Weihs, G., Jennewein, T., Simon, C., Weinfurter, H., Zeilinger, A.: Violation of Bell's inequality under strict Einstein locality conditions. Phys. Rev. Lett. **81**, 5039–5043 (1998)
811. Weinberg, S.: The Quantum Theory of Fields. Cambridge University Press, Cambridge (1995)
812. Welsh, D.J.A.: Complexity: Knots, Colouring and Counting. Cambridge University Press, Cambridge (1993)
813. Werner, R.F.: All teleportation and dense coding schemes. quant-ph/0003070 (2000)
814. Wehrl, A.: General properties of entropy. Rev. Mod. Phys. **50**, 221–260 (1978)
815. Wheeler, J.A., Zurek, W.H.: Quantum Theory and Measurement. Princeton University Press, Princeton (1983)
816. Wikipedia: Amino acid. http://en.wikipedia.org/wiki/Amino_acid (2007)

817. Wikipedia: List of standard amino acids. http://en.wikipedia.org/wiki/List_of_standard_amino_acids (2007)
818. Wiesner, S.: Conjugate coding. SIGACT News **15**(1), 78–88 (1983)
819. Wolfs, T.F., Zwart, G., Bakker, M., Valk, M., Kuiken, C.L., Goudsmit, J.: Naturally occurring mutations within HIV-1 V3 genomic RNA lead to antigenic variation dependent on a single amino acid substitution. Virology **185**(1), 195 (1991)
820. Wong, J.T.F.: A co-evolution theory of the genetic code. Proc. Natl. Acad. Sci. USA **72**, 1909–1912 (1975)
821. Wootters, W.K., Zurek, W.H.: A single quantum cannot be cloned. Nature **299**, 802–803 (1982)
822. Xavier, G.B., Walenta, N., Vilela de Faria, G., Temporao, G.P., Gisin, N., Zbinden, H., von der Weid, J.P.: Experimental polarization encoded quantum key distribution over optical fibres with real-time continuous birefringence compensation. New J. Phys. **11**, 045015 (2009), 5 pp.
823. Yao, A.C.: Quantum circuit complexity. In: FOCS, pp. 352–361 (1993)
824. Yang, J., Bao, X.-H., Zhang, H., Chen, S., Peng, C.-Z., Chen, Z.-B., Pan, J.-W.: Experimental quantum teleportation and multi-photon entanglement via interfering narrowband photon sources. Phys. Rev. A **80**, 042321 (2009)
825. Ye, J., Vernooy, D.W., Kimble, H.J.: Trapping of single atoms in cavity QED. quant-ph/9908007 (1999)
826. Yoshida, K.: Functional Analysis. Springer, New York (1978)
827. Yuen, P.: Anonymous key quantum cryptography and unconditionally secure quantum bit commitment. In: Proceeding of the Fifth International Conference on Quantum Communication, Computing, and Measurement, pp. 285–293 (2000)
828. Yuen, H.P., Ozawa, M.: Ultimate information carrying limit of quantum systems. Phys. Rev. Lett. **70**, 363–366 (1993)
829. Zadeh, L.A.: Fuzzy sets. Inf. Control **8**, 338–356 (1965)
830. Zalka, C.: Grover's quantum searching is optimal. Phys. Rev. A **60**, 2746 (1999)
831. Zanardi, P., Rasetti, M.: Noiseless quantum codes. Phys. Rev. Lett. **79**, 3306–3311 (1997)
832. Zaslavskii, G.M.: Stochasticity of Dynamical System. Nauka, Moscow (1984)
833. Zelenov, E.I.: Quantum approximation theorem. P-Adic Numbers, Ultrametric Anal. Appl. **1**(1), 88–90 (2009)
834. Zhang, Q., Goebel, A., Wagenknecht, C., Chen, Y.A., Zhao, B., Yang, T., Mair, A., Schmiedmayer, J., Pan, J.-W.: Experimental quantum teleportation of a two-qubit composite system. Nat. Phys. **2**, 678–682 (2006)
835. Zhang, Y.-S., Li, C.-F., Guo, G.-C.: Quantum key distribution via quantum encryption. Phys. Rev. A **64**, 024302 (2001)
836. Zubarev, D.N.: Nonequilibrium Statistical Thermodynamics. Consultants Bureau, New York (1974)
837. Zurek, W.H.: Pointer basis of quantum apparatus: Into what mixture does the wave packet collapse? Phys. Rev. D **24**, 1516 (1981)
838. Zurek, W.H.: Decoherence, einselection, and the quantum origins of the classical. Rev. Mod. Phys. **75**, 715–775 (2003)

Index

M. Ohya, I. Volovich, *Mathematical Foundations of Quantum Information and Computation and Its Applications to Nano- and Bio-systems*,
Theoretical and Mathematical Physics,
DOI 10.1007/978-94-007-0171-7, © Springer Science+Business Media B.V. 2011

R